Methods for Evaluating Biological Nitrogen Fixation

Methods for Evaluating Biological Nitrogen Fixation

Edited by

F. J. Bergersen
CSIRO, Division of Plant Industry, Canberra, Australia

A Wiley—Interscience Publication

JOHN WILEY & SONS
Chichester • New York • Brisbane • Toronto

British Library Cataloguing in Publication Data:

Methods for evaluating biological nitrogen fixation.
1. Nitrogen—Fixation—Measurement
I. Bergersen, F. J.
574.1'33 QR89.7 79-41785

ISBN 0 471 27759 2

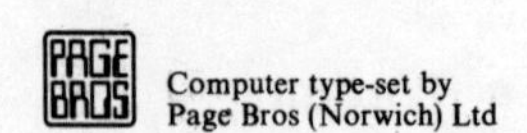

Computer type-set by
Page Bros (Norwich) Ltd

Contributors

DR. C. A. APPLEBY	*Division of Plant Industry, CSIRO, P.O. Box 1600, Canberra City, A.C.T. 2601, Australia.*
DR. F. J. BERGERSEN	*Division of Plant Industry, CSIRO, P.O. Box 1600, Canberra City, A.C.T. 2601, Australia.*
PROFESSOR G. BOND, F.R.S.	*Department of Botany, The University, Glasgow G12 8QQ, Scotland (now retired).*
MR. J. BROCKWELL	*Division of Plant Industry, CSIRO, P.O. Box 1600, Canberra City, A.C.T. 2601, Australia.*
DR. F. C. CANNON	*ARC Unit of Nitrogen Fixation, University of Sussex, Brighton BN1 9RQ, England.*
DR. H. DALTON	*Department of Biological Sciences, University of Warwick, Coventry CV4 7AL, England.*
DR. JOHANNA DÖBEREINER	*Programa Fixação Biológica de Nitrogênio, EMBRAPA, SNLCS-CNPq, Km 47, 23460 Seropédica, Rio de Janerio, Brazil.*
DR. W. F. DUDMAN	*Division of Plant Industry, CSIRO, P.O. Box 1600, Canberra City, A.C.T. 2601, Australia.*
DR. R. R. EADY	*ARC Unit of Nitrogen Fixation, University of Sussex, Brighton BN1 9RQ, England.*
DR. K. J. F. FARNDEN	*Department of Biochemistry, University of Otago, P.O. Box 56, Dunedin, New Zealand.*
DR. A. H. GIBSON	*Division of Plant Industry, CSIRO, P.O. Box 1600, Canberra City, A.C.T. 2601, Australia.*
PROFESSOR R. KNOWLES	*Department of Microbiology, Macdonald College, McGill University, Quebec HOA 1CO, Canada.*
DR. J. G. ROBERTSON	*Division of Applied Biochemistry, DSIR, Private Bag, Palmerston North, New Zealand.*
DR. E. A. SCHWINGHAMER	*Division of Plant Industry, CSIRO, P.O. Box 1600, Canberra City, A.C.T. 2601, Australia.*

PROFESSOR W. D. P. STEWART, F.R.S.	*Department of Biological Sciences, The University, Dundee DD1 4HN, Scotland.*
DR. J. A. THOMPSON	*New South Wales Department of Agriculture, P.O. Box 720, Gosford, N.S.W. 2250, Australia.*
MR. G. L. TURNER	*Division of Plant Industry, CSIRO, P.O. Box 1600, Canberra City, A.C.T. 2601, Australia.*
DR. C. T. WHEELER	*Department of Botany, The University, Glasgow G12 8QQ, Scotland.*

Contents

Preface

This book was originally conceived during discussions in Brazil in 1973. Dr Johanna Döbereiner, Professor W. D. P. Stewart, and I had been engaged in teaching a graduate course in Nitrogen Fixation at the Instituto Agronomico in Campinas for two weeks prior to the IV Global Impacts of Applied Microbiology Congress in São Paulo. After these events, we were enjoying a tropical evening, relaxing in hammocks slung between the supporting pillars of the verandah of the Döbereiner home near Rio de Janeiro. It seemed to us that there was an urgent need for a practical book to complement the many 'state of knowledge' books which were then being prepared. Professor Stewart and I refined our ideas during the long flight from Rio to London a few days later, but the project lapsed as each of us became engaged in our other work. Late in 1975, during a meeting at the UNEP headquarters in Nairobi, Professor C.-G. Hedén passed a note addressed to Dr. Döbereiner and myself, suggesting that we undertake the writing of a book about the evaluation of biological nitrogen fixation systems. However, once again the pressure of work intervened and the matter lapsed until I was approached by Dr H. A. Jones of John Wiley and Sons, at the suggestion of Professor Hedén, in June 1976. I consulted with Dr. Döbereiner and Professor Stewart by letter. They readily agreed with my proposal that we enlarge the concept to an edited book in which they agreed to join as contributors, sharing the preparation of the material with our colleagues from the Division of Plant Industry, CSIRO, Canberra, and from laboratories in the United Kingdom, Canada, and New Zealand. Only two of those originally invited to contribute were unable to accept, because of prior commitments in 1977–78. All contributors have shared in developing the framework of the book and many suggestions have been incorporated in the chapter outlines as a result of consultation by correspondence.

The purpose of this book is to bring together, from very diverse sources, currently used methods related to biological nitrogen fixation. We have not intended to produce a set of standard methods. Rather, we have tried to produce comprehensive information which will provide a starting point for scientists entering the field for the first time, or for those who are changing their research programmes within the overall topic. Most chapters have been prepared by single authors and therefore will inevitably reflect the experience and preferences of each of them. We see this book as being complementary to many books and reviews published about the current state of knowledge of the subject and it will provide a convenient source for authors who submit papers to journals whose Editors have restricted space for the inclusion of detailed methods sections.

F. J. BERGERSEN
1979

Section I

INTRODUCTION

Methods for Evaluating Biological Nitrogen Fixation
Edited by F. J. Bergersen

F. J. Bergersen
*Division of Plant Industry,
CSIRO
Canberra, Australia*

Methods, Accidents, and Design

(a) HISTORICAL RELATIONSHIPS BETWEEN METHODOLOGY AND RESEARCH IN NITROGEN FIXATION

From the discovery of nitrogen as a major constituent of the terrestrial atmosphere by the great English and French chemists of the latter half of the eighteenth century, through the period of the controversy surrounding the 'doctrine of the aerial nutrition of plants' in the early nineteenth century, to the proofs of the fixation of atmospheric N_2 by microbes and nodulated legumes late in that century, progress was almost entirely dependent on the development of suitable methods and experimental techniques. Indeed, on many occasions, development of theories greatly outstripped the capacity of techniques suitable for their critical evaluation. It is interesting to note that, in addition to the great agricultural interest of the subject, part of the stimulus for the development the chemical analysis of nitrogen compounds came from concern in England about the agricultural, industrial, and domestic pollution of waterways and water supplies (see, for example, Warrington, 1878, where reference is made to a Rivers Pollution Commission Report). In France, Boussingault (1853) remarked that river water in Paris resembled a pile of manure! We tend to think that environmental concern is a modern phenomenon.

The development of methods of nitrogen analysis was spectacular; nevertheless, Lawes and Gilbert (1851) remarked that they were not adequate to measure the addition of 100 pounds of ammonia to an acre of soil, even when that soil was greatly depleted in nitrogen by repeated cropping. Progress therefore depended heavily upon further technical and experimental developments.

The history of the controversies of the period 1840–90 which surrounded the study of the nitrogen nutrition of plants, and especially the question of their ability to utilize atmospheric N_2, provides lessons in experimental design, analytical methodology, and a human commentary on the scientists involved. Excellent accounts may be found in Wilson (1940) and Aulie (1970) and it is of great interest to consult at least two of the original papers, by Lawes *et al.* (1861) and Lawes and Gilbert (1892). However, many users of this book may not have access to the publications and therefore a brief reiteration is appropriate.

The value of legumes in agriculture was well attested in the writings of the ancients and this knowledge was conserved through the Dark Ages and Mediaeval times by means of monkish copying of such Roman Writers as Varro, Cato, and Columella. It was not until the late nineteenth century that the basis for this value became apparent. The practice of including a legume in crop rotation systems of farming seems to have been introduced into England by Sir Richard Weston, who was impressed by the benefits of clover in cropping sequences during a visit to Flanders in 1639 (Aulie, 1970). The practice became well established and was described by Jethro Tull in 1731 (Aulie, 1970) with respect to the inclusion of clover, sainfoin, and lucerne in planting sequences. This was found to be beneficial by empirical observation, the criterion being the profitibility of the farms on which it was used. On the basis of this experience, crop rotation spread to Europe, where food production increased by 40–50% during the eighteenth century. The search for the reasons for the benefits of legumes became a subject for scientific study early in the development of agricultural chemistry.

The development of a reliable method for estimating the carbon, hydrogen, oxygen, and nitrogen content of organic matter by Dumas (1834) directly permitted the next stage of understanding. Boussingault conducted experiments on forage crops in 1838–44, from which it emerged that the benefits of legumes and their superior nutritive value were related to their nitrogen content. In 1841 he published the puzzling results of analyses which showed that, in land initially exhausted by repeated cropping of cereals, during a 5-year rotation, an excess of nitrogen accumulated beyond that added in manures. Could the excess of nitrogen have been acquired from the air? (Aulie, 1970).

At about the same time (1840), the Baron von Liebig presented a simple explanation of how plants obtained their nitrogen. He contended that ammonia, released to the atmosphere from putrefaction, was returned to the soil in rain-water and absorbed from the air by the leaves of plants (Aulie,

1970). There was a simple logic to the theory which had been tentatively suggested by other European scientists as early as 1804. Analysis detected ammonia in the atmosphere but it was the great reputation of von Liebig as an organic chemist which gave credence to his theory. There was no real basis for the adequacy of this source of nitrogen for plant growth. As early as 1838, Boussingault was aware of the importance of establishing the concentration of ammonia in the air and, even more important, of the rate at which it was released to the air and returned to living matter. However, the Dumas method was not suitable for this purpose and other methods were inadequate (Aulie, 1970). It was not until 1853–54 that he was able to publish values for the ammonia content of rain-water and river and well waters using a concentration, distillation, and titration method which was accurate to 0.03–0.04 mg of ammonia per litre. At about the same time, at Rothamsted in England, Sir John Lawes and Dr. J. H. Gilbert, who was fresh from obtaining a doctorate under von Liebig, began to measure ammonia and nitric acid in rain-water (Lawes and Gilbert, 1854). They used essentially similar methods, evaluating their results rigorously for accuracy. They concluded that the ammonia dissolved from the atmosphere was inadequate to account for observed crop yields. They were not satisfied enough with their methods for nitrates to draw reliable conclusions. However, Thomas Way (1856) showed clearly that atmospheric nitrates and ammonia could not supply the needs of plants for nitrogen. These methods proved indispensable to the development of understanding of nitrogen transformations and led towards an understanding of the fixation of atmospheric nitrogen.

With the convincing elimination of combined nitrogen in the atmosphere as a significant nitrogen source for plants, the stage was set for critical experiments about nitrogen fixation. Aulie (1970) gives an interesting summary of the controversy, before the French Academy, between Boussingault and Ville during 1850–56. The former concluded from scrupulously controlled experiments in enclosed systems that plants cannot utilize the molecular nitrogen of the air, whereas the latter concluded that they could and that the controversy hinged on the design of apparatus and experimental technique. In spite of Boussingault's precise work, the Academy favoured the conclusions of Ville.

At Rothamsted, between 1844 and 1859, Lawes *et al.* (1861) evaluated the work of Boussingault, Ville, and others, some of whose experiments they drew on when designing their own. The paper begins with summaries of their field experiments with legumes and cereals, clearly showing the effects of legumes in providing nitrogen during rotations. They discussed possible sources for the 'actually large total amounts of combined nitrogen which we know to exist and to circulate in land and water, animal, and vegetable life and in the atmosphere'. During 1857–59, experiments were carried out in an enclosed system, the construction and use of which were carefully described.

Cereals grew well but legumes sometimes grew poorly. Nevertheless, it was concluded the 'results do not indicate any assimilation of free 'nitrogen'. The authors remark, somewhat plaintively, that 'it still remains not very obvious to what actions a large proportion of the existing combined 'nitrogen may be attributed'.

The paper by Lawes *et al.* (1861) effectively negated the possibility that plants could utilize nitrogen from the atmosphere. The experiments were models of excellent design and execution but they were misleading. The microbial agents of nitrogen fixation (then unknown) had been carefully removed because the experience of the previous 10 years or so had emphasized the importance of eliminating all forms of combined nitrogen from the air, water, and soil and from the inner surfaces of the apparatus. No scientist of the period noted the connection between the use of calcined soil and negative results, or remarked on the presence or absence of root nodules, on the legumes. Legume nodules had been studied microscopically at least as early as 1858 by Lachmann (Fred *et al.*, 1932), and were considered to contain microorganisms.

The matter progressed no further for 20 years until W. O. Atwater, working at the newly established Connecticut State Experiment Station (Wilson, 1963), experimented in a glasshouse with peas grown in pots supplied with various amounts of nutrients, including combined nitrogen. The balance obtained at the end of the experiment showed that the plants had acquired an excess of nitrogen above that supplied. Although recognizing that the possibility 'that plants assimilate free nitrogen is contrary to general belief and the results of the best investigators on the subject . . .', Atwater continued to make the point (Atwater, 1885). Although remaining uncertain whether the basis of the reaction was physical or biological, he was certain that the plant was involved (Atwater, 1886).

Meanwhile, in Germany in 1883, two plant chemists, Hellriegel and Wilfarth, began to study the effects of rates of application of nitrogenous fertilizers on yields of crops. After three years they had three well-established findings: (i) in sand cultures with complete nutrient supply except for nitrogen, cereals grew in proportion to amount of added nitrate; (ii) peas often showed gains of nitrogen which came from another undefined source; and (iii) in control treatments, with no added nitrate some peas were yellow and stunted and others green and vigorous. They reasoned that the results could be explained by bacteria dropping on to the sterilized pots from the air, by chance infecting the roots to produce nodules, thus enabling some of the pea plants to utilize free nitrogen from the air. They set out to test this by deliberately infecting sterilized pots of peas with extracts of garden soil. Their first experiment failed because they used an unsuitable sand. The second experiment produced clear results. Peas grown in sterile sand, protected from contamination by sterile cotton, grew poorly without added combined nitrogen; those to which soil extract had been added grew vigorously. In all cases good growth was correlated with the presence on the roots of many root nodules. The connection between agricultural chemistry and microbiology had been made for nitrogen fixation as had already been done for nitrification (Warrington, 1883). The results were presented in 1886 with an impressive demonstration before the experimental agriculture section of the Versammlung Deutscher Naturforscher und Aertze (Hellriegel and Wilfarth, 1888). The Chairman of the meeting was Gilbert of Rothamsted.

Characteristically, the Rothamsted workers remained sceptical (Lawes and Gilbert, 1889) until they had completed their own evaluations of the new experiments. As was to be expected, their experiments were thoroughly described and were as conclusive in proving the nitrogen fixing ability of nodulated legumes (Lawes and Gilbert, 1890, 1892) as had been the earlier negative experiments. In contrast to the German experiments, the Rothamsted work was backed by comprehensive nitrogen analyses and observation of nodule development.

(b) THE DESIGN OF EXPERIMENTS

This brief account of the early controversies has been included here as an example of the interactions which take place in science between experimental techniques, experimental and scientific philosophy, and the personal views and perspectives of the experimenters. The methods to be used in an investigation are important but only within the total experimental approach. The design of an experiment implies the framing of a question in terms which the experiment is capable of answering. The methods certainly place constraints upon the answers which will be obtained. The constraints must be defined by the inclusion of appropriate controls and comparisons and the overall design must allow for the serendipity which so frequently puzzles or inspires the experimenter.

Sometimes methods enjoy periods of fashionable use, with the consequent danger of being used inappropriately. For example, the mere possession of a gas chromatograph is insufficient reason to use the acetylene reduction assay for nitrogenase when measuring nitrogen fixation by a nodulated plant or by a laboratory culture of Azotobacter. In many cases it will be both easier and more suitable to measure the actual increment of total nitrogen during the experiment, using a method such as one of those devised in the last century.

The following principles of experimental design may be discerned in the papers cited in the preceding historical section. To some they may be obvious, but they are listed in the hope that the proper place of methods in overall design may be seen.

1. Find out what is already known and do not confine the survey to familiar

journals or even the same discipline; do not disregard the observations of field workers, even when they are not scientists, and above all, beware of preconceived ideas, especially when they are not supported by data; even when they are supported by data, maintain a healthy scepticism.
2. Define the problem and carefully frame the question which your experiment is to answer. Remember that in doing this, you may not have sufficient information to ask the question properly. Therefore, allow for unexpected results.
3. Obtain statistical advice early. It is easier to modify the design than to repeat the experiment.
4. Determine which controls and comparisons to include in order to safeguard the objectives of the experiment.
5. Select the methods to be used. Test your ability to use them correctly and determine whether they are adequate for the objectives of the experiment.
6. Check to see that your resources are sufficient to deal with the sampling and analysis which are required to be completed within the time limits imposed by the experimental design.
7. Even with all these precautions, do not be disappointed when experiments do not work. Keep on trying!

(c) CURRENTLY INCREASED RESEARCH IN NITROGEN FIXATION

There is no doubt that there has been a recent increase in the number of scientists engaged in nitrogen fixation research and the number continues to increase. For example, regarding studies of the legume symbiosis alone, papers abstracted for the private circulation *Rhizobium Newsletter* rose from 85 in 1968 to 450 in 1976 (Gibson, personal communication). These papers appeared in 300 journals published in 45 countries during 1972–76. Similarly increased research has occurred in other aspects of nitrogen fixation. There are many reasons for the expanding research effort in the subject. The value of biological nitrogen fixation in agricultural systems was re-emphasized by the ‘energy crisis’ of 1973, when the costs of nitrogenous fertilizers rose steeply because of the increased cost of petroleum. Increased resources for research in the subject have been recommended in some countries (e.g. Anon, 1977a). There have also been significant technical and methodological improvements which have enabled new studies to begin. Hardy *et al.* (1973) remarked that the growth of the use of the acetylene reduction assay for nitrogenase activity became exponential just prior to presentation of their paper. This fulfilled the predictions of Wilson and Fred (1935) that between 1965 and 1970 the number of papers on nitrogen fixation would peak at about 100 per year unless some new factor encouraged a further period of exponential growth in research. Other factors have included the coincidence of development of the technical capacities to study such aspects as the genetics, enzymology, and physiology of nitrogenase, as well as the discovery of a number of new nitrogen-fixing systems which may assume practical importance.

Nitrogen fixation research is now current in several spheres. The process may be beneficial or detrimental. The role in agriculture and food production has been emphasized; so also has the place of nitrogen fixation in the nitrogen economy of natural ecosystems, many of which are the subject of efforts in the field of conservation. The role of nitrogen-fixing bacteria in the hind-gut of termites (French *et al.*, 1976) may be crucial in maintaining decomposition cycles in some forest and savannah ecosystems. Nitrogen fixation may be beneficial in some waste disposal systems, but often produces problems, such as the stench which is often characteristic of sugar wastes, where such organisms as *Klebsiella* spp. grow readily. Blue-green algae not infrequently cause problems in storages of domestic and industrial waters and in fish production systems. Jutono (1973) reported that three genera of nitrogen-fixing blue-green algae were primary colonizers of stone surfaces of historic buildings in Java, contributing significantly to surface detoriation. Examples such as these illustrate the broadening technical areas to which nitrogen fixation research is being applied.

(d) OTHER SOURCES OF TECHNICAL INFORMATION

The increase in nitrogen fixation research has generated a need for a compilation of the methods of the subject. This need was seen during the International Biological Programme of 1966–74. A valuable, widely used handbook about methods for use with the legume root-nodule bacteria was produced (Vincent, 1970). Parkinson *et al.* (1971) included nitrogen-fixing bacteria in their handbook about the ecology of soil microorganisms and Allen *et al.* (1973) produced an IBP catalogue of *Rhizobium* collections. In the present work, our intention is that a similar function be fulfilled on a broader basis, with sufficient background information to make the methods easily comprehensible in the context of current knowledge, and with a good bibliography from which other material may be obtained. There have been many recent reviews and books about the current state of knowledge of nitrogen fixation (e.g. Postgate, 1971; Lie and Mulder, 1971; Quispel, 1974; Burns and Hardy, 1975; Hardy 1976–78; Newton and Nyman, 1976; Stewart, 1975; Nutman, 1975; Dalton and Mortenson, 1972; Stewart, 1973; Burris, 1974; Eady and Postgate, 1974; Dilworth, 1974; Evans and Barber, 1977). Only a few of these deal in depth with any aspects of methodology. Older works which did so, such as those by Fred *et al.* (1932) and Wilson (1940), are no longer readily available and many of the methods described therein are now superseded. In contemporary research papers, details of methods are often so abbreviated that it would be difficult or impossible for a novice to repeat the experiments which are described.

Some of the methods described in this book will be available in or adapted from other methods series (e.g. Glick, 1954–64; Paech and Tracey, 1959–64; Colowick and Kaplan, 1955–80; Norris and Ribbons 1969–73; Norris, 1976; Booth 1971). However, most will be scattered through the literature and some may not have been described in this way before. We believe that their value will be evident when they are all gathered together in a single source.

Section II

LABORATORY METHODS

Methods for Evaluating Biological Nitrogen Fixation
Edited by F. J. Bergersen

H. Dalton
Department of Biological Sciences,
University of Warwick, Coventry, England.

1

The Cultivation of Diazotrophic Microorganisms

(a) SURVEY OF TAXONOMY, CLASSIFICATION, AND IDENTIFICATION OF DIAZOTROPHS

The ability to fix dinitrogen is limited to prokaryotic organisms, i.e. bacteria and blue-green bacteria (Cyanophyceae; Cyanobacteria). Unfortunately, apart from the Azotobacteraceae and the genus *Rhizobium* there is no family or genus that is exclusively nitrogen fixing although many families have representatives that can do so. Table 1 lists those organisms that are known to fix nitrogen. Although it includes the photosynthetic bacteria, cyanobacteria are dealt with in detail in Chapter III.5. The identification of the nitrogen-fixing character has been greatly aided by the acetylene reduction test (see Chapter II.3) and has led to a re-evaluation of the ability of several organisms previously thought to possess the nitrogen-fixing character when assayed by less sensitive means. The ease with which this test can be performed has also meant that the occasional new isolate can be found that possesses nitrogenase activity. All organisms so far isolated as nitrogen fixers produce ethylene from acetylene and, provided that the negative control in the presence of NH_4^+ is performed, is a good indicator of the nitrogen-fixing ability. Confirmation of the test should be made using $^{15}N_2$.

Diazotrophs (nitrogen fixers) can be simply classified by dividing them into symbiotic and free-living forms. Apart from some strains of *Rhizobium* the symbiotic forms cannot yet be cultivated outside the host under nitrogen-fixing conditions and discussion of their characteristics will be found in Chapter II.5.

In general, the distribution and physiology of diazotrophs is strongly influenced by their reaction towards molecular oxygen and as such serve as a useful basis for their classification and identification.

Table 1. Nitrogen-fixing Organisms

Order	Family	Genus	Species
Eubacteriales	Azotobacteraceae	*Azotobacter*	*beijerinckii*
			chroococcum
			paspali
			vinelandii
		Azomonas	*insignis*
			macrocytogenes
			agilis
		Beijerinckia	*indica*
			fluminensis
			derxii
		Derxia	*gummosa*
	Rhizobiaceae*	*Rhizobium*	species (cowpea)
			japonicum
			leguminosarum

Table 1, (continued)

Order	Family	Genus	Species
	Bacillaceae	*Bacillus*	*macerans*
			polymyxa
		Clostridium	*butyricum*
			pasteurianum
			saccharobutyricum
			acetobutyricum
			beijerinckia
			tyrobutyricum
			acetobutylicum
			felsineum
			kluyverii
			lactoacetophilum
			madisonii
			pectinovorum
			tetanomorphum
			butylicum
	Enterobacteriaceae	*Klebsiella*	*pneumoniae*
			aerogenes
		Enterobacter	*aerogenes*
			cloacae
			agglomerans
		Erwinia	*herbicola*
		Citrobacter	*freundii*
			intermedius
		Escherichia	*coli*
			intermedia
Actinomycetales	Mycobacteriaceae	*Mycobacterium*	*flavum*
			roseo-album
			azotabsorptum
	Corynebacteriaceae	*Xanthobacter*	*autotrophicus*
Hyphomicrobiales	Hyphomicrobiaceae	*Rhodomicrobium*	*vannielii*
Pseudomonadales	Athiorhodaceae	*Rhodopseudomonas*	*palustris*
			capsulata
			gelatinosa
			spheroides
			'x'
		Rhodospirillum	*rubrum*
	Thiorhodaceae	*Chromatium*	*vinosum*
			minutissimum
			sp.
	Chlorobacteriaceae	*Chlorobium*	*thiosulfatophilum*
	Ectothiorhodaceae	*Ectothiorhodospira*	*shaposhnikovii*
	Methanomonadaceae	*Methylosinus*	*trichosporium*
			sporium
		Methylococcus	*capsulatus*
	Thiobacteriaceae	*Thiobacillus*	*ferro-oxidans*

Table 1, (continued)

Order	Family	Genus	Species
Pseudomonadales	Spirillaceae	*Azospirillum*	*lipoferum*
			brasilense
		Aquaspirillum	*peregrinum*
			fasciculus
		Desulfovibrio	*desulfuricans*
			vulgaris
			gigas
		Desulfotomaculum	*orientis*
			ruminis
Nostocales	Nostocaceae	*Anabaena*	
		Anabaenopsis	
		Aulosira	
		Cylindrospermum	
		Nostoc	
	Rivulariaceae	*Calothrix*	
	Scytonemataceae	*Scytonema*	
		Tolypothrix	
	Oscillatoriaceae	*Trichodesmium*	
		Lyngbya	
		Phormidium	
		Plectonema	
Stigonematales	Stigonemataceae	*Fischerella*	
		Hapalosiphon	
		Mastigocladus	
		Stigonema	
		Westiellopsis	
Chroococcales		*Gloeocapsa*	

*The species of *Rhizobium* listed have been reported to fix N_2 in culture. Other species (*R. trifolii, R. meliloti, R. lupini,* and *R. phaseoli*) fix N_2 in symbiosis with appropriate host legumes.

(i) The aerobes

Azotobacteraceae

The Azotobacteraceae are Gram-negative heterotrophic bacteria that have been the subject of taxonomic wrangles for a few years (see Dalton, 1974). Bergey's Manual (8th Edition) (1974) lists four genera based on morphological characteristics and the ability to utilize certain substrates. They are listed into two groups: Group I comprise *Azotobacter* and *Azomonas* and Group II *Beijerinckia* and *Derxia*. Group I organisms are large ovoid cells that grow rapidly (generation times of about 3 h), form some extracellular slime, and are catalase positive. They can be differentiated at the genus level according to their G + C content and the presence of cysts. Table 2 lists some of the salient differentiating characters of Group I organisms. Group II organisms

Table 2. Identifying features of Group I of the Azotobacteraceae

Feature	*Azotobacter*				*Azomonas*		
	chroococcum	*vinelandii*	*paspali*	*beijerinckii*	*agilis*	*insignis*	*macrocytogenes*
Cysts	+	+	+	+	−	−	−
Mole-% G + C	65–66	66	63–65	66	53–54	57–58	58–59
Growth on:							
Starch	+	−	−	−	−	−	−
Mannitol	+	+	−	−	−	−	−
Rhammose	−	+	−	−	−	−	−
Motility	+	+	+	−	+	+	+
Extracellular slime	+	+	+	+	+	−	+
habitat (limited to fresh water)					+	+	

are small rods that grow slowly (generation times in excess of 3 h), form copious amounts of slime, and possess large internal lipid bodies. The organisms are restricted mostly to tropical environments, particularly acid soils, but some have been isolated in temperate areas. The main distinguishing features between the two genera comprising the group, *Beijerinckia* and *Derxia,* is that the former are catalase positive and possess bipolar lipid bodies whereas the latter lack catalase and have lipid bodies dispersed throughout the cytoplasm.

Little biochemical or physiological information of Group II organisms is available but some of their more distinguishing features are listed in Table 3.

Table 3. Identifying features of Group II of the Azotobacteraceae

Feature	*Beijerinckia*				*Derxia*
	mobilis	*indica*	*fluminensis*	*derxii*	*gummosa*
Catalase	+	+	+	+	−
Mole-% G + C	n.d.	54.7	56.2	59.1	70.4
Growth on:					
Benzoate	+	−	−	−	
Arabinose	+	+	+	−	
Inulin	15% of strains	65% of strains +	−	−	
Nitrate	+	50% of strains +	−	−	+

Rhizobiaceae

This family includes both *Agrobacterium* and *Rhizobium,* but only members of the latter genus have been shown to fix N_2. The organisms normally only fix N_2 when they have entered the host plant (in most cases this is a legume, but there has been one report of *Rhizobium* forming active nitrogen-fixing nodules on the non-legume *Parasponia rugosa,* formerly indentified as *Trema* sp.; Trinick, 1973; Akkermans *et al.*, 1978a), although some strains will fix nitrogen (by the acetylene reduction test) in pure culture (see Gibson *et al.*, 1977) and can therefore be considered as free-living nitrogen fixers also.

Classification of *Rhizobium* species has been based largely on inoculation tests (see Chapters III.1 and III.2) based on host plant specificity; however, only a small percentage of legumes have been examined and an even smaller percentage have been investigated culturally. One convenient preliminary test in the classification of rhizobia is based on their rate of growth on yeast extract medium. The so called 'fast growers' give abundant growth after 3–5 days at 30°C and include *R. phaseoli, R. trifolii, R. leguminosarum* and *R. meliloti,* although the last organism will grow at 37°C unlike other rhizobia. Their G + C content is between 59 and 63%. The 'slow growers', i.e. those which produce little growth after 5 days but moderate growth after 10 days, include *R. lupini, R. japonicum,* the *Lotus* rhizobia and most cowpea rhizobia. The G + C content of the slow growers is between 61.6 and 65.5%. For more detailed aspects of the identification of *Rhizobium* spp., see Chapters II.9 and III.1.

Mycobacteriaceae

Four species of Gram-negative *Mycobacterium* have been reported as nitrogen fixers, namely *M. flavum* 301, *M. roseo-album, M. invisible* and *M. azotabsorptum,* all of which were isolated from Russian soils. They fix N_2 only in the presence of ethanol or organic acids, no fixation being reported when carbohydrates were used as the carbon source. The Mycobacteria tend to be more acid resistant than *Azotobacter* and growth under N_2-fixing conditions is stimulated by factors present in yeast extract. The taxonomic position of the diazotrophic Mycobacteria is still uncertain. No Mycobacteria appear as N_2 fixers in Bergey's Manual (8th Edition) (1974) and, according to Biggins and Postgate (1971a), *M. flavum* 301 does not resemble Jensen's description of *M. flavum* and appears, on the basis of its G + C content, to be either *Nocardia* or *Mycobacterium.* The absence of the iron-chelating growth factor, mycobactin, when grown in an iron-deficient growth medium would tend to exclude the organism from the latter genus and place it among the Nocardias.

Only *M. flavum* 301 has been studied in any detail with respect to N_2 fixation. Nitrogenase can be extracted from the organism in a particulate form, as in *Azotobacter,* but, unlike the *Azotobacter* preparation, is oxygen sensitive. This apparent lack of conformational protection (Dalton and Postgate, 1969b) from oxygen by the particles coupled with an inability to augment respiration when faced with an oxygen stress may account for the extreme micro-aerophilic nature of the organism when grown under nitrogen-fixing conditions.

Xanthobacter autotrophicus

This organism, originally classified as *Corynebacterium autotrophicum,* is a facultative autotroph that fixes N_2 either under autotrophic (H_2 as energy source and CO_2) or heterotrophic (sucrose) conditions (Gototov and Schlegel, 1974).

A number of strains have been isolated (Baumgarten *et al.,* 1974) and give either a negative (strain G2 29) or a positive Gram reaction (14g, 7c), although the cell walls of the latter more closely resemble Gram-negative bacteria. *X. autrophicus* is a non-motile, encapsulated, rod-shaped organism that tends to become coccoid with age. The colonies are yellow and grow on only a few carbohydrates. Good growth is observed on a number of organic acids with doubling times of about 3 h. Like other nitrogen-fixing aerobes, *X. autotrophicus* fixes N_2 only under micro-aerophilic conditions (Berndt *et al.*, 1976) and is readily isolated on mineral media under a 10% air, 10% CO_2 and 80% H_2 gas mixture.

Spirillum *spp.*

There has been a certain amount of debate concerning the nomenclatural status of spirilla and even now the situation has not been completely resolved (see Krieg, 1976). The redefined description of the genus *Spirillum* (Hylemon *et al.*, 1973) only includes *S. volutans,* a non-nitrogen-fixing species. Other members of the genus have been ascribed either to the genus *Aquaspirillum* or *Oceanospirillum,* depending on their habitat. Prior to this a number of nitrogen-fixing spirilla were reportedly isolated by Russian scientists but none were tested for their ability to assimilate isotopic dinitrogen and confirmation of the isolates has not been forthcoming from Western scientists. Beijerinck (1925) first described *Spirillum lipoferum* as being capable of N_2 fixation but did not obtain the organism in pure culture. Becking (1963) re-isolated the organism in pure culture and confirmed N_2 fixation isotopically. Subsequently, Dobereiner and Day (1975) observed *S. lipoferum* in association with *Digitaria decumbens* and *Zea mays* (Von Bulow and Dobereiner, 1975) root systems where it was apparently providing the plants with a fixed nitrogen source. In C_4 plants, malate can be excreted by the root system to provide a good source of carbon for the organism; indeed, malate appears to be the best substrate for growth of the organism in pure culture (Okon *et al.*, 1976a). Growth and N_2 fixation of pure cultures of *S. lipoferum* are best supported by organic acids such as malate, succinate, and lactate. *S. lipoferum* is capable of augmented respiration (Dobereiner, 1977) but does not appear to have the respiratory capability of *Azotobacter* (Dalton and Postgate, 1969a; Drozd and Postgate, 1970b).

All strains of *S. lipoferum* will use nitrate as a nitrogen source in preference to N_2, and a tentative division of over 50 strains has been made according to whether or not gas was produced from ammonium nitrate after growth for 2 days (Dobereiner, 1977). Another scheme has also been proposed, based on growth and acetylene reduction on glucose and time taken for colonies to form a pink pigment (Okon *et al.*, 1976b). Group I strains did not utilize glucose and their colonies on nutrient agar were pigmented after incubation for 1 week; Group II strains would grow on glucose but were pigmented only after incubation for 5 weeks. Unfortunately, it was not possible to relate group characteristics with the plant species from which the organisms were isolated.

As a result of DNA homology studies, Krieg (1977) and Tarrand *et al.*, (1978) have proposed the generic name *Azospirillium* to embrace the diazotrophic strains of *Spirillim lipoferum.* These are then defined as Gram-negative rods that are motile by a single polar flagellum and tend to be helically curved. They are oxidase positive and do not form cysts. The genus can be divided into two species, one of which does not require biotin and does not produce acid from glucose or ribose. This is *A. brasilense* and would be equivalent to the Group I strains. The other species, *A. lipoferum,* had a biotin requirement and formed acid from glucose and ribose being formally equivalent to the Group II strains. Furthermore, two species of Aquaspirillum, *A. peregrinum* and *A. fasciculus,* appear to reduce acetylene when cultured in nitrogen-free semi-solid medium (Von Bulow and Dobereiner, 1975), but only in the presence of a small amount of yeast extract. The nitrogenase activity in these species has been confirmed by Strength *et al.*, (1976). These organisms are covered in more detail in Chapter III.3.

Thiobacillus ferro-oxidans

The thiobacilli are Gram-negative, motile, rod-shaped organisms that grow by oxidizing reduced sulphur compounds. All but one (*Th. perometabolis*) are capable of strictly autotrophic growth, although seven out of the eleven species of the genus do have strains that are facultatively heterotropic. *Thiobacillus ferro-oxidans* is the only species that can derive energy from the oxidation of iron (II). When plated on to a minimal medium, a brown precipitate of iron (III) sulphate is formed in and around the colonies, and this can be used as a diagnostic test for the organism (Mackintosh, 1978). The organism has an optimum pH growth range between 1.8 and 5.8. Certain strains of this organism have only recently been shown to be nitrogen-fixers by the $^{15}N_2$ technique (Mackintosh, 1978), although earlier work had established nitrogenase activity in whole cells by the acetylene reduction technique.

Methane-oxidizing bacteria

A methane-oxidizing organism was first described in 1906 by Sohngen and named *Bacillus methanicus.* Up to 1966 only three species had been isolated and described, Sohngen's isolate being renamed as *Pseudomonas methanica.* One isolate, named *Pseudomonas methanitrificans* by Davis *et al.* (1964), was reported as fixing N_2 and this was later confirmed using $^{15}N_2$ (Coty, 1967). Whittenbury *et al.* (1970) devised enrichment and isolation techniques which led to the characterization of over 100 strains of pure cultures of methane-oxidizing bacteria. The organisms are Gram negative and only use either methane or methanol as carbon source. They were classified into five groups on the basis of morphology, fine structure (membrane type), and type of resting stage formed (exospores or cyst). A more recent classification scheme has been devised (Whittenbury and Dalton, 1979).

The strain *Ps. methanitrificans* from Coty was found to be contaminated with a number of other bacteria and '*Methylosinus trichosporium*', a spore-forming motile organism was isolated from his culture which actively reduced acetylene to ethylene. De Bont and Mulder (1974) reported that an organism resembling *Methylosinus sporium* (based on the classification scheme of Whittenbury *et al.*, 1970) would fix dinitrogen when assayed by the $^{15}N_2$ technique but would not reduce acetylene to ethylene. This observation was explained later by Dalton and Whittenbury (1976), who found that acetylene was a specific and potent inhibitor of methane oxidation in *Methylococcus capsulatus* (strain Bath) and that acetylene reduction assays could be readily performed if an oxidizable substrate other than methane was provided (e.g. methanol, formaldehyde, or formate).

Of the methane oxidizers so far isolated, three 'species' have been found to be nitrogen fixers and some of their diagnostic features are listed in Table 4.

Table 4. The diazotrophic methane oxidizers

Species	Morphology	Growth temperature (°C)	Autotrophic CO_2 fixation	Spore/cyst
Methylococcus capsulatus (strain Bath)	Coccus	45	+	'Immature'
Methylosinus trichosporium (4 strains)	Pear-shaped	30	–	Encapsulated spore
Methylosinus sporium (2 strains)	Kidney-shaped	30	–	No capsule surrounding spore

(ii) Facultative anaerobes

There are only two families that contain diazotrophic facultative anaerobes: the Bacillaceae, of which only the genus *Bacillus* is facultative and the Enterobacteriaceae. In both cases the organisms are capable of heterotrophic growth under aerobic or anaerobic conditions, but dinitrogen is fixed only under anaerobic or extremely micro-aerophilic conditions.

The genus Bacillus

Only two species, *B. polymyxa* and *B. macerans,* are authentic nitrogen fixers. They are rod-shaped organisms that form spores. The Gram reaction is variable and cannot be usefully used as a diagnostic key. Both species are motile and catalase positive. One nitrogen-fixing strain of *B. polymyxa* (strain Hino) was isolated from Japanese soil (Hino, 1955). It did not ferment lactose, glycerol, or arabinose, which is typically utilized by the species; in all other respects, however, it was identical with that species. *B. macerans* differs from *B. polymyxa* in that the former does not produce acetylmethylcarbinol in the Voges Proskauer test. Morphologically, *B. polymyxa* is the most easily distinguished species amongst the bacilli owing to its very characteristic spores, which have heavily ribbed surfaces and are star-shaped in cross-section. The strains have been successfully distinguished from one another by the antigenic nature of the spores and susceptibility of the organism to bacteriophage (Davies, 1951; Francis and Rippon, 1949).

DNA base composition studies of *Bacillus* species reveal that strains of *B. polymyxa* other than strain Hino formed a narrow cluster within the range 43.2–45.6 G + C; strain Hino was 51.4%, which was close to that of the *B. macerans* strains, which were between 49 and 51%.

Although 12 of the 13 strains of *B. macerans* investigated would fix $^{15}N_2$, no successful preparation of a nitrogenase extract has been forthcoming from these organisms. This contrasts with the successful extracts prepared from *B. polymyxa* (Fisher and Wilson, 1970), in which pyruvate would serve as a source of energy and reductant for nitrogenase through the phosphoroclastic system.

Line and Loutit (1971) tentatively classified one of their isolated from soil as being *B. circulans*. Several features of the organism suggest that it may not belong to this species (*viz.* ability to fix N_2, lack of milk coagulation, and variability of spore diameter), but until *B. circulans* has been adequately defined the significance of the anomalous features cannot be known. In several respects the organism resembles *B. macerans,* but, *B. circulans* does not produce gas from carbohydrates. However, this cannot be used as a good differentiating character as several strains of *B. macerans* also do not produce gas. Clearly, the classification of these organisms is still not resolved.

Enterobacteriaceae

A number of Enterobacteriaceae have now been found to be natural nitrogen fixers and have been isolated from a wide variety of environmental situations including various soils, water, plants, urine, the intestines of humans, pigs, guinea pigs, termites, and paper mill effluents.

The recent observations that representatives of five genera of this family can be readily isolated from a single source (paper mill effluent; Neilson and

Sparell, 1976) as nitrogen-fixing organisms and that Bergey's Manual (8th Edition) (1974) has no mention of the nitrogen-fixing character in a single member of the Enterobacteriaceae apart from *Klebsiella pneumoniae* lead one to the following conclusions: either the ability to fix N_2 has been missed in all taxonomic studies of the enteric bacteria, or the nitrogen-fixing character is labile and readily lost or acquired by these organisms.

The acetylene reduction technique has been a recent innovation compared with the establishment of authentic Enterobacteriaceae and may not have been used to screen these organisms systematically for nitrogenase. This may account for lack of appearance of this trait in Bergey's enterics. However, the fact that the genes that code for nitrogenase synthesis can be readily transferred from *Klebsiella pneumoniae* to other organisms, either by conjugation (Dixon and Postgate, 1976; Cannon *et al.*, 1976; Cannon and Postgate, 1976) or transduction (Streicher *et al.*, 1971), may support the second conclusion. Although a systematic investigation of whether the *nif* genes present in these enterics are located within the host DNA or are present on a plasmid has not been made, the observation that not all isolates of a species will fix N_2 suggests that *nif* may be plasmid-borne.

The environments from which diazotrophic enterics have been isolated (see above) have been conducive for their selection because the fixed nitrogen content, in general, has been meagre. This is particularly true of the intestines of those macro-organisms whose diet consists largely of carbohydrates, and of paper mill waste, which is also mainly cellulosic. It is difficult to estimate exactly how much organisms such as *Enterobacter cloacae* contribute to the nitrogen economy of the soil and intestines from which they have been isolated. The very low efficiencies of N_2 fixation by these organisms coupled, in some cases, with their sparse occurrence indicates that their contribution may be low, although some reports (e.g. Raju *et al.*, 1972; Bergersen and Hipsley, 1970) suggest that they may be important in this respect.

A number of nitrogen-free media have been used for enrichment and isolation of diazotrophic Enterobacteriaceae. Neilson and Sparell (1976) used selective media containing galactitol (see Section b) for their isolations, media containing glycerol only gave low selectivity and those based on ribitol only selected for *Klebsiella pneumoniae* and *Enterobacter aerogenes*. French *et al.* (1976) used Hino and Wilson's medium (see Section b) and Bergersen and Hipsley (1970) used Burk's medium (see Section b). In some cases growth factors were required and could be met by addition of small amounts of yeast extract. Colonies were usually observed after several days anaerobic incubation at 37°C, but incubation for 1 week gave better results. Several subcultures were usually necessary to produce pure cultures.

Tribe: klebsielleae. N_2-fixing members of this tribe are found in the genera *Klebsiella* and *Enterobacter*. Confusion over the classification of *Klebsiella* in previous years (*e.g. K. pneumoniae* strains being indistinguishable from *Aerobacter aerogenes* strains) has led to the abolition of the genus *Aerobacter*. Motile aerobacters are now *Enterobacter* and non-motile aerobacters are classified as *Klebsiella*. There are probably only two authentic *Klebsiella* species that fix N_2, namely *K. pneumoniae* and *K. aerogenes*. They have been isolated from many natural ecosystems as Gram-negative, rod-shaped organisms capable of fermenting many cargohydrates with the production of CO_2 and H_2 (Bergersen and Hipsley, 1970; Line and Loutit, 1971; Evans *et al.*, 1972; French *et al.*, 1976; Neilson and Sparell, 1976). *K. aerogenes* and *K. pneumoniae* can be distinguished from one another and from diazotrophic *Enterobacter cloacae* and *E. aerogenes* biochemically (see Table 6), although Bergey's Manual (8th Edition) (1974) classifies *K. aerogenes* with its description of *K. pneumoniae*. The *Enterobacter* isolates have been found in association with plants (Line and Loutit, 1971; Raju *et al.*, 1972; Nelson *et al.*, 1976), where they presumably contribute to their nitrogen economy.

It was generally considered that members of this tribe only fix N_2 anaerobically; however, observations in several laboratories over the past few years has suggested that some of the organisms can grow and fix N_2 at low dissolved oxygen tensions (Klucas, 1972; Hill, 1975; Neilson and Sparrel, 1976). Hill (1976) showed that *K. pneumoniae* is capable of limited respiratory and conformational protection of its nitrogenase from oxygen damage and that this accounted for the ability of the organism to fix N_2 between 2 and 10 mmHg but not at 15 mmHg.

Citrobacter freundii. This facultative anaerobe is a Gram-negative rod that was isolated from the hind gut of Australian termites by French *et al.* (1976) and from paper mill process waters by Neilson and Sparrel (1976). Like other facultative anaerobes it only fixes N_2 anaerobically and dinitrogen fixation has been confirmed with $^{15}N_2$. The organism was isolated by dissection of the intestinal tract of termites and streaking the contents of the hind gut into Hino and Wilson's nitrogen-free agar. Such a procedure generally produced only a single colony type after incubation at 28°C. Three isolates were characterized in detail by French *et al.* (1976). On nutrient agar, irregular creamy-grey colonies were formed. The organisms were motile, reduced nitrate, were catalase and methyl red positive, and fermented sucrose, xylose, arabinose, and galactose with production of gas. They were Kovacs oxidase negative, gluconate negative, phenylalanine deaminase negative and arginine dihydrolase positive.

Neilson and Sparell (1976) investigated eleven isolates from effluents using a selective medium for Enterobacteriaceae containing galacticol as the carbon source. About 70% of the strains tested would reduce acetylene to ethylene. The biochemical characteristics of the strains were similar to the isolate of French *et al.* (1976) except that the effluent organism was H_2S negative.

Table 5. Biochemical characteristics of nitrogen-fixing Enterobacteriaceae from paper mill process waters (after Neilson and Sparell, 1976): percentage of positive occurrence of characters

Characteristic	*Klebsiella pneumoniae*		*Enterobacter cloacae*	*Enterobacter aerogenes*	*Erwinia herbicola*	*Citrobacter freundii*	*Citrobacter intermedius*	*Escherichia coli*
Biotype	A	B						
Number of isolates	48	19	27	1	14	8	1	1
Nitrogenase	50	80	60	+	40	70	+	25
Voges Proskauer	100	100	100	+	40	0	0	0
Decarboxylases:								
Arginine	0	0	100	−	0	80	−	0
Lysine	90	0	0	+	0	0	−	25
Ornithine	0	0	40	+	0	20	+	50

Neilson and Sparell (1976) also isolated *C. intermedius* that fixed N_2; it could be differentiated from *C. freundii* according to the presence of amino acid decarboxylases (see Table 5) and the negative indole test and positive H_2S production test of the latter.

Table 6. Characteristics of diazotrophic Klebsielleae

Characteristic	*Klebsiella pneumoniae*	*Klebsiella aerogenes*	*Enterobacter cloacae*	*Enterobacter aerogenes*
Voges Proskauer	+	−	+	+
Methyl red	−	+	−	−
Malonate	+	−	±	+
Motility	−	−	+	+
Ornithine decarboxylase	−	−	+	+
Arginine dihydrolase	−	−	+	−
Lysine decarboxylase	+	+	−	+
Esculin hydrolysis			−	+

Escherichia. Apart from a strain of *Escherichia coli* into which the *nif* genes have been transferred, the only reported species of this genus to fix N_2 up to a few years ago was *E. intermedia,* which was isolated from the rhizosphere of New Zealand grasses (Line and Loutit, 1971). A naturally occurring strain of *E. coli* was recently isolated from paper mill effluents (Neilson and Sparell, 1976) but there is some doubt as to its validity (Postgate, personal communication). In the most recent edition of Bergey's Manual (8th Edition) (1974), *E. intermedia* no longer exists and is now classified as *Citrobacter intermedius.* It has all the characteristics of the *Escherichia* genus but will utilize citrate as sole carbon source.

Erwinia. A number of nitrogen-fixing strains of *Erwinia herbicola* were isolated from paper mill effluents. The organism was biochemically distinguishable from other Enterobacteriaceae (see Table 5) although it fitted the description of *Enterobacter agglomerans* (Ewing and Fife, 1972), which had also been isolated as a diazotroph from the gut of termites (Potrikus and Breznak, 1977) and decaying fir trees (Aho *et al.*, 1974). *Enterobacter agglom-*

erans is now classified as *Erwinia herbicola* in Bergey's Manual (8th Edition) (1974).

(iii) Anaerobes

Only three genera of obligately anaerobic non-photosynthetic bacteria have diazotrophic representatives: *Clostridium, Desulfovibrio,* and *Desulfotomaculum.*

The genus clostridium

In Bergey's Manual (7th Edition) (1957), ten of the ninety-three species of the genus were listed as diazotrophs based on the $^{15}N_2$ incorporation data of Rosenblum and Wilson (1949). In the 8th Edition (1974), however, only sixty-one species of *Clostridium* are listed, of which only three are given the status of diszotrophs, namely *C. pasteurianum, C. butyricum,* and *C. acetobutylicum.* The 8th Edition still retains *C. kluyveri* and *C. felsineum* but not as diazotrophs. Five of Rosenblum and Wilson's diazotrophs have disappeared completely, *viz. C. butylicum, C. beijerinckii, C. madisonii, C. lactoacetophilum,* and *C. pectinovorum.* The reasons for this are obscure but it is probably because the characters used to distinguish them from other members of the genus in the 7th Edition were not sufficient or too vague to warrant their inclusion in the 8th Edition. The present author is uncertain if all of the original diazotrophic species are still available and so only those diazotrophs as listed in the 8th Edition of Bergey's Manual will be discussed.

The three species are all Gram-positive, spore-forming rods which are capable of fermenting a wide variety of carbohydrates to produce particularly acetic and butyric acids. They have been readily isolated from soil and are particularly abundant in the rhizosphere of plants, where root exudates provide a good supply of fermentable carbon substrates. *C. butyricum* and *C. pasteurianum* are readily distinguished from *C. acetobutylicum* by the ability of the latter to hydrolyse gelatin and produce butane-2,3-diol, no liquefaction or butane-2,3-diol production occurring with the other two.

Table 7 lists some of the more useful characteristics used to distinguish the three diazotrophic species. Optimum fixation of nitrogen generally occurs at acidic pH values in the complete absence of oxygen.

Genera desulfovibrio *and* desulfotomaculum

These organisms are anaerobic sulphate-reducing bacteria which grow in media containing lactate and sulphate, the latter undergoing dissimilatory reduction to sulphide, giving cultures their characteristic smell. The genera are distinguished from one another by the presence of either terminal or subterminal spores in *Desulfotomaculum* species. *Desulfovibrio* species do not form spores.

Table 7. Distinguishing characteristics of diazotrophic clostridia

Characteristics	*C. butyricum*	*C. pasterianum*	*C. acetobutylicum*
Gelatin hydrolysis	−	−	+
Butane-2,3-diol produced	−	−	+
Fermentation of:			
Lactose	+	−	+
Ribose	+	−	−
Cellobiose	+	−	−
Xylose	+	−	+
Melezitose	−	+	−

The genus *Desulfovibrio* consists of five species of which three, *D. desulfuricans, D. vulgaris,* and *D. gigas,* have been shown to be diazotrophic to varying extents (Reiderer–Henderson and Wilson, 1970; Postgate, 1970). Straightforward acetylene reduction tests often fail to show measurable nitrogenase activity for these cultures, whereas the use of Pankhurst tubes does (Campbell and Evans, 1969). Table 8 lists some of the important features used to distinguish the three species. *D. desulfuricans* and *D. vulgaris* have been isolated from polluted waters or waterlogged soils rich in organic substrates, whereas *D. gigas* was isolated near Marseilles, France, from salt water but saline media are not used for its cultivation.

Two of the three species of the spore-forming *Desulfotomaculum* genus reduce acetylene, which was completely inhibited by ammonium chloride (Postgate, 1970). Some $^{15}N_2$ incorporation has been observed but the values were not very high. *Dm. ruminis* was originally isolated from sheep rumen and can also be found in soil. It can be distinguished from the other diazotrophic member of the genus, *Dm. orientis,* on morphological grounds since the latter are fat, curved rods with round central spores showing a 'tumbling and twisting' motility, whereas the former are straight rods with oval spores which show only slight tumbling motility. Furthermore, growth of *Dm. ruminis* can be supported with pyruvate without sulphate whereas *Dm. orientis* requires sulphate for growth on pyruvate.

Desulfovibrio species can be readily distinguished from *Desulfotomaculum* species on the following characteristics: (a) spores; (b) G + C of *Desulfovibrio*

Table 8. Characteristics of *Desulfovibrio* species

Characteristics	*D. desulfuricans*	*D. vulgaris* strain Hildenborough	*D. gigas*
Dimensions (μm)	0.5–1 × 3–5	0.5–1 × 3–5	1.2–1.5 × 5–10
Polar flagella:			
Single	+	+	–
Lophotrichous	–	–	+
Growth in:			
Pyruvate minus sulphate	+	–	–
Malate plus sulphate	+	–	–
Hibitane resistance (mg/1)	10–25	2.5	2.5
E_h for growth at pH 7.2 (mV)	−100	−100	80

species are greater than 46.1% whereas those *Desulfotomaculum* species are less than 45.6%; (c) The presence of c_3 cytochrome and desulfoviridin in *Desulfovibrio;* and (d) polar flagellation in *Desulfovibrio* as opposed to peritrichous flagellation in *Desulfotomaculum.*

(b) MEDIA

(i) For *rhizobium* spp.

Undefined medium

Rhizobia are mesophilic aerobes that can grow on heterotrophic media containing either NO_3^- or NH_4^+ as a nitrogen source. When N_2 is a nitrogen source, the growth conditions and media constituents are more complex than when combined nitrogen sources are used, and will be dealt with separately.

There are a wide variety of strains that differ in their growth requirements; some require amino acids and/or vitamins as supplementary nutrients, in which case the reader will have to check with papers dealing with these strains for details of specific media recipes.

The most commonly used propagation medium is a yeast extract/mannitol broth which has the following composition:

Mannitol	10 g
Yeast extract	0.4 g
K_2HPO_4	0.5 g
$MgSO_4 \cdot 7H_2O$	0.2 g
NaCl	0.1 g
Distilled water	1 l

This medium is autoclaved at 121 °C for 15 min. For the preparation of a solid medium, 15 g l^{-1} of agar is added and if agar slopes are used for maintenance of stock cultures it is often advantageous to add solid calcium carbonate to the medium. This is evenly dispersed throughout the medium and serves as a neutralizing agent for any acid that may be produced on prolonged incubation.

The technique used in our laboratory for preservation of nitrogen-fixing bacteria is outlined in Section h.

Defined medium

Defined medium after Brown and Dilworth (1975):

Glucose	2.5 g
KH_2PO_4	0.36 g
K_2HPO_4	1.4 g

$MgSO_4 \cdot 7H_2O$	0.25 g
$CaCl_2 \cdot 2H_2O$	0.02 g
NaCl	0.2 g
$FeCl_3$	6.6 mg
EDTA	0.15 mg
$ZnSO_4 \cdot 7H_2O$	0.16 mg
$Na_2MoO_4 \cdot 7H_2O$	0.2 mg
H_3BO_3	0.25 mg
$MnSO_4 \cdot 4H_2O$	0.2 mg
$CuSO_4 \cdot 5H_2O$	0.02 mg
$CoCl_2 \cdot 6H_2O$	1.0 μg
Thiamine·HCl	1 mg
Ca Pantothenate	2 mg
Biotin	1 μg
Distilled water	1 l
pH adjusted to 7.0	

In this meduim the nitrogen source can be varied to include either KNO_3 (0.7 g l^{-1}), NH_4Cl (0.7 g l^{-1}) or L-glutamate (1.0 g l^{-1}). The phosphates are usually sterilized separately to avoid precipitation.

Defined media for growth under N_2-fixing conditions. Although a number of strains of *Rhizobium* have been cultivated in the free-living state and demonstrated to fix N_2, none of the media so far described for such cultivation has completely excluded the presence of a small amount of a fixed nitrogen source. Cultivation of the cowpea rhizobia (*Rhizobium* sp.) strains 32HI and CB756 and *Rhizobium japonicum* strains CB1809, 311b83, and 61A76 in the media described below have led to the successful demonstration of nitrogenase activity in whole cells by the acetylene reduction technique.

Liquid Medium
(Bergersen *et al.*, 1976)

Glycerol	43 mM
Inositol	5.6 mM
Na succinate	25 mM
Glutamine	1–2 mM
K_2HPO_4	30 mM
$CaCl_2 \cdot 2H_2O$	0.7 mM
$MgCl_2$	0.14 mM
$MnSO_4$	58 μM
H_3BO_3	82 μM
$ZnSO_4 \cdot 7H_2O$	3.5 μM
KI	6 μM
$CuSO_4 \cdot 5H_2O$	0.8 μM

$Na_2MoO_4{\cdot}2H_2O$	0.4 μM
$CoCl_2{\cdot}6H_2O$	0.4 μM
$FeSO_4{\cdot}7H_2O$	54 μM
Na_2EDTA	54 μM

The pH is adjusted to 6.25 before autoclaving. The two cowpea strains tested, *viz.* CB756 and 32HI, did not respond to addition of vitamins. A similar liquid medium in which the only carbon cource was malic acid was used by Tjepkema and Evans (1975) to cultivate 32HI under nitrogen-fixing conditions. In both media, however, successful demonstration of nitrogenase activity depended on the maintenance of a low dissolved oxygen tension.

Solid Medium CS7
(Pagen *et al.*, 1975)

Na succinate	25 mM
L-Arabinose	25 mM
Myo-inositol	5.6 mM
Glutamine	2 mM
$MgSO_4{\cdot}7H_2O$	0.14 mM
KCl	0.9 mM
$CaCl_2{\cdot}2H_2O$	0.7 mM
KH_2PO_4	2.2 mM
$MnSO_4{\cdot}4H_2O$	58 mM
H_3BO_3	82 mM
$ZnSO_4{\cdot}7H_2O$	3.5 mM
KI	6 mM
$CuSO_4{\cdot}5H_2O$	0.8 mM
$Na_2MoO_4{\cdot}2H_2O$	0.4 mM
$CoCl_2{\cdot}6H_2O$	0.4 mM
$FeSO_4{\cdot}7H_2O$	54 mM
Na_2EDTA	54 mM
Thiamine·HC1	15 mM
Nicotinic acid	41 mM
Pyridoxine·HCl	2.4 mM
Agar (Noble Difco)	1% w/v
pH	5.9

Eight strains of cowpea rhizobia and eight *R. japonicum* strains showed positive acetylene reduction when cultured on the CS7 medium. Arabinose, which is expensive, can be replaced with galactose in the basic medium.

(ii) For *azotobacter* spp.

For both continuous and batch cultures a modified Burk's medium (Dalton and Postgate, 1969a) is used, containing the following in distilled water.

Mannitol	10.0 g l^{-1}
K_2HPO_4	0.64 g l^{-1}
KH_2PO_4	0.16 g l^{-1}
NaCl	0.2 g l^{-1}
$MgSO_4{\cdot}7H_2O$	0.2 g l^{-1}
$CaCl_2$	0.1 g l^{-1}

plus the following trace elements:

$FeSO_4{\cdot}7H_2O$	2.5 mg l^{-1}
H_3BO_3	2.9 mg l^{-1}
$CoSO_4{\cdot}7H_2O$	1.2 mg l^{-1}
$CuSO_4{\cdot}5H_2O$	0.1 mg l^{-1}
$MnCl_2{\cdot}4H_2O$	0.09 mg l^{-1}
$Na_2MoO_4{\cdot}2H_2O$	2.5 mg l^{-1}
$ZnSO_4{\cdot}7H_2O$	2.1 mg l^{-1}
Nitrilotriacetic acid	100 mg l^{-1}

Nitrilotriacetic acid was added to a solution of the trace elememts and the pH adjusted to 7.0 with NaOH before adding to the bulk of the medium.

The final pH of the medium was 7.4 ± 0.2 and batches of the medium were autoclaved at 121 °C for 20 min. In autoclaves where there is no temperature probe for fluids, 20 l of medium are autoclaved at 121 °C for 45 min. After autoclaving a visible precipitate was visible, which disappeared after several days at room temperature. The precipitation could be avoided by autoclaving the phosphates separately and adding to the bulk of the medium when cool.

If a fixed nitrogen source was required, this could be added at a final concentration of 1.5 g l^{-1}, or 0.15 g l^{-1} if an ammonium-limited culture was required.

Solid media were prepared from the above 'mannitol-B_6' medium by the addition of 2% of agar.

(iii) For *azospirillum* spp.

The following medium in distilled water is used (Okon *et al.*, 1976b):

K_2HPO_4	6.0 g l^{-1}
KH_2PO_4	4.0 g l^{-1}
$MgSO_4{\cdot}7H_2O$	0.2 g l^{-1}
NaCl	0.1 g l^{-1}
$CaCl_2$	0.02 g l^{-1}
$FeCl_3$	0.01 g l^{-1}
$Na_2MoO_4{\cdot}2H_2O$	0.002 g l^{-1}

The pH was adjusted to 6.8 with NaOH; 0.5%w/v DL-malic acid, succinic acid, DL-lactic acid, or pyruvic acid were good carbon sources and gave faster growth rates(t_d = 20 h) and acetylene reduction rates than other sources tested. The acids, apart from the pyruvic acid, were prepared as 10-fold concentrated solutions and sterilized separately by autoclaving; pyruvic acid was sterilized by filtration. These were then added to the bulk of the medium when cool. Addition of a small amount of fixed nitrogen (0.005% N as KNO_3 or yeast extract) to the medium generally reduced the lag phase of the cultures. Organisms were grown at 30°C. When sodium salts of the acid substrates were used growth was accompanied by an increase in the pH of the culture. At pH in excess of 7.8 there is a sharp decrease in the acetylene reduction activity. This medium was devised for liquid cultures for laboratory production of cells. Media for isolation from soil and root samples are described in Chapter III.3

The medium could be made semi-solid by adding 0.05% agar. This restricted mixing of the culture and enabled micro-aerophilic conditions to be attained. Alternatively, the medium in 1-l Roux bottles could be gassed with a sterile N_2/O_2 mixture and sealed with a serum stopper.

For growth of larger quantities of cells a 1–2% inoculum of cells grown on medium containing 0.1% of NH_4Cl is sufficient for good growth.

(iv) For *desulfovibrio* spp.

'Berre S' strain modified 'C' in distilled water is used (Postgate, 1966):

K_2HPO_4	0.5 g l^{-1}
Na_2SO_4	3.8 g l^{-1}
$CaCl_2 \cdot 2H_2O$	0.2 g l^{-1}
$MgSO_4 \cdot 7H_2O$	0.7 g l^{-1}
Trisodium citrate	4.5 g l^{-1}
Sodium lactate	6.75 g l^{-1}
$FeSO_4 \cdot 7H_2O$	4 mg l^{-1}
$Na_2MoO_4 \cdot 2H_2O$	1 mg l^{-1}
H_3BO_3	1.16 mg l^{-1}
$CoSO_4 \cdot 7H_2O$	0.5 mg l^{-1}
$CuSO_4 \cdot 5H_2O$	0.04 mg l^{-1}
$MnCl_2 \cdot 4H_2O$	0.035 mg l^{-1}
$ZnSO_4 \cdot 7H_2O$	0.87 mg l^{-1}

The pH of the medium was adjusted to 7.8 with NaOH or HCl and autoclaved in 10-l batches at 121°C for 15 min.

The presence of citrate prevents precipitation during autoclaving, which could remove essential trace elements. The medium is sparged with N_2 to make it anaerobic and is maintained at 30°C during growth.

(v) For *clostridium* spp.

Spore stocks of *Cl. pasteurianum* can be readily revived on a potato medium which is made as follows. Solid calcium carbonate is added to cover the bottom of a test-tube (150 × 18 mm). Fresh potatoes that have been washed, peeled and finely diced are added to give a layer 3 cm deep. To this, 10 ml of 2% sucrose in tap water are added. The tubes are then plugged with non-absorbent cotton-wool and autoclaved for 20 min at 120°C. After cooling, the tubes are inoculated and sealed with a plug of absorbent cotton-wool to which 2 drops of 2 N K_2CO_3 followed by 2 drops of 4 M pyrogallol are added. The tubes are then lightly closed with a rubber stopper and incubated at 30 °C. Vigorous evolution of hydrogen occurs when the culture is growing and the tube may explode if the stopper is on too tightly.

Once a tube culture has grown (*ca.* 24 h) this inoculum may be transferred into a larger (500–1000 ml) culture medium which consists of the following in tap water:

Sucrose	20 g l^{-1}	
$CaCO_3$	3 g l^{-1}	
$MgCl_2 \cdot 6H_2O$	0.15 g l^{-1}	
NaCl	0.1 g l^{-1}	
Na_2MoO_4	0.0 g l^{-1}	
Na_2SO_4	0.07 g l^{-1}	
KH_2PO_4	0.5 g l^{-1}	Sterilized separately and added to the sterilized medium when cool
$K_2HPO_4 \cdot 3H_2O$	0.92 g l^{-1}	
Biotin	5 μg l^{-1}	
p-aminobenzoic acid	5 μg l^{-1}	

The medium is autoclaved and N_2 is bubbled through it to make it anaerobic. Before inoculation add 4 ml of sterile 5% $FeCl_3.6H_2O$ (in absolute ethanol). The temperature is maintained at 30°C. For larger quantities of cells the $CaCO_3$ can be decreased to 1 g l^{-1} and the phosphate concentration doubled.

It is usual with the large-scale cultivation methods (> 100 l) to maintain the pH at 6.0 by the addition of 10 N KOH. The Hino and Wilson medium prescribed below for *Klebsiella pneumoniae* has been used successfully for the growth of *Cl. pasteurianum*.

For chemostat cultures the $CaCO_3$ is omitted and the pH maintained at 6.2 either by increasing the phosphate concentration or by automatic pH control.

(vi) For *mycobacterium flavum*

The following medium (Biggins and Postgate, 1969) in distilled water is used:

Sodium lactate	4 g l^{-1}
K_2HPO_4	1.67 g l^{-1}

KH_2PO_4	0.87 g l^{-1}
$MgSO_4 \cdot 7H_2O$	0.29 g l^{-1}
$CaCl_2$	0.07 g l^{-1}
NaCl	0.48 g l^{-1}
$FeCl_3 \cdot 6H_2O$	0.01 g l^{-1}
$Na_2MoO_4 \cdot 2H_2O$	0.005 g l^{-1}
$ZnSO_4 \cdot 7H_2O$	0.0002 g l^{-1}
$MnSO_4 \cdot 4H_2O$	0.003 g l^{-1}
H_3BO_3	0.005 g l^{-1}
$CoSO_4 \cdot 7H_2O$	2 μg l^{-1}
Biotin	20 μg l^{-1}
Yeast extract	0.08 g l^{-1}
pH	6.8

Bothe and Yates (1976) successfully omitted biotin, Zn, Mn, H_3BO_3, and Co in large-scale cultivation.

(vii) For *klebsiella* spp.

The following medium (Hill, 1976) is used in distilled water:

Sucrose	20 g l^{-1}
Na_2HPO_4	10.4 g l^{-1}
KH_2PO_4	3.4 g l^{-1}
Iron (III) citrate	36 mg l^{-1}
$MgSO_4$	30 mg l^{-1}
$CaCl_2 \cdot 2H_2O$	26 mg l^{-1}
$MnSO_4$	0.3 mg l^{-1}
$Na_2MoO_4 \cdot 2H_2O$	7.6 mg l^{-1}

The phosphates are sterilized separately and added to the bulk of the medium when cool. The organisms are grown at 30 °C at pH 6.7 ± 0.1, which can be maintained by the buffering capacity of the medium.

For large-scale cultivation Hino and Wilson's medium is routinely used. This contains the following ingredients in distilled water:

Sucrose	20 g l^{-1}
K_2HPO_4	12.06 g l^{-1}
KH_2PO_4	3.4 g l^{-1}
$MgSO_4 \cdot 7H_2O$	0.5 g l^{-1}
NaCl	0.01 g l^{-1}
$FeSO_4 \cdot 7H_2O$	0.015 g l^{-1}
$Na_2MoO_4 \cdot 2H_2O$	0.005 g l^{-1}

Growth is at 30 °C and the medium is sparged with N_2.

(viii) For *bacillus polymyxa*

This organism, like *Klebsiella pneumoniae,* is a facultative anaerobe and can be grown in the same medium as used for that organism with added calcium carbonate (10 g), biotin (5 μg) and *p*-aminobenzoic acid (10 μg).

(ix) For *corynebacterium autotrophicum (xanthobacter autotrophicus)*

This mixotrophic hydrogen bacterium fixes N_2 under autotrophic (H_2/CO_2) and heterotrophic (sucrose) conditions. The medium (Berndt *et al.* 1976) for growth under heterotrophic conditions contains the following ingredients in distilled water:

Sucrose	5.0 g l^{-1}
$Na_2HPO_4 \cdot 12H_2O$	9.0 g l^{-1}
KH_2PO_4	1.5 g l^{-1}
NH_4Cl	1.0 g l^{-1}
$MgSO_4 \cdot 7H_2O$	0.2 g l^{-1}
$CaCl_2 \cdot 2H_2O$	0.014 g l^{-1}
Iron (III) citrate·$5H_2O$	0.0022 g l^{-1}
Trace element solution	0.25 ml

The trace element solution contains the following:

$ZnSO_4 \cdot 7H_2O$	100 mg l^{-1}
$MnCl_2 \cdot 4H_2O$	30 mg l^{-1}
H_3BO_3	300 mg l^{-1}
$CoCl_2 \cdot 6H_2O$	200 mg l^{-1}
$CuCl_2 \cdot 2H_2O$	10 mg l^{-1}
$NiCl_2 \cdot 6H_2O$	20 mg l^{-1}
$Na_2MoO_4 \cdot 2H_2O$	30 mg l^{-1}

The pH of the medium is adjusted to 7.1–7.2 if necessary. The ammonium chloride is included to initiate growth and permits the organism to grow at higher oxygen tensions (up to 0.15 atm O_2 in gas phase or 5 μg l^{-1} in the liquid phase) during the early stages of growth. When the ammonium chloride has been completely utilized nitrogenase is derepressed and the aeration is reduced by diluting the air inflow with more N_2 to give oxygen tensions of about 0.02 atm in the gas phase or 0.6 μg l^{-1} in the liquid phase.

(x) For *thiobacillus ferro-oxidans*

Low phosphate medium

This is used for the production of inoculum (Mackintosh, 1978):

$(NH_4)_2SO_4$	1.0 mM
KH_2PO_4	0.2 mM
$MgCl_2$	0.125
$CaCl_2$	1.0 mM
$MnCl_2$	} 0.5 μM
$ZnCl_2$	
$COCl_2$	
H_3BO_3	
Na_2MoO_4	
$CuCl_2$	
H_2SO_4	9.6 mM
$FeSO_4$	180 mM

The final pH of the medium is adjusted to 1.8. This medium considerably reduced the amount of basic iron sulphates formed during growth and hence prevented the loss of bacteria due to adsorption to the precipitate.

Medium for nitrogen fixation

This is the F_2 medium described by Mackintosh (1978):

$MgSO_4$	2.0 mM
KH_2PO_4	0.3 mM
KCl	0.7 mM
$CaCl_2$	0.06 mM
Na_2MoO_4	0.04 mM
$FeSO_4$	180 mM

The pH is adjusted to 2.2 and both media are sterilized by autoclaving, except for the iron (II) sulphate, which is filtered through a 0.2-μm membrane filter and added to the bulk of the medium when cool. For the preparation of solid media, 0.8% w/v agarose is added to the liquid medium.

To test for purity of the culture the organisms must be grown on solid media. Because the organisms grow slowly and only form microcolonies on agar, it is important to observe colony development microscopically and a system based on the slide culture technique (Postgate *et al.*, 1961) has been developed by Mackintosh (1978). After inoculation, the culture is placed inside an anaerobic jar under an air/N_2/CO_2 (1:94:5) mixture and incubated at 30°C for 36 h. Colonies of *Th. ferrooxidans* form a brown precipitate due to iron (III) sulphate and therefore distinguish them from other thiobacilli that may grow on such a medium. These colonies can be seen readily using a plate microscope after a further period of incubation (usually 7 days) and picked off for inoculation into liquid media. The process can be repeated to ensure purity of the culture.

For growth in liquid media, all-glass vessels must be used because of the low pH of the culture. The gas mixture given above is obtained by passing each component through flow meters and forcing the gas through sintered-glass spargers into the liquid. After growth for 48 h in low phosphate medium about 0.5 g of cells can be obtained from 20 l of culture (Mackintosh, 1978).

(xi) For *methyloccoccus capsulatus*

The methane-oxidizing bacteria are aerobic and only a few strains appear to fix N_2. The recipe used here is for *M. capsulatus* strain Bath, which grows optimally at 45 °C (Dalton and Whittenbury, 1976).

The mineral salts medium contain the following ingredients in distilled water:

Na_2HPO_4	0.33 g l^{-1}
KH_2PO_4	0.26 g l^{-1}
$MgSO_4 \cdot 7H_2O$	1.0 g l^{-1}
$CaCl_2$	0.2 g l^{-1}
FeEDTA	5.0 mg l^{-1}
$Na_2MoO_4 \cdot 2H_2O$	2.0 mg l^{-1}
$FeSO_4 \cdot 7H_2O$	500 μg l^{-1}
$ZnSO_4 \cdot 7H_2O$	400 μg l^{-1}
$MnCl_2 \cdot 4H_2O$	20 μg l^{-1}
$CoCl_2 \cdot 6H_2O$	50 μg l^{-1}
H_3BO_3	15 μg l^{-1}
$NiCl_2 \cdot 6H_2O$	10 μg l^{-1}
EDTA	250 μg l^{-1}

The pH of the medium is adjusted to 6.8 before autoclaving. The phosphates are sterilized separately and added to the bulk of the medium when cool. Because the phosphate concentration is low, owing to its inhibitory nature, it is necessary to control the pH by automatic addition of alkali to maintain the pH between 6.6 and 6.8. Gas mixtures are explosive between 5 and 15% of CH_4 in air and so it is preferred to grow the organisms at 20% CH_4 in air at 45 °C, although 4% can be used, but growth then becomes carbon-limited. These organisms are also sensitive to high oxygen tensions so it is necessary to culture them at low DOT values (less than 0.05 atm).

(xii) Media preparation—liquid cultures

Most of the media described in this chapter are relatively simple to prepare since they generally consist of non-heat labile components. The ingredients are dissolved in either distilled water or, when large-scale cultivation is used, tap water. In the latter instance it is essential to test the efficacy of the tap water before inoculating with a seed culture since the presence of different trace metals and various minerals may affect the growth of the organism in question. This can be easily done on a small scale when preparation of the seed inoculum is undertaken.

In some media, precipitation of phosphates occurs on cooling after autoclaving, so it is desirable to sterilize the phosphate component of the medium separately and add it to the bulk of the medium when cool. The pH of the medium should be checked after the phosphates have been added and adjusted with either sterile acid or alkali to bring the pH to the desired value. Any other additions should also be made at this stage, i.e. certain vitamins or amino acids that are heat-labile and are therefore sterilized by filtration. Sterilization of medium constituents by filtration is best achieved by suction through pre-sterilized membrane filters in which the maximum pore size is 0.5 μm; 0.2-μm filters are best (e.g. Metrical GA-8 Gelman membrane).

Media are generally sterilized by autoclaving at 121 °C (15 p.s.i.) for 20 min. If the autoclave is not fitted with thermocouples, by which the exact temperature in a 'dummy' container of medium can be measured and hence the temperature in the vessel to be sterilized can be gauged, then it will be necessary to add sufficient time to the sterilizing cycle to allow the vessel and its contents to reach the sterilizing temperature. In the author's experience 20-l vessels of medium require approximately 1 h of sterilization at 121 °C after an initial free steaming period of 30 min.

Large quantities of medium can be sterilized by filtration, although this method tends not to be as efficient as autoclaving since organisms have been known to penetrate the membrane. The membranes themselves are paper-thin and need to be handled with care. They are usually supported on a perforated 142-mm stainless-steel disc holder (e.g. Gelman 11350 or Millipore YY3014230) and are autoclaved in the supporting apparatus prior to use. It is desirable to have a pre-filter in the system to remove any larger particles which may clog the small-pore filter. This can be achieved by incorporating a pre-filter pad in the filtration device or by using a commercially available filter tube in the line between the medium and the membrane filter (e.g. 0.45 μm Gelman filter cartridge 12505.1, Gelman-Hawkesley).

(c) CULTIVATION METHODS

(i) Batch cultures

In an attempt to understand the detailed mechanism of nitrogenase action by physico-chemical means it has become necessary to produce large quantities of the enzyme and ancillary proteins and enzymes. If one considers that to obtain about 1 g of pure MoFe protein from *A. vinelandii* requires about 2 kg centrifuged cells and that an average yield of cells from a 400-l fermenter is about 0.5 kg, then the necessity for large-scale cultivation becomes obvious.

Preparation of inoculum

Before it is possible to produce large quantities of cells it is necessary to produce an inoculum, and this is usually about 10% of the volume of the culture medium. In most instances it is desirable to start with a pure, single colony isolate of the organism. This can be obtained by selecting a well isolated colony from a plate culture of the organism grown on nitrogen-free agar. The composition of the agar medium should ideally be the same as that of the liquid medium in which the organism is to be grown.

Very often cultures may be obtained either on agar slopes, on glass beads, as lyophilized powders, or, in the case of *Clostridium* and *Bacillus*, as spore stocks. These cultures have to be revived before they can be selected as pure colonies from a plate. The usual method of revival is to transfer them, aseptically, to an appropriate liquid medium (usually NH_4^+-supplemented mineral medium) and incubated at the required temperature (usually 30 °C) and oxygen regime. In the case of *Clostridium* revival in potato medium (see Section b) in stoppered tubes is adequate. When the cultures can then be plated on to nitrogen-free agar medium and colonies selected for further cultivation. The colony is carefully transferred into 25 ml of sterilized nitrogen-free liquid medium in a 250-ml flask and then incubated at the required temperature. For many aerobic nitrogen fixers the cotton-wool plug in the neck of the flask will prevent contaminants from entering and permit the free entry of oxygen to the culture. A number of aerobic nitrogen fixers tend to be adversely affected by ambient oxygen concentrations and grow only at oxygen concentrations less than atmospheric. In these circumstances the inoculated flask can be sealed with a sterile serum stopper and gassed with an oxygen/nitrogen mixture to give pO_2 values of less than 0.2 atm. The gases, which are mixed after passing through flow meters to give the required pO_2 value, are fed through a sterile cotton-wool filter and a hypodermic needle is inserted through the serum stopper. Effluent gas is passed through a second needle in the stopper.

The flasks can then be incubated at the required temperature statically in an incubator or preferably with shaking. At this stage, however, cultures can be started by transferring the colony of organisms into a liquid culture supplemented with ammonium chloride. This has the advantage of obviating the need to provide the correct oxygen tension for growth, since all aerobic and facultatively anaerobic nitrogen fixers grow well in such media and are generally unaffected by atmospheric oxygen concentrations. In some cases it may be necessary to inoculate several flasks with colonies from plates to ensure that the organisms in at least one flask will grow.

The procedure adopted for the growth of anaerobes is essentially the same as that described above for growth in stoppered flasks under reduced oxygen tensions, except that the flasks are thoroughly gassed with high purity nitrogen after inoculation.

Once growth has been established in liquid culture it is possible to transfer the culture to a larger volume. The contents of several flasks can be transferred to 500 or 1000 ml of culture medium. For anaerobes it is sufficient to transfer about 25 ml of an actively growing inoculum into 500 or 1,000 ml of freshly sterilized medium in round flat-bottomed flasks so that addition of the cells almost fills the flask. Under these conditions anaerobic conditions are rapidly attained and the cells will grow well. For aerobic cells it is necessary to use vessels that can be readily aerated. In the author's experience 2-l glass Quickfit vessels (FV2L, Corning Ltd., Stone, Staffs., England) with lids (MAF 2/2) are ideal since they can be fitted with either spargers or air inlet and outlet tubes. The air inlets must, of course, be fed with air through sterile cotton-wool filters. The culture vessel is either kept in a constant-temperature room or fitted with a constant temperature device (Baker, 1968). If spargers are used bubbling is sufficient to keep the culture mixed; if not, then a magnetic bar should be incorporated and the vessel placed on a magnetic stirrer.

Such a system can be readily converted to continuous culture (see Section c.ii) and as such can act as a continuous supply of cells which can be collected from the overflow. If this can be done then it is the preferred method since it eliminates the tedium of having to prepare fresh inocula each time. The purity of the continuous culture should, of course, be checked daily, as well as any inoculum prepared from the chemostat. It is advisable, if other factors have not forced it earlier, to strip down the chemostat every 2 or 3 months and re-inoculate with a fresh inoculum prepared from the stock culture as described earlier. The possibility that the organism grown in the chemostat has mutated from the parent strain can be a real one and should be checked. Periodic stripping down of the chemostat will help to prevent a mutant organism from becoming established.

The size of the inoculum required for growth in a fermenter depends to a large extent on the organism and how well grown it is. Although a 10% inoculum is useful as a general guide, many organisms will grow well from a smaller inoculum, e.g. *C. pasteurianum* (2%), *K. pneumoniae* (5%), *A. chroococcum* (7%), and *M. capsulatus* (8%). A large inoculum tends to reduce any lag phase that may occur when a small inoculum is transferred to fresh medium.

If a 100-l fermenter system is to be used, then the 500–1000 ml of inoculum can be transferred into a similar vessel of 10 l capacity. For 400 and 1000-l systems correspondingly larger inocula will be required and can be prepared either by using a number of vessels of 20-l capacity (Baker, 1978) or by inoculating a seed culture vessel of 100-l capacity.

Large-scale growth of diazotrophs

The design, construction and operation of all-glass fermenters with capa-

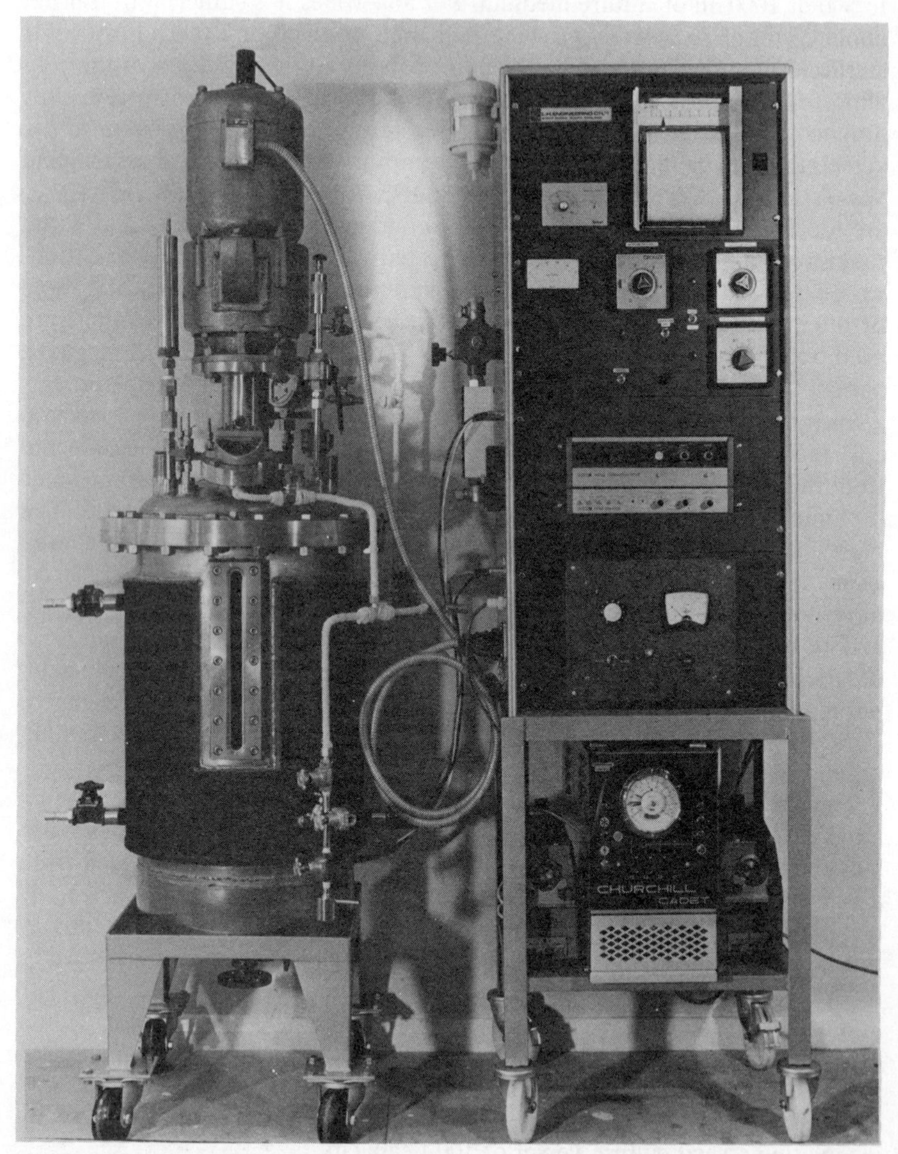

Figure 1. A 60-l batch fermenter with control panel (courtesy of L.H. Engineering)

cities up to 1000 l have been described recently (Baker, 1978) and the reader is referred to that paper for details of such a system. In many instances, however, the time required to build such a system can be a drawback, although in terms of cost it is far superior to commercially available fermenters of comparable size.

A number of manufacturers produce complete pilot-plant fermenter systems that can be tailor-made to suit the customer's requirements. These can include small-scale seed fermenters (from 5 up to 50 l) as well as larger fermenters of capacity up to 500 l. For most biochemical work fermenter systems of 100-l capacity are adequate.

The system used in our laboratory consists of a 100-l jacketed stainless-steel vessel stirred by a 1-h.p. motor placed on the lid of the vessel. The stirrer shaft is connected to the motor via copper and carbon bush seals and has two six-vaned impellors placed 30 cm apart. Gases are fed to the vessel through flow meters and sterile, non-absorbent cotton-wool and cartridge filters to the bottom of the vessel. Four vertically placed baffles help to break up gas bubbles entering the fermenter to give the K_LA values shown in Table 9.

Table 9. Oxygen transfer coefficient (K_LA) values for a 100-l fermenter (L.H. Engineering)

Motor speed (r.p.m.)	Aeration (l. min^{-1})	K_LA(g O_2 h^{-1})
100	25	0.25
100	100	1.38
200	50	2.08
300	25	3.75
300	100	7.89

These values, using air in the gas stream, give a range of oxygen solution rates that are more than adequate to grow aerobic nitrogen-fixing organisms at relatively high (*ca.* 5 g l^{-1} dry weight) cell densities. We have grown organisms in such a culture vessel at cell densities of 10 g l^{-1} dry weight.

Sterilization

The vessel can be sterilized by steam under pressure. We routinely sterilize at 10 p.s.i. for 5 h although a mixture of formaldehyde vapour and steam at 80 °C would suffice (Baker, 1978). The vessel is equipped with a pressure-indicating dial and pressure-relief valve to prevent any excess pressure building up in the vessel. The pH electrode, which is inserted through a port on the side of the vessel, can be sterilized *in situ* without causing it any damage.

Membrane-covered oxygen electrodes cannot be sterilized at this pressure unless the pressures inside and outside the electrode are equal. It would be possible to do this and still sterilize *in situ* by encasing the electrode in a chamber and applying equal pressure to the chamber and the vessel. Normally the oxygen electrode is sterilized by autoclaving separately and inserted later.

Steam, from a boiler, is allowed to enter the vessel and all the parts with silicone tubing connected (i.e. acid/alkali, antifoam, sampling, medium inlet, and air inlet and outlet) are opened slightly and capped with aluminium foil at the open end to prevent back-contamination of the lines. After sterilization the parts are closed and the separately sterilized acid/alkali and antifoam reservoirs are connected aseptically after removing the foil caps. The air inlet and outlet filters are then connected; the inlet filter is a pre-sterilized glass tube (250 × 60 mm) packed with glass-fibre; the outlet filter consists of a water-cooled condenser which is inserted through a rubber bung into a 2-l Erlenmeyer flask, the side-arm being connected to a non-absorbent cotton-wool filter. This arrangement of the gas outlet system prevents any moistening of the filter, which could result in a restriction of gas flow and possible contamination of the culture by condensing moist vapour into the flask. The sampling port is a stainless-steel hood connected to the sampling line of the fermenter fitted with a side-arm air filter. The hood accepts standard screw-necked flasks. During sterilization of the vessel condensed water accumulates, which is run out through the bottom tap of the fermenter before medium is added.

The medium is sterilized by filtration as described earlier. We use the Millipore system in our laboratory, although similar systems by other manufacturers are just as good. The liquid medium is poured in batches into a 40-l dispensing pressure vessel (Sartorius 16602) and connected to the input side of the pre-sterilized membrane filter apparatus. The pressure in the vessel is increased to give a reasonable flow-rate through the filter (*ca.* 40 l in 1 h at 30 p.s.i.). Because of the tendency to form precipitates, the phosphate is always filter-sterilized in one batch separately from the rest of the medium constituents.

Controls

Temperature. This is controlled by water circulating around the jacket of the fermenter via a Churchill Thermocirculator set to the desired temperature. For most nitrogen-fixing organisms 30 °C is the optimum temperature. *Methylococcus capsulatus*, however, is grown at 45 °C on methane under nitrogen-fixing conditions and at high cell densities (greater than 5 mg ml^{-1} dry weight) produces heat so the temperature in the thermocirculator has to be reduced to about 41 °C to maintain the optimum temperature in the fermenter.

Oxygen. The oxygen levels in the culture is an important parameter only with aerobic nitrogen fixers, since the anaerobes are grown under oxygen-free nitrogen. Many aerobic nitrogen-fixers will not grow or fix N_2 if the oxygen concentration in the culture is too high (Dalton and Postgate, 1969b) and so some indication of its level is important.

The dissolved oxygen level in a culture can be conveniently monitored by the autoclavable membrane-covered galvanic cell probe (Borkowski and Johnson, 1967) that is now manufactured by a number of companies. The only problem with the electrode is that the anode metal is gradually dissolved by oxygen and has to be replaced eventually. To prolong the life of the electrode we routinely flush the headspace above the electrolyte with oxygen-free nitrogen when in use and short out the terminals when not in use. Obviously, exposing the electrode to high concentration of dissolved oxygen for prolonged periods will greatly reduce its effective lifetime, but in normal use in a fermenter where the oxygen content is usually low, such electrodes will normally last for up to 4 months in continuous use.

The oxygen electrode measures the oxygen tension in the liquid phase rather than its concentration (Kinsey and Bottomley, 1963), since dissolving solutes in the liquid can reduce the oxygen concentration without affecting the oxygen electrode reading at a given gas pressure. The oxygen tension in liquid is given by $T_o = fc$, where T_o is oxygen tension in the liquid, f is an activity coefficient and c is oxygen concentration in the liquid. When the gas is saturating the liquid, then equating the partial pressure with T_o can give f, which is the reciprocal of Henry's constant (H). However, to calculate the concentration of oxygen in solution from the oxygen electrode readings will involve the assumption of some value of H which depends on the temperature and dissolved solutes in the medium. If a value for the oxygen concentration in air (1 atm)-saturated water is required, then the following values may be used:

Temperature (°C)	Oxygen concentration	
	mg l^{-1}	mM
25	8.10	0.253
30	7.53	0.235
35	6.99	0.218
40	6.59	0.206
45	6.15	0.192

However, these will vary by dissolving solutes in the water.

For most purposes it is convenient to express the electrode readings as dissolved oxygen tension (DOT) in mmHg. Before a fermenter run is commenced the electrode is calibrated by forcing air through the medium in the fermenter at high stirring rates to obtain the 100% reading [equivalent to pO_2 (0.209) × barometric pressure (mmHg)] and adjusting the current output reading from the electrode accordingly. This should be done until the reading is stable and then the zero is obtained by repeating the operation using oxygen-free nitrogen. Full-scale deflection of the meter will be roughly equivalent to 159 mmHg.

For most fermenter runs, however, this degree of accuracy is rarely required and only becomes important when studying the effects of oxygen in chemostat cultures (see Section c.ii).

Control of DOT values in large fermenters during batch growth is usually effected manually either by altering the stirring rate or preferably by altering the pO_2 value of the inflowing gas. Most aerobic nitrogen fixers only grow and fix N_2 well within certain limits of dissolved oxygen (Table 10). It is important particularly during the early stages of growth that the DOT value does not greatly exceed the values quoted in Table 10. Once growth has been established then the respiratory capacity of the culture is sufficient to tolerate minor increases in the DOT value.

The DOT value can be controlled automatically from the current output of the oxygen probe by passing it through a shunt resistance and amplifying the voltage drop across it. This in turn can be fed to a controller which either activates a pneumatic valve to control the air supply to the culture (MacLennan and Pirt, 1966; Harper and Lynch, 1973) to give control of ±5% at 2 mmHg and ±3.3% at 30 mmHg or is used to vary the voltage supply to a d.c.-wound shunt motor, thereby controlling the speed of rotation of the stirrer (Moss and Bush, 1967). In the latter case this may give rise to large variations (±15%) in the culture volume which is probably of little significance in batch cultures. What is more significant, however, is the cost and reliability of stirrer speed-controlled systems, which may be high and prone to mechanical faults. In the author's hands the latter system has proved extremely reliable on small-scale systems (up to 5 l) but problems have been encountered on large-scale systems (>100 l).

Table 10. Optimum dissolved oxygen tension (DOT) values (in mmHg) for aerobic growth of nitrogen-fixing organisms

Organism	DOT	Reference
Corynebacterium autotrophicum	12	Berndt *et al.* (1976)
Spirillum lipoferum	6	Okon *et al.* (1976)
	1.5	Dobereiner (1977)
Azotobacter chroococcum	16	Dalton and Postgate (1969)
Methylococcus capsulatus	6	Dalton (unpublished work)

pH. In general, nitrogen-fixing cultures become acidic during the growth cycle and, because fixation is often inhibited at pH values less than 6.0, it becomes necessary to control this parameter. In media where the phosphate concentration is fairly high the buffering capacity afforded may be sufficient to prevent the pH from fluctuating too drastically. In some cases, however, this may not be so and it is always advisable to monitor the pH of the culture continuously using one of the many steam-sterilizable pH probes that are commercially available. When actively growing, cultures of *Azotobacter*, *Clostridium*, and *Methylococcus* require large amounts of alkali to keep the pH of the culture at its optimum value. A 5 N solution of steam-sterilized KOH is used to adjust the pH of the culture. This can either be added manually, although this is not very convenient if the culture starts growing at an awkward time (as they do very often), or automatically. The automatic control system that we use relies on the EIL 9150 pH controller (EIL, Chertsey, Surrey, U.K.), which can be used for the addition of either acid or alkali when connected to a suitable pump (e.g. HR Flow Inducer, Watson Marlow, Falmouth, Cornwall, U.K.). The accuracy of control of the system is within 0.05 pH unit.

Foaming. At high cell densities (usually in excess of 5 mg ml^{-1} dry weight) many cultures produce large amounts of foam as a result of cell lysis and aeration. This can be reduced by the periodic addition of small amounts of silicone antifoam agent (Antifoam A, Sigma Chemical Co., St. Louis, Mo., U.S.A.). This is done manually when necessary (here again it can be inconvenient) or automatically either by adding a small amount at pre-set time intervals using a timer and pump or by incorporation of a liquid level sensor and relay (Noflote Level Controller, Fielden Electronics Ltd., Manchester, U.K.). In the author's experience the latter system tends to be unreliable unless correctly adjusted, and this adjustment may be different for different conditions.

Harvesting of cells

The only practicable way of separating large quantities of bacterial cells from their growth medium is by continuous centrifugation using a solid-bowl centrifuge (Solomons, 1969). There are a number of different models available (Solomons, 1969), but the types used most by the author has been the Westfalia SA7 and KA2 models. The conical disc-type machine (SA7) is the larger of the two, capable of harvesting 100 l of *Azotobacter*, *Klebsiella*, and *Methylococcus* cultures in 15 min (Baker, 1978). The smaller four-chamber bowl (KA2) takes about three times as long to effect a comparable harvest. Depending on the position of the centrifuge relative to the fermenter, harvesting the culture can be accomplished either by gravity feeding or by pumping. In the latter case we routinely use a Charles Austin Pump (Byfleet, Surrey, U.K.), although any pump capable of flow-rates up to 400 l h^{-1} would be sufficient.

It is often desirable to harvest cells at a lower temperature than the growth temperature to preserve enzyme activity and prevent the action of proteases present in many bacterial cells. With relatively small culture volumes (up to 100 l) we pass the culture prior to centrifugation through a 10 m × 10 mm coil of stainless-steel tubing immersed in a water/ice-bath because the centrifuges described above have no facilities for cooling the bowl. For cultures of *Methylococcus* grown at 45 °C this has the effect of reducing the feed to the centrifuge to about 12 °C. For much larger culture volumes it may be necessary to use a commercial heat exchanger. One such system used in the ARC Unit at Sussex is the Paraflow plate cooler manufactured by A.P.V. Ltd. (Crawley, Sussex, U.K.) which, when used with cold water at 15 °C as the coolant, will reduce the temperature of a culture from 30 to 18 °C. Furthermore, the centrifuge rotor is also cooled with a jet of cold water (Baker, 1978).

The temperature to which the cells are cooled prior to harvesting can be important if good nitrogenase activity is to be preserved. Too sudden a drop in temperature can result in cold shock of the cells, with subsequent loss of of enzyme activity. In the case of *Mycobacterium flavum*, if the temperature of the culture is reduced to 10 °C then there is a marked increase in the viscosity of the culture, which has a deleterious effect on the efficiency of the harvest.

The harvesting of up to 40 l of cells can be readily accomplished using the small continuous-action rotor attachment for the MSE 18 centrifuge (MSE Instruments, Crawley, Sussex, U.K.) or the KSB continuous flow equipment with an RC2B refrigerated centrifuge (Ivan Sorvall, Norwalk, Conn., U.S.A.). The head can be cooled by the centrifuge and there is no necessity to chill the culture before centrifugation. Twenty liters of *Azotobacter* or *Methylococcus* cells can be effectively separated from culture medium in 30 min at maximum centrifuge speed (18000 r.p.m.).

In all cases the wet packed cell mass is removed from the centrifuge by hand. With the Westfalia KA2 and MSE 18 systems this is the only facility available but with the Westfalia SA7 automatic 'desludging' can be used, which releases the contents of the bowl in about 1 l of liquid. Such automatic device is undesirable for some diazotrophs since it leads to rupture of the cells and consequent loss of enzyme (Baker, 1978).

The flow-rates attainable with the SA7 to produce a 'gin-clear' effluent of some diazotrophs has been given by Baker (1978), in which the values range from about 2 l min^{-1} for *Mycobacterium flavum* up to 13 l min^{-1} for a *Clostridium* species. In general, the rates with the KA2 and the MSE 18 are about one third and one tenth, respectively, of the rates obtained with the SA7.

(ii) Continuous cultures

Although batch culture techniques are almost universally used for the production of cells for enzyme purification, there are some disadvantages. Because of the very nature of such techniques cells can vary in many respects from one batch to another; indeed, Munson and Burris (1969) found that only 11 extracts of the 49 prepared from different batches of *Rhodospirillum rubrum* had activities greater than 0.5 nmol N_2 fixed min mg^{-1} and 24 extracts had no activity. When the cells were grown in continuous (chemostat) culture, however, they were able reproducibly to obtain active extracts which in some cases gave specific activities as high as 4.3 nmol N_2 fixed min mg^{-1}. Chemostat-grown cells, however, are rarely used as a source of enzymes for purification but the technique is unrivalled when it comes to growing cells for physiological studies since it can provide the experimenter with a reproducible population in a constant 'physiological state' almost *ad infinitum* (barring accidents, mechanical failure, or contamination). One of the major advantages, indeed a prerequisite, of any chemostat culture experiment, is that the cells are growing in a steady state in which all nutrients except one (the formal growth-limiting nutrient) are in excess of cellular requirements. This means that it is possible simply to perturb a single parameter by either altering the nature of the growth-limiting substrate or the rate at which the cells are growing, provided it is less than the maximum specific growth rate of the population (μ_{max}), and following the subsequent changes to the cells when the new constraint is applied. Consequently, the technique has proved invaluable in the study of how oxygen and nitrogen in particular effect diazotrophs both in physiological and biochemical terms.

Since the purpose of this book is to cover the practical rather than the theoretical aspects of nitrogen fixation, the reader is referred to Herbert *et al.* (1956), Postgate (1965), and Pirt (1975) for basic treatments of the theory and applications of continuous culture. I shall restrict my comments to the practical aspects of the use of the chemostat.

The apparatus

Many companies now offer a range of complete 'off-the-shelf' chemostats which vary in degree of sophistication, e.g. LKB, Braun, Gallenkamp, New Brunswick, Biolafitte, Chemap, and L.H. Engineering. Each unit consists basically of a stirred vessel of capacity up to 10 l, with medium addition pump, pH, temperature and foam control as well as measurement of dissolved oxygen, which may or may not be controlled. The last item is an important asset if aerobic nitrogen fixers are to be grown since the dissolved oxygen tension can have profound effects on cellular metabolism of diazotrophs. In the author's laboratory several commercial chemostats have been fitted with

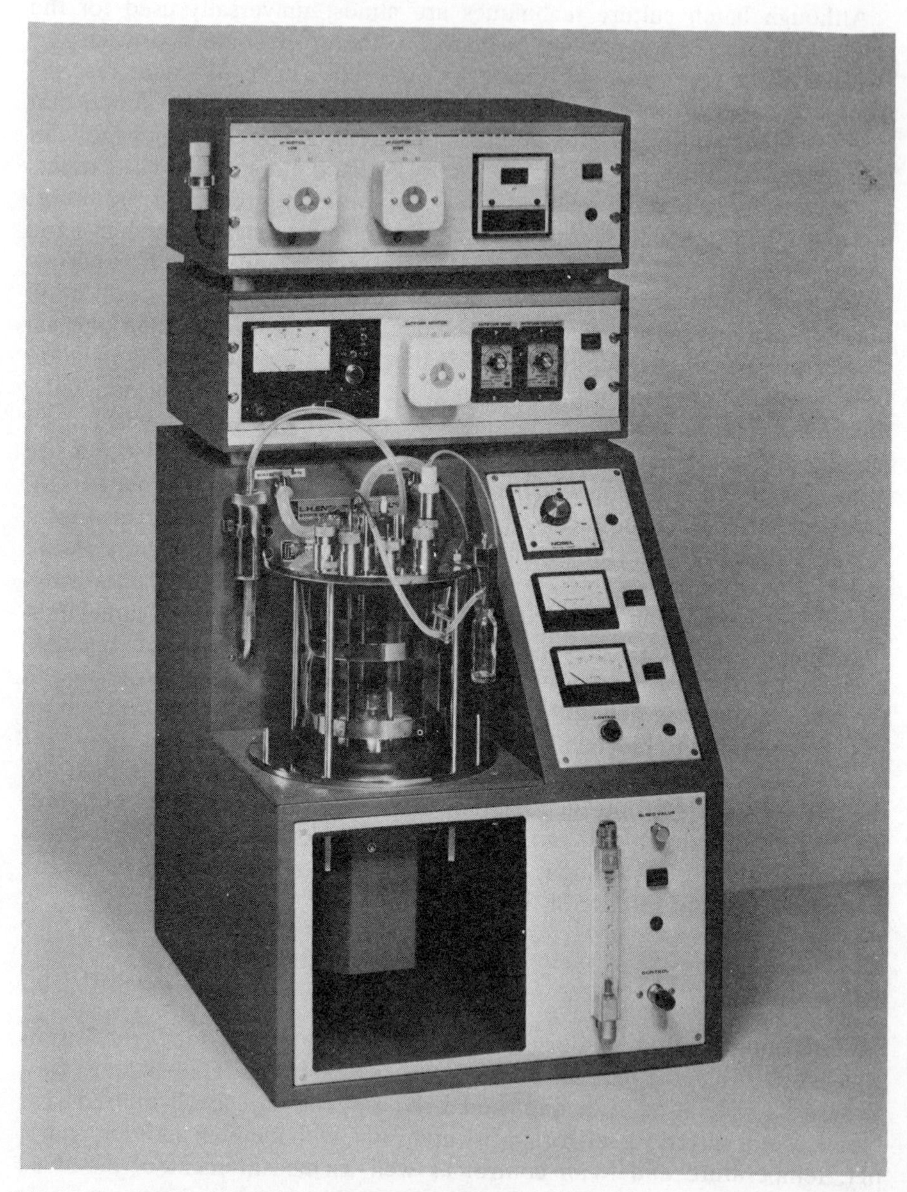

Figure 2. A 2-l chemostat with magnetically coupled stirrer (courtesy of L.H. Engineering)

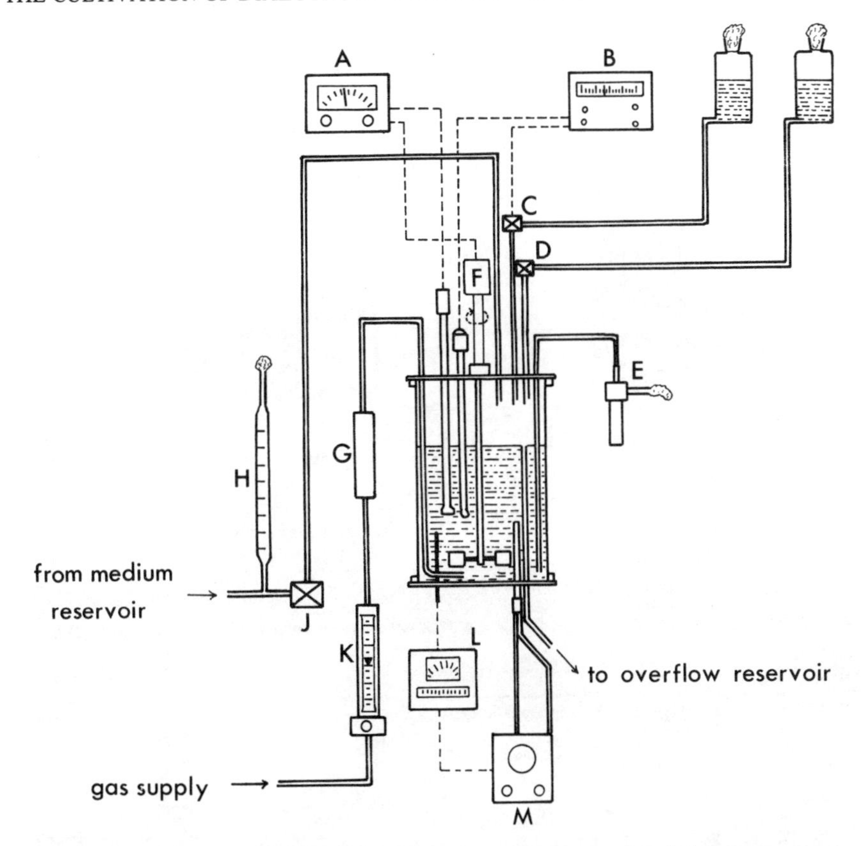

Figure 3. Schematic diagram of a chemostat. For operational details see text. A, Dissolved oxygen controller; B, pH controller; C, pump for acid or alkali addition; D, pump for antifoam addition; E, sample hood and bottle; F, stirrer motor; G, filter; H, pipette for measuring medium flow-rate; J; pump for medium addition; K, gas flow meter and regulator; L, temperature indicator; M, temperature controller

dissolved oxygen controllers which operate by controlling the stirrer speed. Coarse adjustment of the DOT is made by adjustment of the flow-rate of gases into the culture and the controller is adjusted to give fine control. Such a system has been used for several years to effect good control of DOT in cultures of *Methylococcus capsulatus* under diazotrophic conditions. The author has been able to maintain DOT values of 2–20 mmHg with an accuracy of ±8% in steady-state cultures for several months with minimal fluctuation of culture volume between the high and low set points. Other systems offered by Western Biological operate through control of gas flow-rates by solenoid valve switching of gases or a combination of stirrer speed and gas flow-rate.

Figure 4. 'Home-made' 500-ml working volume chemostat. Temperature, pH, gas and medium flow-rates, culture volume and stirrer speed are all controlled; dissolved oxygen tension is monitored but could be controlled by any one of the methods described in the text

For the more adventurous experimenters (or those with limited funds to spend on a commercial system), then it is possible to construct one's own chemostat from 'off-the-shelf' apparatus. Baker (1968) has detailed the construction of such an apparatus and its use has expanded considerably since then (Dalton and Postgate, 1969a b; Drozd and Postgate, 1970 a, b; Hill, 1976; Drozd *et al.*, 1972; Lees and Postgate, 1973; Hine and Lees, 1976; Eady *et al.*, 1978). The vessel and most of the glass components are available from Jobling Laboratory Division, Staffordshire, England. The weir-type overflow from the vessel can either be inserted by a competent glassblower or an 'air-lift' system can be used in which the culture volume is kept constant by the internal pressure forcing excess culture through a glass tube inserted into one of the ports on the top of the vessel, which leads to an overflow reservoir. For the cultivation of aerobes, silicone rubber tubing is used for all of the connections, but because of its permeability to air is unsuitable for use in anaerobic systems. Autoclavable PVC tubing is preferred here provided that the connections are firmly wired on.

Indication and control of pH can be effected using the EIL system described above and the temperature can either be controlled by use of a contact thermometer, relay, and heating lamp (see Baker, 1968), by keeping the vessel in a constant-temperature room, or by circulating constant-temperature water through a jacket built around the vessel. We routinely use the last method in vessels manufactured by T.W. Wingent (Cambridge, U.K.) in conjunction with a constant-temperature water circulator.

Oxygen measurement and control can be achieved using the methods outlined for batch culture. In addition we have also used a system for oxygen measurement based on the diffusion of oxygen into a thin-walled coil of silicone rubber tubing in the vessel. Basically a coil of known length (usually about 3 m of 2.0 mm I.D. × 0.5 mm wall thickness) is carefully wound around a former inside the vessel through which helium is passed at a low (*ca.* 5 ml min^{-1}), but reproducibly accurate, rate via a Norgren RO6 valve. The outlet is connected via a 5-ml sample loop (Gas Sampling Valve 12654, W.G. Pye, Cambridge, U.K.) to a Pye 104 gas chromatograph fitted with a thermal conductivity head on a 3-m molecular sieve 5A (60–80 mesh) column. Helium is used as the carrier gas and the temperature is 50 °C. The response from the chromatograph is fed to a Servoscribe recorder (Smiths Industries Ltd., Cricklewood, London, U.K.) fitted with an integrator. To determine the DOT level of the culture a 5-ml sample of gas from the silicone rubber coil which contains oxygen that has diffused into it from the culture is diverted into the sample loop and the integrated value of the peak due to oxygen is determined. The system can be calibrated by vigorously aerating culture medium in the vessel with known mixtures of oxygen and nitrogen before inoculation of the chemostat. Using this system we have been able to measure accurately down to 0.06 mmHg (*ca.* 0.1 μM O_2 at 30 °C), which is about 10 times better than the accuracy of the oxygen electrode.

Operating the chemostat

The apparatus, which includes the medium reservoir, the culture vessel and its probes, the acid/alkali and antifoam reservoirs, and the overflow vessel, must first be assembled correctly and sterilized with as many, or preferably all, of the tubing connections having been made prior to autoclaving. The culture vessel itself is filled with medium so that the probes are immersed during autoclaving (although this is not absolutely necessary). The medium used to fill the culture vessel should be the same as that in the medium reservoir; it can be supplemented with a fixed nitrogen source if required. This supplementation is particularly useful for aerobic diazotrophs because their initial growth under nitrogen-fixing conditions is markedly influenced by the oxygen tension (a difficult parameter to control in sparse populations) but is not so influenced when grown with a fixed nitrogen source. Once the organisms have grown sufficiently then oxygen control in dense populations is relatively easy.

When the vessel has cooled after sterilization any connections that have to be made must be done aseptically.

It is important, once the vessel is ready for use, to check that the oxygen and pH probes are reading correctly. The medium in the culture vessel is brought to the growth temperature and the oxygen probe tested by forcing oxygen-free nitrogen through the culture medium until the meter records its lowest value (usually after 30–60 min). The zero control knob is then adjusted so that the meter reads zero. Air is then passed through the vessel at maximum stirrer speed until the oxygen meter records its highest value. The 100% control knob is then adjusted to read 100%, which is equivalent to 100% air-saturated medium or a DOT value of $0.209 \times$ barometric pressure (mmHg). If required, further calibration of the oxygen probe can be effected by using a variety of known O_2/N_2 gas mixtures and repeating the process used for the 100% value.

The pH probe is tested by sampling the medium in the vessel, measuring its pH value on an external probe and meter and adjusting the meter reading on the fermenter accordingly.

The air flow-rate (or N_2 flow-rate if an anaerobe is being cultivated), stirrer speed, pH, and temperature controls are finally set to their desired values so that the vessel is ready for inoculation. The inoculum (usually about 20 ml of a well grown culture is sufficient for a culture vessel with a working volume of 3 l) is then added to the culture vessel either through a special inoculation port or through the sampling port, after having transferred the inoculum to a sterile universal bottle and attached it, aseptically, to the sample hood.

The culture is then left to grow until a reasonable cell density is attained (usually 0.5 mg ml^{-1} dry weight is sufficient), at which point the flow of medium to the vessel is started. It is important routinely (2 or 3 times a day) to check the chemostat to make sure that it is functioning correctly and to keep a record of all operations performed on the apparatus and the culture.

One of the prerequisites of continuous culture theory is that one of the components of the medium is growth-limiting. Therefore, the medium in the reservoir must have been so formulated that one component is present at a sufficiently low concentration that it alone determines the biomass concentration at a particular dilution rate. To verify that this is indeed the case there are several criteria that must be satisfied. The first, and purely indicative, criterion is the virtual absence of that particular component from the culture supernatant. The culture is sampled, usually into a bottle containing 1 drop of sulphuric acid or mercury (II) chloride to stop cellular metabolism, and centrifuged to remove bacterial cells, yielding a clear supernatant. The conclusion that the component (carbon, nitrogen, phosphorus, sulphur, magnesium, etc.) is either present or absent clearly depends on the sensitivity of the method used for detecting it. If it cannot be detected then this is indicative of nutrient limitation; however, it must be borne in mind that continuous culture theory predicts that there must be a vanishingly low (but finite!) concentration of that nutrient in the culture supernatant since its concentration ultimately controls the specific growth rate of the culture. At a specific growth rate equal to $1\frac{1}{2}$ times the maximum growth rate the concentration of growth-limiting nutrient in the culture supernatant (i.e. in the growth vessel) for different organisms can be as low as 0.07 mg l^{-1} or as high as 25 mg l^{-1} (Pirt, 1975), even though the concentration in the medium reservoir is 1000 times this value.

Much better criteria depend on observing changes in the cell density when changes are made in the inflowing medium. These are as follows: a decrease in concentration of the limiting nutrient gives a proportionate decrease in bacterial concentration; a doubling in concentration of all other nutrients in the medium has no effect on bacterial concentration. At the same time, several other operations must also be performed. The culture must be in a steady state if results are to be reproducible. By that we mean that there is no change with time in either the biomass or limiting nutrient concentrations. This can easily be checked by periodically sampling the culture and measuring the cell density (either optically, by dry weight determinations, or measurement of some component of the organism) and substrate concentration. If there is no change in these parameters over a 48-h period the culture can be considered to be in the steady state.

Having established a steady state, it is essential to know the specific growth rate of the population in that steady state. This is given by the dilution rate (D) of the system, which is numerically equal to the flow-rate of medium to the vessel (F) divided by the culture volume (V). Only in the steady state does this relationship hold. It can easily be determined by measuring the amount of medium pumped to the vessel via a pipette connected to the medium reservoir. The supply line from the reservoir is clamped off, the clamp on the pipette tubing is removed, and the time taken for a known amount of medium to be pumped to the vessel is measured. The culture volume can be determined by momentarily stopping the stirrer and reading the culture volume from calibration marks made on the side of the vessel.

Any fluctuation in parameters such as dilution rate, pH, temperature, and dissolved oxygen tension can upset a steady state. Hence it is important to ensure that these values are constantly monitored (at least twice a day) and seen to be stable for at least two replacements of the culture volume before a steady state can be declared. Together with these checks, one should also monitor the culture for possible contamination by microscopic inspection and plating samples on to nutrient and nitrogen-free agar plates. At the first sign of contamination it is worth leaving the culture for a further 12 h or so and re-checking in case the contaminant was due to a spurious sampling technique. If the rogue organism still persists then the culture should be stripped down, re-sterilized and inoculated again.

(d) SPECIAL CULTURE TECHNIQUES

Because oxygen can have a profound effect upon the growth of anaerobic and aerobic diazotrophs alike, a number of culture techniques have been developed that enable the oxygen regime to be controlled simply, thereby allowing good growth of diazotrophs in liquid and solid media.

(i) Pankhurst tubes

The use of these tubes for determining estimates of most probable number of diazotrophs by the acetylene reduction technique has been described by Campbell and Evans (1969). They consist basically of a test-tube with a smaller tube connected to the main tube by a side-arm. The side-arm is plugged with cotton-wool, 10 ml of medium are introduced into the main tube and the top is plugged with cotton-wool. The tube is then sterilized by autoclaving and, when cool, the main tube is inoculated and sealed with a sterile No. 33 Suba Seal (William Freeman, Barnsley, Yorkshire, U.K.). A small ball of cotton-wool is placed inside the smaller tube, to which is added 1 ml of saturated pyrogallol and 1 ml of a solution of 10% NaOH and 15% K_2CO_3, capped with a No. 33 Suba Seal and incubated at the required temperature. The alkaline pyrogallol absorbs any oxygen in the tube and also releases carbon dioxide, which is important in initiating the growth of anaerobic diazotrophs. Carbon monoxide may be released by the alkaline pyrogallol solution but usually in such small amounts that it does not affect growth of the organisms.

(ii) Semi-solid agar technique

This technique has proved successful for the cultivation of a large number of aerobic diazotrophs. Its major advantage over many other small-scale batch methods is that only a single tube is used to provide a wide range of oxygen regimes, whereas other methods require incubation of cultures at different oxygen tensions.

A 40-ml volume of culture medium is dispensed into a test-tube (150 × 23 mm) and agar is added to give a final concentration of 0.15%. The tube is plugged with cotton-wool and autoclaved. The tube is then transferred to a water-bath at 30 °C, allowed to cool and then inoculated with a suspension of cells. Repeated pipetting up and down of the medium ensures adequate dispersal of the cells. The cotton-wool plug can either be retained or replaced with a sterile No. 49 Suba Seal if incubation under a specific gas atmosphere is required. After incubation for several days growth is initiated as a band or pellicle below the surface of the semi-solid nitrogen-free medium. The position of the band depends on the oxygen regime that is optimum for the growth of the organism in question. Extremely microaerophilic diazotrophs band much lower in the medium than do those which require higher oxygen tensions for growth. This is because there is a gradient of oxygen concentration in the medium which is high at the surface and low at the bottom.

We have used this technique successfully for the cultivation of methane-oxidizing bacteria in which the gas atmosphere above that agar consists of methane, nitrogen, and oxygen. Because the gas atmosphere can be retained by insertion of a Suba Seal stopper, the technique also lends itself to measurement of acetylene reduction rates of cultures *in situ*.

(iii) Nylon bag technique

When bacterial cultures are grown on solid media in enclosed containers (i.e. desiccators or plastic containers) the gas atmosphere is continuously changing. To maintain a constant gas composition in the atmosphere, Hill (1973) used nylon bags through which a constant stream of gas of known composition was passed at about 500–1000 ml min^{-1}. The gas was bubbled through water via PVC tubing to minimize evaporation from the surface of the plate before entering the bag. The outlet tube was also connected to a Drechsel bottle of water to indicate that there was a positive pressure in the bag. These bags had a working volume of up to 7 l and could accommodate many agar plates—if more were required they could be connected in series. The main advantage of nylon bags is that they are relatively impermeable to gases and can also be used for the growth of anaerobes. In our laboratory, Dr C.S. Dow has cultivated *Desulfovibrio* and a range of photosynthetic bacteria on agar plates in such bags. The inoculated culture plates and a beaker containing 20 ml of saturated pyrogallol (see above) are placed inside the bag, which is then sealed at the open end using a commercial freezer bag sealer. Oxygen-free nitrogen is then blown into the bag via a hypodermic syringe needle inserted into its surface; a second needle of smaller diameter is inserted to vent the gas. After passing the gas through the bag for 20 min (the bag is fully distended at this stage) the smaller needle is removed, 20 ml of alkaline K_2CO_3 are injected into the pyrogallol solution, and the hole is sealed with Sellotape. Finally, the nitrogen supply needle is removed and quickly sealed with Sellotape. The bag and its contents can be transferred, on a tray, to the incubator. Because the nylon bags are transparent, visual inspection of colony development can be made without removing the plates.

(e) SOURCES OF CULTURES

Many diazotrophic microbes can be obtained from National Type Culture Collections, which are usually a fairly reliable source. Many laboratories however, stock their own cultures of a variety of organisms and are usually able to supply researchers with a slope or ampoule of cells with some advice on cultivation techniques if required. This is particularly so if mutants of a particular strain are required, since many of them reside with the primary investigator and only rarely appear in the major culture collections. For those workers who are new to the field and require certain laboratory strains, a number of primary sources of diazotrophs that will usually supply some of their cultures are listed below, although there is no guarantee that all of the organisms will be available.

Source	*Organisms*
Professor J. R. Postgate, ARC Unit of Nitrogen Fixation, University of Sussex, Brighton BN1 9QJ, U.K.	*Klebsiella* (and various mutants), *Desulfovibrio* and *Desulfotomaculum*, *Azotobacter*, *Derxia*, *Mycobacterium*.
Professor R.H. Burris, Department of Biochemistry, College of Agriculture and Life Sciences, University of Wisconsin, Madison, Wisc. 53706, U.S.A.	*Clostridium*, *Azospirillum*, *Bacillus*.

Address	Cultures
Professor H.G. Schlegel, Institut für Mikrobiologie der Universität Göttingen, Grisebachstrasse 8, 3400 Göttingen, Federal Republic of Germany.	Diazotrophic hydrogen oxidizers.
Dr J. Döbereiner, Empresa Brasileira de Pesquisa Agropecuaria, Km 47 23460 Seropédica, Rio de Janeiro, Brazil.	*Azotobacter paspali,* *Azospirillum.*
The Curator of Rhizobia Collection, Microbiology Department, Rothamstead Experimental Station, Harpenden, Herts., U.K.	Principally temperate *Rhizobium* species.
Dr F. J. Bergersen, Division of Plant Industry, CSIRO, P.O. Box 1600, Canberra City, A.C.T. 2601, Australia.	*Rhizobium* species.
Dr J. Burton, Nitragin Company, 2101 West Custer Avenue, Milwaukee, Wisc. 53209, U.S.A.	Many strains of *Rhizobium.*
Head of Department of Soil Science and Plant Nutrition, Institute of Agriculture, University of Western Australia, Nedlands, W.A. 6009.	Many types of Rhizobia, including those indigenous to W. Australia.
Dr H. Dalton, Department of Biological Sciences, University of Warwick, Coventry CV4 7AL, U.K.	Diazotrophic methane oxidizers.

Three major National culture collections stock a number of diazotrophs:

Country	*Collection*	*Address*
U.K.	National Collection of Industrial Bacteria (NCIB)	Torry Research Station, 135 Abbey Road, Aberdeen, Scotland.
U.S.A	American Type Culture Collection (ATCC)	12301 Parklawn Drive, Rockville, Md., U.S.A.
India	National Collection of Industrial Microorganisms (NCIM)	National Chemical Laboratory, CSIR, Poona-8, Maharashtra, India.

The National culture collections will charge a handling fee for any cultures they supply.

The World Federation for Culture Collections (WFCC) has published the 'World Directory of Collections of Cultures of Microorganisms' under the Editorship of Martin and Skerman (1972), which lists all of the organisms held at that time by culture collections in most countries of the world. It is written in six languages and is a very useful reference source if one wishes to obtain a culture of any organism.

The WFCC is an affiliation, with various levels of membership, of culture collections around the world. Past, current, and projected activities and interests of the WFCC are summarized periodically in the *WFCC Newsletter,* which is also a forum for members. The Federation also sponsors international meetings at intervals (e.g. the 3rd International Conference on Culture Collections, Bombay, 1977). Further information can be obtained from the Secretary, World Federation for Culture Collections, Culture Collection of Fungi (QM), Department of Botany, University of Massachusetts, Amherst, Mass. 01002, U.S.A.

It is unlikely that major catalogues, such as those mentioned above, will be published again at regular intervals. The information is now stored in a data bank controlled by the World Data Center for Microorganisms, a body sponsored by UNESCO, the UN Environment Programme, the International Cell Research Organization, the World Federation for Culture Collections, and the University of Queensland. Information about cultures can be obtained from The Director, World Data Center for Microorganisms, University of Queensland, St. Lucia, Queensland 4067, Australia.

The addresses given above are correct as at the time of writing but may change as office-holders change.

(f) CULTURE PRESERVATION

Maintenance of cultures can pose a few problems principally because no single method for preservation has universal applicability to all cultures. It is fair to say, however, that freeze-drying of cells is more widely used successfully than other methods. Even then the length of time that cells can be kept in the freeze-dried state and still retain a reasonable (better than 10%) viability when revitalized varies from one organism to another. The need to keep a stock supply of cultures in the hypobiotic state is obvious when one has experienced the problems that occur in both large- and small-scale cultivation from contamination by unwanted organisms. It is often far simpler to start with a fresh supply of the original culture than attempt to purify the diazotroph from a contaminated culture. Furthermore, it is possible that the organisms in the culture vessel, particularly after prolonged cultivation in the chemostat or repeated subculture in batch culture, may differ significantly from the organism when originally isolated from the environment. It is important, therefore, that cultures should be checked periodically for any phenotypic variation and replaced with a fresh culture from the original isolate if the variation is physiologically or biochemically significant.

A fairly comprehensive survey of methods available for culture preservation has been published (Lapage *et al.*, 1970), so I shall deal only briefly with those methods which have proved successful for the preservation of diazotrophs.

(i) Freeze-drying

This method involves the removal of water from a frozen solution of cells by sublimation from the ice as vapour. Two methods of freeze-drying are usually used—in one the sample is centrifuged to overcome frothing until the material is frozen, and the other requires pre-freezing of the cell sample before connection to the vacuum system.

Essentially both methods are similar, although the former tends to give slightly better recoveries of viable cells. With the centrifugal method a sterile glass ampoule is aseptically loaded with about 2 ml of an actively grown culture of cells re-suspended in a 7% glucose broth solution to give about 10^{10} cells ml^{-1}. The ampoule is attached to the head of the centrifuged freeze-drier (Model 5PS, Edwards High Vacuum Ltd, Crawley, Sussex, U.K.) and centrifuged. Primary drying of the suspension is achieved by applying a vacuum, which also has the effect of freezing the sample by evaporation of some of the re-suspending fluid. At the conclusion of the primary drying about 5–10% of the water still remains in the ampoule. The neck of the ampoule is then constricted and further evacuation effects the secondary drying process, which leaves about 1–2% of water in the sample. The neck of the ampoule is then completely restricted under vacuum. The 1–2% retention of water in the sample is important if good viability is to be achieved upon revitalization, since complete desiccation often results in very low viabilities. The inclusion of 7% glucose in the cell sample before freeze-drying ensures that the residual water level after drying will be at this required value.

This method of culture preservation will preserve cultures of *Azotobacter beijerinckii, A. chroococcum, A. vinelandii, Clostridium acetobutylicum, Desulfovibrio desulfuricans, D. gigas, D. vulgaris, Desulfotomaculum orientis, Dm ruminis, Thiobacillus ferro-oxidans, Bacillus polymyxa, Enterobacter cloacae,* and *Klebsiella pneumoniae* for at least 4 years with a good retention of viability upon revitalization. The ampoules can be stored at room temperature, although with the Azotobacters, *Desulfovibrio gigas,* and *Thiobacillus* it is best to store the ampoules at 0–5 °C.Other *Azotobacter* strains and *Derxia* are poorly preserved by this method.

(ii) Sterile soil

The use of sterile soil as a supporting preservative is particularly valuable in keeping hypobiotic cultures of spore-forming diazotrophs and *Clostridium* species in particular. In essence, a well grown suspension of the organism (in which spores are easily discernible) is added to sterile soil and dried at room temperature. The sample can then be stored in the fridge and revitalized by taking a pinch of soil from the tube and adding to fresh medium.

(iii) Storage in liquid nitrogen

This method has proved reasonably successful for a number of organisms and depends on fairly carefully controlled conditions of freezing and thawing (Bridges, 1966) and on the nature of the suspending fluid.

(iv) Storage on glass beads at −70°C

This method has been used routinely in the author's laboratory for a number of diazotrophs, including *Rhizobium* spp. About 100 glass beads (3 mm diameter) are sterilized in a McCartney bottle with 2 ml of glycerol. When cool, 8 ml of a well grown suspension of organisms are added to the bottle, which is then shaken. The excess liquid is poured off and the contents of the bottle are put into a deep-freeze either at −20 or −70 °C. When a fresh culture is required, one or two beads coated with the organism are aseptically removed and revived in the appropriate medium. Cultures of *Rhizobium* spp., methane oxidizers, *Azotobacter,* and *Klebsiella* have been stored successfully in this way for several years. It has the advantage that it is extremely simple to do and the same sample can be revived reproducibly many times because only one or two beads from the stock are removed at a time.

Methods for Evaluating Biological Nitrogen Fixation
Edited by F. J. Bergersen

F. J. Bergersen
Division of Plant Industry,
CSIRO, Canberra, Australia.

2

Measurement of Nitrogen Fixation by Direct Means

(a) INTRODUCTION

In this chapter, methods will be described which may be used for the measurement of nitrogen fixation by detecting increases in the total nitrogen content of a system, by detecting characteristic changes in the nitrogen-containing components of a system, or by use of isotopically enriched or depleted nitrogen gas. The term 'direct means' in the title to the chapter refers to the measurement of nitrogen itself, as distinct from indirect methods which measure correlated activities such as acetylene reduction or hydrogen evolution. The methods described are basic and applications for some of them will be found in succeeding chapters. The nitrogen analyses which are described are generally useful, but are also those which may be used for ^{15}N analysis. Some of the methods described are not strictly direct. Rather, their main applications will provide the basis of expression of results. An example would be the estimation of protein as a basis for expressing nitrogenase activity.

The fixation of atmospheric N_2 by living systems should lead to a demonstrable net increase on the nitrogen content of these systems. This increase should be detectable by direct analysis of total nitrogen. Early work on biological nitrogen fixation employed such methods with conspicuous success (e.g. Helgriegel and Wilfarth, 1888, and the earlier work of Boussingault, reviewed by Aulie, 1970). However, there is often acute analytical difficulty in measuring increases in nitrogen content which amount to only a small proportion of that which is already present in the system. The system may also be losing nitrogen by denitrification or by leaching, thus diminishing the net gain. Other analytical difficulties include the possibile involvement of nitrogen compounds which resist analysis, problems of variability in nitrogen content of biological material, and analytical errors. These matters have been discussed by a number of authors and will not be reiterated here (see, for example, Wilson, 1940, pp. 95–98; Martin and Skyring, 1962; Parker, 1961). However, nitrogen analysis can be used successfully when simple systems are studied and the product of fixation is readily isolated. An example of this is the use of analysis of NH_3 to measure N_2 fixation by cell-free extracts which do not assimilate the NH_3 into other products (e.g. Koch *et al.,* 1967).

Some special techniques have been devised to minimize the difficulty of measuring gains of total nitrogen in biological systems. For example, Hurwitz and Wilson (1940) used a gasometric method in which the gas pressure was maintained constant by the addition of O_2 and absorption of CO_2. If N_2 was fixed, the pO_2 in the system increased. If the added O_2 simply replaced O_2 consumed in respiration, the pO_2 remained constant. Although this technique was successful, it was not very sensitive and Wilson (1940, p. 108) suggested that isotopic N, used as a tracer in N_2 gas, should prove to be valuable for N_2 fixation studies.

Ruben *et al.* (1940) used the short-lived radioactive isotope of nitrogen ^{13}N to demonstrate apparent N_2 fixation by barley leaves (a finding subsequently found to be false by Burris, 1941) and ^{13}N has been used subsequently (Nicholas *et al.*, 1961; Wolk *et al.*, 1974); however, the technical difficulties impose several restrictions on the use of this isotope. Burris and Miller (1941) were the first to indicate the application of the stable isotope, ^{15}N, as tracer in N_2 fixation studies and many reports employing the method followed quickly (e.g. Burris 1941; Burris *et al.*, 1942; Burris and Wilson, 1957).

These experiments utilized atmospheres in which the N_2 was enriched with ^{15}N. Later, other methods were developed in which ^{15}N added to systems as combined nitrogen was diluted with atmospheric N_2.

(b) ANALYSIS OF TOTAL NITROGEN

There are two methods upon which all commonly used analyses are based. The first is the oxidative method, in which organic material is oxidized in the presence of copper oxide to produce N_2 gas, the volume of which is measured. This method is based on the original Dumas technique, and is usually restricted now to commercially produced apparatus such as the Coleman nitrogen analyzer. The methods for use of such apparatus are usually specified by the manufacturer and will vary with design.

More commonly, a 'wet' digestion based on that attributed to Kjeldahl is used. In this, organic and mineral nitrogen is reduced to NH_3 in hot, concentrated sulphuric acid in the presence of a catalyst. The NH_3 is recovered by distillation or diffusion and estimated by titration or colorimetrically. There are many variations of this method. Those which have been found to be most useful in the author's laboratory will be described and reference made to some variations which may be useful. Other accounts were given by Umbreit and Bond (1936) and Parker (1961); Bremner (1965) gives accounts of nitrogen analysis for soils.

(i) Preparation of material for digestion

Plant material should be oven-dried, ground or milled to a homogeneous mixture, and samples of 200–300 mg weighed for digestion. Samples of other materials should be selected to have similar nitrogen contents (2–10 mg as N). The methods, with care, may be adapted to a micro-scale of 1–14 μg N, and thus are suitable for use with samples of bacterial cultures. However, in all cases it is best to work with samples as large as practicable in order to minimize errors.

(ii) Digestion reagent

Potassium sulphate (100 g) to elevate the boiling point and powdered

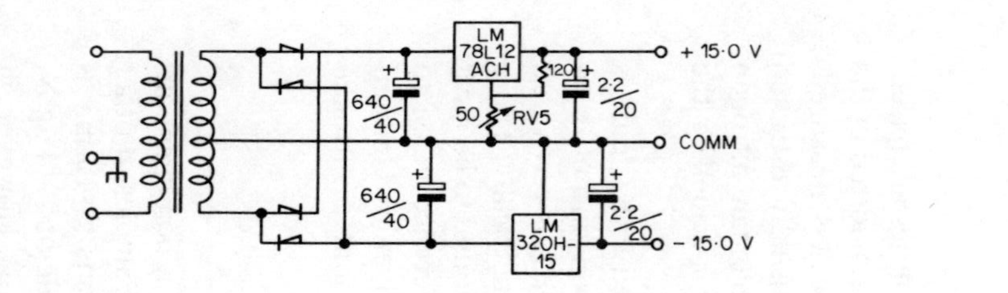

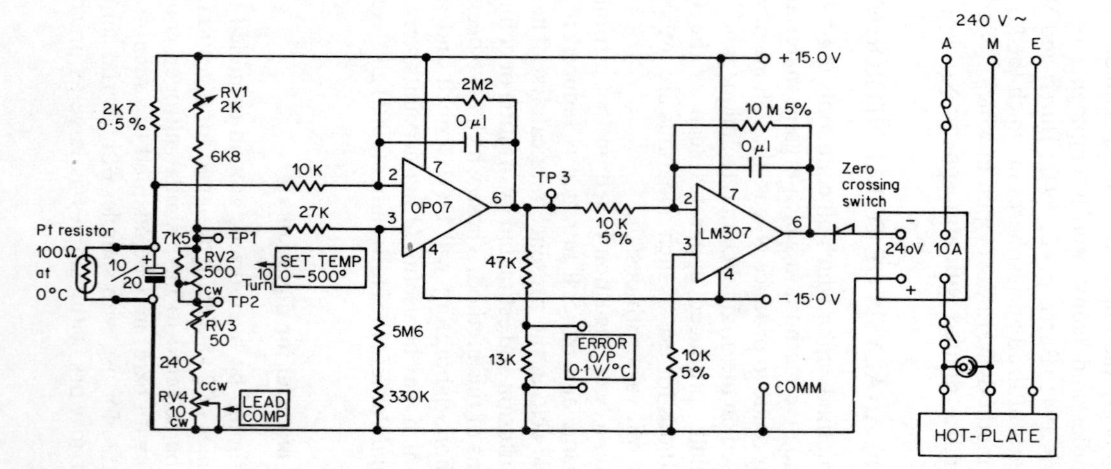

Figure 1. Circuit for control of hot-plate temperature for Erlenmeyer flask digestions. The 100-ohm Pt resistor is thermally connected to the centre of the underside of the hot-plate, using silver connecting leads to avoid oxidation due to heat. All resistors are 1% tolerance unless otherwise stated, and all 1% and 0.5% resistors are metal film types. For initial adjustment: (1) RV2 and RV4, full ccw; connect a 100-ohm, 0.1% resistor in place of the Pt sensor and adjust RV5 for 15 V supply. (2) Adjust RV1 and RV3 for 1.505 V at TP1 and 0.536 V at TP2. (3) RV2 full ccw, and adjust RV1 for 0.00 V at TP3. (4) Connect sensor, cooled in melting ice, with the silver leads as indicated and adjust RV4 for 0.00 V at TP3. The circuit for 240 V a.c. is readily adaptable to other supply voltages

metallic selenium (1 g) are dissolved in concentrated sulphuric acid (1 l) by heating at 250–300 °C for 3–4 h. This reagent has been found to be as good as, or better than, any published formulations which contain other catalysts. Complete recovery of tryptophan and nicotinic acid nitrogen are usual when the digestion is continued for 1–2 h after clearing. A blank digestion should always be included, because the reagent often becomes contaminated with ammonia from laboratory air.

(iii) Digestion

Digestion vessels may be conventional Kjeldahl flasks of nominal capacity 100 ml, with necks about 16 cm long; with advantage they may be graduated at 100 ml. With these, any type of digestion furnace may be used within a well vented fume hood, using a rack to support the neck of the flasks at 45° from the horizontal. The area of the flask surface which is heated should not be greater than the surface exposed to the digestion liquid and losses are minimized when heating is controlled so that refluxing of acid is limited to the lower third of the neck of the flasks. Flasks should be rotated during the digestion and, if necessary, cooled and washed down with distilled water to ensure complete digestion of material carried up the flask during the early stages of digestion. Glass beads or other aids are usually an advantage to

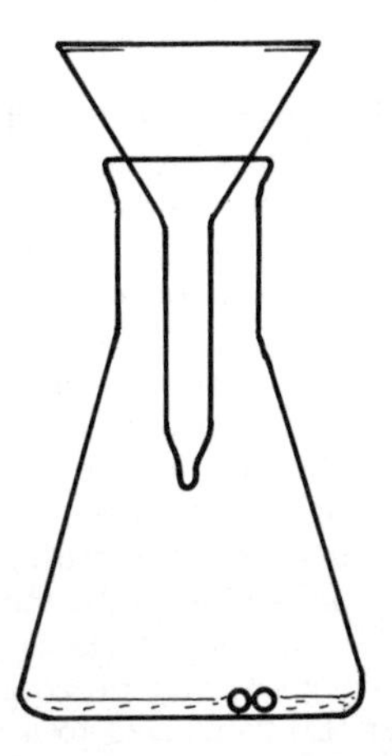

Figure 2. Kjeldahl digestion in Erlenmeyer flask. For plant material (220–300 mg) the flask should be of 100 ml capacity; for micro- and semimicro applications, Pyrex (R) type No. 4980, 25 or 50 ml capacity. The closures are small funnels with the tips sealed and glass beads or other boiling aids are provided

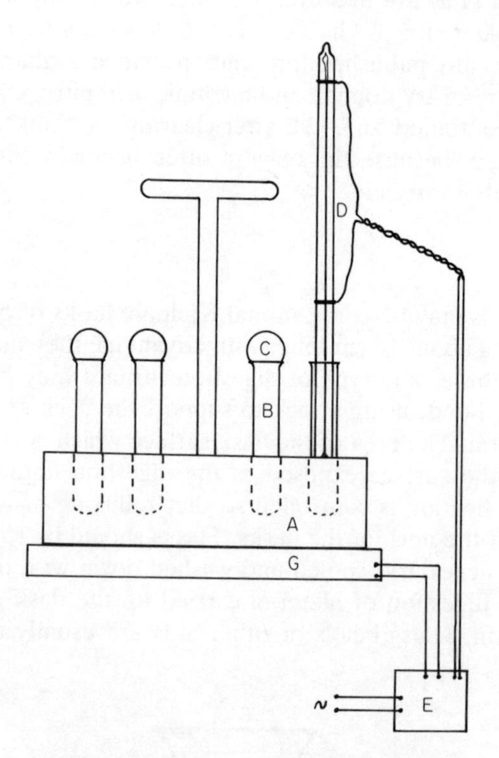

Figure 3. Apparatus for Kjeldahl digestions in test-tubes. A steel block (A) is bored to receive Pyrex test-tubes (B) selected according to the desired scale. The holes should be deep enough to allow good heat transfer. The acid refluxing should rise no higher than half of the distance between the top of the block and the tops of the tubes. After evaporation of water from the sample, the tubes may be closed with glass marbles (C) or with sealed funnels as in Figure 2. One of the tubes carries a mercury-in-glass, electric contact thermometer (D) connected to a relay (E) to control the heating element of the hot-plate (G). The bulb of the thermometer is immersed in digestion reagent and is set to control the tube temperature at 320–330 °C. The number of tubes per block is determined by requirements and hot-plate size

promote steady boiling, especially during evaporation of water early in the digestion. Control of digestion temperature is important since high temperatures promote digestion, but temperatures above 330–350 °C lead to losses of NH_3. We have found that losses are minimized and greater precision achieved by digesting in ordinary Erlenmeyer flasks (capacity 25–100 ml according to scale), heated on an electric hot-plate whose surface temperature is thermostatically controlled (Figure 1). After evaporation of water from the sample, the necks of the flasks are closed with small glass funnels, the tips of which have been sealed and drawn to a point (Figure 2).

A further alternative is to use glass tubes held vertically in a steel block, to promote even heating. One tube can be fitted with a mercury-in-glass contact thermometer to control the hot-plate temperature (Figure 3). This method is well adapted for micro-scale digestions and is widely used, but it offers no advantages over the Erlenmeyer flask method, and is less adaptable in scale.

Amounts of digestion reagent used and digestion times must be adapted to the material being analyzed and to the method of estimation of NH_3 which is to be used. For plant material, 5 ml of reagent are usually sufficient for 200–300-mg dry weight samples. For samples of bacterial cultures, which may contain residual medium components such as sugars, it is sometimes necessary to use proportionally more reagent. For samples of 10 mg (dry weight) of washed bacterial cells, digestion is usually complete with 1 ml of digestion reagent.

Plants usually contain nitrate. When its amount is small, there is usually sufficient carbon present to effect its reduction to ammonia during digestion. However, this may be variable and if there is reason to suspect that significant amounts of nitrate are present it is advisable to modify the digestion procedure. There are several suitable methods—the simplest is to convert the nitrate into nitrosalicylic acid, which digests readily to produce ammonia. The weighed sample in the digestion vessel is well mixed with 1–3 ml of a solution of pure salicylic acid (5% w/v) in concentrated sulphuric acid for at least 20 min. Sodium thiosulphate (0.3–1 g) is then added and the mixture gently heated until fuming. Then it is cooled, digestion reagent is added and digestion is carried out as described previously. Alternatively, nitrate remaining after digestion may be recovered by boiling the digest with Devarda's alloy (BDH Chemicals, Poole, Dorset, U.K.) after the ammonia has been distilled see Section c.ii).

(iv) Recovery of ammonia by distillation and estimation by titration

Several commercially available stills are suitable for recovery of NH_3 from Kjeldahl digests (Figure 4). Their use has been described by Humphries (1956) and Bremner and Edwards (1965). The NH_3 passing from the condenser

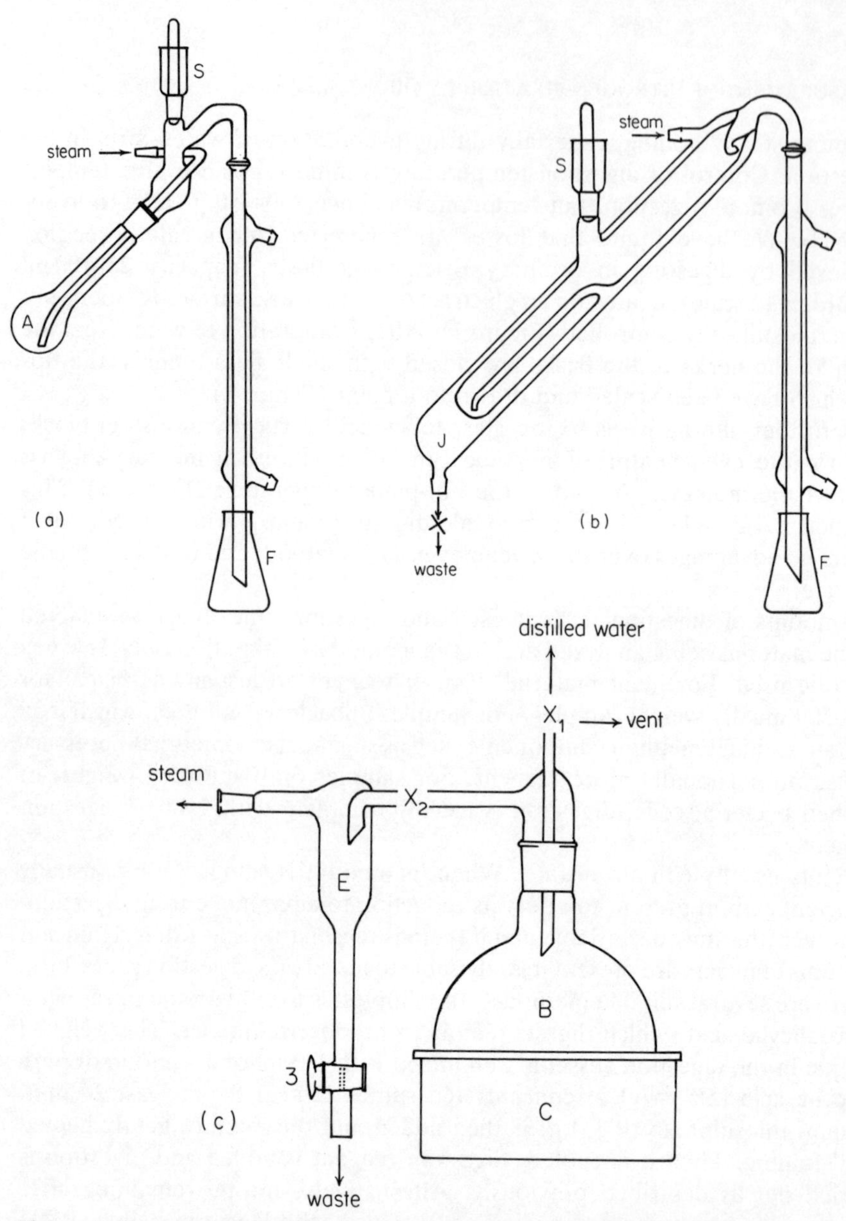

Figure 4. Commercially available stills for recovery of ammonia. (a) Quickfit semi-micro Kjeldahl assembly, similar to Bremner and Edwards (1965). The digestion flasks (A) may be of 50 or 100 ml capacity. (b) Quickfit Markham still. (c) A steam generator assembly. B is a 2–3 l distillation flask heated with an electric mantle (C). A supply of distilled water is connected through a two-way stopcock (X_1). Sample and alkali are introduced through S and distillate collected in boric acid/indicator solutions in Erlenmeyer flasks (F). When used with the still shown in (a), with X_2 open and X_1 closed, a partial vacuum develops in B when the heating is stopped. This facilitates removal of sample residue from A and rinsing of the system prior to connecting a new digestion flask. With the Markham still (b) the same effect is obtained when the steam supply is disconnected and steam condenses in the jacket (J)

is trapped in 2% boric acid containing 5 ml l^{-1} of 0.1% methyl red in 95% ethanol and 20 ml l^{-1} of 0.1% bromocresol green in 95% ethanol. It may be necessary to increase the indicator concentrations with some dye batches. For micro-Kjeldhal distillations, the boric acid concentration should be reduced to about 0.2%.

The water in the steam generator should be acidified with about 1 ml l^{-1} of H_2SO_4 to trap any NH_3 which may be present and the level of water should not be allowed to drop to less than half its initial volume. The rate of boiling and the cooling water flow-rate in the condenser should be adjusted so that the distillate emerges from the condenser at close to room temperature. Before use, the distillation apparatus should be cleaned and steamed out for 10–30 min. The aliquot of the digested sample is added with the steam flowing and rinsed in with distilled water. When steam is passing freely through the samples the tip of the condenser is lowered so that it is immersed in the boric acid/indicator solution and sufficient alkali (usually 40% NaOH) is added slowly to render the sample alkaline to added phenolphthalein. After about 4 min the condenser tip is raised above the boric acid solution. For ordinary Kjeldahl distillations (0.05–5.0 mg N), 20 ml of boric acid are sufficient and the collection of about 30–35 ml of distillate is usually considered to be adequate. The steam line is then closed (venting the generator to atmosphere) and the sample of digest is sucked out to waste as the steam condenses in E. Water is then flushed through the sample inlet port, discharged to waste and the still is ready for the next sample. The NaOH should be freshly prepared and protected from atmospheric CO_2.

When samples containing ^{15}N are being distilled, additional precautions must be observed in order to avoid possible contamination from trace amounts of highly labelled NH_3 remaining in the still. Bremner and Edwards (1965) advised the distillation of about 15 ml of ethanol between samples to remove traces of NH_3 adhering to the glass surfaces of the still shown in Figure 4a. In our laboratory, no cross-contamination has been encountered with the Markham still (Figure 4b), provided a total of about 100 ml of distillate is collected for each sample.

For 1–10 mg N the ammonia in the distillate is titrated with standardized approximately N/28 HCl (31.8 ml of concentrated HCl diluted to 10 l with distilled water) to a faint pink end-point (pH 4.9). In the micro-range, the titrating acid should be 10 times more dilute. A 1 ml volume of N/28 HCl is equivalent to 0.5 mg of nitrogen. The titrating acid is standardized with standard borax, prepared by two successive recrystallizations of borax from a solution saturated at 55 °C. The crystals are washed with cold water. They are finally washed with dry ethanol and diethyl ether and equilibrated for 24 h over water saturated with NaCl and sucrose, before storing in a tightly stoppered bottle kept in a desiccator (Humphries, 1956). About 0.3 g of the standard borax is weighed accurately, dissolved in water and diluted to 50 ml with water. A 10 ml volume of this solution is titrated with the HCl solution to be standardized, using one drop of 0.1% ethanolic methyl red as indicator.

$$\text{Normality of acid} = \frac{\text{wt. borax} \times 200}{190.72 \times \text{ml acid}}$$

For micro-titrations, an AGLA micrometer syringe (Burroughs Wellcome & Co., London, U.K.) proved useful.

(v) Recovery of ammonia by diffusion

Conway (1950) discussed the use of these methods and the Conway diffusion dish has been used by many authors, e.g. Parker (1961) and Freney and Wetselaar (1967). In our laboratory the Conway dish has been replaced by a simple assembly consisting of a 10 ml Erlenmeyer flask, closed with a rubber stopper which bears an etched glass rod protuding from its lower surface (Figure 5). The etched tip of the rod may be prepared by sand-blasting or by treatment with hydrofluoric acid. Its purpose is to retain a film of acid during the diffusion. The concentration of the acid is adjusted according to the amount of ammonia to be diffused. Dilute sulphuric acid (10% v/v) is adequate

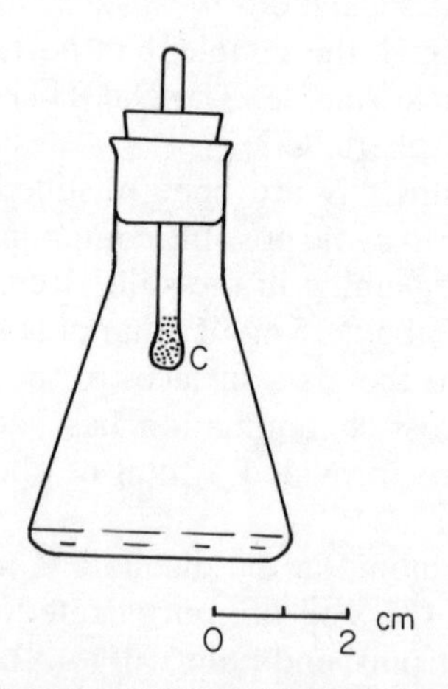

Figure 5. Recovery of ammonia by diffusion. An aliquot of digested sample is placed in the 25-ml Erlenmeyer flask and titrated to pH 4–5 (1 drop of neutral-red indicator just orange). When cool, 1 ml of 20% NaOH is added and the flask quickly closed with a rubber stopper bearing an etched glass rod (C) dipped in 10% H_2SO_4. Overnight, NH_3 diffuses quantitatively into the acid on the rod

for up to 150 μg N per diffusion. Routinely, it is convenient to use an aliquot of digest containing 1–14 μg N, which is in the range of a suitable colorimetric assay for ammonia (see later). Larger sample sizes may be used and aliquots of the diffusate used for the colorimetric assay. The latter procedure is convenient when ^{15}N analysis is to follow.

The procedure is as follows. The acid digest is diluted to a suitable volume (100 ml for 200–300 mg of plant material containing 1–10 mg N). When cool, an aliquot containing 1–14 μg N is carefully pipetted into the 500 ml Erlenmeyer diffusion flask, a drop of neutral red indicator added, and the solution neutralized to pH 4–5 (orange) with NaOH, taking care not to swirl acid up the sides of the flask. After cooling to room temperature, the glass rods are dipped into 10% H_2SO_4, 1 ml of 20% NaOH is added to the flask contents, and the stopper with the glass rod is quickly inserted before gently swirling to mix. Alternatively, the glass rod may be replaced with a piece of Nichrome or stainless steel wire inserted into the rubber stopper. The wire is bent into a hook upon which is hung a small square of acid-soaked filter-paper. The flasks are left overnight on the bench, avoiding positions where temperature gradients may cause condensation within the flask. The next morning the stoppers are carefully removed and the tips of the glass rods soaked in 1 ml of distilled water in a 10-ml tube in which the colorimetric assay may be carried out. If larger quantities of N are required in the diffusate (e.g. for subsequent isotopic analysis), the aliquot of digest used may contain up to 150 μg N, the glass rods may be rinsed into a volumetric flask and the diffusate diluted to a standard volume, a portion of which is used for the colorimetric assay.

(vi) Colorimetric analysis of ammonia

In the author's experience, the most convenient and reliable colorimetric assay for ammonia is that described by Chaney and Marbach (1962). It reliably measures up to 14 μg NH_3-N in 1 ml.

Reagents

A.	Phenol	50 g l^{-1}
	Na nitroprusside	0.25 g l^{-1}
B.	NaOH	25 g l^{-1}
	Na hypochlorite	2.1 g l^{-1}

(28.2 ml of BDH Analar solution; 1 N hypochlorite in 0.1 N NaOH)

Method. One part by volume of each solution is diluted with 4 parts of water and 5 ml of each diluted reagent are added in turn to 1 ml of sample; with agitation colour is fully developed in 30 min at room temperature and is stable for several hours. Each set of estimates should include at least two standards and a reagent blank. Record the optical absorbance at 625 nm.

The methods may be adapted to use with Kjeldahl digests directly, without a distillation or diffusion step as follows. The digest is diluted with water to give a solution 5% in H_2SO_4 (5.0 ml of digestion reagent for 200–300 mg of sample, diluted to 100 ml). An aliquot of this containing 1–10 μg N is diluted to 1 ml with 5% H_2SO_4. Reagent B above is modified by increasing the NaOH concentration to 40 g l^{-1}. In all other respects, proceed as above. Standards should also be prepared in 5% H_2SO_4.

(vii) Estimation of ammonia by AutoAnalyzer

For large numbers of estimations of total nitrogen, automatic analysis is an advantage. After digestion of the samples as described above, ammonia is determined by an automatic procedure, usually using the blue colour formed by reaction with hypochlorite and phenol, but omitting the nitroprusside catalyst. The following method is that described by Williams and Twine (1967) for use with a Technicon AutoAnalyzer.

Reagents

A. Sodium phenate. Add 90 ml of distilled water to 500 g of phenol and, when dissolved, mix with 2.5 l of 20% w/v NaOH.

B. Sodium hydroxide, 0.52%.

C. Sodium hypochlorite. Dilute 50 ml of commercial sodium hypochlorite (12% NaOCl) to 200 ml with distilled water.

D. Diluent. Mix 3 ml of Teepol detergent with 10 l of distilled water.

Standards. Dissolve 9.433 g of $(NH_4)_2SO_4$ in 1 l of distilled water; 1 ml ≡ 2.000 mg N. Dilute to give a range of standards containing 0–0.15 mg ml^{-1} N and 5 ml of digestion mixture per 100 ml.

Samples. Use digestions containing 5 ml of digestion reagent and dilute to 100 ml with distilled water. Mix and stand overnight to allow solids to settle. Decant a suitable aliquot into the sample cup for analysis.

Determination. The manifold and flow sequence is shown in Figure 6. A sampling rate of 40 per hour, a heating temperature of 95 °C, and measurement of colour at 630 nm are used. It is not necessary to use a wash between samples.

Reagent blanks and standards are interspersed with replicated samples. Analyses should be repeated when replicates differ by more than 5–10%. This method is independent of interference from up to 50 p.p.m. of Ca, Mg, or Se, 40 p.p.m. of P or 2 p.p.m. of Mn, Fe, or Al. These are greater than the concentrations likely to be encountered in digests of biological materials.

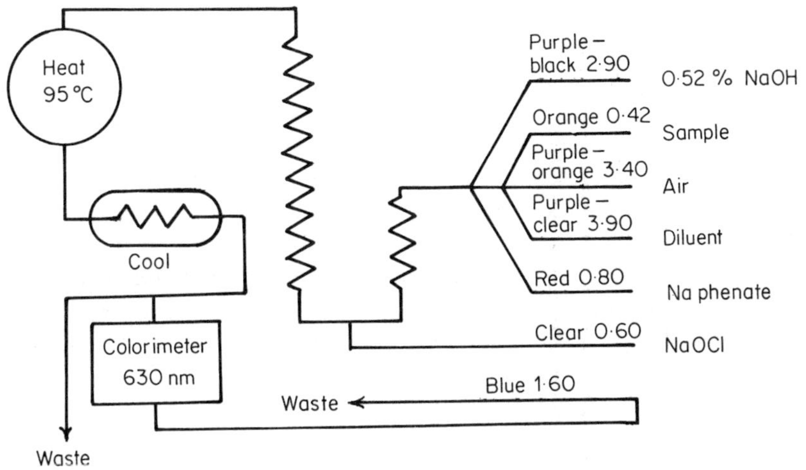

Figure 6. Manifold and flow sequence for the determination of nitrogen in Kjeldahl digests with a Technicon AutoAnalyzer. A 16-mm tubular flow cuvette is used and pump tube sizes are specified in accordance with the Technicon colour code; flow-rates in ml min^{-1} are indicated on each line. (Redrawn from Williams and Twine, 1967, and reproduced by permission of CSIRO, Australia)

However, small variations in the H_2SO_4 concentration influence the results, and care should be taken to measure the digestion reagent accurately and to avoid losses during digestion.

(c) ANALYSIS OF NON-PROTEIN NITROGEN

Apart from analysis of component compounds, estimation of non-protein nitrogen is usually developed from one of several arbitrary fractionations based on the general properties of proteins.

(i) Extraction

The non-protein nitrogen of plant tissues may be extracted by grinding in acid which is strong enough to precipitate all proteins. The methods of choice will depend on preliminary analysis of the tissue concerned, but the following will form the basis from which to develop suitable procedures.

For root nodule material, Aprison *et al.* (1954) used 3 N HCl to extract early products of nitrogen fixation. Routinely, in the author's laboratory 1 g (fresh weight) of nodules or similar tissue is extracted by crushing with a glass rod in 5 ml of 3 N HCl, centrifuging, and repeating the crushing and extraction of the pellet twice with 2 ml of 3 N HCl. The combined supernatants are then digested and analyzed for total nitrogen, as above. The same acid extraction method can be applied to centrifugally separated tissue fractions by adding equal volumes of 6 N HCl to suspensions of particles, or centrifugally clarified buffer extracts. The acidified mixtures, after standing at room temperature for 15 min, are clarified by centrifugation, the pellets re-extracted with 3 N HCl, and the combined supernatants analyzed as before. Suspensions of bacteria may be extracted in a similar fashion, but usually 10% w/v trichloroacetic acid is superior to 3 N HCl. The removal of soluble proteins from a buffer extract is best accomplished by making the solution 0.1 N in H_2SO_4 and 1–2% w/v in sodium tungstate, which effectively precipitates proteins and all but the smallest polypeptides.

Non-protein nitrogen is frequently extracted by treatment of tissues, bacteria, and other biological material, with boiling 80% v/v ethanol. Following clarification of the extracts by filtration or centrifugation, it is often necessary to remove the ethanol by evaporation to dryness, on a rotary evaporator. This applies especially when analysis for ^{15}N is to follow, because traces of ethanol give rise to spurious peaks in the mass spectrometer. Following drying, the sample is redissolved in a suitable buffer.

(ii) Ammonia, amide, and nitrate plus nitrite nitrogen

A relatively crude, but often useful analysis of non-protein nitrogen involves the successive removal of ammonia, amide, and nitrate plus nitrite nitrogen. A micro-Kjeldahl distillation apparatus such as that illustrated in Figure 4a is used.

Reagents

A. Sodium borate, pH 10. A saturated solution of borax ($Na_2B_4O_7.10H_2O$) is adjusted to pH 10 with NaOH and kept tightly stoppered to exclude CO_2.
B. NaOH, 40% w/v.
C. Devarda's alloy (BDH Chemicals, Poole, Dorset, U.K.) sieved to remove fine powder but passing 85 mesh.
D. Boric acid, 2% w/v.

Procedure. The sample size should be adjusted to contain approximately 100 μg of N in the least abundant fraction. The extract is carefully neutralized to about pH 4 with NaOH, and then run into the distillation vessel. An equal volume of saturated borate (pH 10) is added and the free ammonia is distilled and trapped in 5–10 ml of 2% boric acid. When distillation is complete, a fresh boric acid flask is placed in position, 5 ml of 40% NaOH added to the sample, and the amide nitrogen is distilled into 5–10 ml of 2% boric acid. When complete, the steam flow is stopped a further boric acid flask is placed in position, the sample flask is removed, 0.2 g of Devarda's alloy added, and the flask quickly reconnected to the still. The steam is passed, slowly at first to avoid undue splashing due to the reaction, and the nitrite and nitrate nitrogen are collected as distilled ammonia. Ammonia in the distillates is quantified as previously described. Nitrogen remaining after these distillations is likely to be very largely amino and nucleotide nitrogen. It may be estimated by difference or analysed in a separate aliquot of extract.

(iii) Sensitive indirect estimation of ammonia

A specific enzymatic–radiochemical technique for estimating ammonia in the range 0.1–10 nmol has been developed by Kalb *et al.* (1978). Commercially available enzymes and radioactively labelled chemicals are used. The basis of the method is the following reaction:

$$\alpha\text{-}[1\text{-}^{14}C]\text{ketoglutarate} + NADPH + NH_4^+ \xrightleftharpoons{\text{glutamate dehydrogenase}} [1\text{-}^{14}C]\text{glutamate} + NADP^+$$

in which, with the α-ketoglutarate and NADPH in excess, quantitative yields of glutamate are produced from ammonia. The remaining radioacitve α-ketoglutarate is then decarboxylated by treatment with H_2O_2 and the $^{14}CO_2$ removed by heating. Interference in the above reaction due to accumulating $NADP^+$ is avoided by including glucose-6-phosphate and glucose-6-phosphate dehydrogenase in the reaction to regenerate NADPH. Radioactivity in the mixture is then measured by scintillation methods, providing a direct estimation of NH_4^+ originally present in the reaction. The method has an additional advantage in that, by measuring NH_4^+ before and after glutamine hydrolysis, determination of glutamine is also possible.

In principle, this method is widely applicable but it must be adapted to specific experimental requirements because of possible intereference by components present. Kalb *et al.* (1978) give certain modifications for simple or complex reaction mixtures. Those wishing to use this method should therefore consult the original paper and develop the assay for their requirements. Thus, R.N.F. Thorneley (personal communication) has developed an adaptation for the estimation of minute amounts of NH_4^+ produced in rapid stopped-flow experiments with nitrogenase.

(iv) Amino nitrogen

After removal of ammonia and amide nitrogen from extracts, α-amino nitrogen can be analyzed using an adaptation of the ninhydrin method in which the coloured reaction product is decomposed in acid with the production of ammonia (Kennedy, 1965). The use of the ninhydrin colour is inappropriate for the estimation of mixtures of amino acids in unresolved extracts. Its use in the analysis of resolved amino acids is common and will be briefly described later (Section c.v).

Reagents

A. Acetate buffer. Sodium acetate trihydrate (270 g) is dissolved in 200 ml of CO_2-free distilled water, 50 ml of glacial acetic acid are added and the solution is diluted to 750 ml (pH 5.4).
B. Ninhydrin, 3% w/v in distilled water.
C. Hydrochloric acid, 2 N.
D. Hydrogen peroxide, 30% w/v (100 volume).

Procedure. Residues of extracts after distillation of ammonia and amide nitrogen are neutralized with H_2SO_4 and evaporated to dryness in a rotary evaporator. Excess salts are removed by dissolving the residue in hot 80% ethanol, which is transferred to a fresh flask and again evaporated to dryness. This residue, which should contain 50–100 μg of amino nitrogen, is dissolved in 1 ml of water and 1 ml of pH 5.4 acetate buffer and 2.0 ml of ninhydrin are added. The mixture is heated in a boiling water-bath for 10 min. The ninhydrin–ammonia complex is then destroyed by addition of 4.0 ml of 2 N hydrochloric acid and heating for a further 10 min. The released ninhydrin is then destroyed by treatment with 0.25 ml of hydrogen peroxide for 5 min at 100°C. Ammonia is recovered by steam distillation as described above, and estimated by titration or by colorimetric analysis of a measured aliquot of distillate. It may also be used for isotopic analysis (Section e.vi).

Alternatively, aliphatic amino nitrogen may be analysed by the modified Van Slyke method (Kenten, 1956, Bergersen, 1965), but the apparatus is not widely available.

(v) Amino acid analysis

The analysis of amino acids is a subject in itself and detailed treatment of methods is not warranted here. The methods of choice depend upon the subject of the investigation and the equipment available. Sources of well established methods will be cited, and principles to be followed in relation to nitrogen fixation studies will be briefly outlined. The methods available span the range from paper chromatography, for qualitative and semi-quantitative analyses, to sophisticated automatic analysis.

Paper and thin-layer chromatography; paper electrophoresis

Principles of paper chromatography of amino acids were described by Hellmann (1956). Smith (1969) and Zweig and Whitaker (1971) give data about paper selection and preparation of solvents, R_F values, and location reagents. Electrophoresis of amino acids on paper and a combination of chromatography and electrophoresis have often proved useful for resolution of complex mixtures (Dreyer and Bynum, 1967; Smith, 1969; Zweig and Whitaker, 1971). Estimation of amino acids on paper can be effected by direct photometric scanning of the coloured spots which result from treatment with a number of reagents, but more commonly the spots are excised, eluted and after redevelopment of the colour, the amino acid is estimated spectrophotometricallly. If ninhydrin is used, the amino nitrogen may be recovered as described above, and used for isotopic analysis (Kennedy 1965) provided that the spot contains sufficient nitrogen. Alternatively, the amino acid may be digested as previously described, and the total nitrogen analyzed. Wolk *et al.* (1976) and Meeks *et al.* (1977) have described thin-layer chromatographic and electrophoretic methods for studying nitrogen fixation products in ^{13}N experiments.

Thin-layer chromatography on silica gel, alumina, or cellulose offers increased resolution of complex mixtures because of reduced spreading of spots. The methods may be adapted from micro to semi-preparative scales and will be generally applicable to nitrogen fixation studies. Comprehensive accounts of these methods can be found in Brenner and Niederwieser (1967), Pataki (1966), and Stahl and Ashwood (1969).

Ion-exchange chromatography

Following the classic paper of Moore and Stein (1951), the separation of amino acid mixtures by ion-exchange chromatography has become a standard procedure and forms the basis of most systems of automatic analysis. Standard reference works on these methods are Spackman *et al.* (1958), Jacobs (1966), Spackman (1967), and Hare (1977). Most of these methods have been developed for work with protein hydrolysates but are readily applicable to studies of nitrogen fixation products. Newer methods have concentrated on analysis in the nanomole and sub-nanomole ranges. These methods are not suitable for stable isotope analysis and it may be necessary to use commercially available preparative equipment which is fitted with effluent splitting and fraction collection facilities. If no commercially built equipment is available, the above references supply sufficient detail to permit the assembly of equipment which is appropriate to the research problems.

(vi) Ureides

Allantoin and allantoic acid are the principal nitrogen compounds exported from the nodules of some legumes. They are conveniently assayed (Herridge, 1977) in 80% ethanolic extracts of plant tissues (1 g fresh weight of tissue with 10 ml 80% v/v ethanol in water, extracted at −20 °C for 24 h, then homogenized, filtered, and washed twice with 80% ethanol), which have been taken to dryness and redissolved in distilled water. If green tissues are being analysed, interfering pigments should be removed by extraction with light petroleum.

Materials and reagents

A. Cellulose thin-layer plates (e.g. Eastman Chromatogram Sheet with fluorescent indicator).
B. Butanol–acetic acid–water (12:3:5) solvent.
C. Standards prepared from allantoin and allantoic acid in ethanol (0.2 mg ml^{-1}).
D. Ehrlich reagent. Dissolve *p*-dimethylaminobenzaldehyde (10% w/v) in concentrated HCl. Mix 1 volume of this with 4 volumes of acetone immediately before use.
E. Scanning densitometer for reflectance measurements (e.g. the Chromoscan, Joyce Loebl, Middlesex, U.K.)

Procedure. Alliquots (5 μl) of the extracts and standards are spotted on the thin-layer plates and developed with the solvent by the ascending technique at 22 °C. When the solvent front has risen to within 2 cm of the top of the plates, they are air-dried for 1 h, dipped in Ehrlich reagent, air dried again for 30 min and then heated at 75 °C for 5 min to complete development of the colour. The density of the yellow spots is measured at 400–435 nm using the intergrating densitometer in the reflectance mode and concentrations are estimated from values for standards run on the same chromatogram. The calibration graph is linear in the range 0–1.0 μg for both allantoin and allantoic acid.

(d) MEASUREMENT OF PROTEIN

Protein content is a frequently used standard upon which to base measurements of biological activity. In addition, there will sometimes be differences in total protein content, or in content of specific proteins, which express differences in nitrogen fixation. The following are some simple, widely used methods for measuring protein, but for methods involving preparation of purified specific proteins the reader is referred to standard works on ion-exchange chromatography of proteins (e.g. Fasold *et al.*, 1961; Kirkland, 1971; Jakoby, 1971), gel filtration chromatography (e.g. Determann, 1968; Kirkland, 1971; Jakoby, 1971), preparative techniques of electrophoresis (e.g. Jakoby, 1971), and affinity chromatography (Jakoby, 1974).

(i) Protein nitrogen

The residue following extraction of non-protein nitrogen (see above) may be analyzed for total N (Section b) as a crude estimate of protein nitrogen, the trichloroacetic acid and tungstic acid methods being preferred. Precipitation of soluble proteins in the presence of 0.1 N H_2SO_4 and 1–2% sodium tungstate is a standard procedure. Crude estimates of protein content are frequently obtained by multiplying values for total nitrogen obtained by Kjeldahl digestion, by 1/0.16 (6.25), since many protein preparations contain about 16% nitrogen.

(ii) Colorimetric assays for protein

Two methods will be described. The first, widely known as the Lowry method, is the more sensitive (1–100 μg of protein) and is recommended for soluble proteins and enzyme preparations. The second is useful in the range 10 μg–2 mg of protein and, in our laboratory, is preferred for estimating the protein of bacterial cells.

Method 1 (Lowry *et al.*, 1951)

Reagents

A. Na_2CO_3 (2% w/v) in 0.1 N NaOH.
B. $CuSO_4 \cdot 5H_2O$ (0.5% w/v) in 1% (w/v) Na or K tartrate.
C. Alkaline copper solution prepared by mixing 50 ml of reagent A with 1 ml of reagent B; discard after 1 day.
D. Carbonate–copper solution prepared as for reagent C, but omitting NaOH from reagent A.
E. Diluted Folin reagent. Titrate Folin–Ciocalteau phenol reagent (Fisher Scientific Co., New York, U.S.A., BDH Chemicals, Poole, Dorset, U.K.) with standard NaOH, to a phenolphthalein end-point. On the basis of this titration, dilute the Folin reagent to 1 N in acid (usually about 1:2).

Procedure. The samples should contain 1–100 μg of protein in about 0.2 ml. Add 1.0 ml of reagent C, mix, and stand for at least 10 min at 20 °C. Quickly add and mix 0.1 ml of reagent E. Stand for at least 30 min. Read the absorbance at 750 nm for 5–25 μg of protein; for higher protein concentrations, read at 500 nm. Any multiple of the volumes given may be used as required. Standards are conveniently prepared from crystalline bovine serum albumin, but for greatest accuracy the colour produced should be calibrated against a weighed portion of the protein being measured. For example, 1 μg of bovine albumin is equivalent to 0.97 μg of human serum protein (Lowry *et al.*, 1951). Proteins precipitated with tungstate, as described above, will usually dissolve easily in reagent C, but trichloroacetic acid or HCl precipitates and dried preparations dissolve slowly and may have to be heated. If this is done, the standards should be treated similarly.

If samples to be assayed contain significant non-protein nitrogen in the form of amino acids and nucleotides, some of which produce colour, the protein should be precipitated with tungstic acid, centrifuged, washed, and then redissolved as above.

Method 2 (Goa, 1953)

Reagents

A. NaOH, 3% w/v.
B. Trichloroacetic acid, 11% or 26% w/v.
C. Benedict's reagent. Prepared as follows. Dissolve 173 g of sodium citrate and 100 g of Na_2CO_3 in about 500 ml of distilled water, with gentle heating and stirring; do not boil. Filter while warm. Separately dissolve 17.3 g $CuSO_4 \cdot 5H_2O$ in about 100 ml of distilled water, with gentle heating. Mix the two solutions in a 1 l volumetric flask, cool, and make up to 1 l with distilled water. Keep tightly stoppered in the dark.

Procedure A (0.1–2 mg of protein). Centrifuge the bacteria in a glass centrifuge tube, re-suspend in 1 ml of water, add 4 ml of 11% trichloroacetic acid, and stand at room temperature for 10 min. Centrifuge and discard the supernatant. Repeat this at least once to remove non-protein nitrogenous compounds which react with Benedict's solution. Disperse the treated cells in 4 ml of 3% NaOH, warming if necessary until clear. Add 0.2 ml of Benedict's solution and read the absorbance at 330 nm after 15 min, using a reagent blank. Standards of bovine serum albumin should be prepared in 1 ml and treated in the same way. Sometimes a precipitate forms. This may be centrifuged away without affecting the coloured product.

Procedure B (10–400 μg of protein). Suspend the centrifuged bacteria in 0.2 ml of distilled water and add 0.1 ml of 26% trichloroacetic acid. Then proceed as in procedure A but using these smaller volumes of reagents. Finally disperse the cells in 0.5 ml of 3% NaOH and add 25 μl of Benedict's reagent. Record the absorbance at 330 nm, using microcuvettes.

(e) METHODS USING ^{15}N

(i) Preparation, properties, and natural abundance of ^{15}N

The existence of the heavy, stable isotope of nitrogen was recognized by Naude and its mass was measured by Jordan and Bainbridge (1936). The nucleus of this isotope contains one neutron more than the nucleus of the abundant atom and thus has an atomic weight of 15.0049 daltons (relative to O = 16.0000). Thode and Urey (1939) first produced useful amounts of this isotope which was almost immediately used in tracer experiments on the incorporation of dietary amino acids into proteins in rats (Schoenheimer and Rittenberg, 1939). Rittenberg had worked with Urey and then moved to Columbia University to pioneer the use of stable isotopes in biochemistry (reviewed by Hevesy, 1948).

The separation of the atoms of ^{15}N from the more abundant ^{14}N atoms makes use of the small physical and chemical differences which are found between simple compounds that contain them. Early separations employed electromagnetic methods to separate the heavy isotopes of a number of elements but these processes are not easily applicable to the separation of compounds that contain them. For example, the isotopes of N_2 gas can be separated in this way, but efficient conversion of the resultant enriched gas to convenient compounds is difficult. The method pioneered by Urey involved the isotopic separation which occurs when a gaseous compound of an element exchanges with a compound of the same element in solution. At equilibrium, the gas and liquid phases have different isotopic concentrations. The method is used for the separation of isotopes of O, C, N, and S. In the case of nitrogen, NH_3 gas is passed through a counter-flowing solution of an ammonium salt and the enriched NH_3 is easily recovered as an $^{15}NH_4^+$ salt, which can then be readily reacted to produce other chemical compounds. The separation reaction is due to the equilibrium constant of the following reaction being greater than unity:

$$^{15}NH_3(\text{gas}) + {}^{14}NH_4^+(\text{aq.}) \overset{K_{eq} = 1.033}{\leftrightarrows} {}^{14}NH_3(\text{gas}) + {}^{15}NH_4^+(\text{aq.})$$

A great variety of compounds labelled with ^{15}N enrichments up to 99% are now commercially available (Table 1). From all sources, the cost is directly related to the amount of isotopic element purchased.

The above equation illustrates a general property of reactions of isotopically labelled compounds. That is, although the chemical properties of the ^{15}N and ^{14}N atoms are almost identical, small quantitative differences are found owing to their different masses and different activation energies. For example, in addition to differences in chemical reactivity, the diffusion rates of molecules are inversely proportional to the square roots of their molecular weights and therefore $^{15}N_2$ will diffuse slightly more slowly that $^{14}N_2$.

Very soon after the first use of ^{15}N as a tracer, Burris and co-workers at the University of Wisconsin commenced to use $^{15}N_2$ enriched atmospheres to study N_2 fixation. One of the first publications established that there was no detectable discrimination due to mass and no detectable mass fractionation in the course of N_2 fixation experiments with *Azotobacter vinelandii* (Burris and Miller, 1941). It was clearly shown that ^{15}N-labelled N_2 behaved as unlabelled N_2 within the limits of experimental error. Little attention has been given subsequently to mass discrimination effects in N_2 fixation experi-

Table 1. Some commercial sources of ^{15}N labelled compounds

Company	Address	Compounds available
Bio-Rad Laboratories	32nd and Griffin Ave., Richmond, Calif. 94804, U.S.A.	NH_3, NH_4 salts, NHO_3, KNO_3, KNO_2, $NaNO_2$, $NaNO_3$, amino acids, and other compounds on request
Monsanto Research Corporation, Mound Laboratory	P.O Box 32, Miamisburg, Ohio, 45342, U.S.A.	NH_3, $(NH_4)_2SO_4$, HNO_3, KNO_3, N_2
Isocommerz GmbH Exportkontor	Permoserstrasse 15, 705 Leipzig, G.D.R.	NH3, NH_4^+ salts, HNO_3, nitrate salts, N_2, amino acids, etc., many compounds on request
Bureau des Isotopes Stables, Centre d'Études Nucléaires de Saclay	B.P No. 2, 91190 Gif-sur-Yvette, France	NH_3, NH_4^+ salts, HNO_3, N_2, and other compounds

ments. This is probably because these experiments are usually of short duration and the effects are small enough to be neglected. However, when N transformation processes occur in nature over prolonged periods, variations in the abundance of ^{15}N in various compounds are readily seen (e.g. Cheng *et al.*, 1964, detected differences between various soils in the enrichment of ^{15}N). These variations are probably almost entirely due to isotopic discrimination effects such as those observed by Wellman *et al.* (1968) in which reduction of nitrate by *Pseudomonas stutzeri* favoured the reduction of $^{14}NO_3^-$ over $^{15}NO_3^-$ by a factor of 1.02. It should be noted that this value is similar to the K_{eq} in the above equation. Long-term studies of N_2 fixation in enclosed soil systems should probably take account of these effects. However, the more immediate practical effect is that the natural isotopic content of various N compounds in the environment is often found to be different from that of atmospheric N_2.

Most text-books give the natural abundance of ^{15}N as 0.365 atoms-% i.e. 0.365% of the atoms of a sample of a natural N compound will be ^{15}N. This value is obtained by analysis of the N_2 of the atmosphere. In our laboratory, such measurements have ranged from 0.362 to 0.368 atoms-%. With this low concentration, effectively all of the ^{15}N is present as molecules of $^{14}N^{15}N$ with a mass of 29 daltons. Equilibrium samples of other compounds derived from atmospheric N_2, should also have the same composition. However, as noted above, mass discrimination effects can alter the abundance of ^{15}N slightly in natural materials, and in laboratory chemicals large discrepancies from 0.365 atoms-% ^{15}N are sometimes found. These matters will be referred to when control material for $^{15}N_2$ fixation experiments is being considered.

(ii) Early uses of $^{15}N_2$

The use of $^{15}N_2$ incorporation as a measure of N_2 fixation was rapidly developed at the University of Wisconsin in 1941–42 and the first full description of an isotopic N_2 fixation experiment was given by Burris (1942). He showed that glutamic acid contained the highest ^{15}N content following a 90-min exposure of a growing culture of *Azotobacter vinelandii* to $^{15}N_2$. This amino acid was therefore an important early product of the N_2 fixation process. In the same year, Burris *et al.* (1942) showed that cultures of *Rhizobium leguminosarum* did not incorporate ^{15}N from $^{15}N_2$, whilst nodulated plants of several legumes did incorporate the isotope when is was supplied as $^{15}N_2$ to the roots, but not when it was supplied to the plant shoots. They also clearly showed that the fixed N was distributed throughout the host plant. Thus, in one short paper, analytical proof was obtained that N_2 was fixed only within the nodules and that it was rapidly used by plants. These findings had been accepted by most biologists for many years, but no direct analytical evidence had previously been obtained.

In these early experiments, the isotopic analysis was carried out with a mass spectrometer built in the Physics Department of the University of Wisconsin. The development of ^{15}N techniques and their use in other laboratories was limited mainly by the restricted availability of mass spectrometers. The techniques for preparing and handling ^{15}N-labelled material were developed quickly and their improvement still continues. Beginning in 1958–60, small mass spectrometers for isotope analysis were developed and became more widely available. With the advent of solid-state electronics and improvements in vacuum technology, adequate mass spectrometers are now within the resources of agricultural and biological scientists in most countries, often as a central analytical facility. For some types of work, optical ^{15}N analyzers may offer even cheaper analytical solutions but they are not suitable for all types of experiments. These instruments will be described later.

Experiments employing ^{15}N have been used for the investigation of many aspects of the biochemistry of N_2 fixation, the study of factors affecting N_2 fixation, confirmation of the ability of various species of microorganisms to fix N_2, studies of the translocation of symbiotically fixed N in plants, many N balance studies, and the calibration of the acetylene reduction technique. With the advent of the last-mentioned technique, the importance of ^{15}N has diminished in the field of N_2 fixation; however isotopic experiments remain the final method for the establishment of many qualitative and quantitative aspects of the subject.

(iii) Preparation of $^{15}N_2$ gas

Sources of ^{15}N-labelled materials are listed in Table 1. The list is not exhaustive but is given for the benefit of readers who do not have ready access to purchasing information. Information about sources may be obtained from the International Atomic Energy Agency, Vienna, Austria.

Nitrogen gas enriched with ^{15}N can be purchased from commercial sources but is is very expensive and may be difficult to handle unless special vacuum lines fitted with magnetic means of breaking seals under vacuum are available. An ammonium salt is a more convenient source of ^{15}N and can be readily converted to $^{15}N_2$ gas by oxidation. This must be done in an evacuated apparatus embodying facilities for transferring the $^{15}N_2$ to a suitable storage vessel. It is desirable to build such apparatus in, or near to, the location in which the experiments are to be conducted. The preferred oxidation is carried out with copper (II) oxide at about 600 °C.

$$2NH_3 + 3CuO = N_2 + 3Cu + 3H_2O$$

In this reaction, traces of oxides of nitrogen may be produced but the reaction is otherwise very clean. Oxidation with hypobromite can also be used, but this method usually produces more impurities and in practice seldom gives quantitative recoveries of N_2.

The apparatus in use in the author's laboratory is illustrated in Figure 7. It consists of an ammonia generator (G), connected through an NaOH-filled tube to an evacuated system for circulating the NH_3 over CuO in a silica combustion tube (C), which can be heated in an electric furnace. The gas is circulated by means of a simplified Toepler pump (TP). There is a mercury manometer (M) and a gas reservoir vessel (R) forms a part of the system.

The procedure is as follows. The CuO is regenerated before each use by heating the tube (C) with the system filled with O_2. While C is cooling, the

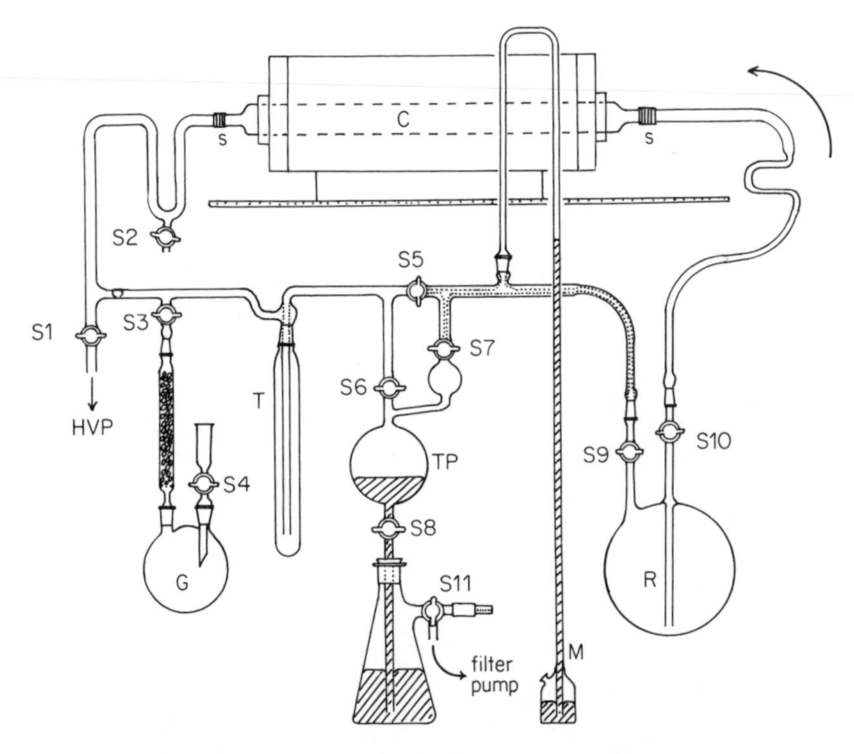

Figure 7. Apparatus for production of $^{15}N_2$ gas from a $^{15}NH_4^+$ salt. It consists of a quartz combustion tube (C), filled with CuO granules, and joined through graded seals (S) to a Pyrex vacuum line as shown; a rotary vacuum pump (HVP) gives a vacuum of about 0.01 torr. G is an NH_4 generator in which the salt is placed and NH_3 liberated by addition of alkali through S4. T is a liquid nitrogen-cooled trap for N impurities, TP a simple Toepler pump for circulation of the gas in the system and R is the gas reservoir into which the $^{15}N_2$ is finally pumped. A mercury manometer (M) is also provided. For details of operation, see text

sample of $^{15}NH_4^+$ salt is placed in G and the entire system evacuated by a vacuum pump (HVP). The vacuum is maintained at 0.01 torr for at least 1 h to ensure that the contents of the system are well out-gassed and a test should be run to see that no leaks are present. The pump is then disconnected by closing S1. The mercury in TP should be raised up to S6 and S7. A solution of NaOH (40%) is then run into G from the funnel and G is then heated gently. The NH_3 gas is then displaced into the system by gentle boiling of the liquid in G until only about 2 cm of the solid NaOH pellets remain in the tube (these prevent water entering the system). The mercury in TP is then lowered to draw the last of the NH_3 into the system and S3 is closed. (The generator should immediately be rinsed because the corrosive contents etch

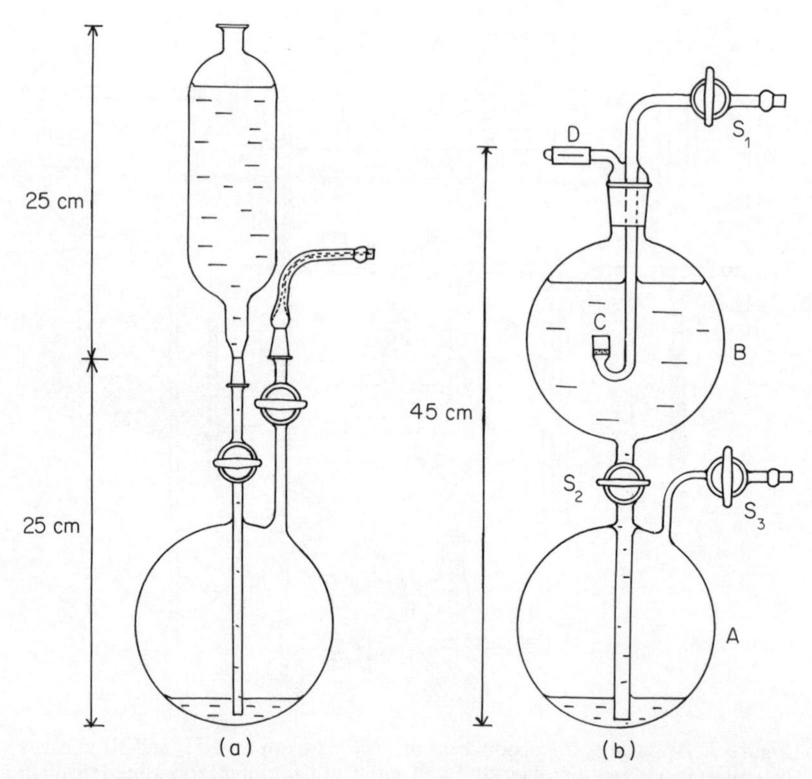

Figure 8. $^{15}N_2$ reservoirs. (a) The gas reservoir shown in Figure 7 has been fitted with a reservoir for displacement fluid and an attachment for a rubber connection to a gas manifold. (b) A gas reservoir for the preparation of ^{15}N gas mixtures and for anaerobic mixtures. The $^{15}N_2$ is kept in the lower bulb (A) under a displacement fluid in B which can be purged by passing argon or helium through C; D is a rubber Bunsen valve. S_1,S_2, and S_3 are high-vacuum stopcocks

the glass.) The furnace is then heated to 600 °C and the NH_3 gas circulated in the system, in the direction of the arrow, by means of the Toepler pump for 1 h after the manometer (M) indicates that a constant volume has been reached. This ensures completion of the reaction. Finally, the trap (T) is cooled in liquid N_2, and the gas is circulated to remove impurities. S10 is then closed and the N_2 gas pumped into R. The Toepler pump is operated as follows. With the mercury lowered to S8, S5 and S6 are closed, S7 is opened, and the mercury raised to S7, which is then closed. The mercury is then lowered and S6 opened, thus again charging the pump with gas from the system. The mercury is raised by connecting S11 to the atmosphere (or a slight positive air pressure) through a capillary, and lowered by connecting S11 to a vacuum provided by a filter pump.

(iv) Storage of $^{15}N_2$ and preparation of gas mixture

After collection of the $^{15}N_2$, the pressure in the container will be less than 1 atm. Figure 8a illustrates a displacement fluid reservoir which can be added to the gas reservoir to provide a useful 'gasometer' from which the $^{15}N_2$ can be dispensed. The displacement fluid should be acidified to trap any residual NH_3 or NH_3 which may dissolve from laboratory air, and it can also be staurated with K_2SO_4 to reduce the solubility of the $^{15}N_2$ in the fluid and thus minimize loss of the isotope. The gas in this apparatus becomes contaminated with O_2 from air dissolved in the displacement fluid. This is of little importance for aerobic systems, but is unacceptable for anaerobic N_2 fixation experiments. For these, a storage container such as that illustrated in Figure 8b should be used. In this, the fluid in the upper chamber can be purged with O_2-free N_2. Further, an O_2-absorbing displacement fluid can be used, such as a solution of a chromium (II) salt. Pyrogallol should be avoided because it produces CO upon ageing; CO not only inhibits N_2 fixation, but also interferes with mass spectrometry.

Gas mixtures for use in $^{15}N_2$ fixation experiments are conveniently prepared by the use of a manifold, mercury manometer, and vacuum pump (Figure 9). For most accurate results, the gas mixtures should be prepared and stored in separate vessels from which a sample can be drawn for analysis (Figure 8) and from which it is transferred into evacuated experimental vessels. However, sometimes it is necessary to prepare the mixtures in the reaction vessels. This may be done using the manifold system (Figure 9) or by injecting $^{15}N_2$ through suitable injection caps (*cf.* the use of C_2H_2—Chapter II.3).

(v) Instruments for measuring ^{15}N concentrations

The use of ^{15}N depends upon the development of suitable instruments for the measurement of its concentration, in the presence of a large excess of

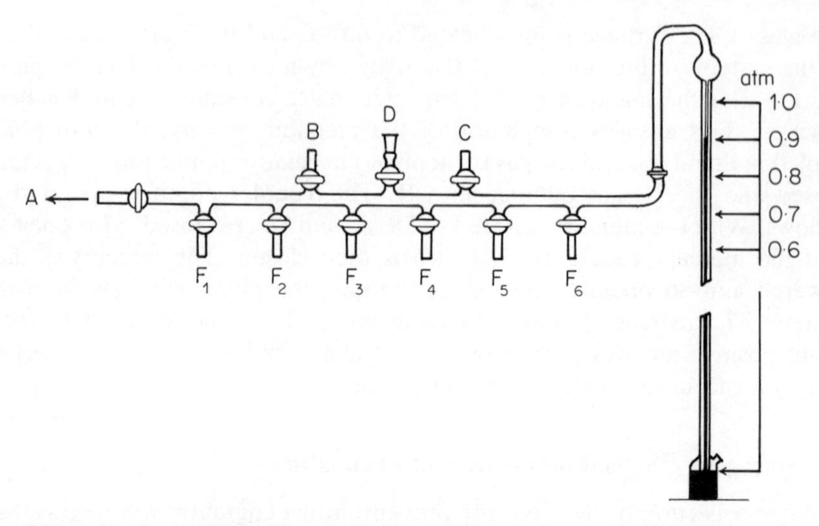

Figure 9. A manifold for preparation of gas mixtures and distribution to reaction vessels. It is constructed of 1–1.5 mm bore Pyrex tubing and stopcocks are of good vacuum quality. A vacuum pump is connected at A. The ^{15}N reservoir is connected to B, and the gas mixture reservoir to C. Sources of other gases are connected at F_{1-6} as required. E is a mercury manometer, calibrated in fractions of 1 atm. First, all connections and the gas receiver (Figure 8b) are pumped out and the system is tested for leaks. Next, the pump is disconnected, $^{15}N_2$ admitted to the desired pressure and then other gases in turn are admitted as required, the diluent gas (usually argon or helium) being added last. A sample of the mixture for analysis may be withdrawn at D. Reaction vessels are then connected at F_{1-6}, evacuated and filled with the mixture to the desired pressure

^{14}N. Similar considerations applied to the use of the stable oxygen isotope ^{18}O and the hydrogen isotope deuterium (D or ^2H). However, in the case of deuterium, the additional neutron doubles the mass of atomic hydrogen, leading to correspondingly large differences in physical properties. For example, relatively simple methods can be used to measure differences in the density of water containing D_2O (see, for example, Frances *et al.*, 1959, pp. 97—101). However, the only suitable instrument for ^{15}N and ^{18}O was the mass spectrometer, which had been developed originally by Thomson (1907) and improved for isotope work by Aston (1927), Nier (1940), and many others. More recently, optical methods have been developed and applied to the measurement of isotope abundance (Broida and Chapman, 1958). In the case of ^{15}N, all methods involve the conversion of a sample of a nitrogen-containing compound to N_2 gas, which is isotopically analyzed by measuring the proportion of molecules of masses 28, 29 and 30 present in the sample.

Mass Spectrometer

There are a great variety of designs of mass spectrometer but all have basically similar features, which are summarized in Figure 10. All parts of the analyzer are situated in a vacuum chamber at a pressure of 10^{-7}–10^{-8} torr (1 torr = 1 mmHg = 0.1333 kPa). Firstly, there is an ion source in which ions

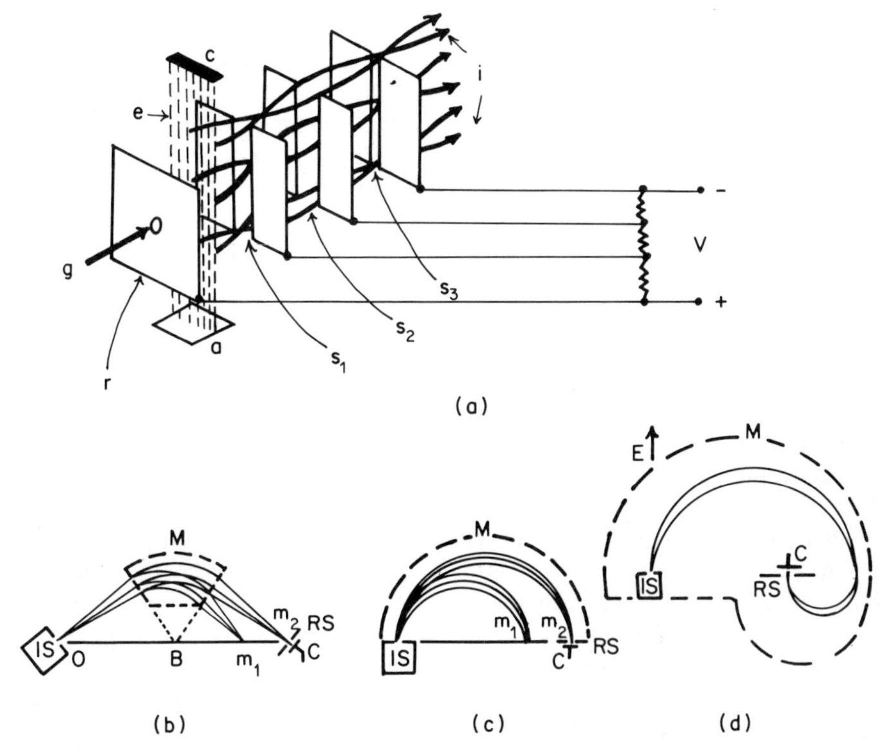

Figure 10. Mass spectrometer principles. (a) Diagram of an ion source. The gas sample (g) enters the source through a hole in the repeller plate (r). This gas is ionized by the electron beam (e) which passes between the heated filament cathode (c) and the anode (a) under the influence of an electric field of about 100 V. The electron beam is often collimated by a small magnetic field. The positively charged ions of the sample gas are accelerated and focused by electrically charged plates (r, s_1, s_2, and s_3). s_3 represents the object slit whose image is focused by the ion optics of the analyser system. (b)–(d) Schematic diagrams of the ion optics of mass spectrometers commonly used for isotope analysis. IS, ion source; RS, resolving slit; C, ion collector. The dashed line (M) indicates the boundary of the magnetic field whose direction is normal to the plane of the diagram. m_1 and m_2 represent beams of ions of different mass. [The diagrams are not to scale. In fact, analysers of type (b) are the largest and cycloidal analysers (d) may be very small.] (b) 60° sector type; (c) 180° sector type; (d) cycloidal type, in which an electric field (E) is superimposed at right-angles to the magnetic field

are produced from the material being studied. Secondly, there is a mass-resolving system in which some characteristic (kinetic energy or position in space) is imparted differentially to the ions of different mass emerging from the ion source. The third component is the ion collector, where ions of different mass can be collected, producing electrically measurable signals as their charges are dissipated.

Most mass spectrometers employ elecron bombardment for the ionization of a stream of vapour or gas. A tungsten or rhenium filament, heated by a stable current, is the source of electrons which are then directed by an electric field across the stream of vapour or gas. Only moderate electron energies (from about 4.5 V for caesium to about 25 V for helium) are required to strip an outer electron from molecules of the sample, producing positively charged ions. By increasing the electron energy to about 100 V, the probability of ionization is increased. The optimum conditions to be used are usually a function of each ion source design. The remainder of the source consists of a series of positively charged plates. The first, the repeller plate, bears the highest charge and is situated immediately behind the ionizing electron beam. The others are divided into segments by slits. This permits focusing of the ion beam by adjustment of the charge on each segment. Each succeeding plate is at a lower positive potential and the result is that the ions are accelerated down a potential gradient. The final pair of plates form the entrance slit through which the focused ion beam enters the analyzer chamber (see Figure 10).

There are a number of different types of mass-resolving systems, employing magnetic and/or electric fields to impart differentially curved paths to ions of different mass. Most instruments used for isotope work employ focusing magnetic analyzers which usually use 60°, 90°, or 180° sector magnets to provide a strong field whose direction is normal to the plane of the ion beam (Figure 10b and c). The magnets may be strong permanent magnets or electromagnets, the field strength of which can be varied by varying the magnet current and/or moving the pole pieces. In such a system, an ion of mass m (daltons), or kinetic energy V (volts), and moving in a magnetic field of strength B (gauss) exhibits a trajectory with a radius curvature (r) according to the expression

$$r = \frac{144}{B}\sqrt{mV}$$

Thus, by varying B or V, r for any given mass can be altered. A further property of ion beams in magnetic fields is that if they enter the field from a focus they will emerge from it to an approximate focus in the same plane as the initial focus (Figure 10). These properties enable a beam of ions of a given mass to be focused at a given point in the analyzer space by varying V or B.

Another type of analyzer system, which is widely used because of its compactness, is the cycloidal analyzer, which employs combined magnetic and electric fields at right-angles (Figure 10d).

A third principle is embodied in the time-of-flight mass spectrometer in which an undeflected, pulsed ion beam is used and the ions arrive at the collector in pulses in order of decreasing mass. This type of instrument is seldom used for isotope abundance work.

Ion-collecting systems usually consist of insulated collecting plates, screened by slits, and placed within deep Faraday screens. The collectors are connected to the grid leak resistance of an electrometer or to other sensitive input devices. When an ion beam is focused on the collecting plate, a small current is produced which is proportional to the ion current and which can be amplified and measured electronically. These currents may be recorded as a mass spectrum when the magnetic field strength or accelerating voltage is varied, causing successive ion beams to fall on the collector. In addition, many isotope mass spectrometers are equipped with devices which facilitate the determination of the ratios of ions of various mass which are present in a sample.

All of the functions outlined above require complex electronic circuits for their function. These are shown diagramatically in Figure 11, which represents a mass spectrometer in use in the author's laboratory. The ion source consists of two parts: (1) a circuit which maintains a very stable current through the filament which is the source of the electron beam, so that the resulting ion beam is very stable; this circuit also controls the intensity of the electron beam; and (2) a circuit which provides the ion accelerating voltage and allows the focusing of the ion beam. The measuring system consists of circuits which amplify the feeble ion currents into measurable signals (e–h) and a measuring device which measures the ion beam or measures the ratio of the beams of ions of different mass (i); a meter or recorder displays the resultant signal (j).

As mentioned earlier, mass spectrometers function at high vacuum, and this is usually provided by vapour diffusion pumps (oil or mercury) backed by a mechanical pump. (Ion pumps or sputter pumps should not be used for isotope ratio mass spectrometers because various gases are differentially pumped, thus affecting sample composition during measurement.) A further essential feature of the mass spectrometer is the sample inlet system. This is constructed of glass or metal with suitable stopcocks or valves and is also a vacuum system. It allows the evacuation of previous samples, the admission of new samples, and their expansion to a suitable low pressure in a sample reservoir which is large enough to provide a source of sample which does not lose pressure significantly during the course of measurement. The sample reservoir is connected to the ion source through a restriction (leak) which enables the sample to pass at a controlled rate into the ion source.

Full but simple descriptions of mass spectrometer principles can be found in Robinson (1960), Beynon (1960), Francis *et al.* (1959), and White (1968).

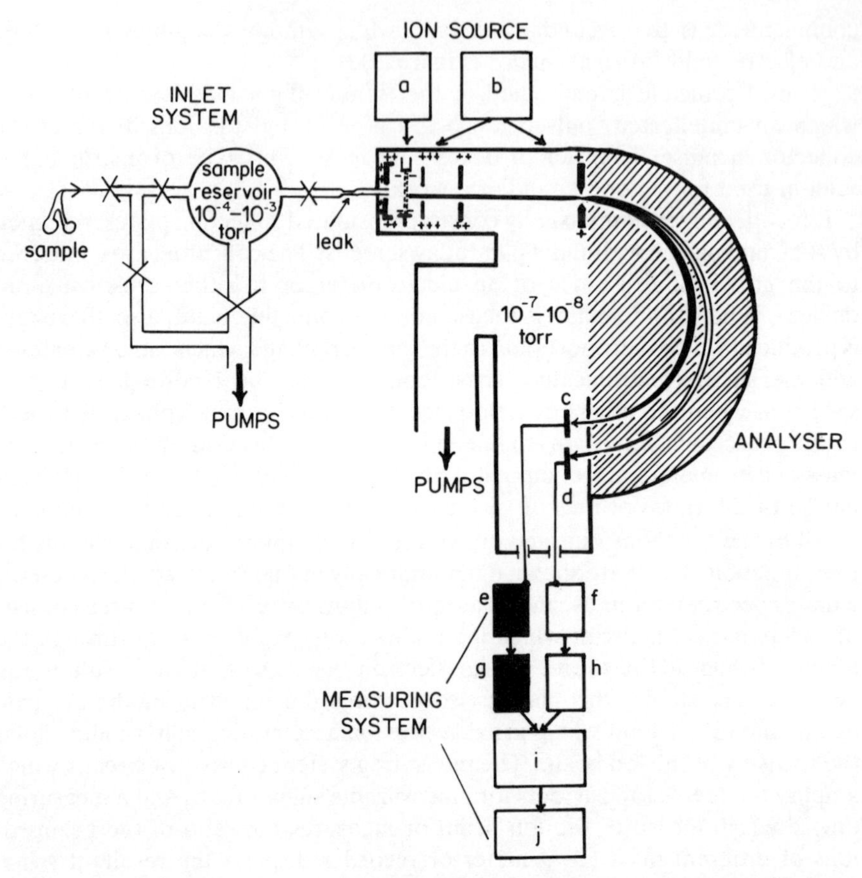

Figure 11. Schematic diagram of a ratio-measuring mass spectrometer used for isotope analysis. The shaded area represents a magnetic field, normal to the plane of the paper, of about 40000 gauss. The inlet system is evacuated by two oil vapour diffusion pumps in series, backed with a two-staged rotary pump. The analyser system is evacuated by two vapour diffusion pumps in series, backed with a two-stage rotary pump. The X symbols in the inlet system represent vacuum-tight valves. The ion source is controlled by a filament current regulator (a) and a high voltage supply (b). The lighter isotope ion beam is collected on c and the signal amplified through e and g. The heavier ion beam is collected on d and the signal amplified through f and h. The ratio of the two signals is measured in i and the result displayed on a recorder (j)

Emission Spectrometer

The emission spectroscopic method for the measurement of ^{15}N enrichment has become widely used and commercial instruments are available (e.g. see *Jena Review,* 1969, No. 3). The instruments make use of the isotope shift in the emission spectra of nitrogen when it is excited by microwave radiation. Different emission lines are produced by $^{14}N_2$, $^{14}N^{15}N$ and $^{15}N_2$ and their intensities corresponds to the concentrations of the various molecular species (see Broida and Chapman, 1958).

The instruments consist of a high-frequency microwave power generator with an antenna or resonant cavity for the transfer of the energy to the gas sample, which is contained in a tube constructed of special glass or quartz. The light emitted by the excited gas is collected and focused by a quartz lens and passed through a UV monochromator whose prism or grating can be driven to enable scanning of the spectrum in the range in which the bands of the most suitable N_2 transitions occur (usually 297.7 nm for $^{14}N_2$, 298.3 nm for $^{14}N^{15}N$ and 298.9 nm for $^{15}N_2$; Proksch, 1972). The intensity of the bands is measured by a photomultiplier and recorded as a mass spectrum. The atoms-% ^{15}N values are determined with reference to sealed tubes containing standard nitrogen samples whose ^{15}N enrichment has been accurately determined by mass spectrometry.

This apparatus is very suitable for measurements where a standard deviation of 3% for replicate samples is acceptable and where the ^{15}N values of the samples are higher than about 0.05–0.10 atoms-% excess. Its advantages are that it is suitable for routine use in experiments where many samples of small size (0.2–10 μg) are to be analyzed in large batches, such as in fertilizer nitrogen or nitrogen balance studies. The instrument is also cheaper than most mass spectrometers. An evaluation of the method in comparison with mass spectrometry was given by Proksch (1972) and an evaluation of a commercially available instrument by Keeney and Tedesco (1973).

(vi) Preparation of samples and mass spectrometric analysis

It is assumed that the sample consists of an acidic solution of NH_4^+ prepared as a distillate or diffusate as described in Sections b, c, and d. The problem is to determine the ^{15}N content of these samples. The first step is the conversion of the NH_4^+ into N_2 gas in an evacuated vessel. A number of different techniques have been described (e.g. Francis *et al.*, 1959; Smith *et al.*, 1963). The most convenient of them employ various modifications of the apparatus attributed to Rittenberg (1946).

Apparatus. Rittenberg vessels are constructed of glass with two limbs joined in an inverted Y-shape; a vacuum-tight stopper-stopcock and a suitable connection to vacuum line and mass spectrometer are located at the top (Figure 12). If adequate nitrogenous samples are available, the size of the vessel is not critical. However, in N_2 fixation experiments, analysis of small samples (say 50–100 μg of N) is sometimes required. This is assisted by the use of small Rittenberg vessels, which give higher sample pressures for a

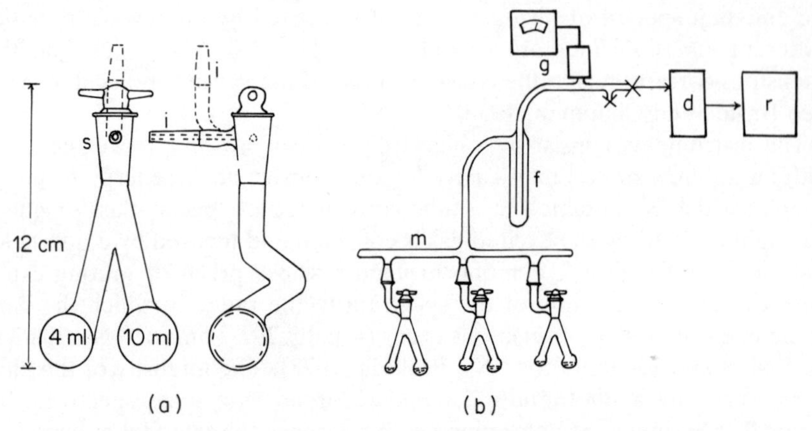

Figure 12. Apparatus for conversion of $^{15}NH_4^+$ to $^{15}N_2$. (a) Rittenberg tube. s is a vacuum-tight stopper/stopcock and i a suitable standard joint for connection to vacuum line and mass spectrometer. (b) A vacuum line for evacuation of Rittenberg tubes. m, manifold; f, trap cooled with liquid N_2; g, vacuum gauge, d, vapour diffusion pump; r, rotary vacuum pump

given amount of nitrogen. The size of sample to be used depends on the mass spectrometer design and sensitivity. Some types have expansion chambers of variable volume to compensate for sample size.

Reagents

A. Alkaline Hypobromite. Dissolve 500 g of NaOH in 688 ml of distilled water. Dissolve 8 ml of bromine in 70 ml of distilled water. To the bromine solution, chilled in ice, slowly add 40 ml of the NaOH with stirring. The bright yellow reagent is stable in a refrigerator for several weeks and, when prepared in this way, does not form the copious precipitate of NaBr encountered with other formulations (Francis *et al.*, 1959). In our laboratory, we have not encountered problems due to the evolution of O_2 from the alkaline hypobromite reagent. Sims and Cocking (1958) noted this problem as interfering with the correction for contamination of N_2 with air in mass spectrometric analysis and suppressed the evolution of O_2 by adding 0.1% w/v of KI to the reagent before use.

B. Sample. The sample should contain the required amount of nitrogen for analysis in 1 ml of acidified solution. If it is too dilute, it should be concentrated by boiling, but care should be taken to avoid crystallization of boric acid.

Procedure

1. The nitrogenous sample is placed in one limb of the Rittenberg vessel

and sufficient alkaline hypobromite to completely oxidize the sample (1.5 ml of reagent will oxidise 100 μg of N) is placed in the other limb.

2. The vessel is then capped, attached to a vacuum line (Figure 12) and evacuated to remove all traces of air, including air dissolved in the liquid. If the pumping rate is fast, the sample may freeze and pumping should continue for at least 15 min to remove adsorbed air.

3. The vessel is removed from the vacuum line after closing the stopcock. The reagent is then tipped into the sample and allowed to react for several minutes. The reaction is

$$3NaOBr + 2NH_3 \rightarrow N_2 + 3NaBr + 3H_2O$$

4. The Rittenberg vessel is attached to the inlet of the mass spectrometer, the limbs immersed in a dry-ice/ethanol mixture (temperature −70 to −80 °C), and the connection pumped out with the inlet vacuum system. This freezing step prevents the passage of corrosive bromine into the mass spectrometer and removes impurities such as the higher oxides of nitrogen. Liquid N_2 should not be used because it causes strong adsorption of the sample on the ice crystals.

5. The vacuum pumps are disconnected and the sample is expanded into the sample reservoir.

6. Admission of the sample into the analyzer chamber is done in such a way that the sample produces a mass 28 signal (peak) of a standardized size (usually near the maximum for the instrument). This is recorded (usually as a signal strength in volts, but sometimes as an ion current in microamps). The focus is then changed to bring the mass 29 on to the collector. Figure 13 shows the mass spectrum of ^{15}N-enriched samples of N_2. For manual determination of mass 29:28 ratios, the mass 29 signal is recorded and the ratio of the two signals is calculated. In double collector ratio instruments, when the mass 29 beam is focused on one collector, the mass 28 peak is focused on the other and the ratio of the two ion beams can be directly determined. In single collector ratio instruments, the mass 28 peak is 'backed-off' to zero and then the mass 29 peak is focused and compared with the 'back-off' voltage to determine the ratio.

(7) Finally, the mass 32 beam is focused on the same collector on which the mass 29 peak was originally measured. This enables a correction to be applied for traces of air which may have contaminated the sample.

(vii) Calculations, controls, and corrections

All calculations are made on the basis that atoms of ^{14}N and ^{15}N in the N_2 gas are in equilibrium. This is attained when NH_3 is oxidized to N_2 in the preparation of gas for N_2 fixation experiments or during preparation of N_2 for mass spectrometric or emission spectroscopic analysis. It is important that

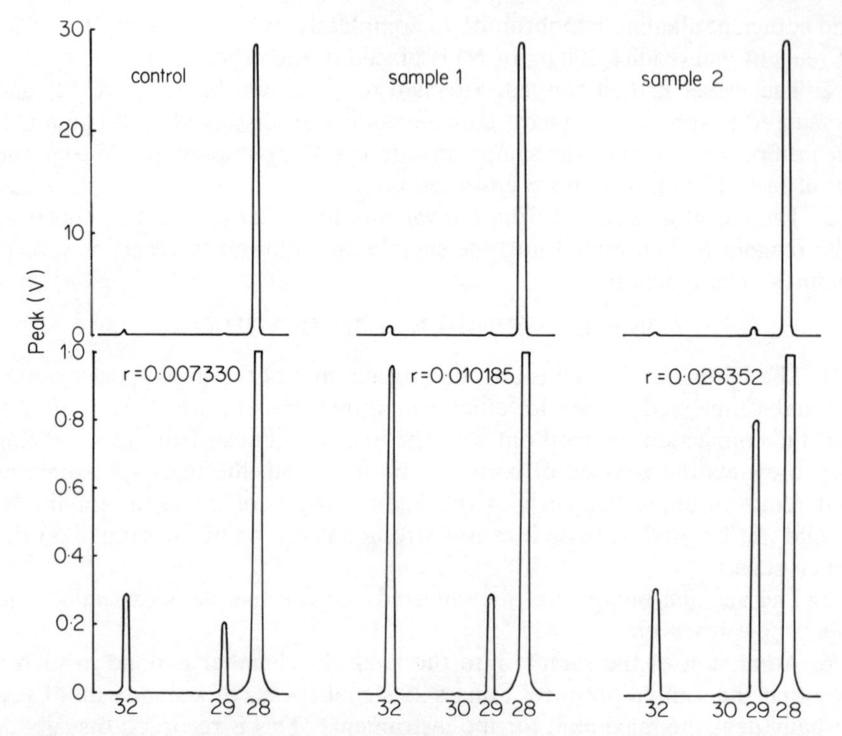

Figure 13. Mass spectra of three samples of N_2, two of which are enriched in ^{15}N. The upper spectra were recorded in the least sensitive range of the mass spectrometer and the lower spectra with a 30-fold greater sensitivity. Sample 1 has suffered an air leakage, as shown by the larger mass 32 peak. Calculation of atom-% excess ^{15}N values for these samples is given in Table 2

diluent or carrier nitrogen be provided before this step to ensure equilibration. A non-equilibrium situation results when unlabelled N_2 is added to a sample containing $^{15}N_2$. Such non-equilibrium mixtures can only be analysed by recording the concentration of all molecular species at mass 28, 29, and 30.

If the relative numbers of atoms of ^{14}N and ^{15}N in an equilibrium isotopic mixture are represented by p and q, the numbers of molecules of masses 28, 29, and 30 in the gas will be in the proportions $p^2{:}2pq{:}q^2$. The ratio measured is for the mass 28 and 29 beams. Therefore.

$$\text{ratios } 29/28 = r = 2pq/p^2 = 2q/p$$

Therefore,

$$\text{atoms-\% } ^{15}\text{N} = \frac{100q}{q+p}$$

$$= \frac{\frac{100q}{p}}{1 + p/q}$$

$$= \frac{\frac{100r}{2}}{1 + r/2}$$

$$= \frac{100r}{2 + r}$$

The value of atoms-% ^{15}N of the sample minus the atoms-% ^{15}N of the control nitrogen gives the 'atoms-% excess ^{15}N' value, which is almost universally used in N_2 fixation studies.

Alternatively, some workers prefer to use the enrichment function, δ, which is calculated as follows:

$$\delta = \frac{r_{\text{sample}} - r_{\text{standard}}}{r_{\text{standard}}}$$

The standard is N_2 of accurately known mass 29:28 ratio, perhaps generated from a standard sample of an ammonium salt.

Reference has already been made to the variability in the ^{15}N content found in naturally occurring nitrogen compounds due to mass-discrimination effects. For accurate results in N_2 fixation experiments, the natural abundance in the compound being studied should be determined for control material which has passed through the same analytical procedures as the ^{15}N-enriched material. Thus, in a legume experiment, a control sample of nodules, kept in air, should be extracted and analyzed in the same way as a sample of nodules exposed to $^{15}N_2$ and the difference between the atoms-% excess value of the two samples used to give the atoms-% ^{15}N excess. When carrier compounds are used, a control should be the biological material before exposure to ^{15}N, to which the carrier compound is added in the same way as it would be to the $^{15}N_2$-exposed material.

It is true that variations in natural ^{15}N abundance are small, but they contribute to discrepancies which may become important, particularly in experiments in which the distribution of fixed N_2 among various components is being determined.

In spite of all precautions, traces of air will be present in most samples when the mass ratio is measured, and nitrogen in this air will lower the isotopic enrichment of the sample. A correction can be applied if the mass 28 peak (which closely approximates the magnitude of the N_2 sample) and the mass 32 peak (O_2) are measured. The correction can be applied as a routine to the isotope enrichment.

Firstly, it is necessary to know the ratio of mass 28 and mass 32 peaks (measured on the same collector used for mass 28 in the isotope ratio measurement) when contaminating air alone is in the system. This factor can be obtained for residual air from a blank determination in which the reagents, but no NH_4^+-N, are present in the Rittenberg vessel. Sometimes an air leak may occur at some stage of the analysis. In this case the correction factor would be derived from the mass ratio 28:32 for a small sample of air admitted to the mass spectrometer. Now, let us suppose that the size of the mass 28 peak of the isotope sample is represented by X and the size of the O_2 peak found to be present in this sample is represented by Y. The mass ratio 28:32 for contaminating air was found to be 5.4 (a figure often found in practice for leaking air). Then the contribution made by contaminating air N_2 to mass 28 in the isotope ratio measurement would be $5.4Y$ or $(5.4Y \times 100)/X = A\%$. The true atoms-% ^{15}N excess is then given by multiplying the apparent value by $100/(100 - A)$. An example of a calculation based on the isotopic samples shown in Figure 13 is given in Table 2.

(vii) Enrichment, depletion, and dilution

Most commonly, ^{15}N is used in nitrogen fixation experiments as an enrichment of the gas-phase N_2. Fixed nitrogen is then traced as ^{15}N excess or enrichment in the total nitrogen of the total biomass, in components of the biomass, or in specific chemical components. Increases in ^{15}N content as a function of time may be sought, or pulse-chase experiments may be used in the investigation of assimilatory pathways.

During the manufacture of ^{15}N-enriched compounds, material depleted in ^{15}N is generated and can be used as isotopic tracer material. However, when completely depleted, the maximum difference in $^{15}N_2$ content due to uptake of $^{14}N_2$ is about 0.36 atoms-%, thus limiting the general usefulness of this material to certain long-term experiments and material with very high specific acitivity and low nitrogen content. Among other sources, 99.99% ^{14}N ammonium sulphate and nitrate (double labelled) is available from Monsanto Research Corporation (Table 1).

Sometimes it is not possible to use $^{15}N_2$ in a contained gas phase, for example in field experiments. In these circumstances techniques of isotope dilution have been devised (Chapter III.4). In these, soil nitrogen is enriched with ^{15}N and the rate at which plant ^{15}N is diluted gives an estimate of N_2 fixation.

(ix) Designing ^{15}N experiments

We have considered the essentials of the technology of ^{15}N experiments in N_2 fixation. However, the performance of these experiments is an art, learned

Table 2. An example of the calculation of atoms-% excess ^{15}N from mass 29:28 ratios. The mass spectra in Figure 13 were used. For other details see text

				Correction for residual air in sample				Corrected atoms-% excess
				Peak height (V)				
Sample No.	Sample ratio (R)	Atoms-%	Atoms-% excess (E)	Mass 28 (X)	Mass 32 (Y)	$\frac{5.4\,Y \times 100}{X} = A$	$\frac{100}{100 - A}$	$E \times \frac{100}{100 - A}$
Control	0.007332 0.007328	0.3652	—	28.56	0.28	—	—	—
1	0.010186 0.010184	0.5066	0.1414	28.62	0.96	18.133	1.221	0.1727
2	0.028356 0.028348	1.3978	1.0326	28.76	0.33	6.196	1.066	1.1007

only by experience. Repeating experiments exactly as done by others is imitative and is therefore less of an art. It may be useful but it is not as creative, even when applied to novel experimental material. Each type of experiment should be designed to give the information required. The following matters should be carefully considered when designing these experiments.

The purpose of the experiments

The most important uses of $^{15}N_2$ have been in short-term experiments. These have ranged from the simple establishment of N_2 fixation by a bacterial culture or symbiotic system (e.g. Bergersen and Hipsley, 1970; Bergersen *et al.*, 1965) to complex experiments about detailed enzyme mechanisms involved in N_2 fixation (e.g. Bergersen and Turner, 1973). In the first type, establishment of ^{15}N enrichment from N_2 gas into the N of the bacteria or plant tissue is all that is required. Quantitative results are not very important. In the latter type experiment, quantitative aspects are of paramount importance and great care must be exercised to account for and correct as many sources of error as possible. Varying needs for quantitative data must be considered when designing experiments. For many purposes, comparative results are sufficient, e.g. 'Does system A or treatment A fix more N_2 than B?' In these cases, provided the absolute amounts of N being analysed are identical in A and B, the relative enrichments (atoms-% ^{15}N excess values) of A and B are sufficient to provide the answer. If the question is 'By how much does N_2 fixation by A exceed that of B?', then additional information is required, *viz.* the total N content of A and B, and the atoms-% ^{15}N of the experimental gas phase; the $^{15}N_2$ concentration in the gas phase should be non-limiting.

Various types of enclosures have been used for $^{15}N_2$ fixation experiments. One of the important points to consider is that the gas space must be sufficient to contain adequate O_2 for aerobic systems. Burris (1974) has described some systems used in his laboratory. Warburg vessels have been used and some special vessels have been designed (e.g. Figure 14). Generally, flasks or bottles closed with injectable caps are adequate.

Long-term N_2 fixation experiments are much more difficult, chiefly because of the difficulty of maintaining constant gas-phase conditions in a closed system. Where such experiments involve soil in which other parts of the nitrogen cycle are active, the only way to measure N_2 fixation is by the construction of a complete ^{15}N balance for the system. Ross *et al.* (1964, 1968) have used gas lysimeters for this type of experiment, recovering 99–100% of ^{15}N added. Modifications to the apparatus were described by Stefanson (1970). Several of these lysimeters, lit by artificial light, have been used in experiments with growing plants over a period of 40 days. Because of the complexity of the equipment necessary and the large amount of analytical work generated, the purposes of long-term N_2 fixation experiments should be very carefully

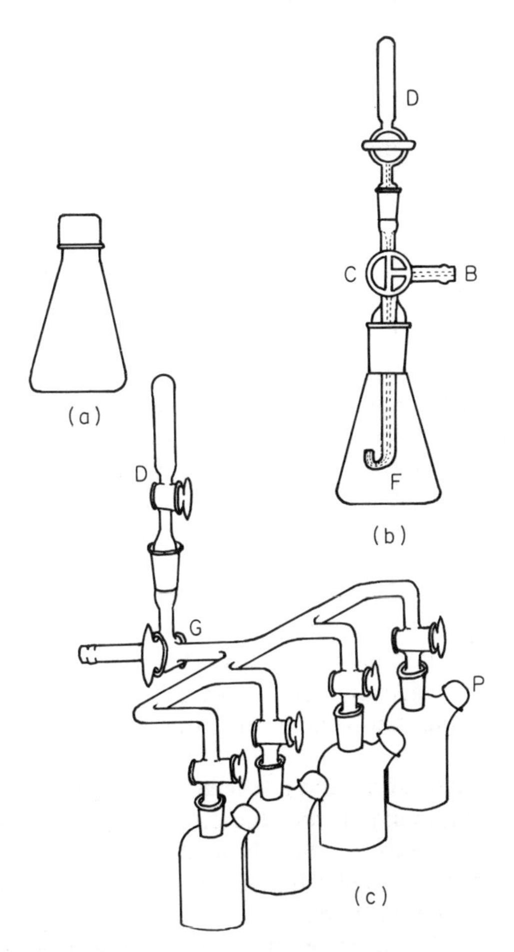

Figure 14. Examples of incubation vessels for $^{15}N_2$ experiments. (a) Flask with rubber injection closure (A). It can be evacuated and filled by connecting to the manifold (Figure 9) where F_{1-6} have been terminated with pieces of rubber tube, each bearing a hypodermic needle. Alternatively, gases may be added to the flask by injection. (b) A flask (F) with gas sampling attachment (Bergersen, 1963) in which evacuation and filling with gas mixture takes place through B. The three-way tap (C) allows evacuation of the sample bulb (D) and collection of a gas sample from F. The assembly is usually mounted on a support for shaking in a thermostat. (c) A manifold for reaction vessels, four of which are shown. The assembly actually bears 16 places and is mounted for shaking. Sampling of gas from each vessel is possible, via the three-way stopcock (G). In addition, each vessel has a rubber-capped injection port (P)

considered before they are initiated. However, long-term experiments about uptake of fertilizer ^{15}N by growing plants are much simpler (see, for example, Martin *et al.*, 1963; Wetselaar *et al.*, 1973).

What level of ^{15}N enrichment should be used?

Preliminary information is needed before deciding on the level of ^{15}N to use in the N_2 for experiments. Firstly, one should know the amount of N_2 needed for an adequate sample to be used in mass spectrometer or ^{15}N emission spectrometer. This will guide in the establishment of sample sizes. The sensitivity and reproducibility of the ^{15}N measurement should also be known before designing the experiment. For example, in our laboratory, duplicate determinations using 50 μg of N are adequate to measure enrichments as low as 0.005 atoms-% excess; however, the variations in biological material make 0.010 atoms-% excess a more prudent lower limit. Other mass spectrometers may require larger samples (e.g. Smith *et al.*, 1963; 250 μg of N). Emission spectrometry requires less sample (1–10 μg of N; Proksch, 1972) but the accuracy of ^{15}N measurement is not as good as with mass spectrometry and generally the lower limit of enrichment is considered to be 0.05 atoms-% excess.

Next, it is necessary to know the nitrogen content or the content of the particular nitrogen compounds to be studied in the experimental material. From this and the sensitivity of the ^{15}N analysis, the sample size can be calculated and the degree of replication decided.

Finally, an estimate should be made of the order of magnitude of the N_2 fixation to be expected in the experiment. In general, it is the lower limit which is important. If we underestimate the N_2 fixation, most methods are adequate to deal with high enrichment. However, it is possible to fail to detect N_2 fixation if we overestimate its magnitude and use an inadequate ^{15}N enrichment in the N_2 of the experimental gas phase.

These matters are best illustrated by an example. Let us suppose that it is desired to measure N_2 fixation into the non-protein nitrogen of samples of detached soybean nodules.

1. Mass spectrometer: sample size, 50 μg N as N_2; ^{15}N detection limit, 0.010 atoms-% excess ^{15}N.
2. Soybean nodules contain about 2.5 mg of non-protein N per gram (fresh weight). Thus, 1 g (fresh weight) of nodules would provide sufficient N for five replicate determinations.
3. Soybean crops have been reported to fix 0.2–2.0 g of N per plant in 90 days. Plants bear an average of 10 g of nodules over this period. N_2 fixation would therefore by 9–90 μg N g^{-1} (fresh weight) nodules per hour. That is, N_2 fixation amounts to between 9/2500 and 90/2500 or 0.36–3.6% of the non-protein N of the nodules in a 1-h experiment.

4. We shall require for this experiment that the lower limit of N_2 fixation should give ten times the minimum enrichment (see 1 above)—that is 0.100 atom-% excess. If the gas-phase N_2 contained 100% of ^{15}N, the enrichment obtained in a 1-h experiment would be expected to be 0.360–3.600 atoms-% ^{15}N excess. Therefore, we should use N_2 gas containing about 30 atoms-% ^{15}N.
5. The experimental gas mixture should contain about 0.2 atm of ^{15}N-labelled N_2. This is about three times the K_m (N_2) value for detached nodules, thus giving close to maximum rates of N_2 fixation.

Special Application

With modern isotope ratio mass spectrometers such as the Micromass 602 or 903 (VG-Micromass, Winsford, Cheshire, U.K.), it is possible to measure with confidence $\delta^{15}N$ values as low as 0.5‰, or 0.00018 atoms-% ^{15}N excess. This level of accuracy permits the measurement of nitrogen fixation of about 70 pmol of NH_3 in a reaction mixture exposed to 95 atoms-% $^{15}N_2$, and to which 600 μg of carrier NH_3-N has been added to promote the recovery of the labelled NH_3 which has been fixed. This method has been tested recently in our laboratory and is perhaps an alternative to the indirect enzymatic–radiochemical estimation of ammonia (section c, iii) for use in rapid stopped-flow reactions or other applications in which measurements in the nmole to pmole range are required.

Economies with $^{15}N_2$

Isotopic N is expensive and it is therefore desirable to avoid wastage. With anaerobic experiments, the gas space can be kept to a small volume which is just large enough to avoid significant changes in pN_2 during the course of the experiment. However, with aerobic N_2-fixing systems, changes in pO_2 influence N_2 fixation rates greatly and gas volumes must be large enough to avoid significant reduction in pO_2 due to respiration during the course of the experiment. For longer term experiments, addition of O_2 by some automatic device is possible, but this may be more expensive than the saving of ^{15}N which is effected by the use of a smaller gas space.

Economy of $^{15}N_2$ can also be obtained by using low pN_2 values with a diluent inert gas such as argon or helium to make up 1 atm pressure. For comparative purposes 0.1 atm $^{15}N_2$ can be used but, for experiments relating results to N_2 fixation in natural conditions, 0.78 atm $^{15}N_2$ would be necessary. Some savings of $^{15}N_2$ can be effected by using 2 or 3 limiting N_2 concentrations and calculating what the fixation would have been at 0.78 atm N_2 (Bergersen, 1970). This technique is not advised unless the experimenter is very experienced.

Some workers have advised the use of low enrichment of ^{15}N as an economy measure and have developed techniques for this purpose. Whilst there are circumstances in which this approach may be justified, it is this author's opinion that, in general, the lower enrichments cause increased analytical restraints, require greater replications for precision, and in several other ways generate additional costs which offset savings generated from lower isotope usage.

Another method of economising on $^{15}N_2$ is the recovery of gas mixtures for re-use. In practice this is rarely done because of the hazard of accumulating gaseous products (e.g. ^{15}N-labelled oxides of nitrogen or NH_3) which can be assimilated by processes other than N_2 fixation and so give spurious results. Other gaseous products may also accumulate, leading to unexpected problems. However, if the $^{15}N_2$ is re-purified after each experiment and then used to make fresh gas mixtures, these problems are minimized. Purification would involve removal of O_2, CO_2, and NH_3 by absorption, and of oxides of nitrogen by trapping at liquid N_2 temperatures, in a suitably designed gas system. Use of this method would perhaps be justified if large volumes of gas mixtures could be recovered over a period of experiments. However, most workers in N_2 fixation accept that the $^{15}N_2$ used in experiments is expendable because of the high cost of recovery.

(f) METHODS USING ^{13}N

(i) Background

Although the short-lived radioactive isotope ^{13}N (half-life 9.96 min) was used as early as 1940 for nitrogen fixation experiments (Ruben *et al.*, 1961), it is still not widely used. However, recently there have been several reports from Michigan State University of its use in experiments tracing products of fixation in blue-green algae (e.g. Meeks *et al.*, 1978). The short half-life of ^{13}N indicates clearly that this isotope must be produced in a facility near to its point of use. It is therefore likely that use of ^{13}N in nitrogen fixation experiments will continue to be restricted, and hence details of the methods will not be given here. Instead, an outline of the preparation and uses of the isotopes will be given. Readers are advised to consult original publications for further information. The techniques available and the precise data which they provide outweigh the daunting prospects of developing a capacity for work with short-lived isotopes.

^{13}N is a positron emitter (β^{+}_{max} 1.19 MeV) which may be detected by means of the 0.511-MeV annihilation radiation. Technology for measuring this high-energy emission is now available and offers opportunities for high resolution. The short half-life of ^{13}N means that samples from double-labelling experiments may be assayed for high-energy emissions and then, after decay of the ^{13}N, for the low-energy emission of ^{14}C, for example.

(ii) Preparation of ^{13}N

Production and use of ^{13}N in biological experiments has recently been reviewed by Straatmann (1977), who cites several recent reviews dealing with problems of ^{13}N production. Some processes have utilized bombardment of various targets with protons or deuterons in cyclotrons. Others have utilized Van de Graff accelerators. The method chosen is obviously dominated by the local availability of these machines. Austin *et al.* (1975) described a batch technique which produces useful yields of labelled N_2 gas with radioactivity of about 20 mCi ml^{-1}. Amorphous carbon targets (17 mg) highly enriched in ^{13}C are used; they are surprisingly cheap. These targets are irradiated in a cyclotron for 20–30 min with 1–4-μA beams of 7-MeV protons which leave the target with an energy of about 4 MeV. After irradiation, the target material is converted to $^{13}N_2$ by the Dumas reaction in a modified Coleman nitrogen analyser. The labelled gas is compressed into a 1-ml vial after passing through a liquid N_2-cooled trap to remove oxides of nitrogen. Typically, 12–15 min elapsed between the end of the bombardment and availability of the labelled gas for use. The gas contained essentially pure $^{13}N^{14}N$. Radioactive impurities in the irradiated target did not pass beyond the Dumas reaction. The only significant impurity was about 0.015% of NO which arose from the KNO_3 added to the target to provide carrier N, before conversion to N_2.

(iii) Rapid Analytical Techniques

In spite of the short half-life of ^{13}N, many useful experiments can be done. Exposure of biological material to the labelled N_2 gas can commence within 15–30 min of the completion of irradiation, leaving 1–15 h for processing and analysis before the radioactivity becomes too low for accurate measurement. With accurate timing, all measurements can be corrected for decay. Accurate measurement of background radiation is also essential.

Wolk *et al.* (1974) used $^{13}N_2$ and autoradiography to localize N_2 fixation products in heterocysts of blue-green algae. The exposure lasted 2 or 15 min, after which the diluted cell suspension was spread directly on to an emulsion-coated microscope slide. The enclosed slide holder was evacuated to dry the preparation. After at least 2 h, the emulsion was developed. Thick layers of emulsion were used and disintegrations recorded as individual tracks.

Wolk *et al.* (1976) and Meeks *et al.* (1978) have described rapid, high-voltage electrophoretic and chromatographic methods using thin layers (0.1 mm) of cellulose on glass plates for identification of the products of nitrogen fixation following short exposures (1–120 sec) to $[^{13}N]N_2$. Following development, the plates were scanned for radioactivity using conventional or specially developed scanning equipment. Identification of radioactive spots was made by inclusion of unlabelled amino acids, etc., whose positions were established by ninhydrin. Alternatively, ^{14}C-labelled standards were included and initial scanning for ^{13}N made with the detector shielded by thin aluminium window to exclude ^{14}C emissions. Later, after decay of the ^{13}N, the plates were again scanned with the aluminium shield removed. In another variation, regions of thin-layer plates corresponding to specific amino acids were localized with ninhydrin, the cellulose was scraped off, eluted and the radioactivity of the elute measured using a scintillation spectrometer.

(iv) Hazards

Because of the short half-life of ^{13}N, radiation hazards are minimal; however, Austin *et al.* (1975) measured whole-body exposures of 2 mrem per batch during preparation of the carbon target for combustion and 4 mrem per batch for the remainder of the operation of $[^{13}N]N_2$ preparation. These exposures were inevitable because of the nature of the operation, but if frequency of experiments warranted it, substantial reductions in exposure could be achieved using normal technology for handling radioactivity of this type.

(g) ACKNOWLEDGEMENTS

The author expresses his thanks to B. N. Condon, Electronic Services, CSIRO Division of Plant Industry, for permission to reproduce the circuit in Figure 1, to Professor P. Wolk and Asssociate Professor J. M. Tiedje, Michigan State University, for advice about ^{13}N methods, and to colleagues who have freely discussed details of personal experience in the use of many methods described in this chapter.

Methods for Evaluating Biological Nitrogen Fixation
Edited by F. J. Bergersen

G. L. Turner and A. H. Gibson
Division of Plant Industry,
CSIRO, Canberra, Australia.

3

Measurement of Nitrogen Fixation by Indirect Means

(a) INTRODUCTION

The activity of nitrogenase can be determined directly by measurement of the fixation product, ammonia, and the further products of its assimilation, as described in Chapter II.2, or by indirect measurements which are based on the ability of nitrogenase to reduce a number of substrates other than molecular nitrogen. Substrates such as nitrous oxide (N_2O) (Hoch *et al.*, 1960), azide (N_3^-) (Schöllhorn and Burris, 1967a), and cyanide (CN^-) and isocyanide (NC^-) (Kelly *et al.*, 1967; Hardy and Knight, 1967; Hardy and Jackson, 1967) are reduced and the requirements for ATP and reductant are similar to those for N_2 fixation. The reduction of protons to molecular hydrogen has been used to measure nitrogenase activity (Bulen *et al.*, 1965).

Few observations in nitrogen fixation research, or any other branch of science, have had as much impact as the observation that nitrogenase can reduce acetylene to ethylene. Independently, Dilworth (1966) and Schöllhorn and Burris (1966, 1967b) observed the inhibitory effect of C_2H_2 on N_2 fixation by extracts of *Clostridium pasteurianum*. Dilworth (1966) further observed by mass spectometry that the C_2H_2 was reduced to C_2H_4. Both gases are readily measured at very low concentration by gas chromatographic techniques. As a consequence of its high sensitivity, low equipment and resource costs, simplicity, and rapidity of use, the acetylene reduction assay quickly became a standard procedure in many laboratories. Its applications have been many and varied, ranging from studies with pure enzyme preparations to natural ecosystems. The availability of this assay has been a major factor in the proliferation of N_2 fixation research and in the numerous important advances made in the subject in recent years.

(b) THE ACETYLENE REDUCTION ASSAY

(i) Preparation for assays

Acetylene is a dangerous gas, forming explosive mixtures with air. The explosion limits are 2.5–80% in air (Perry, 1950). It should not be used in confined spaces without appropriate precautions.

Preparation of gas phase

The source of acetylene varies, some laboratories preferring commercially available compressed gas while others prepare their own from calcium carbide. Industrial-grade commercial gas is dissolved in acetone and often contains trace quantities of H_2, H_2S, PH_3, NH_3, CO_2, and C_2H_4. Acceptable purification can be accomplished by bubbling the gas through two scrubbing (Drechsel) bottles, one containing concentrated H_2SO_4 and the other water. The gas may be used directly from the cylinder without scrubbing, but this procedure should only be adopted after comparing scrubbed and unscrubbed gases in a series of assays. Acetylene has also been prepared by the addition of water to commercial calcium carbide (Postgate, 1972; Tjepkema and Burris, 1976). The reaction also forms small amounts of PH_3, CH_4, and C_2H_4. The level of background C_2H_4 in C_2H_2 cylinders tends to fall as the cylinder pressure falls with use; such cylinders should be kept for assays in which only low levels of C_2H_2 reduction are expected.

The usual source of C_2H_2 in our laboratory is a mixture containing C_2H_2 in argon, commercially prepared with 12.5% or 25% C_2H_2. The upper level of C_2H_4 is specified as 2 v.p.m. The mixture is provided in 920 × 290 mm cylinders containing 3 m^3 of C_2H_2: argon at 1400 kPa or 200 p.s.i. (below the explosion limit). This mixture can be diluted with argon, oxygen, or both, depending upon the types of assay required.

Gas mixtures are made using a rotary pump, manometer and vacuum manifold (Figure 1), with the calculated proportions being metered by manometer pressure directly into the assay vessels, or as we usually prefer, into evacuated mechanical reservoirs (Figure 2). For aerobic work, industrial oxygen is adequate as only trace quantities of N_2 and CO_2, and no CO, are present. For anaerobic work, industrial-grade (welding) argon is generally satisfactory provided there is less than 10 v.p.m. of O_2. Where natural gas is plentiful, it may be cheaper to use helium instead of argon or N_2 to prepare anaerobic gas mixtures.

The use of a mechanical reservoir (Figure 2) ensures a constancy in the gas mixture over a number of assays, a factor which could be important where small differences in O_2, or even C_2H_2, concentration could effect the assays. The mechanical reservoirs have threaded pistons to expel the gas mixture into the incubation vessels as required; in our case, one turn provides 15 ml. The cylinders can be evacuated, and when filled the composition of the gas

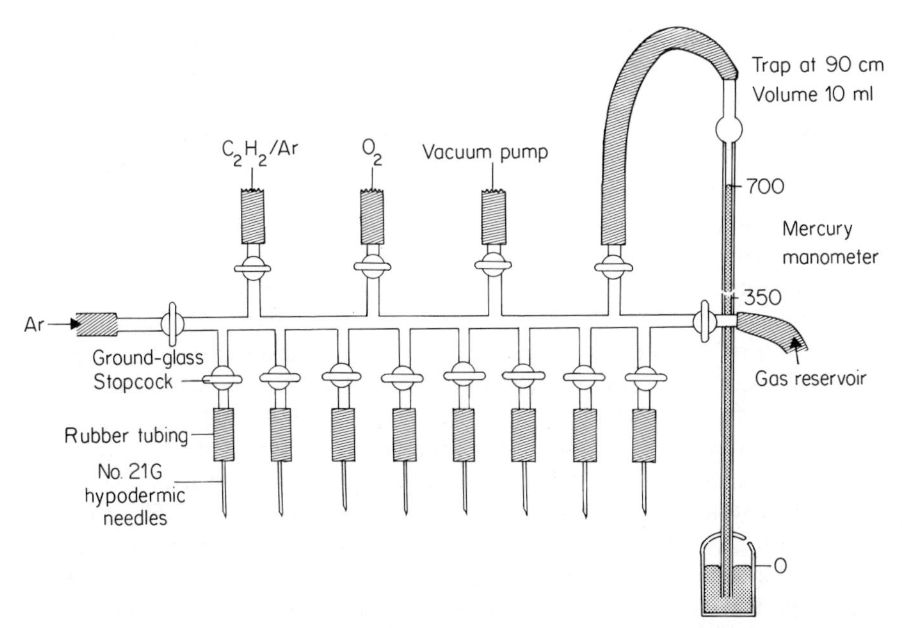

Figure 1. Manifold made from fine-bore thick glass tubing, for use in preparing acetylene reduction assay gas mixtures, either for storage in mechanical gas reservoirs (Figure 2) or for direct use in evacuated assay vessels

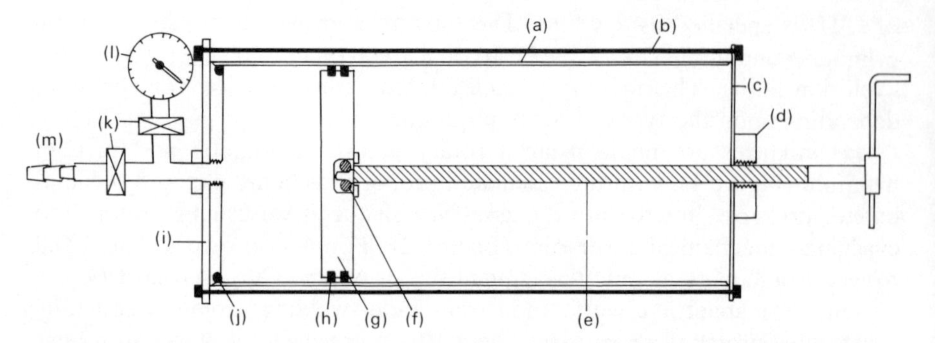

Figure 2. Cross-section of gas reservoirs capable of evacuation to 0.01 atm. (a) Stainless-steel cylinder, 31 cm long, 13 cm diameter (capacity 3 l). (b) Steel rods (4) fitted with threads and holding end plates tight against cylinder. (c) End plate, hexagonal shape, 0.5 mm M/S. (d) Quick-release screw split nut (not essential to have quick release split). (e) Piston-drive screw, threaded to provide 15-ml delivery with 1 turn. (f) Ball-bearing socket attaching screw to piston. (g) Piston, with two grooves for Ludowici fluid seals, (h). (i) Base plate, hexagonal shape, fitted with O-ring, (j), to maintain gas-tight seal. (k) Saunders diaphragm valves. (l) Sphygnomometer pressure gauge (0.7–1.5 atm). (m) Hose fitting

mixture is maintained for at least 7 days. They are fitted with sphygnomometer gauges to determine when the pressure in the container is 1 atm. The incubation vessels are either evacuated on the manifold and filled with the gas mixture, or flushed through with the pre-mixed gas (this requires at least 20 volumes to reduce the O_2 level to 1 v.p.m.). For field assays, football bladders have been used successfully to transport C_2H_2 for injection into assay vessels (Stewart *et al.*, 1971).

Injection of C_2H_2 directly into the incubation vessel(s) is a common practice. A rubber tube is fitted to the outlet of the gas cylinder regulator, or beyond the Drechsel bottles if the gas is being scrubbed, and C_2H_2 drawn into a syringe inserted through the wall of the tubing. If done in a laboratory, or confined space, the outflow from the tubing should be voided into a fume hood, or through a window, as a safety precaution. After injection, an equivalent volume of gas mixture is removed from the incubation vessel to return the pressure to 1 atm. For aerobic assays, this will cause a decline in the oxygen concentration equivalent to the concentration of C_2H_2 achieved, e.g. if 10% of C_2H_2 is added to an air-filled vessel, removal of the mixture will cause a decline in pO_2 from 0.21 to 0.19 atm. Some N_2-fixing systems are very responsive to small changes in pO_2, and this decline could have significant consequences in the level of activity, especially where the assay is used to estimate an absolute rate of N_2 fixation. Hand-operated vacuum pumps are available commercially (e.g. Neward Enterprises, Cucamonga, California,

U.S.A.), and these can reduce the pressure within a vessel to 0.2 atm. The vessel can then be brought to atmospheric pressure with a prepared gas mixture containing the appropriate levels of C_2H_2, O_2, and N_2 or Ar, and the operation repeated. Flushing the vessels with a prepared gas mixture is not recommended as it requires at least 10–15 volumes of the mixture to obtain reasonable replacement of the air. For safety reasons, gas mixtures containing C_2H_2 and O_2 should not be prepared and transported under pressure. However, gas proportionators (e.g. No. 665 with 602 and 603 gas tubes, Matheson Gas Products, East Rutherford, N.J., U.S.A.) have been used to prepare a gas mixture *in situ*, and this is flushed through the assay vessel for 1 min (Hardy and Holsten, 1977).

Preparation of sample

The acetylene reduction assay has been used to determine N_2-fixing activity in many and diverse systems (e.g. Hardy *et al*, 1973; Burris, 1974; Hardy and Holsten, 1977). These range from undefined soil cores, through whole plant systems (both legumes and non-legumes), nodulated roots, aquatic samples, lichen crusts, and bacterial cultures, to enzyme preparations in various stages of purification. Details of various assays are described in Chapters II.4, II.5, II.6, III.3, III.4, and III.5.

Incubation of sample

Almost any type of enclosure may be used for measuring acetylene reduction, provided that the material is not porous or capable of occluding the C_2H_2 or C_2H_4. The incubation vessel used in any assay is determined by the size and composition of the sample. The essential feature is that the volume be adequate to ensure that the gas phase does not change substantially during the course of the assay, especially with regard to O_2 concentration. However, by keeping the volume to a minimum, consistent with maintaining pO_2, the sensitivity of measurement of C_2H_4 is increased. Most systems employ a vessel of fixed volume, ranging from 2–3-ml glass vials through to 4–5-l jars. However, plastic bags fitted with inlet ports and closed by heat-sealing or tape are suitable for use under some field conditions involving rice plants (Lee *et al.*, 1977) or nodulated roots of legumes or non-legumes (Hardy *et al.*, 1973; see also Chapter II.4). Metal cylinders driven into the soil, and sealed over the top with a transparent cover (Balandreau *et al.*, 1978; see also Chapter III.3, Figure 2), have been used for *in situ* estimations of N_2-fixing activity. Gas loss through the bottom is reduced by applying water around the outside of the cylinder. Soil cores have been incubated in large syringes (Hardy *et al.*, 1968), bottles (Silvester and Bennett, 1973), and within specially constructed soil borers (Zuberer and Silver, 1978). Plastic canopies have been used for *in situ* studies of ecosystems (Jordan *et al.*, 1978), and glass canopies for soil samples (Paul *et al.*, 1971).

A number of systems have been used to provide gas-tight sealing of the assay vessels and, at the same time, provide access for introducing and

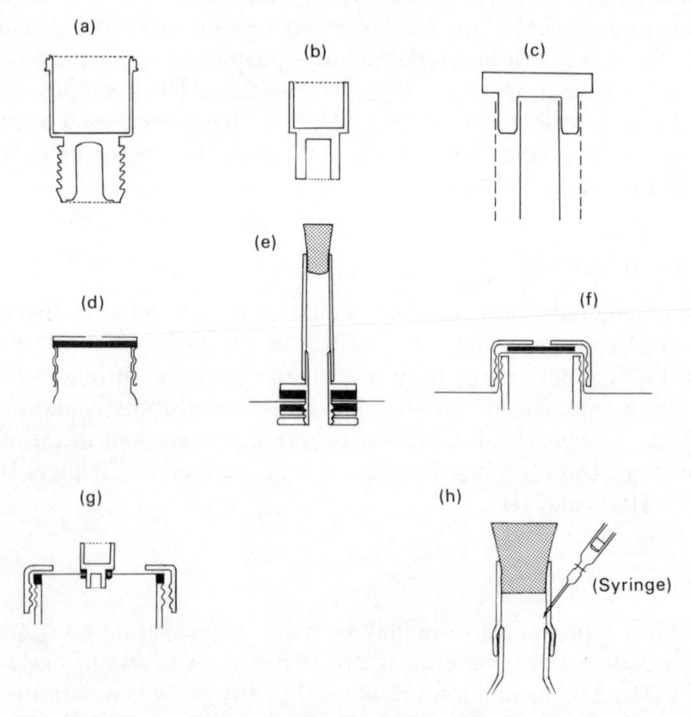

Figure 3. Examples of closures and sampling systems used in acetylene reduction assays. (a) Suba Seals, in a range to fit openings 6.5–25.5 mm. (b) A.H. Thomas serum stoppers, to fit openings 5–16 mm. (c) Blood bottle stoppers, fitting over (20 mm) or within (25 mm) test-tubes. (d) McCartney or other screw-cap bottles, with rubber liner and hole in cap for sampling. (e) Brass filling port, made in our workshops, for use with polyethylene bags. A hole is made in the bag, and a nut tightens against the flange on the port, with protection for the bag provided by rubber washers. A short length of plastic tubing is fitted over the top, and connects to the mechanical reservoir (Figure 2). Sealed by small rubber stopper. Internal diameter 4 mm. (f) Top of plastic wash bottle and Bakelite cap, with 5-mm hole and fitted with rubber liner, also for use with polyethylene bags. (g) Mason, or fruit-preserving, jar with a Suba Seal fitted to a hole in the thin metal lid and sealed with Araldite or Silastic sealing compound. (h) Erlenmeyer flask fitted with wide-bore latex (soft) rubber tubing and closed with a rubber stopper. Injection and sampling through the tubing

removing gas (Figure 3). These range from rubber serum stoppers fitting tightly into glass vessels (e.g. Suba Seals, William Freeman, Barnsley, Yorkshire, U.K.; Arthur H. Thomas, Philadelphia, Pa., U.S.A.) to makeshift arrangements involving latex rubber tubing (Figure 3h). For assays in 29-ml McCartney bottles (or similar type), a hole is made in the metal cap, and the rubber liner inside the screw-cap serves to produce a seal as well as a means of access for injection and sampling (Figure 3d). For large volume assays, Mason (or fruit-preserving) jars up to 5 l in capacity have a gas-tight lid into which can be sealed a serum stopper (Figure 3g). Alternatively, plastic bags can be fitted with a filling port (Figure 3e), with the sample taken directly through the wall of the bag or through a thin rubber sheet taped to the bag if multiple samplings are to be made. Plastic wash-bottle necks and screw lids have also been used, in association with a rubber liner (Figure 3f), although our experience indicates that such bags are hard to fill through a narrow-gauge needle.

Continuous assay of chemostat cultures of bacteria (La Rue and Kurz, 1970) and of nodulated legumes (Fishbeck *et al.*, 1973) has been achieved by the incorporation of a low level of C_2H_2 into a gas stream passing through the incubation system, or by pulsing C_2H_2 through the system at regular intervals (Mederski and Streeter, 1977). The effluent is sampled at regular intervals. The difficulty of continuous exposure is to ensure that C_2H_2 reduction measures only a low, but constant, proportion of nitrogenase activity in order that it does not interfere greatly with normal N_2 fixation. The known growth-regulating effects of C_2H_4 and the recent reports of increased nitrogenase activity following exposure to C_2H_2 (David and Fay, 1977) indicate the need for caution in adopting and/or interpreting these assay procedures.

Disposable syringes have been used successfully for assays of nodulated roots, soil cores, marine samples and *Rhizobium*-tissue culture samples, with samples of the gas phase being collected with other syringes or into evacuated tubes for storage before C_2H_4 determination (e.g. Hardy *et al.*, 1973; Stewart *et al.*, 1971). Within the laboratory, N_2-fixing systems may be cultured in the assay vessel. For example, modified Pankhurst tubes can be used for screening large numbers of potential diazotrophs (Campbell and Evans, 1969; Tubb and Postgate, 1973), while McCartney bottles (29 ml) have been used for the culture and assay of strains of rhizobia (Gibson *et al.*, 1976b). Special chambers for root enclosure have been used for rhizosphere studies (Pan and Schmidt, 1969; Huang *et al.*, 1975).

(ii) Gas-phase sampling

At the completion of the assay period, the gas phase is sampled with a gas-tight syringe, and introduced directly into the gas chromatograph or stored for later determination of C_2H_4. Sample sizes vary from 200 to 1000 μl. We routinely inject 200 μl, removing 250 μl (allowing 10 s for equilibration), and adjusting the volume before injecting the sample quickly, but smoothly, into the gas chromatograph. Injection procedures should be standardised as far as possible. The intention is to provide straight-sided triangular peaks of equal base and constant retention time. Any deviation in peak shape or retention time (even by 1 s) indicates leaky septa, a decline in carrier gas flow-rate, loss of combustion gas pressure, or variation in injection technique (see Section b.v.). Before commencing to use the acetylene reduction assay, operators should experiment with injection procedures to ascertain the most effective technique.

Under laboratory conditions, we use glass barrel syringes of the type made by SGE Pty Ltd (North Melbourne, Victoria 3051, Australia). These syringes have replaceable needles and the plunger is fitted with a gas sealing gland. Teflon tape, as used by plumbers, is wrapped around the upper part of the metal plunger, and should be replaced every 10–20 samplings to ensure that the syringe remains gas tight. Other gas-tight syringes (e.g. Gastight No. 1725, Hamilton and Co. Inc., Whittier, Calif. U.S.A., and many medical disposable syringes) are also suitable for sampling. For enzyme assays, dry, grease-free, glass syringes with accurate graduations are important to ensure accurate sampling and no ethylene carry-over from vial to vial. As part of the sampling procedure, it is important to flush the syringe 3–4 times with air between samples. This applies particularly where samples without nitrogenase activity are assayed, as carry-over within the needle space can give a false-positive determination. When sampling a gas phase, the vessel should be shaken, particularly if there are impediments to the free diffusion of gas (e.g. roots). Where there is doubt regarding the presence of activity, a second determination should be made some time after the first.

Disposable plastic syringes are frequently used when the rate of sampling exceeds the rate at which samples can be processed, or where short-term storage is necessary. The needles of the syringes can be stuck into rubber mats or stoppers and kept for several hours before analysis. Alternatively, the tips of the needles can be placed under water during storage (J. Hoglund, personal communication). New disposable plastic syringes are often used for each sample, although our experience indicates syringes can be 'opened' and 'rested' for several weeks before re-use. Both C_2H_2 and C_2H_4 are adsorbed by the plastic, and could thus provide ethylene peaks in the absence of nitrogenase. Similar considerations apply to rubber closures and again a rest period, with the caps exposed to an open air flow, is recommended. One problem with using disposable syringes is that gauge 21–23 needles cause more rapid deterioration of the septum in the injection port of the gas chromatograph than the very fine needles (0.5 mm) on the gas-tight glass syringes (10–12 compared with up to 100 determinations). Furthermore, the accuracy of sample size is less with disposable syringes.

For storage periods longer than several hours, pre-evacuated blood sampling tubes (Venoject—Jintan Terumo, Tokyo, Japan, or Vacutainer—Becton, Dickinson and Co., Rutherford, N.J., U.S.A.) are suitable. The sample is taken from the assay containers through replaceable double-ended sampling needles which screw into holders. They are available in a range of sizes from 2 to 15 ml. When purchased, these tubes are evacuated to 0.1–0.2 atm, and we routinely re-evacuate immediately before use. Ethylene adsorption on to the rubber caps should also be considered in making the calculations. After the C_2H_4 determination is made, the caps should be removed and 'rested' before re-use.

Consideration must be given to the reduced gas pressure within the syringe when taking the sample, especially where the ratio of syringe volume to assay vessel volume is greater than 1:20. Samples of 250 μl reduce the gas pressure within a 5-ml gas space by 5%, and this should be considered in the subsequent calculations, especially when a number of consecutive determinations are made. Similarly, sampling into evacuated tubes requires that consideration be given to a decline in the C_2H_4 concentration between the assay vessel and the syringe used to inject the sample into the gas chromatograph. Provided that the C_2H_2 response on the gas chromatograph is still linear at the high concentrations used, pressure loss due to sampling can be determined by using the C_2H_2 peak as an internal standard, using the formula

$$\frac{C_2H_2 \text{ peak} \times \text{attenuation (control)}}{C_2H_2 \text{ peak} \times \text{attenuation (sample)}}$$

However, the high solubility of C_2H_2 in water (free liquid, plant tissue) requires that this correction must be used with caution (see section b.v.). With non-rigid containers such as plastic bags, the ratio of volume of gas sample to volume of assay sample is usually small enough for corrections to be neglected.

Sampling the gas phase is preferable to stopping the reaction by the addition of acid (H_2SO_4, CCl_3COOH, $HClO_4$), which may be slow in penetrating the biological material or cause release of C_2H_4 from rubber caps. However, injection of acid to terminate reactions is suitable for enzyme preparations or bacterial suspensions.

(iii) Analysis of C_2H_4 and C_2H_2 by gas chromatography

Although C_2H_4 can be measured colorimetrically (La Rue and Kurz, 1973a) or by mass spectrometry (Dilworth, 1966), the most common and sensitive method of detection employs gas chromatography (Hardy *et al.*, 1968; Postgate, 1972). Unlike some other substrates reduced by nitrogenase, there is only one product (ethylene) from the reduction of acetylene by the enzyme system.

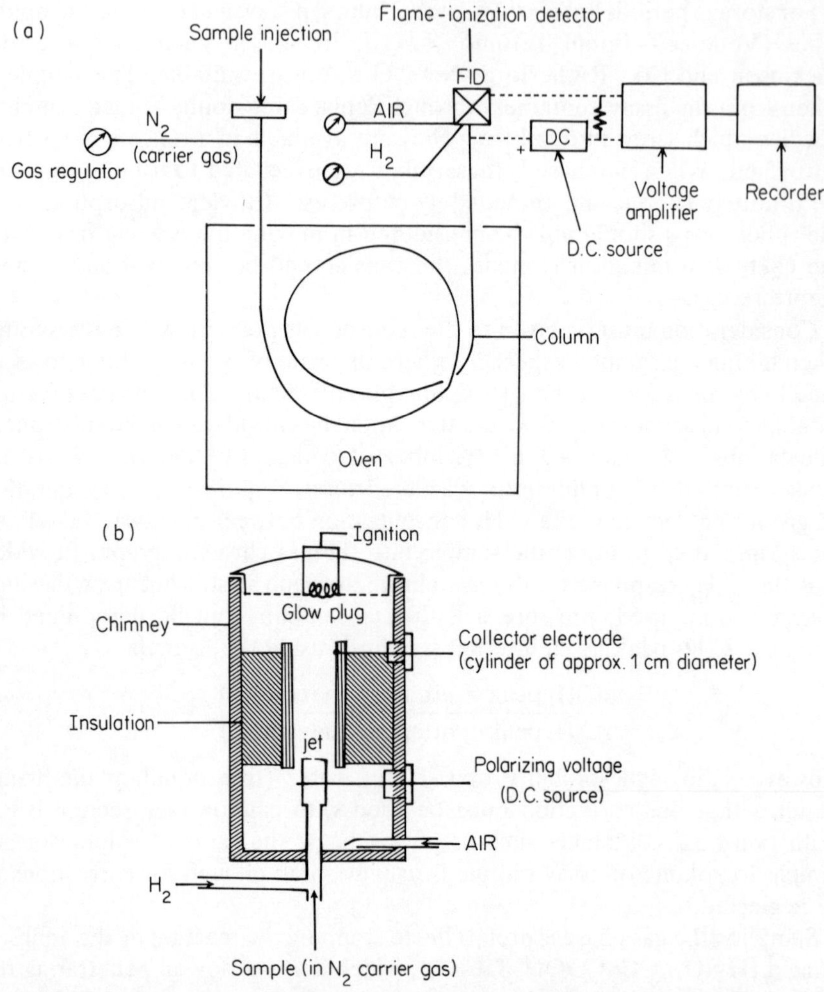

Figure 4. Diagrammatic representation of (a) the operation of a gas chromatograph fitted with (b) a hydrogen flame-ionization detector

Gas chromatographs equipped with H_2 flame-ionization detectors (FID) are used almost universally for the detection of C_2H_4 and C_2H_2. The principle behind the operation of the FID gas chromatograph is that the gases passing through the column in the carrier gas stream (Figure 4a) separate into their different components. To achieve adequate separation, attention to the carrier gas and its flow-rate, column length and packing material, and oven temperature is essential. As the effluent gas (not the inert carrier gas) is burnt in the H_2 flame, the free electrons and ions produced cause an increase in the electrical conductivity between the two electrodes, one at the tip of the flame jet and the other in the cylinder surrounding the jet (Figure 4b). This increase in conductivity lowers the resistance between the electrodes, causing current to flow. This is measured as a drop in voltage and recorded on a strip-chart recorder (Figure 5). The highest sensitivity is achieved with nitrogen as the carrier gas, and helium is the least sensitive; argon produces lower, and slower, C_2H_2 and C_2H_4 peaks than nitrogen but the areas under the peaks are similar for both carriers. Only compounds containing CH groups are burnt in the H_2 flame so that the response is not complicated by non-combustible gases such as oxides of carbon, NH_3, water, NO_x, or N_2. The system is very sensitive. For example, analysis of ^{15}N in our laboratory in an Atlas-Werke M86 mass spectrometer requires 5×10^{-6} mol of available N_2 with an enrich-

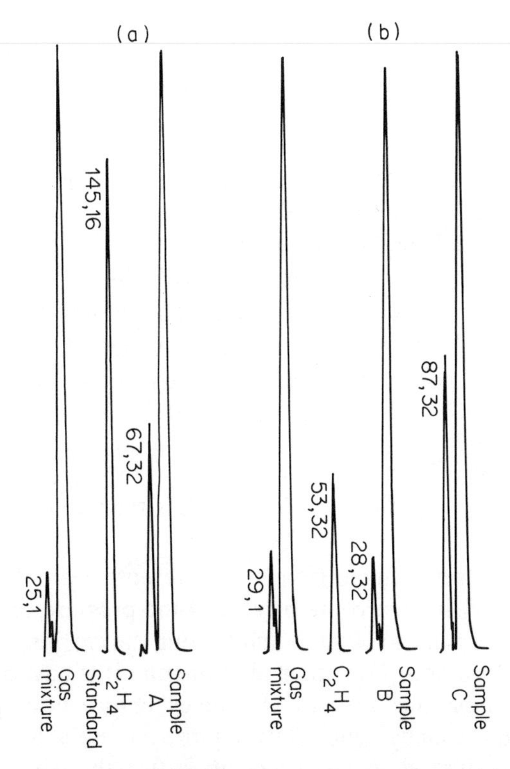

Figure 5. Example of a recorder chart: (a) following injection of C_2H_4 gas mixture, standard C_2H_4 (0.0411%), and unknown sample A, and (b) following injection of C_2H_4 gas mixture, sample from standard assay vessel containing 10 μl of C_2H_4, and two unknown samples

ment of 0.01% (5×10^{-9} mol). Analysis of C_2H_4 with an FID can be accomplished with great accuracy on a sample containing 5×10^{-11} mol of C_2H_4, while amounts as low as 1×10^{-12} mol can be detected. The level of C_2H_2 used in the incubation assays is usually adequate for the C_2H_2 peak to be used as an internal standard (but see precautions). Alternatively, propylene can be used as an internal standard (see Chapter III.4, Section b.iii).

A number of column packing materials can be used to separate C_2H_4 and C_2H_2, including Porapak N,R,T, and Q (Waters Associates, Farmingham, Mass., U.S.A.), and 20% ethyl-*N′N′*-dimethyloxalamide on 100–120-mesh acid-washed firebrick (Supelco Inc., Bellefonte, Pa., U.S.A.) (Hardy *et al.*, 1968). Nitrogen is the commonly used carrier gas, although helium is a suitable alternative.

In our laboratory, the following system has been used for a number of years to provide quick assays with good peak separation.

Detector type	H_2 FID. H_2, 25 ml min^{-1}; air, 200 ml min^{-1} [both supplied at 138 kPa (20 p.s.i.)]. Temperature, 85 °C.
Column	Stainless steel, 1 m × 2 mm I.D., Porapak R (100–120 mesh).
Carrier gas	N_2, 414 kPa (60 p.s.i.), 45 ml min^{-1}.
Oven temperature	Isothermal, 60 °C.
Septum type	Hamilton Tri-layer.
Retention times	C_2H_4, 1.3 min; C_2H_2, 1.8 min.

The retention time is governed by column length and diameter, packing, carrier gas flow-rate, and oven temperature. A suitable alternative to the above is an oven temperature of 65 °C, with a 0.6-m column, providing retention times of 0.7 and 1.1 for C_2H_4 and C_2H_2, respectively. Higher temperatures may provide faster responses, but the separations are poorer and there is the danger that detector signals are produced which are faster than the response time of the recorder, i.e. less than 100% of total peak is recorded.

Ignition of the detector flame frequently presents a problem. With some instruments, this may be overcome by raising the pressure of H_2 to 200 kPa, restricting the air pressure to 60 kPa, and leaving the carrier gas valve closed, before attempting ignition. For other instruments, it is advisable to wait for 2–3 min after opening the air and H_2 valves before attempting ignition. Successful ignition is usually denoted by a ‘pop’; a double ‘pop’ indicates that the flame has extinguished. A good indication that the flame is alight is the fogging of a metal spatula or mirror held over the detector exit. Before re-adjusting the air and H_2 pressures, the carrier gas valve is opened.

The signal is transmitted to a recorder, 1 mV input. Following a pressure peak immediately after injection, there is a slight negative deflection, followed immediately by a double peak (shoulder and peak), regarded in our laboratory as the 'garbage' peak. In order to produce easily read charts, the recorder pen is not positioned until it returns to the baseline after the 'garbage' peak. With the increasing availability of microprocessors, it is likely that they will be used, especially where large numbers of assays are carried out, for recording output data, either as peak heights or as the area under the curve. Although the latter gives a more technically correct estimate, peak height provides a satisfactory value in most circumstances. However, it is vital to standardize injection procedures and to ensure that all gas flows are maintained at the correct rates through the series of assays.

A new type of gas chromatograph, especially useful for field operations, has been developed by Mallard *et al.* (1977). The detector is a Taguchi Model TGS-812 gas sensor (Figaro Engineering Co., Toyonaka City, Asaka 560, Japan; available from many electronic retail outlets), consisting of a small filament heater surrounded by a sintered semiconductor. When the filament is heated in air in the presence of combustible gases, the decrease in the resistance of the semiconductor is measured with an ammeter. A 12-V power source is required, e.g. the cigarette lighter outlet in a motor vehicle. Compressed air, to serve as carrier gas, is obtained from a pump-tank as used for spraying insecticides, to which is fitted an activated carbon filter and two-stage pressure regulator to maintain a pressure of 35 kPa. A conventional column for use with standard gas chromatography is used. The authors estimate the cost at less than U.S.$50, depending on the availability in most laboratories of some materials, and the weight is less than 2 kg. The instrument should be amenable to further modification to permit easier reading. A sensitivity of 10 p.p.m. of C_2H_4 in a 1-ml sample is claimed. The instrument has been used successfully in a field study in the Australian Alps (Carr *et al.*, 1979).

(iv) Calculations

Before using the gas chromatograph for assays, it is necessary to check the linearity of the response over a range of concentrations of both C_2H_4 and C_2H_2. Subsequently one or two different standards should be run each day, provided the response is linear. Changes in detector response to a given concentration may occur due to detector fouling.

Two types of standard are used in our laboratory, one being a pre-mix of ethylene in argon (known concentration) and the other an injection of pure ethylene into a reaction vessel of the same volume as those used for the experiment. The former requires a knowledge of the assay reaction vessel volumes in order to calculate the molar quantities of C_2H_4 produced, and is essential where volumes vary, as is common in a range of McCartney bottles or preserving jars. The latter approach assumes that all vessels have the same volume as it provides a correlation between molar quantity of C_2H_4 added and peak height. Alternatively, a change in peak height following the injection of an accurately determined volume of C_2H_4 may be used to determine vessel volumes.

In the following calculations, it is assumed that the volume of the sample taken is the same as the volume of the standard injected into the gas chromatograph.

C_2H_4/argon pre-mix calibration

An evacuated bottle is flushed with the pre-mix of known C_2H_4 concentration [say $Y\%$ v/v, which is $Y/(22.4 \times 100)$ mmol ml^{-1}, or 0.446 Y µmol ml^{-1}; normally we use a mixture in which Y is about 0.05]. From the peak derived for the standard sample [$p_1 \times a_1$ or peak height (mm) × attenuation], a factor, D, is derived and this is applied to all assays:

$$D = \frac{0.446\ Y}{p_1 \times a_1}$$

(For accurate work, D should be corrected for temperature and pressure.) Assuming that a sample gives a peak $P \times A$, the total C_2H_4 in the reaction vessel is calculated as

$$\text{Total } C_2H_4\ (\mu\text{mol}) = P \times A \times D \times V$$

where V ml is the volume of the reaction vessel gas phase. Correction for background C_2H_4 (i.e. C_2H_4 in the C_2H_2 gas mixture) is made as follows:

$$\text{Total } C_2H_4 = [(P \times A) - (p_2 \times a_2)]D \times V$$

where p_2 and a_2 are the C_2H_4 peak values for the C_2H_2 gas mixtures.

Example. Figure 5a shows an example chart, with readout for the gas mixture ($p_2 = 25$, $a_2 = 1$), the C_2H_4 standard ($Y = 0.0411$, $p_1 = 145$, $a_1 = 16$), and an unknown sample ($P = 67$, $A = 32$) taken from an assay vessel with a gas volume of 125 ml.

$$D = \frac{0.446 \times 0.0411}{145 \times 16}$$

$$= 7.9 \times 10^{-6}$$

Total C_2H_4 produced in the assay

$$= [(67 \times 32) - (25 \times 1)] \times 7.9 \times 10^{-6} \times 125$$

$$= 2.093\ \mu\text{mol}.$$

Injection into reaction vessel

A known volume of pure C_2H_4 (Z μl) is injected into a reaction vessel with the same gas space volume as a regular assay vessel. The amount of C_2H_4 is $Z/22.4$ μmol ($0.0446\ Z$ μmol), which produces a peak $p_1 \times a_1$. If the unknown sample produces a peak $P \times A$, the total C_2H_4 is given by

$$\text{Total } C_2H_4\ (\mu\text{mol}) = 0.0446Z \times \frac{(P \times A)}{(p_1 \times a_1)}$$

For any set of assays, V, p_1, and a_1 are constant, and can be combined to produce a constant, $E\ [= 0.0446\ Z/(p_1 \times a_1)]$. Hence, in calculating the total C_2H_4 produced, and taking into consideration the background C_2H_4 ($p_2 \times a_2$) in the C_2H_2 gas mixture, the equation is

$$\text{Total } C_2H_4\ (\mu\text{mol}) = [(P \times A) - (p_2 \times a_2)]E$$

Examples. Figure 5b shows an example chart for the C_2H_4 gas mixture ($p_2 = 29$, $a_2 = 1$), the C_2H_4 standard obtained following injection of 10 μl C_2H_4 into a vessel of same volume as the assay vessels ($p_1 = 53$, $a_1 = 32$), and two unknown samples. The constant, E, is given by

$$E = \frac{0.0446 \times 10}{53 \times 32}$$

$$= 0.000263.$$

Total C_2H_4 production for the assay B

$$= [(28 \times 32) - (29 \times 1)]2.63 \times 10^{-4}$$

$$= 0.228\ \mu\text{mol}.$$

Total C_2H_4 production for assay C

$$= [(87 \times 32) - (29 \times 1)]2.63 \times 10^{-4}$$

$$= 0.725\ \mu\text{mol}.$$

Nitrogenase (C_2H_2-reducing) activity is normally expressed as nanomoles or micromoles of C_2H_4 per unit of sample (millilitres of culture, milligrams of protein, milligrams dry weight of cells, grams of nodule fresh or dry weight, etc.) per unit time.

(v) Precautions

The apparent simplicity of the acetylene reduction assay can be deceptive.

Broadly, potential problems can be considered as operational or interpretative, although each may have a bearing on the other.

Operational

Operational problems may be mechanical or related to the supply and maintenance of appropriate conditions during the course of the assay. One dilemma facing operators is whether or not to correct for gas leakage when the C_2H_2 peak in samples is lower than that found in prepared gas mixtures or immediately after injection of C_2H_2 into an aerobic gas phase. Leaking caps, poor sealing of the incubation vessel, leaking sample syringes used for storage, or inadequate sealing of the gas syringes used for injecting samples into the gas chromatograph can lead to lower C_2H_2 and C_2H_4 peaks. However, C_2H_2 is very soluble (42.8 mM at 1 atm and 20°C), and a large volume of water in the assay vessel, including that in soil or in plant material, can cause an appreciable decline in the C_2H_2 concentration in the gas phase. Similarly, C_2H_2 and, to a lesser extent, C_2H_4 are readily absorbed by rubber used in sampling stoppers or in gas sampling tubes (e.g. Venoject tubes). Where the lower C_2H_4 peak is due to gas loss, as demonstrated by the use of an internal standard supplied with the C_2H_2 (e.g. methane, propane, propylene—see discussion by Knowles, Chapter III.4) correction is justified. When the pO_2 supplied is less than ambient, care should be taken to prevent leakage of O_2 into the vessels.

The C_2H_2 peaks will be low if there is a blockage in the needle of the syringe used to inject the sample into the gas chromatograph. This problem can arise with very sharp needles (a coring effect) and with particular types of material used in the stoppers and injection septum. We use 0.5-mm diameter needles in our syringes as this preserves the life of the septa and the stoppers. With larger needles (No. 22G or 23G), the septum should be replaced after 10–15 injections, especially when time-course assays are being made and the samples are taken directly from assay vessel to the gas chromatograph.

Any drop in the pressure of the carrier gas, due to a leaky injection septum or declining cylinder supply, or exhaustion of H_2 or air cylinders, will lower both C_2H_2 and C_2H_4 peaks. This can be detected by an increase in the retention times and lower peak heights. In addition to watching retention times and all inlet gas pressures, *frequent* observation of oven temperature should be made to detect problems at an early stage.

Sampling stoppers (e.g. Suba Seals) can be used many times without appreciable gas leakage. However, C_2H_4 and C_2H_2, absorbed by the rubber (Kavanagh and Postgate, 1970) and slowly released to the atmosphere can cause problems in low-activity assays, especially where the assay is designed to detect the presence or absence of nitrogenase activity, or where endogenous

C_2H_4 production is being determined. When assay vessels with stoppers inserted are autoclaved, C_2H_4 may be released (Jacobsen and McGlasson, 1970), e.g. we have found 0.32 ± 0.16 nmol of C_2H_4 in a 13.5-ml bottle. Normally new stoppers are washed in the detergent Teepol and thoroughly rinsed before use. Used stoppers are removed from the assay vessels immediately after the assay is completed, rinsed, and allowed to 'rest' for 2–4 weeks before re-use. Ideally they should be placed in a wire rack and a stream of air directed over them. This latter precaution is not necessary if dealing consistently with highly active systems. The absorption of C_2H_4 on to plastic or other synthetic materials (e.g. Perspex, Plexiglass) should be considered when making incubation chambers, or when using, or re-using, disposable plastic syringes.

New blood sampling tubes, used to store gas samples, are evacuated to only 0.1–0.2 atm. If using such tubes without re-evacuation, this value should be determined accurately in order to amend the sample determination to its correct value. This can be done by measuring the volume of water taken up, relative to the volume of the tube (gravimetrically). Similarly, if the volume of the sample taken in the syringe, relative to the volume of the sampling tube or relative to the gas space volume of the assay vessel, is greater than 3%, the determined C_2H_4 value should be corrected accordingly.

The volume of the incubation vessel should be adequate to minimize changes in the conditions during assay. A fall in pO_2, due to respiration, will cause a decline in the rate of C_2H_2 reduction by aerobic N_2-fixing systems (Sprent, 1969). Little is known of the effect of increasing pCO_2 on nitrogenase activity during assays, although a high pCO_2 in the root environment of pot-grown legumes promotes N_2 fixation (Mulder and Veen, 1960).

Acetylene concentration in the reaction gas phase should be adequate to saturate the enzymes, otherwise the rate of reduction is proportional to pC_2H_2 (Hardy *et al.*, 1973). Acetylene pressures of 0.10 atm *in vivo* and 0.02 atm *in vitro* should produce saturation. However, Spiff and Odu (1972) and MacRae (1977) found an increase in C_2H_2 reduction by *Beijerinckia* spp. with increasing pC_2H_2 up to 0.7 atm, indicating the need to determine the optimum pC_2H_2 for each system studied. In longer assays, C_2H_2 will inhibit growth of bacteria (Brouzes and Knowles, 1971) owing to deprivation of supplies of fixed nitrogen. However, no generalization of the effect on nitrogenase activity should be made as others (David and Fay, 1977; Jordan *et al.*, 1978) have found that *Azotobacter vinelandii, Rhodospirillum rubrum, Anabaena* spp., and epiphytic diazotrophs in natural ecosystems exposed to C_2H_2 for long periods show greatly enhanced C_2H_2-reducing activity. This is possibly a consequence of the diazotrophs reacting to an inadequate nitrogen supply by raising the amount and/or level of activity of their nitrogenase.

Further consideration of the precautions to be observed in assaying particular N_2-fixing systems can be found in other chapters.

Interpretative problems

A major problem in the use of the acetylene reduction assay to estimate absolute rates of N_2 fixation is the calibration of the assay. Theoretically, a C_2H_2 to N_2 ratio of 3:1 can be justified and many researchers have used this ratio to convert assay results without attempting validation for the system under examination. In practice, rates from 1.5:1 up to 25:1 have been recorded (see Hardy *et al.* 1973). With field systems, particularly those showing low levels of nitrogenase activity, great difficulty may be encountered in obtaining a reliable estimate of N_2 fixation by ^{15}N or total N methods (Paul, 1975). However, the variability in C_2H_2 to N_2 ratios within, and between, various diazotrophic systems under the relatively stable conditions in the laboratory requires that caution be exercised in the use of a 3:1 conversion ratio. A possible contributory factor to this variation is the production of H_2 during N_2 fixation, and the observation that C_2H_2 competitively inhibits H_2 production such that the electrons involved are used to reduce C_2H_2 to C_2H_4. Observations with legumes in our laboratory, and elsewhere, indicate that the ratio of H_2 produced to N_2 fixed may be greater than 1:1 in some systems, thus precluding the use of a theoretical 4:1 conversion ratio. Some systems have an active uptake hydrogenase that re-metabolizes the H_2 produced, so that attempts to measure endogenous H_2 production by determining H_2 evolved could be meaningless. In addition, the H_2 to N_2 ratio appears to show diurnal variation, further complicating the derivation of a conversion factor.

As a measure of N_2 fixation in natural habitats, the acetylene reduction method suffers from the same disadvantage of other systems using short-term assays to assess overall activity. Variation in activity due to light intensity, temperature, and moisture levels, be they diurnal, day-to-day, or seasonal, makes it difficult to achieve meaningful integration of a series of short-term assays. The inclusion of a factor derived from a series of diurnal assays has been attempted (Sloger *et al.*, 1975; Halliday and Pate, 1976a,b; Ruegg and Alston, 1978). For example, Sloger *et al.* (1975) suggested that sampling of soybeans at 6 a.m., 8 a.m., 10 a.m., and 12 noon gave a good estimate of 24-h activity. Halliday and Pate (1976a) derived a factor of 0.87, which was used to multiply assay results obtained at 10 a.m. in their monthly assays with field grown *Macrozamia riedlei*, to obtain an estimated daily rate of N_2 fixation. For white clover, assays at 9 a.m. gave a good indication of the estimated daily rate of N_2 fixation (Halliday and Pate, 1976b). Goh *et al.* (1978) found different patterns of diurnal variation in studies with field-grown white clover. From these and other results, they concluded 'that the acetylene reduction assay appears to be unsuitable for long-term quantitative measurements of symbiotic N_2 fixation in the field'. Sinclair *et al*, (1976), also working with white clover pastures, had drawn a similar conclusion, although they believed that the assay could be valuable in making short-term comparisons between treatments.

A further precaution to consider in soil and rhizosphere studies in which nitrogenase activity is low is the recent observation (Nohrstedt, 1976; De Bont, 1976) that C_2H_2 inhibits the metabolism of endogenous C_2H_4. Standard practice in such studies is to measure C_2H_4 production before commencing the assay and to correct the results accordingly. With the inhibition of C_2H_4 metabolism by C_2H_2, evolution of endogenous C_2H_4 will increase, leading to an over-estimation of C_2H_2 reduction (Witty, 1979b).

(vi) Recovery of dissolved C_2H_4 and C_2H_2

Recent experiments (Bergersen and Turner, 1975a; Chapter II.8) on the effects of leghaemoglobin on the consumption of O_2 and the nitrogenase activity of bacteroids and bacteria, have necessitated working without a gas phase. The solutions required for analysis are pre-incubated for 60 min at reaction temperature (28 °C) with a gas mixture containing C_2H_2, O_2, and argon. For accuracy, a 25% C_2H_2 in argon mixture is added to a pre-mix of 3% O_2 in argon to achieve the desired level of oxygen (0.5–1.0%). The technique of measuring N_2 fixation by the reduction of C_2H_2 to C_2H_4 is

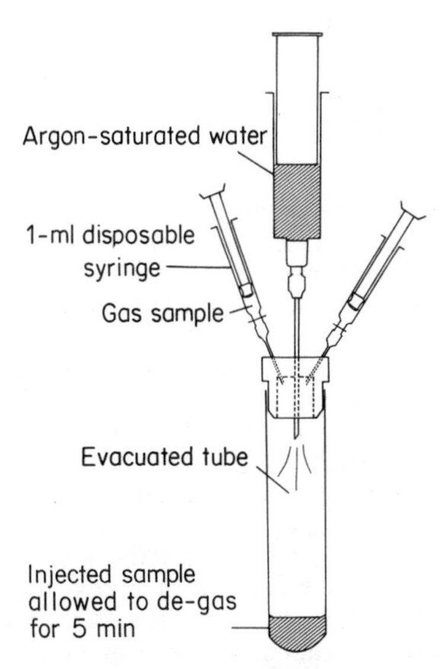

Figure 6. Evacuated blood sampling tube of 15–20 ml capacity. Liquid sample added and allowed to de-gas for 5 min. Argon-saturated water added by large syringe to expel gaseous sample quickly into two 1-ml disposable syringes

particularly suited to solution kinetics because of the high solubility of C_2H_2; in fact, 1 ml of solution at 21 °C prepared as above provides over 200 μl of gas for experiments with 1% O_2 and over 250 μl of gas for those with 0.5% O_2 (barometric pressure for Canberra, Australia, in the range 690–710 mmHg).

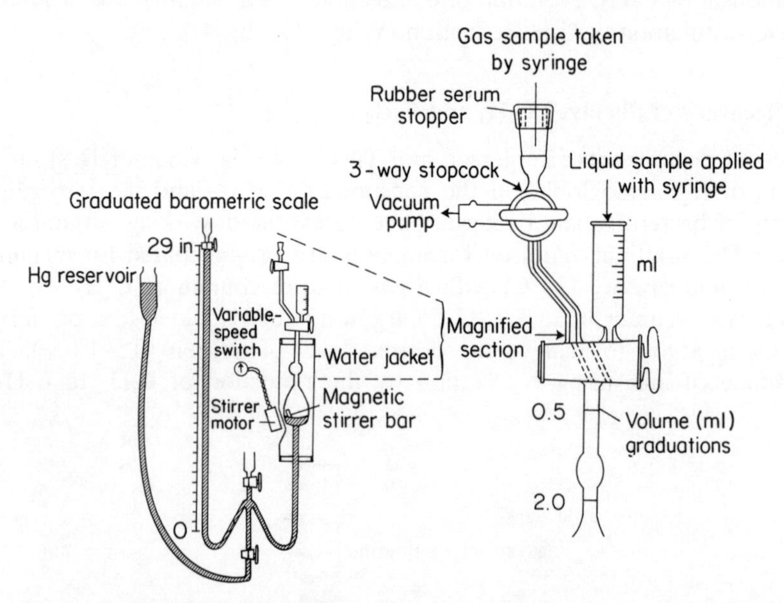

Figure 7. Modified Van Slyke apparatus for de-gassing liquid samples. Water (plus dissolved gases) is drawn into jacketed, stirred chamber by lowering the mercury reservoir. By further lowering the reservoir, the mercury level can be dropped until the stirring bar is in contact with the water. Stirring for 4 min is usually necessary for complete recovery of dissolved gases. The Hg reservoir is then raised until water meniscus is level with the 2-ml graduation mark. The partial pressure of gas is then recorded from the barometric scale (P_1). On further raising the mercury level the gas sample can be directed via the two-way tap into a gas sampling port which has been pre-evacuated. Gas samples (2 × 300 μl) can be taken through a rubber serum stopper. By lowering the mercury level, the water plus remaining gas can be drawn back into the main body of the de-gassing chamber. Again by using the two-way stopcock, the residual gas can be discharged from the chamber by raising the mercury level until the water just touches the stopcock capillary. Lowering of the reservoir will bring the top of the water to the 2-ml graduation mark, at which point the partial pressure can again be read from the scale (P_2). The difference between recorded partial pressures ($P_1 - P_2$) is a measure of the pressure of the sample gas in a volume of 2 ml. The gas volume at ambient temperature and pressure can be calculated from the standard gas laws (at constant temperature), i.e. sample volume (ml) = $(P_1 - P_2) \times 2$ divided by ambient pressure

The reaction in solution is stopped by either injecting a 4-ml sample from a cuvette into 1 ml of 10% trichloroacetic acid (TCA) or, if the experiment is continuous, by pumping 4-ml samples into syringes already filled with 1 ml of 10% TCA (Chapter II.8). The sample is then further treated in either of two ways:

(i) Injecting the sample into a 20-ml evacuated tube (Venoject—Jintan Terumo, Tokyo, Japan) and allowing the sample to de-gas. The evolved gas can be displaced into two 300-μl gas syringes by quickly injecting argon-saturated water into the tube (Figure 6). This method requires calculation of dissolved gases and without careful and quick sampling may produce errors due to the gas redissolving in the displacing water. It is normal to use disposable syringes with No. 23G or 24G needles. After pushing the needles through the tube closure, the syringe plungers are pulled out to the 200-μl graduation and then the argon-saturated water is injected. Disposable syringes are normally made with rubber plunger seals which hold a very good vacuum for the short sampling times encountered.

(ii) De-gassing in a Van Slyke apparatus or a modified version of it (Figure 7). This apparatus provides a direct measure of evolved gas volume by utilizing a graduated de-gassing chamber and a means of reading the sample gas pressure.

Both methods provide 95–96% recovery of dissolved gases, the latter being preferred because of the ease of measurement of gas volumes.

(c) UTILIZATION OF OTHER ALTERNATIVE SUBSTRATES

Nitrogenase is unique in that it is the only isolated enzyme known to reduce the triple bond (Burns and Hardy, 1975). It is capable of reducing a wide range of substrates, including alkynes, isonitriles, nitriles, N_2O, and azide, as well as dinitrogen. In general, the susceptible bonds include those bridging N to N, N to O, N to C, and C to C, although not all such bonds are reduced. In the sequence of decreasing $N \equiv N$ character to increasing $C \equiv C$ character, there is a decline in the extent to which complete cleavage occurs. In the middle of the sequence, multiple reduction products are generated. These compounds are valuable in assessing nitrogenase activity only when the products can be identified readily and quantitatively, and where the proportions of such products are constant.

(i) Alkynes

Alkynes of general formula C_nH_{2n-2}, including acetylene, are reduced to olefins, the final product of such a chain not being realized (e.g. a saturated alkane such as C_2H_6). Reduction of long-chain alkynes occurs only when the acetylenic linkage is at the end of the chain. For analytical work, only substrates that produce a gaseous reduction product at experimental temperatures can be used, e.g. propyne or but-1-yne (Hardy and Jackson, 1967; Kelly, 1967). The product can be analysed for the reduced compound (e.g. propylene, but-1-ene) by gas chromatography. The procedures used are similar to those employed for acetylene reduction assays.

(ii) Cyanides and other nitriles

Although a potent inhibitor of metalloprotein enzymes, cyanides are reduced by the nitrogenase enzyme complex *in vitro*. HCN and KCN are the most commonly used substrates (Hardy and Knight, 1967; Kelly, 1967). The products of the reaction are primarily methane and ammonia, but methylamine and traces of C_2H_4 and C_2H_6 are also found. Methane can be detected by gas chromatography or ammonia analysed for nitrogen reduced in the ionic species $—C \equiv N$.

Linear chain alkyl cyanides are also substrates for nitrogenase, but not the branched compounds (Burns and Hardy, 1975). The overall form of the reaction is

$$RCN \rightarrow RCH_3 + NH_3$$

with no apparent deviation. Ammonia is determined analytically.

Alkenyl cyanides, particularly acrylonitrile, have been used to study the site of reduction of nitrogenase (Fuchsman and Hardy, 1972). The reactivity of the nitrile group of these compounds may be due to a double bond in the olefin group (Burns and Hardy, 1975). Acrylonitrile is reduced to ammonia, propylene, and propane, but not to amines or imines. Either the ammonia or the hydrocarbons can be determined analytically. Both *Azotobacter* extracts and intact N_2-grown *Azotobacter* cells reduce acrylonitrile (Burns and Hardy, 1975).

The more reactive isocyanides, such as methyl, ethyl, and vinyl isocyanides, have been used as substrates for nitrogenase (Hardy and Jackson, 1967; Kelly *et al.*, 1967), with the ethyl compound being reduced more slowly than the others (Kelly, 1967). The products of the reaction consist mainly of methane, with minor amounts of ethane, ethylene, and some C_3 compounds. Isonitrile reductions have been reported for many cell-free extracts and for intact clover nodules (Kelly, 1967).

(iii) Nitrous oxide

Nitrous oxide (dinitrogen oxide) is reduced to N_2 by the nitrogenase complex in intact soybean nodules (Hoch *et al.*, 1960), or by nitrogenase preparations from *Azotobacter* and *Clostridium* (Hardy and Knight, 1966). There is little or no formation of ammonia as N_2O is a specific inhibitor of N_2 fixation

(Repaske and Wilson, 1952). The product of the reaction can be determined by mass spectrometry.

(iv) Cyclopropene

Nitrogenase preparations of *Azotobacter vinelandii* reduce cyclopropene to cyclopropane and propene (McKenna *et al.*, 1976).

$$\text{cyclo-}(\mathrm{CH_2})\mathrm{HC{=}CH} \xrightarrow[\text{ATP, }2e^-]{\text{Nitrogenase}} \text{cyclo-}(\mathrm{CH_2})\mathrm{H_2C{-}CH_2} + \mathrm{CH_3CH{=}CH_2}$$

Propane was not detectable. The ratio of the initial rate constants for the formation of propene and cyclopropane was 2:1, the two compounds being analysed by gas chromatography or ^{1}H NMR spectrometry. Further analysis of the reaction could assist in the elucidation of the site of reduction of N_2 on nitrogenase.

(v) 6-Cyanopurine

When incorporated in agar medium, 6-cyanopurine acts as a qualitative colour indicator for N_2-fixing cultures of *Klebsiella pneumoniae* and for certain classes of *nif* mutations (MacNeil and Brill, 1978). Mutants with defects in the regulatory genes *nifA–nifL* were not coloured, while mutations in six other genes gave colonies with a deeper purple colour than wild-type colonies. Although the pigment formation is not catalysed by a nitrogenase-specific reduction of the $C \equiv N$ bond, 6-cyanopurine will aid genetic studies with anaerobic N_2-fixing bacteria. The compound is added to the medium at the rate of 125 μg ml^{-1} after autoclaving. The colour is unstable in air.

(d) H_2 EVOLUTION FROM NITROGENASE

The relationship between H_2, N_2, and nitrogenase is complex. *In vitro* studies with nitrogenase preparations show that 25–33% of the electrons used in nitrogenase function are donated to protons, with the consequent production of H_2. This reduction of protons is ATP-dependent, and hence the study of H_2 production, as well as providing an understanding of the mechanism of N_2 fixation, has important implications in elucidating the energy costs of N_2 fixation. In a number of aerobic diazotrophs (*Azotobacter*, some *Rhizobium* spp.), the examination of H_2 production is complicated by the presence of unidirectional uptake hydrogenases. In *Clostridium pasteurianum* and other anaerobic diazotrophs, a conventional hydrogenase capable of both producing and assimilating H_2 provides a further complication. Furthermore, H_2 is involved in an exchange reaction when nitrogenase reduces N_2, as shown by the formation of HD when D_2 is supplied under assay conditions; hydrogenase activity also involves an exchange reaction with H_2.

(i) Measurement of H_2

Commonly used methods for determining H_2 production involve gas chromatography, amperometric determinations, and mass spectrometry. A tritium exchange assay has been developed to detect hydrogenase activity.

Gas chromatography

The following system is used in our laboratory.

Detector type	Thermal conductivity (Aerograph 200); temperature, 90 °C; filament current, 115 mA.
Column	Stainless steel, 2 m × 2 mm I.D., molecular sieve 5A, 60–80 mesh.
Carrier gas	Argon, 274 kPa (40 p.s.i.), 15 ml min^{-1}.
Oven temperature	Isothermal, 65 °C.
Septum type	Hamilton Tri-layer.
Retention times	H_2, 0.45 min; O_2, 0.80 min; N_2, 1.55 min.

With this system, 0.002% w/v of H_2 (i.e. 0.2 nmol) can be measured in a 200-μl gas sample, the detector being less sensitive for H_2 than the FID for C_2H_4. Daily standards (500 μl of H_2 in 25 ml of air) are run. The detector provides a linear response at H_2 concentrations up to 10%. Every 6 weeks, the column is reconditioned by heating at 200 °C for 2 h with a flow of argon. When not in use, the oven and carrier gas (reduced to 50 kPa) are left operating but the current to the detector is switched off. See also Chapter II.4 Section e for description of a system for determining H_2, O_2, CO_2, C_2H_4, and C_2H_2.

For assays of H_2 uptake, the appropriate amount of H_2 is injected into the assay vessel and, after shaking, a sample is taken to determine the initial level. Alternatively, 2% of H_2 is provided in the gas mixture prepared in a cylinder (Figure 2). After evacuation of the cylinder, the appropriate pressure of H_2 is passed into the cylinder, which is then brought to atmospheric pressure with air.

By using N_2 as the carrier gas, there is a small decrease in sensitivity but assays can be carried out twice as quickly as there is no N_2 peak. However, this method is not to be recommended owing to the absence of an internal standard.

Amperometric determination

This method (Hanus *et al.*, 1979) employs a Clark oxygen electrode (Yellow Springs Instruments, Yellow Springs, Ohio, U.S.A.) with the polarizing voltage reversed. The probe consists of a platinum cathode and a silver wire anode immersed in 50% KCl solution, all isolated from the external environment by a Teflon membrane permeable to O_2 but not to water or ions. For O_2, the Pt electrode is polarized at -0.8 V with respect to the Ag/AgCl electrode; for H_2, the Pt electrode is polarized at $+0.6$ V. The reaction is assumed to be

$$H_2 \xrightarrow{Pt} 2H^+ + 2e^-$$

and

$$2AgCl + 2e^- \xrightarrow{Ag} 2Ag + 2Cl^-$$

The rate of electron flow is limited by the diffusion rate of H_2 through the membrane, which is a function of the H_2 pressure outside the membrane. The flow of electrons is measured by a nanoammeter, the lower limit of detection being of the order 5×10^{-4} atm partial pressure of H_2.

The system was first devised by Jones and Bishop (1976) for studies with *Anabaena* and has recently been described in detail (Hanus *et al.*, 1979) following extensive use with legume nodules (e.g. Schubert and Evans, 1976). The probe is fitted into a small gas-tight compartment containing nodules or other material. A Keithley 602 solid-state electrometer (Keithley Instruments, Cleveland, Ohio, U.S.A.) is used as the ammeter (5–100 nA) and to drive a strip-chart recorder, although an alternative meter and amplifier can be constructed cheaply. The cuvette should provide for circulation of thermostatically controlled water; glass is preferable to other materials which may adsorb H_2.

New probes must be sensitized as there is insufficient AgCl on the Ag electrode. The probe is be operated as an O_2 probe for 24 h. The polarizing voltage is then reversed and pure H_2 introduced to the chamber. Sensitivity to H_2 is increased by successive flushing and H_2 treatment (0.5–1 min), continuing until a current of approximately 5 μA is obtained with pure H_2. When a decrease in sensitivity in the membrane cannot be rectified, the membrane should be replaced. Hanus *et al.* (1979) advise that it is better to work with reduced sensitivity in order to avoid the increased noise and drift associated with new membranes.

When not in immediate use, the probe should be stored with the polarizing voltage 'on' or it will require re-sensitizing for H_2. To store, place it in a cuvette filled with water. The sensitivity can be increased after storage, or extended use, by drying the chamber and alternating the polarity of the voltage with a frequency of one reversal per second for about 15 s. The probe is then treated with 1% H_2 and flushed with air several times over a 30-min period. The instrument is standardized with 1% H_2 in argon, a procedure which should be repeated every 2–3 h to account for any decrease in sensitivity.

Mass spectrometry

A proportion of the reaction gas is sampled into an evacuated sampling tube fitted with a stopcock. The sampling point is provided with a fitting (B14 male or female cone, etc.) depending on the inlet system of the mass spectrometer. To minimize leakage, samples should be analysed as quickly as possible. Two methods can be used. One is to determine the H_2 to N_2 or H_2 to He (where He is added as an internal standard) ratio, and to compare the results with those for a standard mixture; alternatively, the results are calculated using the 'source factors' (a measure of the ionizability of the molecules) for H_2 and N_2. The second approach involves component analysis of all molecular species, utilizing source factors for each component to calculate the percentage composition of each species. With the first method, the decline in pressure in the sample tube is not relevant as the $H_2:N_2$ ratio of the sample, and the N_2 level of the standard, are used to determine the H_2 level in the sample.

Mass spectrometry is required to measure HD formation in the deuterium exchange reaction involving nitrogenase (Turner and Bergersen, 1969).

Tritium exchange

This technique determines hydrogenase activity, and is based on the exchange reaction between water and tritium (T_2) gas:

$$HOH + T_2 \rightarrow HT + HTO$$

In the system described by Lim (1978), 1 μl of T_2 containing 2.27 mCi ml^{-1} is introduced to the gas phase of a *ca*. 20-ml incubation vessel containing 1 ml of culture suspension, or nodules, and incubated for 1 h. The reaction is stopped with methanol and extracts made into 10 ml of Aquasol-2 (New England Nuclear, Boston, MA., U.S.A.). The amount of HTO is determined in a liquid scintillation spectrometer.

(ii) Applications

The techniques described have been variously used to measure H_2 production by nitrogenase preparations, cultured bacteria, bacteroid preparations and N_2-fixing symbiotic associations, H_2 uptake by bacteria, bacteroids and symbiotic associations, and the hydrogen exchange reactions associated with both nitrogenase and hydrogenase activities. The intention of these investigations has been to examine the mechanism of N_2 fixation (e.g. Bulen *et al.*, 1965; Jackson *et al.*, 1968), the production of hydrogen by cyanobacteria for fuel purposes (Benemann and Weare, 1974; Daday *et al.*, 1977), the utilization of H_2 as an energy source for biological N_2 fixation (De Bont and Leijten, 1976; Emerich *et al.*, 1979), and the energy loss involved in H_2 evolution during N_2 fixation in legume and non-legume nodules (e.g Schubert and Evans, 1976; Carter *et al.*, 1978; Bethlenfalvay and Phillips, 1978).

Measurements of H_2 evolution from some N_2-fixing bacteria are confounded by the activity of an uptake hydrogenase. This has applied particularly to the legume studies, in which the term 'relative efficiency' (*R.E.*) has been used to express the proportion of total electron flow through nitrogenase that is involved in H_2 production

$$R.E. = 1 - \frac{H_2 \text{ evolved}_{(\text{air})}}{C_2H_2 \text{ reduced}_{(\text{air})}}$$

or

$$R.E. = 1 - \frac{H_2 \text{ evolved}_{(\text{air})}}{H_2 \text{ evolved}_{(\text{argon}/O_2)}}$$

Hence, where an uptake hydrogenase exists, H_2 evolution is a measure of the difference in H_2 production and H_2 uptake. In some systems we have observed high hydrogenase activity, such that little or no H_2 is evolved under an argon/oxygen atmosphere; in other systems, hydrogenase activity is only sufficient to assimilate that H_2 produced during normal N_2 fixation under air.

Carbon monoxide is an effective inhibitor of N_2 reduction, C_2H_2 reduction, and the HD exchange reaction of nitrogenase, but does not retard H_2 evolution, at least in *Azotobacter chroococcum* (Smith *et al.*, 1976); with legume nodules, 2% of CO inhibits H_2 evolution but this is probably a function of CO inactivation of leghaemoglobin. Acetylene has been reported to inhibit hydrogenase in *Azotobacter* (Smith *et al.*, 1976) but with the bacteroids of *R. japonicum* (Emerich *et al.*, 1979), and the methane-oxidising *Methylosinus* (De Bont, 1976), exogenously supplied H_2 promoted C_2H_2 reduction. In considering hydrogenase activity, attention must be given to the conventional hydrogenases (ATP-independent) which can produce, or assimilate, H_2 depending on the conditions.

(e) CONCLUDING REMARKS

Although numerous substrates are reduced by nitrogenase, and could be used for assaying activity, the simplicity of the acetylene reduction assay has meant that most laboratories involved in nitrogen fixation research use this technique. As indicated here, the apparent simplicity of the assay can be deceptive. Apart from operational problems, interpretation of the results requires great care. For the different N_2-fixing systems, interpretative problems are discussed in the relevant chapters of this book. In essence, the major problem concerns the precise relationship between C_2H_2 reduction and N_2 fixation, and the appropriate conversion ratio to be used in determining actual rates of N_2 fixation. In addition to the relationship between C_2H_2 and N_2 reduction at the time of assay, there is the difficulty of effectively integrating a number of assays, to achieve a meaningful estimate of N_2 fixation over time. As a generality, the further one moves from *in vitro* studies towards the field situation, the greater are the operational and interpretative problems. These problems, plus those associated with variation in H_2 production that may, or may not, be directly related to N_2-fixing activity, require that further detailed analysis of the factors affecting C_2H_2 reduction, and N_2 reduction, be made as a matter of some urgency if the acetylene reduction assay is to be used to provide accurate estimates of nitrogen fixation.

Methods for Evaluating Biological Nitrogen Fixation
Edited by F. J. Bergersen

A. H. Gibson
*Division of Plant Industry,
CSIRO, Canberra, Australia*

4

Methods for Legumes in Glasshouses and Controlled Environment Cabinets

(a) INTRODUCTION

Culture methods for experiments with nodulated legumes are almost as varied as the number of authors contributing to the subject. Despite this variability, there are only two major principles to apply in selecting a system to use. The degree of bacteriological control must be adequate to ensure that there is no ambiguity in the results. That is, depending on the purposes of the experiments, contamination by other bacteria must be minimal and not influence the results or their interpretation. The second principle is that the environmental conditions, including mineral nutrition, should be optimal for plant growth. The basis of these principles is to permit the maximum opportunity for the plant–bacteria associations under study to express fully their capacity for symbiosis. Undue limitations on growth may minimize the expression of differences between associations of different symbiotic capacity.

Frequently the experimental conditions applied meet some requirements, but fail, in varying degrees, to meet other requirements. Both symbionts must be considered. Microbiologists sometimes neglect the welfare of the plants in their efforts to impose strict bacteriological control, forgetting the need for adequate levels of CO_2, light, and moisture. Similarly, plant-orientated scientists may not appreciate the problems associated in achieving and maintaining bacteriological control while at the same time providing adequate conditions for the bacteria, and the nodules, to function properly. In this chapter, factors to be considered in selecting and using culture techniques for different purposes will be examined, and a number of systems described.

(b) PRINCIPLES OF ENVIRONMENTAL CONTROL

The term 'environment' is used here in its broadest context—physical, nutritional, and growth medium pH. The selected conditions should not restrict plant growth, nor should they impose any restrictions on the interpretation of the study. The exception to this generalization occurs when the effects of an environmental variable are being investigated, but even under those circumstances the other environmental conditions should be optimal or comparable to those likely to apply when the particular stress is encountered in the field. Recent reviews of the influence of the physical environment (Dart, 1977; Gibson, 1976a, 1977; Lie, 1974) and mineral nutrition (Munns, 1977a, b) on nodule formation and nitrogen fixation provide access to detailed information on the subject. In this section, those features of the environment which should be considered in selecting experimental conditions are discussed.

(i) Light

The three important aspects of light are the irradiance, the spectral composition, and the photoperiodic daylength. Within glasshouses, the conditions are similar to those found in the field, with the exception that the irradiance is lowered, especially where shading is required to maintain reasonable temperatures. The most common form of shading is 'whitewash'. Various commercial preparations are available (e.g. Parasol, Phytochemical Laboratories, A. Gysalinck, Corme, Belgium; Stay-wite, Taubmans Paints, Birmingham Ave., Villawood, N.S.W., Australia), containing finely ground limestone, an organic glue, and a vegetable oil. It is appled in mid-Spring and scrubbed off in mid-Autumn. Slatted blinds over the glass (as used at Rothamsted Experimental Station, Harpenden, Herts., U.K.) allow greater flexibility in coping with cloudy periods. A modification of this approach is the use of aluminium louvres, attached to a frame over the glasshouse and operated by levers at ground level (Hely, 1959). The angle of the louvres is readily changed as the light conditions vary, and the system is amenable to automation, using a photocell and electronic circuits. None of these procedures is necessary where adequate cooling is available, but this is very expensive to establish and to operate.

The flowering of many legumes is determined by the length of the photoperiod and, in some instances, by temperature. For some species, the maintenance of long days (e.g. greater than 14 h) will keep them in the vegetative phase, while other species grown under the same conditions will flower prematurely. In general, plants in the Fabeae (e.g. *Vicia, Pisum, Cicer*) and Genisteae (e.g. *Lupinus*) are long day plants (flowering when daylength lengthens) or indifferent, whereas those in the Phaseoleae (e.g *Glycine, Vigna, Phaseolus*) are short day plants or indifferent (Evans and King, 1975). Regardless of these generalizations, care should be taken to ascertain the photoperiod characteristics of a species, or cultivar of a species, before commencing investigations. Many species show a change in N_2-fixing ability during pod-fill (e.g. Mague and Burris, 1972, soybeans; LaRue and Kurz, 1973b, peas; for *Vigna* studies, see Doku, 1970, and Summerfield *et al.*, 1974), an effect attributed largely to competition between the nodules and the developing pods for photosynthate (Lawn and Brun, 1974). Moderately low light intensity, as provided by 60–100-W incandescent globes, is adequate to extend the photoperiod. Under glasshouse conditions, the photoperiod can be shortened by removing the plants to a darkened environment, either by hand or by various mechanical devices. Examples of the latter include controlled-temperature cabinets fitted with light-tight panels and tops that rise over the cabinets at pre-set times (Morse and Evans, 1962) or trolleys that are pulled into, and out of, dark chambers at pre-determined times (e.g. Summerfield *et al.*, 1976).

Supplementary illumination is frequently required during the winter in glasshouses located in regions of high latitude. High-pressure mercury vapour lamps (400–1000 W) provide a moderate level of irradiance. Such lamps internally silvered on the upper part (Reflectalite types) provide less shading of natural illumination than lamps fitted beneath external reflectors. Care should be taken to achieve uniform irradiance over the work bench by careful spacing of the lamps.

Controlled-environment cabinets have both advantages and disadvantages. The irradiance level, spectral composition, and daylength are readily controlled, and it is possible to repeat the same conditions for different experiments. Unfortunately, the light intensity provided by most units is relatively low in relation to normal field conditions (rarely greater than 40%). To some extent, the uniform irradiance level throughout the photoperiod compensates for lower intensities in the field during the early and late periods of the day. Of concern is the reduction in light intensity, by approximately 8%, for each pane of glass between the lights and the plants. More importantly, accumulation of dust on the glass seriously lowers the irradiance level, and periodic cleaning of all glass surfaces is essential.

Depending on the nature of the experiments, consideration must be given to the spacing of the plants. For symbiotic effectiveness studies, it is common to grow the plants so that they are separated and mutual shading is minimized. Growth rates are generally retarded where the leaf area index (LAI, or ratio of total leaf area to experimental area) exceeds 4. If attempting to simulate a crop situation, the plants around the perimeter of the experimental area should be treated as 'buffer' plants and not be included in the experimental material; in the glasshouse these plants receive more light than those inside the area.

Legumes are particularly sensitive to spectral composition, and both nodule formation and plant growth are affected by far-red wavelengths. The inclusion of tungsten filament lamps in the light bank, either as conventional globes or as incandescent strip lamps (as used in decorative lighting or to illuminate mirrors), can overcome some of the imbalance problems found with certain types of fluorescent tubes. A common problem is the failure of cabinet-grown plants to have the same appearance as field-grown plants, mainly due to excessive etiolation of the former. At the Plant Growth Laboratory, Reading, Berkshire, U.K., all species to be examined are grown under a range of light sources before commencing experimentation. Subsequent studies utilize that light combination which produces plants most similar to those grown with natural light (Summerfield *et al.*, 1977).

Exposure of roots to light may influence nodule formation. Lie (1969a) has shown that exposure of shoots or roots to far-red light alone retards the nodulation of *Pisum sativum* and *Vicia faba,* indicating an involvement of phytochrome in nodulation. In this laboratory, we have observed nodule formation on tube-grown plants whose roots received light, whereas such plants grown with darkened roots failed to nodulate (Gibson, 1968). With other species, illumination of the roots may retard, or even inhibit, nodulation. For plants grown completely within test-tubes, the placement of the tubes in close-fitting holes in wooden blocks up to 7 cm deep reduces, but does not eliminate, illumination of the roots. Paper cylinders wrapped around the tubes, especially for systems in which only the roots grow inside the tube (Section c.v), provide an effective cover. Alternatively, the tubes can be placed in close-fitting holes in the top of a box, or in the cover on a water-bath (Figure 5).

(ii) Temperature

Both nodule formation and the level of nitrogen fixation are influenced strongly by temperature, especially that in the root zone. And just as host species differ in their response to temperature, plants nodulated by different strains of rhizobia also vary in their N_2-fixing capability (Gibson, 1965; Dart *et al.*, 1976).

For most purposes, such as the provision of nodule material for biochemical or physiological studies on bacteroids or nodule enzymes, the determination of general symbiotic effectiveness of a range of host × strain combinations, the study of plant nutrition, or studies on the infection process, optimal temperatures should be provided. For most temperate species (e.g. *Trifolium, Pisum, Phaseolus*), this will be 18–22 °C during the light period, and down to 14 °C during the dark period. Tropical species may fail to nodulate at these temperatures, and regimes in the range 28–32 °C/23–25 °C are the most favourable for nodulation and growth. Some of these species have a very narrow optimum temperature range (e.g. *Glycine wightii*), whereas others, such as *Macroptilium atropurpureum,* have a broad optimum (Gibson, 1971). There are a number of species whose temperature optima are intermediate to the ranges quoted above; these include *Glycine max, Medicago sativa,* and *Lupinus* spp. The temperature range that encourages most rapid nodule formation may differ from that which supports the highest rates of N_2 fixation. For example, *Trifolium subterraneum* nodulates rapidly at 30 °C but few strains of *R. trifolii* fix significant levels of N_2 above 26 °C. At the lower end of the temperature scale, nodulation of *T. subterraneum* is retarded below 18 °C and shows a minimum around 7 °C; this too is strain dependent (Roughley and Dart, 1970). Nodulated plants transferred to lower temperaures can achieve appreciable rates of N_2 fixation (Gibson, 1967a, b).

For glasshouse studies, black or dark pots should be avoided, as the temperatures in these pots may be up to 5 °C higher than ambient temperatures. In one recorded instance (Meyer and Anderson, 1959), the use of dark pots severely restricted N_2 fixation by *Trifolium subterraneum* plants. Trays of flowing water can be used to prevent abnormally high temperatures (Rogers, 1969), provided that the pots are sealed. For providing supplementary heating in the root zone, Bergersen and Turner (1970) used insulated metal boxes fitted with copper-covered heating elements (as used in heating soil), two

Figure 1. Heated boxes for promoting the nodulation of soybeans grown in pots in the glasshouse. Note the fan (another is located inside the near end of the box), the heating wire and thermostat in the open box. Aluminium louvres for glasshouse shading can be seen in the background. A window in the glass wall is covered with fine mesh to exclude insects

small fans, and a thermostat. A moulding around the top of the pots rests on the edge of holes cut into the top of the box so that the shoots are freely exposed to the glasshouse conditions (Figure 1). This system is used with soybeans during the first 2–3 weeks after planting. The temperature in the box is regulated at 26–28 °C whereas the glasshouse air temperature is 25/20 °C. Another system of localized heating uses large diameter (120–150 cm) transparent plastic tubing extending the length of a bench or series of benches. Heated, or cooled, air is circulated continuously through the tubing, thus saving on general glasshouse heating (E. W. Jaworski, personal communication).

The temperature of the water or nutrient solution used to irrigate the plants may have a marked effect on plant growth. The use of tap water or nutrient below 20°C affected the form and growth of *Glycine max, Lablab purpureus,* and *Vigna unguiculata,* although the effect on soil temperaure was small and persisted for less than 1 h (Brockwell and Gault, 1976). The growth of *Lupinus* spp. and *Phaseolus vulgaris* was not affected by the temperature of the water. For these reasons, nutrient solutions should be kept at glasshouse temperatures, even if it requires the construction of a heated mobile tank and pumping system (Gault *et al.,* 1977).

(iii) Water

Excess or insufficient water will affect adversely both nodulation and nitrogen fixation. The ideal water content for pot-grown plants is 'field capacity', i.e. where all free water will drain from the pots. Excess water lowers the O_2 tension in the root zone, with a consequent fall in the metabolic activity of both the rhizobia and the plant roots. Lowered O_2 levels in the root zone retard the nodulation of clovers (Bond, 1950; Loveday, 1963) and cowpeas (Minchin *et al.,* 1978). Water stress reduces the level of root hair infection and nodule formation by clovers (Worral and Roughley, 1976), and when applied to nodulated plants, lowers the level of N_2 fixation (Sprent, 1972, 1976; Huang *et al.,* 1975; Gallacher and Sprent, 1978). Provided that the nodules do not lose more than 15% of their fresh weight, full N_2-fixing activity can be recovered on re-wetting; otherwise, the nodules die and new nodules must be developed.

Species differ in their ability to tolerate water-logging conditions. In this laboratory, most clovers will nodulate well when grown with the roots under water (as in tube culture), but with *Medicago sativa* and a number of tropical species, care must be exercised to keep the level of free water 5–8 cm below the top of the root zone.

Two problems associated with water levels are the maintenance of an adequate supply and the measurement of the water tension. For most experimental systems, the principal concern is maintenance of optimal conditions. Provided the pots are free-draining, watering once or twice per day is adequate, although this will depend on the porosity and water-holding capacity of the medium, the age of the plants, and the environmental conditions. For example, Harper (1971) uses gravel beds for growing soybeans, and they are sub-irrigated from a central reservoir every 2 h. When plants are grown in Perlite (a granulated, porous Ca–Mg silicate preparation used for pot culture work in many countries), watering twice per day is adequate except where the temperature exceeds 25°C or large plants transpire at a high rate. Placing the pots in shallow trays of water partially alleviates the problem but can cause cross-contamination problems. Mixing one part of Vermiculite (a micaceous material commonly used for plant culture) with 2 parts of perlite increases the water-holding capacity but retains the free-draining character of perlite. Acid-washed sand as a basic medium offers many advantages but if the texture is too fine water-logging problems may be experienced.

Ideally, the water-holding capacity of a soil or medium should be determined prior to sowing (soil sample air-dried, weighed, and then re-weighed after water has been allowed to drain through) and the pots weighed daily to restore the initial weight. In order to reduce labour costs, automatic watering devices have been developed. The simplest involves trickle irrigation for 60–90 s at regular intervals during the day, controlled by a regulator clock. Alternatively, with a system developed in the CSIRO Division of Tropical Crops and Pastures, Brisbane, Australia, the pots can be placed on a conveyor belt which delivers the pots to a balance and watering device; water is added to bring the pot to a pre-determined weight 2–3 times each day (Andrew and Cowper, 1973).

Maintenance of pre-determined soil moisture levels is always a problem and various systems have been developed. One such system involves soil columns of different length, each standing in free water, so that the water potentials at the top of the column decrease with increasing column length. Hack (1971) developed a system with a number of auto-irrigator cells around each plant, each being supplied from barostat reservoirs to provide pre-determined hydrostatic pressures in all parts of the pot. More recently, polyethylene glycol (PEG) has been used to regulate water supply. As PEG is toxic to many roots, it must be separated from the growth medium by a semipermeable membrane; such a system has been described by Johnson and Asay (1978). None of these systems has been used to study moisture stress effects on legume nodulation and N_2 fixation, but should be adaptable for that purpose.

The above discussion involves 'open' pots whose use involves considerable contamination risks. For small seeded legumes, particularly those grown for a limited period, tube culture (Section c.ii and c.v) provides satisfactory bacteriological control. Sterile water, or nutrient solution, is added as required. For larger seeded legumes, the Leonard jar assemblies (Section c.vii), or modifications to that approach, are commonly used.

Measurement of water potential

Various techniques have been used to measure the water potential in soils and plant tissues. For leaves, the easiest method involves a pressure chamber (Scholander *et al.*, 1965) into which a leaf is placed with the petiole extruding. Pressure in the chamber is increased slowly until free water returns to the cut surface of the petiole. Other systems involve commercial dew point hygrometers or thermocouple psychrometers (Nelsen *et al.*, 1978); the psychrometer can also be used for determining water potential of soil or inoculants. For further reading, see Begg and Turner (1976), Rawlins (1976), Boyer (1976), and Sprent (1976).

Begg and Turner (1976) comment on the relationship between water stress induced in pots and that found in the field. Whereas stress under laboratory conditions is induced more rapidly, often due to small pot size and lack of reserve water supplies, in the field the stress usually develops more slowly, allowing for a degree of adaptation by the plants.

(iv) Carbon dioxide

Carbon dioxide is important in legume nodulation studies for two reasons. The close relationship between photosynthesis, N_2 fixation, and plant growth is widely accepted. Assuming a normal CO_2 level of 320 v.p.m., the rate of photosynthesis will decline as the CO_2 concentration falls to *ca.* 75 v.p.m.; at this level, the respiration rate of legumes equals the rate of photosynthesis. This has important considerations with controlled-environment chamber design, and it has been suggested that the fresh air flow into a cabinet should be 4–6 times greater than the maximum CO_2 demand of the plants (Morse and Evans, 1962) in order to maintain ambient CO_2 levels. The growth, and hence the N_2 fixation, of many plants in cotton-plugged test-tubes (Section c.ii) is frequently retarded owing to CO_2 limitation (Gibson, 1967c). The plug prevents turbulent mixing of air outside the tube with the air space within the tube, such that the growth is limited by the rate of CO_2 diffusion into the tube. This limitation can effectively eliminate real bacterial strain differences in symbiotic effectiveness tests. The use of loose-fitting metal 'bacteriological' caps, or foam push-in caps, restricts diffusion 6–8 times more than cotton plugs and should be avoided. Legumes will show a marked response to increased CO_2 levels above ambient (Wilson *et al.*, 1933; Hardy and Havelka, 1976; Havelka and Hardy, 1976), but this is a characteristic of the plant rather than of the symbiotic association. Although marked increases in N_2 fixation are found with >800 v.p.m. CO_2 in the air around the leaves, such treatment also promotes the growth of plants supplied with combined nitrogen.

Enrichment of the rhizosphere with CO_2 promotes N_2 fixation, especially in soils of low pH (<5.5) (Mulder and Veen, 1960). Lupin nodules can 'fix' CO_2 through the enzyme phosphoenolpyruvate carboxylase (Christeller *et al.*, 1977), but the significance of this enzyme activity in promoting N_2 fixation has not been evaluated fully. The high level of air turbulence around pots in controlled-environment cabinets could lower the pCO_2 in the rhizosphere, hence reducing growth compared with that found under comparable conditions in the field.

(v) Mineral nutrition

Although legumes are often considered to require a higher plane of mineral nutrition than many non-legumes, Munns (1977a) regards this generality as fallacious. There is as much variation in the nutritional requirements for maximum production among non-legumes as there is among the legumes. Many *Lupinus* spp. are well adapted to a poor nutritional environment (Gladstones, 1970), while some *Stylosanthes* species are more productive on a low level of mineral nutrition than on a 'normal' level.

For any experimental system, all the necessary nutrients, with the exception of combined nitrogen, must be provided at levels adequate to ensure that nutrition does not limit the expression of symbiotic capacity. Many elements have been implicated in the successful formation and function of legume nodules. Calcium, phosphorus, molybdenum, boron, cobalt, copper, and iron each have a particular role in the symbiosis (Munns, 1977a). However, in formulating any medium, consideration must be given to the adequacy of other elements such as sulphur, potassium, sodium, chloride, and the trace element zinc.

Table 1 provides the formulation for N-free mineral nutrient solutions used in a number of legume nodulation studies. There is a basic similarity between the formulations, the differences lying in the compounds used rather than in their concentration. In general, most elements are provided in excess amounts and any nutrient disorders are likely to arise from toxicity rather than deficiency. While any of the formulations would probably be satisfactory for most legumes, it is advisable for those starting work with any species to compare the growth of their plants on two or more formulations. Downward adjustment of phosphate levels may be necessary although this will also reduce the pH-buffering capacity of the medium. Obvious symptoms include leaf margin scorch, leaf wrinkle, deformed leaves, and chlorosis, but in many cases nutrient disorders are only reflected in reduced yield. Tisdale and Nelson (1966) discuss the diagnosis of such disorders for a number of legumes.

Caution should be exercised in interpreting deficiency symptoms in the early stages of plant growth. Such effects may be due to delayed commencement of N_2 fixation, due in turn to the use of sub-optimal temperature conditions.

For tube culture work, it was the practice in this laboratory to provide plants with 1/4-strength nutrient solution at the first and subsequent waterings. More recently, we have found that if the subsequent waterings are made with sterile distilled or demineralized water plant growth is as good, or better, than that obtained with 1/4-strength nutrient. For pot-grown plants, only one watering per day is with nutrient solution, any other watering being made with water. This minimizes the possibility of 'salt' injury to the plants.

Table 1. Formulation of N-free mineral nutrient solutions used to culture nodulated legumes. A common trace element solution is provided*. All quantities in mg l^{-1}

Compound	Tube culture†				Hydroponics	Pot culture		
	Thornton (1930b)	Jensen (1942b)	Brockwell (unpublished)	Fåhraeus (1957)	Munns (1968)	McKnight‡ (1949)	Gibson (unpublished)	Bergersen (unpublished)
K_2HPO_4	500	200	500	—	—	—	130	—
Na_2HPO_4	—	—	—	150	—	—	—	—
KH_2PO_4	—	—	—	100	140	200	—	30
$Ca_3(PO_4)_2$	2000	1000	2000	—	—	—	—	—
$CaHPO_4$	—	—	—	—	—	—	—	—
$CaCO_3$	—	—	100	—	—	—	—	—
$CaCl_2$	—	—	—	100	—	—	6	1
$CaSO_4 \cdot 2H_2O$	—	—	—	—	1088	1500	—	—
$MgSO_4 \cdot 7H_2O$	200	200	200	120	230	200	110	30
NaCl	100	200	100	—	—	—	—	—
KCl	—	—	—	—	0.7	300	110	40
$FePO_4$	1000	—	500	—	—	—	—	—
$FeCl_3$	10	140	100	—	3§	140	8§	3§
Fecitrate	—	—	—	5	—	—	—	—
pH	6.6	6.6	6.6	6.5	6.5	7.0	6.8	6.8

*Trace element stock solution: H_3BO_3, 2.86 g; $MnSO_4 \cdot H_2O$, 2.08 g; $ZnSO_4 \cdot 7H_2O$, 0.22 g; $CuSO_4 \cdot 5H_2O$,0.08 g; Na_2MoO_4, 0.11 g; all in 1000 ml. Use at a rate of 1.0 ml l^{-1}.
†Agar 1.2–1.6%, depending on batch and firmness required. Use 1.6% for semi-enclosed tube culture.
‡McKnight's solution also used in vermiculite tube culture.
§Sequestered with ethylenediaminetetraacetic acid: 22.8 g of EDTA, 250 ml of 1 N KOH, plus 10 g of $FeCl_3$ (or 17.2 g of $FeSO_4 \cdot 7H_2O$) in 1000 ml of water. Aerate vigorously overnight.

(vi) pH

The majority of species nodulate most profusely in the pH range 6.0–7.0, but there is wide variability between species in their ability to form nodules outside this range. For example, *Medicago* species show poor nodulation below pH 6.0, and develop few nodules below pH 5.2. Clovers are more tolerant of low pH than medics, but pH 4.7 is the lower limit for nodulation of most *Trifolium* species. The so-called tropical legumes may nodulate at a pH as low as 4.0 but, as shown by Munns *et al.* (1977), many show a strong response to the increase in pH obtained by liming; others, such as *Stylosanthes* may show a decline in nodulation above pH 5.5, while even those species responding overall to lime may have depressed nodulation early in their growth (Munns and Fox, 1976).

The effect of alkaline pH in media and soils has received less attention than acidic pH. However, caution must be exercised when raising the pH of any medium above 6.0–6.5 owing to the lowered availability of such nutrients as Fe, B, Zn, and phosphate. Conversely, low pH promotes the availability of Al and Mn, and this could lead to toxicity syptoms.

With artificial media, the pH is controlled primarily by the pH of the nutrient solution. Problems have been encountered with the micaceous substrate vermiculite, and thorough overnight washing with acidified water (pH 5) is recommended. Vermiculite from some sources may have a pH in excess of 9.0. Furthermore, the use of vermiculite in media intended to provide a range of pH values is not recommended, owing to its inherent buffering capacity. For example, media established with a pH range of 5.0–8.7 reverted to a range of 6.0–7.0 within 2 days (D. C. Jordan, personal communication). With agar media, pH buffering is achieved by the inclusion of K_2HPO_4 and KH_2PO_4 at levels higher than necessary to provide adequate K and P. In hydroponic culture systems, reasonable control of the pH, especially where different sources of nitrogen are used, has been achieved by the use of the cation-exchange resin Amberlite IRC 50 (Harper and Nicholas, 1976) and a nutrient recirculating system. Many legumes cause a marked drop in the pH of the medium, be it 'solid' or liquid, due to H^+ excretion during cation uptake and to the excretion of organic acids under some conditions. Buffering the nutrient reduces this effect, but care should be taken to ensure that the fall in pH does not retard nodulation or interfere with the uptake of essential nutrients. With the uptake of NO_3^- from the medium, the pH tends to rise, whereas with NH_4^+ uptake, the pH falls. The use of NH_4NO_3 as a source of nitrogen in control treatments will minimize the pH change.

By controlling the pH of the medium, it should be possible to control the number of nodules formed. For example, if growing a termperate legume that forms many small nodules, and the intention is to obtain a large amount of nodule tissue for studies on the bacteroids, dropping the pH 7 days after nodules start to develop should both limit the number formed and increase their size (Munns, 1968; Lie, 1969b). Care must be taken to minimize any adverse effects of low pH, such as reducing the availibility of some nutrients. Lie and Brotonegoro (1969) have used Fe(III) EDTA with peas to achieve the same effect on nodule number and size.

(vii) Combined nitrogen

The literature on the inhibitory effects of combined nitrogen, particularly nitrate, on root hair infection, nodule initiation, nodule development, and N_2 fixation is very extensive (see Dart, 1977; Munns, 1977a; Gibson and Pagan, 1977). The extent of these effects is influenced by the host plant, the bacterial strain, and the form and concentration of the combined nitrogen source. Combined nitrogen can also promote nodulation and N_2 fixation, but this is usually found after an initial period of inhibition during which plant growth is promoted by the combined nitrogen; the higher levels of nodulation and N_2 fixation are a consequence of the larger plant size. Combined nitrogen can also alleviate inhibitory effects of high light intensity on the nodulation of young seedlings (Orcutt and Fred, 1935).

When growing plants in soil, there is little that can be done to control available combined N. Growing a non-legume will lower the level of available combined N (Simpson and Gibson, 1970), but continuing nitrification, particularly following regular watering and drying, will mean that nodulated plants will be receiving two forms of N, *viz.* nitrate and N_2.

With sand (especially acid-washed), agar, or artificial solid substrates such as perlite and vermiculite, there will be little or no available combined N. The danger arises in the source of water used to prepare the agar medium or nutrient solution. Distilled or deionized water is satisfactory, but tap water from some sources may contain up to 7 p.p.m. of NO_3–N, which is sufficient to retard, then promote, nodulation (Gibson and Nutman, 1960). Before starting glasshouse experiments which utilize large amounts of water, NO_3–N levels in the tap water should be determined. Depending on the nature of the experiment, and the likely effects of this combined N on N_2 fixation, it may be necessary to supply distilled or deionized water.

Nitrogen-free media are commonly used for most types of legume nodulation experiments, be they designed to examine N_2 fixation by different host × strain combinations, the effect of environmental or nutritional variables, or to produce material for biochemical or physiological studies on nodules or bacteroids. This approach permits a more ready assessment of N_2 fixed. Furthermore, there are no complications in interpreting the results due to differential inhibition of nodulation by combined N. However, for experiments designed to determine the most effective strains of rhizobia, the use of N-free media may be questioned. Most soils have low to moderate levels of available N at the time of sowing. Strains of rhizobia show differences in their response to combined N (Gibson, 1976b), and the performance of some less effective strains is improved, relative to known effective strains. For this reason, consideration should be given to including 10–20 p.p.m. of combined N in the initial medium when conducting strain tests in the laboratory or glasshouse.

Controls containing combined N are a vital feature of many experiments, although such controls are all too frequently omitted. They provide a standard with which to compare the inoculated treatments (Section f.iv). They also give an indication of host response; too many papers appear in which the influence of a particular treatment is attributed to an effect on nodulation or N_2 fixation, whereas it could be an effect on the host.

In concluding this section on principles of environmental control, it should be stressed that individual workers should determine the conditions required to give the best plant growth, bearing in mind the species being used, the nature of the experimental work, and the principles enunciated above. The variation between species in their response to environmental and nutritional conditions mitigates against presenting details for individual species. In undertaking these preliminary studies, experience in handling the material, and in understanding the complexities of the symbiotic associations, will be obtained.

(c) MICROBIOLOGICALLY CONTROLLED SYSTEMS

The degree of microbiological control to be exercised in any culture system depends on the purpose and nature of the experiments. For studies involving strains of rhizobia whose performance with a given host is not known, or where those strains are slow to nodulate a host, or the level of symbiotic effectiveness is low, strict microbiological control is essential to ensure that each test strain is the only microorganism, or at least the only strain of rhizobia, present in the system. At the other end of the scale, where a strain is known to nodulate a host promptly and form a highly effective symbiosis, and the intention is to obtain bulk nodule material for biochemical or physiological studies, some relaxation of the levels of bacteriological control exercised *may* be tolerated. Such relaxation is acceptable only when the experimenter has a sound knowledge of the symbiotic attributes of the host and bacterial strain(s) being used, and is aware of the dangers of contamination when interpreting the results. In many glasshouses and laboratories, a high load of rhizobia may exist in the atmosphere and extreme difficulties may be experienced in keeping uninoculated control cultures clean. When there is any doubt, it is advisable to adopt a more cautious approach.

Many techniques have been devised to achieve microbiological control, ranging from the use of completely enclosed systems (e.g. cotton-plugged test-tubes) to systems in which the plant shoots are freely exposed to the atmosphere but with which every effort is made to keep the root system free from contamination. The most easily handled system involves plugged test-tubes, primarily because of the need for minimum watering and hence a reduced risk of contamination. However, such systems are only suitable for small-seeded legumes, especially where the intention is to make comparisons of the symbiotic effectiveness of different strains of rhizobia.

For some purposes (e.g. counting populations of soil and inoculant populations of rhizobia), small-seeded species with the same nodulating characteristics as larger seeded species can be used. These include *Ornithopus sativus* for counting *R. lupini, Vicia hirsuta* for *R. leguminosarum, Glycine ussuriensis* for *R. japonicum,* and *Macroptilum atropurpureum* for the cowpea-type rhizobia, *Rhizobium* sp. (Chapter III.1, Table 1). Where the intention is to determine the number of rhizobia able to nodulate a particular host species, tests should be made to ensure that the surrogate host gives the same results as the species under examination. For example, *M. atropurpureum* nodulates with most strains of *Rhizobium* sp., but a number of other 'tropical' legumes are nodulated only by specific strains. Furthermore, symbiotic effectiveness (Section f.iv) is frequently more specific than nodulation, and extreme caution should be used in extrapolating from surrogate hosts to other hosts.

(i) Seed sterilization

The seed sample should be clean and free of broken seed. If the purpose of experimentation is a symbiotic effectiveness study, plant-to-plant variation can be reduced by selecting seed of relatively uniform size (e.g. by sieving, visual grading, or weighing).

There is no universally adopted method for surface sterilizing seeds, and frequently the method used will depend on seed hardness. With seeds that have a very soft coat (e.g. *Arachis hypogea*), a 2–5 min exposure to a 2% solution of sodium hypochlorite (commonly obtained as bleach) is adequate, provided it is followed by 5–6 rinses with sterile water. A more common method uses $HgCl_2$ [4 ml of a cooled saturated solution (28 g of $HgCl_2$ per 100 ml of water at 80 °C) in 100 ml of water plus 0.5 ml of concentrated HCl]. The seeds in a flask or tube are immersed for 10 s in 95% ethanol, then in the $HgCl_2$ solution for 3–5 min. Care should be taken to ensure that the sides of the container, and the rubber stopper surfaces in contact with the container, are wetted by the solution. After decanting, the seeds should be thoroughly rinsed 8–10 times with sterile water and allowed to stand in the last washing. Two further changes of water should be made at 15-min intervals.

For seeds with hard coats, concentrated H_2SO_4 is a suitable sterilant and softener. Ideally the seeds (up to 600 small seeds in a 50-ml flask) are desiccated overnight to remove any traces of surface moisture. After brief exposure to ethanol (10 s), the seeds are covered with concentrated H_2SO_4 for 5–20 min, the time being determined by experimentation with each batch of seed. The acid is rinsed over the sides of the flask and the base of the rubber stopper. The acid is drained *thoroughly* into an acid-waste flask. The first rinse is with 30 ml of sterile water, with thorough shaking to minimize localized heating, and is followed by a further 9 rinses before allowing the seeds to imbibe in the last wash. Fine emery- or sand-paper scarification of seed is also used to soften hard seed coats. Scratching each seed with a pin also serves to permit entry of water to the seed.

The above procedures are adequate to kill most surface microorganisms but contaminants under the seed coat are difficult to remove. They are unlikely to be rhizobia, but can cause problems in rhizosphere studies (Evans *et al.*, 1972a). Plating the seeds on YMA provides a ready test for detection of rhizobia, but contaminating anaerobes and microaerophiles (e.g. *Klebsiella pneumoniae*) are unlikely to be detected.

After allowing the seeds to imbibe, they may be sown directly into the culture medium (twice the number required, and thinned after germination) or plated on 20 ml of sterile 0.9% agar in Petri dishes. If plating, the number of seeds is restricted to 60–100 per plate, as the capacity of the agar to supply water for germination is limited. The plates are inverted during germination to encourage the development of straight radicles. For clovers, and other species with a cold requirement, 48 h at 4 °C prior to incubation at 20–25 °C produces more even germination than immediate incubation.

For large seeded legumes sown into vermiculite/perlite medium, and to be inoculated with strains known to nodulate the host promptly and effectively, adequate sterilization is achieved by rinsing the seeds with 95% ethanol (10–20 s), followed by 5 rinses with sterile water. This should only be used after trials to check its efficacy, and in all such experiments the use of uninoculated controls is vital.

(ii) Cotton-stoppered tubes

Thornton (1930b) developed a tube culture technique by which the plants were grown completely within cotton-plugged test-tubes on agar slopes (Figure 2a). For legumes with seeds up to 8 mg, 20 × 150 mm tubes are commonly used, with 8 ml of seedling nutrient agar (Table 1) added. The tubes are stoppered with a firmly rolled (not tight) non-absorbent cotton plug (not push-in plugs as the plugs must be removed and replaced several times), autoclaved (120 °C, 20 min), and sloped with the top of the agar half way up the tube. Experience has shown that it is preferable to melt the agar (1.2–1.6%, depending on the ability to provide a firm slope) in the required volume of *distilled* water before adding nutrients as stock solutions. The pH of the medium is adjusted and it is dispensed. Automatic-filling, hand-operated syringes (e.g. Cornwall Syringes, Becton-Dickinson, Rutherford, N.J., U.S.A.) or, for large numbers of tubes, automatic pipetting machines, greatly facilitate the task.

The time-consuming and boring task of plug rolling is eased by cutting 4.5 cm wide strips from a roll of cotton-wool with a band-saw. A motor-driven, flat, tapered spindle can be used to roll the plugs from the strips (Gault *et al.*, 1980). The variable-speed motor runs at 75 r.p.m (rolling phase) or 220 r.p.m. (smoothing phase), and is controlled by a foot switch. Annoyance from cotton-wool dust is avoided by locating the machine in a box with a fan exhausting outside the building. A fume hood should not be used as dust accumulation will constitute a fire hazard.

Table 1 shows four commonly used nutrient solutions for tube culture. For clovers, Jensen's medium is most satisfactory although trial experiments should be conducted when working with a species not used previously. Reduction in the phosphate levels may reduce leaf scorch with some species (see also Sections b.v and b.vi).

Vermiculite is a better medium than agar for the cultivation of some *Medicago* species (Brockwell and Hely, 1966). Vermiculite (*ca.* 1 g) is added to a depth of 5 cm and 6–7 ml of nutrient solution (McKnights solution, Table 1) are run into the tube before plugging.

With these tubes, the seeds are germinated *in situ,* near the top of the slope, or seedlings with radicles 0.7–1.5 cm long are transferred to the slopes. The sowing of seedlings is facilitated by using an inoculation needle bent to the shape of a shepherd's crook (S-shaped), or a pair of long-armed, lightly sprung forceps. For vermiculite culture, the seedling is buried 2 mm beneath the surface, with the radicle against the wall of the tube; the top of the vermiculite is pressed with a sterile, flat-ended glass rod after sowing.

For periods of growth in excess of 21 days, the addition of sterile water or nutrient solution (1/4 strength of that used in the agar) is recommended, with further additions as required (Section b.v). Pipettes or hand-syringes can be used, although we find that a burette, fitted with a three-way stopcock and a bell-shaped hood around the tip of the pipette, minimizes contamination. The burette is filled from an overhead reservoir (Figure 2b). The plugged burette and reservoir fitting, and the stopcock, are autoclaved separately and assembled in a clean, still atmosphere. Flasks of sterile water or nutrient can be replaced as needed.

For either system, metal bacteriological caps must be avoided owing to restrictions on CO_2 diffusion. With cotton plugs, the restricted diffusion of CO_2 limits growth to less than 1.0 mg dry weight per day with 20 mm diameter tubes, and to less than 1.5 mg per day with 25 mm diameter tubes. This is of little concern when only the ability to form nodules is being examined, as with plant dilution tests, but caution should be exercised when looking for

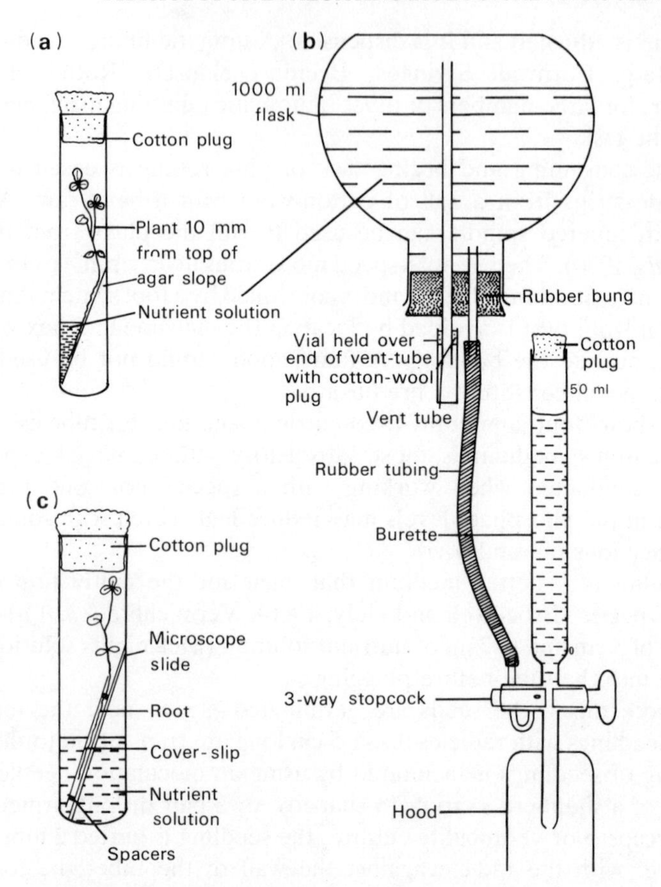

Figure 2. (a) Cotton-stoppered tube with short agar slope (Section c. ii). (b) Watering burette with hood and nutrient reservoir. (c) Fåhraeus slide method (Section c. iii) for observation of root hairs

differences in symbiotic effectiveness between strains of rhizobia (Gibson, 1967c). As a guide to using 20 mm diameter tubes, the exponential growth of *T. repens* (seed weight, 0.7 mg) is retarded at 28 days, while for *M. sativa* (2.7 mg) and *T. subterraneum* (10 mg), the corresponding periods are approximately 14 and 8 days. After these times, the increase in dry weight, and total N, is linear, being restricted to the rate of CO_2 diffusion, and differences between strains are minimized.

(iii) Fåhraeus slide technique

In order to examine root hair development and infection, Fåhraeus (1957) developed a system in which the roots grow down the surface of a microscope slide, underneath a rectangular cover-slip covering the lower two thirds of the slide (Figure 2c). The cover-slip is held away from the slide by 'spacers' of 2–4 chips of cover-glass cemented together. Nutrient solution (Table 1) is added (20 ml) to 25 × 150 mm cotton-stoppered tubes and the assembly autoclaved. Before sowing germinated seeds between the slide and the cover-slip, sterile soft agar (0.3% w/v) is run into the space and allowed to set (in a sterile Petri dish). At various intervals, the slide can be removed from the tube for microscopic examination of the roots (Nutman, 1959) (see Section f.i).

(iv) Petri dish method

For genetic studies with rhizobia, it is essential that many isolates be tested for loss of ability to form nodules, or to fix N_2, in a rapid screening procedure. To facilitate testing, B. G. Rolfe and P. M. Gresshoff (personal communication) have developed a Petri dish method for use with small seeded legumes such as *Trifolium repens.* Germinated seedlings (3–6) are transferred to the mid-line of a plastic dish (99 mm) containing 30 ml of agar, with Fåhraeus nutrient solution (Table 1). Prior to sowing, the lower half of the agar is streaked with the appropriate culture, using a sterile, disposable wooden applicator stick. The plates are left horizontal for 60 min after sowing, to enable the seedlings to adhere to the agar, and a 100 × 25 mm strip of Nesco-film (Nippon Shoji Kaisha Ltd, Osaka, Japan) is stretched around the join of the dishes before putting them in a vertical position. Two slits are made in this film to permit gas exchange (sterile forceps are run across the top of a group of plates). Many plates can be placed in rows beneath the light bank in a controlled environment cabinet (e.g. 22°C, 18 h light period; 19°C, 6 h dark period; 300 μEinstein $m^{-2} s^{-1}$). Assessment is made progressively up to 6 weeks. Owing to the heterozygous nature of *T. repens,* a plate and hence a 'mutant' colony is discarded as soon as one plant shows definite evidence of N_2 fixation. Where all plants fail to nodulate, the 'mutant' is regarded as non-nodulating. The technique will have very limited value for examining differences in degree of symbiotic effectiveness owing to the restrictive conditions under which the plants are grown but, within the aims outlined above, should provide a useful screening procedure.

(v) Semi-enclosed tubes

With this tube culture system, the roots are grown under bacteriologically controlled conditions within a tube, and the shoots are free above the tube (Figure 3), removing the restriction on growth due to reduced CO_2 levels within cotton-stoppered tubes. The system also permits the imposition of separate shoot and root temperatures (Gibson, 1965).

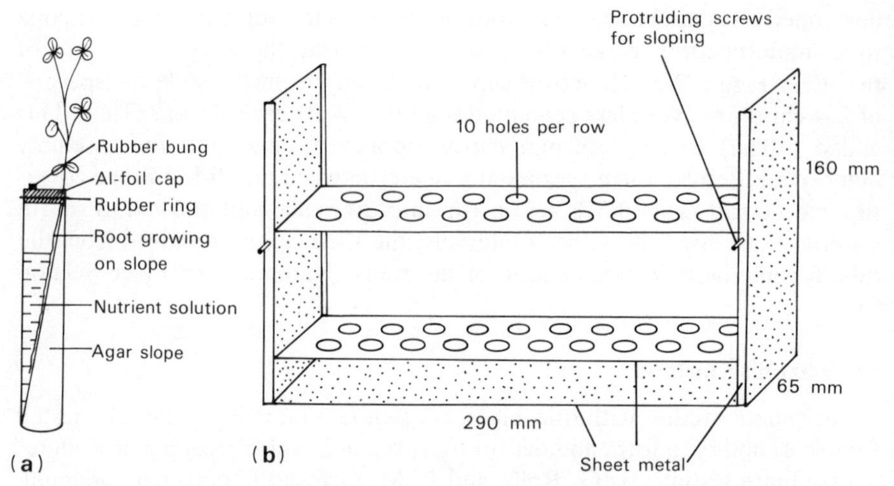

Figure 3. (a) Test-tube with roots growing on long agar slope and shoots exposed to atmosphere (Section c. v). (b) Sloping rack for tubes (Section c. v)

Plants from seed weighing 1.0–15.0 mg grow very well in this system. They include *Trifolium subterraneum*, *T. pratense*, *Medicago sativa*, *M. truncatula*, *Macroptilium atropurpureum*, *Stylosanthes humilis*, *Terramnus uncinatum*, *Desmodium* spp., and *Glycine ussuriensis*.

Agar medium is prepared as in Section c.ii, and 12.5 ml are dispensed in to 20 × 150 mm tubes (25 ml into 25 × 150 mm tubes). Circles of aluminium foil (45 mm diameter) are stamped from sheets of foil folded to 50 mm width and interleaved with newspapar (this aids separation of the discs). Thin foil may have to be doubled before folding. The disc is held in place over the tube top by a rubber band, or with an Elastrator ring normally used for marking and tailing lambs. It is applied with Elastrator pliers (Elastrator Pty. Ltd, 414 Collins Street, Melbourne, Australia; similar products can be obtained from other companies) and, by using a ring dispenser, two operators can cap and ring 1000 tubes in 2 h. After autoclaving, the tubes are sloped so that the agar just reaches the foil cap. Metal racks, containing two rows of ten holes in plates 6 and 12 cm above the base, and fitted with two screws protruding 7 mm from the side of the rack, 13 cm above the base (Figure 3b), greatly facilitate the sloping operation.

Seeds are surface sterilized and germinated in inverted Petri dishes (Section c.i). A sterile needle, of the same diameter as the radicles, is used to hole the foil 3 mm from the edge and immediately above the agar slope. Using sterile forceps to hold the cotyledons, the radicle is inserted through the hole *away from the slope*. The cotyledons are then moved towards the centre of the tube so that the radicle lies *on* the agar surface.

Care must be taken to ensure that the seed coat does not dry, otherwise the cotyledons will not separate. One method involves covering the tubes with an inverted 25 × 50 mm tube containing moistened cotton-wool in the base, and resting on the rubber ring. Wire mesh covers are placed over a closely packed rack of tubes to keep the covers in place. Alternatively, moistened paper tissues (e.g. Kleenex or Scotties) are laid gently over the tubes in the rack, which is then placed in a plastic bag containing free water in the bottom. The bag is closed. In the author's experience this system is slightly less effective than the tube system, although easier to use. Two experienced operators can sow 1000 tubes in 2 h provided that the seedlings are well germinated (radicles 10–15 mm long).

After 3 days, the covers are removed and any adhering seed coats gently removed with forceps or a needle. A hole is made in the foil cap on the opposite side to the seedling, using a sterile sharpened probe of 5 mm diam-

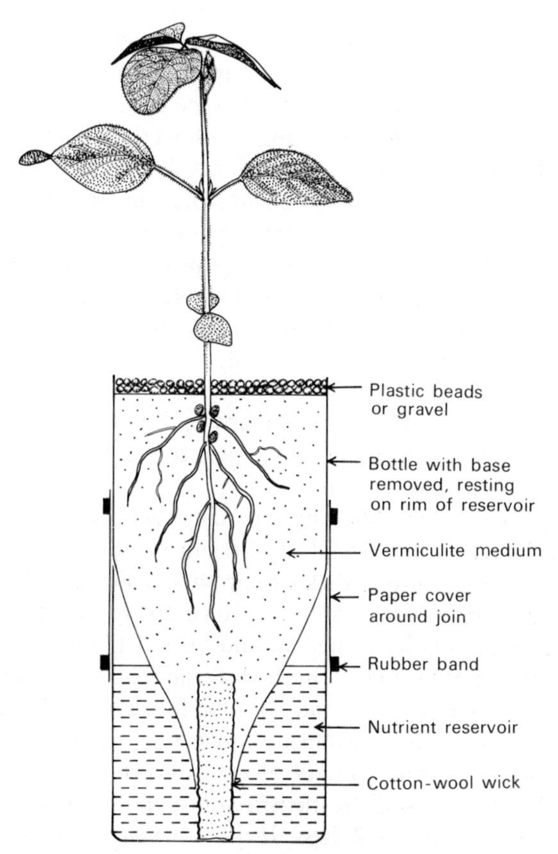

Figure 4. Leonard jar assembly (Section c. vii)

eter. Sterile nutrient solution (1/4 strength of that used in the agar) or sterile water is added to within 15 mm (*Trifolium* spp.) or 40 mm (*M. sativa,* tropical species) of the top of the tube, the lower level being essential for those species whose nodulation is retarded under water. The hole is filled with a sterile (rinsed in ethanol, dried at 80 °C) rubber stopper, 4–7 mm diameter, 14 mm long. Inoculum may be added in the watering solution, or several drops of a suspension containing 10^7–10^8 organisms/ml added by pipette; the nutrient is gently agitated over the roots at the top of the slope to provide uniform dispersion of the bacteria. The tubes are then transferred to the growth conditions. Ideally the roots should be darkened, either by a thick paper sleeve or by suspending the tube through holes into a darkened box (or water bath) (Figure 5).

The penalty for the improved plant growth is the need for more frequent watering than with the closed tube technique. Normally watering is not required for 14–16 days after setting-up (depending on the growth temperature), but it becomes more frequent until harvest at 28–35 days. Watering of a large number of plants is done most effectively with an automatic pipetting machine operated by a foot switch. The interior of the syringe and tubing, and the exterior of the tubing to be placed in the watering solution, is sterilized by immersion in 70% ethanol for 5 min, followed by passing 2 l of very hot sterile water through the syringe and tubing. For longer term experiments, the plants should be well spaced to ensure that the leaf area index (Section b.i) does not exceed 4.0 over the experimental area.

Contamination of uninoculated control plants by rhizobia is very rare (less than 1 in 100 tubes) and little or no trouble is encountered with other contaminants. However, the technique is not recommended for use where absolute bacteriological control is essential, as in ascertaining whether a strain will nodulate a particular host or when doing plant dilution counts. Where such control is not essential, as in many effectiveness tests with known nodulating strains, or environmental studies, the technique has many advantages over the closed tube system.

The technique can be modified by replacing the agar with a 1:1 mixture of vermiculite and perlite, plus 15 ml of nutrient solution. The tubes are filled completely with the mixture in order to provide a substrate and water supply for the seedlings when sown as above. Alternatively, surface-sterilized seeds are inserted through a hole in the cap and buried 15 mm below the surface. A sterile rubber stopper is placed in the hole and 2–3 days later the emerging shoot is pulled through the hole and the tube inoculated. The rubber stopper is replaced gently. Subsequent watering is effected through this hole. Provided that the occasional contaminant is of no concern (standard uninoculated controls give the most pessimistic measure of this), the stopper can be omitted after the first watering at 10–12 days. If the tubes are to be placed in a water-bath, the addition of one or two glass marbles on top of a shallow layer of vermiculite/perlite at the bottom of the tubes during preparation will reduce the tendency for the tubes to float. This modification is used only where the plants grow poorly on agar. The disadvantages are that nodulation is difficult to observe on a daily basis, and there is a need for more frequent watering.

(vi) Plastic pouches

A disposable pouch system was proposed by Porter *et al.* (1966) and modified by Weaver and Frederick (1972) as another method for plant dilution counts for rhizobia in soils and in legume inoculants (Chapter III.1, Section b). The method is particularly useful for larger seeded legumes, e.g. *Glycine max.* When properly stacked, as in a gramophone record holder, many pouches can be located in a controlled-environment cabinet.

The pouches contain a sheet of sterile absorbent paper, with a V-shaped trough at the top. Surface-sterilized seeds are placed in this trough, and 40 ml of nutrient solution containing the appropriate dilutions of rhizobia are added to the bag through a plastic drinking straw located at one side of the bag. The system should not be used for assessing the level of symbiotic effectiveness. There should also be included a number of uninoculated pouches to ascertain the level of any contamination. Its use with strains of bacteria whose nodulating behaviour is not known, or which are slow to nodulate a particular host, is not recommended. The basis of the technique is the assumption that one viable cell of rhizobia added to a pouch will lead to nodule formation.

These pouches may be purchased (American Hospital Supply Corp., Moline, Ill., U.S.A.) or produced in the laboratory, using polythene sandwich bags and sheets of blotting paper cut to the size of the bag. Paper folded to have the V-trough is autoclaved in sealed foil (30 min, 121 °C) to prevent wetting, and inserted into the bags immediately before use.

(vii) Leonard jar assemblies

The most common method for growing large-seeded legumes under bacteriological control is the Leonard jar technique (Leonard, 1944) or some modification thereof. The assembly consists of two parts, a wide-mouthed container to serve as a reservoir of nutrient solution (preserving jars are ideal owing to their weight and stability), and an upper vessel containing sand or other supporting medium (Figure 4). The upper vessel is conveniently made from a round, brown beer or spirits bottle (750 ml capacity) with the bottom removed. Removal of the bottom can be achieved in various ways but L. Diatloff (personal communication) recommends that a mark be made around the bottle 10 mm from the base with a diamond-point glass knife. The bottle is then rotated slowly in a cradle, with the very flat flame from a porcelain-tipped fishtail burner directed at the mark. After four turns, the bottle is plunged into cold water while covering the neck opening with a finger. Sharp edges are ground off with an emery wheel. Alternatively, the bottoms of the bottles can be removed with a glassworker's abrasive wheel.

The neck of the bottle is plugged with a cotton-wool strip, or with a lamp wick (thoroughly washed in detergent and rinsed) before filling with well washed medium–coarse river sand amended with K_2CO_3 (1 g kg^{-1}) to within 50 mm of the top. Vermiculite or perlite, or mixtures, can also be used. With vermiculite, the wick should extend 50 mm into the bottle, whereas with sand it should extend nearly to the top of the medium, to ensure adequate distribution of the solution by capillarity. As indicated previously, it may be essential to wash the vermiculite if the pH is high. The medium is moistened with nutrient solution (Table 1). The same nutrient is used to fill the nutrient reservoir bottle (preferably 1–2 l capacity). The medium bottle is then inverted into the reservoir so that the shoulder of the former rests securely on the mouth of the latter and the bottom of the bottle is within 50 mm of the base of the reservoir. The top is covered with a Petri dish half, and either the assembly is placed in a paper bag or the junction of the vessels is wrapped with paper held in place by rubber bands. The assembly is autoclaved for 2 h at 121 °C (150 kPa).

Surface-sterilized seed, or seedlings, are sown below the surface of the medium, inoculated, and the Petri dish cover is replaced until the cotyledons emerge. At this time, the sand is covered with dry-sterilized gravel or Alkathene (ICI, Australia) polyethylene beads (sterilized in 95% ethanol, dried, and allowed to cool) to a depth of 20 mm.

The units, including uninoculated controls, are then transferred into a glasshouse or growth cabinet. If it is necessary to refill the reservoir during the course of the experiment, the two containers are separated and normal bacteriological precautions used in adding sterile nutrient to the depleted reservoir or in exchanging empty reservoirs for freshly sterilized full reservoirs. If sound bacteriological procedures are used, and the assemblies are placed in a clean environment, contamination by other rhizobia is rare. No water should enter the growth medium through the surface (the gravel or beads should remain dry). However, it is extremely difficult to maintain complete bacteriological control, and other bacteria will be found in root washings. This is an important point to remember if using the technique for purposes other than observing nodulation or measuring symbiotic response, e.g. grass–rhizosphere studies. The likely source of contamination is from beneath the seed coat.

An interesting modification to the above technique utilized high-density polyethylene bottles (2 l) able to withstand steam sterilization (Wacek and Alm, 1978). These bottles are cut with a knife just below the shoulder; after inserting a wick, the top is filled with sand and the base is used as the reservoir. The bottles will withstand numerous autoclavings.

(viii) Open trays, capilliary watering

For genetic studies on the host plants, it is common to inoculate a number of plants with the one strain of rhizobia. Devine and Reisinger (1978) have modified the Leonard jar technique for larger scale operations. Plexiglass trays (58 × 63 × 15 cm) are fitted with two rows of three Plexiglass tubes (6.4 cm deep, 4.4 cm diameter) protruding from the bottom and fitted with 18-mesh wire screen. Two trays and tubes are filled with sterilized sand, perlite, or as they prefer, vermiculite, and lowered into nutrient reservoirs constructed of black Plexiglass (61 × 130 × 11.4 cm) and fitted with a draining spigot. Water supply to the plants is achieved by capilliary action through the medium in the tubes. Thirty-six soybean plants can be grown per tray, and there are two trays per reservoir (80 l capacity). Nutrient renewal is made every 2 weeks, and plants normally grow for 8 weeks. The trays are disinfected with 50% ethanol in water, or other disinfectants, between uses.

A similar method for growing large seeded legumes was described by Vincent (1970), although his apparatus requires nutrient addition every 2–3 days.

(ix) Open pots

Where the host species and bacterial strain form nodules promptly, good bacteriological control can be achieved by growing the plants in open pots, *provided* that the aerial load of bacteria is low. The pots are thoroughly washed and rinsed with disinfectant (e.g. Dettol or some other bacteriocide) immediately prior to filling with the growth medium. A perlite/vermiculite mixture (2:1) provides good aeration and reasonable water-holding capacity. Under some circumstances, this medium can be autoclaved in polyethylene bags able to withstand such treatment, and the contents poured into the pots immediately before use, although it is most unlikely that the media will contain a population of rhizobia at the time of purchase. The vermiculite may require neutralization treatment as indicated in Section b.vi. The medium should be firmly, but not tightly, packed. The pots should have a drainage hole, covered by mesh to prevent loss of medium.

Other materials used as base substrate include pumice and sand/vermiculite mixtures, both providing good aeration and drainage. For their soybean nodule studies, Bergersen and Turner (1970) use a 1:1 mixture of washed vermiculite and washed river sand in 250-mm diameter pots (10 plants per pot). The nutrient solution is McKnights solution (Table 1), diluted 1:1 with tap water. After germination, the pots are placed in the tops of heated boxes (Figure 1) at 26–28 °C for 2 weeks to encourage prompt nodulation. The glasshouse is maintained at 25/20 °C on a 12 h/12 h cycle. Robertson *et al.* (1975a) prefer sterile pumice as their substrate for lupin culture. The plants are grown in sterile stainless-steel trays in a growth cabinet (23/20 °C, 12 h daily light period).

When sowing, holes are made in the medium and inoculated with 1 ml of a suspension of the appropriate bacteria (10^7–10^8/ml) before adding the surface-sterilized seed. The hole is made, and covered in, using a sterile glass rod. Soybean and lupins are best sown at 4 cm depth, and smaller seeds closer to the surface. The pots are then watered with sterile water or, as is frequently adequate, with demineralized water. As soon as the cotyledons emerge, the surface is covered with a 20 mm depth of sterile (ethanol-rinsed) polyethylene beads.

Nutrient solution (Table 1) is added through a watering tube placed in the side of the pot and with the top covered with a specimen tube or test-tube. Alternatively, nutrient is added through the beads once daily, but this should only be done if the precautions are adequate to minimize the bacterial contamination in the air. As conditions demand, additional demineralized water is supplied once or more daily. Great care should be exercised to prevent flow through the pots, and so minimize contamination of the atmosphere (especially in controlled-environment cabinets). The beads serve a double function of minimizing moisture loss from the system and minimizing upward movement of the bacteria. Similarly, inoculating the bottom of the hole prior to sowing minimizes the bacterial load carried above the surface during germination. This procedure also provides heavy nodulation in the crown region of the plant (especially useful for those experiments designed to provide nodule tissue) and minimizes the possibility that any contaminating rhizobia will form a significant proportion of the nodule population. Experience is the best teacher in determining how much water or nutrient solution to add to maintain adequate water supplies without the risk of flow-through contamination.

Provided that the local water supply is free of mineral salts (especially nitrates and carbonates) and organic materials, tap water can be used to water the plants; this includes preparation of the nutrient solution. Installation of an in-line UV water sterilizer (e.g. Aquafine Corporation, Burbank, California, U.S.A.) will minimize the risk of contamination by rhizobia, particularly where the supply is drawn from underground wells.

Obviously, the open pot system has its limitations and must be used with care. There needs to be a clear recognition of the experimental aims and the problems of interpretation which arise if even a low proportion of the nodules are formed by contaminating rhizobia. The efficacy of the system depends on the establishment of a high rhizosphere population of inoculant bacteria and their influence on minimizing nodulation by contaminating bacteria. Each experiment should contain 5–8% of the pots as uninoculated controls, well spread through the experimental area. As soon as such plants show any nodules, or greening due to N_2 fixation, the experiment should be terminated. If sufficient time has elapsed since establishment (Section f.iv), harvest may be appropriate. While generalization can be dangerous, in many cases nodules would appear on uninoculated controls contaminated by a given load of rhizobia before such contamination would effect nodulation of inoculated plants. Checks of identity, by serological or other means (Chapter II.9), should be made of nodule populations where doubts exist.

(x) Hydroponics

Large-vessel hydroponic culture has rarely been used for nodulation studies, owing primarily to difficulties in maintaining bacteriological control and for achieving good nodulation in water. Munns (1968) successfully developed a system for examining pH and combined N effects on the nodulation of *Medicago sativa*. Polyethylene bags were used as liners for 3-l glass beakers or 22-l steel drums. The nutrient solution (Table 1) was aerated continuously with filtered air, using fish-tank aerators to disperse the air. The pH of the solution was adjusted daily with H_2SO_4 or KOH, and varied no more than 0.2 unit from the nominal value. The seeds were germinated on cotton gauze over nutrient solution, and transferred to the containers when the roots were 5–6 cm long (7–8 days). The containers had a thin plywood, or polyurethane, cover with 2–3 mm diameter holes, through which the roots were inserted into the nutrient solution 25 mm from the top of the container. The seedlings were held in place by sterile cotton-wool.

A simple, inexpensive aeroponic system for the culture of nodulated legumes has been described by Zobel *et al.*, (1976); it is particularly useful for plants susceptible to water-logging.

(xi) Other systems

The foregoing systems provide a range of techniques that cover the common requirements in most laboratories studying various aspects of the legume–*Rhizobium* association under microbiologically controlled conditions. However, there are a number of specialized techniques that have been developed for particular purposes. The principles behind these techniques are described but intending users should consult the references cited for further details.

Tissue culture

Several systems have been used to examine the relationship between plant cells and rhizobia in attempts to elucidate the nature of host factors responsible for the induction of nitrogenase in rhizobia. Holsten *et al.* (1971) inoculated actively growing soybean cell culture suspensions with *Rhizobium japonicum*, continued the culture on a klinostat for 3–7 days, then plated the cells on semi-solid agar. The liquid Murashige and Skoog medium was supplemented with coconut milk (15%) and 2 mg l^{-1} of 2,4-dichlorophenoxyacetic acid, but highest nitrogenase activity was found when these amendments were omitted from the semi-solid agar medium. Phillips (1974a,b) used a similar approach but Child and LaRue (1974) modified the system by plating the cells on auxin-free agar for 7–14 days prior to inoculation. Induction of nitrogenase occurred when some strains of *Rhizobium* sp. were cultured adjacent to, but not in contact with, cultured cowpea or tobacco cells on agar (Scowcroft and Gibson, 1975), but further interest in this system declined when it was shown that these strains could fix N_2 on a suitably amended defined medium (e.g. Pagan *et al.,* 1975).

Reporter and Hermina (1975) adopted a different approach by culturing plant cells and bacteria in separate containers, connected by a tube with 1–3 bacterial filters installed to allow only the passage of chemical compounds. With this 'trans-filter' system, they were able to show that a host factor could induce nitrogenase in certain 'slow-growing' strains of rhizobia. The method has now been refined so that the bacterial cells are cultured within a dialysis bag within the plant cell culture medium (Bednarski and Reporter, 1978). Their results indicate that the bacterial cells are essential to stimulate the plant cells to produce the factor which leads to induction of nitrogenase in the rhizobia. Using this modification, or using an extract from the culture medium in which plant cells have been stimulated to produce the 'inducer', nitrogenase has been induced in many strains of rhizobia, both 'fast growers' and 'slow growers', that have not been shown to fix N_2 on a defined medium.

For tissue culture studies, the need to maintain absolute bacteriological control is paramount. Similarly, it is essential that appropriate checks be made at all stages of operations, especially where manipulative procedures are involved, to ensure that the culture is a pure strain of rhizobia. In essence, it should be possible to show close correspondence between total counts on the population done by conventional plating and plant dilution tests (Chapter III.1, Section b.ii). Media favouring the growth of other diazotrophs (e.g. *Klebsiella, Azotobacter*), but not rhizobia, should be used. Serological examination of isolates, or preferably immunofluorescence examination of populations to ensure homogeneity, should also be made (Chapter II.9).

Excised root culture

Excised root culture techniques for nodulation studies have been developed in an attempt to define host factors that are important in nodule formation and function. Raggio *et al.* (1957) described two systems. With one, the cut end of an excised root is inserted into a small vial containing organic nutrients, KNO_3, and an agar solution. This vial is located in a recess in an N-free inorganic nutrient agar in a large Petri dish, and the root grows on the agar surface. With the second method, the vial plus excized root is suspended from the top of a test-tube, and the root grows into an inorganic nutrient solution. The former method was preferred in most of the subsequent studies on the effect of carbohydrates, combined nitrogen (Raggio *et al.*, 1965; Cartwright, 1967), and temperature (Barrios *et al.*, 1963). Bunting and Horrocks (1964) modified the technique by including a portion of the hypocotyl when excising the root, and claimed that this greatly increased the number of roots nodulating and the number of nodules formed. Of interest is the fact that only *Phaseolus vulgaris* has been used in these studies, and this is one of the more difficult to nodulate species under field conditions.

Propagules

Stolon and stem cuttings have been used to minimize host genotype variation in the heterozygous species *Trifolium repens* and *Medicago sativa.* Joshi *et al.*, (1967) used tips of young, actively growing stolons from *T. repens,* each tip bearing a single apical growing point and an incipient root. The cut end was sealed in paraffin wax and the whole tip (25–50 mm) surface sterilized in 80% ethanol for 60 s, followed by 0.1% w/v $HgCl_2$ for 3 min. After thorough washing, the 'ramets' were transferred to agar slopes in cotton-plugged test-tubes (Section c.ii); inoculation was made after 15 days. Mytton (1976) modified the procedure by reducing sterilization to 5 s in 80% ethanol and growing the 'ramets' so that only the roots were within test-tubes (Section c.v, Figure 3a). The most effective propagation was achieved by including one folded and one expanded leaf on each stolon tip.

Stem cuttings of *Medicago sativa* were used by Lindsay and Jordan (1976) to produce populations of genetically identical plants. Donor plants were severely cut every 4 weeks to increase the yield of new stems. Three weeks after the fourth cutback, stems were cut to produce pieces 30 mm long (20 mm below the nodes, 7 mm above) and the nodal leaves were removed. The lower end of the cutting was dipped in talc containing 0.1% w/v of 3-indolebutyric acid (Rooting Powder I, May and Baker, Bramalea, Ontario, Canada), and the cutting placed in a box of sterile, fine quartz sand with the nodes just above the surface. After watering with a complete nutrient solution, the box was covered with transparent plastic film. The cuttings were transferred to pots 21 days later and, after 14 days (this timing was critical), inoculated with rhizobia. Upper stem segments gave the most vigorous propagules during the first 3 weeks, but subsequently were no better than cuttings taken lower on the stems. A similar technique has been used by Seetin and Barnes (1977) for their host genetic studies on the nodulation of *M. sativa.*

Grafting

Differential effects of shoots and roots on various aspects of the symbiosis between legumes and rhizobia have been studied with shoot–root grafts (Hely, Bonnier, and Manil, 1953; Tanner and Anderson, 1963). Lawn and Brun

(1974b) made grafts between soybean cultivars to study the influence of the shoots, and the roots, on total nodule development, total nitrogen fixation and the specific nitrogen-fixing activity of the nodules (C_2H_2 reduction, g^{-1} nodule fresh weight). V-grafts in the hypocotyl region are used most commonly. The excised shoot is pared to expose 3–4 mm of cambial tissue on either side of the hypocotyl. When inserting this into a vertical cut in the top of the root stock, care should be taken to ensure that the cambial tissue of scion and stock are in contact on at least one side of the graft. The join is covered by a 1–2 cm section of plastic drinking straw (slit vertically, rolled to root diameter and the roll maintained by a very small rubber ring; this is put over the stock before making the slit and pulled over the graft after inserting the scion). The join is sprayed with water and the pot placed in a plastic bag to retain a high humidity. Carbon dioxide is injected into the bag daily until the graft takes. The plants are hardened slowly by cutting a corner from the bag on one day, opening the top the next day, and removing the bag the following day. In a recent experiment, R. J. Lawn and the author achieved over 95% success in grafting together four different *Vigna* species.

(d) NON-STERILE SYSTEMS

There are situations when the soil is essential to meet experimental requirements, but for reasons of reliability in reproducing environmental conditions, time saving in travel, etc., the experiments are best made in the laboratory or glasshouse. These include the comparison of inoculation responses on different soil types, comparison of differences in the populations of indigenous rhizobia, the sampling of populations of rhizobia in soil at a time when it is not practical to grow plants in the field, and the assessment of fertilizer responses.

(i) Pot culture

A common approach is to collect surface soil to a depth of 10–15 cm, allow it to air-dry, and pass it through a 5-mm screen to remove stones, large clods, and root pieces. The use of a finer screen can lead to loss of structure in the soil and subsequent aeration problems. A weighed amount of soil is put directly into the pots, or into polyethylene bags which act as liners for the pots. To facilitate later watering, all pots are tared to the same weight by adding gravel or sand to the botton of the pots. Where applicable, basal dressings of superphosphate, K_2SO_4, lime, etc., are made before sowing, either as surface incorporation or thoroughly mixed through the soil before adding to the pots. Disturbing the profile promotes mineralization of soil N, although this effect can be alleviated partially by growing oats or barley for 4–6 weeks, and removing them at soil level prior to sowing the legume

(Simpson and Gibson, 1970). For experiments intended to examine inoculation effects, any form of soil sterilization (steam, chemical, irradiation) is likely to increase nitrogen availability and grossly disrupt the proportions of the different elements of the natural soil microflora (e.g. V–A mycorrhiza). Demineralized or distilled water is preferred to tap water for daily watering, and care should be taken to use water at the same temperature as that in the glasshouse (Brockwell and Gault, 1976).

(ii) Soil cores

Vincent (1970) has described a soil core technique used successfully in his laboratory for many years, especially when examining inoculation effects where there are different populations of rhizobia in different soils. Sealed, but empty, food cans (80 mm or more diameter) are opened with a can opener that turns the cut edges of the can inward and so reinforces it. A metal cap protects the can's shape when using a heavy hammer to drive the can into the soil to within 20–30 mm of its full depth. The can and soil core are lifted with a spade and placed in a plastic bag to prevent soil movement and desiccation during transport. In the glasshouse, the base of the plastic bag is pierced to allow free drainage, and the bag and can are placed in a wide-mouth drainage jar such that the can is 60–90 mm above the base of the jar. If the fit is tight, as with a flexible plastic jar with a mouth opening just greater than the can (i.e. the can is suspended), an air escape hole is made in the jar 2 cm below the bottom of the can. The top of the plastic bag is left over the core during the early stages of plant growth, then opened to provide protection from lateral wind-borne contamination and desiccation. As required, sterile (boiled) water is applied to the top of the cores until a slight excess drains into the jar.

(e) ENVIRONMENTAL CONTROL SYSTEMS

The principles of environmental control were considered in Section b; the following descriptions apply to systems that have been developed for legume nodulation studies. They are selective, but provide a range of controls that researchers may feel are useful for their studies, either as indicated or modified.

(i) Glasshouse

The principal considerations in any glasshouse operation are the maintenance of hygiene and the provision of suitable light and temperature conditions. Hygiene must be practised at two levels. The aerial load of contaminating rhizobia must be kept to a minimum. This can be achieved by minimizing run-through from pots, and by having a floor that can be hosed down daily. Drainage saucers under the pots should be avoided as the provide a continuing source of contamination; they cannot be rinsed out regularly without great difficulty and without the risk of contaminating the pots during their movement.

Legumes are particularly susceptible to may fungal pathogens, to predatory and/or virus-spreading insects and to some arachnids. As far as possible, only those people not working in other glasshouses should be permitted entry to the glasshouse. Where there is any danger from this source, footwear should be cleaned or changed before entry, a clean laboratory coat or overall put on at the entrance, and for the more hirsute members of the population, the hair combed to remove insects. Fine mesh should be placed over evaporative cooler entry ports and over ventilation louvres in the roof or side of the glasshouse (Figure 1). Under extreme circumstances, insect-proof vestibules may have to be installed at the entrance to the glasshouse. At the first indication of aphids or red spider, fumigation (by experts) should be carried out, although a high degree of success with biological control can be achieved under some circumstances. For example, red spider (*Tetranychus urticae*) may be controlled by the predacious mites *Phytoseiulus persimilis* and *Typhodromus occidentalis* (or *Amblyserus fallacis* under higher temperatures), while white fly (*Trialeurodes vaporariorum*) is controlled by the parasitic chalcid wasp *Encarsia formosa* (Perifleur Ltd., Woodlands Avenue, Littlehampton, Sussex, U.K.).

Louvres, blinds, or whitewash help keep the heat load to acceptable levels in hot environments (Section b.i). Evaporative coolers may be used under low humidity conditions. Flowing water over the roof and/or regular hosing of the glasshouse floor may be necessary under very hot conditions. As important as cooling is the need to provide adequate heat for the species being grown. Many tropical legumes perform poorly at temperatures below 25 °C and heating is frequently necessary, especially at night.

Photoperiodic effects on plant development were discussed in Section b.i. Light intensity is also important. For example, with soybeans grown during the winter months in Canberra (latitude 35 °S; skies normally clear), bacteroid preparations show greatly lowered nitrogenase activity (F. J. Bergersen, personal communication), and normal experimentation is usually suspended during June–August. This problem has not been overcome by providing normal supplementary lighting.

(ii) Controlled-environment conditions

A wide range of commercial 'growth cabinets' are available, and details should be obtained from suppliers. The principal consideration is that the light source should be adequate in terms of quality and quantity (Section b.i).

Humidity control is not a luxury if it reduces the need for watering, with its associated contamination risks.

For tube culture work (Sections c.ii and c.iii), fluorescent tubes located beside, rather than above, racks of test-tubes provide an efficient and effective form of illumination (Gibson and Nutman, 1960; Vincent, 1970). Pairs of 1.3 m long 40-W tubes, or single 1.5 m long 80- or 120-W tubes are arranged on an open stand, alternating with double-row racks of plant tubes. The main consideration is to prevent overheating within the test-tubes and ideally the equipment is located within a temperature-controlled room with a good air flow. Light intensity is less important than temperature control as CO_2 diffusion into the tubes is the main limiting factor to growth. However, irradiance should exceed 75 μEinstein m^{-2} s^{-1} within the tubes. If flourescent tubes with adequate spectral composition to promote normal growth cannot be obtained (a wide range is now available), supplementation with two or three 40-W incandescent bulbs above each light row may be necessary. However, this will double the heat load. Similarly, the ballasts for the flourescent tubes, which produce 65% of the heat load, should be located in a well ventilated position outside the room. The need for adequate dark period temperatures should not be ignored when establishing the system.

Root temperature control

In the field, soil temperature shows less variation than air temperature and it is logical that root temperature control, independent of shoot temperature control, be exercised. The following description applies to a cabinet in which both shoot and root temperature are controlled independently; the irradiance level and duration are also controlled. This cabinet has been used for studies involving the effects of root temperaure variation as well as experiments under standard conditions (Gibson, 1965, 1969; Gibson and Pagan, 1977). The system is adapted for use with the semi-enclosed tube technique (Section c.v), but a modified version has been developed for pot use.

The cabinet comprises four copper tanks (41 × 33 cm, 22 cm deep), illuminated by a bank of twenty 1.5 m long 120-W fluorescent lamps plus four 60 mm long 20-W incandescent tubes (the total irradiance is 300 μEinstein m^{-2} s^{-1}) (Figure 5). A low level of cooling is provided to each tank from a cooling coil located beneath a thermostatically controlled heater-pump which, in conjunction with a pipe to the opposite end of the tank, maintains vigorous circulation of water within the tank. Eight aluminium racks each hold 23 tubes; the total capacity of the cabinet is 736 plants. The light bank is located 45 cm above the top of the tanks, and separated from the plant space by double glazing. Filtered air is drawn continuously past the flourescent tube ballasts (transformers) located within a large box on the front of the cabinet,

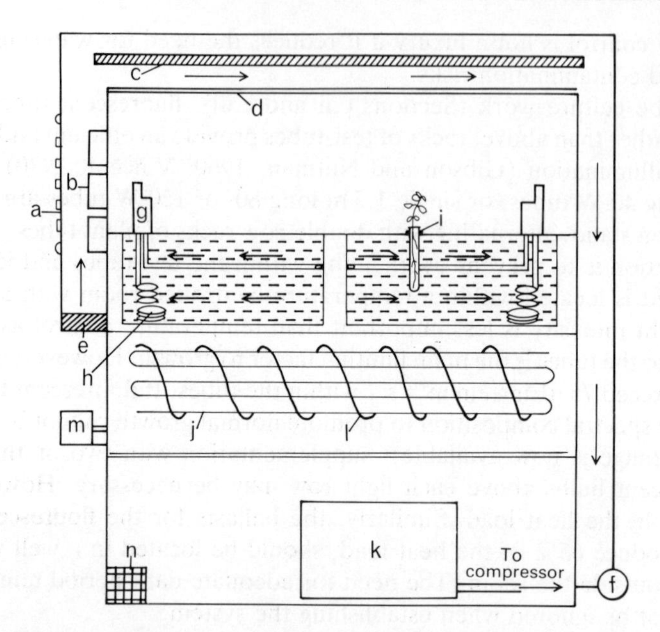

Figure 5. Cross-sectional view of controlled-environment cabinet with root temperature tanks. (a) Control panel (light, fan, and heater switches; time switch); (b) bank of ballasts (transformers) for fluorescent lights; (c) light bank (fluorescent plus incandescent); (d) double glazing; (e) air filter; (f) exhaust fan; (g) pump to heat and circulate water; (h) cooling coil for water-bath; (i) racks holding culture tubes rest on top of water-bath; (j) cooling condensers, connected to 'brine tank' (k); (1) copper-sheathed (soil) heating wire; (m) motors to drive air-circulating fans (fans not shown); (n) pumps to circulate coolant to (h) and (j)

and past the fluorescent tubes before exhausting to the outside air. This ensures that the tube envelope temperature is kept around 35–40 °C, the temperature for optimum operation. Air temperature is controlled by fans on each side of the cabinets, and below the level of the tanks. These fans draw air past cooling coils and heating elements (copper-sheathed soild heating wires), and circulate it up the outside of the insulated root tanks and across the top of the tanks. The cabinet is divided into two halves, and within 5–8 °C limits, different shoot temperature regimes can be applied to each half. The shoot temperature is controlled so that the air is being cooled or heated when the temperature rises of falls 0.5 °C from the selected temperature. A 5 h.p. compressor cools a 1:2 polyethylene glycol/water mixture in a coolant tank; this mixture is pumped through the air cooling coils (on demand) and the water-baths (continuously). This 5 h.p. compressor serves two such cabinets.

For larger plants, tanks to fit in LB growth cabinets (Morse and Evans, 1962) have been constructed. Four insulated tanks, 25 cm width, are located in a cabinet. Drain plugs are fitted to the base of 8 cm diameter pots, and these screw into one of three drainage lines running the length of the tank. An automatic watering system delivers sterile nutrient solution every 2 h (the growth medium is perlite). The nutrient is circulated through stainless-steel pipes within the tanks for temperature equilibration before being added to the pots. The principle of temperature control in the tanks is the same as that described above except that the coolant is pumped from a refrigerated water-bath. This equipment can also be used in a glasshouse.

(f) EXPERIMENTAL OBSERVATIONS AND THEIR ASSESSMENT

Within this section, consideration will be given to procedures commonly employed in studying various aspects of legume nodulation. Light and electron microscopy techniques used are too specialized and extensive for consideration here, but access to suitable source material will be found in the excellent review by Goodchild (1977).

(i) Root hair infection

For continuing observations on root hair infection, the Fåhraeus slide technique (Section c.iii) is ideal for small-seeded legumes. Depth of focus problems, and the greater length of root, restrict its usefulness with medium and large seeded species. Nutman (1959, 1962) found that bright field illumination, using a 30× objective and 8× eyepieces, gave optimal magnification for routine searching. Doubtful infections were checked at higher magnification and/or with phase contrast illumination.

Several staining procedures have been recommended for studying infection threads. Worral and Roughley (1976) fixed roots in 3% v/v acrolein (which is highly toxic) in tap water at ice-bath temperature for 12 h, and stored them at 3°C in 0.025 M phosphate buffer, pH 6.8, before staining lightly with toluidine blue. Phillips (1971) fixed the roots for 24 h in Carnoy's solution ($C_2H_5OH:CHCl_3:CH_3COOH$, 6:3:1), then stored the roots in 70% ethanol after thorough rinsing. To stain, the roots were rinsed in water then dipped in 0.05% toluidine blue at pH 6.8 for 3 s. The infection threads stained red and the root hairs blue.

'Curling' of root hairs is frequently used to describe a response to inoculation. Some curling is non-specific, and Yao and Vincent (1969) have proposed three categories as a rational basis for describing this phenomenon:

(i) branched, but no curling;
(ii) moderately curled, with the tip of the root hair curled through an angle of rotation of at least 90° but less than 360°;

(iii) markedly curled, with the tip curled at least 360°; such hairs are usually deformed markedly and tightly curled. This condition is invariably restricted to host plants inoculated with rhizobia able to nodulate that host.

(ii) Time to initial nodulation

The time elapsed from inoculation until the first visible nodule appears has been used as a measure of environmental effects (physical, pH, nutritional) on the establishment of the symbiosis (Gibson, 1967a; Munns, 1968; Gates, 1974) and of the symbiotic compatibility between particular hosts and strains of rhizobia (Nutman, 1967). It is the summation of a number of physiological processes, starting with the attachment of the bacteria to the root hairs and finalized when the cell division initiated by infection has progressed to an identifiable structure. Distinction between young nodules and lateral root initials requires experience but the former tend to be rounded compared with the pointed lateral initials. Where continuing observations are possible, as with agar culture, plastic growth pouches, and hydroponics or with vermiculite in tubes where the root grows next to the glass (Section c), doubtful nodules can be recorded and the observations verified or discarded the following day. Observations are facilitated by using a light box and by reducing the amount of $CaCO_3$ or Ca phosphate precipitate in the agar.

Determinations of nodule number have been used for various reasons but care must be used in the interpretation of the results. Successful infections frequently cause the cessation of further infection in their immediate vicinity (Nutman, 1959), so that nodule number is not a good indication of the number of infections. Delaying inoculation will lead to many, small nodules while the application of combined N delays nodulation but can lead to more nodules and an increase in total nodule weight.

'Pseudo-nodules' are found on some plants, especially when they are inoculated with rhizobia from heterologous 'species'. They should not be confused with true nodules and, where doubts exist, sample nodules should be examined under the microscope. On *Medicago sativa* and *Trifolium repens* these 'nodules' usually have a white, fluffy exterior. The reason for their development has not been ascertained, i.e. it is not known whether there is root hair infection or whether they are a wound response. They frequently appear at the point of emergence of a lateral root, but this is a common feature of the nodulation of many 'tropical' legumes and it should not be assumed that all such structures are 'pseudo-nodules'.

(iii) Dry weight and total nitrogen

Plant material should be dried in a fan-equipped oven until the material has attained a constant weight. The airflow in the oven should not be restricted by over-loading. Forty-eight hours at 75–80 °C should be adequate. Higher temperatures can lead to volatilization of some N compounds in the tissue. The material should be stored in a desiccator or in a plastic bag containing a desiccant such as silica gel or anhydrous calcium chloride until weighed.

The Kjeldahl digestion for N analysis outlined by Bergersen (Chapter II.2) can handle up to 300 mg of plant material, depending on flask size. For large quantities, the material must be ground and subsampled. The Wiley-type hammer-mill breaks the material to a size able to pass through screens located around the outside of the chamber. This tends to stratify the ground material in the collector and careful mixing is required before subsampling. The preferred alternative is a coffee-grinder type of mill in which the blades, located in the base of the chamber, rotate at 26000 r.p.m. The material in the chamber is covered with a wooden plunger, and the ultimate fineness of the ground material is determined by the degree to which it is confined by the plunger and the length of the grinding. Where possible, nodules should be separated from the roots, as the high level of N in these discrete structures can lead to unevenness in the distribution of N through the material. Ground material should be kept in a desiccator.

After grinding and mixing, two subsamples are digested, each of approximately 200 mg (accurately weighed). The digestion mixture is diluted to 5% acid and allowed stand overnight before two replicates (*ca.* 2.5 ml) are taken for analysis with a Technicon AutoAnalyzer (Chapter II.2).

(iv) Determination of symbiotic effectiveness

The term 'symbiotic effectiveness' provides an indication of a nodulated plant's ability to fix nitrogen. Effectiveness is expressed in qualitative terms—high, moderate, or intermediate, or ineffective—or in quantitative terms, by which growth and/or N_2 fixation of the nodulated plants is related to the growth of plants receiving adequate combined nitrogen or to the growth and N_2 fixation of plants nodulated by a standard strain. Occasionally the term 'efficiency' is used instead of 'effectiveness', but this is incorrect. Efficiency is properly expressed as a qualified rate, e.g. mg N_2 fixed per g nodule weight, per mg bacteroid protein or weight, or per mg carbohydrate utilized by the nodules in fixing N_2. Although 'highly effective' associations are usually, but not always, 'efficient' (Gibson, 1966b, 1969), the two terms have a different basis of expression. Each should be used in its correct context.

As a general principle, tests for symbiotic effectiveness should continue until the best strains have fixed *at least* five times, and preferably more, nitrogen as was initially present in the seed. Under favourable conditions, this should occur within 28–35 days from inoculation. Subtraction of dry weight or total N parameters for uninoculated control plants from the test or standard plants will do little to affect the results, and in the case of dry weight may be misleading owing to the lower level of N in the tissues of the uninoculated plants.

Nitrogen controls

Theoretically, nitrogen controls provide a measure of the potential growth of the species under the experimental conditions used. Expressing the growth of the inoculated plants as a percentage of that made by the controls provides an indication of 'absolute' symbiotic effectiveness. Two precautions must be observed. One is that the addition of N (KNO_3, or NH_4NO_3 which reduces the pH shift) is delayed until the nodules appear; only a low level (50% of seed N) is then supplied while nitrogenase activity in the nodules of the inoculated plants is developing. Further N is added when nitrogenase activity commences. The second precaution involves prevention of damage to the plants by excessive N additions. Weekly additions should be made, based on expected growth in that period, plus 50%. For example, plants of subterranean clover at 21 days may weigh 40 mg, and with a 14% relative growth rate (see below), would increase to 100.1 mg by 28 days (i.e. 60.1 mg increase). With 4% N in the tissue, this will require 2.4 mg of N; hence 3.6 mg of N should be added. Similar considerations apply to other species, and to other systems, the only requirement being an estimate of dry weight increase during each weekly period. In field studies, very high rates of N fertilizer applications may be required to ensure that nitrogen supply is not limiting growth. A soybean crop yielding 5000 kg ha^{-1} will remove approximately 350 kg of N in the seed, and a further 100–150 kg of N will be required for vegetative growth (including roots). In addition, some correction for leaching and denitrification losses should be made. Obviously there must be a number of applications during the growing season if toxicity problems are to be avoided, and this can cause technical problems as the crop develops.

Owing to the ability of the N controls to take up, but not necessarily assimilate, more N than they require, the %N in the tissues of the control plants may be unnecessarily high. Hence the comparison of inoculated plants with the N controls is most appropriately made on a dry weight basis.

$$\text{Symbiotic effectiveness (S.E.),\%} = \frac{\text{Dry wt. inoculated}}{\text{Dry wt. N controls}} \times 100$$

Standard strains

The data for plants inoculated with strains under test, be they known strains (Gibson and Brockwell, 1968) or field isolates of undetermined symbiotic capacity (Gibson *et al.*, 1975), can be compared with that for plants inoculated with a 'standard' strain. In the above examples, the standard strain was

Rhizobium trifolii, TAl, the strain in Australian commercial inoculants at the time.

$$\text{S.E.,\%} = \frac{\text{Dry Wt., inoculated test strain}}{\text{Dry Wt., inoculated standard strain}} \times 100$$

In this calculation, dry weight can be replaced by total plant N.

Relative growth rate

The above methods of determining symbiotic effectiveness depend on one harvest and the results integrate effects due to nodulation and N_2 fixation in the early stages of plant growth. Another approach is to make two harvests, one soon after N_2 fixation commences in most plants, and a second 14–21 days later. This provides information on three characteristics:

(i) symbiotic effectiveness, as discussed above;
(ii) the relative speed of establishment of N_2-fixing associations;
(iii) the relative growth rate between the two harvests. Such a value predicts subsequent growth and N_2 fixation, and in many ways provides a better indication of symbiotic effectiveness than results obtained from a single harvest.

For such an approach, the plants within a treatment are 'paired' on the basis of similarity of size, leaf development, and root and nodule development, at the time of the first harvest. One of each pair is then harvested. After the second harvest, the data from the two plants in a pair are used to calculate relative growth rate (R_W) as follows:

$$R_W, [\text{mg mg}^{-1}\,\text{day}^{-1}] = \frac{\log_e W_2 - \log_e W_1}{t_2 - t_1}$$

where W_1 and W_2 are total plant weights at harvest times t_1 and t_2, respectively. The weight values may be replaced by total plant N values in the calculation to provide relative nitrogen assimilation rates (R_N) in mg N per mg N per day (Gibson, 1965). A typical R_W value is 0.12 mg mg^{-1} day^{-1}, indicating that the plant increases in weight by 12% per day. The doubling time is 6 days (for any situation, $R_W \times$ doubling time = 72).

Experimental design and analysis of results

For determining the symbiotic effectiveness of particular strains of rhizobia with a given host, 8–10 replicates are required for genetically homogeneous plant material (e.g. *Trifolium subterraneum, Glycine max*), whereas 15–20 replicates are required for genetically heterogeneous material (e.g. *Trifolium repens, Medicago sativa*). Where possible, the plant material should be thoroughly randomized before inoculation (e.g. in tube culture experiments) in order to minimize the effects of small differences at planting or resulting from a location effect in cabinets or glasshouses during the pre-inoculation period. After inoculation, the plants should be randomized into n blocks, where n is the number of replicates. Location effects in the growth cabinet or, more important, in the glasshouse, can then be removed as a block effect in the statistical analysis.

Full randomization into blocks is often impossible when examining environmental effects, owing to the problem of arranging more than one cabinet, or tank within a cabinet, for each condition under examination. Within the condition, any other variables under examination should be randomized in blocks. Those undertaking this type of research are strongly advised to seek expert statistical advice before setting up such experiments. Basically, factorial analysis of variance with a single error term is not applicable, and a second error term, based on 'blocks within environmental conditions', must be used to examine environmental treatment effects (see Gibson, 1966a). Unless there is sufficient reason to relate the individual blocks within a treatment to corresponding blocks in other treatments, the 'block sum of squares' should not be partitioned from the 'blocks within environmental condition sum of squares'. All other treatment effects, including interactions with the environmental conditions, are tested against the residual error mean square.

Crop legumes

The use of dry weight and total N values as a measure of symbiotic effectiveness is readily justified for pasture species. With crop species, for which grain yield is the principal yield parameter, difficulties arise in culturing plants to maturity, under bacteriologically controlled conditions, in the laboratory or glasshouse. There is also the problem that plants with the highest dry weight and N content do not always outyield apparently poorer symbiotic combinations in terms of grain yield. This is usually associated with non-optimal conditions for the reproductive phase, e.g. inadequacies of the imposed daylength, inadequate spacing of pots, or deficiencies or excesses in water supply. These are questions few researchers in the area of legume nodulation have attempted to answer. At this stage, it can only be recommended that the best strains of rhizobia from glasshouse experiments be further examined under field conditions before incorporation into inoculants. Such a practice should be adopted with all strains being tested for inclusion in inoculants, including those for pasture species, but for crop legumes the practice appears essential (see also Chapter III.1).

Red leaf markers

As an early guide to symbiotic effectiveness of strains of *Rhizobium meliloti* with *Medicago truncatula,* and of *R. trifolii* strains with *T. subterraneum,* Brockwell (1956, 1958) selected cultivars of these species that have a high level of anthocyanin in the foliage of young plants (Chapter III.1). Nodulation by the most effective strains of rhizobia leads to early loss of this pigmentation. Such material could be useful where many strains are to be screened on these hosts, provided that these cultivars have the same symbiotic characteristics with respect to bacterial strains as the hosts for which strain selection is being made.

(v) The acetylene reduction assay

Whereas measurements of dry weight and nitrogen in plants provide an integrated measure of activity since germination or a previous harvest, the acetylene reduction assay provides an estimate of nitrogenase activity at the time of the assay. How well these estimates indicate N_2 fixation in absolute terms is variable and caution must be exercised in the interpretation of the results. The general methodology of the assay is described in Chapter II.3, and in this section aspects relevant to the legume will be discussed.

Selection of material

Assays may be carried out on whole plants *in situ* (in test-tubes or in pots), on nodulated roots (*in situ* or removed from the growth medium), or with excised nodules. *In situ* assays on whole plants are preferred but difficulties may arise in finding containers in which to make the assay. Plastic buckets have been used for pea plants (LaRue and Kurz, 1973b). For pasture species, Sinclair (1973) put mini-swards (15 × 15 × 10 cm) of white clover into shallow 5-l polyethylene boxes fitted with snap-on lids, while Ruegg and Alston (1978) placed pots of *Medicago truncatula* beneath a bell-jar on a base-plate; rubber seals prevented leakage.

When removing nodulated roots from the growth medium, care should be taken to avoid nodule loss, and desiccation of the material after removal. The effect of disturbing the root system has not been investigated thoroughly. Fishbeck *et al.* (1973) found that activity was higher following removal of roots from soil, especially if it was moist, but unpublished observations in several laboratories indicate that the reverse may be true in many instances. Washing the roots should be avoided if possible as it may reduce activity markedly—a 90% decline with lupin nodules has been observed in this laboratory due to temperature effects and the blockage of nodule aeration pathways. Shoot removal will cause a decline in activity, but this *may* not occur for 30 min and *may* not exceed 20% in the short term (e.g. Mederski and Streeter, 1977). Despite these problems, it is the most common type of assay in both laboratory and field, albeit largely due to convenience.

Excised nodules were used in many early experiments involving C_2H_2 reduction (Sprent, 1969; Schwinghamer *et al.*, 1970; Dart and Day, 1971), but it was quickly obvious that nitrogenase activity was greatly reduced compared with that of nodulated roots. For example, Bergersen (1970) found that the nitrogenase activity of excised nodules was 10–20% of that by nodules attached to roots when assays were carried out during the light period; for material taken during the dark period, the specific activities were similar. The method of detachment is important. Nodules cut from roots at the point of attachment show higher activity than those removed with forceps; those removed with the fingers have the lowest activity (unpublished results).

Polyethylene bag assay vessels

Most assay vessels have a fixed volume (Chapter II.3 and above) but plastic (polyethylene) bags have some advantages under field conditions (see also Chapters III.3 and III.4). The polyethylene should be tested for impermeability to C_2H_2 and C_2H_4. The bags may be purchased or prepared from flat tubing (circumference 20 cm) by cutting into desired lengths and heat-sealing one end. A filling port (Chapter II.3, Figure 3e) is fitted through a hole punched in the tube, and a 15 × 10 mm strip of 2-mm thick rubber taped to the bag (for later sampling). Nodulated roots are placed in the bag and the open end is heat-sealed or, if in the field, sealed with 25-mm width tape. Air is forced from the bag through the port, and the plastic tubing on the port clamped (artery forceps are ideal) until the bag is filled from the mechanical gas reservoir (Chapter II.3, Figure 2). The volume of gas mixture is measured (e.g. one revolution ≡ 15 ml) and the plastic tubing closed with a rubber stopper. The pressure in the bag should not exceed atmospheric, as indicated by the reservoir gauge. After incubation, the gas is sampled through the rubber strip, which is then re-sealed with another strip of tape. To check for leaks, the volume of the bag is ascertained by water displacement at the beginning and end of the assay, using the apparatus described in Figure 6.

Another 'bag' system in which the shoots of plants growing in pots protrude through the top of the bag (1% agar is the sealant) is described by Döbereiner (Chapter III.3, Figure 3).

Incubation of samples

For field operations, a common procedure is to replace the assay vessel in the soil from which the sample was taken. Alternatively, it may be left in a shaded place. In the laboratory, the temperature of incubation should be as close as possible to the temperature under which the plants are being grown. The termperature–response curves for different legumes vary (Hardy *et al.*, 1968; Dart and Day, 1971; Gibson, 1971; Waughman, 1977), some having a

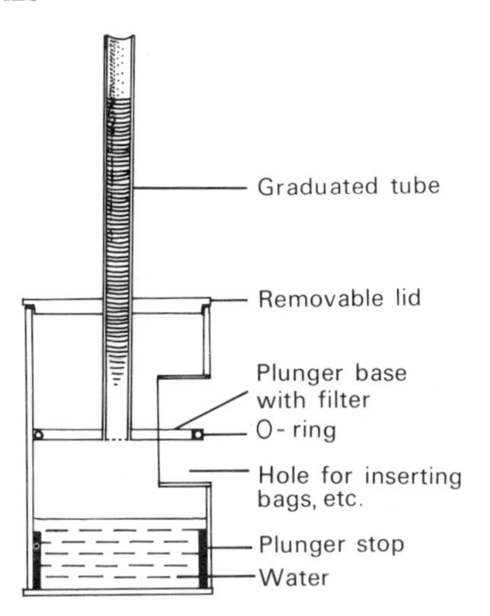

Figure 6. Apparatus for determining root and plastic assay bag volumes by water displacement. A Perspex cylinder (A), 14.5 cm in diameter and 27.5 cm high, has an access port in the side wall and a removable lid. The piston (B) consists of a Perspex tube (graduated at 2-ml intervals) connected to a flat disc with an O-ring partly recessed into the circumference to provide a water-tight seal. Water is added to 10 mm over the 'stops' on the inner wall of A. When the piston is depressed to the 'stops', water rises through a coarse filter on the bottom of the disc into the tube; the level is recorded. The procedure is repeated with a nodulated root, or an assay bag containing gas and root, in the water. The volume is determined by difference

broad optimum and others a narrow optimum. Despite this, care must be exercised in selecting the incubation temperature as the response is dependent on the growth temperature (Gibson, 1976b).

Before taking a sample, the assay vessel should be shaken to achieve thorough mixing of the gas, and to minimize gradients. In a confined space, gas diffusion is relatively slow and the C_2H_4 concentration at the top of the vessel may be lower than that in the immediate vicinity of the nodules. Samples may be taken for immediate injection into the gas chromatograph, or stored in syringes or evacuated blood-sampling tubes (Chapter II.3).

Special systems

Several systems for the continuous assay of nodulated legumes have been developed, and in some it is possible to measure C_2H_4 production and respiration simultaneously. Some of these systems are described briefly here, but further details should be obtained from the cited sources.

Huang *et al.* (1975) developed a double chamber system, with the lower chamber for the C_2H_2 reduction assays and the upper chamber for CO_2 assimilation measurements. A potted plant was suspended in the lower chamber, which was sealed with a lid, the stem protruding through a rubber stopper. Acetylene (10% gas volume in lower cylinder) was injected through a port and the excess gas volume was taken up in two gas-receiving bags attached to the chamber. These were used to mix the gas before being clamped shut. A small fan in the base of the chamber facilitated mixing during the assay. Samples taken every 30 min indicated that linearity of C_2H_4 production was achieved for at least 5 h.

This type of system was developed further by Mederski and Streeter (1977) so that continuous, automated sampling was possible. A nodulated root system was suspended in a chamber and continuously sprayed with nutrient solution from a reservoir in the base. A gas train was developed to mix C_2H_2 with air flowing continuously through the chamber (200 ml of C_2H_2 in 2 l of air per minute). On the outlet line, an automatic sampling valve injected a sample into the gas chromatograph every 3.5 min. The rate of C_2H_4 production, as detected in the air stream, reached a maximum within 10 min; the assays normally lasted 20 min, after which the C_2H_2 supply was shut off and the flow-rate of air increased to 10 l min^{-1} for the next 40 min; neither C_2H_2 nor C_2H_4 could be detected 15 min after the fast flush began. The 'pulsed assay' plants maintained a high rate of nitrogenase activity for 6 days, with some evidence of a 15% decline in activity towards the end of the period. Where plants were exposed to C_2H_2 continuously for 6 days, the decline was much faster, probably reflecting the absence of N uptake by such plants since the commencement of the assay.

To facilitate measurements on root respiration, Minchin *et al.* (1977) developed a chamber for measuring CO_2 production continuously over most of the growth period of a plant (e.g. cowpea). The effluent gas is passed through an infrared gas analyser. Irrigation is manual or automatic, and precautions are taken to collect any overflow. This system is being modified to include C_2H_2 reduction assay facilities (F. R. Minchin, personal communication).

Haystead *et al.* (1979) incorporated some of the principles behind the above systems to produce an apparatus capable of simultaneously measuring shoot, and root, CO_2 exchange, and C_2H_2 reduction by nodulated roots. A number of plants can be assayed concurrently. The basis of the system is that the plants are grown in vermiculite, or perlite, in nylon mesh bags which fit into a wide mouth preserving jar (volume 1350 ml) with an open screw lid. An aluminium plate seals the jar, with holes for the shoots (sealed with Silastic 9161 RTV from Dow-Corning) and for irrigation. For CO_2 assimilation assays, the top of the jar is pushed through the Neoprene rubber base of a 3-l Perspex chamber, the rubber acting as a seal. For C_2H_2 reduction assays, holes in the side of the jar are closed with Suba Seals, and C_2H_2 is injected. Samples are taken by syringe for C_2H_4 determination in a gas chromatograph, and for CO_2 determination in an infrared gas analyser. The CO_2 analysis depends on the injection of the sample into a CO_2-free air stream (see Atkins and Pate, 1977, for a detailed description of this method), but a gas chromatographic procedure could be used (Section f.vi). Although the system is not automated, it would be amenable to such modification if the facilities were available.

Interpretation

The principal value of the C_2H_2 reduction assay is to determine changes in nitrogenase activity as a consequence of imposed experimental treatments. As discussed in Chapter II.3, the assay provides a measure of total electron flow through the nitrogenase enzyme system, including those electrons normally utilized in H_2 production. Attempts to express the assay in terms of actual amounts of N_2 fixed have met with variable success. On some occasions, a C_2H_2 to N_2 ratio of 3:1 has been found, but the range extends from 0.5:1 up to 6:1 in well conducted experiments (e.g. Hardy *et al.*, 1973; Sloger *et al.*, 1975; Bezdicek *et al.*, 1978). The reason for this variation is not understood at present. In part it is due to the inclusion of electron flow normally producing H_2 in the total C_2H_4 values. Species, moisture effects, and 'time of year' differences in the ratio have been observed (Sinclair *et al.*, 1976, 1978), and in this laboratory differences due to the strain of *R. trifolii* nodulating sub clover have been found (unpublished results). Regardless of the reason for these effects, it is obvious that interpretation of the C_2H_2 reduction assay, in terms of actual N_2 fixed, should be undertaken with extreme caution.

(vi) Measurement of H_2, O_2, and CO_2

With increasing interests in energy costs and respiration, there is a need to determine CO_2, O_2, and H_2. These can be measured with a gas chromatograph equipped with a thermal conductivity detector (TCD or Katharometer). For CO_2, a Porapak N column, as used for C_2H_2 and C_2H_4 determinations, is satisfactory, with '$O_2 + N_2$', then CO_2, eluting ahead of the C_2H_4 and C_2H_2. By using a gas chromatograph fitted with both a flame-ionization detector (FID) and a TCD, and a column splitter, half the effluent may be passed through the TCD and half through the FID. Two recorders are necessary, unless a multi-channel integrator is installed. With a Spectra-

Physics Minigrator (Spectra-Physics, Santa Clara, Calif., U.S.A., and Darmstadt, W. Germany) fitted with channel changer, the CO_2 peak can be recorded before switching to the C_2H_4 and C_2H_2 peaks. Alternatively, it is possible to pass the effluent through the TCD, and as it is not combusted or changed, then to pass it through the FID. For CO_2, it is necessary to change the carrier gas from N_2 (the common carrier for C_2H_2 reduction assays) to He to achieve maximum sensitivity for CO_2.

H_2, O_2, and N_2 are also detected by the TCD (see Chapter II.3, Section d), but a Porapak column is less sensitive than a column of molecular sieve 5A. Such a column can also be installed in the above instrument, but the splitter is closed (if conducting concurrent C_2H_2 reduction assays) and the carrier gas changed to argon. However, note that very little H_2 should be produced during C_2H_2 reduction assays, as all the electrons passing through nitrogenase should be used to reduce C_2H_2 to C_2H_4. Hydrogen evolution during C_2H_2 reduction indicates incomplete saturation of the enzyme with C_2H_2.

The key to the successful laboratory or glasshouse culture of plants is awareness. This includes the shortcomings of the system used (contamination risks, restrictions imposed by growth conditions) and the extent to which any shortcomings will affect the interpretation of results within the context of experimental intentions. Adherence to the principles outlined in this chapter should minimize potential problems, but at all stages it is necessary to retain an awareness of any limitations inherent in the chosen system.

Methods for Evaluating Biological Nitrogen Fixation
Edited by F. J. Bergersen

G. Bond* and C. T. Wheeler
Department of Botany,
The University, Glasgow, Scotland.

5

Non-legume Nodule Systems

(a) INTRODUCTION

Although some evidence of the presence of nitrogen-fixing root nodules on certain non-legume angiosperms became available around the end of the 19th century, for a long time afterwards these nodules received only spasmodic study, chiefly by workers interested in the identity of the endophytes. Between the two World Wars there was, however, some appreciation by foresters, especially in Germany, of the possibility of exploiting the fixation associated

*Now retired.

with one particular non-legume, the alder tree. Since 1950 a more sustained study—assisted now by the ^{15}N technique, the electron microscope, and latterly the acetylene assay—has led to a rapid expansion of knowledge of non-legume nodules, which is still gathering momentum.

At present almost 170 non-legume angiosperm species, all trees or shrubs and distributed through 16 genera, are known to bear root nodules tenanted by a nitrogen-fixing endophyte, and with the exception of *Parasponia* spp. (see Section f.iii) the nodules of all these plant species are of the same basic type in their morphology and the nature of their endophytes. Since the nodules of *Alnus* have received far more attention than those of other genera it is convenient, as an interim measure pending the invention of an improved terminology for root nodules in general, to describe the nodules of all the other genera, with the one exception indicated, as being of *Alnus*-type (but see also Section e.).

It is certain that the presence of *Alnus*-type nodules on further plant species still awaits discovery, especially in the more remote regions of the world. Problems of high scientific interest await investigation in the already known nodulating species, particularly in genera other than *Alnus*, and the results of such studies may contribute significantly to the attempts now being considered or made to inculcate the habit of nodule formation into additional crop plants. Also, in view of the cost of fertilizer nitrogen there is an urgent need for study on how the fixation associated with non-legume nodules can be exploited.

(b) PLANT SPECIES KNOWN TO BEAR *ALNUS*—TYPE NODULES

An investigator desiring to enter the non-legume nodule field of work will wish to know which plant species bear such nodules. The list provided in Table 1 of species bearing nodules of *Alnus*-type, which as noted is by far the commonest type of non-legume nodule, is believed by the authors to be complete up to the end of 1979. The original papers in which nodulation was first recorded were cited for most of the species listed in Table 1 by Rodríguez-Barrueco (1968) or by Bond (1976a), while the latter also reported a number of new records made by collaborators in the IBP survey of non-legume nodulation. Citations for species recorded to be nodulated since that survey are supplied in footnotes to Table 1. The lack of authorities for some specific names in Table 1 indicates that none were provided by the authors of the original reports. It is probable that in consequence of a revision of the taxonomy of the genus *Casuarina* which is in progress, there is some synonymity between the species listed under that genus.

The genus *Arctostaphylos*, which for some years has been included in lists of nodulating non-legumes, has been omitted here for two reasons. Firstly, the original report that hypertrophies observed on the roots of *A. uva-ursi*

Table 1. Plant species known to bear *Alnus*-type nodules

CORIARIA

C. angustissima, C. arborea Lindsay, *C. intermedia* Matsum., *C. japonica* A. Gray, *C. kingiana, C. lurida, C. myrtifolia* L., *C. nepalensis* Wall., *C. plumosa, C. pottsiana, C. pteridoides, C. sarmentosa, C. thymifolia* Humb. & Bonpl.

ALNUS

A. cordata (Lois.) Desf., *A. crispa* (Ait.) Pursh., *A. fauriei* Lev. et Vnt., *A. firma* Sieb. et Zucc., *A. formosana* Mak., *A. fruticosa, A. glutinosa, A. hirsuta, A. incana, A. inokumai* Murai & Kusaka, *A. japonica* Sieb. et Zucc., *A. jorullensis* Kunth, *A. maritima* Nutt., *A. matsumurae* Callier, *A. maximowiczii* Callier, *A. mayrii* Callier, *A. mollis, A. multinervis* Matsumura, *A. nepalensis, A. nitida, A. oblongifolia*[a] Torr., *A. orientalis, A. rhombifolia*[b] Nutt., *A. rubra* Bong., *A. serrulata, A. serrulatoides* Callier, *A. sieboldiana* Matsum., *A. sinuata* Rydb., *A. tenuifolia* Sarg. var. *glabra* Call., *A. tinctoria* Sarg., *A. trabeculosa* Handel-Mazzett, *A. undulata, A. viridis* Regel, *A. rugosa*

MYRICA

M. adenophora Hance, *M. asplenifolia* (= *Comptonia peregrina*), *M. brevifolia* E. May ex C.DC., *M. californica* Cham. & Schlecht., *M. carolinensis, M. cerifera* L., *M. cordifolia* L., *M. diversifolia* Adamson, *M. esculenta* Ham., *M. faya* Ait., *M. gale* L., *M. humilis* Cham. & Schlecht., *M. integra* (A. Chev.) Killick, *M. javanica* Blume, *M. kandtiana* Engl., *M. kraussiana* Buching. ex Meisn., *M. microbracteata* Weim., *M. pensylvanica, M. pilulifera* Rendle, *M. pubescens* Willd., *M. quercifolia* L., *M. rubra* Sieb. et Zucc., *M. salicifolia* Hochst., *M. sapida* var. *longifolia, M. serrata* Lam.

CASUARINA

C. cunninghamiana Miq., *C. decussata* Benth., *C. deplanchei* Miq., *C. distyla* Vent., *C. equisetifolia, C. fraseriana, C. glauca, C. huegeliana, C. lepidophloia, C. littoralis* Salisb., *C. montana, C. muellerana* Miq., *C. muricata, C. nana* Sieb. ex Spreng, *C. nodiflora, C. papuana* S. Moore, *C. pusilla* Macklin, *C. quadrivalvis, C. rigida* Miq., *C. rumphiana* Miq., *C. stricta, C. sumatrana, C. tenuissima, C. torulosa* Ait.

HIPPOPHAË

H. rhamnoides L.

ELAEAGNUS

E. angustifolia L., *E. argentea* Pursh., *E. commutata* Bernh., *E. conferta* Roxb., *E. edulis, E. glabra* Thunb., *E. latifolia* L., *E. longipes* A. Gray, *E. macrophylla* Thunb., *E. matsunoana* Makino, *E. multiflora, E. murakamiana* Thunb., *E. pungens, E. rhamnoides, E. triflora*[c] Roxb., *E. umbellata* Thunb., *E. yoshinoi* Makino

SHEPHERDIA

S. canadensis Nutt., *S. argentea*

CEANOTHUS

C. americanus L., *C. azureus, C. cordulatus* Kell., *C. crassifolius* Torr., *C. cuneatus* (Hook.) Nutt., *C. delilianus, C. divaricatus* Nutt., *C. diversifolius* Kell., *C. fendleri, C. foliosus* Parry, *C. fresnensis* Dudley, *C. glabra, C. gloriosus* Howell var. *exaltatus, C. greggii* Gray, *C. griseus* (Trel.) McMinn, *C. impressus* Trel., *C. incana* T. & G., *C. integerrimus* H. & A., *C. intermedius, C. jepsonii* Greene, *C. leucodermis* Greene, *C. microphyllus, C. oliganthus* Nutt., *C. ovatus, C. parvifolius* Trel., *C. prostratus* Benth., *C. rigidus* Nutt., *C. sanguineus* Pursh., *C. sorediatus* H. & A., *C. thyrsiflorus* Esch., *C. velutinus* Dougl.

Table 1, (continued)

DISCARIA

D. americana[d] Gill. et Hook., *D. nana*[d] (Clos) Weberb., *D. serratifolia*[d] (Vent.) Benth. et Hook., *D. toumatou* Raoul., *D. trinervis*[d] (Gill.) Reiche

COLLETIA

C. armata[e] Miers (= *C. spinosa* Lam.), *C. paradoxa* (Spreng.) Escal., *C. spinosossima*[d] Gmel.

DRYAS

D. drummondii Richardson, *D. integrifolia* Vahl., *D. octopetala* L.

PURSHIA

P. glandulosa Curran, *P. tridentata* (Pursh.) DC.

CERCOCARPUS

C. betuloides, *C. ledifolius* Nutt., *C. montanus* Raf., *C. paucidentatus* Britt.

RUBUS

R. ellipticus[f] J. E. Smith

DATISCA

D. cannabina[g] L., *D. glomerata*[h] Baill.

[a]Nodulation recorded by Askew and Lane (1979).
[b]Nodulation observed by C.T. Wheeler in Oregon.
[c]Dr J. H. Becking (personal communication) observed nodules on plants of this species growing in Western Java.
[d]Nodulation in these species was recorded by Medan and Tortosa (1976).
[e]Nodulation observed by G. Bond in glasshouse-grown plants.
[f]Nodulation was recorded by Soemartono, reported in Bond (1976a), and later observed also by Dr J. H. Becking (personal communication).
[g]Nodulation was recorded by Trotter (1902) and re-discovered by Chaudhary—see Calvert, Chaudhary, and Lalonde (1979).
[h]Nodulation recorded in Calvert, Chaudhary, and Lalonde (1979).

were *Alnus*-type nodules mentioned only the external appearance, and has not been followed by a fuller examination. Secondly, one of the present authors (G.B.) noted recently that in some other species of this genus the formation of hypertrophies as a normal feature in development, unrelated to any invasion by a micro-organism, had been reported in older literature (Wieslander and Schreiber, 1939). Tiffney, Benson, and Eveleigh (1978) have described how, in *A. uva-ursi* itself, clusters of vegetative buds sometimes occur on the rhizome close to the soil surface, and show some superficial resemblance to *Alnus*-type nodules.

For several reasons it is better for a worker to elect to investigate the nodules of a species which is native in his region. Thus he will then have access to nodule material, while if he wishes to culture nodulated plants a source of seed or of other propagules will be to hand. Institutional or commercial sources of seed of some of the species listed in Table 1 do exist (see Section d.i), but as regards nodules, which are usually required for the preparation of inocula, the person working on non-native species is likely to be dependent on some helpful individual abroad, and on rapid conveyance of the material by air with due compliance with regulations governing the import of living plant material.

Thus it is recommended that the tiro investigator, if necessary with the help of a plant taxonomist, will discover from an appropriate flora whether any of the species listed in Table 1 are native to his region, and will then proceed to locate them. It is fortunate that some of the genera are very widely distributed, and that few if any of the larger countries have no non-legume nodulating species. In addition, it should be noted that in some of the genera listed there are many species which have not yet been examined for the presence of nodules, and it is very likely that these, when inspected, will also prove to be nodulated; thus in *Myrica* there are some 10 such species, in *Casuarina* about 20, in *Elaeagnus* about 25, in *Ceanothus* 20, in *Discaria* 5, in *Colletia* 14, and in *Cercocarpus* about 16 species. Now that Chinese forest scientists have become interested in non-legume nodules it is probable that the number of unexamined species in some of the genera under discussion will soon be markedly reduced.

Opportunities for the discovery of additional genera of nodulating plants appear to exist particularly in the family Rhamnacae. In that family *Colletia* and *Discaria*, according to Engler's classification, are placed in the tribe Colletieae, which also includes the genera *Talguenea*, *Trevoa*, *Retanilla*, and *Adolphia*. The first three of these are to be found in South America, while the fourth is represented in Mexico and California. It was clearly possible that species of some or all of these genera would prove to be nodulated, and in fact the presence on *Trevoa trinervis* of hypertrophies resembling *Alnus*-type nodules in external appearance and in the ability to reduce acetylene has now been reported (Rundel and Neel, 1978).

(c) CHARACTERISTICS OF *ALNUS*-TYPE NODULES

(i) Position on root system

In some genera, including *Alnus*, nodules occur just below the soil surface and so can be found after slight excavation with a trowel or similar tool. More difficulty may be experienced with other genera. Thus, in Alaska *Dryas* nodules cannot be found near the soil surface, and are mostly at a depth of 15 cm or more (Lawrence et al., 1967), while at a Canadian site few nodules were found above a depth of 50 cm (Lalonde, quoted in Bond, 1976a). At some *Casuarina* sites in Australia, where the upper layers of soil were extremely dry, Coyne (quoted in Bond, 1976a) was unable to find nodules by hand excavation, but believed that they were present in deeper, moister strata which he could not reach.

Data on the regularity of nodulation under field conditions were obtained for a few species during the IBP survey (Bond, 1976a). For example, in Europe *Alnus glutinosa* proved to be almost unfailingly nodulated, and *Myrica gale* not so regularly, although *Myrica* species native to North America were usually found to be nodulated. As noted already there is uncertainty on the regularity of nodulation in *Casuarina*, and there is the puzzling situation that on present evidence nodulation in *Dryas* is restricted to North America.

(ii) External appearance of the nodules

Following invasion of a root by the appropriate soil organism, *Alnus*-type nodules begin as simple, swollen structures, but as a result of quickly repeated branching, usually in more than one plane, coralloid nodule clusters are formed, as indeed happens in some legumes. In the non-legumes the clusters, which are often roughly spherical in shape, continue growth for some years and may reach a diameter of several centimetres where the host species attains tree size, such as species of *Alnus*, *Elaeagnus*, and *Casuarina*, but are usually smaller in hosts of shrub habit. The external colour of the nodule clusters is usually some shade of brown or fawn, the outer, younger lobes being typically less deeply pigmented than the inner, older parts. The interior of a lobe is usually cream or fawn in colour; haemoglobin, if present at all in non-legume nodules (see Bond, 1974, for review of literature), is never in sufficient quantity to impart a red colour to the nodule tissue, although a faint pink colour is sometimes seen in *Casuarina* nodules. In *Myrica*, *Casuarina*, and *Rubus* a rootlet eventually grows from the tip of each nodule lobe, giving a whiskery appearance to the cluster; at least in the first two genera, in plants growing in water culture or in loose media such as peat, these rootlets grow in an upwards direction, although in denser media this is less obvious.

It should be noted carefully that in many plant species the formation of structures which in the present context are pseudo-nodules can be induced by a considerable variety of agents, including nematodes and other animal organisms that deposit eggs in root tissues, the crown gall organism, and sometimes a mycorrhizal fungus, as in *Aesculus* spp. Vigilance is required in this respect. Investigation of the internal structure (see below) will quickly indicate the category to which any nodules that have been found belong. Actually it has been reported that some pseudo-nodules induced by egg-laying insects show nitrogenase activity (Farnsworth, 1976), but this requires confirmation.

(iii) Internal structure of the nodules

This information is provided for the benefit of an investigator who has found what appear to be true nodules on some plant species not previously reported to be nodule-forming, and who wishes to ascertain whether their internal structure conforms to the *Alnus*-type, to the *Parasponia*-type, or to a type not hitherto described. In the *Alnus*-type each lobe of a nodule cluster when sectioned shows a gross structure akin to that of a young normal root, except that there is now a superficial periderm and a distended cortex, some cells of which contain dense growths of the endophyte; these infected cells, which are often considerably enlarged, are in some host species scattered through most of the cortex, but in others are confined to the mid-cortex and form a ring as seen in transverse section. The congested contents of the infected cell can be elucidated satisfactorily only under the electron microscope, when it is revealed that in most recently infected cells the endophyte (now agreed to be an actinomycete—see Section e.) has the form of fine, septate hyphae, usually of diameter 0.5–1.0 nm. The tips of these hyphae soon enlarge to form vesicles which may be spherical, pear-shaped, or cylindrical, depending on the host-plant genus. The vesicles, which are believed to be mainly or wholly responsible for the nitrogen fixation, are commonly disposed peripherally within the infected cell so that they lie near the cell wall, although in *Coriaria* and probably in *Datisca* also (Calvert, Chaudhary, and Lalonde, 1979) this arrangement is reversed, with the vesicles directed inwards towards the vacuole. In other infected cells the hyphae may give rise to large numbers of small unicellular bodies often termed granulae, which appear to be resting spores; unlike the hyphae and vesicles these spores persist in the nodule cells and merge into the soil on nodule decay, possibly serving for the infection of fresh plants. In *Alnus glutinosa* a distinction has been drawn between nodules with abundant granulae and those in which they are infrequent (van Dijk, 1978).

A preliminary study of the nature of the endophyte in newly discovered nodules can be made by means of squash preparations from young nodule lobes, stained with cotton blue or another dye; usually, skeins of fine hyphae and the vesicles will be visible under the light microscope in the case of an *Alnus*-type nodule.

Further information on nodule structure and cytology can be found in the following papers which, collectively, cover most of the genera: Gardner (1976); Bond (1974, 1976a, 1976b); Becking (1970, 1975); Lalonde and Knowles (1975).

(iv) Nitrogen fixation—methods for detection and measurement

Fixation appears to be an unfailing characteristic of *Alnus*-type nodules, except for certain ineffective nodules which may arise under laboratory conditions as a result of the inoculation of host plants with unusual endophytes (e.g. Mian *et al.*, 1976). Although the nodules of only about 60 of the species listed in Table 1 have been tested for fixation, this total includes positive evidence for several species from each of the genera with numerous recorded species, and at least one species from each of the remaining genera except *Rubus*. The methods by which this evidence has been obtained are essentially those that are used for legumes, so that here it will only be necessary to mention any modifications which have proved advantageous with non-legumes.

Long-term growth

The long-term growth method carried out in a glasshouse or growth room, is particularly instructive. In it the ability of a nodulated plant to grow and accumulate nitrogen in a rooting medium devoid of combined nitrogen is tested, with non-nodulated plants as controls. The experiment is usually begun with seedlings and continued for several months; methods for plant culture and nodule induction will be described later. Especially if water culture is used, such a trial yields information on the rate of nodule development, on the external features of the nodules at different stages, and on how the onset of fixation—which will be signified by an increased greening of the leaves—relates to nodule development. Also provided is visual evidence on whether fixation in the nodules is sufficiently rapid to allow vigorous plant growth in the nitrogen-free rooting medium. After the plants have been harvested, Kjeldahl analyses on the dry matter will allow calculation of the amount of nitrogen fixed during the growth period. As a guide it may be noted that in experiments in Glasgow lasting about 5 months from transplanting, nodulated plants of *Alnus glutinosa* accumulated up to 490 mg of nitrogen, *Myrica gale* up to 158 mg and *Myrica cerifera* up to 85 mg per plant; the nitrogen content of corresponding non-nodulated plants in no case exceeded 1 mg per plant. As shown by a review of the literature (Bond, 1974), evidence of fixation has been obtained by this method in respect of about 40 of the species listed in Table 1. Figure 1 shows the visual results of experiments with two species.

For demonstration purposes the plants can be grown on for longer periods, and nodulated alders, casuarinas, or myricas attain very impressive dimensions after 2 years' growth in nitrogen-free rooting media.

Usually experiments of this type have been carried out under non-sterile conditions, and there is the theoretical possibility that any fixation observed has been at least partly due to free-living contaminant bacteria which gained access to the cultures, perhaps in the inoculum applied to induce nodulation, and established themselves on the surface of roots and nodules. In the past, owing to the lack of pure cultures of the non-legume endophytes, the conduction of long-term growth trials under conditions in which the endophyte is the only microorganism associated with the root system has been much more difficult than in the case of legumes. However, using methods to be described later, Quispel (1954) carried out such a trial with alder and found

Figure 1. Plants of *Myrica gale* (upper) and *Casuarina cunninghamiana* (lower) after growth for 5 months in culture solution free of combined nitrogen, the plants on the left with and those on the right without nodules. The edges of the teak supports are 7 inches (17.5 cm) long. (Reproduced from Bond, 1963, by permission of the Society for General Microbiology Limited)

that fixation still occurred. Also, by means of the ^{15}N technique (see below) it can be proved that fixation is exclusively a nodular process.

^{15}N methods

The ^{15}N technique (Chapter 2), because of its much greater sensitivity than the Kjeldahl process, and its dependence on testing for an increase in ^{15}N content over a base or datum level which is practically constant in all normal biological samples, proved a very useful adjunct to the method described above. Bond *et al.* (1954) sealed the root systems of young, intact alder plants into glass bottles partly filled with culture solution, the remaining space being charged with an atmosphere labelled with $^{15}N_2$. After exposure for 6 days, analysis showed that all parts of the plants were enriched with ^{15}N, the enrichment in the nodules being about twice that in the roots and shoots. Later experience (Bond, 1956a) showed that a similar result could be obtained after exposure for only 6 h, except that the disparity between the labelling in the nodules and elsewhere in the plant was now much greater. These findings pointed strongly to the nodules as the site of fixation.

By means of the isotopic technique it became possible to detect and measure fixation in nodules detached from the plant. Virtanen *et al.* (1954) detached nodules from a 4-year-old alder plant and exposed them to excess of $^{15}N_2$ for periods varying from 0.5 to 2 h. No significant enrichment was detected when the nodule nitrogen as a whole was subsequently assayed, but clear enrichment was shown by the acid-soluble nitrogen fraction, even after exposure for only 0.5 h. In Glasgow, tests with detached nodules taken from 1–2-year-old plants of various non-legume genera showed highly significant enrichment on the whole-nodule-nitrogen basis after exposures of a few hours, the nodules of *Hippophaë* and *Casuarina* being particularly active (Bond, 1957). Samples of roots alone, without nodules, showed no enrichment, proving that fixation is confined to the nodules. Calculation indicated, however, that the rate of fixation in detached nodules was much below that prevailing in the nodules prior to detachment, and although the reason for this is not clearly established, a lack of respirable substrate or a partial inactivation of nitrogenase as a result of the entry of air into the nodule tissues through the wound may be responsible.

A study of the effect of oxygen tension on fixation in detached nodules from various genera (Bond, 1961) showed that fixation was very small at low levels of oxygen, but rose with increasing supply and attained a maximum in the range 12–25% oxygen, depending on the plant species. At higher oxygen levels fixation was inhibited. Other tests showed that fixation in these non-legume nodules was inhibited by low levels of carbon monoxide and by higher levels of hydrogen (Bond, 1960). In these experiments nodule samples of 0.5–1.0 g fresh weight were enclosed in 28-ml glass tubes charged with a gas mixture containing 10% nitrogen, 20% oxygen, and 70% argon, the nitrogen usually having a 34 atom-% excess of ^{15}N. Incubation was carried out at 24 °C for a few hours. In order to reduce sample error, samples for the experiments were usually drawn from bulked, well mixed nodules taken from several plants. All nodules were from glasshouse-grown plants.

The ^{15}N method has also been used to test for fixation in samples of nodules brought in from the field, from species hitherto not tested by any method. Thus, Ziegler and Hüser (1963) tested *Myrica asplenifolia* nodules, Sloger and Silver (1965) those of *Myrica cerifera*, Delwiche *et al.* (1965) those of several species of *Ceanothus*, and Van Ryssen and Grobbelaar (1970) nodules of several species of *Myrica*. Bond (1956b) was able to demonstrate, by the use of ^{15}N, the occurrence of fixation in alder nodules still attached to the tree in the field.

Bond and Mackintosh (1975a) used the isotopic method in a study of diurnal changes in fixation in *Casuarina* plants 1 m tall, growing in a glasshouse subject to the usual variations in light intensity and temperature. At different times of day nodules were detached and incubated for a short time under excess of $^{15}N_2$ at the prevailing glasshouse temperature. The more rapid fixation observed over the mid-day period could be only partially explained by the higher temperature, and light intensity was considered to be a contributory factor. A further finding was that when nodules detached at different times of day were all incubated at the same, favourable temperature, the potential for fixation thus revealed was highest in the early morning and late evening, and probably overnight also. It appears that in this tropical genus, which has a high temperature requirement for fixation (optimum above 36 °C), cool nights in the glasshouse restrained fixation, leading to a potential building up in the nodules.

The ^{15}N technique has also been used, e.g. by Bond and Mackintosh (1975b), to evaluate the extent to which fixation persists when nodulated non-legumes are grown for a period in a rooting medium containing combined nitrogen. Nitrate or ammonium nitrogen carrying a low label with ^{15}N is fed to the plants, and after a suitable growth period the assay of plant nitrogen for ^{15}N content enables the amounts of nitrogen gained by uptake of labelled nitrogen and by fixation of ordinary atmospheric nitrogen to be calculated.

^{15}N has also been used as a tracer in attempting to identify the substances into which newly fixed nitrogen passes in *Alnus* and *Myrica* nodules (Leaf *et al.* 1958, 1959).

From the above account it will be obvious that the isotopic technique has been of the greatest assistance in the study of non-legume nodules. It is, however, arduous, exacting, and expensive, and because of the cost of ^{15}N itself, the experimenter is always tempted to use an undesirably low container to sample volume ratio, except in a well organized laboratory where there are facilities for recovering and purifying residual ^{15}N from incubation vessels.

Acetylene reduction assays

The acetylene reduction assay (Chapter 3), by virtue of its relative simplicity combined with high sensitivity, has found considerable use in non-legume studies. In their pioneering work, Stewart *et al.* (1967) found that the nodules of two non-legume species, which had been shown previously by the methods described above to be nitrogen fixing, also reduced acetylene to ethylene; this, reinforced by the experience of subsequent workers, confirms that the nitrogenase of non-legume nodules resembles in this respect that in other biological material.

The method is well adapted for use at field sites remote from a laboratory for screening newly discovered non-legume nodules for nitrogenase activity. For this purpose, the assay of nodules detached from the roots is usually satisfactory, for although it was shown by the ^{15}N method that detachment considerably reduces nitrogenase activity, some survives for a few hours. It would, however, be even better to assay nodules still attached to short lengths of root. Under field conditions acetylene can be generated from calcium carbide, although in the U.S.A. portable 1-l bottles of acetylene are available from Airco, Montvale, N.J. Samples of non-nodulated root material from the same plant species should be assayed, to allow of a correction for any ethylene present originally as a contaminant in the acetylene, or formed in the general metabolism of the tissues. Becking (1976a) used the acetylene technique to test various materials in the field for nitrogenase activity, and although non-legume nodules were not included, his method for storing gas samples for later analysis is of interest. Soft-glass test-tubes are adjusted to contain exactly 2 ml by pushing a rubber septum down to the appropriate distance from the base, the part of the tube above the septum being then drawn out into a capillary. Before the gas sample is introduced the pressure in the storage tube is reduced to 0.5 atm by inserting the needle of a syringe through the septum and withdrawing the plunger to the 2-ml mark on the barrel. The latter is then replaced with another syringe containing a 1-ml gas sample from an assay tube, this sample being injected into the storage tube. After withdrawing the syringe the capillary end of the storage tube is sealed in a Primus stove or similar appliance. Alternatively, good results are reported from the storage of gas samples in blood sampling tubes (e.g. Vacutainer, 10 ml, Becton, Dickson & Co., Rutherford, N.J.; see Chapter II.3) for later analysis.

The assay has also found uses in the laboratory. Waughman (1972) examined the effect of oxygen tension on nitrogenase activity in detached nodules of *Alnus* and *Hippophaë*, and obtained results similar to those secured previously in the ^{15}N method (above). Wheeler (1969, 1971), using de-topped nodulated root systems of young plants, demonstrated a well marked diurnal variation in nitrogenase activity in *Alnus* and *Myrica* which was ascribed to the effect of changes in light intensity. Mian and Bond (1978) used the assay to determine the stage at which nitrogenase activity commences in young, nodulating alder plants in relation to endophyte development within the nodules. The assay has also been used to evaluate nitrogenase activity in nodule homogenates (Section f.i).

The following additional practical points may be noted:

1. Nodule clusters from field plants may in part be several years of age, and nitrogenase activity will be much greater in the peripheral, young lobes than in the inner, older parts where, as noted earlier, the endophyte has mostly disappeared from the nodule cells. Thus, if detached nodules are being used, and it is not specially desired to obtain a mean value for the cluster as a whole, young lobes should be selected.
2. In laboratory experiments, nodule samples from temperate plant species have usually been incubated, during exposure to acetylene, at a temperature in the range 20–25 °C, within which optimal temperatures have been shown to lie (Akkermans, 1971; Wheeler, 1971; Waughman, 1972, 1977).
3. As with the ^{15}N method, respiratory processes during the incubation period may change the composition of the atmosphere within the sample container to an extent which will limit further nitrogenase activity. To obviate this, a reasonably high container to sample volume ratio should be provided, while to remove any further doubt the atmosphere within the container can be analysed for oxygen and carbon dioxide at the end of incubation; also, in planning experiments, it is useful to know the rate at which particular nodular or other material respires.
4. In laboratory experiments where it is desired to maintain nitrogenase activity during assay at the level prevailing in the nodulated plant in its normal growth environment, it is necessary to assay whole plants. The root system can be sealed into suitable glass or plastic containers, or into plastic bags of an appropriate type, while small plants can be wholly enclosed. In all cases there must, of course, be provision for gas injection and sampling. These preparations for assay should be made without any rough handling of the plant or deformation of the root system, otherwise nitrogenase activity is likely to be reduced (Wheeler *et al.*, 1978).
5. In view of the considerable number of situations in which measurable fixation by rhizosphere organisms has been recorded, users of the assay, particularly on field material, should bear in mind that a proportion of any nitrogenase activity detected might be due to such organisms. The effect on the assay result of a preliminary treatment of the material with a surface sterilisant could be studied.

The significance in the larger context of nitrogen fixation of observations made by acetylene assay depends on the existence, in a given experiment, of a constant relation between the measured rate of ethylene formation and that at which nitrogen reduction would proceed in similar material under similar conditions apart from the absence of acetylene. Most workers have assumed that such a relation existed with their material. The further question of what this relation is in quantitative terms for non-legume nodules has received some attention. Fessenden *et al.* (1973) exposed samples of detached nodules of *Myrica asplenifolia* to excess of $^{15}N_2$ for 1 h and then to acetylene for a further 1 h. From 24 samples so treated they obtained a mean ratio of 3.14 for moles of C_2H_2 reduced to moles of N_2 reduced. Values in the range 2.0–2.8 were reported for *Alnus rubra* nodules by Russell and Evans (1970), whilst Akkermans (1971) provided data which appear to indicate ratios ranging from 1.3 to 3.3, with a mean of 2.3, for *Alnus glutinosa* nodules.

(d) CULTURE OF NODULATED PLANTS

It is very likely that a study of particular non-legume nodules will necessitate the raising of nodulated plants in a glasshouse or growth room. Thus, it may be desired to carry out a long-term growth experiment, or to have available a large number of young nodulated plants for laboratory experiments of various kinds or a considerable bulk of nodules for some biochemical study, and although the latter can perhaps be procured from the field it can be considered unethical to raid trees or shrubs growing in the wild for such a purpose. Moreover, delay during transport to the laboratory may reduce the value of the nodule material.

The authors' cultural work has mostly been carried out in glasshouses lit by daylight, with temperatures usually in the range 18–30 °C from April to October and lower over the rest of the year. During the latter period some use has been made of fluorescent lighting in the glasshouse in order to keep evergreen species in growth. Large crops of *Alnus glutinosa* in particular have been raised in growth rooms operating a 16-h light period each day, the light being provided by fluorescent tubes supplemented by tungsten filament bulbs, giving a light intensity of about 14000 lux at plant level. The temperature during the light period is maintained at 19 °C, and during the dark period at 15 °C.

(i) Sources of seed

It has been recommended already that where possible investigators should work with native species, in which case a source of seed will be to hand. If for some reason it is desired to work with exotic species, possible sources of seed are as follows. The species may happen to be growing in a botanic garden in the worker's region; also, many such gardens participate in a seed exchange scheme, and seed for research can sometimes be obtained in this way, although usually in small amounts only. Alternatively, if the species is already under study by a worker in some other country to which it is native, that worker might be able to provide seed. Government forestry institutes maintain seed stocks of many native trees and shrubs, some of which may be nodule-forming non-legumes. Commercial seed firms' tree and shrub lists often include such plants, and on several occasions the authors have obtained seed of species of *Alnus*, *Hippophaë*, *Elaeagnus*, *Ceanothus*, and *Casuarina* from the seed firm of Vilmorin-Andrieux, Paris, France.

(ii) Pre-treatment and sowing of seed

In the authors' experience, *Alnus* and *Casuarina* seed needs no pre-treatment, although a thick sowing of *Alnus* seed is advisable since percentage germination is usually low, due in part, according to McVean (1955), to a failure of embryo formation in many seeds. Seed of *Myrica gale* and *Hippophaë* often germinates more freely if it is previously subjected to cold treatment by being mixed with slightly moistened peat or sand and stored at about 2°C for a month or longer. In the particular case of *Ceanothus* spp. it is recommended that heat treatment should first be given by suspending the seed in a muslin bag in 1 l of water pre-heated to 90°C, and leaving them immersed until the water has cooled to room temperature; the seed is then given a cold treatment as above. The seed of most other *Myrica* species is different from that of *M. gale*, and is usually enclosed in an outer woody coat, which may be wax covered. Chipping the coat with a sharp knife often helps germination, but the authors have had difficulty in securing germination uniform in respect of time with these species. Tiffney *et al.* (1979) secured fair germination in *M. pensylvanica* after seed treatment involving de-waxing, mechanical scarification and soaking in 0.1% gibberellic acid. The germination of *Purshia* seed is improved by pre-immersion in a 3% aqueous solution of thiourea for 15 min (Hubbard and Pearson, 1958).

The authors' experience is that the seed of these non-legumes, if taken straight from the parent tree or bush without any contact with the soil below, is rarely, if ever, contaminated with the nodule organism, so that for routine experiments the surface sterilization of seed is usually unnecessary.

Seed of most species can be sown in shallow plastic trays of horticultural type, previously swabbed with a disinfectant solution as a precaution against the possible presence of the nodule endophyte, followed by ample rinsing; the trays can be filled with Peralite moistened with a very dilute nitrogen-free culture solution such as that of Crone (see below) at 1/8th of normal strength. The grade of Peralite that the authors use requires the addition of an amount of solution equal to 3.5 times the weight of the Peralite for the latter to be moistened sufficiently. The weight of the filled and sown tray is noted, and is restored to that weight from time to time by the addition of distilled water. On germination the seed of *Hippophaë* and of *Purshia* quickly produces a deep-growing tap-root, and this seed should be sown in a container 15 cm or more in depth.

The authors' early trials with species of *Ceanothus* and *Coriaria* suggested that crushed-nodule inocula (see below) were not effective in inducing nodulation in these species, and habitat soil, i.e. soil collected from around the roots of the species in question growing in the wild, was procured and the seed sown in this. However, a recent re-examination of the situation in *Coriaria* (Cañizo and Rodríguez-Barrueco, 1976) suggested that the original difficulties were due to the pH of the culture solution being too low for the endophyte.

(iii) Alternative propagules

Although seeds provide the most readily transportable propagules, these plants can also be raised in other ways. Plants of *Myrica asplenifolia* were grown in the authors' laboratory from 10-cm lengths of underground stem sent from Canada by Dr R. J. Fessenden, while plants of *Colletia paradoxa* were raised from leaf cuttings taken from a plant in the Glasgow Botanic Gardens. Clonal plants of species of *Alnus* have been raised from single internode cuttings by Gordon and Wheeler (1978), Huss-Dannell (1978) and other workers, while the petioles of detached leaves of *A. glutinosa* occasionally form roots (Wheeler and McLaughlin, 1979).

(iv) Subsequent culture of plants

The authors' custom has been to transplant seedlings of *Alnus* and *Myrica* species from the seed trays when one to two leaves have emerged, of *Casuarina* when the epicotyl has a height of about 0.5 cm, while those of *Hippophaë*, *Elaeagnus* and *Purshia* have been moved at the cotyledon stage because of the risk of breaking the long tap-root if further growth is permitted.

Water culture

The transplants have mostly been set up in water culture, and although any of the standard culture solutions in their nitrogen-free versions would probably suit these non-legumes, the present authors, like Krebber (1932) and Roberg (1934), have used the culture solution devised by Crone (1902). Its nitrogen-free form contains the following: KCl 0.75 g, $CaSO_4 \cdot 2H_2O$ 0.5 g, $MgSO_4 \cdot 7H_2O$ 0.5 g, $Ca_3(PO_4)_2$ 0.25 g, $Fe_3(PO_4)_2 \cdot 8H_2O$ 0.25 g, and distilled water 1 l. One millilitre of a minor element concentrate should be added to each litre of Crone's medium, the recommended recipe for this concentrate being as follows: 0.62 g H_3BO_3; 0.40 g of each of Na_2SiO_3, $MnCl_2 \cdot 4H_2O$, and $KMnO_4$; 0.055 g of each of $CuSO_4 \cdot 5H_2O$, $ZnSO_4 \cdot 7H_2O$, $Al_2(SO_4)_3$,

$NiSO_4 \cdot 6H_2O$, $CoCl_2$, and TiO_2; 0.035 g of each of $Li_2SO_4 \cdot H_2O$, $SnCl_2 \cdot 2H_2O$, KI, KBr, and $Na_2MoO_4 \cdot 2H_2O$; and distilled water 1 l. This recipe is similar to the well known Hoagland A–Z solution.

An unusual feature of Crone's medium is that three of the salts, in the amounts shown, do not completely dissolve in 1 l of water, the effect being that in respect of five major essential elements the medium is for a period self-regenerating, since the initially undissolved fraction will dissolve as plant uptake of ions proceeds. Thus the culture solution can be renewed less frequently than with a fully soluble salt mixture. It is convenient to prepare 10 l of the culture solution at a time in an aspirator bottle, which can be rolled to and fro on a bench for several minutes to dissolve as much as possible of the salt mixture. The solution should also be well shaken before it is dispensed into each culture vessel. The natural pH of the Crone solution is about 6.3, which is a favourable level for infection and nodulation in most species (Bond *et al.*, 1954). Subsequently the plants become more pH tolerant, although for best results some regular control of pH is necessary (Bond, 1951; Ferguson and Bond, 1953; Bond *et al.*, 1956).

Transplanted seedlings grow better if diluted forms of the Crone solution are supplied, such as one-quarter or one-third of the strength indicated in the recipe above. As growth proceeds the solutions are strengthened. A small, once-only addition of ammonium or nitrate nitrogen (1–2 mg of nitrogen per litre of culture solution) can be made to newly transplanted seedlings to prevent extreme nitrogen deficiency pending nodule development. With most species, seedlings are inoculated with the nodule organism within a few days of transplanting, but with *Hippophaë* and *Casuarina* better results have been obtained by allowing the seedlings to grow on for 3–4 weeks in the presence of 10 mg of nitrate nitrogen per litre of culture solution before inoculation. Methods for inoculation are described in the next sub-section.

Containers for water cultures have consisted of glazed straight-sided earthenware jars 14 cm in diameter, 17 cm high, with a capacity of about 2 l. They were heat-sterilized before each occasion of use. The covers for these jars have been waxed teak squares 1 cm thick, 17.5 cm edge (Figure 1), replaced latterly with squares of black polyethylene of similar size. The squares are bored with holes (13 mm diameter) for six plants and, to ensure that the roots of the often very diminutive seedlings make proper contact with the solution, the seedlings have been supported on rings (3 mm thick) cut from rubber tubing of suitable diameter and bore, inserted into the holes in the squares, as shown in Figure 2. The covers have been washed with disinfectant before each occasion of use. The number of plants per jar can be reduced as growth proceeds, while supports for the plants can be provided by slender stakes inserted into additional holes bored in the covers. Where plants need to be grown for a few weeks only, 800-ml beakers wrapped in black paper and covered with squares smaller than those indicated above are

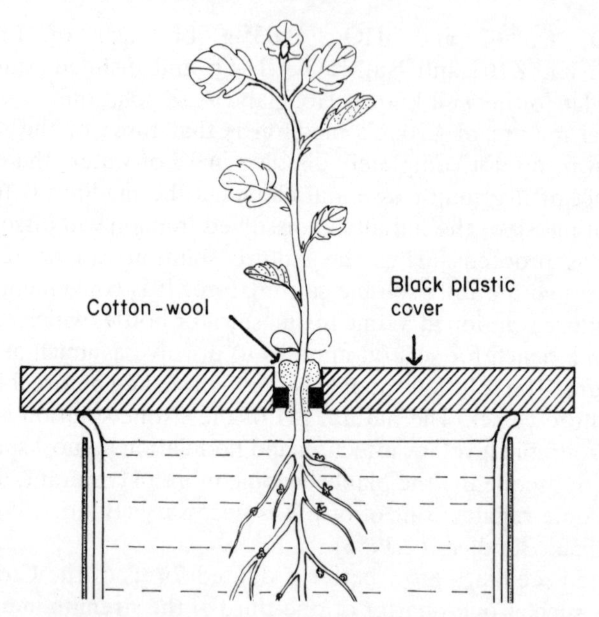

Figure 2. Details of covers employed for water culture in jars or beakers. The rubber ring supporting the plant is shown in solid black. Actually there were five holes for plants in the cover. Approx. 3/5 actual size. Drawing by Dr S. Mian

more economical of bench space and culture solution. Beaker culture was employed in studies of the importance of molybdenum and cobalt in the growth of nodulated non-legumes (Hewitt and Bond, 1961, 1966).

Plants to be grown into a second year can, if necessary, be transferred into larger containers such as 5- or 8-l Doulton pots, fitted with larger covers of the type already described, but bored now for a single plant and with a loose fillet leading to the hole to allow insertion of the plant.

Experience has shown that with the arrangement described above the growth of most species is satisfactory without forced aeration, at least until a considerable size has been attained. A comparison in the case of *Alnus glutinosa* showed no benefit from forced aeration until the plants had attained a height of 20 cm, while excellent plants of *Myrica gale* 80 cm in height have been grown without aeration. Species which tend to form massive root systems, such as *Hippophaë*, *Purshia*, and *Colletia*, have been provided with aeration, for which the oil-free Reciprotor pump (Reciprotor A/S, Copenhagen-Bagsvaerd, Denmark) has been found to be very satisfactory, although care has to be taken not to disperse the undissolved part of the Crone's medium lying at the bottom of the culture jar.

The frequency with which the culture solution in the 2-l jars is changed depends on the size of the plants. Initially every 2–3 weeks suffices, but with larger plants a weekly change is recommended.

Peralite culture

Considerable use has also been made of Peralite culture. This medium, obtained from British Gypsum Co. Ltd., replaced the sand originally used for filling seed trays and for providing an inert basis for continued plant growth, and because of its low specific gravity much labour has been saved. The authors' bioassay, using wheat plants, showed that Peralite has a negligible content of combined nitrogen. It is also free of non-legume endophytes.

Rigid plastic pots and also collapsible Transpots have been used as containers for culture in Peralite. Crone's solution has again been employed, while distilled water has been added daily, or as required, to restore the weight of the culture to a previously calculated figure (see Section d.ii). Plants from seed sown in habitat soil have usually been transplanted into Peralite after they have nodulated.

Compared with water culture, Peralite culture has the advantages that the plants usually need no staking and that the cultures are easily moved around the glasshouse or laboratory. Aeration of the root system is presumably adequate, while the slight alkalinity of the medium favours the growth of species of *Coriaria*, *Ceanothus*, and *Hippophaë*. On the other hand, nodule development cannot easily be followed on plants in Peralite, while the nutrient status and the pH within the rooting medium are less easily controlled than in water culture; where such control is desired, frequent flushing is necessary (Bond and Mackintosh, 1975b). Plants of *Alnus glutinosa* tend to grow more slowly in Peralite than in water culture.

It has been recommended above that equipment used in plant culture should be sterilized by appropriate methods prior to use. This is particularly important if uninoculated control plants, designed to remain free of nodules, are being grown, while additional precautions should be taken to prevent contamination between adjacent cultures.

Zobel *et al.* (1976) reported that good results had been obtained by growing certain non-legumes aeroponically, i.e. with their roots constantly bathed in a nutrient mist.

(v) Inoculation of young plants

Since, as in legumes, the microorganism responsible for the induction of nodules in non-legumes is not carried within the seed, under natural conditions nodulation in young plants is dependent on invasion of the roots by the appropriate organism from the soil, while in cultured plants inoculation is necessary. Because until very recently (see below) none of the non-legume endophytes has been available in pure culture, other ways of infecting the roots have been used, and may have to continue in use for some time in certain genera.

The commonest way, dating back to Hiltner in 1896, has been to apply a crushed-nodule inoculum to the roots. In the present state of knowledge it will be best if the nodules that are required for this purpose are taken from field plants of the same species as the plants that are to be inoculated. Although, by the use of crushed-nodule inocula (Bond, 1974) and recently of pure cultures (see Section e.) there has been considerable study of the possibility of satisfactory cross-inoculation between different species and genera of host plants, information is still very incomplete, especially in connection with genera more recently discovered to be nodule-bearing. Also largely unexplored is the question (which is now acquiring practical importance) of whether in a species such as *Alnus glutinosa* the endophyte, as in legumes, exists as strains which differ in the rate at which they fix nitrogen, although all could be classed as reasonably effective.

A crushed-nodule inoculum is prepared by grinding the appropriate nodules, well washed and surface dried, in distilled water in a sterilized mortar and pestle or other device, a small amount of sharp sand being added to aid disintegration. The proportions are not very critical, but commonly we have used 2 g of nodule per 100 ml of water, although in many instances more dilute inocula would probably be just as satisfactory: in the particular case of alder an inoculum only 1/20th of the above strength gave good results, while on the basis of their own trials Lalonde and Fortin (1972) calculated that as little as 1 μg of nodular matter, suitably suspended, sufficed for the inoculation of an alder seedling. If desired the inoculum can be passed through muslin to remove the larger tissue fragments, and perhaps it is a worthwhile precaution to avoid its direct exposure to sunlight. The inoculum can be applied to the seedling root system by means of a small, previously disinfected brush or, for more critical work, a measured amount (say 0.5 ml) can be applied with a pipette which has a coarse orifice. Seedlings to be grown on in Peralite culture can be dipped into an appropriate inoculum prior to planting. For particular purposes, e.g. cross-inoculation studies, the surface sterilization of the nodules to be used in inoculum preparation may be contemplated, but as will be explained in Section e, this can rarely be achieved by ordinary methods.

As a guide it may be noted that under the present authors' conditions, in response to inoculation carried out as above nodules first appear in *Alnus glutinosa* after 9–14 days from inoculation, in *Myrica gale* and *Hippophaë* after 2–3 weeks, and in *Casuarina* after 3–4 weeks.

As reported already, in some genera nodulation has been more reliably secured by sowing the seed in habitat soil. Cañizo and Rodríguez-Barrueco

(1976) reported satisfactory results from the application of an extract of habitat soil in *Coriaria* sp.

(e) METHODS RELATING TO THE ISOLATION OF NON-LEGUME ENDOPHYTES

From the early years of this century, periodic attempts have been made to isolate the endophytes from *Alnus*-type nodules, especially that from the nodules of alder itself, and the literature contains over 30 accounts of such attempts. However, it is only recently that a well substantiated claim of success has emerged.

The earlier attempts were reviewed by Bond (1963, 1967) and by Quispel (1974), and only a brief statement of methods employed is necessary now. The nodules, after a deemed surface sterilization, followed in some cases by excision of the outer tissues as a further precaution against contaminants, were often crushed in sterile water and the suspension was plated out. Sections of nodules were plated out by some workers, while others picked out endophytic structures with a micro-manipulator. A wide variety of media and cultural conditions were employed. Through such procedures organisms which showed actinomycete-like structure and some resemblance to the endophyte as seen in the nodule cells were sometimes obtained, but many of the workers concerned were unable to obtain nodulation in test seedlings following inoculation with the isolated organism. Other workers claimed success in such tests, but their claims were open to criticism on the grounds that the nodules took far too long to appear, or that the number of uninoculated, control plants was too small, while one isolation which seemed to have promise died out before confirmatory tests could be made.

The studies of Quispel (1954), although they did not result in a confirmed isolation, have been of much value to later workers. He showed that the removal of surface contaminants from alder nodules was very difficult to achieve, since when nodules were planted out on a rich medium after being shaken with standard sterilizing solutions such as mercury (II) chloride, only a small proportion remained free of obvious contaminant growth after incubation for several weeks. By applying an inoculum prepared from such uncontaminated nodules to alder seedlings grown under aseptic conditions, he raised for the first time nodulated alders which appeared to have no microorganism associated with them other than the endophyte and, as noted in Section c.iv, he observed that nitrogen fixation still occurred.

Lalonde and Fortin (1972) devised another means of obtaining an inoculum free of contaminants. Three generations of young *Alnus crispa* plants, all aseptic prior to inoculation, were grown, nodulation in the first generation being induced by an inoculum prepared from field nodules partially surface sterilized with sodium hypochlorite solution, while the succeeding generations were inoculated from the nodules present on the previous generation, again after surface sterilization. In all cases the inocula were made as dilute as possible, consistent with an ability to induce satisfactory nodulation. Bacteriological tests showed that the nodules of the first, and to some extent those of the second generation bore contaminants, but that those of the third generation were contaminant-free. Thus, an inoculum prepared from these last nodules could serve in the raising of further generations of aseptic nodulated plants.

Lalonde *et al.* (1975) isolated from the nodules of aseptic plants of *A. crispa*, grown as above, and also from field nodules, identical slow-growing actinomycetes which were morphologically similar to the endophyte within the original nodules. The proportions of the various isomers of diaminopimelic acid—increasingly used as a means of characterizing actinomycetes—present in the isolates were similar to that in the endophyte, while immunological tests also appeared to confirm their identity. Full details of these techniques may be found in the original paper. However, in tests carried out on a large scale, involving about 2000 aseptically grown plants, inoculation with these isolates failed to induce any nodulation. The authors remained convinced that they had isolated the alder endophyte, although in a non-infective state. A number of earlier workers suspected this of their own isolations, but were unable to provide the evidence furnished by Lalonde *et al.* (1975).

Callaham *et al.* (1978) reported the isolation of the endophyte of the nodules of *Comptonia peregrina* (also known as *Myrica asplenifolia*) without loss of infectivity. The following is an outline of their multi-stage procedure; full details are given in the original paper. Following Quispel's (1954) procedure (see above), surface-sterilized nodules were incubated on a bacteriological medium for 6 weeks, when nodules free of contaminant growth were sliced and incubated for several hours in a nutrient medium containing also cellulase and pectinase, with the object of hydrolysing the cell walls. The nodule tissues were then teased apart in a similar medium but lacking the enzymes, and the resulting suspension was filtered to remove the coarser cell debris. The endophyte material was washed by centrifugation, suspended in the same liquid medium and incubated. A slow growth of a filamentous organism producing colonies up to 2 mm in diameter after several months was observed. Subsequently, faster growth was obtained on other media. The organism, which appears to be microaerophilic, was regarded by Callaham *et al.* (1978) as actinomycetous in nature, although its relation to the known groups of actinomycetes required further study. The hyphae did not produce the club-shaped vesicles typically present in the infected cells of *Myrica* species (though their formation has since been observed in other culture media—see Lalonde and Calvert, 1979), but they did form spores which appeared to correspond to the so-called bacteroids or granulae observed in the nodules of other *Myrica* species (Becking, 1970; Gardner, 1976) and well known in alder nodules. The ability of the isolate to induce nodulation in *C. peregrina* was tested in three experiments, each involving a considerable number of young plants; practically all inoculated plants formed nodules, which showed nitrogenase activity when tested by acetylene assay, and, although in one experiment some nodules appeared on control plants, a satisfactory explanation was available. Other findings reported by Callaham *et al.* (1978) were that their isolate also induced nodulation in *Myrica gale* and *M. cerifera*, and that Dr. M. Lalonde, to whom cultures were sent, found in his laboratory that the isolate caused nodulation in *Alnus glutinosa* and *A. crispa*, later extended to other *Alnus* spp. (Lalonde and Calvert, 1979).

Berry and Torrey (1979) reported the isolation of the endophyte from the nodules of another host species, namely *Alnus rubra*; the nodules induced to form on young plants of that species by inoculation with the isolate were shown to be active in acetylene reduction. This isolate again showed the characters of an actinomycete and also a degree of promiscuity, since it induced nodulation in other species of *Alnus* and also in species of *Myrica*, though not of *Ceanothus*. In a valuable review of methods available for the isolation of the endophytes from *Alnus*-type nodules, Baker and Torrey (1979) stressed the importance of using culture media selective for actinomycetes, and of providing micro-aerobic conditions during culture; they also reported some further isolations and the results of various cross-inoculation tests.

That these nodule endophytes are truly actinomycetes has now been accepted (Lechevalier and Lechevalier, 1979), and following the proposal of Becking (1970) the organisms have been placed provisionally in the single genus *Frankia*, a name originally created in the nineteenth century for these nodule organisms in honour of a German botanist, well known for his pioneering studies of associations between higher plants and microorganisms. This acceptance of identity has encouraged the search for a better designation for these nodules, more precise than '*Alnus*-type nodules' and less cumbersome than 'actinomycetous nodules'. The term 'actinorhizae' has been proposed (Torrey and Tjepkema, 1979) and has gained some acceptance, but it appears that the term 'actinorhizal nodules' would be preferable, since these particular associations are not the only examples of the invasion of roots by actinomycetes. A further point is that those authors who see good reason for the spelling 'mycorrhiza' will doubtless adopt the corresponding rendering of this new term.

(f) OTHER METHODS AND TOPICS

(i) Preparation of nodule homogenates retaining nitrogenase activity

In unpublished trials carried out in the early 1970s the present authors sought to find whether nitrogenase activity in non-legume nodules could survive nodule disintegration. Using methods which had succeeded with legumes (Koch *et al.*, 1967), nodules of various non-legumes were, by means of a pestle and mortar, ground up finely in phosphate buffer containing polyvinylpyrrolidone (PVP) and ascorbate or sodium dithionite (20 mM), the operations being carried out in a glove-box through which a current of argon was passing, with monitoring of the inner atmosphere to confirm that anaerobic conditions prevailed. The nodule slurries were assayed for acetylene-reducing activity in incubation media containing dithionite and ATP or an ATP-generating system, with a gaseous phase of 20% acetylene and 80% argon. The activities found were usually less than 1% of that in intact nodules, and only occasionally rose to 5%. Thus, it appeared that a modified technique was necessary for non-legume nodules, perhaps partly because of the relatively high level of phenolases and phenolic compounds in the nodules; in the particular case of alder, *o*-diphenol oxidase activity in the nodules is 2–5 times higher than in the roots (Wheeler, unpublished data).

After much exploratory work, van Straten *et al.* (1977) and Akkermans *et al.* (1977) arrived at a technique which yielded preparations of disintegrated nodules of *Alnus*, *Hippophaë*, *Myrica*, and *Shepherdia* that retained 25–40% of the nitrogenase activity of the intact nodules, on the basis of acetylene assays. While the original papers should be consulted for full details, an outline of procedure is as follows. The nodules to be used were stored for 20 min between wet filter-paper since, for an unknown reason, this considerably enhanced the eventual nitrogenase activity. The nodules were then disintegrated in a Virtis homogenizer together with an appropriate amount of a buffer (0.05 M Tris–HCl, pH 8.0) containing also 0.3 M sucrose, 1,4-dithioerythritol, 100 mM sodium dithionite, and PVP. The chamber of the homogenizer was flushed with argon and the homogenization time was usually 4 min at 3000 r.p.m. The homogenate, still under anaerobic conditions, was filtered to remove coarser particles, yielding a filtrate containing intact and disrupted vesicle clusters, hyphal fragments, and plant cell debris. The medium for acetylene assay contained the same constituents as that used in homogenization, plus $MgCl_2$ and ATP, while the gas phase was 10% acetylene in argon. By further anaerobic sieving of the homogenate, preparations containing a higher proportion of intact vesicle clusters could be obtained, and showed enhanced activity.

In an alternative method (Benson and Eveleigh, 1979), which gave good results with *Myrica pensylvanica*, nodules were crushed in liquid nitrogen in a pestle and mortar and the resulting homogenates was transferred into tubes filled with liquid nitrogen, which were sealed with a serum stopper vented with a hypodermic needle. When the liquid nitrogen had boiled off the tubes were gassed with purified argon, appropriate volumes of degassed 0.1 M Hepes buffer (pH 7.8) and 0.1 M sodium dithionite added, and the suspension was used for assays of acetylene-reducing activity. High concentrations of dithionite (150 mM), ATP, and $MgCl_2$ (both at 30 mM) were required for maximal activity, while the addition of insoluble polyvinylpyrrolidone extended the period over which acetylene reduction persisted. Experience showed that activity was not enhanced by the inclusion of an osmoticum such as sucrose in the incubation medium.

Benson, Arp, and Burris (1979a,b) ground nodules of *Alnus glutinosa* in liquid nitrogen, suspended the powder and centrifuged it four times in 100 mM potassium phosphate buffer, pH 7.4, containing 10 mM sodium dithionite. In order to free the homogenate from inhibitory phenolic compounds the medium was then changed to 20 mM Tris-HCL, pH 7.2, with 20 mM ascorbate and 2 mM dithionite, and the nodule material washed four additional times. The sedimented material was re-suspended in twice its volume of the second medium, mixed with an equal weight of solid polyvinylpyrrolidone and subjected to mild sonication for 1.5 to 2.0 min. The supernatant from centrifugation of the sonicated material at 20000**g** for 20 min was cell free and would reduce acetylene in the presence of ATP, Mg^{2+}, and sodium dithionite. Anaerobic conditions were maintained during the above operations.

(ii) Estimation of non-legume fixation in the field in terms of Kilograms of N per Hectare per year

Confirmation of the inference drawn from laboratory results that considerable amounts of nitrogen will be fixed by nodulated non-legumes under field conditions is obviously desirable, although the tree or shrub habit of these plants makes the securement of such confirmation considerably more difficult than in the case of herbaceous legume crop plants.

A number of investigators seeking information on this matter have used the classical method of Kjeldahl analysis of soil and biomass. Crocker and Major (1955), in a study of the colonization of recently deglaciated areas in Alaska, deduced from analyses of the soil under *Alnus crispa* thickets of increasing age that there was an annual accretion to the soil of about 62 kg ha^{-1} of nitrogen, most of this being credited to nodular fixation, although a contribution from free-living fixers could not be excluded. Youngberg and Wollum (1976), after observations for 10 years involving analyses of soil, litter, and biomass, on plots colonized by *Ceanothus velutinus*, concluded that nitrogen accretion for the 10-year period ranged from 715 to 1080 kg ha^{-1} at different sites. These authors also give references to other studies of the same type.

Stewart and Pearson (1967), in a study of nitrogen accumulation in stands of *Hippophaë rhamnoides*, again used the above method but supplemented it by direct measurement of fixation rates in samples of nodules by the ^{15}N method. Pieces of root about 12 cm long bearing nodule clusters were enclosed in glass tubes and exposed to excess of ^{15}N for 2 h in the field. The results confirmed that a considerable proportion of the observed accretion of nitrogen in plants and soil which, for example, amounted to 133 kg ha^{-1} annually in stands of 11–13-year-old bushes, could be attributed to nodule fixation.

Akkermans (1971), and Akkermans and van Dijk (1976) tried to determine nodular fixation in a large stand of alder (*Alnus glutinosa*) trees, chiefly by means of acetylene assays on nodule samples. At regular intervals over a year, nodule clusters still attached to root pieces 2–5 cm long were quickly enclosed in glass vessels and assayed for acetylene reduction in the field at 21 °C. Preliminary tests had indicated that the severance of the root from the tree had no early effect on nitrogenase activity within the nodules borne by the root. The observed activity was suitably corrected depending on the actual soil temperature at the time. Much variation in activity was experienced between different nodule samples on a given occasion, and in order to obtain a fairly reliable mean figure about 50 nodule samples were assayed on each occasion. No nitrogenase activity was detected over the period between autumn leaf fall and bud opening in the following spring, a finding corroborated by Pizelle and Thiery (1977). The weights of nodules present on sample excavated trees of various age groups were then determined, and from the total number of trees of each age group in the stand it was possible to estimate the mean weight of nodules as in kilograms per hectare. On the basis of that figure and the result of the acetylene assays, together with an assumption that the conversion ratio for this material was 3 : 1, it was calculated that 58 kg of nitrogen had been fixed per hectare during the year of observation.

Fessenden *et al.* (1973) found much variation in nitrogenase activity as determined by acetylene assay in detached nodules of *Myrica asplenifolia* brought in from the field, and commented on their unsuitability on that account for use in laboratory experiments.

(iii) Nodules of *Parasponia* species

Trinick (1973) reported that root nodules had been detected on saplings of a non-legume growing as weeds in tea plantations in New Guinea. He believed that the non-legume in question belonged to the genus *Trema*, which falls within the family Ulmaceae, but Akkermans *et al.* (1978a) have now established that the plant should be placed in the closely related genus *Parasponia*, the actual species being *P. rugosa* Bl. Further accounts of the nodules were provided by Trinick (1975) and Trinick and Galbraith (1976). In external appearance they show some similarity to nodules of *Alnus*-type, since the lobes branch and clusters are eventually formed, but they resemble legume nodules in the ease with which they become detached from the roots. Internally, the gross structure of the nodule lobe is rather similar to that of the *Alnus*-type, with a central stele and broad cortex in which the majority of cells contain the endophyte, except for an outer layer about 10 cells deep which remains uninfected. Differences from the typical *Alnus*-type nodule are that there is no superficial periderm, and that the infected cells do not completely surround the stele, but lie in a horse-shoe shaped area as seen in a transection, although this feature is also shown in one genus listed in Table 1, namely *Coriaria*. Surprise was aroused when the nature of what is now to be called the *Parasponia rugosa* endophyte was investigated (Trinick, 1975; Trinick and Galbraith, 1976), for it proved to be a bacterium of *Rhizobium* type, with a notable difference from the legume nodule in that in over two thirds of the infected cells the rhizobia remain confined within infection threads, and only in the remaining infected cells do they escape and develop into branched bacteroids. The organism proved to be easily isolated, and was found to induce nodulation in some tropical legumes that are normally nodulated by cowpea-type rhizobia. Long-term growth trials showed that the normal host plant, if nodulated, could thrive in a nitrogen-free rooting medium, indicating that the nodules are nitrogen fixing, while acetylene assays confirmed the presence of nitrogenase.

Akkermans *et al.* (1978b) intimated that they had discovered an old report of nodulation in Indonesia in another species of *Parasponia*, namely *P. parviflora* Miq., and that they had confirmed this by personal observation. Becking (1979) showed that the endophyte within these Indonesian nodules was again a *Rhizobium*, though with peculiar features similar to those reported above for the endophyte of *P. rugosa*; isolates from the nodules of *P. parviflora* induced nodulation in several legumes. It is possible that other examples of non-legume nodules containing rhizobial endophytes will eventually be found, either in other species of *Parasponia* or in other genera.

Methods for Evaluating Biological Nitrogen Fixation
Edited by F. J. Bergersen

R. R. Eady
ARC Unit of Nitrogen Fixation,
University of Sussex, Brighton, England

6

Methods for Studying Nitrogenase

Table 1. Organisms from which active nitrogenase has been extracted

Organisms	Comments	References*
Anaerobic bacteria:		
Clostridium pasteurianum (Cp)	Cp1 and Cp2 characterized	Zumft and Mortenson (1973) Tso *et al.* (1972)
Desulphovibrio desulphuricans (Dd)	Extracts	Sekiguchi and Nosoh (1973)
Chromatium vinosum (Cv) (photosynthetic)	Cv1 characterized	Evans *et al.* (1973)
Rhodospirillum rubrum (Rr) (photosynthetic)	Rr1 and Rr2 purified; activating factor present	Ludden and Burris (1976), Emerich and Burris (1978)
Rhodopseudomonas palustris (Rp) (photosynthetic)	Rp1 and Rp2 separated; putative activating factor present	Zumft and Castillo (1978)
Facultatively anaerobic bacteria:		
Klebsiella pneumoniae (Kp)	Kp1 and Kp2 characterized	Eady *et al.* (1972), Smith *et al.* (1976a)
Bacillus polymyxa (Bp)	Bp1 and Bp2 purified	Emerich and Burris (1978)
Escherichia coli C-M74 (Ec) (genetically constructed hybrid, carries Kp *nif* genes)	Ec1 immunochemically identical with Kp1	Cannon *et al.* (1974)

Aerobic bacteria:		
Azotobacter vinelandii (Av)	Av1 and Av2 characterized	Burns and Hardy (1972), Shah and Brill (1973)
Azotobacter chroococcum (Ac)	Ac1 and Ac2 characterized	Yates and Planqué (1975)
Corynebacterium autotrophicum (Ca)	Ca1 and Ca2 characterized	Berndt *et al.* (1978)
Spirillum lipoferum (S1)†	Extracts; activating factor present	Okon *et al.* (1977)
Mycobacterium flavum (Mf)	Mf1 and Mf2 separated	Biggins and Postgate (1971b)
Blue-green algae:		
Anabaena cylindrica (Acy)	Acy1 and Acy2 separated	Tsai and Mortenson (1978)
Gloeocapsa sp LB795 (G sp)	Extracts	Gallon *et al.* (1972)
Plectonema boryanum 594 (Pb)	Extracts	Haystead *et al.* (1970)
Symbiotic bacteria:		
Rhizobium japonicum (Rj) bacteroids	Rj1 and Rj2 separated; Rj1 characterized	Bergersen and Turner (1970) Israel *et al.* (1974)
Rhizobium lupini (R1) bacteroids	Rl1 and Rl2 characterized	Whiting and Dilworth (1974)
Frankia alni (Fa) (an actonomycete)	Extracts	Van Stratten *et al.* (1977)

*The references given are to those reports of extracts of the highest specific activity, or where the nitrogenase components have been separated, to the purification procedure.
†Also known as *Azospirillum lipoferum* or *A. brasilense* (see Chapter III.3).

(a) PREPARATION OF CELLS WITH ACTIVE NITROGENASE

Extracts with nitrogenase activity have been prepared from 19 different organisms. These range from the free-living anaerobic and aerobic bacteria, through both legume and non-legume symbiotic systems, to the blue-green algae (see Table 1). The media and growth conditions for the production of cells containing active nitrogenase are described in Chapter II.2. In order to provide sufficient material for the isolation and characterization of nitrogenase from free-living organisms, culture volumes of 80–1000 l have been used. For the Rhizobium-legume system 150–450 g fresh weight of nodules are required.

(i) Preparation of bacteroids from root nodules

In the Rhizobium-legume N_2 fixing system, the bacteroids of the root nodules are the site of nitrogen fixation. Purification of nitrogenase from these sources involves maceration of the nodule and separation of the intact bacteroids from plant debris. During this procedure nitrogenase has to be protected from oxygen inactivation, and also from the action of inhibitory phenolic compounds of plant origin. This protection is achieved by carrying out the steps in the isolation with the minimum exposure to air using anaerobic buffers and syringe transfer techniques (see below), and by including polyvinylpyrrolidone in the medium used for preparation of the bacteroids.

The tissue is blended in an Omni-mixer (Ivan Sorvall and Co.), fitted with gassing vents to allow the contents to be flushed with nitrogen for 10 min (with shaking) before the tissue is disrupted. Filtration steps are carried out in a glove-box flushed with nitrogen (Evans *et al.*, 1972b). More rigorous anaerobic techniques can be obtained using a modified Omni-mixer (Bergersen and Turner, 1973). An outlet tube at the bottom of the vessel allows the homogenate to be passed through a modified sterilizing filter. The filter (11.5 cm diameter) is formed by two layers of nylon cloth (mesh 200) made air-tight with Neoprene O-rings and gaskets. After the tissue has been macerated it is forced under a slight gas pressure through the filter.

Both soluble polyvinylpyrrolidone (Kollidon 25, average molecular weight 25 000; BASF) and insoluble polyvinylpyrrolidone (Polyclar AT, General Aniline Co., New York, U.S.A.) have been used in isolation procedures. Before use the insoluble polyvinylpyrrolidone is prepared by soaking in a 5-fold (w/w) excess of 3 M HCl for 12 h. It is then filtered, washed with water, and equilibrated with 20 mM phosphate buffer (pH 7.4).

In a typical preparation of *R. japonicum* bacteroids (Evans *et al.*, 1972), fresh or frozen nodules (150 g), 50 g of solid polyvinylpyrrolidone, 300 mg of dithionite, and 400 ml of 20 mM potassium phosphate buffer (pH 7.4) containing 0.2 M sodium ascorbate are combined in an Omni-mixer flushed with nitrogen. After gassing for 10 min the mixture is macerated for 5 min. The gas vents are then closed and the Omni-mixer is transferred to a glove-box flushed with nitrogen. The macerate is squeezed through four layers of cheesecloth to remove large debris. Bacteroids are recovered by centrifuging the filtrate at 5000 *g* for 10 min. The supernatant is discarded and the bacteroid pellet re-suspended in 50 mM Tris–HCl buffer (pH 8.0) and washed by centrifuging and re-suspending in the same buffer.

For *R. lupini* bacteroids, the buffer used to macerate the nodule tissue is 50 mM phosphate buffer (pH 7) containing 0.2 M sucrose and 2% w/v of soluble polyvinylpyrrolidone and 0.5 mM dithiothreitol (DTT) (Kennedy, 1970; Whiting and Dilworth, 1974).

(ii) Storage of organisms

The method of storage of cells can affect both the efficiency of cell breakage and the activity of purified nitrogenase components. With the majority of organisms, storage at liquid nitrogen temperatures is used routinely.

When *A. vinelandii* is disrupted by French pressure treatment, cells stored at −20 °C for 1 month or −70 °C for 14 months result in extracts as active as those obtained with freshly harvested cells (Burns and Hardy, 1972). However, cells stored at −20 °C for longer than 1 month do not lyse readily (Emerich and Burris, 1978), and if this method of cell breakage is to be used cells stored for longer periods should be kept in liquid nitrogen. With *A. chroococcum*, freshly harvested cells are used since organisms which have been frozen produce acid (presumably β-hydroxybutyrate) on thawing, which results in nitrogenase preparations of decreased activity.

C. pasteurianum is normally stored as a dried cell powder. Cultures are harvested and the cells vacuum dried in a rotary evaporator at 40 °C (Carnahan *et al.*, 1960). Freeze-drying cannot be used because of the cold lability of nitrogenase in this organism. The dried cells are stored anaerobically in bottles at −20 °C (Mortenson, 1972). It has been suggested that this treatment results in the formation of an inactive de-molybdo species of the Mo–Fe protein since cells stored in liquid nitrogen do not contain the inactive species (Orme-Johnson and Davis, 1977).

Extracts of blue-green algae are prepared from freshly harvested cells, and those from nodules of soybean and lupin either from fresh material or from nodules frozen in liquid nitrogen and subsequently stored at −20 °C.

(b) PREPARATION OF CELL-FREE EXTRACTS

For some organisms the use of a particular method of cell disruption may be critical in obtaining active extracts. It cannot be emphasized too strongly that except in the case of Azotobacter, which contains a particulate, oxygen-tolerant nitrogenase in crude extracts, air must be rigorously excluded during manipulative procedures if activity is to be preserved.

Table 2. Conditions used to prepare crude extracts of various organisms by pressure disruption

Organism	Weight of cell paste (g)	Volume of buffer (ml)	Buffer used for extraction	Disruption pressure (lb in^{-2})	Conditions of centrifugation	Specific activity of crude extract [nmol C_2H_4 produced min^{-1} (mg of protein)$^{-1}$]	Reference
A. vinelandii	1.5 ml g^{-1} of buffer		25 mM phosphate buffer (pH 7.5)	16000–20000	20000 *g*, 16 h	54	Burns and Hardy (1972)
A. chroococcum	1000	500	75 mM Tris–HCl (pH 7.8) containing dithionite and DTT	3000	40000 *g*, 30 min, 5°C	51	Yates and Planqué (1975)
C. autotrophicum	800	800	50 mM Tris–HCl (pH 7.8) containing dithionite and DTT	—	23000 *g*, 2.5h, 5°C	180	Berndt *et al.* (1978)

M. flavum	—	—	25 mM Tris–HCl (pH 7.4)	9000–12 000	28 000 *g*, 30 min	8	Biggins and Postgate (1971a)
R. japonicum bacteroids	Equal weights of bacteroids and buffer		100 mM TES (pH 8.5) containing 1.2 mM dithionite	16 000	48 000 *g*, 60 min, 2°C	30	Israel *et al.* (1974)
S. lipoferum	4 ml g^{-1} of buffer		300 mM Tris–HCl (pH 8.7) containing dithionite and DTT	8000	48 000 *g*, 2 h, 4°C	8	Okon *et al.* (1977)
K. pneumoniae	425	300	25 mM Tris–HCl (pH 7.4) containing dithionite and DTT	15 000	25 000 *g*, 90 min, 5°C	50	Eady *et al.* (1972)
C. vinosum	400	1200	20 mM Tris–HCl (pH 8.5)	10 000	23 000 *g*, 30 min	—	Evans *et al.* (1973)
Gloeocapsa sp.	—	—	20 mM HEPES (pH 7.5)	10 000	—	0.4	Gallon *et al.* (1972)

A French pressure cell was used in all cases except for *C. vinosum*, where a Manton–Gaulin Homogenizer was used. Dithiothreitol (DTT) at 0.1 g l^{-1} was included in the buffers where indicated.

(i) Pressure disruption

In this method it is comparatively easy to exclude air and to scale up to deal with large volumes of material. The cells are suspended in buffer (usually containing dithionite) as an even slurry and then disrupted in a pre-cooled French press. The barrel of the pressure cell is flushed continuously with oxygen-free gas during the filling procedure to maintain anaerobiosis. After passage through the pressure cell the suspension is collected in suitable centrifuge tubes continuously flushed with oxygen-free gas and sealed with an air-tight cap before centrifuging. Table 2 shows the experimental conditions used for different organisms and the scale on which this method has been applied.

(ii) Ultrasonic disruption

This method has been used for the preparation of extracts of *R. lupini* bacteroids (Whiting and Dilworth, 1974), *Rhodopseudomonas palustris* (Zumft and Castillo, 1978), and *Anabaena cylindrica* (Smith and Evans, 1970). During sonication air is excluded by flushing nitrogen or nitrogen containing 10% of hydrogen over the cell suspension.

The method for *R. lupini* involves re-suspension of the isolated bacteroids (see above) in 50 mM TES buffer (pH 7.4) containing 2 mM $MgCl_2$ and 5 mM each of dithionite and DTT. The suspension is sonicated for 15 min at full power in a cooled Raytheon 9-kHz sonicator (Raytheon Co., Mass., U.S.A.). Centrifugation at 80 000 *g* for 30 min gives a crude extract with a specific activity of 20–30.

For *R. palustris*, washed cells are re-suspended in 50 mM TES buffer (pH 8.5) containing 1 mM dithionite, and disrupted by four cycles of 15-s exposure to 90 W with 45-s cooling intervals. Anaerobic centrifugation at 14 500 *g* for 15 min gives a crude extract with a specific activity of 1–1.5. Acetylene reduction is non-linear and is stimulated by Mn^{2+}, providing presumptive evidence for the presence of an activating factor in this organism (see Section d.v).

(iii) Lysis of dried cells

This method has been used only with *C. pasteurianum*. Typically the dried cells (100 g) are lysed by mixing, under hydrogen, with a 12–20 volume excess of 50 mM Tris–HCl buffer (pH 8) containing 150 mg of lysozyme (Mortenson, 1972). Any oxygen introduced during this manipulation is removed by 10 cycles of evacuation and flushing with hydrogen. The mixture is shaken at 30 °C for 1 h and then centrifuged at 37 000 *g* for 15 min at 15 °C. Typical crude extracts have a specific activity of 50–80.

(iv) Lysis of freshly harvested or frozen cells

This method has been used for the large-scale extraction of *B. polymyxa* (Emerich and Burris, 1978). The method involves mixing the cell paste (typically 150 g) with 2.5 volumes of 40 mM cacodylate buffer (pH 7.8) containing 1 mM dithionite, and shaking vigorously at 45–55 °C in a water-bath. During this process any oxygen trapped in the cell paste is removed by evacuation and flushing with hydrogen. When an even slurry of cells has been obtained, 50 ml of the same buffer containing 150 mg of lysozyme, 1 mg of DNase I and 0.1 g of $MgCl_2$ is added. The mixture is then maintained at 21–23 °C and stirred under hydrogen for 45–50 min. The crude extract is obtained by centrifuging anaerobically at 22000 *g* for 1 h at 5–10 °C, and typically has a specific activity of 25–30.

(v) Lysis by osmotic shock

This method is used for the large-scale preparation of extracts of *A. vinelandii* (Shah *et al.*, 1972) and *R. rubrum* (Emerich and Burris, 1978).

For *A. vinelandii*, frozen cells are washed in 25 mM Tris–HCl buffer (pH 7.4), re-suspended in 4 volumes of the same buffer containing 4 M glycerol for 30 min, and then centrifuged at 12000 *g* for 10 min at 40 °C. The supernatant is discarded and the cells are lysed by vigorous shaking of the pellet in 3–4 volumes of 25 mM Tris–HCl buffer (pH 7.4) containing 1 mg ml^{-1} of dithiothreitol and 5–10 μg ml^{-1} of DNase under nitrogen. Crude extracts obtained by centrifuging at 27000 *g* for 30 min have a specific activity of 36.

For *R. rubrum*, freshly harvested or frozen cells (40–50 g) are re-suspended anaerobically in 4 volumes of 4M glycerol containing 10 mg of lysozyme and 1 mg of DNase I. After incubation at 0 °C for 15 min the suspension is centrifuged at 8000 *g* for 15 min and the supernatant discarded. The cells are then lysed by the rapid addition of 50 mM triethanolamine–20 mM Tris–HCl buffer (pH 8.15) at the level of 3 ml per gram of cell paste. The lysate is centrifuged for 10 min at 10000 *g* to remove debris, and then at 45000 *g* for 2 h. The specific activity of the supernatant at this stage is 7–8 but is stimulated up to 5-fold when the activating factor for Rr2 is added to the assay (Ludden and Burris, 1976) (see Section d.i).

(c) PURIFICATION AND PROPERTIES OF NITROGENASE AND ITS COMPONENT PROTEINS

Nitrogenase from a number of these sources has been purified and separated into two proteins, both of which are required for enzymic activity. Both are redox proteins containing Fe–S clusters and the larger protein also contains Mo.

In this chapter, the Mo- and Fe-containing protein is referred to as Mo–Fe protein and the Fe-containing protein as the Fe protein. Components from a particular organism are denoted by a capital letter indicating the genus and a lower-case letter indicating the species of the organism (see Table 1), and the number 1 for the Mo–Fe protein and number 2 for the Fe protein. Some organisms contain an activating factor for the Fe protein which may be present predominantly in an inactive form in crude extracts (see Section d.v).

The strategic approach to the problem of the structure and mechanism of nitrogenase by separation and characterization of the individual component proteins and studies of their properties when recombined has been very successful. Work in a number of laboratories has resulted in the assignment of roles to both proteins and the development of a general scheme for the events during catalysis (see Section c.vi). Only in the case of *A. vinelandii* have extensive studies on the unresolved complex been reported, and recent work has allowed the oxygen tolerance of these preparations to be rationalized.

The increasing application of EPR and Mössbauer spectroscopy and stopped-flow techniques to structural and mechanistic studies requires large amounts of purified enzyme. This need arises not only from the inherent lack of sensitivity of these methods, but also because physicochemical rather than catalytic properties are often being measured.

(i) Anaerobic technique

A major obstacle to the purification of the nitrogenase proteins is their extreme sensitivity to oxygen. On exposure to air the Fe protein is irreversibly inactivated, losing half its activity after exposure for 45 s. The Mo–Fe protein is more stable but typically the activity has a half-life of 10 min in air.

Protection from oxygen inactivation during purification is discussed in this section, and the more rigorous techniques which must be used when dithionite-free or oxidized proteins are manipulated are described in Section e. At all stages during purification deoxygenated buffers containing dithionite are used, and as far as possible procedures should be carried out under an oxygen-free atmosphere. Transfer of nitrogenase components should be made rapidly using a large-scale syringe fitted with a wide-bore tube or needle. Disposable 50-ml plastic syringes are suitable for this purpose. Before use they should be rinsed with anaerobic buffer containing dithionite.

Preparation of buffers

Adjust the required buffer to the desired ionic strength and pH value and transfer it into a suitably sized conical flask to allow the liquid to be bubbled vigorously without spillage. Bubble oxygen-free gas through the liquid for 15 min at a rate sufficient to produce small trapped bubbles in the liquid.

During this process the neck of the flask should be closely covered with aluminium foil to prevent eddying of air into the flask. Then solid dithionite is added to give the desired concentration (usually 1 mM), and dithiothreitol (DTT) is added if required. Once dithionite has been added, exposure to air should be minimized since dithionite oxidation results in acid formation. The use of dithionite to scavenge residual oxygen and to provide a measure of protection against inadvertent exposure to air places a limit on the lower pH range which can be used. At pH values above 7.6 solutions of dithionite are stable for several weeks under anaerobic conditions; at pH 6, decomposition occurs at a rate of 10% min^{-1} (Dixon, 1971).

Anaerobic chromatography

Commercially available chromatographic columns with adjustable end adaptors are suitable for anaerobic chromatography. These provide closed systems and anaerobic conditions are obtained by using buffer containing dithionite fed from a reservoir continuously flushed or bubbled with oxygen-free gas. The connecting tubing to the column should be nylon or thick-walled PVC tubing to reduce the rate of diffusion of oxygen into the buffer. Silicone rubber is too porous and is unsuitable in this application. Samples can be loaded either directly to the inlet line by temporarily stopping the buffer flow and transferring the inlet to a flushed flask containing the sample, or alternatively via a three-way tap in the inlet line connecting either to the reservoir or to the sample vessel.

Columns containing Sephadex (Pharmacia Fine Chemicals AB, Uppsala, Sweden) can be prepared aerobically and subsequently made anaerobic by passing 2–3 bed volumes of buffer containing dithionite through the column. Substituted celluloses (carboxymethyl or diethylaminoethyl) are best treated with dithionite before the column is prepared. The cellulose is adjusted to the required pH and ionic strength and degassed using a vacuum pump. Solid dithionite is then added until the supernatant will reduce methyl viologen paper (see below). During preparation of the column nitrogen is flushed above the settling absorbant. When the column is packed it is equilibrated with 2 bed volumes of the required starting buffer containing dithionite.

The presence of dithionite in reservoir buffers and column effluents can be checked rapidly by using papers impregnated with methyl viologen. These are prepared by soaking strips of chromatography paper (approximately 1 cm wide) for 2 min in 0.25 M Tris–HCl buffer (pH 7.4) containing 1% of methyl viologen. The papers are then dried and cut into approximately 1.5-cm lengths. When the paper is wetted with a few drops of column effluent or reservoir buffer a transient blue colour indicates the presence of dithionite. The UV absorbance of buffers containing 1 mM dithionite is too high to monitor protein elution at 254 or 280 nm and the chromatography is followed visually. Column effluents can be collected either in conical flasks flushed continuously with oxygen-free gas, or via a hypodermic needle into bottles capped with rubber seals. The sealed bottles are flushed with oxygen-free gas and vented by a hypodermic needle during collection.

Precipitation procedures

Precipitation steps used in the purification of various nitrogenases are polyethylene glycol or protamine sulphate precipitation and crystallization. Polyethylene glycol precipitations are carried out in two suction flasks connected together (Tso *et al.*, 1972) by a tube 10 mm in diameter. The weighed polyethylene glycol solid is contained in one flask and the extract in the other. Using this technique, the contents of the flasks can be degassed, flushed with hydrogen, and mixed together without exposure to air during the transfer. Protamine sulphate precipitation and subsequent re-solubilization with cellulose phosphate are carried out in centrifuge tubes using magnetic stirrers, so as to minimize the number of transfer steps (Mortenson, 1972).

Concentration of protein solutions

Ultrafiltration using Diaflo membranes (Amicon Corp., Lexington, Mass., U.S.A.) and a suitable vessel pressurized with nitrogen or argon has been widely applied to the concentration of column effluents. Suitable membranes are XM-50 and XM-100, which are sturdy and have high flow-rates. Recently, on-line concentration of Sephadex column effluents using an Amicon eluate concentrator has been introduced (Emerich and Burris, 1978).

As an alternative to ultrafiltration, batch elution from small DEAE-cellulose columns (2.5 × 4 cm) can be used to concentrate nitrogenase components with known chromatographic behaviour. The protein to be concentrated is adsorbed on the DEAE-cellulose and the eluted at a concentration of salt sufficient to produce a narrow elution band.

Storage of nitrogenase components

Nitrogenase and its component proteins are routinely stored in bead form at liquid nitrogen temperatures with no decrease in activity over many months. Solutions are frozen by dripping them into liquid nitrogen. On a large scale this is conveniently done by dropping the solution from a separating funnel into the liquid nitrogen. The tap should be adjusted to give a flow-rate such that each drop freezes as a discrete bead (approximately 20–40 ml min^{-1}). During this process nitrogen is flushed over the surface of the solution. Preparations from some organisms are stable for short periods at higher temperatures. For example, when frozen rapidly in an ethanol–CO_2 bath and subsequently stored at −20°C, Av1 retains 93% and Av2 76% of its initial activity for 20 days (Shah and Brill, 1973).

(ii) Purification techniques used for nitrogenase components

In this section various purification techniques and their application to nitrogenase proteins from different sources will be considered. This approach, rather than a detailed description of the purification procedures, allows the utility of a particular technique to be assessed more easily. At the end of this section three different but typical purification procedures for highly purified, intensively studied proteins are summarized.

Chromatography on DEAE-cellulose

Chromatography on DEAE-cellulose is a widely used method to resolve nitrogenase into the Mo–Fe protein and Fe protein (see Table 3). In most purification procedures it is used at an early stage, the crude extract being applied to the column. Anaerobic chromatography is normally carried out at room temperature and the column developed with buffers containing increasing salt concentration, either NaCl or $MgCl_2$ (usually applied by syringe transfer to the open top of a column which is flushed continuously with gas). The Mo–Fe protein is eluted first, as a dark brown band by 0.15–0.25 M NaCl, and the Fe protein as a dark yellowish brown band eluted at higher salt concentrations (see Table 3). Both proteins migrate as diffuse tailing bands. The Mo–Fe protein fractions are normally free of contamination by the Fe protein and consequently have no activity when assayed alone. The Fe protein fraction is invariably contaminated with residual Mo–Fe protein and consequently has considerable nitrogenase activity when assayed alone.

In addition to its application in the separation of the nitrogenase proteins, DEAE-cellulose chromatography is applied at a later stage in the purification of the separated proteins. In the final stage in the purification of Cp1 and Kp1, preparations of these fractions, which contain essentially homogeneous proteins, are further purified by elution from DEAE-cellulose with a linear salt gradient. Fractions are obtained which have a higher metal content and specific activity and which do not exhibit the EPR signal at $g = 1.94$ shown by the preparations with lower activity.

The conditions used for Cp1 protein are elution with a 0.15–0.3 M NaCl linear gradient from a DEAE-cellulose column at pH 7.5 equilibrated with 50 mM Tris–HCl buffer initially containing 0.1 M NaCl (Zumft and Mortenson. 1973). A pale yellow band of inactive Cp1 is eluted before the dark brown band of active Cp1.

The conditions used for Kp1 protein are elution from DEAE-cellulose at pH 8.7 (Smith *et al.*, 1976a). A linear gradient of 30–90 mM $MgCl_2$ in

Table 3. Nitrogenases which have been resolved into component proteins by DEAE-cellulose chromatography

Source of nitrogenase	Chromatography buffer	Conditions for elution		Reference
		Mo–Fe protein	Fe protein	
C. pasteurianum	20 mM Tris–HCl (pH 7.4) containing 0.15 M NaCl	0.25 M NaCl	0.4 M NaCl	Tso *et al.* (1972)
K. pneumoniae	25 mM Tris–HCl (pH 7.4) containing DTT	0.23 M NaCl	0.09 M $MgCl_2$	Eady *et al.* (1972)
*B. polymyxa**	25 mM Tris–HCl (pH 7.4) containing 0.15 M NaCl	0.25 M NaCl	0.35 M NaCl	Emerich and Burris (1978)
*A. chroococcum**	25 mM Tris–HCl (pH 7.4) containing DTT	0.15–0.2 M NaCl	0.09 M $MgCl_2$	Yates and Planqué (1975)
*A. vinelandii**	20 mM Tris–HCl (pH 7.2)	0.25 M NaCl	0.35 M NaCl	Burns and Hardy (1972)
C. autotrophicum	25 mM Tris–HCl (pH 7.4) containing DTT	0.15 M NaCl	0.09 M $MgCl_2$	Berndt *et al.* (1978)
S. lipoferum	50 mM Tris–acetate (pH 7.7)	0.2 M NaCl	0.45 M NaCl	Ludden *et al.* (1978)
*R. japonicum** bacteroids	25 mM TES (pH 7.9) containing DTT	0.15 M NaCl	0.05 M $MgCl_2$	Israel *et al.* (1974)
R. lupini bacteroids	50 mM Tris–HCl (pH 7.4) containing 0.1 M NaCl and DTT	0.1 M NaCl	0.1 M $MgCl_2$	Whiting and Dilworth (1974)
*C. vinosum**	20 mM Tris–HCl (pH 7.5) containing 0.1 M NaCl	0.25 M NaCl	0.6 M NaCl	Evans *et al.* (1973)
R. rubrum	50 mM Tris–acetate (pH 7.6)	0.2 M NaCl	0.45 M NaCl	Emerich and Burris (1978)
R. palustris	25 mM TES (pH 7.5)	0.21 M NaCl	Continuous elution	Zumft and Castillo (1978)
*A. cylindrica**	50 mM Tris–HCl (pH 7.5)	0.1–0.6 M NaCl gradient		Tsai and Mortenson (1978)

*DEAE-cellulose chromatography was carried out after preliminary purification using other techniques.

25 mM Tris–HCl buffer is used to elute the column. The distribution of fractions of increased activity is variable and no clear visual separation as occurs with Cp1 is observed.

DEAE-cellulose chromatography is also used as the final step in the purification of Cp2 (Zumft and Mortenson, 1973). A 25 × 2.5 cm column is used in conjunction with a linear gradient (500 ml) of 0.2–0.4 M NaCl.

Gel filtration

Chromatography on Sephadex gels is a widely used purification technique, usually employed after the nitrogenase components have been separated. Table 4 shows the variety of conditions which have been used for proteins from different organisms. An upwards flow of buffer is often used, but care must be taken with samples with high salt and protein concentration to avoid tailing of the loading zone over the edge of the lower supporting bed. This can be overcome by loading the column with a downwards flow until the sample has completely entered the gel, and then rapidly inverting the column. Before the column is loaded the sample should be centrifuged (10000 *g* for 10 min) to remove any precipitated protein. In the case of partially purified Fe protein preparations, several coloured bands may be observed, residual Mo–Fe protein (elutes first) and ferredoxin or pink paramagnetic protein behind the yellow–brown Fe protein.

Gel filtration has been used to resolve nitrogenase of *C. pasteurianum* (Zumft and Mortenson, 1973) and *A. cylindrica* (Tsai and Mortenson, 1978). In both cases a Sephadex G-100 column (50 × 7.5 cm) equilibrated with 50 mM Tris–HCl buffer (pH 7.5) containing 0.1 M NaCl was used.

Precipitation with polyethylene glycol or polypropylene glycol

This procedure has been applied to nitrogenase from *C. pasteurianum* and Cp1 (Tso *et al.*, 1972), nitrogenase from *R. japonicum* bacteroids (Israel *et al.*, 1974), and Bp1 (Emerich and Burris, 1978).

For *C. pasteurianum* nitrogenase, crude extracts are adjusted to pH 6 by the addition of 0.1 M MES and solid polyethylene glycol 6000 (Union Carbide Corp.) added at 15 °C to a final concentration of 10% w/w. The precipitated nucleic acids and high molecular weight material are removed by centrifugation at 27000 *g* for 20 min and discarded. Nitrogenase is then precipitated by adding more polyethylene glycol to 30% w/w and collected by centrifugation at 27000 *g* for 1.5 h, ferredoxin remaining in the supernatant. Nitrogenase is redissolved either directly, or after storage in liquid nitrogen, in 20 mM Tris–HCl buffer (pH 7.4) and centrifuged to remove any insoluble material.

Precipitation with polyethylene glycol is also used at a later stage in the

Table 4. Application of Sephadex gel filtration to the purification of separated Mo–Fe and Fe proteins

Protein	Grade of Sephadex	Column dimensions (cm)	Chromatography buffer	Reference
Mo–Fe proteins:				
Cp1	G-200	85 × 5	50 mM Tris–HCl (pH 8.0)	Tso *et al.* (1972)
Kp1	G-200	80 × 5	25 mM Tris–HCl (pH 8.7), DTT	Eady *et al.* (1972)
Bp1	G-200	80 × 5	50 mM Tris–HCl (pH 8.0)	Emerich and Burris (1978)
Ac1	G-200	40 × 5	25 mM Tris–HCl (pH 8.0), 0.1 M NaCl, DTT	Yates and Planqué (1976)
	G-100	40 × 5	25 mM Tris–HCl (pH 7.4), 0.05 M $MgCl_2$, DTT	
Ca1	G-200	40 × 5	25 mM Tris–HCl (pH 7.4), 0.1 M NaCl, DTT	Berndt *et al.* (1978)
Rj1	G-200	83 × 1.5	20 mM TES (pH 7.5)	Israel *et al.* (1974)
Cv1	G-200	40 × 4.5	40 mM Tris–HCl (pH 8.5), 0.1 M NaCl	Evans *et al.* (1973)
Fe proteins:				
Cp2	G-100	80 × 2.5	50 mM Tris–HCl (pH 8)	Tso *et al.* (1972)
Kp2	G-100	37 × 5	25 mM Tris–HCl (pH 7.4), 0.05 M $MgCl_2$, DTT	Eady *et al.* (1972)
Bp2	G-100	85 × 2.5	20 mM Tris–HCl (pH 7.4), 0.05 M $MgCl_2$	Emerich and Burris (1978)
Ac2	G-100	40 × 5	25 mM Tris–HCl (pH 7.4), 0.05 M $MgCl_2$, DTT	Yates and Planqué (1976)
Ca2	G-100	40 × 5	25 mM Tris–HCl (pH 7.4), 0.05 M $MgCl_2$, DTT	Berndt *et al.* (1978)

Dithiothreitol (DTT) at 0.1 g/1 is included in buffers where indicated.

purification of Cp1 protein. The pH of the Cp1 fraction from a DEAE-cellulose column is adjusted to 6 by the addition of 0.1 M MES. The fraction precipitated at 15 °C in the concentration range 5–14% w/w of polyethylene glycol 6000 is purified Cp1 protein and, after collection by centrifugation at 20 000 g for 30 min, is redissolved in 50 mM Tris–HCl buffer (pH 8.0).

A similar procedure is used to purify Bp1 protein from a DEAE-cellulose column except that polyethylene glycol 4000 is used as precipitant. The protein [in 20 mM Tris–HCl buffer (pH 7.4) and 0.18–0.2 M NaCl] that is precipitated at 15 °C within the concentration range 15–25% w/w of polyethylene glycol is Bp1. The precipitate is recovered by centrifugation at 20 000 g for 20 min and then redissolved in 20 mM Tris–HCl buffer (pH 7.4).

Polypropylene glycol is used to precipitate the nitrogenase of *R. japonicum* bacteroids from crude extracts. Cold polypropylene glycol (P-400, Matheson, Coleman & Bell Co.) is added dropwise to a vigorously stirred crude extract. Nitrogenase is obtained in the fraction precipitated between 21 and 41% v/v of polyethylene glycol. After collection by centrifugation at 30 000 g for 15 min at 2 °C the precipitate is dissolved in 0.1 M MES buffer (pH 7.9).

Preparative gel electrophoresis

Anaerobic preparative polyacrylamide gel electrophoresis is used for the final step in the purification of Av2 (Shah and Brill, 1973), Sl2 (Ludden *et al.*, 1978), and both Rr1 and Rr2 (Emerich and Burris, 1978). In all cases an Ortec 4100 pulsed power supply (Ortec Inc., Oak Ridge, Tenn., U.S.A.) is used to minimize heating effects during electrophoresis. Anaerobic conditions are achieved by pre-electrophoresis of dithionite into the gel for 3 h at 60 V 100 cps, 1 μF before the protein is loaded, and by flushing the buffer reservoirs with oxygen-free gas during electrophoresis.

Av2 is purified by electrophoresis in a Fractophorator (Buchler Instruments, Fort Lee, N.J., U.S.A.), using a 1-cm stacking gel (6% w/v acrylamide) and a 4-cm separating gel (8% w/v acrylamide) 1.3 cm in diameter. The reservoir buffers used are 65 mM Tris–borate buffer (pH 9) containing 0.3 mg ml^{-1} of dithionite. A sample containing 20 mg of Av2 from a DEAE-cellulose column [in 25 mM Tris–HCl buffer (pH 7.4) with 0.35 M NaCl] is loaded under the reservoir buffer and electrophoresed for 10–12 h (overnight). The voltage is then increased to 150 V (10 mA) and the Av2 eluted into 65 mM Tris–HCl buffer (pH 7.4) containing 0.1 mg ml^{-1} each of dithionite and DTT. The time taken for the electrophoresis can be decreased to 6 h by increasing the voltage to 200 V (12 mA).

Sl2 from a DEAE-cellulose column is desalted by passage through a Sephadex G-25 column equilibrated with 50 mM Tris–acetate buffer (pH 7.7) before electrophoresis. A 3-cm diameter polyacrylamide gel consisting of a 2-cm deep stacking gel (4% w/v acrylamide) and a 4-cm deep separating gel

(8.3% w/v acrylamide) is used. The gel and reservoir buffer used in 100 mM Tris–borate (pH 8.6) and the eluting buffer is 100 mM Tris–acetate (pH 8.0). The protein is electrophoresed at 50 V (150 pulses s^{-1}) overnight, and run into the elution buffer at 150 V (150 pulses s^{-1}).

Rr1 from a polyethylene glycol 4000 precipitation step is dissolved in 50 mM Tris–HCl buffer (pH 8) containing 1.5% w/v of sucrose. The 3-cm diameter gel used for electrophoresis is made up of a 1.5-cm stacking gel (4% w/v acrylamide) and a 3-cm separating gel (7.5 w/v acrylamide). The reservoir buffers are 100 mM Tris–borate (pH 8.6) and the eluting buffer is 100 mM Tris–HCl (pH 8.0). The protein is then loaded and run overnight, and the voltage is increased to 150 V (20 mA) to elute Rr1 into the elution buffer.

Rr2 from a DEAE-cellulose column is electrophoresed under similar conditions except that it is first desalted into 50 mM Tris–acetate buffer (pH 7.6) containing 1.5% w/v of sucrose, and an 8% separating gel is used.

Precipitation with protamine sulphate

In addition to its use in the precipitation of nucleic acids from crude extracts, protamine sulphate forms insoluble complexes with the anionic Mo–Fe and Fe proteins of nitrogenase. This property has been used in the purification of nitrogenase from *A. vinelandii* (Burns and Hardy, 1972) and *C. pasteurianum* (Zumft and Mortenson, 1973) and in the subsequent purification of Cp1 and Cp2 (Zumft and Mortenson, 1973) and Cv1 (Evans *et al.*, 1973).

In this method, precipitated material is recovered by centrifugation and the complexed proteins are displaced and re-solubilized by adding an excess of cellulose phosphate. Some variation in the concentration range of protamine sulphate within which these proteins precipitate can occur. This difficulty can be overcome either by monitoring the precipitation of activity from the supernant (Evans *et al.*, 1973), or the molybdenum content of the supernatant by atomic-absorption spectroscopy (Dalton *et al.*, 1971). Protamine sulphate (50 mg ml^{-1}) is added dropwise from a separating funnel (or, for small volumes, from a syringe) to solutions in a centrifuge tube containing a magnetic stirrer bar.

For *A. vinelandii*, protamine sulphate (20 mg ml^{-1}) is added slowly to a magnetically stirred crude extract at pH 6 and 0°C until the $A_{260\,nm}:A_{280\,nm}$ ratio is 0.9:1 (approximately 3 ml per gram of protein). After 20 min precipitated nucleic acids are removed by centrifugation at 20000 *g* for 50 min. A heat treatment (see below) is used to partially purify nitrogenase before it is precipitated with protamine sulphate. The supernant from the heat treatment is adjusted to pH 6.5 with HCl before protamine sulphate (half the volume required to precipitate the nucleic acids) is added. The precipitate of the nitrogenase–protamine sulphate complex is recovered by centrifugation at 27000 *g* for 1 min. The supernatant is discarded and the pellet re-suspended in the minimum volume of 100 mM Tris–HCl buffer (pH 7.2). Dry cellulose phosphate (Cellex-P, Bio-Rad Laboratories), previously equilibrated with phosphate buffer (pH 7), is added at the level of 0.25 g per millilitre of heat-treated supernatant. The pH is adjusted to 7.2 and the suspension stirred. If necessary, additional buffer is added to give an even consistency. Nitrogenase can be recovered from the supernatant by centrifugation.

The concentration range of protamine sulphate over which nitrogenase from *C. vinosum* precipitates (usually 5–15% w/w relative to the initial protein concentration) is determined from activity measurements on the supernatant. Protamine sulphate (50 mg ml^{-1}) is added to the polyethylene glycol-treated crude extract and initially material that precipitates without loss of nitrogenase activity from the supernatant is discarded. When activity is precipitated, protamine sulphate is added until no activity remains in the supernatant. Because Cv2 and Cv1 are precipitated differentially during this process, an excess of Cv2 is used in the assays. The precipitated nitrogenase is recovered by stirring for 30 min with a 7-fold (w/w) excess (over the protamine sulphate used) of cellulose phosphate in 20 mM Tris–HCl buffer (pH 8.0). Cellulose phosphate and insoluble residues are removed by centrifugation at 40000 *g* for 1 h.

Nitrogenase from *C. pasteurianum* is precipitated from crude extracts at pH 8.5 by protamine sulphate (1–11% of the initial protein concentration). The precipitate is collected by centrifugation and nitrogenase redissolved by treatment with a 5-fold (w/w) excess of cellulose phosphate (P-11, Whatman) over the protamine sulphate used. Cp1 and Cp2 are subsequently separated using Sephadex G-100 before further purification using protamine sulphate. Cp1 in 50 mM Tris–HCl buffer (pH 7.5) containing 0.1 M NaCl, is precipitated by 0.5–5% w/w of protamine sulphate, and Cp2 in the same buffer by 1–4% w/w of protamine sulphate.

Heat treatment

A heat treatment to denature contaminating proteins has been applied to the unresolved nitrogenase of *A. vinelandii* (Burns and Hardy, 1972) and *R. japonicum* (Israel *et al.*, 1974), and to Av1 protein (Shah and Brill, 1973). The procedure is conveniently carried out in a suction flask fitted with a two-holed stopper (Burns and Hardy, 1972). Oxygen-free gas is fed into the flask via the side-arm and flows through an outlet hole in the stopper. A thermometer to measure the temperature of the protein solution during heating is fitted to the second hole. The flask should be sufficiently large to allow the contents to be mixed by swirling. To obtain rapid heating a boiling water-bath can be used to heat the contents of the flask to the desired temperature; when this has been reached the flask is transferred to a constant-temperature bath for the required period.

As applied to *A. vinelandii* nitrogenase, crude extracts at pH 7 under nitrogen are heated at 60 °C for 7 min. Denatured protein is then removed by cooling the solution on ice to 20 °C and centrifuging at 27 000 *g* for 30 min (Burns and Hardy, 1972). Partially purified Av1 [in 25 mM Tris–HCl buffer (pH 7.4) containing 0.25 M NaCl] has been treated by heating at 52 °C for 5 min under hydrogen (Shah and Brill, 1973).

Partially purified nitrogenase from *R. japonicum* bacteroids has been treated at 55 °C for 2.5 min to denature contaminating proteins (Israel *et al.*, 1974)

Crystallization

Av1 protein crystallizes from solutions of low ionic strength, and this property is routinely used in the purification of this protein. The method is not generally applicable to other Mo–Fe proteins which do not crystallize under these conditions. The original method (Burns *et al.*, 1970) has been modified (Shah and Brill, 1973) because of difficulty in obtaining crystals. Since other workers have reported success with the original method (Yates and Planqué, 1975; Swisher *et al.*, 1977), both methods are described below.

In the procedure of Burns *et al.* (1970) Av1 from a DEAE-cellulose column [in 15 mM Tris–HCl buffer (pH 7.2) containing 0.25 M NaCl] is concentrated by ultrafiltration to 45 mg ml^{-1} of protein. This solution is then diluted with 6–10 volumes of 20 mM Tris–HCl buffer (pH 7.2). A greyish brown precipitate is formed which, after standing for 60 min, is collected by centrifugation at 20 000 *g* for 5 min. The almost colourless supernatant is discarded, and the dark brown pellet washed by re-suspension in 50 ml of 20 mM Tris–HCl buffer (pH 7.2) and collected by centrifugation. The pellet is then redissolved in 15 ml of the same buffer containing 0.25 M NaCl and centrifuged to remove insoluble amorphous material. The supernatant (33 mg ml^{-1} of protein) is then diluted 8-fold with 20 mM Tris–HCl buffer (pH 7.2), which results in the formation of a dense population of white crystals. These are collected by centrifugation and recrystallized by dissolving in buffer containing 0.25 M NaCl and subsequently diluted as described above. For dilute DEAE-cellulose column effluents the yield can be increased by lowering the pH to 5.8 by the addition of 0.2 M acetic acid at 0 °C before dilution (Burns and Hardy, 1972).

In an alternative method (Shah and Brill, 1973), Av1 protein from a DEAE-cellulose column [in 25 mM Tris–HCl buffer (pH 7.4) containing 0.25 M NaCl] is concentrated to 40 mg ml^{-1} of protein by ultrafiltration, and then diluted 5-fold with degassed 25 mM Tris–HCl buffer (pH 7.4). The solution is concentrated until crystallization starts, then transferred to a centrifuge tube and maintained for 1 h at 38 °C under hydrogen. The dense population of needle-shaped crystals is collected by centrifugation at 20 000 *g* for 10 min and washed by re-suspension in 25 mM Tris–HCl buffer (pH 7.4) containing 42 mM NaCl.

Both of the procedures described above give short, needle-shaped crystals of dimensions approximately 3 × 50 μm. Static diffusion has been used to obtain crystals which are 1–3 mm long and aggregate to 40 μm wide (Bulen, 1976).

Outlines of three typical purification procedures

Detailed description can be found in the references cited.

C. pasteurianum (Zumft and Mortenson, 1973)

Crude extract:
- 0–1% protamine sulphate to remove nucleic acid;
- 1–10% protamine sulphate to supernatant to precipitate nitrogenase;
- Sephadex G-100 chromatography of redissolved precipitate to separate Mo–Fe from Fe protein.

Mo–Fe protein:
- 0–0.5% protamine sulphate, discard precipitate;
- 0.5–5% protamine sulphate to precipitate Mo–Fe protein;
- Sephadex G-200 chromatography of redissolved precipitate;
- repetitive DEAE-cellulose chromatography using gradient elution. (Specific activity 2000).

Fe protein:
- 0–1% protamine sulphate, discard precipitate;
- 1–4% protamine sulphate, to precipitate Fe protein;
- Sephadex G-100 chromatography of redissolved precipitate;
- DEAE-cellulose chromatography using gradient elution. (Specific activity 2200).

A. vinelandii (Shah and Brill, 1973)

Crude extract:
- DEAE-cellulose chromatography to separate Mo–Fe from Fe protein.

Mo–Fe protein:
- Heat treatment at 52°C;
- DEAE-cellulose chromatography;
- crystallization.
- (Specific activity 1638).

Fe protein:
- DEAE-cellulose chromatography;
- preparative gel electrophoresis.
- (Specific activity 1815).

K. pneumoniae (Eady *et al.*, 1972; Smith *et al.*, 1976b)

Crude extract:
- DEAE-cellulose chromatography to separate Mo–Fe from Fe protein.

Mo–Fe protein:
Sephadex G-200 chromatography;
repetitive DEAE-cellulose chromatography using gradient elution. (Specific activity 2250).

Fe protein:
Repetitive Sephadex G-200 chromatography. (Specific activity 1200).

Criteria of purity

Electrophoresis under denaturing conditions in SDS polyacrylamide gels is the most reliable and sensitive criterion of the protein purity of nitrogenase preparations. This technique (see Chapter II.10 for details and the conditions of electrophoresis) avoids the problems encountered with native proteins when multiple bands can arise if oxidative damage occurs.

The extent of contamination with inactive material can be gauged from a comparison of the specific activity of the preparation with the values given for the outline purifications at the end of Section c.ii. The absence of an EPR spectrum at $g = 1.94$ in preparations of the Mo–Fe protein indicates that the inactive demolybdo species of this protein has been removed (Zumft and Mortenson, 1973). For the Fe protein the enhancement of the chelation of Fe with bathophenanthroline by ATP has been suggested as a measure of the amount of active Fe protein (Ljones and Burris, 1978).

Estimation of protein

Protein concentration can be estimated by the Lowry method (Chapter 2). With dried bovine serum albumin as a standard, no correction factor is necessary. Dithionite present in the sample or buffer blank must be decomposed by vigorous vortexing of the diluted samples in air before assay. The biuret and micro-biuret methods have also been used.

(iii) Purification of the nitrogenase complex

The nitrogenase complex has been purified from extracts of *A. vinelandii* prepared by pressure disruption (Bulen and LeCompte, 1972).

All manipulations are carried out under anaerobic conditions. The extract is centrifuged at 205 000 *g* for 30 min and the protein concentration of the supernatant adjusted to 35 mg ml^{-1} by dilution with 25 mM Tris buffer and the pH to 7 by the addition of 0.5 M KOH. Nucleic acids are precipitated with 2% w/v of protamine sulphate (4.6 mg per gram of protein), stirred for 30 min, and removed by centrifugation at 205 000 *g* for 10 min. $MgCl_2$ [2 M in 10 mM TES buffer (pH 7.1)] is added to the supernatant to 10 mM and the pH adjusted to 6.5 with 0.5 N acetic acid. The nitrogenase complex is then precipitated by the addition of 2% w/v of protamine sulphate (1.4 ml per gram of protein) and collected after 15 min by centrifugation at 110 000 *g* for 20 min. The pellet is redissolved by the addition of phospho-cellulose (see above) in 20 mM TES buffer (pH 6.8). Small contaminating proteins are precipitated from the redissolved material by the addition of 2 M $MgCl_2$ to 20 mM, and adjustment of the pH to 7 with 0.5 N KOH. After standing for 30 min with occasional stirring, the precipitated proteins are removed by centrifugation at 30 000 *g* for 15 min. The supernatant is then centrifuged at 205 000 *g* for 2 h to sediment the nitrogenase complex. The pellet is redissolved in 20 mM TES buffer (pH 6.8) containing 0.1 mg ml^{-1} of DTT, and any insoluble material is removed by centrifuging at 60 000 *g* for 10 min. The precipitation of contaminating proteins by $MgCl_2$ (10 mM final concentration) and collection of the nitrogenase complex by sedimentation are repeated. The complex is contaminated with several NAD(P)H-linked enzymic activities and has a nitrogenase specific activity of 125. The oxygen tolerance of this complex is due to the interaction of nitrogenase with an Fe–S protein present in stoichiometric amounts (Haaker and Veeger, 1977).

(iv) Properties of the Mo–Fe proteins

Preparations of the Mo–Fe protein from seven different organisms have been obtained which are homogeneous proteins with very similar physico-chemical properties. The problem of establishing the extent to which these preparations are contaminated by inactive material is discussed below. All common amino acids are present with acidic residues (*ca.* 20%) predominating over basic (*ca.* 10%) residues (see Eady and Smith, 1978).

Subunit structure

The proteins are tetrameric with molecular weights near to 220 000 (see Table 5). The subunit structure of all purified Mo–Fe proteins has been investigated by SDS gel electrophoresis, but it has been shown that the resolution of the two subunit types by this method depended upon the commercial source of the SDS used (Kennedy *et al.*, 1976). The subunits of Cp1 (Huang *et al.*, 1973) and Av1 (Swisher *et al.*, 1977) have been separated by other techniques, and in the case of Kp1 the tryptic peptide maps of the two subunit types have been shown to be different (Kennedy *et al.*, 1976). There is also genetic evidence for two separate structural genes for the Kp1 subunits. The similarity of the amino acid compositions of the Mo–Fe proteins makes it highly likely that all have an $\alpha_2\beta_2$ structure with the α subunits of molecular weight approximately 50 000 and the β subunits of 60 000 (see Eady and Smith, 1978).

Table 5. Molecular weights and metal and sulphide contents of highly purified Mo–Fe proteins

Protein	Mol. wt.	Mo (g-atom mol^{-1})	Fe (g-atom mol^{-1})	S^{2-} (g-atom mol^{-1})	Reference
Cp1	212000	2	24	24	Huang *et al.* (1973)
Kp1	219000	2	32	—	Smith *et al.* (1976a)
Ac1	223000	1.9	22	20	Yates and Planqué (1975)
Av1	234000	2	34–38	26–28	Burns *et al.* (1970), Shah and Brill (1977)
Ca1	232000	2.2	23	20	Berndt *et al.* (1978)
Rj1	197000	1.3	29	26	Israel *et al.* (1974)
Cv1*	N.D.	1.4	19	15	Evans *et al.* (1973)

*The data for Cv1 were recalculated on the basis of an assumed mol. wt. of 220000; N.D. = not determined.

Metal content

The Mo–Fe proteins contain both molybdenum and iron but a definitive composition for these oxygen-sensitive proteins has been difficult to establish. The correlation between activity and metal content is not direct (Orme-Johnson and Davis, 1977); activity can presumably be lost with the retention of metal. Nevertheless, preparations with the highest specific activity contain approximately 2 Mo and 24–30 Fe atoms and an approximately equivalent number of acidlabile S^{2-} atoms per tetramer (see Table 5). The approximate equivalence of Fe and S^{2-} is suggestive of the presence of Fe–S centres similar to those found in simpler Fe–S proteins, and a considerable amount of work has been done in an attempt to characterize such centres in the Mo–Fe proteins.

Optical spectroscopy

Solutions of the Mo–Fe proteins are brownish yellow, but the absorption spectra are very broad, with features at 525 and 557 nm evident in some spectra of crystalline preparations of Av1, which have been attributed to contamination with c-type cytrochrome (Shah and Brill, 1973). There is a general increase in absorption above 450 nm on oxidation.

Electron paramagnetic resonance spectroscopy

At temperatures below 30 K the Mo–Fe proteins exhibit a characteristic rhombic EPR spectrum, of a type not shown by other Fe–S proteins, with g values near 4.3, 3.7, and 2.01. This EPR spectrum has recently been shown to be associated with a novel cluster containing Mo, Fe, and S^{2-}, which can be resolved from the Mo–Fe protein (FeMo-co, see below). For Av1 and FeMo-co the EPR signal integrates to 1 ± 0.1 spin/Mo (Rawlings *et al.*, 1978).

The precise g values of Kp1 (Smith *et al.*, 1973), Bp1 (B. E. Smith and D. J. Lowe, personal communication), and Ca1 (Berndt *et al.*, 1978) are pH dependent owing to the interconversion of a high pH and a low pH form. For Kp1 a pK_a of 8.7 at 0°C was determined for this interconversion. The equilibrium between these two species is displaced by acetylene in favour of the high pH form for Kp1 and Bp1 and the low pH form of Ca1. The reasons for these differences are not known but the perturbation of the equilibrium provides evidence for a substrate binding site on the isolated Mo–Fe proteins.

In turnover, in the presence of Fe protein, MgATP, and dithionite, the EPR spectrum of the Mo–Fe protein is 90% bleached. Under some conditions, e.g. in the presence of the inhibitor CO, or the product ethylene and substrate acetylene, EPR signals characteristic of 4Fe 4s clusters at the -1 and -3 oxidation levels are observed. Since the linewidth of the signals is broadened when ^{57}Fe-substituted Mo–Fe protein is used, these signals provide additional evidence for substrate and inhibitor binding sites on this protein.

The presence of conventional 4Fe 4S clusters in Av1 is further supported by studies on the effect of thiophenol (10 mM) in 80% *N*-methylformamide on the EPR spectrum. Under these conditions the EPR spectra of Av1 are the sum of the spin $S = 3/2$ signal associated with FeMo-co (see below) and at least three 4Fe 4S clusters at the −3 oxidation level (Rawlings *et al.*, 1978). These experiments do not exclude the possibility that 2Fe 2S clusters at present since they are very labile and show no EPR spectra under these conditions.

Mössbauer spectroscopy

Detailed studies have been made on the ^{57}Fe-enriched Kp1 (Smith and Lang, 1974) and Av1 (Münck *et al.*, 1975). Despite the differences in metal composition between the proteins used in these studies (1 Mo and 18 Fe and 1.45 Mo and 17 Fe, respectively, per 220000 mol. wt.) the data are very similar. These proteins may have an all-or-none complement of metal (Zimmerman *et al.*, 1978), as has been reported for the $(4Fe\ 4S)_2$ ferredoxin from *C. acidi urici* (Hong and Rabinowitz, 1970). The Mössbauer spectrum of dithionite-reduced Kp1 at 195 and 77 K consisted of three overlapping doublets, the areas under which corresponded to 2, 8 and 8 Fe atoms per molecule. Very similar data were obtained with Av1 except for an additional weaker doublet observed at 1.5 K corresponding to about 1 Fe atom per molecule. In both Av1 and Kp1 one of the doublets observed at 77 K corresponding to 8 Fe atoms collapsed at 4.2 K, indicating magnetic character. This allowed it to be correlated with the species giving rise to the EPR signal at $g = 4.3, 3.7,$ and 2.01, which has subsequently been shown to reside in FeMo-co. During turnover when the EPR signal is bleached, the Mössbauer parameters of this group of Fe atoms change in a manner consistent with reduction rather than oxidation having occurred (Smith and Lang, 1974).

The iron and molybdenum cofactor

A low-molecular-weight iron- and molybdenum-containing cofactor (FeMo-co) has been isolated from a number of purified Mo–Fe proteins (Shah and Brill, 1977). FeMo-co from Av1 contains 8 Fe and 6 S^{2-} atoms per Mo atom, and the same ratio of Fe to Mo is obtained with the cofactor from Cp1. Biological activity is measured in an assay based on the ability of FeMo-co to restore activity to extracts of a *nif*$^-$ mutant of *A. vinelandii* (strain UW-45) that produces normal levels of Av2 but an inactive form of Av1. On a molybdenum basis, FeMo-co from Cp1, Kp1, Bp1, and Rr1 is as effective as that prepared from Av1 in reconstituting activity in *A. vinelandii* UW-45 extracts.

Solutions of FeMo-co as isolated in *N*-methylformamide (NMF) can be stored anaerobically for 24 h at 25°C or 10 days at −20°C without loss of activity. Activity is unstable in aqueous buffered solutions (20% loss of activity after storage for 18 h at −20°C) even under anaerobic conditions. All activity is lost on exposure to air for 60 s.

Solutions of FeMo-co are brown, but the visible spectrum is featureless. The EPR spectrum is very similar to that of the spin $S=3/2$ signal of dithionite-reduced native Mo–Fe proteins except that the features are broader (Rawlings *et al.*, 1978). Mössbauer spectroscopy of FeMo-co prepared from ^{57}Fe-enriched Av1 shows the majority (at least six) of the Fe atoms to be magnetic (Rawlings *et al.*, 1978) and allows the EPR signal to be assigned to them. Since mercurials destroy both the EPR signal and activity it has been suggested that the six S^{2-} atoms are associated with these Fe atoms. FeMo-co does not contain conventional 4Fe 4S clusters since treatment with thiophenol does not affect the activity or change the EPR spectrum significantly, and $\alpha\alpha'$-bipyridyl complexes less than 5% of the total Fe.

CO binds reversibly to FeMo-co under photoreducing conditions with deazaflavin and EDTA (Rawlings *et al.*, 1978). Binding results in the loss of the EPR signal but full activity is retained. In the presence of sodium borohydride and dithionite at pH 9.6, FeMo-co catalyses the reduction of acetylene to ethylene (Shah *et al.*, 1978). The activity is 8% that of nitrogenase with an equivalent amount of Mo and is strongly inhibited by CO. Nitrogen is not reduced. In contrast to the extreme oxygen sensitivity of FeMo-co when measured in the reactivation assay, acetylene reduction by FeMo-co is only decreased 20% by exposure to air for 18 h.

Purified FeMo-co does not reactivate extracts of nitrate reductase mutants of *Neurospora crassa* and is distinct from the molybdenum cofactor (Mo-co) of this enzyme (Pienkos *et al.*, 1977). Crude extracts of nitrogen-grown *A. vinelandii* contain both FeMo-co and Mo-co, which probably accounts for earlier reports that acid-treated partially purified Mo–Fe protein would reactivate extracts (Nason *et al.*, 1971).

X-ray absorption spectroscopy

This recently developed technique is the only method currently available for studying the molybdenum site of nitrogenase. In this method the absorption of X-rays (from a high-intensity tunable X-ray line derived from a synchrotron source) by Mo atoms in nitrogenase is measured, and compared with that of the molybdenum complexes of known structure.

Studies using this technique have been made on freeze-dried samples of Cp1 (Cramer *et al.*, 1978a), crystals of Av1, and FeMo-co (derived from Av1) in NMF (Cramer *et al.*, 1978b). These samples have the same single inflection point in the absorption edge, characteristic of Mo with ligated S. The presence of a double-bonded oxy-species on the Mo is excluded since this species is invariably associated with a double inflection point on the absorption edge. Mo=O features are observed in spectra of air-oxidized Cp1.

The extended X-ray absorption fine structure (EXAFS) of Av1, Cp1, and FeMo-co are almost identical, implying that in the NMF extraction procedure used in the preparation of FeMo-co the ligation of the Mo has not changed (Cramer *et al.*, 1978a). The strongest EXAFS feature is assigned to Mo–S and is not compatible with the presence of Mo = O, i.e. the molybdenum is ligated only to S in these samples. A second EXAFS component was assigned to Mo–Fe interaction. The complex $[Mo_2Fe_6S_9(SEt)_8]^{3-}$ has been synthesized and the structure determined by X-ray crystallography shows two $MoFe_3S_4$ cubic structural fragments to be present. The EXAFS of this complex is very similar to that of the Mo–Fe proteins.

Redox properties

Most of the potential measurements have been made for the dithionite-reduced and dye-oxidized redox couple (see Orme-Johnson and Davis, 1977, and O'Donnell and Smith, 1978), using the EPR signals at $g = 4.3$, 3.7, and 2.01 to monitor the reaction.

The midpoint potential (E_m) for this process depends on the source of the Mo–Fe protein. Values range from 0 mV for Cp1 through −42 mV for Av1 and Ac1, −95 mV for Bp1, to −180 mV for Kp1 (O'Donnell and Smith, 1978). Cp1 and Bp1 showed non-ideal behaviour, but the value above for Cp1 is close to that of −20 mV as found by Zumft *et al.* (1974). For Cv1 two redox processes at −60 and −260 mV were observed (Albrecht and Evans, 1973) attributed to the same centre being in different environments.

Amperometric measurements on oxidized Av1 gave two processes with E_m values of −320 and −450 mV (Watt and Bulen, 1976). The process occurring at −320 mV is probably the same as that observed at −40 mV, displaced because only methyl viologen ($E_m = -440$ mV) was used as a mediator and would be unable to transfer electrons to Av1 at high potentials (O'Donnell and Smith, 1978). The process occurring at −450 mV has been assigned to the Fe centres which Mössbauer studies show are oxidized with Lauth's violet (Orme-Johnson *et al.*, 1977).

(v) Properties of the Fe proteins

Fe proteins from five different organisms have been obtained as homogeneous protein preparations. As with the Mo–Fe proteins the oxygen sensitivity and instability of the Fe protein make uncertain the extent to which these preparations are contaminated by inactive material. The Fe proteins from *R. rubrum* and *S. lipoferum* are not typical in that they can exist in inactive forms which are activated by a factor which requires ATP (see Section d.v).

All purified Fe proteins have very similar physico-chemical properties. They are very unstable in air, and typically the activity has a half-life of less than 1 min; Cp2 is cold-labile and loses activity when stored at 0°C (Moustafa and Mortenson, 1969). The amino acid compositions are very similar with acid residues (*ca.* 20%) predominating over basic residues (*ca.* 10%); tryptophan is absent. The complete amino acid sequence of Cp2 has been determined and shows no sequence homology with other Fe–S proteins (Tanaka *et al.*, 1977a,b,c).

Subunit structure

The Fe proteins are dimeric and have molecular weights in the range 57 000–72 000 (Table 6). They show a single band on SDS gel electrophoresis which, together with the sequence data for Cp2, makes it likely that these proteins have a γ_2 structure, although Rr2 is resolved into two bands of molecular weights 30 000 and 31 500 on SDS electrophoresis (Ludden and Burris, 1978).

Table 6. Molecular weight and metal and sulphide contents of highly purified Fe proteins

Protein	Mol. wt.	Fe ($g\text{-atom mol}^{-1}$)	S^{2-} ($g\text{-atom mol}^{-1}$)	Reference
Cp2*	57 647	3.8	3.8	Nakos and Mortenson (1971)
Kp2	66 800	4	3.8	Eady *et al.* (1972)
Ac2	65 400	4	3.9	Yates and Planqué (1975)
Av2	64 000	3.4	2.8	Kleiner and Chen (1974)
Ca2	72 600	3.8	2.4	Berndt *et al.* (1978)

*The data for Cp2 have been corrected to the mol. wt. determined from the amino acid sequence.

Metal content

Approximately four Fe and four S^{2-} atoms per dimer are present in the Fe proteins (see Table 6). The Mössbauer and EPR properties as discussed below are consistent with the presence of a single 4Fe 4S cluster, and core extrusion studies on Cp2 have shown that a 4Fe 4S cluster can be extracted (Averill *et al.*, 1978). The sequence data for this protein show that the cysteine groups are not clustered as is the case with ferredoxins containing 4Fe 4S clusters.

Optical spectroscopy

Solutions of Fe proteins are straw yellow but the spectra are broad and featureless. A general increase in absorption occurs on oxidation. This change has been used to measure pre-steady-state electron transfer reactions in turnover, and the kinetics of reduction of the dye-oxidized Fe protein by dithionite. Estimates of the molar absorptivity for this oxidation for Cp2 range from 3900 (Walker and Mortenson, 1973) to 6600 mol l^{-1} cm^{-1} (Ljones and Burris, 1978).

Electron paramagnetic resonance spectroscopy

At temperatures below 30 K the dithionite-reduced Fe proteins are characterized by a rhombic EPR spectrum with $g = 1.95$, similar to reduced ferredoxins. The EPR spectrum of Ac2 is bleached on oxidation by the redox dye phenazine methosulphate, and is fully restored by the uptake of one electron from dithionite (Yates *et al.*, 1975). These properties, together with the extrusion data for Cp2, are consistent with the presence of a 4Fe 4S centre operating between the −3 and −2 oxidation levels. Integration of the EPR signal gives (in electrons per molecule) 0.2 for Cp2, 0.24 for Ac2, 0.28 for Ca2, and 0.45 for Kp2. It has been suggested that this is a consequence of the interaction of the EPR active centre with an as yet unidentified rapidly relaxing paramagnetic centre (Lowe, 1978).

Mössbauer spectroscopy

The Mössbauer spectra of ^{57}Fe-enriched Kp2 shows a doublet at 195 K, which collapses as the temperature is decreased, indicating magnetic character, and allows the EPR signal to be assigned to clusters of Fe atoms. The single doublet which is observed at 195 K is characteristic of reduced ferredoxins with 4Fe 4S centres, and contrasts with the spectra of 2Fe 2S ferredoxins, where in the reduced state the Fe atoms are not equivalent, and the Mössbauer spectrum consists of two overlapping doublets (Smith and Lang, 1974).

Redox properties

Estimates of the number of electrons transferred, and the redox potential of the dye-oxidized/dithionite-reduced redox couple of the Fe protein, have utilized changes in the visible absorption of the protein or changes in the intensity of the EPR signal to monitor the extent of oxidation or reduction.

For Cp2, absorbance changes during the reduction of oxidized protein by dithionite gave a potential of −240 mV and 1.42–2 electrons per 55 000 molecular weight (Walker and Mortenson, 1973). EPR spectroscopic measurements on solutions of known potential gave a potential of −295 mV (Zumft *et al.*, 1974) and −270 mV (Burris and Orme-Johnson, 1976). Both MgATP and MgADP decreased the midpoint potential by 100–200 mV.

Kinetic measurements of these processes have been made on dye-oxidized Ac2 using rapid-freeze EPR techniques in conjunction with stopped-flow spectrophotometry to measure dithionite oxidation at 315 nm or protein reduction at 425 nm (Yates *et al.*, 1975; Thorneley *et al.*, 1976). These studies showed that the actual reductant is $SO_2^{\cdot -}$ produced by dissociation of dithionite ion, and that 1 e was taken up rapidly ($\tau < 1$ ms) with the full reappearance of the EPR signal. MgADP but not MgATP inhibited the reduction ($\tau \approx 92$ ms at 9 mM MgADP). Three slower phases of reduction then occurred, which together accounted for 1 e, but did not result in any change in EPR signal intensity. Oxygen-damaged protein showed only the slow effects which may indicate that preparations of dye-oxidized Ac2 which have not been deliberately exposed to oxygen nevertheless contain some inactive material.

As an alternative to dye oxidation, the Fe protein can be oxidized enzymically. In assays containing only a small amount of Mo–Fe protein and limiting dithionite, oxidized Fe protein accumulates when dithionite is exhausted. Re-reduction of Cp2 by the addition of standardized dithionite solutions gave 1 e (Ljones and Burris, 1978). In this method, presumably only enzymically active Fe protein becomes oxidized.

Interaction with ADP and ATP

Gel filtration techniques have shown that [^{14}C]ATP binds to Cp2, Kp2, and Ac2 (see Eady and Smith. 1978). Quantitative estimates of the binding constants have been obtained for Cp2 using a rapid gel equilibration technique (Tso and Burris, 1973). Two non-cooperative sites for MgATP (K_D = 16.7 μM) and a single MgADP binding site (K_D=5.2 μM) which results in cooperativity of MgATP binding were reported, Later work using this technique gave a value of K_D=53 μM for MgATP (Emerich, 1978), which compares with the value of 85 μM from the ATP-enhanced reactivity of Fe towards the chelator bathophenanthroline disulphonate. Problems associated with the EPR method for MgATP binding studies are discussed by Orme-Johnson and Davis (1977) and Eady and Smith (1978). In the presence of ATP or ADP the midpoint potential of Cp2 decreases by approximately 100 mV and the oxygen sensitivity of Ac2 and Kp2 increases. The reactivity of thiol groups of Kp2 towards DTNB also increases (Thorneley and Eady, 1974). These data are consistent with a gross conformational change occurring on nucleotide binding.

(vi) The nitrogenase system

In addition to the reduction of nitrogen, nitrogenase will reduce acetylene and a number of other substrate analogues. In the absence of an added reducible substrate, protons are reduced and hydrogen is evolved. The reduction of acetylene to ethylene provides a simple and sensitive assay which has been widely applied in biochemical, genetic and field studies of nitrogen fixation (Chapter II.3).

For activity nitrogenase requires ATP, a divalent cation (Mg^{2+} is the most effective), a reductant (usually dithionite *in vitro*), and an anaerobic environment. Studies on the isolated proteins, described above, have shown that MgATP and dithionite interact with the Fe protein, and reducible substrates with the Mo–Fe protein. EPR and Mössbauer spectroscopy and stopped-flow spectrophotometry of the functioning system have resulted in the development of a mechanism in which electrons are transferred from dithionite to the Fe protein, and from the Fe protein to the Mo–Fe protein in an ATP-dependent reaction (see Zumft and Mortenson, 1975; Orme-Johnson and Davis, 1977; Eady and Smith, 1978; and Mortenson and Thorneley, 1979). More recently, rapid quench studies have shown that a pre-steady-state burst of ATP hydrolysis occurs during electron transfer from the Fe to the Mo–Fe protein, indicating that ATP hydrolysis is coupled to electron transfer (Eady *et al.*, 1978). Other studies indicate that during hydrogen evolution the two proteins probably dissociate after the transfer of the electron which results in the bleaching of the g=4.3, 3.7, and 2.01 signal of the Mo–Fe protein (Hageman and Burris, 1978). Different forms of the enzyme are responsible for the reduction of protons, acetylene, and nitrogen (Smith *et al.*, 1976b, 1977; Thorneley and Eady, 1977; Lowe *et al.*, 1978). During the reduction of nitrogen an intermediate which on acid or alkali treatment releases hydrazine has been detected (Thorneley *et al.*, 1978).

(d) ASSAYS

(i) Assay of nitrogenase

In addition to the reduction of nitrogen, nitrogenase will reduce acetylene and a number of other substrate analogues. The reduction of acetylene to ethylene provides a simple and sensitive assay which has been of particular importance in enzymological, genetic, and field studies. In the absence of an added reducible substrate, protons are reduced and hydrogen is evolved.

For activity, nitrogenase requires ATP, a divalent cation (Mg^{2+} is most effective), a reductant (usually dithionite *in vitro*), and an anaerobic environment. Since ADP is a potent inhibitor of activity, assays for nitrogenase usually include an ATP-regenerating system. The basic assay conditions are similar for the different substrates, and are usually carried out in vaccine or serum bottles with a nominal volume of 7.5 ml. Different laboratories use slightly different assay conditions but typical concentrations in the assay are as follows: buffer (pH 6.6–7.5), 40 mM; ATP, 8 mM; $MgCl_2$ 20 mM; dithionite, 20 mM; creatine kinase, 0.1 mg; and acetylene, 0.1 atm. These assay conditions are used for the analytical procedures described in this section.

Requirements

Shaking water-bath, thermostated at 30°C.
Gas-tight syringes, 0.1 and 0.5 ml.
Serum bottles, nominal capacity 7.5 ml.
Rubber closures (Suba Seal No. 25, Griffin and George, Wembley, Middx., U.K.).
Nitrogen (high purity).
Argon (high purity).
Acetylene (from a cylinder, or can be conveniently prepared from calcium carbide and water; store over water in a suitable container to allow gas removal by syringe transfer via rubber closure).
Trichloroacetic acid (TCA), 30% w/v.

The components of the assay are conveniently prepared as two solutions, reaction mixture and creatine kinase solution, which are stable indefinitely at −20°C. Sodium dithionite solutions must be prepared daily.

Reaction mixture: composition for 4 ml, sufficient for 16 assays:

1 ml of 80 mM ATP (pH 7.4);
1 ml of 200 mM $MgCl_2$;
2 ml of 200 mM HEPES buffer (pH 7.5);
55 mg of creatine phosphate.

Creatine kinase solution: 10 mg in 5 ml of 25 mM HEPES buffer (pH 7.5) containing 50 mg of bovine serum albumin.

Sodium dithionite solution (200 mM). Prepare fresh by weighing the dry solid (174 mg) into a bottle, flush with argon, add 5 ml of 25 mM HEPES buffer (pH 7.5), previously bubbled with O_2-free gas, and seal with a rubber closure.

Assay procedures

Pipette the reaction mixture (0.25 ml), creatine kinase solution (0.05 ml), and water (0.6 ml minus the volume of nitrogenase to be added; the final liquid volume is 1 ml) into the serum bottle. Flush with nitrogen (or with argon if acetylene reduction or hydrogen evolution is to be measured), and seal with a rubber closure. Efficient gassing of the bottle at this stage is essential. This may be facilitated using a wide-bore needle (16 gauge) inserted between the rubber closure and the bottle neck, and flushing for 1–2 min. If acetylene reduction is to be measured, inject 1 ml of gas (0.1 atm) through the rubber closure. Inject dithionite solution (0.1 ml) into the mixture and incubate at 30°C for 5 min to allow temperature equilibrium and removal of the residual oxygen by reaction with dithionite. Equilibrate the pressure in the bottle to atmospheric by momentarily piercing (2–3 s) the rubber closure with a hypodermic needle. This equiiibration minimizes loss from the sampling syringe when gas samples are later removed for analysis. Transfer frozen beads of nitrogenase from low-temperature storage to a serum bottle, flush with argon, and seal with a rubber closure. When thawed, a suitable amount of nitrogenase is then injected and the mixture incubated at 30 °C for the required time (usually 5–10 min). During the incubation the bottles are shaken through 4 cm at 90–120 stokes min^{-1} to ensure adequate gas exchange. This is particularly important with nitrogen as a substrate, where inadequate shaking results in undersaturation with nitrogen and the diversion of electrons into hydrogen evolution. Inject 0.1 ml of TCA into the serum bottle to stop the reaction after the required reaction period.

(ii) Product analysis

Ammonia

Ammonia is estimated colorimetrically after microdistillation from the stopped assay mixture. Samples (1 ml) of the stopped reaction mixture are added to 1 ml of saturated potassium carbonate in a serum bottle. The bottle is immediately sealed with a rubber bung of a size to seal the reaction vessel and the test-tube used subsequently carrying a glass rod with a roughened end on which a drop of 0.5 M H_2SO_4 is suspended (Burris, 1972). Microdiffusion is carried out and ammonia in the diffusate is estimated as described in Chapter II.2, Sections b.v and b.vi.

The method described is widely used but has two disadvantages. The microdistillation step, even if the sample is shaken, is time consuming, and considerable manipulative skill is required to avoid the loss of the acid drop from the glass rod. The following method (P. Maryan and W. T. Vorley, personal communication), which avoids microdistillation and gives better reproducibility, can be used, provided that certain buffers (including Tris) are avoided in the assay.

The stopped reaction mixture is gently bubbled with air for 2 min to oxidize the acid decomposition products of dithionite, which otherwise irreversibly inhibit colour formation. The contents are then transferred to a centrifuge tube and precipitated protein removed by centrifuging at 10000 *g* for 10 min. An aliquot is removed for colorimetric estimation of ammonia as described. Colour development using this method is inhibited by different buffers to different extents, so that a calibration graph must be constructed under the experimental conditions used. At ammonia concentrations up to 1 μM linear calibrations are obtained, except that colour development is inhibited to the following extents with 25 mM buffers in the stopped reaction mixture: Bicine 4%, BES 10% and HEPES 35%. Tris, Tricine and TEC inhibit colour formation completely, even when present only as the buffer in the nitrogenase preparation.

Ethylene

Ethylene is readily separated from acetylene by GLC and detected with high sensitivity using a flame-ionization detector. The lower limit of detection is approximately 0.01 nmol ml^{-1}. Methods of analysis are described in Chapter II.3.

Hydrogen

Hydrogen evolution is conveniently measured by GLC using a katharometer as detector, or amperometrically in a specially designed reaction vessel. These methods are described in Chapter II.3.

Phosphate

The hydrolysis of ATP to ADP and Pi can be measured directly by estimation of Pi colorimetrically in the stopped reaction mixture. The ATP-regenerating system is omitted from the assay and is replaced by ATP (15 mM) and $MgCl_2$ (20 mM). Under these conditions the reaction is linear for a reasonable time, and avoids the problems with the acid lability of creatine phosphate. The method described below has high sensitivity (Ottolenghi, 1975).

Requirements

(a) 10% w/v ammonium heptamolybdate.
(b) 3% w/v ascorbic acid in 0.5 M HCl. Keep at 0°C; use within 48 h.
(c) 2% w/v trisodium citrate, plus 2% w/v sodium arsenite plus 2% acetic acid. Dissolve the citrate and arsenite in water, add the acetic acid, and make up to volume with water. **Caution:** arsenite is carcinogenic.
(d) This reagent is prepared by mixing rapidly 20 volumes of ice-cold solution (b) with 1 volume of solution (a), with vigorous mixing. Keep at 0°C and use within 4 h.

Stop the nitrogenase assay with 0.1 ml of TCA (30% w/v) and keep on ice. Samples for gas analysis (hydrogen or acetylene) may be made at this stage. Transfer the stopped assay mixture into tubes and centrifuge at 10 000 *g* for 5 min to remove precipitated protein. Transfer 0.5 ml to test-tubes pre-cooled on ice. Add 1 ml of ice-cold reagent (d), and after 8–10 min on ice add 1.5 ml of solution (c). Remove the tubes from the ice and incubate at 37°C for 12–15 min. Read the absorbance at 850 nm in a 1-cm cell; $\varepsilon \approx 2.5 \times 10^4$ l mol^{-1} cm^{-1}. The colour is stable for 24 h.

(iii) Dithionite oxidation assay

A continuous spectrophotometric assay for nitrogenase is based on the decrease in absorbance which occurs when dithionite is oxidized (Ljones and Burris, 1972). The overall reaction for the oxidation of dithionite to hydrogen sulphite is

$$S_2O_4^{2-} + 2H_2O \rightleftharpoons 2HSO_3^- + 2H^+ + 2e^-$$

but kinetic studies have indicated that many reactions proceed via the pre-dissociation of $S_2O_4^{2-}$ to give $SO_2^{\cdot -}$, which can then undergo a 1e oxidation to hydrogen sulphite:

$$SO_2^{\cdot -} + H_2O \rightleftharpoons HSO_3^- + H^+ + e^-$$

The kinetics of reduction of dye-oxidized Ac2 protein (Thorneley *et al.*, 1976) and the dependence of Av nitrogenase activity on dithionite concentration (Watt and Burns, 1977) are consistent with $SO_2^{\cdot -}$ being the active reductant for nitrogenase.

Requirements

Recording spectrophotometer.
Silica cuvettes (3 ml nominal volume, 10 mm light path with tapered tops).
Rubber closures to fit cuvettes.
Argon (high purity).
Reaction mixture and creatine kinase solutions as given under assays of nitrogenase above.
30 mM sodium dithionite solution (56 mg in 10 ml of degassed 25 mM HEPES buffer, pH 7.5).

The cuvettes are capped with rubber closures and made anaerobic either by flushing with oxygen-free argon or nitrogen for 20 min or by three cycles of evacuation with a vacuum pump and subsequent flushing with gas via a needle. Before transfer to the cuvette, water (to give a final volume of 3 ml), the reaction mixture (1.2 ml) and creatine kinase (0.15 ml) solutions are made anaerobic by flushing, but not bubbling, with oxygen-free gas. The dithionite (0.1 ml) is added by syringe and the initial absorbance at 350 nm is recorded before nitrogenase (100 μg ml^{-1} final concentration is added). The dithionite oxidation assay was described originally for Cp nitrogenase which has an apparent K_m of <15 μM (Ljones and Burris, 1972). The low concentration required to saturate this system allowed the absorbance changes to be monitored at the maximum at 315 nm ($\varepsilon_{315\,nm} = 8000$ $l mol^{-1} cm^{-1}$; Dixon, 1971) providing a very sensitive assay. Nitrogenase from other sources requires higher concentrations of dithionite for saturation (e.g. Av and Kp nitrogenase have apparent K_m values of 0.5 and 1 mM, respectively, and measurements are made at 350 nm ($\varepsilon_{350\,nm} = 1200$ or 1300 $l mol^{-1} cm^{-1}$; Smith *et al.*, 1976b, and Davis *et al.*, 1975, respectively) with a corresponding decrease in sensitivity.

(iv) Various Effects

Effect of Divalent Cation

Mg^{2+} is the most effective divalent cation. In assays from which the ATP-regenerating system is omitted, the efficiency of substituting for Mg^{2+} is Mn > Co > Fe and Ni. Inhibition was caused by Cu and Zn (Burns, 1969; Eady *et al.*, 1972). The optimum ratio of cation to ATP is 0.5 and in general the activity is only 25–33% of that with Mg^{2+}. The importance of Mg^{2+} concentration under standard assay conditions is discussed in the following section on ATP supply.

Effect of ATP supply

ATP is obligatory for nitrogenase function; GTP, UTP, ITP, and CTP are inactive. The substrate is MgATP and the apparent K_m of 0.4 mM is substantially higher than the equilibrium binding constant of MgATP to the Fe protein. MgADP is a potent inhibitor ($K_i' = 20$ μM), and the ratio of MgATP to MgADP controls nitrogenase activity (Davis and Orme-Johnson, 1976). The accumulation of ADP is minimized by incorporating creatine kinase and creatine phosphate as an ATP regenerator in the nitrogenase assay. In routine assays as described below, the creatine kinase activity is sufficient to maintain low ADP levels. Problems can arise at high nitrogenase concentrations such as are used in EPR and Mössbauer studies and if the pH or Mg concentrations are changed (Davis and Orme-Johnson, 1976). When the assay conditions are changed it is advisable to establish that the regenerating system is able to meet the ATP demand of nitrogenase, either by increasing the creatine kinase concentration or by measuring the creatine kinase activity. In addition to its effects on the ATP-regenerating system, Mg is important in nitrogenase function. At concentrations above 30 mM it is inhibitory and also inhibits by sequestering ATP as the inactive Mg_2ATP complex (Thorneley and Willison, 1974). Under most circumstances a 1 mM excess of Mg^{2+} over the nucleotide is adequate.

Effect of nitrogenase concentration

Nitrogenase activity shows a dilution effect, i.e. a disproportionately low activity at low protein concentrations (for purified nitrogenase, below 20 μg ml^{-1}). This has been attributed to the dissociation of the active nitrogenase complex at high dilutions and a $K_{complex}$ of about 3×10^7 M^{-1} obtained for Kp and Ac nitrogenase (Thorneley *et al.*, 1975). The dilution effect is more pronounced at temperatures below 20°C, at low ATP levels, and at high salt concentrations. When nitrogenase activity is first assayed under new conditions a range of protein concentrations should be assayed to establish the linear concentration range. The dilution effect can be overcome and the activity consequently increased by the addition of Fe and Mo–Fe protein (Shah *et al.*, 1972; Thorneley *et al.*, 1975). However, at higher total protein concentrations Kp1 at an eight-fold excess has been shown to produce inhibition (Thorneley *et al.*, 1975).

(v) Assay of separated nitrogenase components

When completely separated, neither the Mo–Fe nor the Fe protein has any enzymic activity alone, and neither protein has been shown to catalyse any partial reactions involving ATP or reducible substrates. A prerequisite for the assay of one component protein is that a supply of the complementary protein is available.

When a fixed concentration of the Fe protein is assayed by titration with increasing amounts of Mo–Fe protein, the activity increases initially to an optimum and then decreases (Figure 1). The optimum ratio of the two proteins is dependent both on the absolute concentration and on the source of the protein. Nitrogenase from some sources shows a sharp inhibition of substrate reduction, but not ATP hydrolysis, beyond the optimum. To measure the optimum activity of the Fe protein a series of assays are set up containing 30–40 μg of Fe protein and a range of Mo–Fe protein (40, 60, 80, 100, 150, and 200 μg per assay). The assays are most easily started by the addition of Fe protein.

When a fixed amount of the Mo–Fe protein is assayed with increasing amounts of Fe protein a normal saturation curve is obtained (Figure 1). To determine the activity of the Mo–Fe protein, duplicate assays containing 50 μg of Mo–Fe protein and a saturating level of Fe protein (600 μg) are set up. If the Fe protein has residual activity due to contamination with Mo–Fe protein, this must be subtracted from the total activity obtained.

Time course of substrate reduction

In the assays described above, substrate reduction proceeds without a detectable lag when nitrogenase is added to the assay. Conditions which result in non-linear rates of substrate reduction are an increasingly powerful tool in the study of nitrogenase enzymology and are discussed below.

The time course of nitrogenase activity in crude extracts of *R. rubrum* is non-linear with a lag of 10–15 min before linear rates of substrate reduction are obtained (Munson and Burris, 1969). The behaviour, first reported with *R. rubrum*, has subsequently been observed with *S. lipoferum* (Okon *et al*, 1977) and *R. palustris* (Zumft and Castillo, 1978). In all three organisms activity is stimulated by Mn^{2+}. As isolated, Rr2 and Sl2 are inactive when recombined with their respective Mo–Fe proteins in the nitrogenase assay

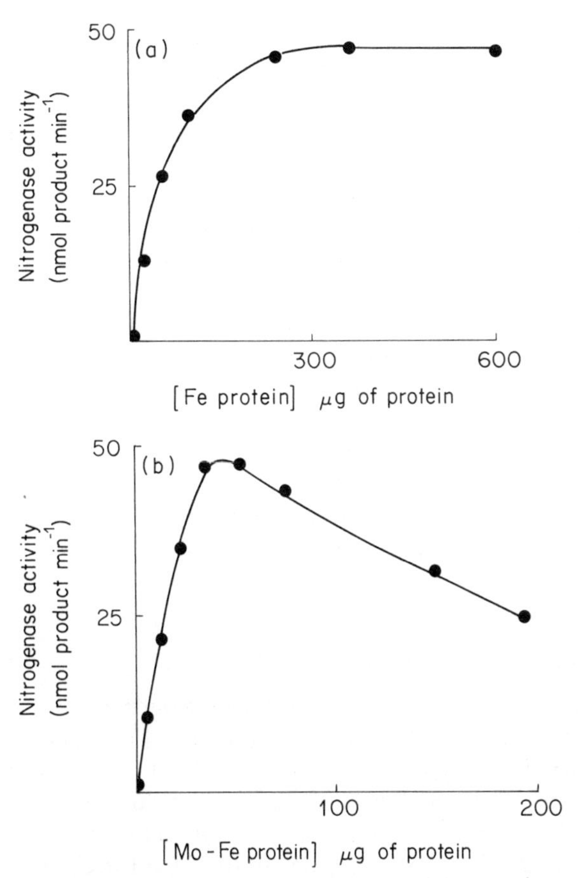

Figure 1. Dependence of nitrogenase activity on the ratio of the two component proteins. Each point is the activity obtained in separate assays as the ratio of proteins is varied. Specific activities are calculated from the maximum activity obtained and the protein concentration of the protein held constant during the titration. (a) Constant Mo–Fe protein ratio; (b) constant Fe protein concentration

system. Activity is restored when these Fe proteins are pre-incubated in the assay system containing Mn^{2+} (0.5 mM) and an activating factor isolated from crude extracts. With Sl2, linear rates of substrate reduction are obtained when Sl1 is added to such pre-incubated assays. It has been established that activating factor requires Mn^{2+} and ATP for activity. Although crude extracts of *R. palustris* show similar behaviour with respect to lags, isolated Rp2 is active when recombined with Rp1 and shows no stimulation with added Mn^{2+} and has linear rates of substrate reduction.

The heterologous nitrogenase formed between Cp2 and Kp1 exhibits a linear rate of hydrogen evolution but lags of 10 and 35 min at 30 °C for the reduction of acetylene and nitrogen, respectively (Smith *et al.*, 1976b, 1977). Similar behaviour for hydrogen evolution and acetylene reduction is observed with the homologous Kp nitrogenase at 10 or 30 °C with a 20-fold excess of Kp1 (Thorneley and Eady, 1977). For Kp nitrogenase, the total electron flow remains constant during the lag, since under these conditions hydrogen evolution shows a burst and is subsequently inhibited as acetylene is reduced. This behaviour has been attributed to different forms of the enzyme being responsible for the reduction of different substrates. It also illustrates a shortcoming of the continuous dithionite oxidation assay for nitrogenase, since the temporal redistribution of electrons to different nitrogenase substrates is not detectable in this assay.

Using the continuous amperometric assays for hydrogen evolution, a lag was observed with Cp nitrogenase at 30°C under conditions of a 500-fold excess of Cp1. This lag correlated with the turnover time in the steady state and is consistent with Cp1 and Cp2 dissociating in each catalytic cycle (Hageman and Burris, 1978).

(vi) Electron donors to nitrogenase

Electrochemical studies with purified nitrogenase and viologen dyes as mediators indicate that the potential required for half maximum activity is -460 ± 5 mV for Av1 (Watt and Bulen, 1976) and Cv nitrogenase (Evans and Albrecht, 1974). Sodium dithionite will reduce nitrogenase directly and efficiently (Bulen *et al.*, 1965) and is widely used *in vitro* assays of nitrogenase. The apparent K_m values range from 10 to 15 μM for Cp, 0.5 mM for Av, and 1 mM for Kp nitrogenase.

Natural electron donors to nitrogenase are reduced ferredoxins and flavodoxins. Ferredoxins from nitrogen-fixing organisms have redox potentials below -390 mV but range in type from 2Fe 2S through 4Fe 4S to $(4Fe\ 4S)_2$ (see Yates, 1977). They are not usually species specific in their ability to donate electrons to nitrogenase of other organisms. Flavodoxins have three available oxidation states, the quinone/semiquinone couple and the semiquinone/hydroquinone couple. The effective couple to Ac and Kp nitrogenase is the fully reduced hydroquinone/semiquinone couple (Yates, 1972), which for proteins from various sources has midpoint potentials ranging from -372 to -450 mV. Flavodoxins vary in their efficiency in transferring electrons to nitrogenase isolated from a different source.

Photoreduction of Ferredoxin and Flavodoxin

Illuminated chloroplasts can be used to transfer electrons from ascorbate-reduced dichlorophenolindophenol ($DPIPH_2$) via photosystem I to ferredoxins or flavodoxins. This system has been used as the basis of an assay to identify natural electron carriers to nitrogenase in a number of organisms; it is more effective for ferredoxins since some flavodoxins are only reduced to the semiquinone level in this system (Benneman *et al.*, 1969).

Chloroplast preparation

Chloroplasts are prepared from fresh spinach or Swiss chard leaves. Soak the leaves in iced water in the light for 2 h and remove the midribs. Blend the leaves (20 g) for 20–40 s in a Waring blender in 65 ml of ice-cold 50 mM Tris–HCl buffer (pH 7.8) containing 0.01 M NaCl and 0.4 M sucrose (Yoch, 1972). Strain the macerate through cheese cloth and then centrifuge at 2500 *g* for 1 min. Discard the supernatant and re-suspend the pellet in 50 ml of the above buffer diluted 10-fold to disrupt the chloroplasts and wash them free of ferredoxin. Collect the chloroplast fragments by centrifuging at 10000 *g* for 5 min and re-suspend in a minimum volume of 5 mM Tris–HCl buffer (pH 7.8). The oxygen-evolving capacity of these preparations is destroyed by heating at 55°C for 5 min. The chlorophyll content can be estimated spectrophotometrically (Whatley and Arnon, 1963). A sample is diluted 1 to 200 in acetone/water (4:1) and, after filtration to remove debris, the absorbance at 652 nm is measured. In a 1-cm cell multiplication of the absorbance by 5.8 gives the concentration of chlorophyll in the original suspension in mg ml^{-1}.

Coupling to nitrogenase

A typical assay system for the electron donor contains the components necessary for nitrogenase activity and to generate photoreductant, but lacks the electron carrier to mediate the transfer of electrons to nitrogenase (Benneman *et al.*, 1969).

The reaction mixture used contains the following in 1.5 ml:

10 mg of chloroplast suspension (300 μg of chlorophyll);
10 μmol of sodium ascorbate;
0.05 μmol of dichlorophenolindophenol;
5 μmol of $MgCl_2$;
40 μmol of creatine phosphate;
4 μmol of ATP;
50 μmol of HEPES buffer (pH 7.4);
0.05 mg of creatine kinase;
0.2–1 mg of nitrogenase preparation.

The reaction is carried out at 30°C in a 7.5-ml rubber-capped serum bottle under 0.2 atm of acetylene in argon illuminated at 9000 ft-candles. Nitrogenase is added by injection after the system has been made anaerobic by illumination for 10 min. Acetylene reduction is dependent on the addition of carrier and follows saturation kinetics.

An alternative method of photoreduction applicable to flavodoxins uses the photochemical reduction of deazaflavin. In the system described below, flavodoxins are rapidly reduced to the hydroquinone level. Reduced deazaflavin alone is 10% as effective as dithionite as an electron donor to nitrogenase. Under continuous illumination, Av flavodoxin reduced by deazaflavin is as effective a donor as is dithionite to Av nitrogenase in lysozyme–toluene-treated *A. vinelandii* cells (Haaker and Veeger, 1977).

Photoreduction of deazaflavin

The assay system contains glucose and glucose oxidase and catalase to remove residual oxygen from the assay components. The reduction can be carried out in the nitrogenase assay system. The additional components required are as follows:

40 μM deazaflavin (3,10-dimethyl-5-deazaisoalloxazine);
10 mM glucose;
10 u of glucose oxidase;
40 $\mu g\ ml^{-1}$ of catalase;
50 mM tricine buffer (pH 7.0).

Illuminate at 20 $mW\ cm^{-1}$ from an 8-W fluorescent tube. Both the illuminated chloroplast system and the photochemical system above will reduce methyl viologen, which can denote electrons to nitrogenase.

(e) SPECIAL TECHNIQUES

In this section, techniques necessary for the anaerobic manipulation associated with sample preparation for physical studies of nitrogenase proteins and the preparation and assay of FeMo-co are described.

(i) Vacuum line techniques

The ease with which the dithionite-reduced Mo–Fe and Fe proteins are inadvertently oxidized or oxidatively inactivated has led to the wide application of vacuum line handling techniques in physical studies on nitrogenase. The double manifold vacuum and gassing system described below, together with various accessory fittings, allows EPR tubes and Mössbauer and ultracentrifuge cells to be filled with the minimum exposure to oxygen.

The system has a number of manifold outlets fitted with sockets (Sl9) to allow flexibility in coupling to the accessories. The outlets can be connected

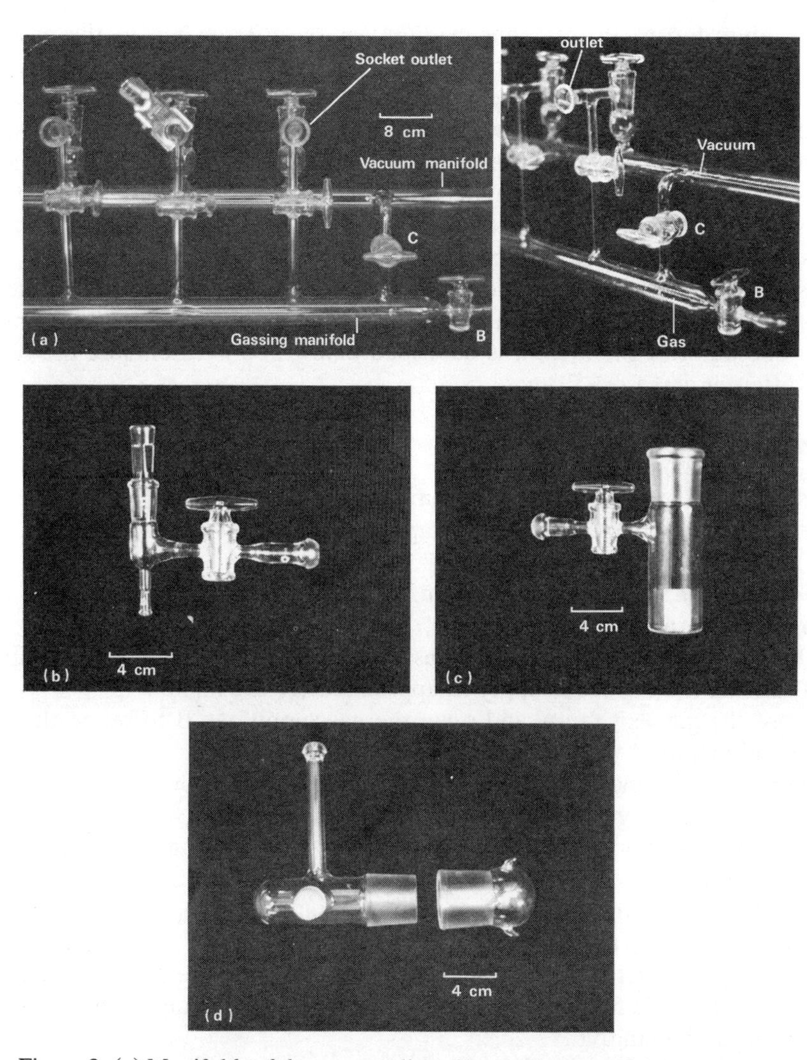

Figure 2. (a) Manifolds of the vacuum line. The gassing manifold is connected to the gas supply train (Figure 3). The socket outlets fit the attachments shown in (b)–(d), the link containing tap C allows the gassing manifold to be de-gassed initially. The gas supply tap B regulates the flow of deoxygenated gas to the gassing manifold. (b) Attachment for filling EPR tubes. A single EPR tube fits on to the attachment by a B5 cone, and is filled using a 30-cm needle passed through a rubber closure that replaces the glass stopper. (c) Attachment for de-gassing Mössbauer cuvettes. A similar but longer (25 cm) attachment is used for de-gassing a number of EPR tubes at the same time. (d) Attachment for de-gassing ultracentrifuge cells. The B40 cone allows the cell to pass through and is filled by syringe through the rubber closure

by a 2-mm tap to the vacuum manifold or by a 2-mm tap to the gassing manifold. The vacuum manifold (2 cm diameter) is connected to the pump via two wide-bore traps (1.75 cm diameter) cooled with liquid nitrogen with a vacustat pressure gauge between. The traps protect the pump from water vapour and allow the pressure in the system to be measured free of water vapour pressure. The gassing manifold is supplied with gas scrubbed by passage through two traps containing photoreduced viologen and a final water wash. The two manifolds are connected by a link with a tap (C) to allow the gassing manifold to be evacuated initially to remove oxygen. Figure 2a shows the arrangement of the two manifolds and Figure 3 the gas scrubbing train. The operation of the system is described below and is essentially the same for different accessory fittings.

Setting-up procedure

Isolate the vacuum manifold from the gassing manifold by closing the interconnecting tap C, and close the taps from the socket outlets to the vacuum manifold.

Turn on the vacuum pump, open the tap to the vacuum manifold and immerse the traps in liquid nitrogen.

Open a socket outlet tap to the gassing manifold, open the gas exit tap A and turn on the nitrogen supply to approximately 250 ml min^{-1}.

Close the gas exit tap A and open the gas supply tap B to purge the gas train and manifold with nitrogen.

Illuminate the methyl viologen traps with an 8-W fluorescent tube until they become dark green, then continue illumination.

Close the socket outlet tap, open the gas exit tap A, and close the gas supply tap B.

Check that all of the socket outlet taps are shut, open the manifold interconnecting tap C, and pump down to a pressure of 5×10^{-3} mmHg.

Close tap to isolate the pump and check that the pressure does not increase significantly over a 5-min period, indicating a leak. Open tap from the pump to the vacuum manifold.

Filling the gassing manifold

Close the manifold interconnecting tap C.

Open tap B very slowly. Check that the gas supply rate is sufficient to prevent liquid being sucked from the gas exit trap.

When the pressure has equalized, indicated by the bubbling of gas through the gas exit trap, close tap B.

Repeat the procedure of interconnecting the manifold pumping down to 5×10^{-3} mmHg and filling the gassing manifold three times. Check that the viologen traps remain reduced. Open the gas supply tap B.

Operation of the system

A number of attachments for EPR, Mössbauer, and ultracentrifuge work are shown in Figure 2b–d. They are connected to the socket outlets by cups (S19) and clips. The socket outlet tap to the vacuum manifold is opened and the attachment evacuated. The attachment is then isolated from the vacuum manifold and opened to the gassing manifold to fill with oxygen-free gas. The evacuation and gassing procedure is repeated four times, and finally the attachment is left filled with gas. The contents are then accessible using a 30-cm long stainless-steel syringe needle or, if hand access is required, the procedure is as follows.

With the socket outlets open to the gassing manifold, close the gas exit tap A and fully open the gas supply tap B. After 1–2 s remove the stopper or rubber closure from the attachment. This manipulation must be done rapidly to prevent excessive build-up of gas pressure in the manifold. This arrangement of taps provides a stream of deoxygenated gas out of the attachment and, although providing protection from ingress of air, operations such as removal of EPR tubes or plugging of an ultracentrifuge cell should be done rapidly. Using this arrangement a number of EPR tubes can be degassed in a single container and filled by a syringe fitted with a 30-cm needle after the stopper has been removed. Removal of a tube is facilitated if the needle is kinked slightly to grip on the inside of the tube. EPR tubes can be degassed singly using the attachment shown in Figure 2b and filled by syringe through the rubber septum. This attachment is particularly useful if a number of additions are to be made to a single tube, or for withdrawal of protein for measurement of enzymic activity after the spectrum has been recorded.

Buffers are degassed in a three-necked round-bottomed flask (500 ml) containing a magnet stirring bar. When 300 ml of buffer are degassed in this size of flask through four cycles of 5 min gassing and evacuation with rapid stirring, the oxygen concentration is 10^{-6} M. The degassed buffer is accessible through a rubber septum in one neck for rinsing syringes, etc. From a dip tube into the buffer in the other neck, buffer can be supplied directly to the head of a Sephadex G-25 column by balancing the gas exit tap A, and the gas supply tap B, to provide a positive pressure on the gassing manifold.

(ii) Removal of oxygen from gases

Trace amounts of oxygen can be removed from gases to be used for anaerobic work either by passage over BASF catalyst at 120 °C or hot copper turnings. They can also be conveniently removed using a gas train filled with a scrubbing solution of photoreduced methyl viologen (see Figure 3). This system additionally provides a visual indication that oxygen is being removed since the solution changes from dark green to yellow on oxidation. The

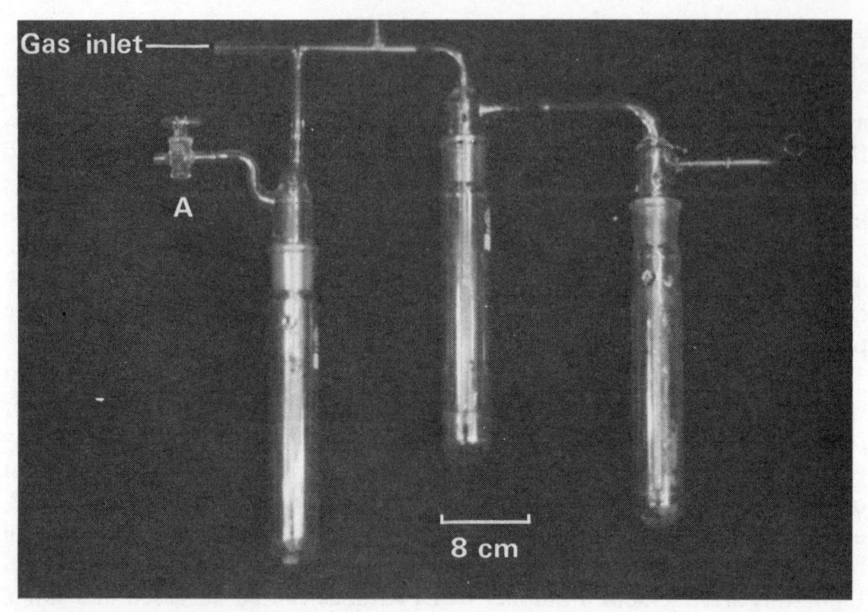

Figure 3. The gas supply train. The left-hand trap contains water and normally the gas escapes through the gas exit tap A. The two remaining traps contain a solution of photoreduced methyl viologen and connect to the gassing manifold by tap B

composition is the same as that of the stock solution of Sweetser (1967) (see below).

(iii) Measurement of trace amounts of oxygen in gases

The method described below is simple and will measure oxygen levels down to 0.2 p.p.m. It is based on the oxidation of photochemically reduced methyl viologen by oxygen (Sweetser, 1967).

Requirements

Gas reaction vessel (see Figure 4).

Spectrophotometer and a heavy black cloth of sufficient size to cover the gas reaction vessel and the sample compartment.

solution (a): 1 mM proflavin hydrochloride (56 mg in 100 ml of water).

solution (b): 10 mM methyl viologen (0.54 g in 200 ml of water).

solution (c): 0.25 M EDTA (18.6 g of the disodium salt in 200 ml of water; pH 6.5).

solution (d): 0.2 M phosphate buffer (pH 6.5).

Prepare a stock solution by mixing 80 ml of solution (a), 200 ml of solution (b), 160 ml of solution (c), and 200 ml of solution (d) and diluting with water to 2 l. Store in a dark bottle in a refrigerator; the solution is stable for several months.

Procedure

Pipette 15 ml of stock solution into the gas reaction vessel and connect the ball-joint to the manifold of the vacuum line. Evacuate the vessel, close the tap, and remove from the manifold. Illuminate with an 8-W fluorescent lamp until the solution is a deep green colour. Shake the vessel vigorously for 1 min to spread the stock solution over the walls to remove absorbed oxygen. It is important to cover the vessel with a dark cloth from this point in order to prevent further reduction of the methyl viologen. Read the absorbance at 570 nm in a spectrophotometer; it will probably be necessary to cover the vessel and cell compartment with a heavy black cloth during this measurement. Take care to avoid trapping bubbles in the cuvette. Repeat the shaking until

Figure 4. A gas reaction vessel (500 ml)

a constant absorbance is obtained. Reconnect the ball-joint of the vessel to the vacuum line, flush, and evacuate the connecting link three times before finally filling the vessel with gas by opening the tap. When the pressure in the vessel is atmospheric, close the tap and remove the vessel from the vacuum line. Shake the vessel vigorously for 30 s, read the absorbance at 570 nm and repeat this process until a constant absorbance (to within 0.005 absorbance unit) is obtained. Reconnect the vessel to the vacuum line and evacuate the gas, then repeat the determination twice more. There is no necessity to replace the stock solution since the reductant, EDTA, is present in excess.

The oxygen content of the gas in parts per million is given by $[20\,500/(V-15)]\delta A$, where V is the volume of the gas reaction vessel and δA is the total decrease in absorption observed on shaking the vessel containing the gas.

(iv) Preparation of dithionite-free protein

Dithionite is removed from protein solutions by anaerobic Sephadex G-25 chromatography. The column is arranged so that a glass dip tube in the buffer degassing flask on the vacuum line can be used to supply degassed buffer to the column head. Residual oxygen in the column is scavenged by loading 5 ml of 100 mM dithionite on to the column and developing it with degassed buffer. When the effluent will no longer reduce methyl viologen papers, the protein sample is loaded on to the column. The column is developed with degassed buffer from the vacuum line, and the protein collected into a glass syringe connected to the column via a thick-walled plastic tube. The syringe is connected with the plunger fully depressed, i.e. with no gas headspace and the effluent containing the protein displaces the plunger. A glass bead in the syringe is used for mixing the contents.

(v) Stopped-flow spectrophotometry

This technique has been used to study pre-steady-state electron transfer from the Fe protein to the Mo–Fe protein and the reduction of dye-oxidized Fe protein by dithionite. A commercial stopped-flow apparatus (American Instrument Co.) modified to achieve a high degree of anaerobicity has been developed (Thorneley, 1974). The modifications described below result in a 100-fold improvement and the rate of oxidation of reduced methyl viologen in the observation chamber shows that less than 5 μM oxygen leaks in over a 30-min period.

Modifications

The observation and mixing chamber is surrounded by a solution of sodium dithionite [0.1 M in 1 M Tris–HCl buffer (pH 7.4)] enclosed in a black acrylic plastic jacket. The dithionite solution is pumped through a closed thermostated circulation loop.

The PTFE tubing connecting the drive and stopping syringes is shrouded with vinyl tubing through which nitrogen is passed.

The Kel-F pistons of the drive syringes have nitrogen flushing through a hole drilled in the centre which emerges through four holes drilled at 90° just behind the rubber O-ring. The gas stream decreases the rate of diffusion of oxygen past the O-rings and absorption on to the syringe walls when the piston is depressed.

When not in use the reaction chamber is filled with a buffered solution of dithionite.

(vi) Electron paramagnetic resonance spectroscopy

Spectra are measured on a spectrometer fitted with facilities for helium cooling (10 K) and accessories for low-temperature measurement. Samples (0.25 ml) are contained in the bottom of a quartz capillary tube (internal diameter 3–4 mm) which are degassed and flushed with purified gas before the sample is introduced, using a gas-tight syringe and a 30-cm stainless-steel needle. The sample is frozen by immersion of the tube in liquid nitrogen (freezing time 20–30 s) or, for more rapid freezing, by complete immersion of the sample in an isopentane bath chilled to −140 °C (freezing time 1 s). If liquid nitrogen is used the tip of the tube should be frozen first in order to minimize tube breakage. Approximate protein concentrations required to obtain EPR spectra are 2 mg ml^{-1} for the Mo–Fe protein and 1 mg ml^{-1} for the Fe protein. The tubes are stored at liquid nitrogen temperature.

(vii) Mössbauer spectroscopy

Protein samples for Mössbauer spectroscopy are enriched in ^{57}Fe by growing the organism on medium from which the trace metals are removed (see below) before ^{57}Fe is added. Spectra are obtained on samples (0.7 ml) in flat-faced polyethylene holders with internal dimensions of $3 \times 14 \times 20$ mm. The holders are degassed on a vacuum line, filled with sample, and frozen by plunging into liquid nitrogen. Forceps are used to transfer the holder to the liquid nitrogen and to hold it upright during the freezing. Samples are stored at liquid nitrogen temperature. Protein concentrations required are 50 mg ml^{-1} for the Mo–Fe protein and 30 mg ml^{-1} for the Fe protein.

(viii) Ultracentrifugation

Assembled ultracentrifuge cells are evacuated and gassed three times on a vacuum line using the attachment shown in Figure 2c, and the cell is filled using a syringe with the needle passed through the rubber closure. The vacuum line is then set to provide a positive gas pressure within the attachment (as described above in the section on operation of the system), the rubber stopper removed, and the cell plugged. Conventional Schlieren interference optics and UV/visible absorbance using a photoelectric scanning system have been used in ultracentrifuge studies. Data must be interpreted with caution since prolonged storage of Cp1 in liquid nitrogen results in the formation of aggregates (Dalton *et al.*, 1971), and with Av1 freeze–thawing produces dissociated species (Swisher *et al.*, 1977).

(ix) Removal of trace metals from growth media

Isotopic enrichment with ^{57}Fe is essential for Mössbauer studies of nitrogenase and is useful in EPR studies to enable an EPR signal to be assigned to a particular protein. ^{99}Mo has also been used to probe the extent of the participation of molybdenum in the EPR active centre of native Mo–Fe protein. Enrichment is achieved by growing the organism on a medium enriched ($>90\%$) with the appropriate isotope.

The cultures are grown in glassware previously cleaned in nitric acid (50% v/v) for 24 h and rinsed exhaustively with deionized water. Trace metals in media constituents are removed by the following treatments.

Dissolve the buffer components (sufficient for 20 l of medium) in 1 l and add 100 g of alumina (aluminium oxide). Transfer to a 2-l flask and autoclave for 15 min at 15 lb in^{-2}. Remove from the autoclave while still hot, and swirl the alumina through the contents of the flask. Stand overnight before filtering through an acid-washed sintered funnel covered with Whatman No. 42 paper. Stand for several days, and remove any precipitate of alumina which forms during this time by centrifuging at 30 000 *g* for 1 h. It is important to remove this precipitate as it will complex the ^{57}Fe added subsequently.

Dissolve the carbon source and other media constituents (sufficient for 20 l, but excluding trace metals) in 1 l of water. Extract twice with a saturated solution of 8-hydroxyquinoline in chloroform, sufficient to form a 3 cm layer in the bottom of the vessel. During extraction (4–6 h), stir the mixture rapidly with a magnetic stirrer. Extract with chloroform alone to remove the colour from the medium. Add the isotopically enriched metal to the medium and sterilize normally. With Hino and Wilson medium (Chapter II.1) this procedure decreases the Fe content of the complete medium to 0.15 mg l^{-1}. To decrease this further, *K. pneumoniae* was grown on one quarter the normal buffer concentration and the pH of the culture regulated automatically (Smith and Lang, 1974).

(x) Isolation and assay of iron and molybdenum cofactor (FeMo-co)

FeMo-co is isolated from the Mo–Fe protein by precipitation and extraction with *N*-methylformamide (Shah and Brill, 1977). The assay of activity is based on the ability of the cofactor (isolated from any Mo–Fe protein) to restore nitrogenase activity to crude extracts of *A. vinelandii* which contain active Av2 but Av1 which is inactive because it lacks FeMo-co.

Requirements for isolation of FeMo-co

Mo–Fe protein [*ca.* 15 mg ml^{-1} in 25 mM Tris–HCl buffer (pH 7.4) containing 0.25 M NaCl].
0.2 M Na_2HPO_4.
1 M citric acid.
0.1 M sodium dithionite.
NN-Dimethylformamide (DMF).
N-Methylformamide (NMF).

NMF and DMF are vacuum distilled before use and contain 1.2 mM dithionite added as a 0.1 M aqueous solution. Both DMF and NMF are neurotoxins and should be used with care in a fume cupboard.

Procedure for isolation of FeMo-co. All manipulations are carried out under hydrogen at 0–4 °C in a centrifuge tube. Dilute the Mo–Fe protein solution (4 ml) with 8 ml of water, and add citric acid solution to a final concentration of 15 mM. Mix thoroughly and stand for 3 min. Add Na_2HPO_4 solution (25 mM final concentration), mix and stand for 25 min. The precipitated protein is recovered by centrifuging at 8000 *g* for 10 min, washed twice by re-suspension in DMF (8 ml) and recovered by centrifuging at 8000 *g* for 10 min. The FeMo-co is extracted into NMF by re-suspending the precipitated DMF-washed pellet and washing with three 4-ml aliquots of NMF. The precipitate is removed by centrifuging at 8000 *g* for 10 min.

Requirements for assay of FeMo-co

FeMo-co in NMF.

Cell-free extracts of *A. vinelandii* mutant UW-45 or *K. pneumoniae* mutant UN 109 from cultures derepressed in medium containing molybdate (see Chapter II.10). Alternatively, wild-type *A. vinelandii* derepressed in medium containing tungstate (5 μM) in place of molybdate can be used.

Procedure for assay of FeMo-co. Incubate various amounts of FeMo-co with the crude extract (6–7 mg of protein) for 30 min at room temperature. Remove an aliquot and assay for acetylene reduction in a standard assay containing excess of Fe protein. Reactivation of extracts is dependent on the concentration of FeMo-co and follows saturation kinetics (Shah and Brill, 1977).

(f) ACKNOWLEDGEMENTS

Many of the techniques described in this chapter have been modified in the light of experience gained over a number of years. The contributions of my colleagues to this process, in particular Drs B. E. Smith, R. N. F. Thorneley, and R. C. Bray, and Messrs. P. Maryan and W. Vorley, are gratefully acknowledged.

Methods for Evaluating Biological Nitrogen Fixation
Edited by F. J. Bergersen

K. J. F. Farnden
Department of Biochemistry,
University of Otago,
Dunedin, New Zealand
and
J. G. Robertson*
Division of Applied Biochemistry,
DSIR, Palmerston North,
New Zealand.

7

Methods for Studying Enzymes Involved in Metabolism Related to Nitrogenase

*Written while on leave at the Charles F. Kettering Research Laboratory, Yellow Springs, Ohio, U.S.A.

(a) INTRODUCTION

In this chapter we present and discuss methods for assaying some of the enzymes involved in metabolism relating to nitrogenase, and also procedures for assaying other enzymes likely to be of interest to workers in the area of biological nitrogen fixation (e.g. hydrogenase, nitrate reductase, and nitrite reductase). In addition, the first section gives methods for the preparation of cell-free extracts from various fractions of legume nodules, free-living nitrogen-fixing bacteria, and cyanobacteria. We have also included assays for enzymes which can be used as markers for subcellular fractions. The last section gives a brief outline of procedures for the measurement of respiration (O_2 uptake and CO_2 evolution) and details of procedures which can be used for the measurement of ATP, ADP, and AMP.

Assays for enzymes relating to nitrogenase require the preliminary preparation of microorganisms or plant tissue. Although growth of free-living microorganisms under carefully controlled conditions (see Chapter II. 1) is taken for granted, such conditions for the growth of plants are often more difficult and expensive to achieve. Growth cabinets do not always provide optimum conditions for growth of plants owing to low light intensity and unnatural spectral quality, uneven heat distribution, inadequate humidity control, or possibly unsatisfactory growth media. Although growth cabinets provide opportunities for research not possible with field-growth plants, care should be taken to ensure that plants grown under artificial conditions are similar to plants grown in the field.

(b) PREPARATION OF CELL-FREE EXTRACTS

Assays for enzymes in fractions from symbiotic associations or nitrogen-fixing organisms should be carried out, at least initially, on freshly grown tissue or cells which have been ruptured and fractionated as quickly as possible after harvesting without freezing or drying. The medium used when rupturing cells will depend on what enzymes are being assayed. The most commonly used buffers are *N*-2-hydroxyethylpiperazine-*N'*-2-ethanesulphonic acid (HEPES), imidazole, phosphate, piperazine-*NN'*-bis(2-ethanesulphonic acid) (PIPES), 2-{[2-hydroxy-1,1-bis(hydroxymethyl)ethyl]amino}ethane sulphonic acid (TES), and tris(hydroxymethyl)aminomethane (Tris). Homogenizing media often contain such enzyme protectants as reducing agents, sulphydryl protectants, protease inhibitors, metal complexing agents, and substances such as polyvinylpyrrolidone, which complexes with phenolics present in extracts of some plants (Loomis, 1969, 1974). In some instances it may not be possible to use a medium, when rupturing cells, which is also suitable for use in the enzyme assays. This problem may be overcome by passing cell extracts through Sephadex G-25 equilibrated with a suitable buffer. Sucrose and other low molecular weight materials of plant or bacterial origin which might interfere with the enzyme assays are also removed by this treatment.

An incubation temperature of 25 °C has been specified for a number of assays when the enzyme source is primarily plant tissue. In other cases, it would be appropriate to incubate at the temperature at which the source material was produced. The International Union of Biochemistry, however, currently recommends 30 °C as the measuring temperature.

(i) Preparation of bacteroid and plant cytosol fractions from legume root nodules

Fractionation of legume nodule tissue into bacteroid and plant cytosol fractions is an essential prerequisite to elucidating the biochemical basis of the symbiosis (Scott *et al.* 1976). Ideally, such fractionations should yield bacteroids which are ultrastructurally intact and metabolically active and a plant cytosol fraction which is free from proteins arising from ruptured bacteroids or plant organelles. Perhaps the most critical method for examining the efficiency of any fractionation is to determine whether enzymes of bacteroid origin are present in the plant cytosol fraction and *vice versa* (Robertson *et al.* 1975b; Planqué *et al.* 1977). A satisfactory separation is most likely to be achieved if relatively young nodules are used for studies of enzymes in different fractions. Nodules undergo a phase of development following infection of legume roots by rhizobia, in which the ratio of nitrogen fixation to nodule weight increases to a maximum over a period of 3–5 weeks, remains approximately constant for 3–5 weeks, and finally declines over a period of 3–8 weeks. Following the initial period of development the nodules are observed to contain increasing amounts of senescent tissue (Syōno *et al.*, 1976) which arises from the death of the plant cells first infected by rhizobia. Nodules subjected to stress are likely to contain increased amounts of senesced tissue. Whenever possible, nodules which have just reached the maximum rate of nitrogen fixation and which have been grown under the best possible environmental conditions should be used for studies of levels of enzymes in various fractions. When using such nodules, the method described below yields bacteroid and plant cytosol fractions which are relatively free from cross contamination.

Apparatus

Pestle and mortar.

Centrifuge (Sorvall RC-5 with SS-34 or HB-4 rotors or equivalent).

Centrifuge tubes, 50 ml.

Miracloth (Chicopee Mills Inc. New York, U.S.A.), or cheesecloth.

Gel filtration column (Pharmacia K9/30 containing Sephadex G-25 or equivalent).

Materials and reagents

Phosphate buffer (0.05 M) adjusted to pH 7.2 with KOH and containing 0.4 M sucrose and 10 mM dithiothreitol (homogenizing medium).

Procedure. All steps in the preparation of plant and bacteroid cytosol fractions are carried out at 0–4 °C. Nodules (approximately 20 g) are picked directly into homogenizing medium, quickly blotted dry on filter-paper to remove excess of liquid, weighed, and then crushed in 2 volumes (w/v) of fresh homogenizing medium using a pestle and mortar. The brei is filtered through two layers of Miracloth, previously wetted with homogenizing medium, and then centrifuged at 250 *g* for 5 min to yield a pellet containing starch granules and plant cell wall debris. The supernatant is removed and centrifuged at 8000 *g* for 5 min to yield a pellet of bacteroids. The post-8000 *g* supernatant is decanted and then centrifuged at 30 000 *g* for 30 min to give a pellet containing a small proportion of bacteroids and also plant mitochondria and membranes and a supernatant containing the soluble proteins from the plant cytoplasm (plant cytosol fraction).

The plant cytosol fraction can be assayed directly or, preferably, passed through Sephadex G-25 equilibrated with a buffer suitable for use in subsequent enzyme assays. The protein fraction from the Sephadex column is assayed as soon as possible or, if necessary, frozen in liquid nitrogen and stored at −20 °C.

The bacteroid pellet is suspended in 10 volumes (v/v) of homogenizing medium and centrifuged at 250 *g* for 5 min to sediment most of the remaining starch granules. The supernatant is centrifuged at 8000 *g* for 5 min to give a pellet of bacteroids which are only slightly contaminated with starch granules and plant organelles and membranes. The bacteroids can be re-suspended in homogenizing medium and the two centrifugation steps repeated if necessary. The final bacteroid pellet is re-suspended in a buffer suitable for the intended enzyme assays and the bacteroids ruptured and fractionated as described for free-living nitrogen-fixing bacteria (see Section 7.b.iii).

General comments. The plant cytosol fraction should be a deep red colour. Browning or development of cloudiness on standing is probably due to reactions caused by the presence of phenols (Loomis, 1969, 1974), which can be removed from nodule breis by including polyvinylpyrrolidone in the homogenizing medium (Evans *et al.*, 1972). Extracts of senescent nodules are also likely to be brown owing to the presence of haem degradation products.

It is advisable to test the effect of different buffers and concentrations of reducing agents in the homogenizing medium on enzyme activity if this is low or unstable (see General comments on glutamate synthase assay, for example).

The bacteroid fraction may contain some bacteroids still surrounded by peribacteroid membranes (Robertson *et al.*, 1978). The numbers of membrane-enclosed bacteroids present will depend on how vigorously the bacteroids have been re-suspended during the washing and differential centrifugation stages. Preparations of bacteroids obtained by crushing nodules using a pestle and mortar are likely to contain more membrane-enclosed bacteroids than are bacteroid preparations from nodules homogenized with a Waring blender. It must be emphasized that bacteroid preparations do not necessarily contain a homogeneous population of organisms. Bacteroid preparations contain organisms which behave differently when subjected to gradient fractionation techniques (Ching *et al.*, 1977: Sutton and Mahoney, 1977). Levels of enzymes relating to nitrogenase in preparations of bacteroids should therefore be interpreted with caution.

Bacteroids capable of fixing nitrogen can be isolated from legume nodule tissue by breaking and fractionating nodules under anaerobic conditions and in the presence of dithionite (Bergersen and Turner, 1967; Evans *et al.*, 1972; Planqué *et al.*, 1977). Bacteroids showing a high degree of viability have been obtained from nodules using protoplasting techniques (Gresshoff *et al.*, 1977). It seems likely that this approach has considerable potential for studies of the biochemistry of the legume nodule symbiosis.

(ii) Subcellular fractionation of legume nodules

The method described above for preparing bacteroid and plant cytosol fractions from legume nodules is suitable for comparing levels of enzymes in these fractions. However, nodules consist of both infected and non-infected tissue and contain various organelles and membranous compartments which are likely to play significant roles in nodule function. Furthermore, the bacteroids are not a homogeneous population of microorganisms. A more detailed understanding of the biochemistry and physiology of legume nodule function requires the correlation of nodule ultrastructure with the localization of enzymes *in vivo* and in isolated fractions. A fractionation scheme is presented for lupin nodules (Robertson *et al.*, 1978) from which a variety of soluble, membranous, organelle and bacteroid fractions can be prepared. In particular, this fractionation has been devised to isolate peribacteroid membranes and bacteroid envelope inner membranes. Although this protocol requires modification for other legume nodules it gives a framework upon which nodule fractionations can be built. It must be emphasized that the fractionation schemes have been worked out using electron microscopic techniques to identify the various membranous components, the metabolic activity of which has not been studied in depth (Robertson *et al.*, 1978).

Apparatus

Pestle and mortar.

Centrifuge rotors and tubes:

Sorvall HB-4 rotor with 50-ml polypropylene tubes.
Beckman 30 rotor with Oak Ridge-type, 30-ml, polycarbonate, screw-cap tubes.
Beckman SW 25.1 rotor with 1 × 3 in cellulose tubes.
Beckman SW 27 rotor with 1 × 3.5 in cellulose tubes.
Refractometer.
Miracloth (Chicopee Mills Inc.).
French pressure cell (Aminco or equivalent).

Materials and reagents

Tris–HCl, 10 mM (pH 8.0).
Sucrose, 16%, containing 50 mM Tris–HCl (pH 8.0) and 2 or 10 mM dithiothreitol.
Sucrose, 30, 35 or 55%, containing 50 mM Tris–HCl (pH 8.0).
Sucrose, 8, 12, 17, 20, 35, 40, 45 or 47%, containing 10 mM Tris–HCl (pH 8.0).
Sucrose, 12%, containing 10 mM Tris–HCl (pH 8.0), 0.1 mM $MgCl_2$ and 20 μg ml^{-1} DNase.

Procedure. The procedure for preparing various bacteroid, plant organelle, membrane, and soluble-protein fractions from the nodules of *Lupinus angustifolius* is summarized in Figures 1 and 2 and is presented in greater detail elsewhere (Robertson *et al.*, 1978). For each 7 *g* wet weight of nodule tissue one centrifuge tube is used at each stage of the fractionation. Throughout the procedures sucrose concentrations, expressed as grams per 100 g of solution, are monitored at 20 °C using an Abbé refractometer. The preparations are carried out at 0–4 °C. Centrifugal forces are given for R_{max}. The HB-4 rotor is used for all centrifugations at 16000 *g* or below. The SW 27 rotor is used for all centrifugations over 16000 *g* involving step or linear gradients, except for the fractionation of SN_4 (Figure 1) at 60000 *g* for 30 min and for the fractionation of SN_6 (Figure 2) at 73000 *g* for 75 min, when the SW 25.1 rotor is used. The Beckman 30 rotor is used for all other centrifugations.

The times between picking nodules and completing the isolation of the membranes is approximately 24 h. The yield of peribacteroid membranes is approximately 20 μg of protein per gram of nodule (wet weight) from 18-day plants. The yield of bacteroid envelope inner membranes is approximately 15 μg of protein per gram of nodule (wet weight) from 18-day plants.

General Comments. The isolation of membrane-enclosed bacteroids depends on gentle crushing of nodule tissue, on gentle re-suspending of the bacteroids during washing, and on maintenance of appropriate osmotic conditions. Using the techniques described above we have also isolated membrane-enclosed bacteroids from a tropical legume, *Dolicos lab lab*. Membrane-

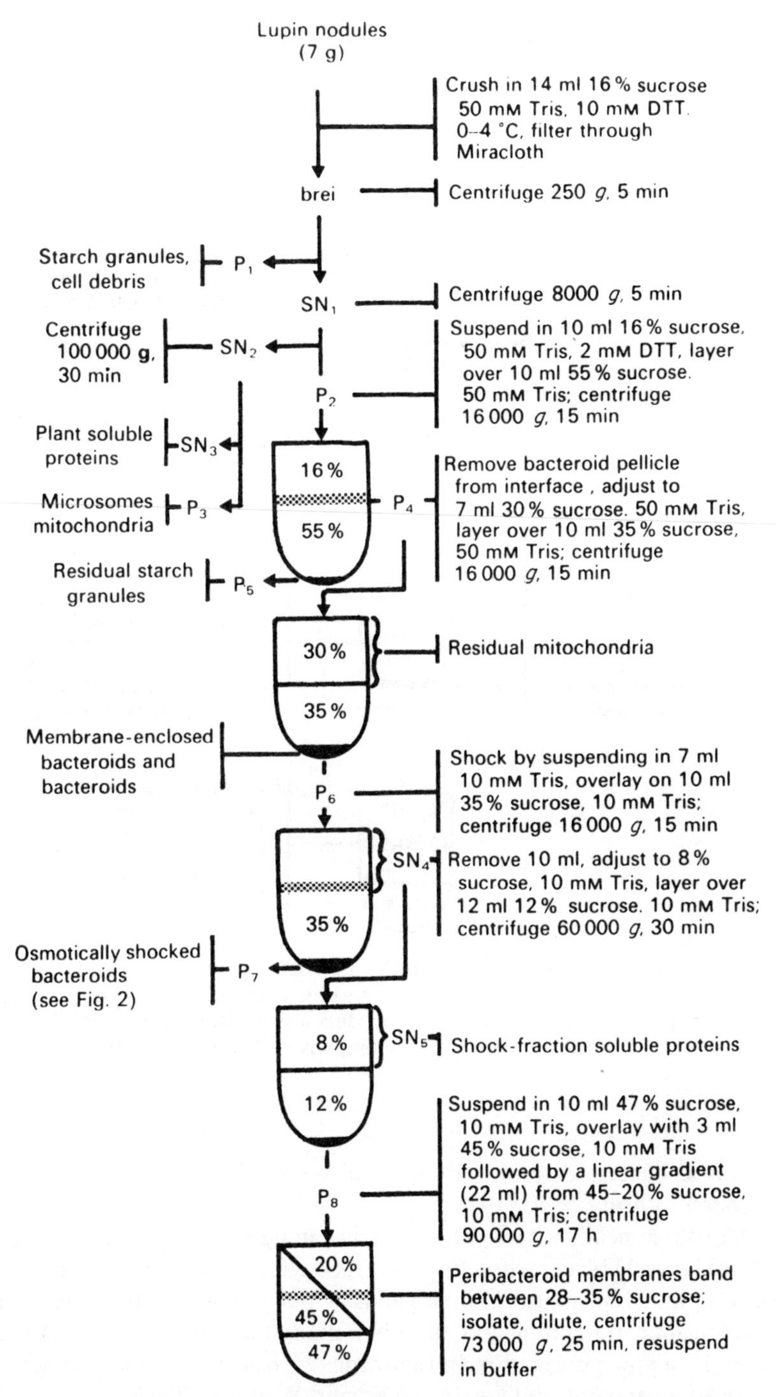

Figure 1. Schematic outline of the protocol for the preparation of various cell fractions, in particular the peribacteroid membranes from lupin nodules. (Reproduced from Robertson *et al.*, 1978, by permission of the Company of Biologists Ltd)

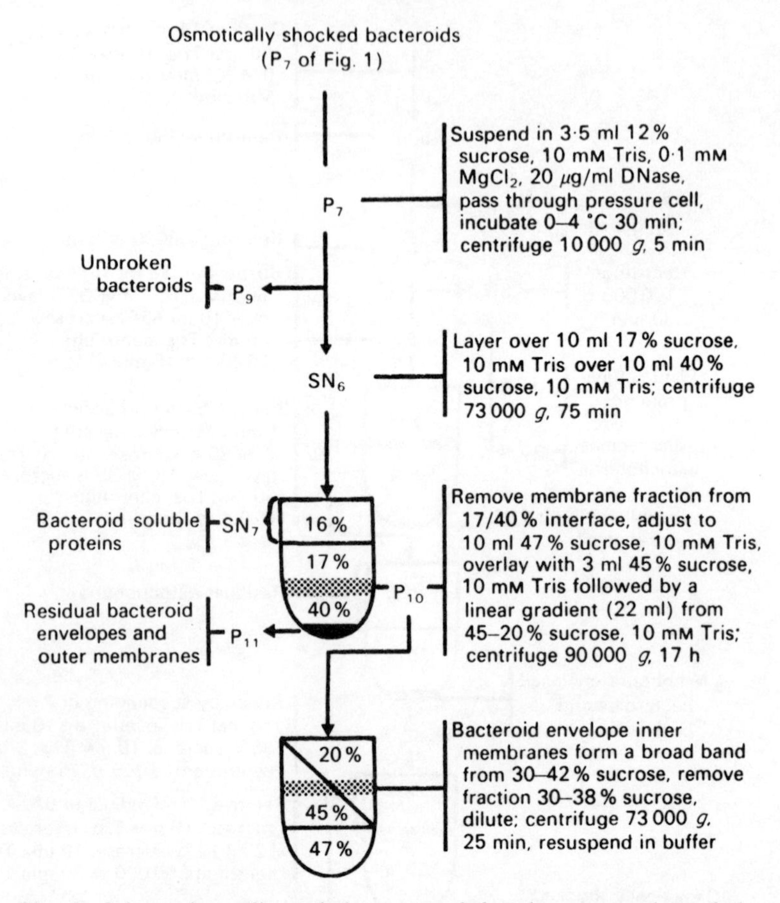

Fig. 2. Schematic outline of the protocol for the preparation of the bacteroid-soluble protein fraction and the bacteroid envelope inner membranes from lupin nodules. (Reproduced from Robertson *et al.*, 1978, by permission of the Company of Biologists Ltd)

enclosed bacteroids and peribacteroid membranes have been isolated from soybean (Verma *et al.*, 1978).

Morphologically intact mitochondria can be prepared from the supernatant SN_2 (Figure 1) by centrifugation at 12000 *g* for 10 min. Further centrifugation of SN_2 at 50000 *g* for 15 min yields a pellet containing vesicles of rough endoplasmic reticulum and of other plant membranes also. Mitochondria capable of phosphorylation and showing respiratory control have been isolated from soybean root nodules (Muecke and Wiskich, 1969).

Bacteroids prepared using the sucrose gradient techniques described are plasmolysed (Robertson *et al.*, 1978). Although we have not examined the effect of lysozyme on these bacteroids, it seems likely that the yield of inner membranes could be improved by such treatment (Planqué *et al.*, 1977).

(iii) Fractionation of free-living nitrogen-fixing bacteria

Nitrogen fixation occurs in a large number of different microbial species (LaRue, 1977; Mulder and Brotonegoro, 1974). Within these species, nitrogenase and enzymes largely associated with ammonia assimilation have been most extensively studied in the anaerobe *Clostridium pasteurianum*, the facultative anaerobe *Klebsiella pneumoniae*, and the aerobes *Azotobacter vinelandii* and *Rhizobium* species (Winter and Burris, 1976; Yates, 1977; Shanmugam *et al.*, 1978). Pathways of carbon metabolism relating to nitrogenase activity and membrane-bound enzyme systems have not been widely studied in these organisms. This situation may soon change as a result of interest in energy requirements for biological fixation and in mechanisms whereby microorganisms establish intracellular anaerobic conditions (Benemann and Valentine, 1972; Postgate, 1974; Haaker and Veeger, 1977; Andersen *et al.*, 1977).

Methods for preparing cell fractions from microorganisms (Coakley *et al.*, 1977) and in particular from nitrogen-fixing microorganisms vary considerably. It must be emphasized that different methods may lead to different conclusions regarding the properties of individual enzymes (Ljones, 1974; Haaker *et al.*, 1977; Schneider and Schlegel, 1977). For example, an enzyme may appear membrane bound or soluble depending upon whether a pressure cell or ultrasonic treatment has been used to break the cells.

Aerobic conditions are commonly employed for preparing cell extracts of nitrogen-fixing bacteria. Provided that cells are broken quickly and efficiently in the presence of an adequately buffered medium at 0–4 °C most enzymes associated with nitrogenase are active in crude extracts. In some cases it may be advantageous to prepare cell extracts under anaerobic conditions so that nitrogenase and hydrogenase activities can be assayed in the same preparation (Planqué *et al.*, 1977; Schneider and Schlegel, 1977). Such conditions may also uncover important facts relating to the properties of other enzymes. Although a discussion of anaerobic techniques is outside the scope of this chapter (see Chapters II. 5 and II. 6), these techniques are becoming more flexible and applicable to general rather than specific use. The following method for the isolation of bacterial cell envelope and cytosol fractions can be carried out under aerobic or anaerobic conditions.

Materials and reagents. The following buffers can be used for preparing cell extracts from nitrogen-fixing and related organisms. The selection of any particular buffer will depend upon what enzymes are to be assayed.

HEPES, 0.05 M, adjusted to pH 7.5 with KOH (Upchurch and Elkan, 1978).
Imidazole, 0.01 M, adjusted to pH 7.0 with HCl and containing 1.0 mM $MnCl_2$ or 20 mM $MgCl_2$ for *Azotobacter vinelandii* (Tronick *et al.*, 1973).
Imidazole, 10 mM, adjusted to pH 7.15 with HCl and containing 10 mM $MnCl_2$ and 2 mM 2-mercaptoethanol (Brenchley, 1973).
Phosphate, 50 mM, adjusted to pH 7.4 with KOH and containing 2 mM 2-mercaptoethanol (Bergersen and Turner, 1978).
Tris, 50 mM, adjusted to pH 8.0 with HCl (Okon *et al.*, 1976a).
Buffer solution containing 0.1 mM $MgCl_2$ and 20 μg ml^{-1} DNase.

Procedure. Organisms are harvested from log or stationary phase cultures under aerobic or anaerobic conditions at 0–4 °C by centrifugation at 10 000 *g* for 10 min. The pellets are washed once by re-suspending in buffer and re-centrifuging. The final pellet is re-suspended in 2–5 volumes of buffer containing 0.1 mM $MgCl_2$ and 20 μg ml^{-1} DNase and passed through a pre-cooled French press at 16 000–20 000 lb in^{-2} (1125–1406 kg cm^{-2}) once or twice, depending upon the organism. The preparation containing broken organisms is held at 0–4 °C for 30 min and then centrifuged at 10 000 *g* for 10 min. Alternatively, the suspension of cells is sonicated at maximum output with alternating 30-s intervals of sonication and cooling for a total sonication time of 2 min and then centrifuged. The supernatant is carefully removed without disturbing the pellet and re-centrifuged at between 30 000 and 100 000 *g* for 30 min to yield a supernatant containing the cytoplasmic proteins and a pellet containing the cell envelopes. The cell envelopes are washed by re-suspending in buffer, then re-centrifuged, initially at 10 000 *g* for 10 min to remove a small proportion of unbroken cells which are discarded, and finally at between 30 000 and 100 000 *g* for 30 min. The pellet of cell envelopes is re-suspended in buffer at approximately 1 mg ml^{-1} of protein and used for enzyme assays. The supernatant containing the cytoplasmic proteins can be assayed directly or, preferably, passed through Sephadex G-25 equilibrated with a suitable buffer for the subsequent enzyme assay. If necessary, the eluate from Sephadex G-25 can be concentrated using ultrafiltration and then stored at −20°C following freezing in liquid nitrogen. It is important to determine whether freezing under these conditions affects enzyme activity. Slow freezing of cell fractions in a deep-freeze is not recommended.

General Comments. Nitrogen-fixing organisms commonly produce large amounts of extracellular polysaccharides which give rise to soft, gummy pellets when cultures are centrifuged. These polysaccharides can be partially removed by washing the cells repeatedly in buffer of low ionic strength, possibly containing up to 10 mM EDTA, or by very brief homogenization using a Waring Blender (Waring Products Division, New Hartford, Conn, U.S.A.) or Polytron (Kinematica GmbH, Lucerne, Switzerland). The conditions of homogenization should be sufficient to disperse the polysaccharide without causing damage to the cells. Blending for up to 45 s appears to have no effect on the viability of some bacteria (Weppner *et al.*, 1977). Alternatively, polysaccharide production by bacteria can be manipulated by altering the conditions of growth (Courtois *et al.*, 1979).

The efficiency of breaking the cells using sonication or French pressing can be determined using light microscopy or by following the release of protein in the 10000 *g* (10 min) supernatant. Total protein can be determined using the method of Lowry *et al.* (1951; Chapter II. 2) following precipitation of proteins in both samples and standards using 5% trichloroacetic acid.

The rate of release of enzymes from cells broken under different conditions should also be determined. The choice of buffers in the breaking medium and in the subsequent assay will depend upon the effect of those buffers on the activities of various enzymes of interest. Imidazole–HCl buffer is commonly used in assays of glutamine synthetase activity (Stadtman and Ginsberg, 1974). Brown and Dilworth (1975) have reported that phosphate and HEPES buffers give higher activities than imidazole and Tris for assays of glutamate synthase from *Rhizobium leguminosarum*. The strength and pH of the breaking buffer is also important since a fall in pH can occur during breaking of microorganisms, resulting in the subsequent sedimentation of proteins which might otherwise have been regarded as soluble (Haaker and Veeger, 1977; D. Keister, personal communication). DNase may not be required to be added to extracts of all organisms and generally not at all where sonication is used to break the cells. The use of protamine sulphate or streptomycin sulphate is not recommended when preparing crude extracts since, some enzymes may coprecipitate with DNA. The inclusion of reducing agents in breaking buffers does not appear to be as important with bacteria as it is with plants, although it is worthwhile testing the effect of 2-mercaptoethanol or dithiothreitol on enzyme activities if these are low or the enzymes are unstable. EDTA may help to stabilize enzyme activities in some cases (Rebello and Strauss, 1969).

Preparation of cell envelopes should be carried out under conditions of centrifugation which do not greatly exceed the minimum required to sediment the envelopes. Where a pellet is difficult to re-suspend or where further fractionation of cell walls and cytoplasmic membranes is to be attempted, it is advisable to sediment the cell envelopes on to a cushion of 55% sucrose rather than to the bottom of the tube. Much of the material in cell envelope preparations will consist of closed cytoplasmic vesicles partially surrounded by cell walls. Further sonication of these vesicles or treatment with low concentrations of detergents may be required to give substrates access to enzymes.

Although not widely explored, cytoplasmic membranes can be prepared from nitrogen-fixing and closely related organisms using lysozyme or lysozyme/EDTA to produce spheroplasts which are lysed by osmotic shock (Kaback, 1971; Brill *et al.*, 1974; Van't Riet and Planta, 1975; Haaker and Veeger, 1977). The membranes could then be isolated using sucrose gradient techniques (Osborn and Munson, 1974; see also Section a.i on isolation of bacteroid membranes). As with cell envelope preparations the possibility that substrates may not have access to the enzymes must be considered (Coakley *et al.*, 1977).

(iv) Fractionation of cyanobacteria

The cyanobacteria (blue-green algae) are divided into three main groups: the filamentous heterocystous forms, the filamentous non-heterocystous forms, and the unicellular forms (Stanier and Cohen-Bazire, 1977). Although virtually all of the heterocystous forms appear to fix nitrogen, it is not certain whether all the non-heterocystous forms do also (Stewart, 1977). Heterocysts are considered to be the major loci of nitrogen fixation in aerobically grown, although not necessarily of anaerobically grown, cyanobacteria (Rippka and Stanier, 1978). The ammonia produced by nitrogen fixation appears to be assimilated by a glutamine synthetase–ferredoxin-dependent glutamate synthase enzyme system (Lea and Miflin, 1975, Wolko *et al.*, 1976). Enzyme assays on extracts of vegetative cells and heterocysts indicate that glutamate synthase activity is present in the vegetative cells whereas glutamine synthetase occurs in both heterocysts and vegetative cells (Dharmawardene *et al.*, 1973; Thomas, *et al.*, 1977).

Several different approaches have been adopted to determine the relative distribution of enzymes in vegetative cells and heterocysts. These involve either sonicating or pressing under mild conditions to rupture the vegetative cells. The contents of these cells are separated from the heterocysts, which are subsequently ruptured using more vigorous conditions of sonication or pressing (Dharmawardene *et al.*, 1973; Peterson and Burris, 1976a; Thomas *et al.*, 1977). A problem associated with these techniques is that the heterocysts are susceptible to damage during the rupturing of the vegetative cells. The degree of damage can be decreased by incubating the filaments in the presence of lysozyme to make the vegetative cells more susceptible to breaking (Fay and Lang, 1971; Tel-Or and Stewart, 1976; Tel-Or *et al.*, 1978; Peterson and Burris, 1978).

Preparation of cell-free extracts from whole filaments and also from vegetative cells and heterocysts following treatment of filaments with lysozyme and EDTA will be described (Thomas *et al.*, 1977; Peterson and Wolk, 1978; R. B. Peterson and C. P. Wolk, personal communication). The method for breaking whole filaments of heterocystous cyanobacteria could also be applied to non-heterocystous organisms. Methods for culturing cyanobacteria are described in Chapter III.5.

Breakage of whole filaments of cyanobacteria

Apparatus

Sonifier (Model S-125, Heat Systems-Ultrasonic Inc., New York, U.S.A., of equivalent) or French pressure cell.

Materials and reagents

TES buffer, 5mM, pH 7.2.
Sodium dithionite.
Dithiothreitol.

Procedure. Cultures of *Anabaena cylindrica* containing organisms at a density of 1.4 μg ml^{-1} of chlorophyll (determined by extracting filaments with 80% acetone and applying the relationship that $\varepsilon_{663\ nm} = 82$ for chlorophyll *a*; Biggins, 1967; Sestak, 1971) are harvested by centrifugation at 5000 *g* for 5 min. The cells are washed in 5 mM TES buffer (pH 7.2) by re-suspending and re-centrifuging, and finally re-suspended at a concentration of 90–125 μg ml^{-1} of chlorophyll using the same buffer. The suspension is degassed three times to 50 μmHg with re-gassing to 1 atm with argon. Further processing of the suspension is carried out under these strictly anaerobic conditions and reagents are degassed as described above. The anaerobic suspension is adjusted to 2.5 mM dithionite and, if found to be necessary for enzyme activity, 1.0 mM dithiothreitol.

Breakage of vegetative cells and heterocysts in the algal suspension is achieved by sonication at 12°C under argon using the Model S-125 Sonifier at 3.0-A output for 60 s ml^{-1} of suspension. Alternatively, the filaments are ruptured by passage (twice) through a French press at 16000–20000 lb in^{-2} (1125–1406 kg cm^{-2}). The suspension of ruptured cells is centrifuged anaerobically at 5000 *g* for 5 min to produce a pellet which contains a small percentage of unbroken heterocysts together with heterocyst and vegetative cell wall debris. The supernatant is taken for assay of nitrogenase and associated enzymes. The pellet is likely to contain most of the hydrogenase activity (Fujita and Myers, 1965; Peterson and Burris, 1978).

General comments. Anaerobic conditions and the use of dithionite are required when preparing cell extracts for assaying nitrogenase and hydrogenase, which are extremely oxygen labile. Breaking filaments under aerobic conditions can be used for most other enzymes. Assays for ferredoxin-linked glutamate synthase should be carried out under anaerobic conditions to prevent oxidation of dithionite required for reduction of ferredoxin or methylviologen (Thomas *et al.*, 1977). It is probably advisable to break and fractionate the filaments under anaerobic conditions rather than to subject solutions of proteins to evacuation prior to assay of glutamate synthase activity.

Isolation of vegative cell and heterocyst fractions

Apparatus

Sonic cleaning bath (Cole-Palmer Model 8845–3).

Materials and reagents

HEPES, 30 mM, PIPES, 30 mM, buffer adjusted to pH 7.2 with KOH, containing 1.0 mM $MgCl_2$ (HP buffer).

HP buffer containing 10 mM Na_2EDTA, with or without 1.0 mg ml^{-1} of lysozyme.

Procedure. Cultures (e.g. *Anabaena variabilis* ATCC 29413 derived from UTEX 1444) are harvested under argon by centrifugation at 10000*g* and 10 °C in a Sorvall SS-34 rotor equipped with a continous flow attachment. The filaments are re-suspended in about 30 ml of culture supernatant to a chlorophyll concentration of 200–300 μg ml^{-1}. The suspension is degassed twice to 50 μmHg with regassing to 1 atm with H_2. This suspension is transferred under H_2 via a degassed syringe into a 50-ml polyethylene centrifuge tube fitted with a rubber serum stopper. All subsequent operations are performed under strictly anaerobic conditions under H_2 and buffers are thoroughly sparged with N_2. $Na_2S_2O_4$ is not present.

Filaments are sedimented at 500*g* for 5 min, then washed twice by re-suspending the pellet in 20 ml of HP buffer containing 10.0 mM EDTA, with re-centrifuging at 500 *g* for 5 min. The filaments are finally re-suspended in 40 ml of HP buffer containing 10 mM EDTA and 1.0 mg ml^{-1} of lysozyme. This suspension is shaken at 200 r.p.m. in a reciprocating water-bath at 30 °C for 30 min. The filaments are fragmented and some heterocysts are detached by this procedure, which renders the vegetative cells more susceptible to destruction by cavitation. The lysozyme-treated material is sedimented at 500 *g* for 5 min and re-suspended in 30 ml of HP buffer. The suspension, in a stoppered 125-ml Erlenmeyer flask under H_2, is immersed in the sonic cleaning bath and subjected to cavitation for 7–10 min to destroy the vegetative cells. The bath is continuously cooled by an immersed copper coil containing circulating tap water at 12–13 °C. Heterocysts present in the sonicated suspension are sedimented at 500 *g* for 5 min and the supernatant (vegetative cell soluble fraction) is removed and used for enyzme assays. The heterocysts are washed free of cell debris by twice re-suspending in 20 ml of HP buffer with centrifugation at 30 *g* for 5 min. After final re-suspension in HP buffer, the heterocysts can be used in studies of photosynthetic and nitrogenase activities or ruptured and fractionated as described for whole filaments.

General comments. Washing the filaments in 10 mM EDTA is helpful for freeing cyanobacteria of extracellular polysaccharides and for making the vegetative cells more susceptible to lysozyme. The use of lysozyme and EDTA, coupled with gentle sonication in the cleaning bath, appears to cause little damage to the heterocysts which contain over 50% of the total *in vivo* nitrogenase activity present in the intact filaments (Peterson and Wolk, 1978). Bath cavitation is an empirical process affected by bath water volume, suspension volume, and the positioning of the flask. The number of heterocysts per millilitre of suspension of whole filaments and isolated heterocysts can be determined by counting heterocysts in a Speirs-Levy counting chamber (C.A. Hausser and Son, Philadelphia, Pa., U.S.A.).

(c) ASSAYS FOR ENZYMES OF AMMONIA ASSIMILATION

(i) Glutamine synthetase

Glutamine synthetase (E.C. 6.3.1.2) has been purified from a wide variety of prokaryotic and eukaryotic sources. In plants and bacteria these enzymes are considered to have a primary role in the assimilation of ammonium into amino acids (Miflin and Lea, 1977). In addition to catalysing the synthesis of glutamine by the reaction shown in equation (1), purified glutamine synthetase will also catalyse the formation of γ-glutamyl hydroxamate by the 'biosynthetic' reaction [equation (2)] or the 'transferase' reaction [equation (3)].

$$\text{glutamate} + \text{ATP} + \text{NH}_4^+ \xrightarrow[\text{(ii) Mn}^{2+}]{\text{(i) Mg}^{2+}} \text{glutamine} + \text{ADP} + \text{P}_\text{i} \qquad (1)$$

$$\text{glutamate} + \text{ATP} + \text{NH}_2\text{OH} \xrightarrow[\text{(ii) Mn}^{2+}]{\text{(i) Mg}^{2+}} \gamma\text{-glutamyl hydroxamate} + \text{ADP} + \text{P}_\text{i} \qquad (2)$$

$$\text{glutamine} + \text{NH}_2\text{OH} \xrightarrow[\text{ADP + arsenate}]{\text{(i) Mg}^{2+}\text{, (ii) Mn}^{2+}} \gamma\text{-glutamyl hydroxamate} + \text{NH}_4^+ \qquad (3)$$

Procedures for measuring each of these reactions are given here. Glutamine synthetase activity [equation (1)] can be assayed by measuring ADP or P_i formation or the formation of [^{14}C]glutamine from [^{14}C]glutamate. Difficulties may be encountered, however, if crude extracts are assayed by these three methods (see general comments on each procedure). Hydroxamate formation [equations (2) and (3)] is measured spectrophotometrically after complexing the hydroxamate with $FeCl_3$. Glutamine synthetase assays based on the formation of γ-glutamyl hydroxamate by the biosynthetic reaction [equation (2)] are of particular value in that they can be used in crude extracts; however there are still some difficulties (see general comments on biosynthetic hydroxamate assay). On the other hand, use of the transferase assay [equation (3)] to measure glutamine synthetase activity has been critized on the grounds that the rates observed are not clearly related to glutamine synthetase activity and that transferase activity is also catalysed by other enzymes in crude extracts (see general comments on transferase assay for further discussion). The transferase assay is included here, however, since it can be used to estimate adenylylated and unadenylylated glutamine synthetase activity.

Biosynthetic assay, P_i measurement

This method is based on that of Shapiro and Stadtman (1970).

Materials and reagents

Imidazole–HCl buffer, 0.2 M, pH 7.0.
$MgSO_4$, 0.3 M.
Sodium glutamate, 0.4 M, pH 7.0.
ATP, 40 mM, pH 7.0.
NH_4Cl, 0.2 M.
Ferrous sulphate reagent: 0.8%w/v $FeSO_4 \cdot 7H_2O$ dissolved in 7.5 mM H_2SO_4, prepared just prior to use.
Ammonium molybdate reagent: 6.6%w/v $(NH_4)_6Mo_7O_{24} \cdot 4H_2O$ in 3.75 M H_2SO_4.
$Na_2HPO_4 \cdot H_2O$, 5 mM.

Procedure. A reaction mixture is prepared by mixing the following volumes of the above solutions: 1.0 ml of imidazole buffer, 0.4 ml of $MgSO_4$, 0.4 ml of glutamate, 0.4 ml of ATP, and 0.6 ml of water.

Glutamine synthetase assays contain 0.14 ml of the reaction mixture and 50 μl of enzyme. Incubate at 25 °C for 5 min. The reaction is started by the addition of 10 μl of the ammonium chloride solution and the reaction mixtures are incubated at 25 °C for 15 min. The reaction is stopped by the addition of 1.8 ml of the ferrous sulphate reagent. This is followed by the addition of 0.15 ml of the ammonium molybdate reagent. Control assays minus ammonium chloride are included. The absorbance of the solution is determined at 660 nm. Absorbances are converted into micromoles of phosphate formed using a calibration graph prepared by adding 50 μl of 0–5 mM $Na_2HPO_4 \cdot H_2O$ to 0.15 ml of the assay mixture. Glutamine synthetase activity is expressed as (μmol phosphate formed)min^{-1} (mg protein)$^{-1}$.

General comments. The inclusion of control assays to measure phosphate in the reaction mixtures and phosphate produced by ATPase activity or non-enzymatic ATP hydrolysis is essential. It is unlikely that this procedure could be used successfully in crude extracts.

Biosynthetic assay, ADP measurement in a coupled assay

The method is based on that of Wellner *et al.*, (1966), in which the ADP formed by the glutamine synthetase reaction is used together with phosphoenolpyruvate in the pyruvate kinase-catalyzed reaction to produce ATP and pyruvate. Pyruvate is then reduced in a reaction catalyzed by lactate dehydrogenase with the concomitant oxidation of NADH.

Materials and reagents

Imidazole–HCl buffer, 0.25 M, pH 7.0.
KCl, 0.5 M.
ATP, 40 mM, pH 7.0.
Glutamate, 0.4 M, pH 7.0.
$MgSO_4$·0.5 M.
$NADHNa_2$, 1.4 mM, freshly prepared.
Phosphoenolpyruvate cyclohexamyl salt (PEP), 15 mM, freshly prepared.
Lactate dehydrogenase–pyruvate kinase solution (4 mg ml^{-1}) (Boehringer-Mannheim). A preparation of crystallized enzymes in 3.2 M ammonium sulphate. The ammonium sulphate is removed by dialysis before use (three periods of 4 h against 1000 volumes of 0.02 M potassium phosphate buffer, pH 7.0).
NH_4Cl, 0.2 M.

Procedure. A reaction mixture is prepared by mixing the following volumes of the above solutions: 2 ml of imidazole buffer, 0.2 ml of KCl, 1 ml of ATP, 1 ml of glutamate, 0.6 ml of $MgSO_4$, and 2.0 ml of water. Glutamine synthetase assays are carried out in 1 ml cuvettes in a constant-temperature cuvette holder at 25 °C. Add to the cuvette 0.68 ml of the reaction mixture, 0.1 ml of the NADH solution, 0.1 ml of PEP, 20 μl of the lactate dehydrogenase–pyruvate kinase solution, and 50 μl of the glutamine synthetase enzyme preparation. The solutions in the cuvette are mixed and the absorbance recorded at 340 nm. Any ADP present will cause a decrease in absorbance. When the ΔA min^{-1} value has returned to zero, the glutamine synthetase reaction is started by the addition of 50 μl of the ammonium chloride solution. The total reaction mixture is immediately mixed and the rate of decrease in absorbance at 340 nm is recorded. A change in absorbance of 0.1–0.5 per minute is an acceptable rate. ΔA_{340nm} min^{-1} is calculated from the linear part of the graph. The number of moles of ADP formed by glutamine synthetase equals the number of moles of NADH oxidized. Glutamine synthetase activity, expressed as (μmol ADP formed)min^{-1}(mg protein)$^{-1}$, is calculated using $\varepsilon_{340nm}(NADH) = 6.22 \times 10^3$ l mol^{-1} cm^{-1}.

General Comments. It may be found that the ΔA min^{-1} value before the addition of NH_4Cl never reaches zero. This is probably due to contaminating

ATPase activity in the glutamine synthetase preparation. Levels of ATPase activity in purified glutamine synthetase should be less than 2% of the biosynthetic activity.

A lag phase is commonly seen after the addition of NH_4Cl. Shapiro and Stadtman (1970) have suggested that pre-incubation of the enzyme with 5 mM $MgCl_2$ for 15 min will overcome this problem. It has been our experience, however, that this treatment will not remove the lag phase from assays of glutamine synthetase purified from lupin nodules (D. K. McCormack and K. J. F. Farnden, unpublished results.).

Pyruvate kinase is inhibited by sodium ions and stimulated by potassium ions. The use of sodium compounds in the preparation of reagents should therefore be avoided wherever possible. It is not essential, however, to make the reaction mixtures completely sodium free.

It is recommended that the PEP and NADH solutions are prepared just prior to use. Phosphoenolpyruvate is unstable in solution. Inhibitors of lactate dehydrogenase have been reported to be formed in solutions of NADH on storage (Fawcette *et al.*, 1961).

Radiochemical Biosynthetic Assay

This procedure is a modification of the biosynthetic P_i measurement assay of Shapiro and Stadtman (1970) in that [^{14}C]glutamate is used as substrate and the reaction rate is determined by the rate of formation of [^{14}C]glutamine.

Apparatus

High-voltage paper electrophoresis equipment.

Materials and reagents

Imidazole–HCl buffer, 0.2 M, pH 7.0.
$MgSO_4$, 0.3 M.
Sodium glutamate, 0.4 M, pH 7.0.
ATP, 40 mM, pH 7.0.
NH_4Cl, 0.2 M.
[U-^{14}C]glutamate (265 mCi mmol^{-1}; 5 μCi ml^{-1}).
Whatman 3MM chromatography paper.
Marker amino acids, glutamine and glutamic acid (2.5 mg ml^{-1}).
Electrophoresis buffer, 0.04 M sodium acetate, pH 5.0.
Ninhydrin (0.2% w/v in acetone).
Scintillation fluid (0.03% w/v POPOP, 0.3% w/v PPO in toluene).

Procedure. A reaction mixture is prepared by mixing the following volumes of the above solutions: 0.25 ml of imidazole buffer, 0.1 ml of $MgSO_4$, 0.1 ml of glutamate, 0.1 ml of ATP, 0.1 ml of [^{14}C]glutamate, and 0.05 ml of water. Glutamine synthetase assays contain 0.14 ml of reaction mixture and 50 μl of enzyme. Incubate at 25 °C for 5 min. The reaction is started by the addition of 10 μl of the ammonium chloride solution. Samples of 20 μl are removed at zero time and at various other times during a 20-min incubation period at 25 °C and applied to a Whatman 3MM paper. Spots are rapidly dried using a hair-dryer and marker amino acids (5 μl) are spotted on top of the samples. Glutamine is separated from glutamic acid by high-voltage paper electrophoresis at pH 5.0 and the radioactivity in each amino acid is determined in a liquid scintillation spectrometer (for details see radiochemical asparaginase assay, Section e.iii). Background counts per minute are subtracted before calculation of results.

Calculation of results

$$\text{Glutamine produced } (\mu\text{mol min}^{-1}) = \frac{\text{cpm Gln}}{\text{cpm Gln} + \text{cpm Glu}} \times \frac{\mu\text{mol Glu in assay}}{\text{incubation time}}$$

Results are expressed as (μmol glutamine) min^{-1} (mg protein)$^{-1}$.

General comments. It is not possible to use this assay to determine glutamine synthetase activity in crude extracts if the extracts used contain glutaminase activity (Robertson *et al.*, 1975a) or if labelled glutamate is metabolized to other compounds which co-electrophorese with glutamine.

The electrophoretic technique used here will rapidly separate glutamate from glutamine, but many other amino acids will co-electrophorese with glutamine. Identification of the product of the reaction as glutamine is therefore essential. This can be achieved by separating aliquots of the supernatant by two-dimensional electrophoresis or chromatography (see Scott *et al.*, 1976, and Streeter, 1973, for methods). In addition, the labelled glutamine can be eluted from the electrophoregram, hydrolyzed with acid, and the product identified as glutamate by the same procedures.

An alternative technique for separating glutamine from glutamate using small columns of Dowex 1 (chloride form) ion-exchange resin has been described by Prusiner and Milner (1970). This technique allows a large number of samples to be analysed for [^{14}C]glutamine, but confirmation that the column eluates contain only labelled glutamine is necessary.

Biosynthetic hydroxamate assay

This is based on the method of Elliott (1955).

Materials and reagents

Imidazole–HCl buffer, 0.25 M, pH 7.0.

Sodium glutamate, 0.3 M, pH 7.0.
ATP, 30 mM, pH 7.0.
$MgSO_4$, 0.5 M.
Hydroxylamine solution: (mix equal volumes of 1 M $NH_2OH.HCl$ and 1 M NaOH just prior to use.
$FeCl_3$ reagent: mix equal volumes of 10% w/v $FeCl_3 \cdot 6H_2O$ in 0.2 M HCl, 24% w/v trichloroacetic acid and 50% v/v HCl.
γ-Glutamyl hydroxamate, 5 mM (standard).

Procedure. Glutamine synthetase hydroxamate assays contain 0.3 ml of imidazole buffer, 0.2 ml of glutamate, 0.2 ml of ATP, 0.1 ml of $MgSO_4$, enzyme, and water to give a total volume of 1.4 ml. Incubate at 25 °C for 5 min. The reaction is started by the addition of 0.1 ml of the hydroxylamine reagent and the reaction mixtures are incubated at 25 °C for 15 min. The reaction is stopped by the addition of 0.4 ml of the $FeCl_3$ reagent. Control assays minus hydroxylamine are included. Assay mixtures are centrifuged at 5000 *g* for 10 min and the absorbance of the supernatant is measured at 540 nm. Absorbances are converted into micromoles of γ-glutamyl hydroxamate using a calibration graph prepared using 0–0.6 ml of the 5 mM γ-glutamyl hydroxamate solution in the assay mixture as above. Glutamine synthetase activity is expressed as (μmol γ-glutamyl hydroxamate formed)min^{-1}(mg protein)$^{-1}$.

General comments. We have found that different batches of commercial γ-glutamyl hydroxamate differ considerably in purity. γ-Glutamyl hydroxamate solutions have therefore been standardized using $\varepsilon_{540\,nm} = 430\,l\,mol^{-1}\,cm^{-1}$ for the hydroxamate–$FeCl_3$ complex.

It may be useful to include EDTA in the glutamine synthetase assays. O'Neal and Joy (1974) have reported a 54% activation of the enzyme from pea leaves by 4 mM EDTA. The activity of glutamine synthetase from the plant fraction of lupin nodules is similarly stimulated 60% by the inclusion of 4 mM EDTA in the assay (D. K. McCormack, unpublished results).

The biosynthetic hydroxamate assay appears to be the assay best suited to measuring glutamine synthetase activity in crude extracts. Studies on purified glutamine synthetase have shown that the apparent K_m values for ATP and glutamate are similar in the presence of NH_4^+ or NH_2OH and that the K_m values for these latter compounds are also similar (Meister, 1974). Furthermore, activites for glutamine synthetase measured by the biosynthetic hydroxamate assay compare favourably with activities determined by the ADP coupled assay and P_i measurement (O'Neal and Joy, 1973).

The problem of glutaminase activity in crude extracts is minimal, since rates of hydrolysis of γ-glutamyl hydroxamate by this enzyme are only 2% of the rates observed with glutamine (Robertson *et al.*, 1975a). Glutaminase activity is inhibited by 5-diazo-4-oxonorvaline. This compound has no effect on γ-glutamyl hydroxamate formation (Robertson *et al.*, 1975a).

Glutamate tRNA synthetase will catalyse the formation of α-glutamyl-hydroxamate from glutamate, NH_2OH, and ATP with the liberation of AMP and PP_i (Eigner and Loftfield, 1974). Mammalian and bacterial glutaminases have been reported to form hydroxamates from glutamate and hydroxylamine in the absence of ATP (Meister *et al.*, 1955). The rate of hydroxamate formation by these reactions should, however, be low compared with the rate observed with glutamine synthetase.

Transferase hydroxamate assay

This assay is based on the method of Shapiro and Stadtman (1970).

Materials and reagents

Imidazole–HCl buffer, 0.5 M, pH 7.0.
Glutamine, 0.15 M.
$MnSO_4$, 0.1 M.
ADP, 0.01 M, pH 7.0.
Sodium arsenate, 1.0 M, pH 7.0.
Hydroxylamine solution: mix equal volumes of 1 M $NH_2OH{\cdot}HCl$ and 1 M NaOH just prior to use.
$FeCl_3$ reagent: mix 10% w/v $FeCl_3{\cdot}6H_2O$, 24% w/v trichloroacetic acid, 6 M HCl and water in the proportions 8:2:1:13.
γ-Glutamyl hydroxamate, 10 mM (standard).

Procedure. A reaction mixture is prepared by mixing the following volumes of the above solutions: 2 ml of imidazole buffer, 4 ml of glutamine, 0.6 ml of $MnSO_4$, 0.8 ml of ADP, 0.4 ml of arsenate, and 2.2 ml of water.

Transferase assays contain 0.5 ml of the reaction mixture, enzyme, and water to give a total volume of 0.9 ml. Incubate at 25 °C for 5 min. The reaction is started by the addition of 0.1 ml of the hydroxylamine solution and the reaction mixtures are incubated at 25 °C for 15 min. The reaction is stopped by the addition of 2 ml of the $FeCl_3$ reagent. Control assays minus hydroxylamine are included. Assay mixtures are centrifuged at 5000 *g* for 10 min and the absorbance of the supernatant is measured at 540 nm. Absorbances are converted into micromoles of γ-glutamyl hydroxamate using a calibration graph prepared using 0–0.4 ml of the 10 mM γ-glutamyl hydroxamate solution in the assay mixture as above (see comment on the purity of commercial γ-glutamyl hydroxamate in the general comments on the previous assay). Transferase activity is expressed as (μmol γ-glutamyl hydroxamate formed)min^{-1} (mg protein)$^{-1}$.

General comments. The use of this assay to measure glutamine synthetase activity has been criticized for two reasons (Miflin and Lea, 1977). Firstly, there is no clear relationship between synthetase and transferase activity, so the rates obtained have no physiological meaning. Secondly, the transferase reaction is catalysed by other enzymes in addition to glutamine synthetase in crude extracts (see Miflin and Lea, 1977, for references). The assay is useful, however , in that it can be used to measure the average state of adenylylation of glutamine synthetase (Ginsburg and Stadtman, 1973). Assays carried out in the presence of 0.3 mM Mn^{2+} measure transferase activity of both adenylylated and unadenylylated glutamine synthetase. Assays carried out in the presence of 0.3 mM Mn^{2+} and 60 mM Mg^{2+} measure the transferase activity of the unadenylylated enzyme only. The average state of adenylylation can then be calculated if the number of subunits adenylylated is known (Ginsburg and Stadtman, 1973) or a relative adenylylation value can be given (Bishop *et al.*, 1976). Inhibition of transferase activity by 60 mM Mg^{2+} and the release of this inhibition by treatment of the enzyme with snake venom phosphodiesterase (removes covalently bound AMP from the enzyme) is taken as evidence of the presence of adenylylated glutamine synthetase.

(ii) Glutamate synthase

It is now considered that the assimilation of ammonium into amino acids occurs via glutamine synthetase and glutamate synthase in higher plants and bacteria (see Miflin and Lea, 1977, for review). $NADP^+$–Glutamate synthase (E.C. 1.4.1.13) has been purified from bacteria (Miller and Stadtman, 1972). NAD^+–Glutamate synthase (E.C. 1.4.1.14 provisionally) has been purified from the plant cytosol fraction of *Lupinus angustifolius* nodules (Boland and Benny, 1977). Wallsgrove *et al.*, (1977) have reported the isolation and purification of a ferredoxin-dependent glutamate synthase (E.C. 1.4.7.1) from the leaves of higher plants.

The glutamate synthase assay procedure described here is based on that of Boland and Benny (1977).

Materials and reagents

HEPES–KOH buffer, 0.2 M, pH 8.6, containing 2% of mercaptoethanol.
2-Oxoglutarate, 10 mM (neutralized with KOH).
NADH (or NADP), 1 mM.
L-Glutamine, 100 mM.

Procedure. The reagents listed above are added to a 1-ml cuvette in the following amounts: 0.5 ml of HEPES buffer, 0.1 ml of 2-oxoglutarate, 0.1 ml of NADH, 0.05 ml of enzyme solution, and 0.2 ml of water. The rate of decrease in the absorbance at 340 nm is monitored at 25 °C before and after the addition of 0.05 ml of the glutamine solution. The former rate serves as a measurement of background NADH oxidation, which, if significant, is subtracted from the rate of the glutamine-dependent activity. Glutamate synthase activity is expressed as (μmol NADH oxidized) min^{-1} (mg $protein)^{-1}$. ε_{340nm} (NADH) = 6.22×10^3 l mol^{-1} cm^{-1}.

General comments. To prepare crude extracts containing glutamate synthase activity it is essential to include high levels of reducing agents in the extraction buffers. Boland *et al.*, (1978), for example, have used buffers containing 10 mM dithiothreitol and 1% v/v 2-mercaptoethanol. The use of Tris buffers in glutamate synthase assays or purification procedures should be avoided (Boland and Benny, 1977; Brown and Dilworth, 1975; Wallsgrove *et al.*, 1977).

Dilworth (1974) has pointed out that it is essential to establish that the oxidation of NAD(P)H that occurs when glutamine is added to the assay system is in fact due to glutamate synthase and not to the coupled action of a glutaminase and glutamate dehydrogenase. This problem has been re-emphasized by Miflin and Lea (1977). If the extracts used to assay for glutamate synthase do contain glutaminase and glutamate dehydrogenase then it may still be possible to demonstrate glutamate synthase activity by adding NH_4Cl (100 mM final concentration) to the cuvette before the glutamine solution in the assay above. The rates of NADH oxidation are measured before and after the addition of glutamine. The former rate is a measure of glutamate dehydrogenase activity and background NAD(P)H oxidation, which can be subtracted from the rate of the glutamine-dependent glutamate synthase activity.

(iii) Asparagine synthetase

Asparagine synthetases from prokaryotes (E.C. 6.3.1.1) convert aspartate to asparagine in the presence of ammonium, ATP, and Mg^{+2} with the stoichiometric production of AMP and pyrophosphate (Cedar and Schwartz, 1969). The enzyme from eukaryotes (E.C. 6.3.5.4) catalyses a similar reaction but utilizes glutamine rather than ammonium as the N donor (Rognes, 1970). Glutamine-dependent asparagine synthetase has been found in the plant fraction of *Lupinus angustifolius* nodules (Scott *et al.*, 1976) and has been purified from *Lupinus luteus* cotyledons (Rognes, 1975). The assay procedure described here is based on that of Scott *et al.* (1976).

Apparatus

High-voltage paper electrophoresis equipment.

Materials and reagents

L-[4-^{14}C]Aspartate (50 mCi $mmol^{-1}$; 50 μCi ml^{-1}), Calatomic.
L-Aspartate, 0.1 M (adjusted to pH 7.8 with KOH).
L-Glutamine, 0.1 M (or NH_4Cl, 0.1 M).
ATP, 0.05 M.
$MgCl_2$, 0.05 M.
Potassium phosphate buffer, 50 mM, pH 7.8, containing 5 mM dithiothreitol and 0.5 M sucrose (sucrose is included as this buffer has been designed for the preparation of the plant cytosol fraction of legume nodules).
Whatman 3MM chromatcgraphy paper.
Marker amino acids, asparagine and aspartic acid (2.5 mg ml^{-1}).
Electrophoresis buffer, 0.04 M sodium acetate, pH 5.0.
Ninhydrin (0.2% w/v in acetone).
Scintillation fluid (0.03% w/v POPOP, 0.3% w/v PPO in toluene).

Procedure. Asparagine synthetase assays contain 20 μl of each of the following solutions listed above: [^{14}C]aspartate, L-aspartate, L-glutamine, ATP, and $MgCl_2$, and 100 μl of enzyme in the phosphate buffer. Samples of 20 μl are removed at zero time and at various other times during a 20-min incubation period at 25 °C and applied to a Whatman 3MM paper. Spots are rapidly dried using a hair-dryer and marker amino acids (5 μl) are spotted on top of the samples. Asparagine is separated from aspartic acid by high-voltage paper electrophoresis at pH 5.0 and the radioactivity in each amino acid determined in a liquid scintillation spectrometer (for details see radiochemical asparaginase assay, Section e.iii). Background counts per minute are subtracted before calculation of results.

Calculation of results

Asparagine produced (μmol min^{-1})

$$= \frac{\text{cpm Asn}}{\text{cpm Asp} + \text{cpm Asn}} \times \frac{\mu\text{mol Asp in assay}}{\text{incubation time}}$$

Results are expressed in (μmol asparagine) min^{-1}(mg protein)$^{-1}$.

General comments. Asparagine synthetase in crude plant extracts is stabilized by mercaptoethanol or dithiothreitol and either glutamine, or $MgCl_2$ and ATP (Scott *et al*., 1976; Rognes, 1975).

[4-^{14}C]Aspartate is used here in preference to [U-^{14}C]aspartate since the latter compound will give rise to labelled alanine (which co-electrophoreses with asparagine) if aspartate-4-decarboxylase activity is present in the crude extract (see Rognes, 1970, for further discussion).

Identification of the product of the reaction as asparagine can be confirmed by separating aliquots of the supernatant by two-dimensional electrophoresis or chromatography (see Scott *et al.*, 1976, and Streeter, 1973, for methods). In addition, the labelled asparagine can be eluted from the electrophoregram, hydrolysed with acid, and the product identified as aspartate by the same procedures.

An alternative technique for separating asparagine from aspartic acid using small columns of Dowex 1 (chloride form) ion-exchange resin has been described by Prusiner and Milner (1970). Confirmation that the column eluate contains only labelled asparagine is essential.

Hydroxamate assays (Rognes, 1975) which measure the formation of β-aspartyl hydroxamate catalysed by purified asparagine synthetase cannot be used in crude extracts as aspartate kinase and aspartyl–tRNA synthetase will catalyse the formation of β- and α-aspartyl hydroxamates, respectively (Wong and Dennis, 1973; Moustafa and Petersen, 1962).

(iv) Aspartate aminotransferase

Aspartate aminotransferase (AAT) (E.C. 2.6.1.1) activity has been demonstrated in the plant fraction of *Glycine max*, *Vicia faba*, *Pisum sativum*, and *Lupinus angustifolius* nodules (Grimes and Masterson, 1971; Ryan *et al.*, 1972; Reynolds and Farnden, 1979). The enzyme has also been found in the bacteroids of *G. max* and *L. angustifolius* nodules and in free-living *R. japonicum* and *R. lupini* (Ryan *et al.*, 1972; Reynolds and Farnden, 1979). A role for the plant AAT in ammonium assimilation in lupin nodules has been proposed by Scott *et al.* (1976).

AAT activity can be conveniently assayed in the direction of aspartate utilization by measuring the rate of oxaloacetate formation with malate dehydrogenase and NADH. The method given below is based on that of Bergmeyer and Bernt (1974).

Materials and reagents

NADH, 12 mM.

Malate dehydrogenase (transaminase activity <0.01% and ammonium free; Sigma, St. Louis, Mo., U.S.A.). Dilute to 600 U ml^{-1} with 0.1 M phosphate buffer (pH 7.5) just prior to use. If malate dehydrogenase is purchased as a suspension in ammonium sulphate, ammonium ions are removed by dialysis (three periods of 4 h against 200 volumes of 0.01 M phosphate buffer, pH 7.5). The contents of the dialysis sac are then diluted with an equal volume of glycerol.

Phosphate buffer, 0.1 M, pH 7.5, containing 0.25 M aspartate.

2-Oxoglutarate, 0.2 M.

Procedure. The reagents listed above are added to a 3-ml cuvette in the following amounts: 0.05 ml of NADH, 0.05 ml of the diluted malate dehydrogenase preparation (30 U per assay), 0.5 ml of enzyme preparation, and 2.3 ml of phosphate buffer containing 0.25 M aspartate. After equilibration at 25 °C in a constant-temperature cuvette holder, the reaction is started by the addition of 0.1 ml of the 2-oxoglutarate solution. Mix, and record the decrease in absorbance at 340 nm over a period of about 5 min. The rate of NADH oxidation equals the rate of conversion of aspartate to oxaloacetate by aspartate aminotransferase. AAT activity, expressed as $\mu mol\ min^{-1}$(mg protein^{-1}), is calculated using ε_{340nm} (NADH) $= 6.22 \times 10^3\ l\ mol^{-1}cm^{-1}$.

General comments. Reaction mixtures must be 'ammonium free' to prevent the oxidation of NADH by glutamate dehydrogenase if it is present in the samples to be assayed.

AAT activity can be assayed in the reverse direction (aspartate formation) by measuring either the disappearance of oxaloacetate at 260 nm (Hatch, 1973) or the appearance of 2-oxoglutarate using $NADP^+$–glutamate dehydrogenase as an indicator enzyme (Huang *et al.*, 1976). The latter assay cannot be used if extracts contain $NADP^+$–malate dehydrogenase activity.

AAT isoenzymes have been detected in the plant fraction of legume nodules using starch and polyacrylamide gel electrophoresis (Grimes and Masterson, 1971; Ryan *et al.*, 1972; Reynolds and Farnden, 1979). A simple, one-step staining procedure for detecting AAT activity in gels following electrophoresis has been described by Decker and Rau (1963).

The enzyme requires pyridoxal phosphate as coenzyme but inclusion of this compound in AAT assays of crude extracts is not usually necessary. Pyridoxal phosphate (25–100 μM) should, however, be included in the buffers used both during the purification of the enzyme and in the assays of the partially purified enzyme.

(v) Glutamate dehydrogenase

Glutamate dehydrogenase (NAD^+, E.C. 1.4.1.2; $NAD(P)^+$, E.C. 1.4.1.3; and $NADP^+$, E.C. 1.4.1.4) activity is commonly found in plants, animals, and bacteria, the coenzyme specificity being dependent on the source of the enzyme (see Davies and Teixeira, 1975, for references). It is now considered unlikely that glutamate dehydrogenase plays a major role in ammonium assimilation in bacteria and higher plants, except in situations of ammonium excess (see Miflin and Lea, 1977, for discussion).

The glutamate dehydrogenase assay described here is based on that of Duke and Ham (1976).

Materials and reagents

Tris–HCl buffer, 0.1 M, pH 7.5.

2-Oxoglutarate, 0.33 M (adjusted to pH 7.5 with NaOH).

NH_4Cl, 3.0 M.
NADH, 1 mM.

Procedure. Glutamate dehydrogenase reaction mixtures contain 1.6 ml of Tris buffer, 0.1 ml of NH_4Cl solution, 0.2 ml of NADH solution, enzyme, and buffer to give a total volume of 2.9 ml. After equilibration in a constant-temperature cuvette holder at 25 °C, the reaction is started by the addition of 0.1 ml of the 2-oxoglutarate solution and the decrease in absorbance at 340 nm is monitored. The difference in the rate of NADH oxidation in the presence and absence of added ammonium chloride is taken as enzyme activity. Glutamate dehydrogenase activity is expressed as (μmol NADH oxidized) min^{-1}(mg protein)$^{-1}$. $\varepsilon_{340nm}(NADH) = 6.22 \times 10^3\ l\ mol^{-1}\ cm^{-1}$.

General comments. Glutamate dehydrogenase from the plant fraction of lupin nodules is inhibited by EDTA, KCN, and pyridoxal phosphate (Stone *et al.*, 1979). EDTA should not be included in the buffers used to prepare cell-free extracts containing this enzyme. The addition of Ca^{2+} has been found to overcome the inhibition of glutamate dehydrogenase activity by both EDTA and EGTA in extracts prepared from several lupin tissues (K. S. Chang and K. J. F. Farnden, unpublished results).

(d) ASSAYS FOR ENZYMES OF CARBON METABOLISM

This section has been restricted to include details and comments on only a few enzyme assays that were considered to be of particular interest to workers examining carbon (or nitrogen) metabolism in fractionated legume root nodules. The probable pathways of hexose catabolism in the plant fraction of legume nodules, bacteroids, and freeliving rhizobia has been reviewed by Robertson and Farnden (1980). Methods for assaying the individual enzymes of these pathways of carbon metabolism in plants or bacteria can be found in *Methods in Enzymology* (Academic Press) or *Methods of Enzymatic Analysis* (Ed. H. U. Bergmeyer, 2nd English Edition, Academic Press).

(i) Phosphoenolpyruvate carboxylase

Phosphoenolpyruvate (PEP) carboxylase (E.C. 4.1.1.31) activity has been detected in the plant cytosol fraction of *Lupinus angustifolius* nodules (Christeller *et al.*, 1977). These authors have proposed that PEP carboxylase provides the oxaloacetate required in the pathway for ammonia assimilation outlined by Scott *et al.* (1976). High PEP carboxylase activity has also been found in extracts of *Vicia faba* nodules (Lawrie and Wheeler, 1975).

PEP carboxylase activity is determined in the presence of NADH and malate dehydrogenase by following either the rate of incorporation of

[^{14}C]HCO_3^- into malate (acid-stable radioactivity) or the rate of NADH oxidation spectrophotometrically. The assay procedures described here are based on the methods of Christeller *et al.* (1977) and Lane *et al.* (1969).

[^{14}C]HCO_3^- Fixation assay

Materials and reagents

[^{14}C]$NaHCO_3$ (55 mCi $mmol^{-1}$); Radiochemical Centre, Amersham, Bucks., U.K.

Phosphoenolpyruvate, 20 mM.

Reaction mixture containing 100 mM Tris–HCl (pH 8.0), 20 mM $MgCl_2$, 10 mM dithiothreitol, 2 mM NADH, 20 mM [^{14}C]$NaHCO_3$ (4 μCi ml^{-1}) and malate dehydrogenase (40 units ml^{-1}).

Scintillation fluid, containing 30 *g* of *p*-terphenyl, 6 l of xylene, and 3 l of Triton X-100.

Procedure. The reaction is carried out in scintillation vials and initiated with 10–100 μl of enzyme extract in 50 mM Tris–HCl (pH 8.0) containing 10 mM $MgCl_2$ and 5 mM dithiothreitol. Each assay contains 0.25 ml of the reaction mixture, 50 μl of PEP, and water to give a final volume of 0.5 ml. The vials are incubated at 25°C for 5 min and the reaction is stopped by the addition of 100 μl of 2 M HCl. Background vials containing the reaction mixture and enzyme but no PEP are treated identically.

The vials are dried at 95°C, cooled, and the dried material is solubilized in 1 ml of water. Scintillation fluid (10 ml) is added. The aqueous phase is suspended by capping the vials and shaking vigorously. The acid-stable ^{14}C activity (as [^{14}C]malate) is determined with a liquid scintillation spectrometer.

Total counts per minute per millilitre of the reaction mixture is determined by adding 5-μl aliquots of the reaction mixture to 1 ml of 10% ethanolamine in scintillation vials. Scintillation fluid (10 ml) is added and the vials are counted.

Counts per minute should be converted into disintegrations per minute (dpm) before calculating the results.

Calculation of results.

$$\text{Conversion factor (CF)} = \frac{\text{total moles of } HCO_3^- \text{ per ml of reaction mixture}}{\text{dpm per ml of reaction mixture}}$$

$$HCO_3^- \text{ fixed (mol min}^{-1}) = [(\text{dpm of sample}) - (\text{dpm of background})] \times \frac{CF}{5\text{ min}}$$

Reaction rate is expressed as μmol min^{-1} (mg protein)$^{-1}$.

General comments. [^{14}C]HCO_3^- is incorporated into oxaloacetate by PEP carboxylase. Oxaloacetate is unstable and highly inhibitory of PEP carboxylase activity. Both of these problems are overcome by the addition of excess of malate dehydrogenase and NADH to convert rapidly all of the oxaloacetate formed into malate.

A range of assay times and enzyme concentrations should be carried out to check that a linear response is always obtained. It is important to mix the sample well at the end of the assay to ensure that all of the reaction mixture in the vial has been acidified. All procedures which involve the release of [^{14}C]CO_2 should be performed in a fume hood.

Spectrophotometric assay

The reaction mixture and conditions are modified from those described for the [^{14}C]HCO_3^- fixation assay in that unlabelled $NaHCO_3$ and less NADH (0.15 μmol ml^{-1} of assay solution) are used. The rate of NADH oxidation is followed at 340 nm (1-cm light path; 25 °C) after initiating the reaction with phosphoenolpyruvate.

(ii) Invertase

Invertase (E.C. 3.2.1.26) occurs in some *Rhizobium* species (Graham, 1964) and also in the plant cytosol of legume nodules (Kidby, 1966; Robertson and Taylor, 1973). The enzyme plays an important role in legume nodules in the initial metabolism of photosynthate for reactions associated with nitrogenase present in bacteroids. The assay procedure described here involves the incubation of sucrose with cell fractions followed by the determination of reducing sugar by the method of Nelson (1944). The sensitivity of the assay is in the range 0–100 μg of glucose.

Materials and reagents

Citric acid, 0.2 M, adjusted to pH 4.5 with NaOH.
Phosphate, 0.2 M, adjusted to pH 7.5 with NaOH.
Sucrose, 2.5 g dissolved in distilled water to a final volume of 50 ml.
Glucose, 10 mg dissolved in distilled water to a final volume of 50 ml.
Copper reagent A. Dissolve 25 g of anhydrous Na_2CO_3, 25 g of Na,K tartrate·$4H_2O$, 20 g of $NaHCO_3$, and 200 g of anhydrous Na_2SO_4 in 800 ml of distilled water and adjust the final volume to 1 l. Filter into a rubber-stoppered bottle and store at 25 °C.
Copper reagent B. Dissolve 15 g of $CuSO_4·5H_2O$ in distilled water to a final volume of 100 ml of solution containing 1–2 drops of concentrated H_2SO_4.

Mixed copper reagents A and B: Immediately before use filter 25 ml of reagent A and add 1 ml of reagent B.

Arsenomolybdate colour reagent. Dissolve 25 g of $(NH_4)_6Mo_7O_{24}\cdot H_2O$ in 450 ml of distilled water, add 21 ml of concentrated H_2SO_4 with mixing, followed by 25 ml of a solution containing 3 g of $Na_2HAsO_4\cdot 7H_2O$. The final solution is stored in a brown bottle for 48 h at 37 °C before use and then stored at 25 °C.

Procedure. Assay mixtures consist of 1.0 ml of 5% sucrose solution, 0.5 ml of 200 mM citrate or phosphate, and 0.5 ml of enzyme. The mixture is incubated at the temperature at which the organism or plant is grown for up to 1 h, boiled for 3 min, then centrifuged at 10 000 *g* for 10 min. To 0.5-ml aliquots of the supernatant is added 0.5 ml of mixed copper reagents A and B. The tubes are covered with aluminium foil or with a glass marble, heated in a boiling water-bath for 20 min, and then cooled in a cold water-bath. To each tube is added 0.5 ml of arsenomolybdate colour reagent followed by 3.5 ml of distilled water. The solutions are mixed and the absorbance read at 520 nm. The absorbance is compared with that of a standard series prepared using 0–0.5 ml of glucose solution (0.2 mg ml^{-1}) adjusted to a final volume of 0.5 ml with distilled water.

(iii) Isocitrate dehydrogenase

$NADP^+$–isocitrate dehydrogenase (E.C. 1.1.1.42) is found in the plant cytosol and bacteriod fractions of *Pisum sativum* and *Lupinus angustifolius* nodules (Kurz and LaRue, 1977; Reynolds and Farnden, 1979). NAD^+–isocitrate dehydrogenase (E.C. 1.1.1.41) occurs in the mitochondrial fraction of *L. angustifolius* nodules (Reynold and Farnden, 1979). Kurz and LaRue (1977) reported that the activity of $NADP^+$–isocitrate dehydrogenase in the bacteriods of *P. sativum* nodules is highest at the time of maximum nitrogen fixation and suggested that this enzyme is a likely source of reductant for nitrogen fixation.

Isocitrate dehydrogenase activity is determined by following the formation of NAD(P)H as measured by the increase in absorbance at 340 mn. The method described below is based on that of Cox (1969).

Materials and Reagents

HEPES–NaOH buffer, 50 mM, pH 7.6.

HEPES–NaOH buffer, 50 mM, pH 7.6 containing 5 M glycerol (used in the preparation of mitochondrial extracts containing NAD^+–isocitrate dehydrogenase).

$MnSO_4$, 30 mM.

$NAD(P)^+$, 20 mM.
Trisodium DL-isocitrate, 120 mM.

All reagents are dissolved in HEPES–NaOH buffer.

Procedure. The reagents listed above are added to a 3 ml cuvette in the following amounts: 2.6 ml of HEPES buffer (pH 7.6), 0.1 ml of manganese sulphate, 0.1 ml of $NAD(P)^+$, and 0.1 ml of enzyme. After equilibration at 25 °C in a constant-temperature cuvette holder the reaction is started by the addition of 0.1 ml of the isocitrate solution and the increase in absorbance at 340 nm is monitored. $NAD(P)^+$–isocitrate dehydrogenase activity is expressed as [μmol NAD(P)H formed]min^{-1}(mg protein)$^{-1}$. ε_{340nm}[NAD(P)H] = 6.22×10^3 l $mol^{-1}cm^{-1}$.

General comments. Mitochondria are re-suspended in HEPES buffer containing 5 M glycerol prior to sonic or French pressure cell disintegation. This buffer stabilizes NAD^+–isocitrate dehydrogenase and is also used for diluting mitochondrial extracts before assay.

The assay given above is suitable for use during purification procedures or in assays aimed at checking mitochondrial and bacteriod fractions of nodules for significant cross-contamination. Rokosh *et al.*, (1973) have pointed out, however, that this assay may be complicated in crude extracts by the endogenous oxidation of NADH and NADPH. These authors modified the assay by including phenazine methosulphate to prevent the accumulation of NADPH or NADH. Dichlorophenolindophenol is then used as a terminal electron acceptor to allow the continuous spectrophotometric measurement of activity. Using this procedure, Rokosh *et al.* (1973) obtained significantly higher (50%) activities for bacteriod $NADP^+$–isocitrate dehydrogenase in crude extracts. It has been our experience (P.H.S. Reynolds and K. J. F. Farnden, unpublished results) that the problem of endogenous NADH and NADPH oxidation in crude extracts can be largely overcome by passing the extracts through a Sephadex G-25 column. This observation is in agreement with the work of Cox (1969). Planqué *et al.* (1977) have reported differently, however.

(iv) 3-Hydroxybutyrate dehydrogenase

3-Hydroxybutyrate dehydrogenase (E.C. 1.1.1.30) activity has been reported to be 'unusually high' in the bacteriods of *Glycine max* nodules (Wong and Evans, 1971), but the role of the enzyme in bacteroid metabolism is not clear (see Bergersen, 1977, for discussion). 3-Hydroxybutyrate dehydrogenase is a useful marker for bacteroid cytoplasmic proteins (Godfrey *et al.*, 1975; Planqué *et al.*, 1977). The assay procedure described here is based on the method of Wong and Evans (1971).

Materials and reagents

Tris–HCl buffer, 50 mM, pH 8.0.
$MgCl_2$, 30 mM.
NAD^+, 12 mM.
Sodium DL-3-hydroxybutyrate, 0.1 M.

Procedure. Tris buffer (2 ml), $MgCl_2$ solution (0.1 ml), NAD^+ solution (0.1 ml), enzyme, and water are added to a cuvette to give a total volume of 2.8 ml. After equilibration at 25 °C, the reaction is started by the addition of 0.2 ml of the 3-hydroxybutyrate solution, and the increase in absorbance at 340 nm is monitored. 3-Hydroxybutyrate dehydrogenase activity is expressed as (μmol NADH formed) min^{-1}(mg protein)$^{-1}$. $\varepsilon_{340nm}(NADH) = 6.22 \times 10^3\ l\ mol^{-1}cm^{-1}$.

General comments. Endogenous NADH oxidation in crude extracts may make accurate estimations of 3-hydroxybutyrate dehydrogenase by the above procedure difficult (see general comments on isocitrate dehydrogenase assays). In this case, it may be preferable to assay 3-hydroxybutyrate dehydrogenase in the reverse direction using acetoacetate as substrate and NADH. A correction for non-specific NADH oxidation can then be made (Planqué *et al.*, 1977).

(v) Fumarase

Fumarase (E.C. 4.2.1.2) is present in bacteriods from legume nodules but occurs in only very low levels in the nodule plant cytosol fraction, probably as a result of leakage from mitochondria (Robertson *et al.*, 1975b). It is therefore a useful marker for determining the degree of contamination of the plant cytosol fraction by enzymes from ruptured bacteriods. The assay procedure described here is based on the methods of Racker (1950) and Hill and Bradshaws (1969).

Materials and reagents

Phosphate buffer, 100 mM (0.68 g of KH_2PO_4 dissolved in distilled water to a final volume of 50 ml after adjusting to pH 7.3 with NaOH).
Malate buffer, 250 mM (0.67 g of L-malic acid dissolved in distilled water to a final volume of 50 ml after adjusting to pH 7.3 with NaOH).

Procedure. Assay mixtures consist of 0.5 ml of phosphate buffer, 0.2 ml of malate, and 0.3 ml of enzyme. The reaction, which is carried out at 25 °C, is started by addition of enzyme. The increase in absorbance at 240 nm is compared with that of a reference cuvette which contains no malate. The reaction is linear for several minutes and the amount of fumarate produced is determined on the basis that a change of 2.4 absorbance units at 240 nm and 25 °C is equivalent to the production of 1 μmol ml^{-1} of fumarate (Bock and Alberty, 1953).

(e) MISCELLANEOUS ENZYME ASSAYS

(i) Hydrogenase

Hydrogenases from *Rhizobium* spp. and *Azotobacter* spp. catalyse the unidirectional uptake of H_2 in the presence of suitable electron acceptors and have been referred to as 'uptake hydrogenases' to distinguish them from the ATP-dependent hydrogen-evolving capacity of nitrogenase (Chapter II.3, Section d) and the reversible hydrogenases common in anaerobic bacteria (see Smith *et al.*, 1976, and Bothe *et al.*, 1977, for discussion). Uptake hydrogenases have been proposed to have an important role in increasing the efficiency of nitrogen fixation owing to their ability to recycle the hydrogen lost by the action of nitrogenase (Dixon, 1972; Smith *et al.*, 1976; Bothe *et al.*, 1977; Tel-Or *et al.*, 1977). Both uptake and reversible hydrogenases can be conveniently assayed by the simple and sensitive tritium exchange assay outlined below. This assay is based on the observation that these hydrogenases will catalyze an exchange reaction between tritium gas and water (Lim, 1978; Mortenson and Chen, 1974). Uptake or reversible hydrogenases have also been assayed manometrically or spectrophotometrically (see Mortenson and Chen, 1974, for discussion), amperometrically (Wang, *et al.*, 1971), or gas chromatographically (Tel-Or *et al.*, 1977).

Apparatus

Disposable culture tubes (16 × 125 mm) and rubber stoppers (serum style) (A. H. Thomas Co., Philadelphia, Pa., U.S.A.).
Gas-tight syringes (Hamilton Co., Reno, Nev., U.S.A.).

Materials and reagents

Tritium gas (2.27 mCi ml^{-1}) (Lawrence Radiation Laboratory, Berkeley, Calif., or New England Nuclear, Boston, Mass., U.S.A.).
PIPES buffer, 50 mM, pH 6.8.
Aquasol-2 (New England Nuclear).
Argon.

Procedure. Tritium (1 μCi, 0.5 μl) is injected into a disposable culture tube (sealed with a serum stopper and under an atmosphere of argon) containing 1 ml of a suspension of either free-living bacteria or bacteroids (about 10^8 cells ml^{-1} in 50 mM PIPES, pH 6.8). Culture tubes are incubated at 25 °C for

1 h with gentle rotary shaking (150 r.p.m.). The reaction is stopped by the addition of 1 ml of methanol. Aliquots (0.1 ml) are removed with a syringe, dissolved in 10 ml of Aquasol-2 and counted in a liquid scintillation spectrometer. Hydrogenase activity determined by this tritium-exchange procedure is expressed as cpm h^{-1}(mg protein)$^{-1}$.

General comments. Hydrogenase from *Rhizobium* spp. has been found to be readily inactivated by oxygen (S. T. Lim, personal communication) and anaerobic procedures should therefore be used when performing the assay. Transfers should be made using gas-tight syringes and buffers deoxygenated on a gassing manifold by evacuating and flushing several times with argon. More than 99% of the radioactivity in the cell suspension at the end of the reaction is found in water, so a measure of the radioactivity in the suspension will give a reliable measurement of the rate of reaction.

This tritium exchange assay has been used (Lim 1978) to survey ten *Rhizobium japonicum* strains for hydrogenase activity (synthesized in free-living cultures only under micro-aerophilic conditions; oxygen concentrations less than 0.5%). Strains which exhibited tritium exchange were found to have H_2 uptake activity (measured by gas chromatography) and relative efficiencies of 0.94–1.00, calculated by the method of Schubert and Evans (1976), using root nodules from soybeans inoculated with the same strains. Strains which had no detectable hydrogenase activity in the tritium exchange assay had no hydrogen uptake activity and had relative efficiencies of 0.6–0.7 (Lim, 1978).

The procedure described above for measuring uptake hydrogenase in free-living cultures of *R. japonicum* has also been used successfully to demonstrate hydrogenase activity in intact legume root nodules (S. T. Lim, personal communication).

(ii) Adenosine Triphosphatase

Adenosine triphosphatase (ATPase) activity is commonly detected in soluble and membrane fractions from nitrogen-fixing organisms and from legume nodules. The assay procedure described here involves the incubation of ATP with cell fractions followed by the determination of the released phosphate using the method of Boyer *et al.* (1959) as modified by Woolfolk *et al* (1966). This method is particularly useful for assaying ATPase activity because of the low rate of ATP hydrolysis during colour development. The assay can be used as a general technique for phosphatases in the range 0–0.25 μmol of phosphate.

The luciferin–luciferase assay (Section f.iii, for measuring ATP) can also be used to measure ATPase activity (determined as removal of ATP with time).

Materials and reagents

Tris base, 250 mM, adjusted to pH 5.5–8.5 using maleic acid.
$ATPNa_2$ or ATP–Tris, 30 mM.
$MgCl_2$, 15 mM.
KCl, 0.5 M.
$Na_2HPO_4 \cdot H_2O$, 1.25 mM.
$FeSO_4 \cdot 7H_2O$: 0.8 g dissolved in 15 ml of 0.05 M H_2SO_4 and diluted to 100 ml with distilled water.
$(NH_4)_6Mo_7O_{24} \cdot 4H_2O$; 6.6 g dissolved in 50 ml of 7.5 M H_2SO_4 and diluted to 100 ml with distilled water.
Trichloroacetic acid, 20% w/v.

Procedure. Assay mixtures consist of 0.2 ml Tris–maleate buffer of the desired pH, 0.1 ml of 30 mM $ATPNa_2$ or ATP–Tris where sodium ion-free systems are required, 0.1 ml of 15 mM $MgCl_2$, 0.1 ml of 0.5 M KCl, and 0.5 ml of enzyme. The $MgCl_2$ and KCl are optional, depending on whether enzyme activity is stimulated by these salts. The mixture is incubated for 30 min at the temperature at which the microorganisms or plants are grown. The reaction is stopped by the addition of 0.4 ml of ice-cold 20% trichloroacetic acid and the tubes are centrifuged at 10000 g for 10 min. To 0.2-ml aliquots of supernatant is added 1.8 ml of freshly prepared $FeSO_4$ reagent followed by 0.15 ml of ammonium molybdate reagent. The solutions are mixed and the absorbance is read at 660 nm after 5 min. The absorbance is compared with that of a standard series prepared using 0–0.20 ml of the 1.25 mM phosphate solution, adjusted to a final volume of 0.20 ml with distilled water.

General comments. Where neutral detergents such as Triton X-100 are present in fractions being assayed for ATPase activity, the detergents are likely to interfere with determinations for phosphate and protein. This problem can be overcome by addition of sodium dodecylsulphate to the systems of assay (Dulley, 1975; Dulley and Grieve, 1975).

(iii) Asparaginase

Asparaginase (E.C. 3.5.1.1) activity has been reported to occur in the maturing seeds of *Lupinus albus* (Atkins *et al.*, 1975) and the enzyme has now been purified from maturing *Lupinus polyphyllus* seeds (Lea *et al.*, 1978). Asparaginase activity has been detected in the plant fraction of *Lupinus angustifolius* nodules prior to the onset of N_2 fixation (Scott *et al.*, 1976). This plant enzyme was different from the bacteroid or free-living *Rhizobium lupini* asparaginase in terms of electrophoretic mobility on polyacrylamide gels and apparent K_m for asparagine (7 mM for the plant enzyme and 6 μM for the rhizobial enzyme) (Scott *et al.*, 1976). K_m values of 10 and 12 mM have been reported for the *L. albus* and *L. polyphyllus* maturing seed enzymes (Atkins *et al.*, 1975; Lea *et al.*, 1978). The enzyme in *Glycine max* bacteriods has a K_m for asparagine of 5 μM (Streeter, 1977).

Asparaginase activity can be assayed radiochemically by following the conversion of [^{14}C]asparagine to [^{14}C] aspartate or by the direct nesslerization of the ammonia produced. The latter assay is based on the procedure used by Castric *et al.*, 1972 to assay β-cyanoalanine hydrolase.

Radiochemical asparaginase assay

Apparatus

High-voltage paper electrophoresis equipment.

Materials and reagents

L-Asparagine, 60 mM.
L-[U-^{14}C]Asparagine (133 mCi mmol^{-1}; 10 μCi ml^{-1}).
Potassium phosphate buffer, 50 mM, pH 8.0.
Ethanol.
Whatman 3MM chromatography paper.
Marker amino acids, asparagine and aspartic acid (2.5 mg ml^{-1}).
Electrophoresis buffer, 0.04 M sodium acetate, pH 5.0.
Ninhydrin (0.2% w/v in acetone).
Scintillation fluid (0.03% w/v POPOP, 0.3% w/v PPO in toluene).

Procedure. Asparaginase assays contain 100 μl of 60 mM asparagine, 20 μl of [U-^{14}C]asparagine, 50 mM phosphate buffer (pH 8.0), and enzyme in phosphate buffer to give a final volume of 300 μl. Samples (20 μl) are removed at zero time and at various other times during a 30-min incubation period and added to 40 μl of ethanol in a centrifuge tube. The tubes are centrifuged at 6000 *g* for 10 min and 20-μl aliquots of the supernatant are spotted on to Whatman 3MM paper (2 cm apart, midway from the ends). Marker amino acids (5 μl) are spotted on top of the samples. Asparagine is separated from aspartic acid by electrophoresis for 45 min at pH 5.0 in 0.04 M sodium acetate at a potential gradient of 30 V cm^{-1} in a tank under Shell-Sol T kept at 5–15 °C. The amino acids are located by spraying with ninhydrin and heating the paper at 50 °C for 5 min. Amino acid spots are cut out from the paper, immersed in scintillation fluid in a counting vial, and the radioactivity on the paper is determined in a liquid scintillation spectrometer. Background counts per minute are subtracted before calculation of the results.

Calculation of results

$$\text{Aspartate produced } (\mu\text{mol min}^{-1}) = \frac{\text{cpm Asp}}{\text{cpm Asp} + \text{cpm Asn}} \times \frac{\mu\text{mol Asn in assay}}{\text{incubation time}}$$

Results are expressed as (μmol aspartate) min^{-1}(mg protein)$^{-1}$.

General comments. Excessive ninhydrin colour development on the paper will result in significant quenching of counts. The paper should therefore be lightly sprayed with ninhydrin and heated only until the spots are just visible.

Identification of the product as aspartate and a check that other amino acids which co-electrophorese with asparagine have not become labelled can be carried out by separating aliquots of the supernatant by two-dimensional electrophoresis or chromatography (see Scott *et al.*, 1976, and Streeter, 1973, for methods).

An alternate method for the separation of aspartic acid from asparagine involving small columns of ion-exchange resin has been used by Streeter (1977).

Spectrophotometric (Nesslerization) assay

Materials and reagents

L-Asparagine, 100 mM.

Potassium phosphate buffer, 50 mM, pH 8.0.

Trichloroacetic acid, 20% w/v.

Nesslers reagent (formula of Bock and Benedict, as described by Hawk, 1965). Place 100 g of mercuric iodide and 70 g of potassium iodide in a 1-l volumetric flask, add about 400 ml of water, and stir until dissolved. Dissolve 100 g of NaOH in about 500 ml of water, cool, and add it with constant stirring to the mixture in the flask; then dilute with water to the 1-l mark. Allow any sediment formed to settle and decant the supernatant for use.

Procedure. Reaction mixtures contain 0.5 ml of 100 mM L-asparagine, 50 mM phosphate buffer (pH 8.0), and enzyme in phosphate buffer to give a final volume of 1.5 ml. Mixtures are incubated for various times (0–30 min) and the reaction is stopped by the addition of 20% trichloroacetic acid (0.1 ml). The samples are centrifuged (6000 *g* for 5 min) and the supernatant solution is added to 6 ml of distilled water. Nesslers reagent (1 ml) is added and the absorbance read at 480 nm after 10 min. Absorbances are converted into micromoles of ammonia from a calibration graph prepared using 0–1.5 ml of 2 mM $(NH_4)_2SO_4$ (4 mM ammonium). Activity is expressed as (μmol ammonia produced) min^{-1}(mg protein)$^{-1}$. Alternatively, the Chaney and Marbach reagent may be used (Chapter II.2, Section b.vi) to estimate the ammonia formed.

General comments. This assay is less sensitive and requires more enzyme than the radiochemical assay. It is useful, however, for the assay of column fractions produced during purification procedures.

(iv) Glutaminase

Glutaminase (E.C. 3.5.1.2) is present in the bacteroids of *L. angustifolius* and *Glycine max* nodules (Robertson *et al.*, 1975a; Streeter, 1977). Extracts of free-living *Rhizobium lupini* have been found to hydrolyse both glutamine and, to a lesser extent, γ-glutamylhydroxamate. The hydrolysis of both of these compounds can be inhibited by L-5-diazo-4-oxonorvaline (Robertson *et al.,* 1975a).

Glutaminase activity can be determined radiochemically or spectrophotometrically using the procedures outline for asparaginase assays (Section d.iii), except that L-glutamine is used as substrate instead of L-asparagine. If the nesslerization assay is used, the tubes must be read exactly 10 min after adding the Nesslers reagent as a small, but significant, time-dependent, non-enzymatic release of ammonia from glutamine will occur. Asparagine is stable under the same conditions.

(v) Nitrate reductase

The most sensitive method for assaying nitrate reductase (E.C. 1.6.6.1) in plant and bacterial extracts is by the spectrophotometric measurement of the nitrite produced. Nitrite is diazotized with sulphanilamide and then reacted with N-(1-naphthyl)ethylenediamine dihydrochloride to produce an azo dye which is measured spectrophotometrically at 540 nm. Procedures for both the *in vitro* and *in vivo* determination of nitrate reductase activity in plants are described here. In addition, assays for the soluble and particulate nitrate reductase of *Rhizobium japonicum* are also presented.

Nitrite determination

This method is based on the method of Nicholas and Nason (1957).

Reagents and materials

N-(1-Naphthyl)ethylenediamine dihydrochloride, 0.02% w/v.
Sulphanilamide, 1% w/v in 3 M HCl.
Sodium nitrite, 1 mM; diluted to 40 μM prior to use.

Procedure. The concentration of nitrite in a sample is determined by adding an equal volume of a 1:1 (v/v) mixture of the naphthylethylenediamine and sulphanilamide reagents. After 10 min, the absorbance is determined at

540 nm. Absorbances are converted into micromoles of nitrite from a calibration graph prepared using 0–1 ml of 40 μM sodium nitrite.

In vitro *Assay for plant nitrate reductase*

This method is based on that described by Neyra and Hageman (1975).

Materials and reagents

Potassium phosphate buffer, 0.1 M, pH 7.5.
Potassium nitrate, 0.1 M.
NADH, 2 mM.
Zinc acetate, 0.5 M.
Phenazine methosulphate, 0.15 mM.

Procedure. The assay mixtures contain 0.5 ml of potassium phosphate buffer (pH 7.5) 0.2 ml of potassium nitrate, 0.2 ml of NADH, enzyme, and distilled water to give a total volume of 2 ml. The reaction is initiated by the addition of enzyme. A zero time and a minus NADH reaction mixture are used as controls. After incubation at 30°C for 15 min the reaction is stopped by the addition of 0.2 ml of 0.5 M zinc acetate. Phenazine methosulphate (0.2 ml) is added to oxidize the excess of NADH, which would interfere with colour development (Scholl *et al.,* 1974). After 20 min, the reaction mixtures are centrifuged at 1000 *g* for 10 min, and the concentration of nitrite in the supernatants is determined spectrophotometrically as described above. Nitrate reductase activity is expressed as (μmol nitrite formed) min^{-1}(mg protein)$^{-1}$.

General comments. Nitrate reductase requires sulphydryl protection, and agents such as cysteine, glutathione, or dithiothreitol should be included in the extraction buffer (Hageman and Hucklesby, 1971). For tissues known to contain high protease activity, protection can be afforded by the inclusion of casein or phenylmethylsulphonyl fluoride in the homogenizing buffer. It is common in the preparation of leaf extracts to use a 25 mM phosphate buffer adjusted to a pH of up to 8.8 so that the homogenate will have a final pH of 7.5 after grinding (Hageman and Hucklesby, 1971). Extracts should be passed through a Sephadex G-25 column before assay to remove inhibitory low molecular weight compounds.

In vivo *Assay for plant nitrate reductase*

The method described here has been widely used for assaying nitrate reductase in leaf tissue and is based on the procedure of Jaworski (1971) as modified by Nicholas *et al* (1976). Nitrate is reduced to nitrate in the dark using endogenous NADH as the reductant. The reaction is carried out in the dark to prevent the further reduction of nitrite to ammonia. A surfactant such as propan-1-ol is used to increase the permeability of the tissue to nitrate and nitrite.

Materials and reagents

Potassium phosphate buffer, 0.2 M, pH 7.5.
Propan-1-ol.
Potassium nitrate, 0.5 M.

Procedure. Incubations contain 5 ml of potassium phosphate buffer, 0.1 ml of propan-1-ol, 1 ml of potassium nitrate, and water to give a final volume of 10 ml. Weighed leaf samples (200 mg; whole, sliced, or punched discs) are placed in the incubation medium, vacuum infiltrated twice (2 min each time), and incubated in the dark in a shaking water-bath at 30 °C. Following incubation, the concentration of nitrite released into the medium is determined as described above. Nitrate reductase activity is expressed as (μmol nitrite formed) h^{-1} (g fresh weight)$^{-1}$.

General comments. Optimum conditions of assay (nitrate, phosphate, and propan-1-ol concentrations and pH) need to be established for different plant tissues (see Nicholas *et al.*, 1976). This assay cannot be used for assaying nitrate reductase in roots as nitrite reduction in this tissue is not light dependent.

Rhizobium nitrate reductase

R. japonicum is known to contain both a soluble (Lowe and Evans, 1964) and a particulate (Cheniae and Evans, 1959) nitrate reductase. The particulate enzyme can utilize viologen dyes (reduced by dithionite), succinate, or NADH as electron donors. Reduced viologen dyes are the only suitable electron donors for the soluble enzyme.

Materials and reagents

Potassium phosphate buffer, 0.2 M, pH 7.0.
Potassium nitrate, 0.1 M.
Methyl viologen or benzyl viologen, 1.5 mM.
Dithionite reagent (40 mg of sodium dithionite, 40 mg of sodium hydrogen carbonate dissolved in 10 ml of water just prior to use).

Procedure. Reaction mixtures contain 0.2 ml of phosphate buffer, 0.2 ml of potassium nitrate, 0.1 ml of viologen dye, enzyme, and water to 0.8 ml. The reaction is started by the addition of 0.2 ml of the dithionite reagent.

Following the incubation of the mixtures at 30 °C for up to 10 min, the reaction is stopped by vigorously shaking the mixture until the dithionite is completely oxidized and the dye colour has disappeared. The concentration of nitrite in aliquots of the reaction mixture is determined as described above. Nitrate reductase activity is expressed as (μmol nitrite produced) min^{-1}(mg protein)$^{-1}$.

General comments. Succinate (0.2 M, 0.1 ml) or NADH (2 mM, 0.1 ml) can be used instead of the viologen dyes and dithionite reagent in assays for the particulate enzyme.

Viologen dyes will also act as electron donors for the reduction of nitrite by nitrite reductase, so the assay described above can only be used in extracts free of nitrite reductase. Nitrite reductase has been found in anaerobic-nitrate-grown *Rhizobium japonicum* cells and not in bacteroids or aerobically grown cells (Daniel and Appleby, 1972). If nitrite reductase is present, nitrate reductase can be assayed by the rate of nitrate disappearance (Daniel and Appleby, 1972).

It is important to ensure that all the dithionite is oxidized at the completion of the reaction, since dithionite will interfere in the nitrite determination. Kennedy *et al.* (1975) have, for example, stopped the reactions by agitation in a stream of oxygen. They also demonstrated that the practice of using dithionite with sodium hydrogen carbonate to maintain anaerobiosis slightly depresses the nitrate reductase activity. However, for routine assays, the method above is more convenient than carrying out the reaction in flasks under nitrogen or argon.

The inclusion of reducing agents (mercaptoethanol, dithiothreitol, or glutathione) in the buffers used to prepare crude extracts of the soluble nitrate reductase from *R. japonicum* has been found to inhibit the enzyme by 50–80%. Potassium cyanate however, has been found to stabilize the enzyme during assay if this compound is added to the assays at a final concentration of 2 mM (Villalobo *et al.*, 1977).

(vi) Nitrite reductase

Nitrite reductase can be assayed by following the rate of removal of nitrite. The assay described below is based on the procedures of Miflin (1967). The enzyme has been found in anaerobic-nitrate-grown cells of *Rhizobium japonicum* (Daniel and Appleby, 1972). In plants the enzyme is thought to be located in the plastids of leaves and roots (Miflin, 1974).

Materials and reagents

Potassium phosphate buffer, 0.2 M, pH 7.0.

Sodium nitrite, 5 mM.
Methyl viologen, 1.5 mM.
Dithionite reagent (see nitrate reductase assay above).

Procedure. Reaction mixtures contain 0.2 ml of phosphate buffer, 0.1 ml of sodium nitrite, 0.1 ml of methyl viologen, enzyme, and water to 0.8 ml. The reaction is started by the addition of 0.2 ml of the dithionite reagent. Following the incubation of the mixtures at 30 °C for 10 min, the reaction is stopped by vigorously shaking the mixtures until the dithionite is completely oxidized and the dye colour has disappeared. The concentration of nitrite remaining in a suitably diluted aliquot of the reaction mixture is determined by the procedure given above for the nitrate reductase assays. Nitrite reductase activity is expressed as (μmol nitrite removed) min^{-1}(mg $protein)^{-1}$.

General comments. Nitrite is very susceptible to reduction so it is important to include controls such as minus enzyme and boiled enzyme.

Dithiothreitol cannot be used as a sulphydryl-protecting agent in the homogenizing buffer, as a non-enzymatic reduction of nitrite will occur in the presence of this compound, viologen dyes, and dithionite.

(f) MEASUREMENT OF RESPIRATION

(i) Uptake of O_2

Manometry

Manometric techniques are based upon the gas law $PV = RT$, where P is pressure, V is volume, T is absolute temperature, and R is the gas constant. The Warburg constant-volume manometer is probably the most commonly used manometric method. In this procedure the reaction is carried out in a small flask which is attached to a manometer. The gas in the flask is maintained at constant volume and temperature, and changes in the quantity of gas (e.g. O_2 uptake) are measured by a change in pressure. Respiratory uptake of O_2 by cells or tissues results in a release of CO_2. If these two gases (CO_2 and O_2) are the only ones involved, the respiration (O_2 uptake) can be measured by absorbing the liberated CO_2 in alkali. This method is known as Warburg's 'Direct method' and depends on the assumption that the absence of CO_2 in the reaction flask will not affect the rate of oxygen uptake. For measurements in the presence of CO_2, the Warburg 'indirect method' or the Pardee method may be used to measure O_2 uptake. Details of each of these methods (apparatus, principle, operation, and applications) are given by Laser (1961), Wilson (1975) and Umbreit *et al.* (1972), the last being a comprehensive text on manometric techniques.

The oxygen electrode

Electrode measurements of O_2 consumption (or evolution) offer several advantages over manometric procedures: (i) rapid response; (ii) greater sensitivity; (iii) multiple additions of substrates and/or inhibitors or effectors are readily made within a single experiment of short duration; and (iv) measurements are made without the removal of CO_2.

Several different O_2 electrodes have been designed, but the Clark electrode (Yellow Springs Instrument Company, Yellow Springs, Ohio, U.S.A.) or variations of the Clark electrode, e.g. Rank electrode (Rank Bros., Bottisham, Cambridge, U.K) are the most commonly used. A diagram of a Rank electrode and reaction chamber is shown in Figure 3. The electrode consists of a platinum cathode and a silver anode in saturated KCl. When a polarizing voltage is applied across the cell, oxygen is reduced electrolytically:

Anode $4Ag + 4Cl^- \rightarrow 4AgCl + 4e$
Cathode $4H^+ + 4e + O_2 \rightarrow 2H_2O$

Sum $4Ag + 4Cl^- + 4H^+ + O_2 \rightarrow 4AgCl + 2H_2O$

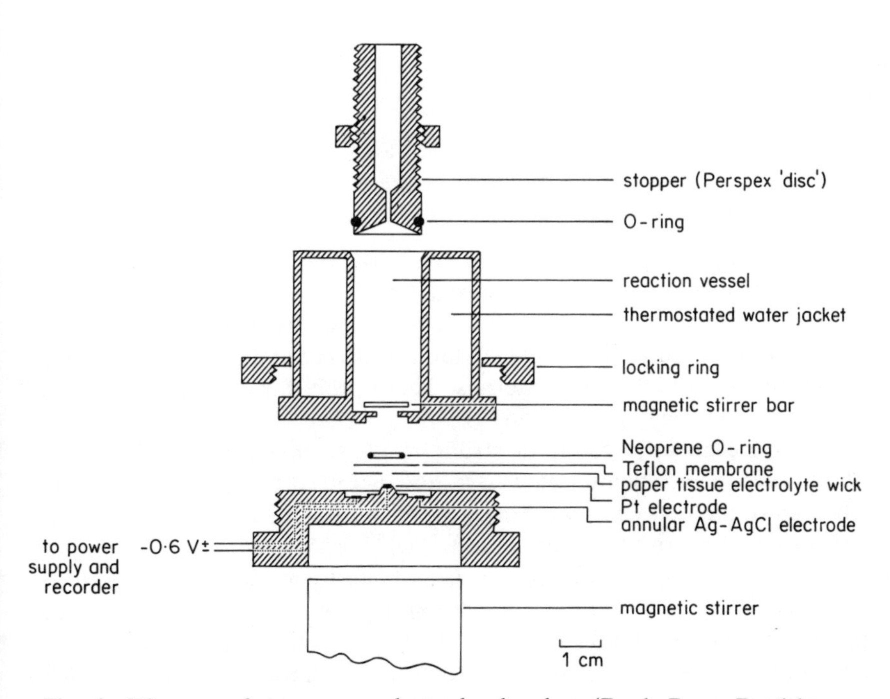

Fig. 3. Diagram of an oxygen electrode chamber (Rank Bros, Bottisham, Cambridge, UK)

At the constant applied voltage, the current generated is proportional to the amount of oxygen diffusing across the membrane which protects the platinum cathode. The membrane is commonly Teflon (PTFE), polyethylene, or polypropylene, but sometimes polycarbonate films are used. The current depends on the temperature of the Teflon membrane and the partial pressure of oxygen in the sample chamber. The reaction vessel is maintained at a constant temperature and the solution to be measured is stirred to prevent oxygen-depleted regions occurring around the cathode. The electrode is attached to a chart recorder and calibrated with air-saturated solutions and zero-oxygen solutions (the latter obtained by flushing with nitrogen, addition of $Na_2S_2O_4$, or the addition of a biological oxygen-consuming system). These solutions are then replaced with the test solutions (temperature equilibrated) and changes in oxygen content are recorded directly on the chart recorder. Small additions of reagents to the reaction vessel can be made with a syringe during the course of the experiment through the bubble escape slit in the stopper.

The rate of oxygen consumption is calculated as follows. If the chart recorder was set at zero for the anaerobic solution and at 100 for the air-saturated solution, then

$$\text{Rate of } O_2 \text{ consumption per minute} = \frac{X\,C\,V}{100}$$

where

X = pen deflection min^{-1};
V = volume of reaction system that was air saturated;
C = O_2 concentration per millilitre of the air-saturated solution at the temperature of assay.

Tables giving the solubility of O_2 in a buffered medium or water at various temperatures were given by Lessler and Brierly (1969) and Cooper (1977). The oxygen content of solutions will vary with ionic strength and more precise methods of calibration are available to take into account the effects of complex buffers containing biological materials. These procedures are presented and discussed by Beechey and Ribbons (1972).

Equally, the linearity of the electrode response should be checked, especially if reactions are being followed at low dissolved O_2 concentrations (the likely area of interest when working with bacteroids). Bergersen and Turner (1975a) found linearity from air-saturated water (244 μM O_2, at 25°C and 700 mmHg) to 1.6 μM O_2, but departures from linearity at lower concentrations. They constructed calibration graphs of log mV *versus* log O_2 concentration, which were checked by injecting small volumes of water containing O_2 concentrations into a sample of reaction solution equilibrated at low pO_2. With these methods, O_2 concentrations down to 0.05 μM could be measured reliably and estimates made down to 0.02 μM. However, Lundsgaard *et al.* (1978)

have shown that substantial inaccuracies are introduced at very low concentrations of dissolved O_2, especially when the K_m for O_2 consumption is below 1–0.1 μM.

Since the commercially available instruments are supplied with detailed operating instructions, these will not be covered here. This information is available, however, in Cooper (1977). Examples of oxygen electrode measurements were given by Lessler and Brierly (1969), Beechey and Ribbons (1972), and Cooper (1977).

Inhibitors of respiration

Inhibitors of electron transport and oxidative phosphorylation have played an important role in developing our understanding of cellular respiration. In general, the same inhibitory agents that have been used in investigations of the respiratory chain in mitochondria can also be used in studies of electron transport and oxidative phosphorylation in bacterial membrane systems. Information on the inhibitors, their site of action, and their use in microbial studies is given by Heinen (1971).

(ii) CO_2 Evolution

Manometry

The 'direct method' of Warburg allows the simultaneous manometric determination of O_2 uptake and CO_2 evolution. The method uses two reaction flasks. One contains alkali in the centre well to absorb CO_2, and therefore records O_2 uptake. The other flask lacks alkali, and therefore records the balance between O_2 uptake and CO_2 evolution. From these two measurements it is possible to calculate the volume of CO_2 evolved (Laser, 1961; Wilson, 1975; Umbreit *et al.*, 1972). This method suffers from the difficulty that one tissue, respiring in the absence of CO_2, is compared with another, respiring in the presence of CO_2. Methods are available to determine whether the presence or absence of CO_2 effects the rate of respiration. If an effect of CO_2 is observed then alternate procedures (outlined in the references cited above) must be used to measure CO_2 formation and O_2 uptake.

Gas chromatography

CO_2 evolution by bacteroid suspensions has been determined by McCrae *et al.* (1978) using a Carle 8500 gas chromatograph. A 45.7 cm × 3.2 mm column was packed with Porapak Q (Waters Associates, Framington, Mass. U.S.A.). Helium was used as the carrier gas at a flow-rate of 17 ml min^{-1}, and the column temperature was 72°C.

Radiorespirometry

The term 'radiorespirometry' (Wang, 1967) refers to a type of experiment in which the rate and extent of the evolution of respiratory $^{14}CO_2$ by a biological system metabolizing substrates uniformly or specifically labelled with ^{14}C are determined.

Radiorespirometry can be used (i) to determine whether a given compound can be utilized by a microorganism, (ii) to identify the occurrence of certain catabolic pathways in the system, and (iii) to estimate the relative participation of concurrent catabolic pathways (e.g. pathways of glucose catabolism). The apparatus required, the experimental design, and the applications of radiorespirometric methods are given in detail by Wang (1967, 1972).

Glucose catabolism in *Rhizobium japonicum* has been investigated radiorespirometrically by Keele *et al.* (1969) and Mulongoy and Elkan (1977).

(iii) Measurement of ATP, ADP, and AMP

Enzymatic procedures for the determination of ATP, ADP, and AMP employing pyridine nucleotide linked reactions are given below. In addition, the determination of ATP with luciferase is also included. This procedure is more sensitive than the pyridine nucleotide linked assays and can be adapted to measure ADP and AMP or ATPase activity (see general comments on luciferase procedure).

Extraction of ATP, ADP, and AMP

For cell suspensions (bacteria, algae) the method of Cole *et al.* (1967) can be used. Inject the cell sample rapidly into ice-cold 30% w/v $HClO_4$ to give a final concentration of 6% $HClO_4$. After 10 min on ice, the extract is shaken well and then neutralized with 1 M KOH. The $KClO_4$ precipitate is allowed to settle (10 min on ice) and the extract is then centrifuged at 6000 *g* for 10 min at 4 °C. These deproteinized and neutralized extracts can be stored for several months at −20 °C without loss of adenine nucleotides. For plant tissues the sample can be frozen in liquid nitrogen in a mortar and then ground to a fine powder with a chilled pestle. Transfer the powdered tissue to a pre-cooled, weighed centrifuge tube and record the weight of the sample. Up to this stage the tissue must be kept frozen. Add 4 volumes of 6% w/v $HClO_4$ to the sample and homogenize with a motor-driven PTFE pestle until the sample has thawed evenly into the acid. Centrifuge at 30 000 *g* for 15 min at 4 °C. The supernatant is neutralized with 1 M KOH and the $KClO_4$ precipitate removed as described above.

Some workers have experienced difficulties with the enzymatic estimation of adenine nucleotides in leaf extracts owing to high rates of endogenous

NADH oxidation and inhibition of coupling enzymes. In this case, further pre-treatment of the extract is necessary (see Macnicol, 1972, for the procedure).

Recoveries should be checked by adding internal standards of ATP, ADP, and AMP during the homogenization in the above procedures. Recoveries of 85% or higher should be achieved.

Measurement of ATP with hexokinase and glucose-6-phosphate dehydrogenase

The method is based on that of Lamprecht and Trautschold (1974) and involves the following reactions:

$$\text{ATP} + \text{glucose} \xrightarrow{\text{hexokinase}} \text{glucose-6-phosphate} + \text{ADP}$$

$$\text{Glucose-6-phosphate} + \text{NADP}^+ \xrightarrow{\text{glucose-6-phosphate dehydrogenase}} \text{6-phosphoglucono-}\delta\text{-lactone} + \text{NADPH} + \text{H}^+$$

NADPH formation is measured spectrophotometrically. For each mole of ATP present, 1 mol of NADPH is formed.

Materials and reagents

Tris–HCl buffer, 0.1 M, pH 7.4.
$NADP^+$, 10 mM.
$MgCl_2 \cdot 6H_2O$, 0.1 M.
Glucose, 0.5 M.
Glucose-6-phosphate dehydrogenase, from yeast: crystalline suspension in 3.2 M ammonium sulphate, specific activity >140 U mg^{-1} (1 mg ml^{-1}).
Hexokinase, from yeast: crystalline suspension in 3.2 M ammonium sulphate, specific activity >140 U mg^{-1} (2 mg ml^{-1}).
Before the start of the experiment, bring these solutions and the ATP samples to 25 °C.

Procedure. Prepare a reaction mixture containing 10 volumes of buffer, 1 volume of $NADP^+$, 2 volumes of $MgCl_2$, 0.5 volume of glucose, and 0.05 volume of glucose-6-phosphate dehydrogenase. Add 1 ml of the reaction mixture to a reference cuvette and 1 ml to a sample cuvette. Add 1 ml of water to the reference cuvette. The ATP sample (up to 0.1 μmol; deproteinized and neutralized) and water are added to the second cuvette to give a total volume of 2 ml. Mix and read the absorbance (A_1) at 340 nm at 5 and 10 min, or until the values are constant. Add 5 μl of the hexokinase solution to each cuvette, mix, and read the absorbance (A_2) at 5 and 10 min, or until the values are constant. The concentration of ATP in the sample is then calculated as follows [ε_{340nm}(NADH) = 6.22×10^3 l mol^{-1} cm^{-1}]:

$$\text{ATP concentration } (\mu\text{mol ml}^{-1}) = \frac{(A_2 - A_1) \times 2}{6.22 \times \text{sample size (ml)}}$$

This value must be multiplied by an appropriate factor to take into account any dilutions that occurred due to the pre-treatment of the sample (e.g. in the deproteinization and neutralization).

General comments. Contamination of the glucose-6-phosphate dehydrogenase with hexokinase (even low levels of 0.1% or less) will initiate a slow but detectable conversion of ATP. This problem may be overcome by not adding the glucose solution until the absorbance of the reaction mixture (now minus glucose) plus the ATP sample is constant. The glucose solution is now added, the absorbance recorded (A_1), followed immediately by the addition of the hexokinase. A_2 is then recorded as above.

Measurement of ADP and AMP with myokinase, pyruvate kinase, and lactate dehydrogenase

This method is based on that described by Jaworek *et al.* (1974). The reactions involved are:

$$\text{AMP} + \text{ATP} \xrightleftharpoons{\text{myokinase}} 2\text{ADP}$$

$$2\text{ADP} + 2\text{phosphoenolpyruvate} \xrightleftharpoons{\text{pyruvate kinase}} 2\text{ATP} + 2\text{pyruvate}$$
$$2\text{Pyruvate} + 2\text{NADH} + 2\text{H}^+ \xrightleftharpoons{\text{lactate dehydrogenase}} 2\text{lactate} + 2\text{NAD}^+$$

ADP is measured first using pyruvate kinase and lactate dehydrogenase. For each mole of ADP present, 1 mol of NADH is oxidized (measured spectrophotometrically at 340 nm). AMP is then measured by the addition of myokinase to the same assay system. One mole of AMP will bring about the oxidation of a further 2 mol of NADH.

Materials and reagents

Tris–HCl buffer, 0.1 M, pH 7.4
$MgCl_2 \cdot 6H_2O$, 0.1 M.
KCl, 1.0 M.
Phosphoenolpyruvate, 0.02 M.
ATP, 0.01 M.
NADH, 7.5 mM.
Lactate dehydrogenase, from rabbit muscle: crystalline suspension in 3.2 M ammonium sulphate, specific activity >500 U mg^{-1} (5 mg ml^{-1}).
Myokinase, from rabbit muscle: crystalline suspension in 3.2 M ammonium sulphate, specific activity >360 U mg^{-1}(5 mg ml^{-1}).

Pyruvate kinase, from rabbit muscle; crystalline suspension in 3.2 M ammonium sulphate, specific activity 200 U mg^{-1} (10 mg ml^{-1}).

Bring the solutions and samples to 25 °C before the start of the experiment.

Procedure. Prepare a reaction mixture containing 7.5 volumes of buffer, 1 volume of $MgCl_2$, 0.5 volume of KCl, 0.5 volume of phosphoenolpyruvate, 0.6 volume of NADH, 0.05 volume of ATP and 0.1 volume of lactate dehydrogenase. Add 0.5 ml of the reaction mixture and 1.5 ml of water to a reference cuvette. Add 1 ml of the reaction mixture to the assay cuvette plus the sample (up to 0.2 μmol of AMP + ADP) and water to give a total volume of 2 ml. Mix and read the absorbance (A_1) at 340 nm until the values are constant. Add 10 μl of the pyruvate kinase solution, mix, and read the absorbance (A_2) at 5-min intervals until the values are constant. Add 10 μl of the myokinase solution, mix, and again read the absorbance (A_3) at 5-min intervals until the values are constant.

The concentrations of ADP and AMP in the sample are then calculated as follows [$\varepsilon_{340nm}(NADH) = 6.22 \times 10^3\ l\ mol^{-1}\ cm^{-1}$]:

$$\text{ADP concentration } (\mu mol\ ml^{-1}) = \frac{(A_1 - A_2) \times 2}{6.22 \times \text{sample size (ml)}}$$

$$\text{AMP concentration } (\mu mol\ ml^{-1}) = \frac{(A_2 - A_3)}{6.22 \times \text{sample size (ml)}}$$

These values must be multiplied by an appropriate factor to take into account any dilutions that occurred due to the pre-treatment of the sample.

Measurement of ATP with luciferase

The ATP-dependent emission of photons by the luciferin-luciferase enzyme system extracted from firefly tails provides a means by which it is possible to assay ATP levels as low as 10^{-12} mol. The number of photons produced over a definite time interval is directly proportional to the ATP concentration of the sample and can be measured using a liquid scintillation spectrometer, or other suitable photometer. The following method is based on those described by Stanley and Williams (1969) and Cole *et al.* (1967).

Apparatus. Packard Model 3330 liquid scintillation spectrometer or similar.

Materials and reagents

Sodium arsenate, 0.1 M, pH 7.4.

Glycylglycine buffer, 50 mM, containing 5 mM $MgSO_4$, pH 7.4.

ATP standard solution, 1 mM (diluted as required just prior to use).

Luciferase: suspend 75 mg of firefly tails (Sigma) in 5.0 ml of sodium arsenate buffer. Homogenize at 4°C using a glass–PTFE homogenizer periodically over a 3-h period. Centrifuge at 2000 *g* for 2 min to remove the cell debris. The supernatant containing luciferase can be stored for up to 3 days at 4 °C without excessive loss of activity.

Procedure. Equilibrate all solutions at 30 °C. Add 5.0 ml of the glycylglycine buffer, 0.1 ml of the ATP sample or standard (2–25 pmol), and 25 μl of the luciferase preparation. Mix, and initiate the counting procedure 15 s after the addition of luciferase. Samples are counted for 5 s.

General comments. The rate of photon production decreases with time. It is essential, therefore, that all determinations are made at corresponding times after the addition of the luciferase preparation. The count recorded in a defined time interval following luciferase addition is proportional to the ATP in the sample.

The sensitivity of the assay can be increased or decreased by altering the gain and discriminator settings on the spectrometer (with a Packard Model 3330 spectrometer, a 50% gain and 50–1000 discriminator divisions have been found to be satisfactory). The coincidence gate must be switched off.

The luciferase is sensitive to ageing and, in addition, alterations in the ionic composition and pH of the samples to be assayed will also affect the count recorded. Standard ATP solutions and the samples to be assayed must therefore be prepared in an identical way and counted at the same time.

ADP and AMP concentrations can be determined by first converting the ADP in an aliquot of the sample into ATP with phosphoenolpyruvate and pyruvate kinase (see above). The total ATP in this sample is then determined with luciferase. In a second aliquot, both AMP and ADP are converted into ATP with phosphoenolpyruvate, pyruvate kinase, and myokinase (see above), and total ATP is measured. Knowing the concentration of ATP in the sample originally, the ADP and AMP concentrations can be calculated from these last two assays by subtraction.

The luciferase assay can also be used to measure ATPase activity (determined as removal of ATP with time).

(g) ACKNOWLEDGEMENTS

We thank members of the Department of Biochemistry, University of Otago, and the Charles F. Kettering Research Laboratory for much helpful criticism and assistance. We also thank R. B. Peterson and C. P. Wolk, MSU-ERDA, Plant Research Laboratory, East Lansing, Michig., U.S.A., M. Boland, J. T. Christeller, and W. A. Laing, DSIR, Palmerston North, New Zealand, D. B. Scott and S. T. Lim, University of California, Davis, Calif., U.S.A., and F. J. Bergersen, Division of Plant Industry, CSIRO, Canberra, Australia, for helpful advice and unpublished information.

Methods for Evaluating Biological Nitrogen Fixation
Edited by F. J. Bergersen

C. A. Appleby and F. J. Bergersen
Division of Plant Industry,
CSIRO, Canberra, Australia.

8

Preparation and Experimental Use of Leghaemoglobin

(a) PREPARATION AND PROPERTIES

Nitrogen-fixing root nodules of legumes are the only place in which a haemoglobin is found among higher plants. Leghaemoglobin, as this haemoprotein is known, has a close relationship with root nodule activity. Therefore, methods for measuring, preparing, and experimentally using leghaemoglobin are of interest to researchers who study nitrogen fixation in legumes. There is also much current research on the chemistry and biochemistry of leghaemoglobin, but much of this work occurs in very specialized fields. We will be describing methods of general application. For specialized aspects, such as the environment of the haem pocket and ligand binding sites of leghaemoglobin, the reader should consult papers such as Aviram *et al.* (1978), Johnson *et al.* (1978), Nicola and Leach (1977a,b), Wittenberg (1978), and Wright and Appleby (1977). The amino acid sequence and molecular structure of leghaemoglobins have been discussed by Ellfolk (1972) and Vainshtein *et al*. (1977), and leghaemoglobin biosynthesis by Verma *et al*. (1974) and Verma and Bal (1976).

The following procedures have been developed in our laboratory to prepare oxyleghaemoglobin for use as an oxygen carrier for the study of nitrogen fixation by *Rhizobium* spp. and other bacteria, at defined, low concentrations of dissolved oxygen. We use soybean leghaemoglobin exclusively, since its properties, including affinity for oxygen, are well known. Also, soybean nodules contain a minimum of polysaccharide and polyphenol oxidase, which may cause difficulty during extraction and purification. The methods will be readily adaptable to the preparation of leghaemoglobins from other legumes.

Soybean nodules (and those of other legumes) contain several leghaemoglobin components (Appleby, 1974; Dilworth, 1980). We routinely use the purified, unfractionated oxyleghaemoglobins (Section a.ii) for our work. These preparations are more resistant to autoxidation than are the isolated components. Also, the unfractionated oxyleghaemoglobin may have a spread of oxygenation and other properties, making it most suitable for physiological studies. For those wishing to compare the functions of individual soybean leghaemoglobin components, a separation procedure is described (Section a.iii). Dilworth (1980) has summarized these procedures for other leghaemoglobins.

(i) Nodule growth and leghaemoglobin extraction

The procedures have been described in several publications (e.g. Appleby, 1974; Bergersen *et al*, 1973; Wittenberg *et al*, 1972; Bergersen and Turner, 1979). Soybeans (*Glycine max* Mer. *cv*. Lincoln) seeds are planted 2–2.5 cm deep in pots (diameter 25–30 cm) of a mixture of washed river gravel and vermiculite (1:1 v/v) which should be sterilized if necessary. The seeds are inoculated at planting, with a suspension from an agar culture of a suitable strain of *R. japonicum* (we use CB1809, syn. USDA 3I1b136, or CC705. syn. Wisconsin 505). The plants are grown in an unshaded glasshouse in which the maximum daytime temperature is controlled at 26–30 °C and the night temperature does not fall below 20 °C. For prompt nodulation, it is an advantage if the root temperature of the plants is maintained at 26 °C from the opening of the cotyledons for 10–15 days. After this period, when crown nodules are well developed the pots are placed in the ambient glasshouse conditions. The pots are either watered twice daily with tap water (which should be at glasshouse temperature) and once weekly with McKnight's nutrient solution (Chapter II.4), or once daily with dilute McKnight's solution (diluted 1:10 with water) and once with tap water. Figure 1 shows data for nodule growth, nitrogenase activity, and leghaemoglobin content for plants grown under these conditions. It can be seen that the best yields of leghaemoglobin are obtained when nitrogen-fixing activity is near the maximum, that is, when the

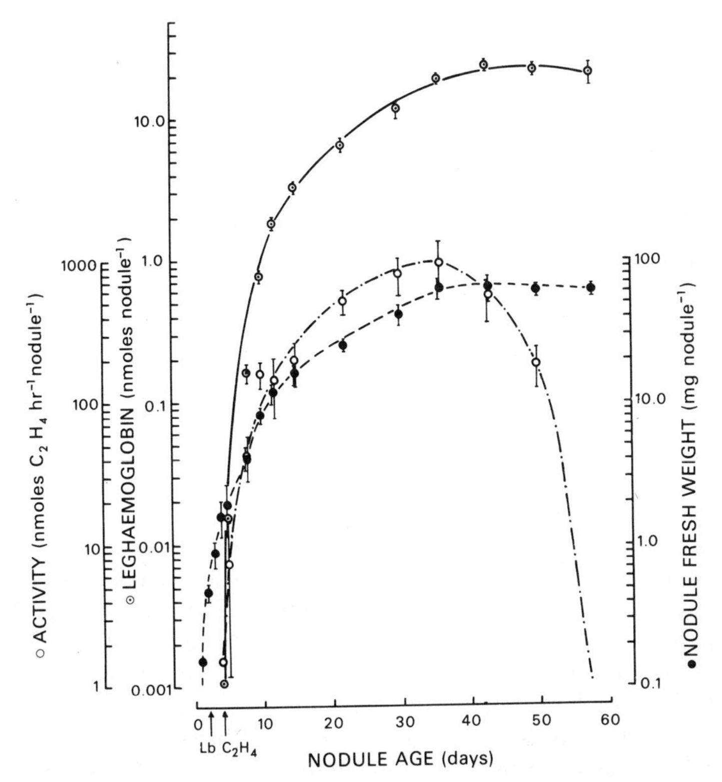

Figure 1. Time course of nodule development, nitrogenase activity and leghaemoglobin accumulation for soybeans grown as described in the text. (Reproduced from Bergersen and Goodchild, 1973, by permission of the authors)

nodules are 30–35 days old (35–45 days from planting). It is very important that nodules are harvested before any have developed green–brown centres. Failure to observe this will result in inferior preparation of leghaemoglobin, which are easily autoxidized. Other workers (e.g. Verma *et al.*, 1974) have obtained satisfactory preparations from plants grown in artificially lit growth rooms at 27–30 °C. If field-grown plants are used it is essential to use a friable soil that is low in available nitrogen and heavily inoculated. Soil temperatures in excess of 30 °C must be avoided.

At harvest, the plants are washed free of adhering growth medium and picked into liquid nitrogen for prolonged storage. Alternatively, they are picked into ice-cold 0.1 M sodium/potassium phosphate buffer (pH 7.4). These nodules may be used immediately or stored frozen for short periods.

All subsequent steps are performed at 0–4 °C and it is essential to use metal-free, glass-distilled water and highest quality metal-free phosphates, ammonium sulphate, and other reagents. If heavy metals, particularly copper, are not excluded, purified oxyleghaemoglobin will be rapidly autoxidized and become useless for physiological studies.

Fresh or thawed nodules are mixed with 1–3 volumes of air-equilibrated 0.1 M sodium/potassium phosphate (pH 7.4) containing 1 mM EDTA, and macerated in a Waring blender, Sorvall Omnimixer, or a similar device. The brei is strained through a double layer of Miracloth and cheesecloth or organdie. The nodule debris is discarded. The turbid red–brown filtrate is clarified by centrifugation at 10000 *g* for 10–30 min. The red–brown supernatant is brought to 55% saturation with ammonium sulphate (388 $g l^{-1}$ at 0 °C), adding the salt slowly with gentle stirring. When dissolved, gentle stirring is continued for at least 15 min, and then the liquid is centrifuged at 10000 *g* for 30 min. The supernatant is brought to 80% saturation by adding 177 $g l^{-1}$ of ammonium sulphate with stirring as before. The mixture is again centrifuged and the red precipitate of leghaemoglobin is dissolved in a minimum volume of 50 mM phosphate buffer (pH 7.4) containing 1 mM EDTA. This is dialysed overnight against the same buffer, using pre-washed Visking dialysis tubing. The next day, the solution is centrifuged at 100000 *g* for 30 min and the clear red supernatant is concentrated to *ca*. 5 mM (assayed by pyridine haemochrome, section b.ii) by pressure filtration over a Diaflo UM10 or PM10 membrane (Amicon, Lexington, Mass., U.S.A.) or other membrane with a 10000 molecular weight exclusion limit. At this stage these preparations contain 80–90% leghaemoglobin, which may have suffered 10–50% autoxidation. The degree of autoxidation depends on the variety of soybean used, growth conditions, nodule age, and especially the degree of heavy metal contamination of the water and reagents used. The preparations may be stored for long periods in liquid nitrogen or for shorter periods in dry ice or deep-freeze cabinets below −20 °C, or used immediately for purification by either procedure (ii) or (iii) below.

(ii) Purification of unfractionated oxyleghaemoglobin

The product of this procedure is routinely used for physiological experiments in which stable supplies of oxygen are required at defined, low concentrations of free oxygen (Sections c and d). The rationale is that high molecular weight impurities are removed from the partly oxygenated leghaemoglobin resulting from the procedure described above (i), by column chromatography on Sephadex G-75, Sephacryl 200 (Pharmacia, Uppsala, Sweden) or similar material. Next, the unfractionated leghaemoglobin is converted to its 100% reduced (ferrous) form by treatment with dithionite and re-chromatographed on Sephadex G-25 or G-15 (Appleby, 1974), Bio-Gel P-10 (Bio-Rad Laboratories, Richmond, Calif., U.S.A.; Bergersen and Turner, 1979), or an equivalent gel. This removes low molecular weight impurities, including flavins, porphyrins, excess of dithionite, and displaced ligands such as nicotinic acid. The ferrous leghaemoglobin becomes oxygenated as it moves down and is eluted from this second column.

The following is a description of a large-scale preparation in which 60 × 50 cm gel columns are loaded to their limit with partially purified leghaemoglobin. If less than 25% of this amount is available, it is better to use smaller diameter columns of the same length. A lesser purification may be achieved if dilute leghaemoglobin is used.

About 50 ml of the dialysed preparation from (i) above, containing 100–200 μmol of leghaemoglobin, is loaded on to a 60 × 5 cm column of Sephadex G-75 or Sephacryl 200, equilibrated with 50 mM phosphate buffer (pH 7.4) containing 1 mM EDTA and eluted with the same buffer at a flow-rate of 2–3 ml min^{-1}. A broad, turbid band of green-brown impurities is eluted first and discarded. The following clear, red fractions containing leghaemoglobin are pooled and concentrated to about 5–10 mM (Section b.ii). The 20–50 ml of concentrate contains a mixture of ferrous oxy and ferric leghaemoglobin, which must be converted to ferrous leghaemoglobin by adding 4 molar equivalents (i.e. 4 mol for each mole of haem) of pure sodium dithionite (e.g. May & Baker, Dagenham, Essex, U.K.), dissolved in a minimum volume of de-aerated phosphate buffer (0.1 M) (pH 7.4). The dithionite solution is added to the leghaemoglobin under N_2 and aeration of the mixture is avoided, so as to prevent the formation of hydrogen peroxide with consequent degradation of the leghaemoglobin.

Meanwhile, a 60 × 5 cm column of Sephadex G-25 or Bio-Gel P-10 is equilibrated with air-saturated 50 mM phosphate buffer (pH 7.4) containing 1 mM EDTA. The 20–50 ml of ferrous leghaemoglobin, containing excess of dithionite, is immediately layered on the top of this column, under N_2, and allowed to run into the surface until no free liquid remains above the gel. Alternatively, the dense solution of ferrous leghaemoglobin may be pumped below free buffer remaining on the top of the column, or applied directly at a rate of 2–3 ml min^{-1} using a top-flow adapter. The column is eluted with the equilibration buffer at a flow-rate of 2–3 ml min^{-1}. The ferrous leghaemoglobin becomes oxygenated as it moves ahead of the dithionite and other low molecular weight impurities into the air saturated column buffer. Complete oxygenation is achieved as later fractions drip into fraction collector tubes. Representative samples of effluent fractions are diluted in the same buffer and examined spectrophotometrically, to ensure the absence of autoxidized leghaemoglobin. The presence of ferric leghaemoglobin is indicated by a characteristic absorption band/shoulder at 626 nm (Figure 2) and by a ratio of α to β absorption bands ($A_{575\,nm}/A_{541\,nm}$) of less than 1.02. Usually, preparations with α/β ratio of > 1.0 are satisfactory. All fractions containing high-quality oxyleghaemoglobin are pooled and, if necessary, concentrated to approximately 2 mM over a Diaflo UM10 membrane using air, not N_2, to pressurize the filtration cell. The solution is then dispensed in aliquots of convenient volume, into small plastic tubes, frozen, and stored in liquid N_2, or in a deep-freeze at -80°C. Preparations are stable for at least 1 year. Only sufficient leghaemoglobin for one experiment should be thawed and kept in ice for immediate use.

(iii) Purification of sub-components as ferric leghaemoglobin

The methods described in this section were developed for the isolation of homogenous sub-components of leghaemoglobin for physico-chemical studies. The basis of the method is gradient elution from columns of Whatman DE-52 cellulose. Since each valence state, or ligand complex of each subcomponent, will behave differently, a confused series of bands will result, unless a single valence state and ligand form of the unfractionated leghaemoglobin is chromatographed. It is most convenient to convert everything to unligated ferric leghaemoglobin by treatment with an excess of potassium hexacyanoferrate (III) (ferricyanide), followed by gel chromatography at pH 9.2 to strip the leghaemoglobins of dissociated ligands such as nicotinate. If needed in the oxygenated form for physiological studies, the isolated subcomponents may then be reduced and oxygenated as described above (ii) for the unfractionated leghaemoglobin. The following procedure is adapted from Appleby *et al.* (1975).

A sample of dialysed, concentrated leghaemoglobin from (i) above (25 ml, 5 mM) is diluted to 65 ml with 0.1 M Tris–HCl buffer (pH 9.2) at 2°C and treated with 4 equivalents of potassium hexacyanoferrate (III) (176 mg), with gentle stirring until O_2 evolution has ceased. This solution is then chromatographed on a 60 × 5 cm column of Sephadex G-25, equilibrated with 10 mM Tris–HCl (pH 9.2) at 2°C, and eluted with the same buffer at a flow-rate of 2 ml min^{-1}. Fractions of 20 ml are collected. A red band of ferric leghaemoglobin is eluted first, followed by a compact band containing excess of potassium hexacyanoferrate (III) and displaced nicotinic acid, which is usually followed by a slowly moving purple band. The pooled fractions containing ferric leghaemoglobin, free of nicotinic acid (Appleby *et al.*, 1975), are concentrated to 20 ml over an Amicon Diaflo UM10 membrane and dialysed for 18 h against two changes of acetate buffer (10 mM, pH 5.2, 4 l) at 2°C. The dialysed leghaemoglobin is clarified by centrifugation and diluted with the same buffer to give approximately 50 ml of 2 mM solution, which is chromatographed on a 25 × 5 cm column of DE-52 cellulose (Whatman). This column is packed with DE-52 cellulose which had been pre-equilibrated with sodium acetate buffer (0.5 M, pH 5.2) and washed with glass-distilled water until the effluent conductivity and pH approached that of 10 mM acetate buffer (pH 5.2), which was finally used for exact equilibration of the packed column at 2 °C. The leghaemoglobin is fractionated and eluted with a linear gradient of sodium acetate buffer (10 mM, pH 5.2, 4 l, to 100 mM, pH 5.2, 4 l) at a flow-rate of 2 ml min^{-1}. If possible, the eluate is monitored with a Uvicord III photometer (LKB-Producter AB, Stockholm, Sweden) at 280 and 546 nm. Peak fractions containing leghaemoglobins *a*, *b*, c_1, c_2, and *d* are concentrated in Amicon Diaflo cells over UM10 membranes, exchanged into 10 mM phosphate buffer (pH 7.4), and finally concentrated to 5–10 ml per component. The concentrations of the components are determined by the pyridine haemochrome method (Section b.ii) and they are stored in liquid N_2.

Analytical and preparative scale flat-bed isoelectric focusing procedures (Fuchsman and Appleby, 1979a) now permit the demonstration of the homogeneity of chromatographically purified soybean leghaemoglobins *a*, *b*, and c_1 and the further separation of leghaemoglobin 'c_2' into the components c_2, and c_3 and leghaemoglobin '*d*' into the components d_1, d_2, and d_3. While this new procedure is recommended for ultimate studies on the structures and functions of leghaemoglobin components, it is too tedious to be considered as a routine purification procedure.

(iv) Properties of soybean leghaemoglobin

The physical and chemical properties of leghaemoglobin have been summarized by Appleby (1974) and Dilworth (1980). Properties which are likely to be related to uses of leghaemoglobin in physiological experiments are reiterated here.

Multiple components

Soybean leghaemoglobin contains four major (*a*, c_1, c_2, c_3) and four minor (*b*, d_1, d_2, d_3) molecular species which differ in molecular weight, isoelectric point, and amino acid sequence. Their optical, ligand-binding, and redox properties are very similar, and all contain protohaem IX as the prosthetic group.

Ligands

Soybean leghaemoglobin has different ligand-binding properties according to whether the haem iron is ferrous or ferric (Table 1). Recently, Aviram *et al.* (1978) have described a higher oxidation state, leghaemoglobin (IV). The most notable features are the high affinity of leghaemoglobin for O_2 and CO (Appleby, 1962; Imamura *et al.*, 1972; Wittenberg *at al.*, 1972) and the ability of ferric and ferrous leghaemoglobin to bind nicotinate (Appleby *et al.*, 1973).

General

In some respects, leghaemoglobin resembles mammalian myoglobin (Antonini and Brunori, 1971). For example, the spectra of the ferrous, ferrous-oxy and ferric forms are similar, and there are similarities in amino acid sequence, molecular weight, and the degree of helical structure (Appleby, 1974). However, there are some important differences, for example in the kinetics of ligand binding (Wittenberg *et al.*, 1972; Imamura *et al.*, 1972). Nevertheless, myoglobin, and leghaemoglobin have been used complementarily in physiological experiments in nitrogen fixation (e.g. Wittenberg *et al.*, 1974; Bergersen and Turner, 1975a,b, 1979).

Antibodies produced in rabbits against purified leghaemoglobin components cross-react strongly, indicating that the immunogenic sites are similar (Hurrell *et al.*, 1976). Consequently, immunological methods are of limited use in identifying the separated components of leghaemoglobins.

(b) SPECTROPHOTOMETRY

In common with other haemoproteins, leghaemoglobins have characteristic optical absorption spectra which play a major role in their analysis and structural studies. In this section, spectral methods of general utility will be described. Data on other aspects may be located from accounts by Appleby (1974), Dilworth (1980), and Fuchsman and Appleby (1979b).

(i) Spectra of common forms of leghaemoglobin

The optical absorption spectra of ferrous, ferrous-oxy, and ferric leghaemoglobin from an unfractionated sample prepared as described above (Section a.ii) are shown in Figure 2. The molar absorptivities (ε) of the various peaks in the visible range agree with those given for leghaemoglobin *c* by Appleby (1974), when adjusted for the approximately 8% overestimate indicated in that publication. However, the molar absorptivities in the Soret region are about 5% below those given by Appleby (1974). During deoxygenation of oxyleghaemoglobin, isosbestic points develop at 527, 552, 569, and 587 nm, provided that no other changes occur.

(ii) Optical estimation of leghaemoglobin concentration

The molar absorptivities measurable from Figure 2 may be used to estimate concentrations of leghaemoglobin. However, for greater sensitivity it is preferable to use the spectrum of the haemochrome produced when pyridine reacts with leghaemoglobin in strong alkali. This reaction also includes free haem and haem from other haemoproteins, so suitable precautions should be taken.

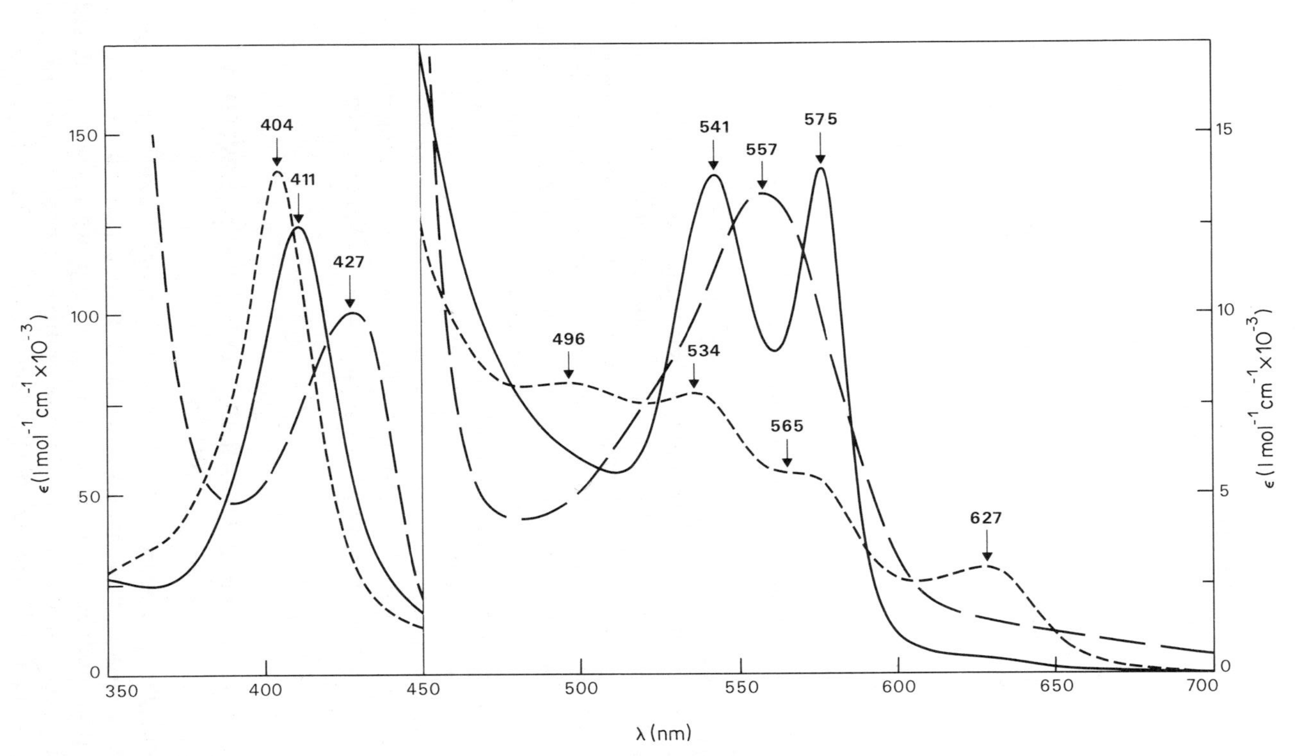

Figure 2. Absorption spectra of purified, unfractionated leghaemoglobin from soybean nodules. ———, Iron (II) oxyleghaemoglobin, prepared as described in the text; – – – –, iron (II) (deoxygenated) leghaemoglobin (reduced with 100 μM sodium dithionite); -----, iron (III) leghaemoglobin [oxidized with potassium hexacyanoferrate (III)]. The spectra were obtained with a solution of 58 μM leghaemoglobin in 25 mM potassium phosphate, pH 7.4

Reagents

Diluent buffer: usually the same as used for the sample of leghaemoglobin being analysed.

Alkaline pyridine reagent: 4.2 M pyridine in 0.2 M NaOH, prepared by dissolving 0.8 g of NaOH in 50 ml of water; when cool, add 33.8 ml of pyridine (33.2 g), dissolve, and dilute to 100 ml with water.

Sodium dithionite (May & Baker, Dagenham, Essex, U.K.), ground finely and stored (in aliquots of 1 g) in small, filled, stoppered tubes in a desiccator.

Procedure. To a suitable dilution of a solution containing leghaemoglobin add an equal volume of alkaline pyridine reagent and mix well. The solution becomes greenish yellow, owing to formation of the ferric haemochrome of protohaem IX. The absorbance is measure at 556 nm against a reagent blank 2–5 min after adding a few crystals of sodium dithionite and stirring without aeration, to reduce the haemochrome. $\varepsilon = 33.9 \times 10^3 \, \mathrm{l \, mol^{-1} \, cm^{-1}}$ (Ohlsson and Paul, 1976).

$$\text{Leghaemoglobin concentration (mM)} = \frac{A_{556\,\mathrm{nm}} \times 2D}{33.9}$$

where D is the initial dilution.

This method should only be used for purified samples or for those in which interfering materials are known to be absent. For less purified preparations, such as clarified extracts of nodules, it is better to use a difference-spectral method as follows. The haemochrome, prepared as above is divided between two cuvettes. The sample cuvette is reduced with dithionite, as above, and the contents of the reference cuvette are oxidized by stirring in a few crystals of potassium hexacyanoferrate (III). The reduced minus oxidized difference spectrum is shown in Figure 3. It is important that the peak should be near 556 nm; a peak at a lower wavelength indicates degradation of the sample. The value of $A_{556\,\mathrm{nm}}$ minus $A_{539\,\mathrm{nm}}$ (trough) is recorded and the leghaemoglobin concentration calculated as above using $\Delta\varepsilon = 23.4 \times 10^3 \, \mathrm{l \, mol^{-1} \, cm^{-1}}$ (Bergersen *et al.* 1973).

(c) USES OF LEGHAEMOGLOBIN IN EXPERIMENTS WITH A GAS PHASE

Leghaemoglobin, prepared as described in Sections a.i and a.ii has been used routinely in our laboratory for experiments with suspensions of bacteroids prepared from nodules, with suspensions of agar-grown bacteria, and with samples from continuous cultures of various nitrogen-fixing bacteria. In some experiments, mammalian myoglobin has been substituted for leghaemoglobin.

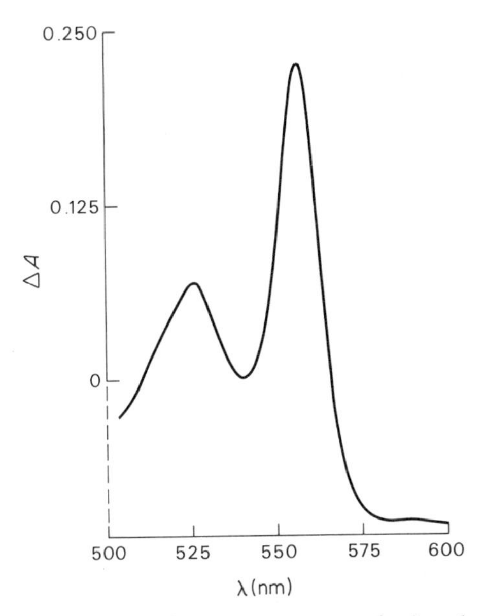

Figure 3. Difference spectrum (reduced minus oxidized) of a pyridine haemochrome prepared from a sample of leghaemoglobin. Leghaemoglobin solution (1.00 ml) was mixed with 2.00 ml of buffer and 3.00 ml of alkaline pyridine reagent; 1.3 ml was placed in each of two cuvettes (10 mm) and the contents of the sample cuvette reduced with a few crystals of sodium dithionite and of the reference cuvette oxidized with a few crystals of potassium of hexacyanoferrate (III). For calculations, see text

Purified lyophilized myoglobin can be obtained from commercial sources (e.g. Sigma), reduced with dithionite and oxygenated on a Bio-Gel P-10 column as described in Section d.ii above (Bergersen and Turner, 1979).

In introducing these gas-phase assays, it must be emphasized that they have limited application. They are convenient for measuring nitrogenase activities under standardized conditions at low concentrations of free, dissolved O_2. However, they are not readily used for relating gas-phase concentrations of O_2 to nitrogenase activities. Nevertheless, these assays are useful when stable, low concentrations of free, dissolved O_2 are required, because the reservoir of O_2 present in the oxygenated carrier is much greater than can be accommodated in a gas phase and consequently reactions are prolonged. This arises

Table 1. Properties of soybean leghaemoglobin, with special reference to binding of O_2 and other ligands. All values refer to an experimental temperature of 20°C except for Lb*a*(Fe^{2+})O_2 at pH 6.5, where 25°C was used

Leghaemoglobin (Lb) species	Molecular weight	Valence state of haem Fe	Ligand	pH	Kinetics					Reference
					$P_{\frac{1}{2}}$ (mmHg)	K' (equilibrium) (M)	k' (association) ($M^{-1} s^{-1}$)	k (dissociation) (s^{-1})	$K' = k/k'$ (M)	
Lb*a*		Fe^{2+}	O_2	7.0	0.040	73×10^{-9}	—	—	—	Appleby (1962)
				6.8	—	—	118×10^6	4.4	37×10^{-9}	Wittenberg *et al.* (1972)
				6.5	0.047	85×10^{-9}	150×10^6	(11.0)*	73×10^{-9}	Imamura *et al.* (1972)
		Fe^{2+}	CO	6.8	—	—	12.7×10^6	—	—	Wittenberg *et al.* (1972).
				6.5	0.00074	0.99×10^{-9}	13.5×10^6	0.012	1.13×10^{-9}	Imamura *et al.* (1972)
		Fe^{2+}	nicotinate	5.3		33×10^{-6}	—	—	—	Appleby *et al.* (1973)
		Fe^{3+}		5.3		1.36×10^{-6}	—	—	—	
		Fe^{2+}		6.8		1.8×10^{-3}	—	—	—	
		Fe^{3+}		6.8		52.6×10^{-6}	—	—	—	
	15775	—	—	—	—	—	—	—	—	Ellfolk and Sievers (1971)
Lb*c*†		Fe^{2+}	O_2	7.0	0.068	123×10^{-9}	—	—	—	Appleby (1962)
				6.8	—	—	97×10^6	4.9	50×10^{-9}	Wittenberg *et al.* (1972)
		Fe^{2+}	CO	6.8	—	—	11.8×10^6	—	—	Wittenberg *et al.* (1972)
	15950	—	—	—	—	—	—	—	—	Sievers *et al.* (1978)
Lb*d*†		Fe^{2+}	O_2	7.0	0.063	114×10^{-9}	—	—	—	Appleby (1962)

*Doubtful value, owing to reduction of some ferric Lb during discharge of LbO_2 in presence of dithionite.
†In this table, 'Lb*c*' refers to an unresolved mixture of Lbc_1, c_2, and c_3, and 'Lb*d*' refers to an unresolved mixture of Lbd_1, d_2 and d_3 (cf. Section a.iii).

because of the high affinity of leghaemoglobin (Lb) for O_2 (Table 1). The reaction

$$LbO_2 \underset{k'}{\overset{k}{\leftrightharpoons}} Lb + O_2$$

has an equilibrium constant $K' = k/k' = 37 \times 10^{-9}$ M. Thus, considering a solution of 100 μM leghaemoglobin with a relative oxygenation ($Y = LbO_2$/total Lb) of 0.5, the concentration of free, dissolved O_2 is 37×10^{-9} M, but the total concentration of O_2 present (free + bound) is more than a 1000-fold greater, namely 50×10^{-6} M.

The presence of a gas phase is convenient for acetylene reduction, or ^{15}N assays of nitrogenase activity. With oxyleghaemoglobin, the gas-phase assays have proved useful for routine assays of bacteroids, and N_2-fixing cultures of rhizobia, *Klebsiella* spp., *Azospirillum lipoferum*, and *A. brasilense*. They were of no advantage for *Azotobacter vinelandii*, in which nitrogenase activity is greatest at higher concentrations of gas-phase O_2.

The following assay is used for samples of cultures and for samples of bacteroids for comparative tests. The conditions must be optimized for each use, therefore ranges are given.

Materials

Bottles or vials, 2 cm internal diameter, 4 cm high, with necks suited to Suba Seals, or other septum caps. Bottles of other dimensions may be used, but optimal experimental conditions will need to be re-determined.
Oxyleghaemoglobin, 2 mM.
Syringes for culture sample and gas sampling.
Reciprocating water-bath shaker at desired temperature, set for 120–150 excursions min^{-1}.
Vacuum pump, manifold, and gases (O_2, argon, C_2H_2 or $^{15}N_2$) (Chapter II.2, Figure 9).
Solution of substrate if required.
Sample of culture or bacterial suspension in a suitable buffer.

Procedure. Place in the bottle 0.25 ml of oxyleghaemoglobin and an added substrate in a 0.1-ml volume, insert the cap, and evacuate for 10 min to remove dissolved air, then fill the bottles with the desired gas mixture. For guidance, with leghaemoglobin try 0.1–1% O_2 in C_2H_2 (10%) and argon (to 100%). With oxymyoglobin try 1–5% O_2. The ideal concentration of O_2 in the gas phase is that which allows the carrier protein in the liquid phase to remain partially oxygenated during the course of the experiment. This can be checked with a manual spectroscope. The bacterial suspension is best prepared under anaerobic conditions, to prevent any inactivation of nitrogenase which may result from contact with air. Samples from cultures may be collected in capped, evacuated bottles. Bacterial suspension (1.65 ml) is injected into the assay bottles and the pressure equalized with the atmosphere by momentarily piercing the cap with a narrow-gauge hypodermic needle. Immediately place the bottles in the thermostat and start the shaker. Samples of the gas phase may be collected at intervals, analysed for ethylene, and the

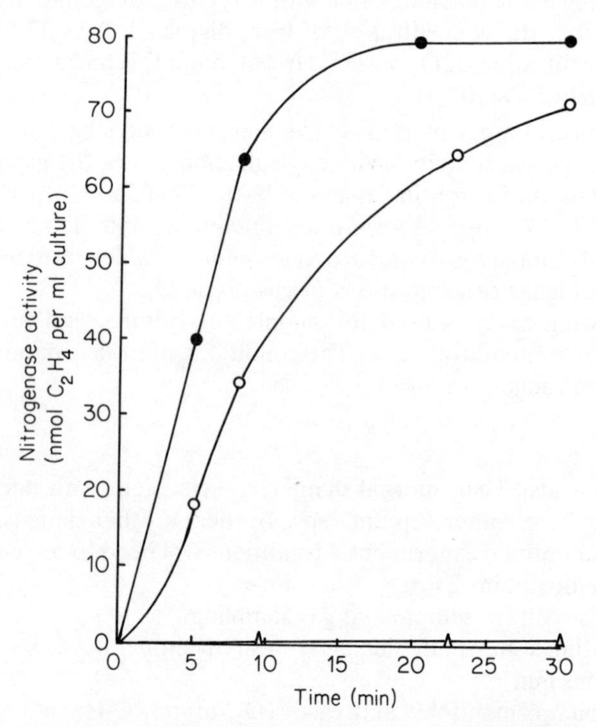

Figure 4. Time course of nitrogenase activity by *Azospirillum brasilense* Sp7 in a reaction with oxyleghaemoglobin, or with oxygen from a gas phase. ●, 0.25 mM oxyleghaemoglobin and no added gas-phase O_2; △, 5 and 10% O_2 in the gas phase, no leghaemoglobin; ○, 2% O_2 in the gas phase, no leghaemoglobin. All assays contained 1.75 ml of culture (0.74 mg ml^{-1}, dry weight)

time course of acetylene reduction plotted. Experience has shown that 2.0 ml is the optimum reaction volume for 2-cm diameter bottles. For bottles with different dimensions, the optimum reaction volume should be established. Figure 4 illustrates results obtained with samples of *A. brasilense* from a continuous culture.

(d) LEGHAEMOGLOBIN IN NO-GAS-PHASE EXPERIMENTS

In addition to providing a source of oxygen, oxyleghaemoglobin and other similar O_2-carrying haemoproteins provide a means of measuring concentration of free O_2, because their absorption spectra indicate their fractional oxygenation (Y). Consequently, rates of oxygen consumption can be calculated from rates of deoxygenation in systems containing known concentrations of leghaemoglobin and myoglobin.

Bergersen and Turner (1975a,b, 1979) have described systems for utilizing these properties of oxygenated leghaemoglobin and myoglobin. The techniques employ a range of apparatus from capped optical cuvettes completely filled with a solution of leghaemoglobin, in which bacteria are suspended, to complex, flowing systems in which various steady states can be established. Most of the latter will be too complex for general use. Therefore, we shall describe the simplest system in detail, with sample calculations and give a brief account of the more complex systems, with an explanatory diagram.

(i) Assays in optical cuvettes

These assays are very suitable for measurements of respiration at concentrations of free O_2 well below the limit of the oxygen electrode. By using reaction solutions which are initially equilibrated with acetylene, nitrogenase activity may be estimated by stopping reactions in a series of cuvettes at intervals, recovering the dissolved gas and analysing the ethylene produced as described in Chapter II.3.

Apparatus and reagents

Spectrophotometer. A good double-beam instrument with a thermostatable sample chamber is required. Preferably, it should be capable of being programmed to record the absorbance (A) at two selected wavelengths alternately. Such an instrument is the Varian Techtron Model 635, fitted with a wavelength programmer and zero offset facility. However, data can be obtained adequately with a manual or scanning instrument provided that spectral data are recorded against a continuously running, accurate time base. Spectra or wavelength pairs can be recorded sequentially, relating A values to the mean time between each pair.

Optical cuvettes, 1-cm light path, fitted with a conical neck to accept a suitable injectable seal. For example a Hellma No. 194 cuvette, fitted with a sleeve-type rubber stopper No. 8826, plug diameter 11 mm (A. H. Thomas, Philadelphia, Pa., U.S.A.) is satisfactory. Alternatively, glass cuvettes, with a 1-cm internal light path and a cone at the top to fit any available rubber closure, can be made by a competent glassblower.

The volume of the assembled cuvettes must be known, and should be in the range 4–5 ml.

Erlenmyer flasks, 250 ml, closed with gas-tight injectable stoppers for equilibration of reaction solutions.

Reaction solutions: these should be prepared in morpholinopropanesulphonic acid (MOPS) or phosphate buffer (25 mM)(pH 7.0–7.4), and should contain oxyleghaemoglobin, prepared as described in Section a.ii, and substrates, such that the desired concentrations are attained following addition of the bacterial suspensions. The final concentration of the oxyleghaemoglobin should be about 100 μM which gives an α absorption peak in the range 1–1.5 absorbance units (Figure 2).

Stopping solution: 10% w/v trichloroacetic acid.

Syringes: 10 ml for injecting reaction solution; a syringe for injecting the chosen volume of bacteroid suspension; 5-ml syringes for stopping reactions; hypodermic needles.

Bacterial suspension. This may be a sample of a culture, or a suspension of washed cells. The concentration should be adjusted so that the sample used in the reaction uses all the oxygen present in 15–30 min.

Glass beads, 2–3 mm diameter.

Procedure. Bacterial suspensions should be collected, processed, and stored anaerobically at the selected reaction temperature, if nitrogenase activities are to be measured. The reaction mixture is shaken gently under an atmosphere containing 1% oxygen in argon, or argon containing a final concentration of 10–20% acetylene, at the selected reaction temperature. (If myoglobin is used, 10–15% O_2 should be used). The cuvettes contain two glass beads to promote mixing. They are capped and flushed with argon. The reference cuvette is completely filled with buffer solution from a syringe or with sufficient buffer so that it is completely full after adding the chosen volume of bacterial suspension. The sample cuvettes are likewise filled with equilibrated reaction solution. It is essential to remove all gas bubbles through a hypodermic needle inserted through the cap. This needle is removed when injection is completed. Reference and sample cuvettes are then placed in the spectrophotometer to ensure temperature equilibrium, and while this is happening preliminary adjustments to the spectrophotometer and recorder are made. For most purposes we use the α-peak of oxyleghaemoglobin (576 nm with bacteria present) and the adjacent trough (561–562 nm) as suitable wavelengths at which to record the deoxygenation of leghaemoglobin. (If myoglobin is used, absorbances at 582 and 564 nm are recorded). With 1% O_2 in the pre-equilibration gas mixture, the fractional oxygenation (Y) of leghaemoglobin will be 0.997. The absorbance difference ($\Delta A = A_{576nm} - A_{562nm}$) is recorded and adjusted to ΔA for $Y = 1.0$, to give ΔA_{oxy} (this can be done before injection of the bacterial suspension; however, if more than 1 mg (dry weight) of bacteria is used in each cuvette, the spectrum may be altered and ΔA_{oxy} should be obtained after addition of the bacteria, using an air-saturated reaction solution in a separate determination). The bacterial suspension is then injected and the cuvettes shaken to disperse the bacteria. Immediately commence recording of the changes in the spectra which accompany deoxygenation.

If the nitrogenase activity is to be measured, several cuvettes can be sequentially placed in the light path and the reaction terminated at various stages in the deoxygenation, by withdrawing the cuvette contents into a 5-ml syringe, containing 0.5 ml of 10% trichloroacetic acid and two glass beads. This should be done with a pressure-releasing needle inserted through the cuvette cap, in such a way as to avoid any degassing of the solution during transfer. The syringe is shaken to mix the reaction mixture and the trichloroacetic acid, and then the dissolved gas is recovered and analysed for ethylene as described in Chapter II.3. Rates of nitrogenase activity are determined from increments of ethylene content between successive samples and these are related to the state of deoxygenation of leghaemoglobin which prevailed during that time interval.

At the end of the series, a value for the fully deoxygenated reaction (ΔA_{red}) is obtained by adding a few crystals of sodium dithionite to the reaction mixture and recording the absorbance difference. The concentration of leghaemoglobin is determined also.

The results of an experiment are illustrated in Figure 5. The calculations are as follows:

1. At time t the fractional oxygenation is given by

$$Y_t = \frac{\Delta A_t - \Delta A_{red}}{\Delta A_{oxy} - \Delta A_{red}}$$

2. The concentration of free dissolved O_2 is given by

$$[O_{2\ free}] = \frac{K'Y}{1 - Y}$$

where $K' = k/k'$ (Table 1). Using the average values for Lb*a* and Lb*c*, $K' = 43.5 \times 10^{-9}$ M (for sperm whale myoglobin, $K' = 786 \times 10^{-9}$ M; Antononi & Brunori, 1971).

3. At time t the total O_2 present in the reaction is given by

$$O_{2\,total} = V_f\,(YC + [O_{2\ free}])$$

where Vf is the reaction fluid volume and C is the concentration of leghaemoglobin.

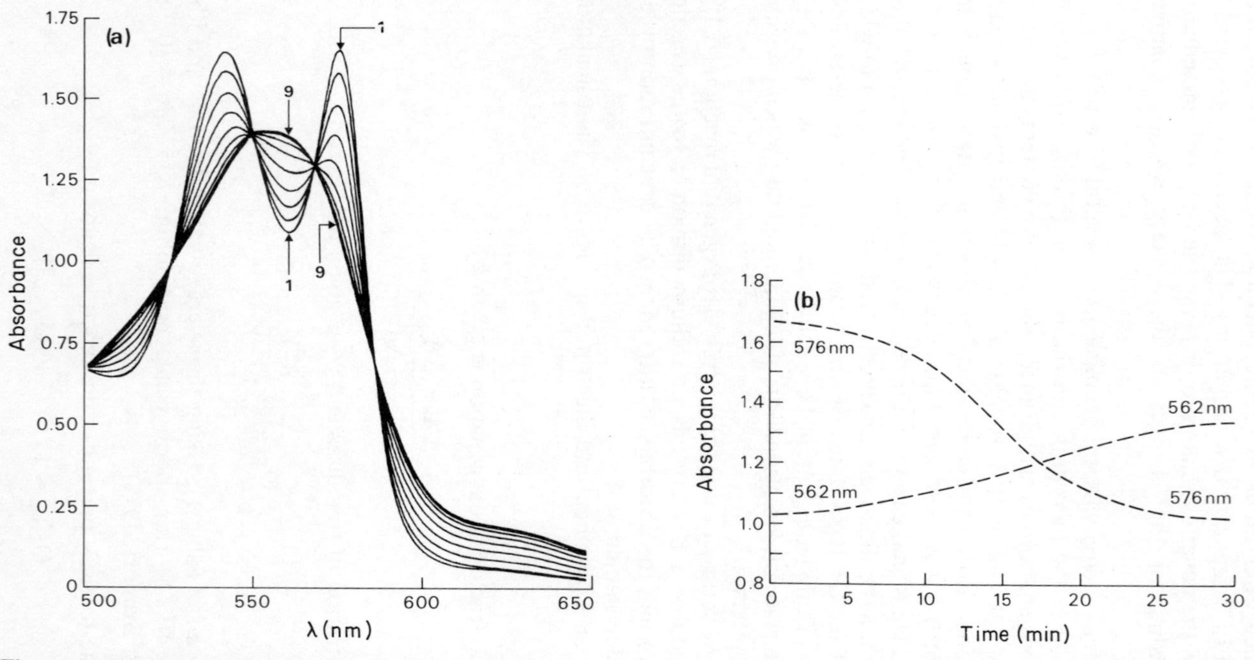

Figure 5. Deoxygenation of leghaemoglobin in a cuvette reaction. (a) Spectra obtained in a reaction in which 4.5 ml of leghaemoglobin (100 μM) was deoxygenated by 0.25 mg (dry weight) of the cowpea *Rhizobium* sp. CB756 from an O_2-limited continuous culture. Scan 1 began 0.5 min after adding the bacteria and there were 3.5 min between the beginning of each scan. (b) Plot of absorbance at 576 and 562 mm as a function of time. ΔA is given by the difference between the two lines, becoming negative after 17 min

4. Rates of O_2 consumption (r_{O_2}) are given by

$$r_{O_2} = \frac{[O_{2\ \text{total}\ 1}] - [O_{2\ \text{total}\ 2}]}{(t_1 - t_2)w}$$

where $t_1 - t_2$ is the time interval in minutes between observations and w is the basis of bacterial activity (mg dry weight or mg of bacterial protein present in each reaction).

5. The concentration of free O_2 prevailing during any time interval is taken as the average:

$$\frac{[O_{2\ \text{free}}\ t_1] + [O_{2\ \text{free}}\ t_2]}{2}$$

Comments

The above method is very satisfactory for measuring O_2 consumption by bacteria at very low concentrations of free, dissolved O_2. It can also be used to relate nitrogenase activity broadly to various ranges of O_2 concentration. The apparatus is simple, requiring only a spectrophotometer of the type found in most modern biochemical laboratories.

The main sources of error arise from the possible presence of ferric leghaemoglobin in the preparations, or to autoxidation occurring during the experiments. Provided that ferric leghaemoglobin does not exceed 5% of the total at the end of the experiment, the effect is minimal (Bergersen and Turner 1979). Autoxidation is minimized by using metal-free distilled water and washed bacteria rather than culture samples. It is also important to avoid using any chemical which may react with leghaemoglobin. Such effects are usually easily seen from examination of the spectra of reaction solutions.

(ii) Flow-cell reactions

Cuvette assays as described above, while useful for relating respiration to free O_2 concentration, are of limited usefulness for nitrogenase measurements. Therefore, two other experimental systems have been devised (Bergersen and Turner 1979). The principles involved are briefly described. However, most workers would probably devise their own modifications to suit the needs of particular experiments.

Sampled reactions

One of the difficulties with cuvette assays is the variability which is sometimes encountered. This is overcome by a system in which a spectrophotometer

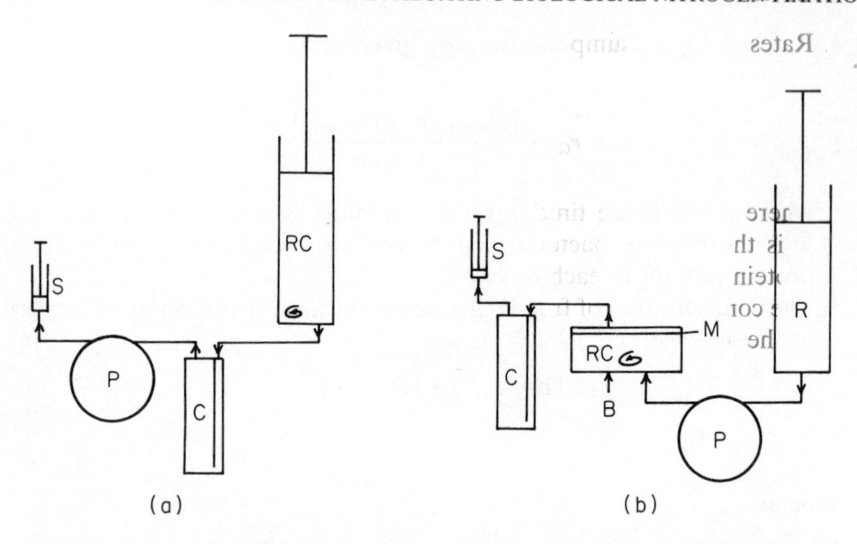

Figure 6. Schematic diagrams of apparatus for use with oxyleghaemoglobin as oxygen source for respiring, nitrogen-fixing bacteria. (a) A sampled reaction system. RC is a magnetically stirred 50-ml syringe reaction chamber containing bacteria suspended in a reaction solution containing about 100 μM leghaemoglobin. C is a spectrophotometer flow cell, optical path length 1 cm. A peristaltic pump (P) is capable of pumping 5 ml of liquid into the sample syringe (S) in 30 s (b) Apparatus for establishment of steady reaction rates. The reaction chamber (RC) is magnetically stirred. Bacteria are injected at B and confined behind a membrane filter (M). Fully oxygenated leghaemoglobin in a reaction solution is supplied from a 50-ml reservoir syringe (R) with a peristaltic pump (P), displacing reaction solution through the spectrophotometer flow cell (C) into a sample syringe (S). For other details, see text and Bergersen and Turner (1979)

flow cell is connected to a stirred reaction chamber, from which samples can be pumped at intervals for determination of accumulating ethylene (Figure 6a). The reaction chamber is conveniently a 50-ml hypodermic syringe with an off-set outlet to allow a small magnetic stirrer to operate. The flow cell and reaction chamber are at the same temperature and are connected in series through a peristaltic pump to a consecutive series of smaller syringes, each containing 0.5 ml of 10% trichloroacetic acid and two glass beads to promote mixing.

The reaction syringe is filled with pre-equilibrated solution, as described for cuvette assays, and the reaction is initiated by injecting bacterial suspension from a small syringe, via a long, fine needle inserted through the larger needle of the reaction syringe. The reaction syringe is then connected to the argon-flushed flow cell and the entire apparatus quickly filled with reaction mixture by means of the pump. If all is correctly done, and no gas bubbles are entrapped anywhere in the system, the reaction in the flow cell is representative of the entire reaction. At timed intervals, samples of the reaction mixture are pumped into the sampling syringes, where the reaction is terminated. Incremental rates of ethylene accumulation may be related to rates of O_2 consumption and concentrations of free O_2 prevailing between samples.

Continuous flow systems

It is sometimes necessary to examine steady reaction rates, rather than reactions in which the conditions are continuously, but slowly, changing. Figure 6b illustrates an experimental system in which steady states can be established. The pre-equilibrated, fully oxygenated reaction solution is supplied from a 50-ml syringe, as before, and the oxygenation of leghaemoglobin in the reaction is monitored by the spectrum of the reaction solution continuously pumped at variable measured rates through a spectrophotometer flow cell and collected in 5-ml aliquots in syringes for determination of dissolved ethylene. The reaction chamber, surrounded by a water-jacket for temperature control, contains a magnetically rotated stirrer, and the bacteria are confined behind a porous membrane filter such as an Amicon Diaflo DPO45 or a Millipore HA (0.45 μm). Stirring rates in the reaction chamber must be sufficient to minimize occlusion of bacteria in the filter surface. Again, it is essential to remove all gas bubbles. Various deoxygenation states of leghaemoglobin may be established at different pumping rates. The oxygenation of leghaemoglobin in the reaction filtrate during a steady state represents the oxygenation in the reaction. The oxygen consumption rate is calculated from the difference between the total oxygen concentration of the reservoir fluid and that of the effluent fluid and the flow-rate; this may be directly related to nitrogenase activity, calculated similarly from ethylene concentrations. The fractional oxygenation (Y) of the effluent leghaemoglobin gives the prevailing concentration of free oxygen.

This system has been used for sustained reactions with soybean bacteroids at concentrations of free O_2 in the range 4–15 nM over periods up to 1.5 h. It seems to us to be applicable to a wide variety of reactions involving nitrogen-fixing bacteria.

Methods for Evaluating Biological Nitrogen Fixation
Edited by F. J. Bergersen

E. A. Schwinghamer and W. F. Dudman
Division of Plant Industry,
CSIRO Canberra, Australia.

9

Methods for Identifying Strains of Diazotrophs

(a) INTRODUCTION

This chapter is concerned with the non-taxonomic identification of strains already classified within a particular species of diazotrophic microorganism. Taxonomic classification and methods of demonstrating nitrogen fixation are described in other chapters (Chapters II.1, II.2, and II.3). In this chapter, it is assumed that the identity of the organism is known at the genus or species level but that it requires further, marker-type means of identification to distinguish it from other closely related organisms, usually within the same species. Recognition of a microbial strain at this level of resolution is a prerequisite for many experiments in the laboratory and in the field, where a few cells may have to be separated from large, unknown background populations.

It is advantageous for the research worker to have a wide range of marker characters to draw upon, since there is a wide spectrum of requirements extending from genetic–biochemical experiments in the laboratory to ecological experiments in the field. This chapter therefore considers briefly markers ranging from serological characters and other naturally occurring properties (e.g. lysogeny or bacteriocinogeny) to selected or induced mutant characters. In view of the breadth of the subject in relation to available space, we have attempted a compromise approach in which experimental detail is limited to a few, key procedures, to allow us to evaluate the suitability or potential of different marker systems for laboratory or field use. The choice of a marker system and awareness of its limitations are as important to the success of the experiment as the isolation and application of the markers. The reader is therefore referred to appropriate reviews or research papers from which detailed procedures suited to the particular diazotroph can be developed. We have concentrated on the genus *Rhizobium* as the 'prototype' diazotrophic organism because of our experience with the rhizobia and because these bacteria demonstrate most of the requirements for strain identification in both laboratory and field situations.

(b) SEROLOGICAL MARKERS

Antisera are the most specific reagents for identifying many strains of microorganisms. Exceptions arising from permanent or acquired sharing of antigenic similarities are known, e.g. the loss of antigenic specificity by strains of *Rhizobium meliloti* after prolonged culture in the laboratory (Humphrey and Vincent, 1975; Wilson *et al.*, 1975). Among the diazotrophs, serological methods have been applied extensively only to *Rhizobium* and *Klebsiella* species, presumably because only these have required detailed strain identification, the former for ecological and the latter for epidemiological reasons.

Space does not allow coverage of the background and the reader is referred to texts dealing with this extensive subject (e.g. Kabat, 1968; Humphrey and White, 1970; Day, 1972; Hobart and McConnell, 1975; Roitt, 1974; Wilson and Miles, 1975; Glynn and Steward, 1978). Results obtained by the application of serological methods to diazotrophs have been briefly summarized (Dudman, 1977) but details of the methods were omitted; the present account deals only with the latter aspect.

(i) Preparation of antigens for immunization

Certain principles are important when considering the growth of microorganisms for use as antigens for immunization. It is necessary to understand the antigenic structure of the organism and the requirements of the serological technique for which the antiserum is to be prepared.

In agglutination reactions, it does not appear to be as important to use a defined medium as it is in the immunodiffusion reaction (Vincent, 1970). However, it is important, for agglutinating antisera, to ensure that the organism is in the correct antigenic phase, if this is subject to variation. *Bacillus polymyxa*, for example, possesses three types of antigens of different specificities: the flagella are strain-specific, the somatic antigens are group-specific, while the spore antigens are common to all strains of the species (Davies, 1951; Wolf and Barker, 1968). The organism is grown in broth culture when spore antigens are to be avoided; when spores are wanted, the organism is cultured on nutrient agar slopes, which favour spore formation. Flagellar antigens of motile cells are prepared from 18-h broth cultures by centrifugation, suspending the cells in saline and treating them with formalin to inactivate the somatic antigens. To obtain somatic antigens, subcultures are selected from smooth colonies on agar, grown in broth, suspended in saline, then heated to destroy the flagellar antigens (Davies, 1951; Wolf and Barker, 1968). *Rhizobium* species are less complex as antigens because they do not have spores, but to ensure the presence of flagellar as well as somatic antigens in immunogenic preparations of *R. meliloti* and *R. trifolii*, Vincent (1941, 1942) used mixtures of young (2-day) and older (5-day) cultures to ensure the presence of actively motile cells in the suspensions to be injected.

On the other hand, in immunodiffusion reactions, antigen mixtures of any degree of complexity do not cause complications because of the resolving power of the technique, but it becomes more important to consider whether or not to use a defined medium on which to grow the organism. If constituents in the medium are themselves antigenic they may give rise to the production of antibodies which may interfere in the subsequent serological analysis by giving rise to falsely positive results. Thus, whenever possible, it is preferable to grow organisms on defined media. We find the defined medium of Bergersen (1961) useful for growing strains of rhizobia for use as antigens for preparing immunodiffusion antisera. In most serological studies with *Azotobacter* species, cultures were grown in defined media (Jensen and Petersen, 1955; Norris, 1960; Holme and Zacharias, 1965; Zarnea *et al.*, 1966; Tchan and de Ville, 1970). Although it is good practice to use a defined medium, this does not mean that growth of immunizing antigens on complex media will necessarily lead to complications. Dazzo and Milam (1975) immunized goats with *Spirillum lipoferum* and *Azotobacter paspali* grown on a trypticase soy broth without the medium causing any apparent interference when the antisera were used in a range of serological reactions (agglutination, immunodiffusion, immunoelectrophoresis, immunofluorescence, and several forms of precipitation).

However, it is our conviction that, on principle, defined media should be used whenever possible to grow the cells to be injected into animals, to ensure that the antisera thus obtained will contain only antibodies against bacterial antigens. The strains that are to be identified serologically may then be grown on any convenient complex medium (such as yeast extract–mannitol agar in the case of *Rhizobium* strains) without fear of interference from the medium, because antibodies against constituents of the medium will be absent from the antisera.

Usually intact cells are injected, although for special purposes disintegrated cells have been used (e.g. Dudman, 1964; Humphrey and Vincent, 1973). Zarnea *et al.* (1966) found that suspensions of *Azotobacter* cells disrupted by repeated freezing and thawing produced superior antisera.

(ii) Preparation of antisera

Investigators wishing to make an antiserum against any antigen must decide upon the animal species to be used, the nature and form of antigen to be injected, the dosage, the route of immunization, the injection schedule, and the manner of bleeding. Readers are referred to other sources for background information regarding these aspects of antiserum production (Evans, 1957; Kabat, 1961; Cruickshank, 1965; Kwapinski, 1965; Williams and Chase, 1967; Nowotny, 1969; Campbell *et al.*, 1970; Vincent, 1970; Oakley, 1971; Kingham, 1971).

In practice, rabbits have been used almost exclusively as the animal species for production of antisera against diazotrophs and all of the various immunization procedures fall into one of two general classes: (1) primary immunization by repeated intravenous injection of the antigen without an adjuvant, at intervals of one or more days, followed by a number of booster injections prior to bleeding, and (2) primary immunization with a single intramuscular injection of the antigen emulsified with an adjuvant, followed by one or more booster injections before bleeding. No systematic comparison has been made of the diazotroph antisera obtained by these procedures, but the first has been widely used in producing antisera for agglutination reactions and the second for antisera used in immunodiffusions. The titres of the antisera obtained by the two methods, when they have been recorded, were similar.

It must be remembered that an antiserum is a complex mixture not only of serum proteins but also of antibody molecules, the relative concentrations of which can be profoundly affected by the immunization procedure. As the ability of various antibody types to participate in different serological reactions is unequal (e.g. IgM antibodies are more efficient in agglutination reactions with particulate antigens but less effective in precipitations with soluble antigens than IgG antibodies; Pike, 1967), it is important for users of serological methods to be sure that they have an appropriate antiserum before reaching conclusions from negative results (see Humphrey and Vincent, 1975).

Our immunization procedure begins by mixing equal volumes of culture suspension (1 ml for each rabbit) and the adjuvant (Freund's complete adjuvant, Difco Laboratories, Detroit, Mich., U.S.A.) in the chamber of a Sorvall

'Omni-mix' homogenizer fitted with a micro-attachment for handling small volumes, and operating this at high speed for 1 min while cooling the chamber in ice. This emulsion is prepared shortly before it is to be used. It is drawn into the hypodermic syringe without the needle in position and 1-ml amounts are injected through a large-bore needle (18 or 20 gauge) into the large thigh muscles of the hind legs of rabbits. Whenever possible, more than one rabbit is used with each antigen because of the variability in the response of individual animals. Additional amounts of the culture suspension, without the adjuvant, are stored frozen. One month after the first injection, the culture suspension (0.5–1 ml per rabbit) is injected subcutaneously without adjuvant into the rabbits to boost their antibody levels. Five days after the second injection a small volume of blood is taken and the serum tested for activity; if satisfactory, the rabbits are bled as described below, on the seventh, ninth, and twelfth days after the second injection. The rabbits may then be rested for a month and again boosted and bled in the same way. This procedure has been described in detail because we have used it successfully to prepare antisera not only against various *Rhizobium* species (Dudman, 1964, 1971; Dudman and Brockwell, 1968), but also against purified nitrogenase Mo–Fe protein (Bergersen *et al.*, 1976) and leghaemoglobins (Dudman and Appleby, unpublished results).

Different methods of bleeding rabbits have been employed by various workers, regardless of the immunization procedure. Some kill the animal and collect as much blood as possible from the jugular vein into a large beaker; others use heart puncture. Our preference is for bleeding from the marginal ear vein, taking 10 and 20 ml of blood alternately on successive bleedings. The blood samples, collected in centrifuge tubes, are incubated for 1 h at 37 °C and the clots loosened from the walls of the tubes with a glass rod. The tubes are held at 4 °C overnight to shrink the clots and then centrifuged. Sometimes a second centrifugation is necessary to remove cells from the supernatant sera. The antisera from the separate bleeds and from all the rabbits immunized with the same antigen should be compared in appropriate serological tests and, if equally satisfactory, pooled to give a larger volume of uniform antiserum which should be divided into smaller samples (no more than 5 ml) for storage.

It is not necessary to add preservatives to serum samples if they are stored frozen (in the deep-freeze compartment of a refrigerator). Repeated thawing and re-freezing does not appear to harm the activity of such sera, but care should be taken to keep thawed sera chilled in ice during manipulations. Alternatively, sera may be stored unfrozen at 4 °C (in the main compartment of a refrigerator), if preservatives are added; phenol (5% solution in saline added with stirring to give a final concentration of 0.25%) and merthiolate (sodium ethylmercurithiosalicylate, final concentration 0.01%) are the most commonly used reagents for this purpose. An alternative procedure is to lyophilize the serum samples and to reconstitute them to their original volume by addition of water but such sera, although active, become extremely turbid and difficult to clarify; this does not interfere with immunodiffusions but may be troublesome for other serological techniques.

(iii) Serological reactions

Reactions between antigens and antibodies may be detected in different ways. Some methods (agglutination and immunodiffusion) have been and continue to be widely applied to diazotrophs. Others, such as immunofluorescence, haemagglutination (Cloonan and Humphrey, 1976), and enzyme-linked immunosorbent assays (Kishinevsky and Bar-Joseph, 1978; Berger *et al.*, 1979), have been less used but show promise of being extremely useful, whereas yet other serological methods (complement fixation, immunoelectrophoresis, and quantitative precipitation) have been tried but have not proved to be so useful or convenient for this purpose (Vincent, 1970; Dudman, 1977). The agglutination test is sensitive and provides information about somatic and flagellar antigens but it is more complex to use and does not give such clear distinctions between reactions of antigenic identity and cross-reactions between closely related strains as does the immunodiffusion test. On the other hand, the immunodiffusion reaction is less sensitive (Marrack, 1963), but this may be rectified by using undiluted antisera and antigen preparations sufficiently concentrated to ensure that negative reactions are truly negative and not positive reactions that are too weak to be detected.

Agglutination

The agglutination reaction is based on the aggregation of bacterial cells or particulate antigens caused by antibody molecules binding to their surfaces and cross-linking them. Such reactions may be performed in different ways, for example with drops of antigens and antisera on glass slides for rapid results. However, for greater accuracy agglutinations are generally performed in small test-tubes by mixing constant amounts of a cell suspension with a dilution series of an antiserum; the highest dilution of antiserum which gives a detectable agglutination is taken as the end-point (the 'titre') of that reaction. Thus, the comparison of bacterial strains which are agglutinated by the same antiserum is made quantitatively by comparing their respective titres.

The cells to be used in agglutinations are grown on a convenient medium and suspended in physiological saline (0.85% w/v NaCl). *Rhizobium* cells possess two types of surface antigen—flagellar (H) and somatic (O)—which may react independently in agglutinations. Heating the cells for 30 min at 100°C destroys the flagellar proteins, leaving only the O antigens to react with the antiserum. Dilute formalin (0.04% formaldehyde) is widely used to inactivate the O antigens of enteric Gram-negative bacteria, but does not do this so well with rhizobia (Vincent, 1970). If the cells have been given these treatments, they are centrifuged and suspended in fresh saline. Because of the tendency of *Rhizobium* cells to form clumps, it is generally advisable to centrifuge the final cell suspension lightly (1000 r.p.m. for 5 min) and to use the evenly turbid supernatant. Strains that tend to auto-agglutinate may form suspensions of greater stability in 0.5% instead of 0.85% saline solution.

The antiserum is diluted with physiological saline in a series of doubling dilutions to cover the range required. The volumes chosen will depend on the number of strains to be examined, allowing 0.5-ml volumes of diluted antiserum for each tube. Thus, for example, a volume of 0.5 ml of undiluted antiserum may be diluted with saline to obtain 5 ml of 1:10 antiserum and 2.5 ml of this then used to obtain a series of doubling dilutions as illustrated in Table 1. The dilutions may be made with a single pipette, provided that it is rinsed thoroughly after each transfer by drawing the diluted serum into the pipette and blowing it out, repeating the process several times.

Table 1. A typical dilution series of antiserum for an agglutination test

	Tube								
	1	2	3	4	5	6	7	8	9
Saline (ml)	4.5	2.5	2.5	2.5	2.5	2.5	2.5	2.5	2.5
Serum (ml)	0.5								
Mix and transer to next tube (ml)	2.5	2.5	2.5	2.5	2.5	2.5	2.5	2.5	2.5*
Reciprocal of final dilution when serum is mixed with equal volume of antigen	20	40	80	160	320	640	1280	2560	5120

*Discard from final tube of series.

It is important in all agglutinations to include an antigen control, omitting antiserum, in which the antigen suspension is combined with an equal volume of saline. The tubes are mixed by swirling and, in a suitable rack, incubated in a water-bath. An incubation temperature of 52°C has been recommended for agglutinations of rhizobia (Vincent, 1970) because flagellar reactions may be read after 1–2 h and somatic reactions after 4 h at this temperature. When incubations are performed at 37°C these times are extended to 4 h and overnight, respectively. Flagellar agglutinations are recognized by the formation of large, flocculent, slow-settling aggregates, whereas somatic reactions are more granular. In suspensions with both types of antigen present, a combined H and O reaction may be detected. The two kinds of agglutination are best seen against a dark background with indirect illumination. Positive reactions, especially near the end-point, may require careful comparison with the control suspension. Strains that react to the same titre with the same serum are regarded as being closely related serologically, if not identical.

It may be necessary to compare such strains further, by absorption experiments. Antisera are absorbed by mixing them with washed suspensions of the absorbing organism, incubating them for periods corresponding to those required for agglutination tests, and centrifuging the mixture. The clear supernatants represent absorbed antisera and may be suitably diluted, making allowance for any dilution arising from the addition of the absorbing strain. The completeness of the absorption should be confirmed by testing the absorbed antiserum against a fresh suspension of cells of the absorbing strain. When absorption is complete, the antiserum fails to agglutinate the absorbing strain. Readers are referred to the classic paper of Vincent (1941) for a description of the agglutination reaction with rhizobia.

Gel immunodiffusion

The immunodiffusion test is the simplest serological reaction to set up and to interpret, and it possesses the important merit of allowing the direct comparison of antigens. This technique has been found very useful in the identification of *Rhizobium* strains (Dudman, 1977). For this purpose, we use gels 4 mm thick, made with 0.75% agar in physiological saline (0.85% NaCl and 0.025% sodium azide). Wells (4 mm diameter) are cut in the gel in a hexagonal array around a central well, all spaced 8 mm between centres (see Figure 1). The Petri dishes are placed over a pattern drawn on paper which serves as a template and the wells are cut with a cork borer; the plugs within the wells are removed with the aid of a wire flattened at one end, and the wells sucked dry with a Pasteur pipette if any liquid is present. Alternatively, the plug of agar may be removed with a narrow tube attached to a pump. The bottoms of the wells are then sealed with drops of molten agar. Gels in plastic dishes do not appear to need to be sealed because the plastic at the base of the wells may be sufficiently water-repellent to prevent antigens and antisera from spreading beneath the agar.

The antiserum is placed in the central well and suspensions of the standard bacterial strain put in two of the outer wells, diametrically opposite each other. The unknown strains are placed in the four remaining wells; each of them is thus adjacent to a standard. Antigens of unknown strains are examined as dense suspensions of cells washed from yeast extract–mannitol agar cultures. Nodules from soybean and cowpea plants may be crushed individually in small volumes of saline and these suspensions used directly in the immunodiffusions; diffusion of the antigens from these slow-growing organisms (from cultures or nodule extracts) is enhanced by heat treatment, wherein material

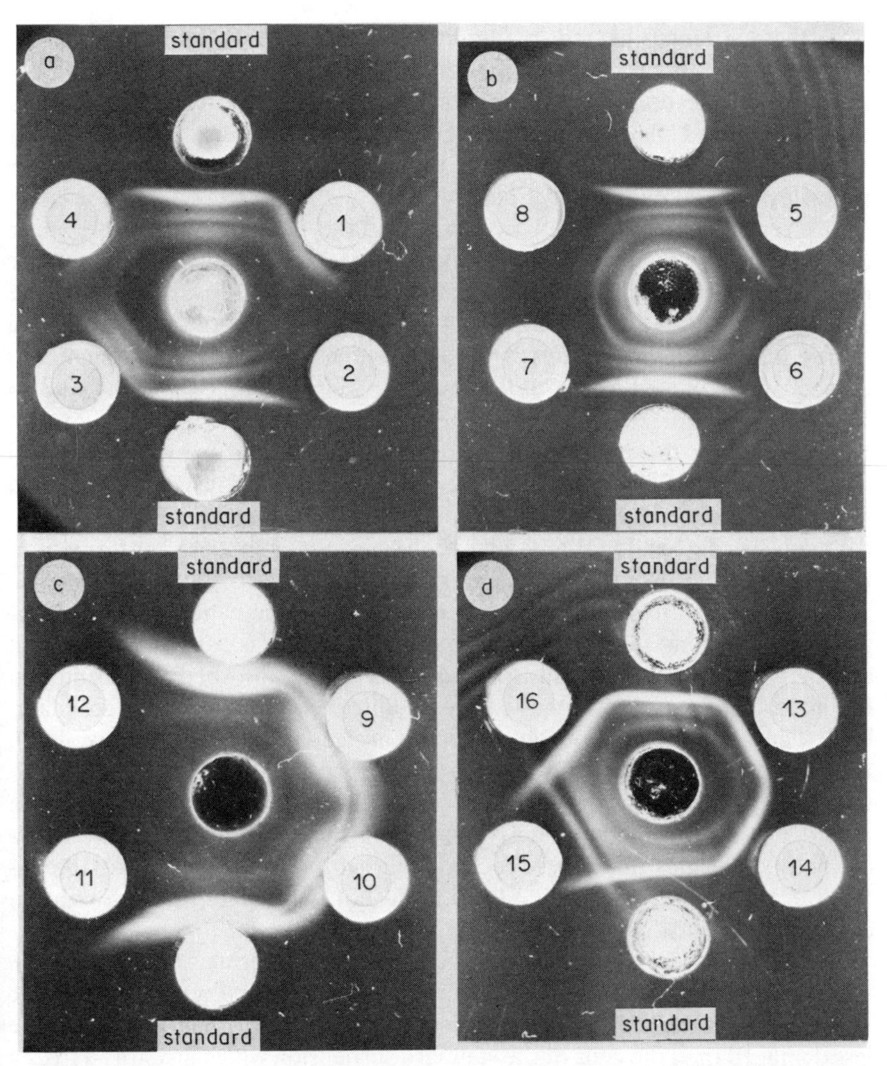

Figure 1. Identification by immunodiffusion of *Rhizobium trifolii* strains isolated from clover nodules during field experiments (Brockwell and Dudman, 1968). The results shown here (photographed after storage of the plates for 3 days at 4 °C) illustrate typical patterns of precipitin bands and show differences in the patterns obtained with the various strains. (a), (b) strain TA1; (c) strain UNZ29; (d) strain WA67. The same arrangement of reagents was used in all plates: antiserum in the centre, standard cell suspensions (10–20 mg cell dry matter ml^{-1}) in the top and bottom wells of each array, and unknown isolates in the remaining outer wells. Isolates 1, 3, 4, 9, 10, 13, 14, and 16 were rated as giving reactions of identity with the standard suspensions in their respective immunodiffusions. Isolates 2, 11, and 12 gave negative results, whereas isolates 5, 6, 7, 8, and 15 gave cross reactions.

Note the differences in the bands obtained in the various cross reactions from these strains is heated in a boiling water-bath for 20 min. Small pieces of paper, coded for identification, may be placed on the agar; although not sterilized, they cause no contamination. The plates are stored with wet paper tissues in closed plastic boxes and kept at 4°C. They are examined daily against dark-field illumination and the results recorded photographically (Crowle, 1973; Williams and Chase, 1971).

The results are interpreted by examining the junctions of the precipitin bands formed by the standard and unknown strains. Three types of result are possible: (i) antigenic identity, indicated by the smooth fusion of the bands (Figure 1: isolates 1, 3, 4, 9, 10, 13, 14, and 16); (ii) cross-reactivity (reaction of non-identity) indicated by a spur at the junction of the bands (Figure 1: isolates 5, 6, 7, 8, and 15): and (iii) absence of any antigenic similarity, shown by the absence of any precipitin band (Figure 1: isolates 2, 11, and 12). Familiarity with the patterns of bands produced by the known strains will lead to greater confidence in the identification of unknown isolates. For a field isolate to be identified as a known strain, its main band must merge with the main specific band of the standard cell suspension without spur formation; with experience this is readily recognizable without much doubt. Some of the precipitin bands are shared with other strains. Specific bands may be recognized by absorbing samples of antiserum with cells of a cross-reacting strain and then examining the bands formed between the absorbed antiserum and the homologous strain. Antisera may be absorbed directly in the gel plate, by first placing a sample of the absorbing antigen in the antiserum well and allowing it to remain for several hours to allow the antigens to diffuse into the gel. Any residual liquid is then removed from the well and the antiserum placed in the well in the usual way; during the subsequent diffusion, the antibody molecules that cross-react with the absorbing antigens are precipitated in the gel near the well edge and do not participate in the formation of precipitin bands.

Suspensions of freshly grown cells generally do not form as many precipitin bands as the standard suspensions, even when they are undoubtedly the same strain. One reason that has been proposed is that freshly harvested cells are less damaged than the cells of the standard suspension which are stored frozen and thawed repeatedly. It is advisable to use the same medium to grow all strains that are to be compared with each other at any one time. The patterns of precipitin bands that are obtained from cultures of a single strain grown on different media, even the same medium in liquid or gel form, can be very dissimilar.

This account of the immunodiffusion technique has been written to describe the identification of *Rhizobium* strains because this is its main application to the study of diazotrophic organisms but immunodiffusions have been used also in taxonomic studies of *Rhizobium* and *Azotobacter* species (Vincent and Humphrey, 1970; Vincent *et al.*, 1973; Norris, 1960). For other applications of the technique to diazotrophs, see Dudman (1977).

Fluorescent antibody techniques

In contrast to the foregoing methods, which require the presence of sufficient amounts of antigen and antibody to give reactions visible to the unaided eye, the fluorescent antibody technique allows the antigens on single cells to be investigated. This makes it possible to examine serologically the rhizobia in crushed nodules of even the smallest size (Trinick, 1969) and to conduct autoecological studies of organisms *in situ* in the soil (Schmidt, 1974).

Antibody molecules are covalently labelled with a fluorescent substituent so that their presence at any site to which they are attached can be readily detected by microscopy in ultraviolet light. In the direct technique, the antibacterial antibodies are labelled, making it necessary to have fluorescent antibodies for each of the strains to be studied. This requirement is eliminated by using the indirect technique, which uses a second antiserum prepared in an animal species different from that used to make the antibacterial antibodies; by labelling the antibodies in this second antiserum (e.g. made in sheep or goats) and using them to detect attachment of the antibacterial antibodies (say from rabbits) it is necessary to have only one type of fluorescent antibody. The second antiserum, required for the indirect technique, can conveniently be purchased with or without a fluorescent label, from suppliers of serological products; if purchased unlabelled, it can be made fluorescent in the same way as antibacterial antibodies, described below.

It is necessary to isolate the globulin fraction from the antiserum before proceeding to label it. The simplest method is to use ammonium sulphate precipitation (Hebert *et al.*, 1973; Nairn, 1976; Nowotny, 1969). A measured volume of the rabbit antiserum is placed in a beaker and stirred constantly while saturated ammonium sulphate solution, sufficient to give a final concentration of 35%, is added slowly (one drop per second) from a pipette. The reaction mixture is stirred for 30 min, set aside for 4 h, and centrifuged to sediment the precipitated protein. No difference is found in the product when the manipulations are performed at 4 °C or at room temperature (Hebert *et al.*, 1973). The precipitate is redissolved in buffered saline (8.5 g of NaCl, 1.07 g of Na_2HPO_4, and 0.39 g of $NaH_2PO_4 \cdot 2H_2O$ in 1 l of distilled water, final pH 7.0; Nairn, 1976) to a final volume equal to the original volume of antiserum and the ammonium sulphate addition repeated. The mixture is stirred for 30 min and centrifuged. The precipitate is dissolved and the process repeated a third time. The final precipitate is dissolved in saline and dialysed at 4 °C against changes of buffered saline until sulphate is not detectable in the outer liquid. Comparison of the initial volume with the final volume of the globulin solution after dialysis allows the dilution to be calculated; the protein concentration may be estimated by the biuret reaction (Kabat, 1961; Nowotny, 1969) or by one of the methods in Chapter II.2.

Fluorescein isothiocyanate (FITC), lissamine rhodamine B, and tetramethylrhodamine isothiocyanate have been widely used as fluorochromes and many conjugation procedures have been described. The following procedure is given by Nairn (1976) for labelling 4 ml of 2% globulin solution with FITC to obtain approximately 4 molecules of fluorescein per molecule of globulin. The reagents (0.1 M Na_2HPO_4, 0.2 M Na_2HPO_4, 0.1 M Na_3PO_4) and the globulin solution are brought to 25 °C in a water-bath. Each milligram of globulin to be labelled require 12.5 μg of FITC and the appropriate weight (1 mg) of FITC isomer I is dissolved in 2 ml of 0.1 M Na_2HPO_4 solution by crushing and stirring with a glass rod for 5–10 min; the solution should be used within 1 hr as it is unstable. A 1-ml volume of 0.2 M Na_2HPO_4 is added dropwise over 2–3 min to the well stirred globulin solution in a 20-ml glass-stoppered flask. The FITC solution (also at 25 °C) is added likewise. Using a pH meter, the reaction of the solution is measured and adjusted to pH 9.5 with 0.1 M Na_3PO_4. The volume is made up to 8 ml by adding 0.85% NaCl and the flask is stoppered and swirled briefly to mix the contents. The mixture is left for 30 min at 25 °C without agitation and then placed in an ice-bath, where it is again swirled to cool the mixture. If necessary, the solution is centrifuged, preferably under refrigeration. Unreacted fluorescent material must be removed; this may be achieved by chromatography on Sephadex G-25 using a column volume approximately six times that of the reaction volume. Phosphate-buffered saline solution is used to elute the column; the first coloured fraction to be eluted is the conjugated globulin. It may be stored deep-frozen.

In the direct fluorescent antibody technique, fixed smears are stained with a range of dilutions of the conjugated antiserum at 37 °C for periods varying between 10 min and 1 h, in a moist chamber to prevent drying, followed by washing with buffered saline (10 min) and then water (5 min). In the indirect technique the unlabelled antiserum is used as just described, followed by the labelled globulin specific for the animal species used to produce the first antiserum; the second globulin stage is treated in the same way. Smears fixed with FITC-labelled globulins are mounted in a mixture containing equal parts of glycerol and isotonic barbital-buffered saline (10.3 g of sodium barbitone, 6.2 g of NaCl, HCl to adjust to pH 8.6, diluted to 1 l with distilled water) (Nairn, 1976).

Space does not permit consideration of the controls necessary to distinguish between specific and non-specific fluorescence or of the microscopy involved in using the technique. Readers should follow the procedures given in original papers and in standard texts (e.g. Kwapinski, 1965; Weir, 1967; Walker *et al*, 1971; Williams and Chase, 1976; Nairn, 1976), as well as the recommendations of the manufacturers of their equipment.

Enzyme-linked immunosorbent assay (ELISA)

The sensitivity of serological reactions can be increased greatly by the use of enzyme-conjugated antibody preparations, which have made it possible to detect the binding of minute amounts of antibodies to extremely small quantities of antigens through the formation of highly coloured products from colourless substrates by the action of the conjugated enzymes. Despite their sensitivity, which has been reported to be similar to that of radioimmunoassays, ELISA methods are rapid and simple to perform, and do not require complex equipment. The only relatively troublesome part of the technique is in the preparation of the conjugated antibody reagents but, once prepared, these are extremely stable and economical to use as only small volumes at very low concentrations are required (Engvall and Perlmann, 1972).

As with fluorescent antibody techniques, there are direct and indirect ELISA methods, depending on whether the enzyme is attached to the antibodies that react with the antigen or to secondary antibodies capable of reacting with the first-mentioned antibodies. Both variations of the technique have been applied successfully to the identification of *Rhizobium* strains from cultures and in extracts of crushed root nodules (Kishinevsky and Bar-Joseph, 1978; Berger *et al.*, 1979). The technique has not yet been tried sufficiently extensively with *Rhizobium* for its scope and limitations to be known—for instance, it is not known whether ELISA can distinguish readily between homologous reactions and cross-reactions between closely related strains—but, nevertheless, we believe it to have great promise. In order to encourage more workers in this field to consider using them in their investigations, the principles and experimental details of ELISA methods that have been applied to *Rhizobium* strain identification are given here in some detail. Further useful and explicit information may be obtained from Voller *et al.* (1976) and Clark and Adams (1977).

The direct ELISA technique was used by Kishinevsky and Bar-Joseph (1978) to identify organisms of the cowpea miscellany specific for nodulating peanuts (*Arachis hypogaea*). It depends on the following sequence, with thorough washing between each successive step: (a) non-specific but firm adsorption of specific antibody to the wells moulded in a polystyrene plate; (b) specific binding of the test antigen to the adsorbed antibody; (c) binding of enzyme-linked antibody to the bound antigen; (d) detection of the presence of the enzyme-bound conjugate by addition of a colourless substrate of the enzyme which then converts it to a coloured product; (e) stopping the reaction after suitable time by addition of NaOH; and (f) examining the colour in each well either qualitatively or quantitatively.

Antisera against *Rhizobium* strains are prepared in rabbits and the globulin fraction is obtained by ammonium sulphate precipitation as described above for the preparation of fluorescent antibodies. The globulins may be purified further by chromatography on columns of diethylaminoethyl(DEAE)-cellulose and collecting the protein fraction that is not bound to the column. An enzyme-linked conjugate of each of the globulin preparations is made by adding 0.3 ml (1.5 mg) of alkaline phosphatase (Sigma, type VII) in phosphate-buffered saline (PBS, containing 0.02 M phosphate, 0.15 M NaCl, and 0.02% sodium azide at pH 7.2) to 0.7 ml of the purified globulin solution (1 mg ml^{-1} of protein in half-strength PBS, indicated by an absorbance of 1.4 at 280 nm) and dialysing extensively against half-strength PBS. The conjugating agent, 0.03 ml of 2% glutaraldehyde in half-strength PBS, is added to the proteins, the mixture kept at room temperature for 4 h, and any unbound glutaraldehyde then removed by dialysis against three changes of PBS. Some workers add 1% of bovine serum albumin to their conjugates before storage at 4°C (Clark and Adams, 1977).

The *Rhizobium* antigens used by Kishinevsky and Bar-Joseph (1978) were obtained from cultures grown on solid defined medium (Bergersen, 1961) or by crushing single washed nodules in PBS–Tween (PBS containing 0.05% of Tween 20).

The direct ELISA reaction is performed with the following steps (all incubations are performed in moist chambers, i.e. with the plates in closed boxes or petri dishes over moistened paper). (1) The wells in polystyrene microtitre plates (Cooke M29AR) are treated with globulin (unconjugated) diluted in 0.05 M Na_2CO_3 buffer (pH 9.6) for 3 h at 35°C. The globulin concentration can be varied but is generally within the range 1–10 μg ml^{-1}; 0.2 ml is placed in each well. Polystyrene plates from different sources have been reported to give variable results and it is advisable not to use the outside rows of wells of any plate (Clark and Adams, 1977). (2) The plates are then given three washes, each of 3 min, in PBS–Tween and shaken dry. The washing must be thorough to remove traces of soluble reactants that could cause non-specific reactions. Treated plates can be stored at −18°C for several weeks or dried over calcium chloride for several days without significant loss of efficiency (Clark and Adams, 1977). (3) A 0.2-ml sample of the test antigen is added to each well and incubated overnight at 6°C. (4) The washing step is repeated. (5) A 0.2-ml volume of the globulin–alkaline phosphatase conjugate is added to each well and incubated for 3 h at 35°C. (6) The washing step is repeated. (7) A 0.2-ml volume of enzyme substrate, *p*-nitrophenyl phosphate (0.6 mg ml^{-1} in 10% diethanolamine buffer adjusted to pH 9.8 with HCl), is added and incubated at room temperature. Alternatively, the substrate may be dissolved in 0.05 M Na_2CO_3 (pH 9.8) (Berger *et al.*, 1979). (8) After 30 min, the reaction is stopped by addition of 0.5 ml of 3 M NaOH to each well. Positive results are indicated by a yellow colour in the wells; the results can be made quantitative by measuring the absorbance of the reaction mixtures at 405 nm.

Kishinevsky and Bar-Joseph (1978) found that homologous reactions could be detected using heat-treated (30 min, 100°C) *Rhizobium* suspensions containing 10^4–10^5 cells ml^{-1}; higher cell densities (10^6–10^8 cells ml^{-1}) were required for positive homologous reactions when the suspensions were used without heat treatment. The suspensions of crushed nodules were not heated before analysis; it was found that even 0.4 mg fresh weight of nodule tissue crushed in 1 ml of buffer was sufficient to give a reliable reaction. Root tissues of uninoculated plants and nodules from plants inoculated with serologically unrelated strains did not interfere. It is obviously important to include all of the positive and negative controls and to use a series of dilutions of the antibody and enzyme-conjugated antibody globulin solutions to determine the optimum concentrations of these for the reaction being investigated.

The indirect ELISA technique, used by Berger *et al.* (1979) to identify strains of *R. leguminosarum*, requires a different sequence of operations. It involves (a) fixation of the antigen (cultured cells or crushed nodule suspensions) to the surface of the well, (b) treatment of the reaction site with a protein solution to eliminate subsequent non-specific binding of antibodies, (c) specific binding of rabbit anti-*Rhizobium* antibodies to the antigen layer, (d) specific binding of enzyme-conjugated sheep anti-rabbit antibodies, and (e) detection of the presence of enzyme-bound conjugate by addition of a colourless substrate of the enzyme which converts it to a coloured product.

The rabbit antisera against rhizobia and the sheep antiserum against rabbit immunoglobulins are prepared in the normal way. The conjugation of the sheep globulin with alkaline phosphatase is carried out in the manner described above.

The *Rhizobium* antigens used for identification by ELISA by Berger *et al.* (1979) were grown in yeast extract–mannitol broth, washed three times in distilled water and suspended in distilled water. Nodule preparations were made by crushing single nodules in sufficient distilled water to give suspensions with absorbances of between 0.45 and 1.0 at a wavelength of 600 nm.

The indirect ELISA reaction, as described by Berger *et al.* (1979), is performed in wells in glass agglutination slides in the following sequence of steps. (1) A 0.05-ml amount of cell or nodule suspension is deposited in each well and spread evenly over the bottom half of the surface. The smears are dried with warm air and heat fixed. Slides may be used immediately or stored at room temperature under vacuum for periods up to 1 week. (2) The slides are washed twice for 5 min in PBS–Tween containing 0.001 M $MgCl_2$ and the excess of buffer is removed. (3) A 0.05-ml amount of bovine globulin (5 mg ml^{-1} in PBS) is added to each well and the slides are incubated for 30 min at room temperature. (4) The slides are washed three times for 5 min. (5) A 0.05-ml amount of rabbit anti-*Rhizobium* globulin at a suitable dilution is added to each well and the slides are incubated at room temperature for 2 h. (6) The slides are washed as in step 4. (7) A 0.05-ml amount of suitably diluted enzyme-conjugated sheep anti-(rabbit globulin) globulin is added and incubated at room temperature for 1 h. (8) The slides are washed as in step 4. (9) A 0.05-ml volume of *p*-nitrophenyl phosphate (1 mg ml^{-1} in 10% diethenolamine buffer, pH 9.8, or in 0.05 M Na_2CO_3, pH 9.8) is added to each sample and the slides are incubated at room temperature for 10–20 min until sufficient colour is produced in the positive controls to be seen and measured colorimetrically. (10) A 0.05-ml amount of 3 M NaOH is added to each well to stop the reaction. For colorimetry at 405 nm, the samples are transferred to tubes containing 3-ml volumes of 0.2 M NaOH or 0.05 M Na_2CO_3 (pH 9.8).

Berger *et al.* (1979) examined the effect of variations of the rabbit and sheep antiglobulin solutions and found their best results were obtained using these reagents at 75 and 18.6 $\mu g\ ml^{-1}$, respectively. They also reported that nodules could be stored frozen and that nodule extracts could be heated in boiling water for 5 min without affecting the results. They obtained complete agreement between their ELISA procedure and immunofluorescence in the identification of *R. leguminosarum* strains in nodule extracts from laboratory- and field-grown plants. It is clear that these ELISA methods hold great promise for the rapid identification of *Rhizobium* strains directly from nodules.

(c) OTHER NATURAL MARKERS

In addition to antigenic properties, bacteria not uncommonly show some other naturally occurring strain differences which can aid in their identification in culture and, in rarer instances, in the field. These properties generally tend to be less reliable or more difficult to apply experimentally than serological differences or many induced mutant markers. However, even those properties which are not suitable for initial, positive recognition of a strain can often suffice as secondary, 'back-up' markers. Some of the more commonly occurring or commonly used characters are briefly considered below; others not mentioned here may be unique to certain microorganisms and serve as effective marker systems.

(i) Morphology in culture or in nature

Distinct differences in colony type, when these occur, provide a direct means of strain recognition. For example, a unique red-colony strain of *Rhizobium* which nodulates *Lotononis* (Norris, 1958) can be easily distinguished from other rhizobia and most other bacteria in culture (Chapter III.1). The frequency of occurrence of such contrasting forms among strains of a species is often low, however. Rough *versus* smooth colony type differences can be used properly only when a very small number of strains are involved, since natural variants of this type appear commonly.

On rare occasions a bacterial strain can be directly, if only tentatively, recognized in the field by virtue of some distinctive alteration of the host plant appearance. Examples of these are the black nodules formed on *Dolichos lablab* (Cloonan, 1963) and several other host plants (Chapter III.1; Stanford *et al.*, 1968), and chlorosis of the soybean plant (Erdman *et al.*, 1957b) as induced by some strains of rhizobia.

In general, morphological character differences are of relatively little value for identification of strains within a species but are of some value for recognition of strains from different species or genera. Their usefulness could be enhanced greatly if traits like colony or nodule pigmentation were genetically transferable (e.g. by promiscuous plasmids) to desired strains across species borders.

(ii) Production of antibiotics, bacteriocins, or bacteriophage

The ability of many strains of bacteria to produce these antibacterial agents (Adams, 1959; Hayes, 1964) provides a useful marker (mainly as a secondary or supporting marker) for such strains, if the action of these agents is sufficiently specific. Bacteriocins and bacteriophage, which require specific adsorption sites on bacteria, commonly have a more limited range of bioactivity and are more specific in their action than the low molecular weight antibiotics commonly produced by bacteria.

Production of an antibiotic (Skerman, 1969; Spooner and Sykes, 1972) by a strain is easily detected by simply spotting a loopful of the test bacteria (either cell mass or dense suspension of cells) on double-layer nutrient agar plates containing cells of an 'indicator' strain in the surface layer. Typically, to obtain a uniform indicator layer or 'lawn', *ca.* 0.2 ml of a 1–3 × 10^8 cells ml^{-1} suspension is added to 2 ml of 0.8% soft agar (kept liquid at 48 °C), mixed, and poured over the surface of the base layer of hard (1.5–1.7%) agar. Alternatively an indicator cell suspension can be spread over the hard agar surface with a glass rod, but the resulting lawn of growth may be more uneven, notably with bacteria which have a very mucoid form of growth. Depending on the size of the zone of inhibition surrounding the spot, from 5 to 15 test cultures can be assayed on one agar plate (100 mm diameter) but it may be necessary to test them on more than one indicator strain, including both sensitive and resistant strains, to increase the reliability of identification of the producing strain.

Production of bacteriocins (bacteriocinogeny) can be demonstrated by one of several procedures (Mayr-Harting *et al.*, 1972). The simplest procedure is that given for antibiotic production, but since large bacteriocins may not diffuse more than a few millimetres laterally into the agar from the spot periphery, the narrow ring of inhibition (halo) may be overgrown and masked by producer cells of very mucoid strains. In such cases the use of a greatly diluted cell suspension (*ca.* 10^6 cells ml^{-1}) for the spot, with the loopful applied to a larger area (1–1.5 cm diameter) will often allow the detection of indicator strain inhibition within the spot area. Since some bacteriocinogenic strains require induction by ultraviolet (UV) light or certain chemical agents, the chance of detection can be greatly increased by exposing the test-strain cell suspension or agar culture plates to a predetermined effective dose of UV light (wavelength 254 nm, from a germicidal lamp). The irradiated bacteria are grown for a period of time equal to one or two cell divisions to allow production, then are killed by exposure to chloroform (1–2 drops per millilitre suspension, for about 10 min) before spotting on to the indicator. Strains not requiring induction (i.e. those which produce bacteriocins during normal growth) can be similarly processed, except that the growth period should be extended to at least three cell generations before killing the cells.

In another long-established but less convenient method for detection of bacteriocinogeny, the test cells are applied as a spot or streak directly to the base agar layer and grown for a suitable period of time to allow the bacteriocins to be released and diffuse into the agar. The bacteria are then killed with chloroform vapour from chloroform added to filter-paper on the inside of the cover-plate of inverted Petri dishes. Residual chloroform is removed from the agar by aeration, the cell mass scraped off with a glass slide, and the indicator bacteria are added by surface-agar layering or surface spreading. Zones of clearing or partial clearing above the original spot or streak area indicate bacteriocinogenicity.

Production of bacteriophage by lysogenic bacteria (Adams, 1959; Kay, 1972) can be most conveniently detected by spotting chloroform-treated cell suspensions on appropriate indicator strains (sensitive and resistant) as described for bacteriocins. However, since bacteriophage multiply and form plaques on the indicator bacteria they can be detected at much lower concentrations than bacteriocins, and it is often possible to observe plaque formation by simply using a very small volume of non-incubated, chloroform-treated cell suspension. Prior irradiation and post-irradiation growth of the bacteria is recommended in the case of inducible lysogenic donor strains. Where a strain is not naturally lysogenic and is sensitive to a known temperate bacteriophage it can often be easily made lysogenic for that bacteriophage. A simple procedure is to spot the bacteriophage on the bacterial indicator layer and remove some bacteria from the developed spot area which will appear less opaque than the non-spot area, due to lysis of most of the bacteria, but will show some regrowth of immune cells carrying the prophage. Subculture these cells by streaking out, and test colonies for lysogenicity as defined by bacteriophage production and self immunity. If the frequency of lysogenization is low, the cells from the spot area can be grown out again and the spot test repeated on these bacteria, with the procedure being repeated until a very turbid spot reaction occurs (i.e. a high proportion of the cells then being lysogenic).

(iii) Sensitivity to antibiotics, bacteriocins, or bacteriophage

Tests for sensitivity to these substances can be performed in several ways:

(1) by incorporating them in the agar, spotting cell suspensions (*ca.* 10^8 cells ml^{-1}) on the agar, and comparing growth in the spot area with that of the control plate; (2) by adding the cells of the test strain in the surface agar layer and spotting either a loopful of cell suspensions of the producing strains or a droplet (alternatively, antibiotic-impregnated paper discs) of previously prepared stocks of the antibacterial agents; and (3) by cross-streaking cell suspensions of producer and sensitive bacteria on the agar and looking for suppression of the sensitive strain at the streak intersection. The first method allows a number of strains to be tested for sensitivity on one plate and would also be suitable for replica plating (Lederberg and Lederberg, 1952) of colonies from a master plate, but requires prior preparation of concentrated stocks of the antibiotic, bacteriocin, or bacteriophage. These stocks usually are membrane-filtered supernatants of broth cultures from the producing strains which have been concentrated by rotary evaporation, high-speed centrifugation, or membrane ultrafiltration. The second methods allows only one strain per plate to be tested for sensitivity but is otherwise convenient, especially when a number of producing strains or samples of antibiotics, bacteriocins, or bacteriophage are needed to confirm the identity of the sensitive strain. A simple variation of this method that allows testing of a larger number of strains is to spot a number of different test-strain cell suspensions on a plate (spot area larger than for the other tests) and place a loopful or small droplet of producer cell material in the centre of each spot. The third method (cross-streaking of cells) is the simplest but may not show good inhibition or lysis of the sensitive strain part of the streak intersection when production of bacteriophage or large bacteriocins is involved, or when the cells of either strain produce large amounts of mucoid slime.

In general, the use of sensitivity to such antibacterial agents as a 'natural' strain marker appears to be less definitive than the use of *production* of these substances (Brockwell *et al.*, 1978). Since even the more specific antagonists will usually attack more than one of a group of strains within a species, production is a more rare event than sensitivity and should, in most cases, offer greater specificity as a marker. Greater reliability of identification can sometimes be achieved by typing strains on the basis of both production and sensitivity (Mayr-Harting *et al.*, 1972). The specificity of reaction to bacteriophage can, however, be greatly increased with systems where host-induced modification (HIM) of bacteriophage is known to occur (Arber and Linn, 1969). HIM has long been the basis of an effective system of strain typing (Adams, 1959) in *Salmonella* and deserves more attention than it has received to date for similar typing of strains among diazotrophic bacteria such as *Rhizobium*, where modification-restriction of bacteriophage DNA has been observed (Schwinghamer, 1965). In this strain-typing procedure the strain specificity pattern of the modifiable bacteriophage is controlled by the bacterial strain on which it is last grown vegetatively (external infection, as opposed to induction of prophage). For example, when such a bacteriophage is grown on some strain 'A' it will produce a much larger number of plaques on strain 'A' as indicator than on strain 'B' as indicator; the reverse efficiency of plaque formation may apply when the bacteriophage is last grown on, and adapted to, strain 'B' or some other strain. In this way, a single bacteriophage can provide a number of different typing stocks which will differentiate between clones of a mixed-strain bacterial population.

(iv) Biochemical properties identifiable at the colony level

Biochemical characteristics are more commonly applied to taxonomic classification of microorganisms, but can also be utilized for strain identification when this is possible at the level of colonies, on agar. These are mentioned among natural marker properties because, even though they may normally be species characteristics, intra-specific strain differences (natural variants) also exist and can be employed as markers. The main biochemical properties adaptable to marker purposes are those involving fermentation or specific enzyme functions, whereby colonies or the agar zone surrounding colonies show distinctive colour reactions after treatment with certain reagents (Hopwood, 1970; Aaronson, 1970). Although these methods have not yet been widely used with diazotrophic organisms, they are included here because of their adaptability to most nitrogen-fixing bacteria.

Colonies of fermenting bacteria produce acid in the presence of the appropriate carbohydrate substrates and can be detected by colour differences when an indicator dye is included in the agar. For example, fermenting colonies become dark purple when grown on eosin–methylene blue agar, in contrast to non-fermenters, which show only the colour of the medium. On agar containing triphenyltetrazolium chloride, the non-fermenters acquire a deep red formazan (insoluble reduction product of tetrazolium) colour, whereas the fermenters remain uncoloured because production of acid inhibits reduction of the dye.

Extracellular enzyme activity forms the basis of a more widely applicable identification method whereby degradative enzyme substrates of macromolecular size are incorporated in the surface agar layer. Enzyme activity in colonies is detected by appearance of clear halos around these colonies, against an opaque, coloured, or fluorescing background of the substrate over the remainder of the agar surface. Enzyme activity, e.g. protease, can be detected directly as clear halos, without further detection reactions when the substrate (milk or casein) is sufficiently opaque. In other cases the agar surface is flooded or sprayed with a reagent to produce the desired non-halo area contrast. For example, for detection of nuclease activity the agar surface is flooded with dilute hydrochloric acid to precipitate undegraded nucleic acid incorporated in the agar, or a fluorescing dye such as acridine orange, which binds to DNA with enhanced fluorescence under UV light, can be included in the agar to reveal the background area of non-degraded nucleic acid. Similar procedures are available for cellulase, urease, and some other degradative enzymes (Aaronson, 1970; Hopwood, 1970; Skerman, 1969). When the substrates are of low molecular weight and cannot be revealed by opacity, appropriate colour indicators and non-substrate materials can sometimes be added to provide contrast for colony and halo detection. An example is the detection of penicillinase activity in colonies (Novick and Richmond, 1965) by flooding the agar surface with a solution of an acid–base indicator (*N*-phenyl-1-napthylamineazo-*o*-carboxybenzene), drying the surface, then flooding again with penicillin solution. Penicillinase-positive colonies turn purple (penicillin converted to penicilloic acid by the enzyme), whereas penicillinase-negative colonies or sectors retain the original colour.

(v) Biochemical properties requiring some biochemical analysis for identification

Methods of strain identification requiring some biochemical analysis of cell material can, for practical purposes, be applied only when a small number of cultures are involved or when other markers are unsatisfactory and confirmation of identity is required. Two examples of this type of identification are isozyme analysis and gas chromatographic methods. The former involves electrophoretic separation of enzymes from cell extracts on paper, starch, or polyacrylamide gel, followed by exposure of the gel to appropriate enzyme substrates and indicator chemicals to identify specific enzymes by their activity (Brewer and Sing, 1971). The method has been used extensively for phylogenetic studies of organisms but could also be adapted for strain identification when sufficient intraspecific variation exists. In *Rhizobium*, for example, sufficient strain differences appear to occur in esterase activity (Murphy and Masterson, 1970) and 3-hydroxybutyrate dehydrogenase activity (Fottrell and O'Hora, 1969) to warrant use of these systems as secondary markers. Analyses based on non-enzymic proteins appear to offer little promise for strain identification.

Alternatively, gas chromatography may be used to analyse fermentation products arising from microbial growth in media or the microbial cells themselves after suitable treatment, to yield distinctive 'fingerprints' or 'signatures' characteristic of the particular genus, species, or strain, as the case may be. Such techniques have not yet been applied to diazotrophs but there is no reason why one of the methods already successfully used with other genera should not be applicable.

It is not possible to give more than a brief outline of the alternative gas chromatographic procedures that have been used. Bacterial cultures (especially anaerobic cultures) with volatile products (acids, alcohols, and gases) may be injected directly or they may be first extracted and the products separated into neutral, acidic, or basic components before the gas chromatographic analysis.

If instead of cultural products in the medium the bacterial cells themselves are to be analysed, they may be treated in a number of alternative ways. The cell samples may be hydrolysed or methanolysed and the constituents released by these procedures made volatile by silylation or trifluoroacetylation prior to analysis. Alternatively, the cell samples may be pyrolysed and their macromolecular constituents converted into smaller volatile fragments which may be resolved directly by gas chromatography. All these methods generate large numbers of peaks and it may be necessary to resort to numerical methods of evaluation rather than rely on visual comparisons for discrimination. The identity of the constituents is not important but recent developments include the coupling of gas chromatographs to mass spectrometers, and this has led to identification of some of the constituents responsible for the peaks seen on the gas chromatograms. Readers interested in applying these methods should consult reviews by Drucker (1976), Mitruka (1976) and Gutteridge and Norris (1979). The literature on the topic is growing rapidly.

In another related instrumental technique, pyrolysis products are analysed directly by mass spectrometry without involvement of a gas chromatograph. This technique was found to be less satisfactory than pyrolysis–gas chromatography in differentiating between strains but it was much more rapid, allowing the analysis of one sample per minute (Meuzelaar, *et al.*, 1975; Gutteridge and Norris, 1979).

(d) MUTANT MARKERS

Although some naturally occurring characters, including antigenicity, provide very effective marker systems in microorganisms, mutation is a more convenient source for a wider range of markers (see Hopwood, 1970, for a comprehensive review; Hayes, 1964). Since strains within a species tend to be similar in most properties, mutants often provide the only reliable means of differentiation. The question of spontaneous *versus* induced mutants is largely one of convenience in isolating the former. Where the frequency of mutation is very low the use of mutagens becomes essential, but it must be remembered that excessive mutagenic treatment results in mostly random accumulation of gene alterations for characters other than the desired ones, and it is desirable that pleiotropic changes affecting critical properties such as nitrogen fixation be avoided for all but certain specific biochemical or genetic purposes.

The choice of which mutant types to use depends on the requirements of the individual experiment. For example, auxotrophs (nutritionally dependent variants) form an essential part of microbial genetic experiments in the laboratory where the composition of the culture medium can be carefully manipulated, but are mostly unsuitable for experiments in the field where the mutant would be at a disadvantage in competition for growth and survival. Antibiotic-resistant mutants are valuable for both laboratory and field experiments, particularly the latter because the antibiotic aids in isolation and enrichment of the resistant strains (i.e. most bacterial contaminants are eliminated during isolation) as well as in identification. Knowledge of the mechanisms of action (e.g. inhibition of synthesis or function of nucleic acids, of proteins, or of the cell wall membrance components) of antibiotics or the nature of mutation to resistance can aid in the choice of antibiotic or antimetabolite to use (Benveniste and Davies, 1973). For example, mutation for resistance to aminoglycosidic antibiotics such as streptomycin or spectinomycin which inhibit protein synthesis often involves alteration of a ribosomal protein. In *Rhizobium*, mutants resistant to these antibiotics (Schwinghamer, 1967, 1968) were less likely to have associated defects in symbiosis than mutants selected for resistance to antibiotics which affect cell wall/membrane function. Whatever the marker used, it should where possible fulfil the following criteria: (a) it should be sufficiently different from the wild-type character and be adaptable to a simple but efficient detection or screening procedure; (b) it should be stable, even through repeated culture or prolonged culture or storage (a knowledge of the approximate reversion frequency is helpful) and (c) it should not differ significantly from the parent strain in key properties such as nodulation or nitrogen fixation (mutants with associated metabolic blocks can, however, be valuable for various biochemical studies).

(i) Mutagenic treatment and isolation of mutants

Since mutation for properties such as antibiotic resistance or reversion from auxotrophy to prototrophy (nutritional independence) involves a gain of function which allows even rarely occurring mutants to be easily isolated on a selective medium, it is usually not necessary to use mutagens to obtain them. This contrasts with isolation of loss-of-function types of mutants such as auxotrophs, where mutagens are used to increase the mutation frequency and some further mutant 'enrichment' procedure is often required.

Radiation and chemical mutagens are generally used to induce mutation (Hopwood, 1970). Ionizing radiations (X-rays, γ-rays, neutrons) are effective mutagens (Hutchinson and Pollard, 1961), but the non-ionizing radiation UV light (Smith and Hanawalt, 1969; Buttolph, 1955) is more readily available in most laboratories. Since low-pressure, mercury-vapour germicidal lamps emit about 90% of their energy at 254 nm, near the absorption peak for nucleic acids, these serve as a convenient source of mutagenic UV radiation. Precise dosimetry is not needed for most routine uses. The exposure time at a given distance for cells in a shallow-depth suspension (preferably agitated with a magnetic stirrer) is determined in a preliminary dose–survival trial; a dose allowing 0.1–1% survival is frequently used for mutagenic treatment, followed by a growth period equal to one or more cell generations for delayed phenotypic expression of mutation.

Chemical mutagens provide the most convenient, all-round means for inducing mutants, partly because little specialized equipment is required and partly because of the wide range of chemicals available, with different groups of chemicals reacting with DNA by different mechanisms. It is thus possible to aim for relatively subtle genetic changes such as base substitution, or for more drastic changes such as deletions and frame-shift mutations. Among the more commonly used chemicals are nitrous acid, alkylating agents (e.g. ethyl methanesulphonate or nitrogen mustard), base analogues (e.g. 2-aminopurine), and *N*-methyl-*N'*-nitro-*N*-nitrosoguanidine (NG); NG is perhaps the most widely used and most effective mutagen for microorganisms, particularly when used under optimum treatment conditions which allow high cell survival (Adelberg *et al.*, 1965; this volume Chapter II.10). The mechanism of action and the experimental use of these chemicals, some of which are highly toxic to humans and differ greatly in their lethality to microorganisms, have been reviewed elsewhere (Hayes, 1964; Kihlman, 1966; Drake, 1970; Freese, 1971). In general, treatment involves exposure (or growth) of cells at a pre-determined time and concentration of the chemical, stopping chemical action by dilution, membrane filtration, or washing (centrifugation), and growing the cells for several generations before screening for mutants.

Mutator genes, which can be regarded as biological mutagens, also warrant a brief mention. When present in a bacterial genome such genes can induce genetic change or enhance the frequency of spontaneous mutation (Cox, 1976). Perhaps the best example of a powerful mutagen of the 'genetic' type as used for a diazotroph (*Klebsiella aerogenes*) is that of the bacteriophage Mu (Bachhuber *et al.*, 1976). The host range of this bacteriophage is normally restricted to the enteric bacteria but its genome can be inserted into a promiscuous plasmid and then transferred by conjugation to an unrelated diazotroph such as *Rhizobium* (Denarie *et al.*, 1977). Various DNA 'insertion or transposable elements' (Starlinger and Saedler, 1976) which are becoming an important experimental tool in genetic engineering, appear to hold similar potential for inducing or manipulating characters which might otherwise be difficult to isolate for strain identification purposes.

For example, transposons have been used effectively for transfer of some antibiotic resistance markers or as mutagens with some diazotrophic bacteria (Chapter II.10), but utilization of these for strain marking purposes is contingent on availability of the necessary genetic expertise and on possible biohazard restrictions for non-laboratory use. These limitations also apply to transfer of antibiotic resistance or other useful markers by plasmids in general, but eventually it should be feasible to apply to field studies those transposable genetic elements which are shown to occur freely among diazotrophs in nature.

Isolation of mutants, whether spontaneous or induced, is a simple matter with 'positive' changes such as antibiotic resistance or prototrophy which are suitable for direct isolation. In the case of antibiotic resistance, bacteria are simply surface-spread or layered on agar containing the antibiotic (antibiotic stock solutions, if not sterile, should be membrane filtered) at a concentration which is at least 4–5 times higher than the level of resistance of the parent strain and of other bacteria from which the mutant must later be distinguished. With antibiotics (e.g. streptomycin) to which mutation produces higher levels of resistance, a higher concentration can be used; however, mutants resistant at very high levels sometimes have an increased likelihood of unwanted pleiotropic effects. Resistant colonies should be subcultured and the clones then spot-tested for maximum level of resistance as well as for possible cross-resistance to other antibiotics and for retention of key properties such as nitrogen fixation. It is normally not necessary to maintain the resistant mutants on media containing the antibiotics in question. Indeed, unstable or antibiotic-dependent resistant mutants are unsuitable for strain identification purposes and are best eliminated. Similarly, mutants isolated for resistance to certain antibiotics or other growth inhibitors are sometimes also auxotrophic (Schwinghamer, 1969). It is therefore desirable to compare all mutants with the wild type strain for growth on a minimal medium, to avoid use of marked strains with metabolic defects that would likely affect their growth competitive ability in the field environment.

Prototrophs are isolated as easily as antibiotic resistant mutants by plating auxotrophic cells on a minimal medium which supports growth of the wild type but not the auxotroph.

Isolation of 'loss' mutants requires some method of mutant enrichment or indirect selection (review by Hopwood, 1970). Most enrichment methods for auxotrophs involve growing the cells in a medium in which only non-mutants grow and are killed by agents or treatments which are selectively lethal to replicating cells. In the best known method, the penicillin-enrichment procedure, the bacteria are first grown in a minimal medium (lacking the growth factor in question) for a period of time to starve only the desired mutant cells and stop or retard their replication; penicillin, an inhibitor of cell wall synthesis, is then added to kill the replicating wild-type cells. The method is highly effective when care is taken to prevent or minimize cross-feeding of mutants by non-mutants in the presence of antibiotic. Where an enrichment method is not practicable, probably the most widely used method for screening a fairly large number (thousands) of cells is by replica plating (Lederberg and

Lederberg, 1952). In this method, a velveteen-covered block or a multiple-pin type of inoculator is used to transfer an imprint of, say, 100 colonies from a 'master' agar plate to other 'tester' plates containing an appropriate indicator or selective agent for detection of either loss or gain of function, e.g. enzyme activity, increased sensitivity to antibiotics, or conditional lethal mutation for temperature sensitivity. Mutant colonies can thus be detected and further characterized for their specific requirements (in the case of auxotrophs) without being sacrificed. Auxanographic procedures, whereby biochemicals to be tested are deposited (as a solid or a concentrated solution) on an agar lawn containing an individual auxotroph, can then be conveniently used to confirm growth responses. These tests are especially useful for detection of partial or multiple growth factor requirements since diffusion of the chemicals into the agar establishes a simple concentration gradient and allows for overlapping of gradients from different chemicals, depending on the spot-test pattern used.

(ii) Use of mutant markers

Mutant markers, notably auxotrophy and antibiotic resistance, have been so widely used for genetic or biochemical genetic studies (Chapter II.10), that it is not necessary here to provide detailed examples of application in laboratory experiments. The reader is referred to the following reviews or representative research papers to illustrate the laboratory use of markers in each of the three main groups of culturable diazotrophs: (a) free-living, non-symbiotic bacteria—Brill (1974), Dixon *et al.* (1977); (b) symbiotic bacteria—Beringer and Hopwood (1976), Brill (1974), Kondorosi *et al.* (1977), Meade and Signer (1977), Schwinghamer (1977); (c) blue-green algae—Herdman and Carr (1972), Shestakov and Khyen (1970). In the case of non-symbiotic diazotrophic bacteria, nitrogenase-deficient (*nif*$^-$) mutants are indispensable for study of the genetic or biochemical basis of nitrogen fixation. This differs from most field studies (excepting experiments using *nif*$^-$ mutants as 'controls' for evaluating levels of biological dinitrogen fixation) in which strain competition and survival ability are the main parameters and the *nif* function should remain intact.

Compared with laboratory experiments, little use has been made of available mutant markers for field experiments, especially with non-symbiotic diazotrophs. Judicious application of some markers, particularly antibiotic resistance, in conjunction with available natural markers (antigenicity and others previously mentioned) would facilitate many types of mixed-strain field studies (survival, competition, nitrogen-fixing efficiency) which have been avoided by investigators unfamiliar with strain identification methods. A single, reliable marker such as streptomycin resistance will suffice for some field experiments. Where two or more test strains are involved, or the population of closely similar, 'mimicking' organisms in the soil is high, a second or third marker is needed. Additional back-up markers (e.g. lysogeny, bacteriophage sensitivity, enzyme activity), which are not suitable for initial screening of a large number of cells, can be used to sort out any doubtful isolate identification. Lysogeny appears to be one of the more effective supporting markers of this type because it allows identification by bacteriophage production, as well as by immunity to the temperate bacteriophage. Resistance to a phage as obtained by mutation is subject to greater risk of pleiotropic change in bacteriophage receptor site and symbiotic effectiveness (Gupta and Kleczkowska, 1962).

In most field experiments with rhizobia, the bacteria to be identified are relatively free of microbial contamination if isolated from fresh, surface-sterilized nodule tissue; even native rhizobia which cannot nodulate the host legume plant used are excluded. Non-symbiotic diazotrophic bacteria or algae would, in similar field experiments, be isolated from soil, water, or plant root surfaces and present more problems of contamination. However, most bacterial contaminants would be eliminated or suppressed by incorporating in the agar medium the antibiotic(s) to which the marked strains are resistant. Fungal contamination can be controlled by adding an antifungal antibiotic such as cycloheximide to the medium. In particular, the use of nitrogen-free or low-nitrogen media provides strongly selective growth conditions for non-symbiotic diazotrophs.

The main application of markers for identification of strains recovered from soil or plant material has in the past been with *Rhizobium*. The markers were mainly antigens, but antibiotic-resistant mutants within recent years (Obaton, 1971; Imshenetskii *et al.*, 1976; Brockwell *et al.*, 1977b) have been used very successfully for strain survival–competition studies in which mixed inocula are applied to the legume seed to be sown in non-sterile potted soil or in the field, and the bacteria are later isolated from individual nodules harvested from the plants at different time intervals. Brockwell *et al.* (1978) compared the practicability of a number of marker systems and found serology, antibiotic resistance, lysogeny, or bacteriocinogeny to be more useful than symbiotic specificity, colony, or nodule characteristics, or sensitivity to bacteriophage and bacteriocins. Figure 2 illustrate the procedure for a hypothetical experiment which simulates some experiments already performed. In this example the three strains are distinguishable from each other and from native strains of *R. trifolii* present in the soil on the basis of double markers, namely antibiotic resistance plus either antigenicity or lysogeny as a second marker (details of seed inoculation and isolation of cultures from nodules are given in Chapter III.1). Where mixed infection in one nodule is suggested by gross differences in the amount of bacterial growth appearing on the antibiotic and non-antibiotic plates, more than one colony should be picked from the relevant plates for assay. The results for an actual experiment of this type (Brockwell

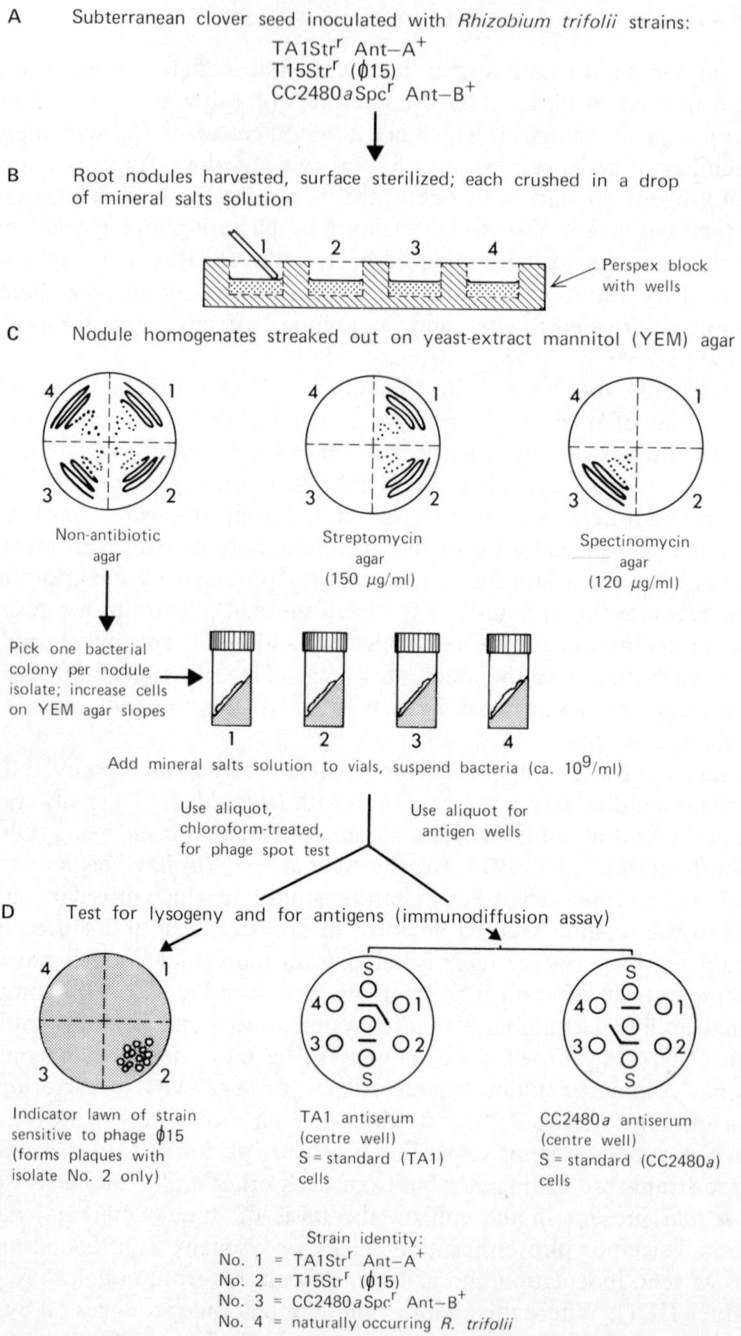

A Subterranean clover seed inoculated with *Rhizobium trifolii* strains:

TA1Str^r Ant–A$^+$
T15Str^r (ϕ15)
CC2480*a*Spc^r Ant–B$^+$

B Root nodules harvested, surface sterilized; each crushed in a drop of mineral salts solution

1 2 3 4

Perspex block with wells

C Nodule homogenates streaked out on yeast-extract mannitol (YEM) agar

4 1 3 2
Non-antibiotic agar

4 1 3 2
Streptomycin agar (150 μg/ml)

4 1 3 2
Spectinomycin agar (120 μg/ml)

Pick one bacterial colony per nodule isolate; increase cells on YEM agar slopes

1 2 3 4

Add mineral salts solution to vials, suspend bacteria (ca. 10^9/ml)

Use aliquot, chloroform-treated, for phage spot test

Use aliquot for antigen wells

D Test for lysogeny and for antigens (immunodiffusion assay)

4 1 3 2

Indicator lawn of strain sensitive to phage ϕ15 (forms plaques with isolate No. 2 only)

S 4 1 3 2 S

TA1 antiserum (centre well)
S = standard (TA1) cells

S 4 1 3 2 S

CC2480*a* antiserum (centre well)
S = standard (CC2480*a*) cells

Strain identity:
No. 1 = TA1Str^r Ant–A$^+$
No. 2 = T15Str^r (ϕ15)
No. 3 = CC2480*a*Spc^r Ant–B$^+$
No. 4 = naturally occurring *R. trifolii*

Figure 2. Outline of an experimental procedure for the identification of marked seed-inoculant strains of *Rhizobium* and naturally occurring rhizobia present in nodules from field grown plants. (Refer to the footnote of Table 2 for the key to abbreviations used for strain identity)

et al., 1977b) are given in Table 2. The experiment had been designed to compare the stability of the two markers shown and to observe the gradual decline of inoculant bacteria in competition with native rhizobia over a period of 3 years in the field. Agreement between these two markers (streptomycin resistance and antigenicity) was at least 95%. Alternatively, two inoculant strains, one carrying Str^r/Ant-1 and the other Spc^r/Ant-2, or four strains each carrying only one of the four markers could have been used in this type of experiment.

Table 2. Identification of *R. trifolii* isolates from clover nodules, using streptomycin resistance and serological characteristics for distinguishing between a doubly-marked (Str^r/Ant^+) inoculant strain and naturally occurring strains

Marker 'phenotype'*	Interpretation of 'phenotype' and type of infection	Isolates in each class			
		Dec. 1969	May 1970	Oct. 1971	Nov. 1972
	Agreement between markers	No. of isolates			
Str^r/Ant^+	Single infection by inoculant strain	78	53	51	50
Str^s/Ant^-	Naturally occurring rhizobia (non-inoculant)	11	9	86	100
Str^r/Ant^+ and Str^s/Ant^-	Mixed infection; inoculant and naturally occurring rhizobia	0	0	6	6
	Disagreement between markers	No. of isolates			
Str^r/Ant^-	Single infection; acquisition of streptomycin resistance by naturally occurring strain, or loss of antigen by inoculant strain	2	1	3	0
Str^s/Ant^+	Single infection; loss of streptomycin resistance by inoculant strain or presence of antigen in naturally occurring strain	0	1	4	2
Str^r/Ant^- and Str^s/Ant^+	Mixed infection; expected to occur only rarely	1	0	0	0
	Total, all classes	Percent of total			
	Isolates identified as Str^r	88	84	40	35
	Isolates identified as Ant^+	86	84	41	37

*Key to 'phenotypes': Str^r = streptomycin-resistant; Str^s = streptomycin-sensitive; Ant^+ = antigenically positive, with serological reaction of identity with inoculant strain; Ant^- = antigenically negative, with no serological reaction of identity.

Methods for Evaluating Biological Nitrogen Fixation
Edited by F. J. Bergersen

F. C. Cannon
ARC Unit of Nitrogen Fixation,
University of Sussex, Brighton, England.

10

Genetic Studies with Diazotrophs

(a) INTRODUCTION

The genetics of nitrogen-fixing systems have developed rapidly during the 1970s. Most of the methods used were adapted from those of conventional microbial genetics. The methods described in this chapter were selected because of their particular relevance to present and future studies. Sources of the methods are cited in each section and in most cases the laboratories from which the publications have come are also the sources of the bacterial strains, phages, and plasmids used. Details of some methods are specific for the organisms used and may need modification for use with others. The genetic symbols used are those of Bachman *et al.* (1976) and Novick *et al.* (1976). A glossary of some genetic terms and abbreviations is given in Table 1. Abbreviations for media and reagents are given at the point of first mention and subsequent uses are cross-referenced to this point in the text.

(b) GENETIC DISSECTION OF *nif* IN *Klebsiella pneumoniae*

A strategy for a comprehensive genetic analysis of *nif* in *K. pneumoniae* is outlined in the flow-sheet below and suitable procedures for carrying out each step are described in the following sections. Useful background information and details of procedures for similar genetic studies of *lac* in *E. coli* are given by Miller (1972).

Isolate *nif* chromosomal point mutants (*nif* ch. pt.).
Isolate *nif* plasmid point mutants (*nif* pl. pt.).
Check point mutants for reversion, suppression, and a transdominant effect on wild-type nitrogenase activity.
Determine linkage of mutations to *his*D by P1 transduction.

↓

ORDER OF *nif* MUTATIONS

Isolate *nif*::Tn and *nif*::Mu chromosomal and plasmid mutants.
Transfer *nif*::Tn mutations to the chromosome by Pl transduction.
Construct the following Nif^-/Nif^- heterogenotes for complementation analysis:

nif pl. pt./*nif* ch. pt. merodiploids ⟶ **_nif_ GENES**

nif::Tn p1. and *nif*::Mu pl./*nif* ch. pt. merodiploids
nif pl. pt./*nif*::Tn and *nif*::Mu ch. merodiploids
nif::Tn and *nif*::Mu pl./*nif*::Tn and *nif*::Mu ch. merodiploids
↓
nif OPERONS

Isolate chromosal *nif* deletions (nif^{∇} ch.): check Nif phenotype by complementation tests, and use as transductional donors and recipients to check for inversions.

Table 1. Glossary of terms used.

allele. One of the alternative mutational forms of a given genetic element.
amplifiable plasmid. A plasmid (usually $< 20 \times 10^6$ daltons) that can replicate in the absence of cell division (usually in the presence of chloramphenicol).
complementation. The process by which two different mutant replicons jointly carry out an action of which neither is capable singly, usually be means of shared gene products.
deletion. A mutation in which a nucleotide sequence is absent.
diploid. Having genes represented twice.
heterogenote. A heterozygous partial diploid.
heterozygote. A bacterium carrying two alleles of one or more genes.
*hsd*R. Gene symbol for host restriction activity.
inversion. A mutation in which the orientation of a nucleotide sequence is inverted.
lysate. Disrupted bacteria. Phage strains are usually preserved and used in the form of lysates of infected cultures.
lysogen. A bacterium that possesses and transmits to its progeny the power to produce phage particles and is therefore prone to lysis.
merodiploid. A partial diploid bacterium.
nif. Genes determining nitrogen fixation.
nif::*Tn*. Denotes a *nif* mutation derived by the insertion of a transposon (Tn) in *nif*.
operon. A gene or cluster of genes which are transcribed into a single messenger RNA molecule.
polar mutation. A mutation that exerts its effects downstream (with respect to direction of transcription) from the altered site.
promotor. A site in DNA at which RNA polymerase binds or initiates transcription.
rec. Recombination deficient.
reversion. Reversal of the effect of one mutation by another, especially when the second occurs at the site of the first.
replicon. A DNA molecule that replicates antonomously.
*rps*E. Gene symbol for 30S ribosomal subunit protein S5. Spectinomycin resistance is a mutant phenotype.
*rps*L. Gene symbol for 30S ribosomal subunit protein S12. Streptomycin resistance is a mutant phenotype.
suppression. Reversal of the effects of one mutation by another when the second occurs at a different site from the first.
transposon (Tn). A DNA segment (usually larger than 2 kilobase pairs) that can insert into several sites in a genome. Drug resistance is a gene function frequently carried on a transposon; e.g. Tn5 confers kanamycin resistance.
transdominant. Two genes on different replicons in the same cell are said to be in the *trans* arrangement with respect to each other. When the product of one gene dominates the phenotype it is said to be transdominant.
transduction. Transfer of bacterial markers from one cell to another by phage particles.
transformation. Transfer of bacterial markers from one cell to another by DNA in solution.
transcription. RNA synthesis guided by a DNA template.
translation. Protein synthesis guided by an RNA template.
zygotic induction. Induction of a prophage on transfer from a lysogenic donor to a non-lysogenic recipient during bacterial mating.

Construct the following Nif^-/Nif^- heterogenotes for genetic mapping:

nif pl. pt./*nif*$^{\triangledown}$ ch.
nif: : *Tn pl*/*nif*$^{\triangledown}$ ch.
nif: : Mu pl/*nif*$^{\triangledown}$ ch.
→ FINE STRUCTURE MAP OF *nif* GENES

SDS–polyacrylamide gel electrophoresis of pulse-labelled extracts of *nif* mutants with and without small amplifiable plasmids carrying parts of the *nif* region.

↓

nif GENE PRODUCTS

Assay nitrogenase and its component activities in crude extracts of *nif* mutants.

↓

FUNCTIONS OF *nif* GENE PRODUCTS

(i) NG (*N*-methyl-*N′*-nitro-*N*-nitrosoguanidine) dose–response curve

A dose–response curve for each strain used in the isolation of mutants is required since the sensitivity of strains to the toxic effect of NG varies. The curve is obtained by determining the viable count of a culture after different times of exposure to NG and plotting the log of the surviving fraction against the time of exposure.

Bacterial strain	*Genotype*	*Important properties*
K. pneumoniae, UNF5023 (Dixon *et al.*, 1977)	*his*D2 *hsd*R1 *rps*L4	$Nif^+His^-Str^r$ Restriction$^-$

Materials

Nutrient broth (NB).
Nutrient agar (NA) is NB + 1.5% agar.
Minimal medium (MM):

K_2HPO_4/KH_2PO_4 buffer, pH 7.2	60 mM
$(NH_4)_2SO_4$	15 mM
$MgSO_4{\cdot}7H_2O$	1 mM
NaCl	0.2 mM
$CaCl_2$	0.01mM
Glucose	0.2% w/v

NG solution: 10 mg ml^{-1} acetone (freshly prepared). NG is dangerous to handle. Solutions should be handled with protective gloves and never pipetted by mouth.

Saline phosphate buffer, pH 7.2:

K_2HPO_4	$7\,g\,l^{-1}$
KH_2PO_4	$3\,g\,l^{-1}$
NaCl	$8.5\,g\,l^{-1}$

Membrane filters, pore size 0.45 μm.

Method. Grow an overnight culture (5 ml) of UNF5023 in nutrient broth on a shaker at 37 °C.

Inoculate 1.0 ml into 50 ml of minimal medium containing histidine (20 μg ml^{-1}) and grow to a population density of 3–5 × 10^8 bacteria ml^{-1} on a shaker at 37 °C.

Add 0.5 ml of NG solution (10 mg ml^{-1} acetone) (this is time zero) and immediately dispense 5-ml aliquots of the culture into eight flasks (25 ml) pre-heated to 37 °C.

Continue incubation on a shaker at 37 °C.

Use one culture for each time interval tested (0, 5, 10, 20, 30, 45, 60, 90 min). Collect the bacteria in 1 ml on a membrane filter, wash with 4 ml of saline phosphate buffer, and re-suspend in 1 ml of the same buffer.

Make serial dilutions in saline phosphate buffer and spread 0.1 ml (in duplicate) of the 10^{-2}, 10^{-3}, 10^{-4}, and 10^{-5} dilutions for the 0, 5, 10, 20, and 30 min points and 0.1 ml (in duplicate) of the 10^{-1}, 10^{-2}, 10^{-3}, and 10^{-4} dilutions for the 45, 60, and 90 min points on nutrient agar plates.

Incubate the plates overnight at 37 °C.

Determine the viable cell count from the colonies on the plates and plot log (% survivors) against exposure time to NG.

(ii) NG mutagenesis and isolation of chromosomal *nif* mutants

A procedure for NG mutagenesis is described because it has been frequently used to generate *nif* mutants in Klebsiella (Streicher *et al.*, 1972; St John *et al.*, 1975; Dixon *et al.*, 1977; and MacNeil *et al.*, 1978b). Procedures for mutagenesis with NG and other mutagens which present fewer problems of secondary mutations are given by Miller (1972).

Bacterial strain. UNF5023, described above in Section b.i., is an example of a suitable strain for isolating Nif^- mutants.

Materials

Nutrient broth, minimal medium, NG solution, and saline phosphate buffer as specified above in Section b.i.

Nitrogen-free Davis and Mingioli medium (NFDM) (Cannon *et al.*, 1976):

K_2HPO_4	12.06 g l^{-1}
KH_2PO_4	3.4 g l^{-1}
$MgSO_4$	0.1 g l^{-1}
$Na_2MoO_4 \cdot 2H_2O$	25 mg l^{-1}
$FeSO_4 \cdot 7H_2O$	25 mg l^{-1}
Glucose	20 g l^{-1}

The phosphate solution should be autoclaved separately.

(Minimal agar is minimal medium as in Section b.i. + 1.5% agar).

Sterile bijou bottles (7 ml capacity).

Suba Seals (William Freeman, Barnsley, Yorkshire, U.K.) or equivalent.

Anaerobic jars, e.g. Gaspak jars (Becton-Dickinson).

Gas chromatograph.

Acetylene.

Serine solution (20 mg ml^{-1} in water), filter sterilized, stored at 4°C.

Method.—Inoculate 5.0 ml of minimal medium containing histidine (20 μg ml^{-1}) with 0.1 ml of an overnight culture of UNF5023 in nutrient broth and grow to a population density of 3–5 × 10^8 bacteria ml^{-1} on a shaker at 37°C.

Add 0.05 ml of NG solution (10 mg ml^{-1} in acetone) and continue incubation on a shaker at 37°C for the time required to decrease the viable cell count of the culture to 10–20% (obtained from an NG killing curve as described above).

Collect the cells on a membrane filter, wash with 5 ml of saline phosphate buffer as in Section b.i., and re-suspend in 50 ml of NFDM + $(NH_4)_2SO_4$(1 mg ml^{-1}) + histidine (20 μg ml^{-1}).

Dispense 4-ml aliquots of the culture into 12 sterile bijou bottles, cap tightly and incubate overnight on a shaker at 30°C. (Dividing the culture into 4-ml portions at this stage increases the probability of obtaining Nif^- mutants from different mutagenic events.)

Make serial dilutions of the cultures in saline phosphate buffer and spread 0.1 and 0.2 ml of the 10^{-6} dilution on NFDM agar plates containing histidine (20 mg ml^{-1}) and serine (50 mg ml^{-1}). Use at least 10 plates for each culture to obtain a minimum of 2000 colonies.

Incubate the plates anaerobically at 30°C for 1 week. Colonies of Nif^- mutants are usually small and opaque in contrast to the large brown colonies of the parental Nif^+ Klebsiella. 1–5% of the colonies should be Nif^-. There is much less contrast between Nif^+ and Nif^- colonies of *E. coli*.

Streak the presumptive Nif^- colonies on minimal agar plates containing histidine (20 μg ml^{-1}), incubate at 37°C, and test those with growth rates comparable to the parental strain for acetylene reduction as follows.

Inoculate 2-ml aliquots of nutrient broth from the minimal agar plates and grow on a shaker at 37°C for 2–4 h. Use the parental strain as a positive control.

Inoculate 4-ml aliquots of NFDM + histidine (20 μg ml^{-1}) + serine (50 μg ml^{-1}) in 7-ml bijou bottles with 0.2 ml of nutrient broth cultures, cap with sterile Suba Seals, equilibrate the gas pressure, and incubate overnight with vigorous shaking at 30°C.

Inject 1 ml of acetylene into each bottle, continue incubation as above, and after 30 min measure the ethylene produced in 0.5 ml of the gas phase with a gas chromatograph.

Measure the absorbance at 540 nm (A_{540}) of the cultures and express the units of ethylene produced per A_{540} unit per hour as a percentage of that produced by the parental strain. Mutants which give less than 20% of wild-type activity are suitable for further studies.

(iii) NG mutagenesis and isolation of plasmid *nif* mutants

The procedure is based on a modification of the method used by Dixon *et al.*, (1977).

Suitable bacterial strains and plasmids	*Genotype or phenotype*	*Important properties*
K. pneumoniae		
UNF107 (Dixon *et al.*, 1977)	*rps*L (*gnd his nif*)	Str^r, *his–nif* deletion
E. coli		
JC5466 (Dixon *et al.*, 1977)	*trp rec*A56 *his rps*E	Spc^r Rec^- His^-
Plasmids		
pRD1 (Dixon *et al.*, 1976)	Km Tc Cb Gnd His Nif ShiA Tra IncP	
pMF100 (Filser and Cannon, unpublished work)	Gnd His Nif ShiA Tra IncP	

Materials

Minimal medium and NG solution as described in Section b.i.
NFDM as described in Section b.ii.

Spectinomycin (Spc) solution (50 mg ml^{-1} in water), filter sterilized, stored at 4°C.

Tryptophan solution (10 mg ml^{-1} in water; add a few drops of 1 M NaOH to dissolve), filter sterilized, stored at 4°C.

Method.—Inoculate 5 ml of minimal medium with 0.1 ml of an overnight nutrient broth culture of UNF107 carrying pRD1 or pMF100 and grow to a population density of 3–5 × 10^8 bacteria ml^{-1} on a shaker at 37°C.

Add 0.05 ml of NG solution (10 mg ml^{-1} in acetone) and continue incubation on a shaker at 37 °C for 10 min.

Collect the cells on a membrane filter and proceed as in Section b.ii to obtain Nif^- mutants.

Transfer the plasmids from these mutants to JC5466 in plate matings as described in Section b.vi and quantify the nitrogenase activities of the transconjugants. [Use minimal agar plates containing spectinomycin (100 μg ml^{-1}) and tryptophan (25 μg ml^{-1}) to select JC5466 transconjugants.]

Mutants which have less than 20% of wild-type activity in both the original and JC5466 backgrounds are suitable for further studies.

(iv) Suppression of *nif* mutations

The effect of a mutation which causes a nonsense codon to appear within a gene can frequently be reversed by suppressor mutations. If the nonsense mutation has a polar effect on promoter–distal gene expression in a multigenic operon, suppression of the mutation relieves the polarity. Useful background information on nonsense mutations and their suppression in *E. coli* is given by Miller (1972).

Suppressibility of *nif* mutations can be tested by transferring plasmids with *nif* mutations to *E. coli* suppressor mutants or by introducing derivatives of the Pl-type plasmid RP4 carrying suppressor genes into chromosomal *nif* mutants.

Method.—To test suppressibility of *nif* mutations on pRD1 or pMF100 (described in Section b.iii), transfer plasmids to His^- *E. coli* suppressor strains (Miller, 1972) in plate matings as described in Section b.vi.

Transfer derivatives of RP4 with suppressor genes to chromosomal *nif* mutants of *K. pneumoniae* in similar plate matings.

Measure the nitrogenase activities of transconjugants using the acetylene reduction test as described in Section b.ii. Use the Nif^- parent and an isogenic Nif^+ strain as negative and positive controls.

(v) Cotransduction of *nif* mutations with *his*D

In P1 generalized transduction approximately 10^{-5} particles in a lysate contain bacterial instead of phage DNA. The derivatives of P1 used for transduction carry kanamycin (Km) or chloramphenicol (Cm) resistance genes and have temperature sensitive repressors (they form lysogens at 30 °C and go into a lytic cycle at 37–43 °C) (Goldberg *et al.*, 1974). Therefore, lysogens of P1-resistant strains can be isolated by selecting for drug resistance and the temperature-sensitive repressor greatly facilitates the production of lysates and minimizes the problem of transductants being killed by active P1 particles.

Consequently, most transductants are also lysogens. The P1 Km*clr*100 or P1Cm*clr*100 prophages can be cured by streaking transductants on plates and incubating them at 37–43 °C. Colonies of the survivors should retain the transduced marker and some should be Km^s or Cm^s.

P1 transduction has shown that most characterised *nif* mutations in *K. pneumoniae* are linked to *his*D. An approximate order of *nif* mutations can be obtained from the frequency of cotransduction since this frequency varies with the distance between a mutation in *his*D and *nif* mutations.

Materials

Luria broth: tryptone 1%, yeast extract 0.5%, NaCl 0.05%; adjust the pH to 7.0.

Luria agar is Luria broth + 1.5% agar.

Soft Luria agar is Luria broth + $CaCl_2$ (5 mM) + glucose (0.2% w/v) + 0.6% agar.

Methods

P1 lysogens of K. pneumoniae (Streicher *et al.*, 1974).—Add $CaCl_2$ (final concentration 5 mM) to a fresh overnight culture in Luria broth, grown on a shaker at 37 °C.

Spread 0.1 ml of the culture and a drop of a P1Km*clr*100 lysate on a Luria agar plate containing kanamycin (20 μg ml^{-1}). Incubate at 30 °C for 24 h.

To check presumptive lysogens, grow overnight cultures of Km^r colonies in nutrient broth at 37 °C, add a few drops of chloroform to each and Whirlimix.

To check for the presence of phage, overlay a Luria agar plate with 3 ml of soft Luria agar containing approximately 10^8 bacteria of a P1-sensitive strain and spot on to its surface drops (10–20 μl) of the chloroform-treated cultures. When the drops have dried, incubate the plates overnight at 37 °C.

Preparation of P1 lysates. The method is based on that of Goldberg *et al.* (1974).

Add 0.5 ml of a fresh overnight Luria broth culture of a P1 lysogen to a 250-ml flask containing 25 ml of Luria broth + $MgCl_2$ (5 mM).

Grow with vigorous shaking at 32 °C to a population density of 3–5 × 10^8 bacteria ml^{-1}.

Add $CaCl_2$ to a final concentration of 5 mM, transfer the flask to a shaker at 43 °C for 15 min, and finally transfer it to a shaker at 37 °C (to achieve changes in the culture temperature rapidly, use shaking water-baths). After 3–4 h at 37 °C the turbidity of the culture should decrease.

Add approximately 1 ml of chloroform and continue shaking at 37 °C for a few minutes.

Whirlimix the culture, remove cell debris by centrifugation and quantify the phage titre in the supernatant as follows.

Make dilutions (10^{-2}, 10^{-4}, 10^{-6}) of the supernatant in a solution of $MgCl_2$ (10 mM) and $CaCl_2$ (5 mM) and place 10 μl of each on a fresh thick Luria agar plate overlaid with 3 ml of soft Luria agar containing 0.3 ml of a log-phase Luria broth culture of a P1-sensitive *K. pneumoniae* strain.

Incubate the plates overnight at 37 °C. P1 titres vary between 10^8 and 10^{10} p.f.u. ml^{-1}.

Transduction of his*D2* nif *mutants* (Kennedy, 1977).—Add 0.1 ml of fresh overnight cultures of the appropriate mutants (e.g. *nif* mutants of UNF5023) to 5 ml of Luria broth and grow on a shaker at 37 °C to a population density of $4–6 \times 10^8$ bacteria ml^{-1}.

Re-suspend the bacteria in 5 ml of a solution of $MgCl_2$ (10 mM) and $CaCl_2$ (5 mM).

Spread 0.2 ml of the bacterial suspension (*ca.* 10^8 bacteria) and approximately 10^9 P1 phage (prepared as above on an appropriate donor strain, e.g. Kp18, Kennedy, 1977) on minimal agar plates. His^+ transductants should appear after incubation for 1–2 days at 30 °C (expect 10^{-6}–10^{-7} per added phage). Since most transductants will also be P1 lysogens, incubation at higher temperatures would cause lysis of the transductants.

Patch the His^+ colonies on to NFDM agar plates (Section b.ii) and incubate anaerobically at 30 °C for 5–7 days. The proportion of His^+ transductants that are also Nif^+ is the cotransduction frequency. The distance between a *his*D (e.g. *his*D2) and *nif* mutations can be estimated from the cotransduction frequencies using the equation $f = (1 - d/L)^3$ (Wu, 1966), where f = cotransduction frequency, L = length of transducing DNA (80 kb for P1) and d = distance in kilobases (kb) of DNA between the *nif* and *his* mutations.

To isolate transductants which have been cured of the prophage, grow overnight cultures in nutrient broth at 30 °C and add sodium citrate to a final concentration of 100 mM before spreading 0.1-ml aliquots on nutrient agar plates containing sodium citrate (100 mM). Incubate the plates overnight at 42 °C. Check colonies for sensitivity to P1 and to Km or Cm.

(vi) Isolation of *nif*::Tn5, *nif*::Tn7, and *nif*::Tn10 plasmid mutants

A transposon inactivates a gene into which it inserts and usually has a strong polar effect on the expression of genes located 'downstream' in the same operon (Kleckner *et al.*, 1977). Mutations produced by the insertion of transposons into *nif* genes are useful for isolating deletions, fine structure mapping, elucidating the operon structure of *nif* genes, and identifying their products. Such studies have so far been confined to Klebsiella *nif*, but now that a transmissible P1 type plasmid with properties suitable for generating insertion mutations with Tn5 has become available, these can be extended to *nif* in other genera (Beringer *et al.*, 1978; Johnston *et al.*, 1978).

The method described for the isolation of *nif*: : Tn plasmid mutants is based on that of Merrick *et al.*, (1978).

Bacterial strains (Merrick et al., *1978)*	*Genotypes or phenotypes*	*Important properties*
E. coli K12		
UNF510	*his trp lys lac rps*L Km^r	Str^r, His^-, Tn5 (Km^r) inserted in the chromosome
J62: : Tn7	*his trp pro*C *lac nal* Tp^r	Nal^r His^- Tn7 (Tp^r) inserted in the chromosome
DU3083	*his trp lys lac rsp*L Tc^r	Str^r His^- Tn10 (Tc^r) inserted in the chromosome

JC5466, UNF107, plasmids PMF100 and pRD1; for details see Section b.iii.
UNF5023: described in Section b.i.

Materials

Kanamycin (Km) solution (20 mg ml^{-1} in water), filter sterilized, stored at 4°C.
Tetracycline (Tc) solution (5 mg ml^{-1} in water), filter sterilized, stored at −20°C.
Naladixic acid (Nal) solution (20 mg ml^{-1} in water), filter sterilized, stored at 4°C.
Trimethoprim lactate (Tp) solution (20 mg ml^{-1} in water), filter sterilized, stored at 4°C.

Methods

nif: : *Tn5 mutations on pMF100*. Construct strain UNF510 (pMF100) by a plate mating of JC5466 (pMF100) × UNF510. Use fresh overnight cultures of JC5466 (pMF100) in minimal medium (Section b.i) plus tryptophan (25 μg ml^{-1}) and UNF510 in nutrient broth. Mark a dried nutrient agar plate into three sections, for donor, recipient, and mating mixture. Apply one drop of donor and recipient to appropriate sections and one drop of each to the mating section. Three spots of confluent growth should appear after overnight incubation at 37°C.

Touch the surface of each spot with the ball-point of a fine platinum wire and streak on to minimal agar plates containing streptomycin (250 μg ml^{-1}), tryptophan (25 μg ml^{-1}), and lysine (20 μg ml^{-1}). These plates are selective for His^+ Str^r transconjugants of UNF510. Use two plates, one for donor and recipient controls and one for transconjugants. Incubate the plates at 37°C for approximately 24 h.

Streak a few of the transconjugant colonies on similar minimal agar plates, incubate overnight at 37°C, and store at 4–10°C for subsequent use. Transposition of Tn5 from the chromosome of UNF510 onto pMF100 can now occur and the derivative plasmids can be isolated by their transfer to another strain.

Cross UNF510(pMF100) with UNF107 in a plate mating as described above with the following modifications.

Grow a fresh overnight culture of UNF510 (pMF100) in minimal medium containing tryptophan (25 μg ml^{-1}) and lysine (20 μg ml^{-1}).

Use at least 10 mating mixtures to increase the probability of obtaining *nif*: : Tn5 mutants from different transposition events.

After overnight incubation at 37°C of the mating mixtures and controls, suspend approximately 2×10^9 bacteria from each patch of growth in 2 ml of saline phosphate buffer (Section b.i), make serial dilutions (10^{-1}, 10^{-2}, 10^{-3}, and 10^{-4}) of mating mixtures in the same buffer and plate as follows: 0.1 ml of neat suspensions of donors and recipients on minimal agar plates with and without kanamycin (20 μg ml^{-1}); 0.1 and 0.2 ml of the 10^{-4} dilution of each mating mixture on minimal agar plates (selective for His^+ transconjugants); 0.1 ml of the 10^{-1} (5 plates) and 0.2 ml of the 10^{-2} (5 plates) dilution of the mating mixtures on minimal agar plates + kanamycin (20 μg ml^{-1}) (selective for His^+Km^r transconjugants).

Incubate the plates at 37°C for approximately 24 h. Expect a transfer frequency of 10^{-2} per donor bacterium for *his* and a transposition frequency of 10^{-2}–10^{-3} per transonjugant.

Determine the numbers of His^+ and His^+Km^r progeny. The ratio of the two numbers gives the frequency of transposition.

To test the His^+Km^r colonies for Nif^- replicate on to NFDM agar plates (Section b.ii) and incubate anaerobically at 30°C for 6–7 days.

Quantify the nitrogenase activity of presumptive *nif*: : Tn5 insertion mutants by the acetylene reduction test as described in Section b.ii. Mutants with less than 20% of wild-type activity are suitable for further studies.

nif: : *Tn7 mutations on pMF100 or pRD1.* The procedure for isolating *nif*: : Tn7 insertion mutants is similar to that described for *nif*: : Tn5 except for the following details.

Construct strain J62: : Tn7 (pMF100) by crossing JC5466 (pMF100) with J62: : Tn7 in a plate mating as described above. Select $His^+Tp^rNal^r$ transconjugants on minimal agar plates (Section b.ii) containing proline (20 μg ml^{-1}), tryptophan (25 μg ml^{-1}), naladixic acid (20 μg ml^{-1}), and trimethoprim (20 μg ml^{-1}).

Isolate transonjugants with Tn7 inserted into pMF100 by crossing J62: : Tn7 (pMF100) with UNF107 in at least 10 independent plate matings.

Suspend approximately 2×10^9 bacteria from each patch of growth in 2 ml of saline phosphate buffer, make dilutions (10^{-2} and 10^{-4}) of the mating mixtures in the same buffer and plate: 0.1 ml of the neat suspensions of donors and recipients on minimal agar plates with and without trimethoprim (20 μg ml^{-1}); 0.1 and 0.2 ml of the 10^{-4} dilution of mating mixtures on minimal agar plates (selective for His^+ progeny of UNF107); 0.2 ml (8 plates) of mating mixture neat suspensions on minimal agar plates + trimethoprim (20 μg ml^{-1}) (selective for His^+Tp^r progeny of UNF107).

The plating protocol is based on an expected transfer frequency of 10^{-2} per donor bacterium for *his* and a transposition frequency of 10^{-4}–10^{-5} per transconjugant for Tn7.

nif::*Tn10 mutations on pMF100. Nif*::Tn10 insertion mutants can be isolated by a similar procedure to that described for Tn5 and Tn7 with the following modifications.

Construct strain DU3083 (pMF100) by crossing JC5466 (pMF100) with DU3083 in a plate mating as described above for Tn5. Select $His^+Tc^rSm^r$ transconjugants on minimal agar plates containing tryptophan (25 μg ml^{-1}), lysine (20 μg ml^{-1}), and streptomycin (250 μg ml^{-1}).

To isolate transconjugants with Tn10 inserted into pMF100, cross DU3083 (pMF100) with UNF107 in at least 10 independent plate matings.

Plate: 0.1 ml of neat suspensions of donor and recipients on minimal agar plates with and without tetracycline (15 μg ml^{-1}); 0.1 and 0.2 ml of the 10^{-4} dilution of mating mixtures on minimal agar plates (selective for UNF107 His^+ progeny); 0.2 ml (8 plates) of mating mixture neat suspensions on minimal agar plates + tetracycline (15 μg ml^{-1}) (selective for His^+Tc^r UNF107 progeny).

Expect a transfer frequency of 10^{-2} per donor bacterium for *his* and a transposition frequency of 10^{-5}–10^{-6} per transconjugant for Tn10.

(vii) Isolation of *nif*::Tn chromosomal mutants

Chromosomal *nif*::Tn mutations can be isolated in *K. pneumoniae* strains carrying ColE1::Tn plasmids which are temperature-sensitive for replication. The method involves selection for the drug resistance specified by the transposon at the non-permissive temperature for plasmid replication. Alternatively, plasmid *nif*::Tn mutations can be transferred on to the chromosome by P1 transduction (Merrick *et al.*, 1978).

Materials. pHM5 is a mutant of ColE1 which is temperature-sensitive for replication and carries Tn5 (H. Meade, personal communication).

Methods

Isolation of nif::*Tn5 insertion mutants of UNF5023 (pHM5)*.—Construct UNF5023 (pHM5) by transformation of UNF5023 with pHM5 DNA as described in Section g.

Add 0.1 ml of a fresh overnight culture of UNF5023 (pHM5) grown at 30 °C in Luria broth (Section b.v) + kanamycin (25 μg ml^{-1}) to 5 ml of the same medium and grow on a shaker at 37 °C for 24 h.

Make 10^{-5} and 10^{-6} dilutions in saline phosphate buffer (Section b.i) and spread 0.1 and 0.2 ml of the 10^{-5} and 0.1 ml of the 10^{-6} dilutions on 10–20 plates of nutrient agar + kanamycin (25 μg ml^{-1}). Incubate overnight at 37 °C.

Replicate colonies on to NFDM agar (Section b.ii) containing histidine (20 μg ml^{-1}) and serine (50 μg ml^{-1}). Incubate the plates anaerobically at 30 °C for 5–7 days.

Quantify nitrogenase activity of presumptive *nif*::Tn mutants by the acetylene reduction test as described in Section b.ii. Mutants with less than 20% of wild-type activity are suitable for further studies.

Transfer of plasmid nif::*Tn mutations to the chromosome by P1 transduction* (Merrick *et al.*, 1978)

Prepare P1Cm*clr*100 lysogens of strains carrying pRD1::Tn or pMF100::Tn as described in Section b.v. with the following modifications.

Select lysogens on Luria agar (Section b.v.) plates containing chloramphenicol at 40 μg ml^{-1} for *K. pneumoniae* and 12.5 μg ml^{-1} for *E. coli.*

Prepare P1Cm*clr*100 lysates as described in Section b.v. The phage titres are usually lower (10^8–10^9 p.f.u. ml^{-1}) from plasmid-bearing lysogens.

Transduce the *nif*::Tn mutation on to the chromosome of UNF5023 using the transduction procedure described in Section b.v.

Transductants which are His^+Nif^- and resistant to the antibiotic appropriate for the transposon being used are suitable for further studies. Check that their complementation patterns are the same as those observed for the parental donors (see Section b.ix for genetic complementation methods).

(viii) Isolation of *nif*::Mu insertion mutants

Bacteriophage Mu integrates randomly into the bacterial chromosome, irreversibly inactivates genes into which it inserts, and exerts a polar effect on promoter distal genes of the same operon (Howe and Bade, 1975). Excision of a heat-inducible Mu prophage frequently deletes neighbouring bacterial DNA (Howe, 1973). These properties of Mu have been used to isolate *nif* insertion and deletion mutations on the chromosome of *K. pneumoniae* and on derivatives of pRD1 (Bachhuber *et al.*, 1976; MacNeil *et al.*, 1978a; MacNeil *et al.*, 1978b; Elmerich *et al.*, 1978).

Bacterial strains	*Genotypes or phenotypes*
K. pneumoniae	
UN729 (Bachhuber *et al.*, 1976)	*rfb*-4001 (Mu sensitive)
E. coli K12	
UQ27 (Mu*cts*Cam4005) (MacNeil *et al.*, 1978a)	*his ara gal*K *mal*A *xyl mtl rps*L *nalA*
UQ28 Muc$^+$M*am*1124) (MacNeil *et al.*, 1978a)	*his ara gal*K *mal*A *xyl mtl rps*L *rps*E
MH812 (Bachhuber *et al.*, 1976)	*thr leu met lac sup*E *hsd*M *hsd*R
M107 (Bachhuber *et al.*, 1976)	*lac sup*D *rps*L
Phages	
Mu*cts*61 (Bachhuber *et al.*, 1976)	
Mu*c*25 (Bachhuber *et al.*, 1976)	
Plasmid	
pTM4010 (MacNeil *et al.*, 1978a)	Derivative of pRD1 which does not confer resistance to Mu or P1.

Preparation of Mu lysates

By lytic infection. Grow a culture of a Mu-sensitive strain to a population density of 1–2 × 10^8 cell ml^{-1} in Luria broth (Section b.v) + $CaCl_2$ (2.5 mM) + $MgCl_2$ (2.5 mM) on a shaker at 37°C.

Add two fresh Mu plaques to 0.2-ml aliquots of the culture and allow 20 min at 37°C without shaking for adsorption.

Dilute with 7.8 ml of the same medium and shake vigorously at 37°C until lysis occurs (usually 3–5 h).

Chill the lysate, add a few drops of chloroform, Whirlimix and remove cell debris by centrifugation at 12000*g* for 15 min (use polypropylene tubes).

Filter the lysate through a membrane filter (pore size 0.45 μm) and store the phage over chloroform at 4°C.

To determine the phage titre, make serial dilutions of the lysate in a solution of $CaCl_2$ (5 mM) and $MgCl_2$ (5 mM) and spot 10 μl of the 10^{-5}, 10^{-6} and 10^{-7} dilutions on a thick Luria agar (Section b.v) plate overlaid with 3 ml of soft Luria agar +$CaCl_2$ (5 mM) +$MgCl_2$ (5 mM) containing 10^8 bacteria of an Mu-sensitive *hsdR E. coli* strain (e.g.MH812).

Incubate the plates overnight at 37°C.

By heat induction of Mucts lysogens. Grow a culture (25 ml in a 250-ml flask) of a Mu*cts* lysogen in Luria broth +$MgCl_2$ (5 mM) with vigorous shaking at 32°C to a population density of 2–4 × 10^8 bacteria ml^{-1}.

Add $CaCl_2$ to a final concentration of 5.0 mM and transfer the flask to a shaking water-bath at 42 °C for 15 min.

Continue incubation at 37 °C with vigorous shaking until lysis occurs (usually 5–7 h).

Chill the lysate, add a few drops of chloroform, Whirlimix and remove cell debris by centrifugation at 12 000 *g* for 15 min.

Filter the lysate through a membrane filter (pore size 0.45 μm), store over chloroform at 4 °C and determine the phage titre as described above.

Isolation of Mu lysogens

Overlay a Luria agar (Section b.v.) plate with 3 ml of soft Luria agar (Section b.v) +$CaCl_2$ (5 mM) +$MgCl_2$ (5 mM) containing approximately 10^8 bacteria of the strain to be lysogenized and spot on to its surface a drop of Mu lysate.

Incubate the plate overnight at 32 °C (Mu*cts*) or 37 °C (Muc^+).

To isolate lysogens of Mu sensitive *E. coli*, suspend the bacteria which survive at the centre of the plaque in Luria broth, streak on to nutrient agar plates and incubate overnight at the appropriate temperature.

To isolate lysogens of a *K. pneumoniae* strain which is moderately sensitive to Mu (e.g. UW729), suspend the bacteria which survive at the centre of the plaque in 2 ml of Luria broth +$CaCl_2$ (5 mM) +$MgCl_2$ (5 mM), infect with two fresh plaques of Mu*c*25, and grow overnight at 30 or 37 °C as appropriate. The latter phage should kill non-lysogenic survivors. Streak the culture on to a nutrient agar plate and incubate overnight at the appropriate temperature.

Check colonies for Mu lysogeny as follows.

Cross-streak. Streak colonies of presumptive lysogens on thick Luria agar plates using a platinum ball-point at 90° angles to a streak of a Mu*cts*61 or Mu*c*25 lysate and incubate the plates overnight at 32 °C. Immune or resistant colonies give streaks of confluent growth, whereas those of sensitive colonies are lysed at the intersection of the streaks.

Phage release. Spread a 3-ml soft overlay of Luria agar +$CaCl_2$ (5 mM) +$MgCl_2$ (5 mM) containing approximately 10^8 cells of a Mu-sensitive *hsdR E. coli* strain (e.g. MH812) on a Luria agar plate.

Patch colonies of presumptive lysogens onto the surface of the overlay using toothpicks or prepare 1-ml Luria broth cultures from the colonies and spot 10-μl aliquots on to the surface of the overlay. Incubate the plates overnight at 37 °C. Phage released by lysogens lyse the surrounding indicator bacteria, giving rise to a halo effect.

Chromosol nif: : *Mu mutants*

Grow an overnight culture of a Mu-sensitive variant of *K. pneumoniae* (e.g. UN729) in nutrient broth on a shaker at 37 °C.

Add 0.2 ml of the culture to 10 ml of Luria broth $+CaCl_2$ (5 mM) $+MgCl_2$ (5 mM) and grow to a population density of 1–2 × 10^8 cells ml^{-1} on a shaker at 30 °C.

Infect at a multiplicity of 5–10 with a lysate of Mu*cts*61 previously prepared by heat induction of a Mu*cts*61 lysogen, e.g. UN729 (Mu*cts*61). Allow 20 min at 30 °C for adsorption, dispense 1-ml aliquots of the culture into bijou bottles, and continue incubation on a shaker at 30 °C for 16–18 h.

Harvest cells by centrifugation, wash with 5 ml of $MgCl_2$ (5 mM)–$CaCl_2$ (5 mM) solution, and suspend in 1 ml of the same solution.

Dilute 0.2 ml of the cell suspensions in 5.0 ml of Luria broth $+CaCl_2$ (5 mM) $+MgCl_2$ (5 mM) and grow to population densities of 1–2 × 10.8 cells ml^{-1} on a shaker at 30 °C.

To eliminate non-lysogenic cells infect each culture with phage from two fresh plaques of Mu*c*25 propagated on UN729 and allow 20 min at 30 °C for adsorption.

Dilute 0.5 ml of the cultures in 5.0 ml of Luria broth $+CaCl_2$ (5 mM) $+MgCl_2$ (5 mM) and continue incubation on a shaker at 32 °C for 5 h.

Re-suspend cells in 4-ml aliquots of NFDM (Section b.ii) $+(NH_4)_2SO_4$ (1 mg ml^{-1}) contained in 7-ml bijou bottles, cap the bottles tightly, and incubate overnight on a shaker at 30 °C.

Centrifuge 1.2 ml of each culture in 7-ml bijou bottles and discard the supernatant.

Add 3.6 ml of NFDM to each bottle, cap with a Suba Seal or similar closure, equilibrate the pressure with a needle, and Whirlimix to suspend the pellets.

Incubate at 30 °C with vigorous shaking for 6 h.

Add 0.2 ml of ampicillin solution (10 mg ml^{-1}) and 0.2 ml of D-cycloserine solution (0.2 M); final concentrations, 500 μg ml^{-1} and 10 mM, respectively.

Continue vigorous shaking of cultures at 30 °C until lysis occurs (usually 2–3 h).

Centrifuge the bijou bottles to pellet the cells, wash with 4 ml of saline phosphate buffer, re-suspend the cells in 4 ml of NFDM $+(NH_4)_2SO_4$ (1 mg ml^{-1}), cap the bottles tightly, and incubate overnight with vigorous shaking at 30 °C.

Make serial dilutions of the cultures in saline phosphate buffer (Section b.ii) and spread 0.1 and 0.2 ml of the 10^{-6} dilution on NFDM agar plates containing serine (50 μg ml^{-1}). Incubate the plates anaerobically at 30 °C for 6–7 days.

Streak presumptive Nif^- mutants on minimal agar plates (Section b.i) and quantify the nitrogenase activities of those with growth rates similar to wild type using the acetylene reduction test as described in Section b.ii.

Mutants with less than 20% of wild-type activity are suitable for further studies and should be tested for Mu lysogeny using the cross streak and phage release tests described above.

Plasmid nif::*Mu insertion mutants*

The procedure is essentially that of MacNeil *et al.* (1978a) with some modifications and involves crossing UQ27 (Mu*cts*Cam4005) (pTM4010) with UQ28 (Muc^+Mam1124) under conditions which induce Mu*cts*Cam4005 in the donor, and selecting His^+ transconjugants which are then screened for spontaneous release of phage. The Muc^+Mam1124 prophage in the recipient prevents zygotic induction in transconjugants which receive a prophage on pTM4010 and complements the Cam4005 mutation of the donor prophage, thus enabling synthesis of mature phage.

Grow overnight cultures of UQ27 (Mu*cts*Cam4005) (pTM4010) in minimal medium and UQ28 (Muc^+Mam1124) in Luria broth on a shaker at 30 °C.

Dilute 0.1 ml of the cultures in 5.0 ml of Luria broth and grow to a population density of 2–4 × 10^8 cells ml^{-1} on a shaker at 30 °C.

Spread 0.1 ml of donor and recipient cultures together and separately on nutrient agar plates and incubate for 6 h at 42 °C.

Suspend approximately 10^9 bacteria from the mating mixture and control plates in 1-ml volumes of saline phosphate buffer, dilute the mating mixture to 10^{-5} in the same buffer, and spread 0.1 ml on each of 50 minimal agar plates containing spectinomycin (200 μg ml^{-1}). Spread 0.1 ml of the undiluted control cultures on five similar plates. Incubate the plates at 37 °C for 36 h. Expect 50–100 transconjugant colonies per plate, assuming a transfer frequency of 10^{-1} per donor.

Replicate or patch transconjugant colonies on to NFDM plates (Section b.ii) containing serine (50 μg ml^{-1}) and also on to the surface of soft Luria agar +$CaCl_2$ (5 mM) +$MgCl_2$ (5 mM) overlays spread on Luria agar +$CaCl_2$ (5 mM) +$MgCl_2$ (5 mM) plates and containing approximately 10^8 bacteria of a freshly grown liquid culture of M107(Su^+). The latter plates test for phage release.

Incubate NFDM plates anaerobically at 30 °C for 6–7 days and the Luria agar plates overnight at 37 °C.

Colonies which grow poorly on NFDM plates and release phage are presumptive *nif*::Mu insertion mutants. Quantify their nitrogenase activities using the acetylene reduction test as described in Section b.ii and retain those with activities less than 20% of wild type for further studies. Also, colonies or patches which have a Nif^+ phenotype and release phage are presumptive pTM4010::Mu lysogens and could be useful for the isolation of Mu induced *nif* deletions.

(ix) Isolation of *nif* deletion mutants

Tn-induced nif *deletions* (Merrick *et al.*, 1978).

The imprecise excision of transposons generates deletions which extend to either or sometimes both sides of the point of insertion. Klebsiella *nif* deletions are isolated by carbenicillin/D-cycloserine enrichment of drug sensitive (Tn7 and Tn10) or His$^-$ (Tn5) derivatives in populations of chromosomal *nif*: : Tn mutants.

Materials

Histidine assay broth: Difco. Use 1% w/v in minimal medium (Section b.i).

Methods

Tetracycline-sensitive deletions of nif: : *Tn10 nutants.* Grow an overnight culture of a *nif*: : Tn10 mutant in 5 ml of Luria broth (Section b.v) on a shaker at 37 °C.

Sub-culture (2%) in 50 ml of Luria broth containing tetracycline (5 μg ml^{-1}) and incubate on a shaker at 37 °C until a population density of approximately 2×10^8 bacteria ml^{-1} is reached.

Add carbenicillin (800 μg ml^{-1}) and D-cycloserine (8 mM) and continue incubation on a shaker at 37 °C until lysis occurs.

Collect survivors on a membrane filter, wash with 5–10 ml of saline phosphate buffer (Section b.i) and transfer the membrane to 10 ml of Luria broth in a 50-ml flask.

Incubate overnight on a shaker at 37 °C.

Plate 0.1 ml of 10^{-5} and 10^{-6} dilutions in saline phosphate on nutrient agar plates (Section b.i), incubate overnight at 37 °C, and replicate colonies on to nutrient agar plates containing tetracycline (10 μg ml^{-1}).

Repeat the enrichment cycle twice more, plating cultures as above at the end of each cycle.

Check the *nif* complementation patterns of tetracycline sensitive colonies using the plate method described in Section b.x. Those with altered Nif phenotypes are presumptive deletion or inversion mutants.

his–nif *deletions of* nif: : *Tn5 mutants.* The procedure is similar to that described for tetracycline sensitive deletions with the following modifications.

Centrifuge and re-suspend an overnight culture of a *nif*: : Tn5 mutant grown in 5 ml of Luria broth in an equal volume of saline phosphate buffer.

Add 1 ml of the cell suspension to 50 ml of histidine assay broth and proceed as above, substituting histidine assay broth for Luria broth containing tetracycline and re-suspending bacteria from Luria broth cultures in saline phosphate buffer before dilution in histidine assay broth at the beginning of each enrichment cycle.

To screen for His^- deletions replicate colonies from nutrient agar plates on to minimal agar plates with and without histidine (20 μg ml^{-1}).

Mu-induced nif *deletions.* UN729(*nif*: : Mu*cts*61) and UQ27 (pTM4010: : Mu*cts*Cam4005) are suitable strains for the isolation of chromosomal and plasmid *nif* deletions, respectively (Bachhuber *et al.*, 1976; MacNeil *et al.*, 1978a).

Grow overnight cultures in Luria broth on a shaker at 32 °C and add sodium citrate to a final concentration of 100 mM.

Spread 10–20 aliquots (0.1 ml) of each culture on separate Luria agar plates containing sodium citrate (100 mM) and incubate overnight at 42 °C.

To purify thermosensitive colonies, streak several hundred on the same medium and incubate overnight at 42 °C.

Check purified clones derived from UN729(*nif*: : Mu*cts*61) for phage release on a Mu-sensitive *hsd*R *E. coli* strain (e.g. MH812) and those from UQ27(pTN4010: : Mu*cts*Cam4005) on M107(Su^+) as described in Section b.viii.

Test clones which do not produce phage for sensitivity to Mu*c*25 using the cross-streak test as described in Section b.viii.

Check the *nif* complementation patterns of sensitive clones using the plate method described in Section b.x. Those with altered Nif phenotypes are presumptive *nif* deletions and are suitable for fine structure mapping.

(x) Complementation analysis of *nif* mutants

Nif mutants of *K. pneumoniae* have been isolated in recombination-proficient strains and the frequency at which Nif^+ recombinants occurred in merodiploids gave rise to ambiguity in plate complementation tests (Dixon *et al.*, 1977). Two different strategies were adopted to solve this problem. Dixon *et al.*, (1977), Elmerich *et al.* (1978), and Merrick *et al.* (1978) used the acetylene reduction test of liquid cultures to quantify the nitrogenase activities of freshly constructed merodiploids. MacNeil *et al.* (1978b) introduced an *E. coli rec*A mutation into Klebsiella Nif^- recipients and were therefore able to use a much less laborious plate complementation procedure.

Bacterial strains	*Genotype*
E. coli K12	
JC5466 (Dixon *et al.*, 1977)	*trp rec*A56 *his rps*E
K. pneumoniae	
UNF107 (Section b.iii)	*rps*L (*gnd his nif*)
CK263 (Dixon *et al.*, 1977)	*his*D2 *nif*A2263 *hsd*R1

Liquid method. Suitable strains for a typical complementation test are:

UNF107 (total *nif* deletion)	−negative control
UNF107 (pRD7, *nif*A2007) CK263 (*nif*A2263)	−haploid controls
CK263 (pRD7)	−Merodiploid
CK263 (pRD1)	−positive control

To construct fresh merodiploids, grow log phase cultures of donors [e.g. JC5466 (pRD7)] and recipients (e.g. CK263) in 1 ml of nutrient broth (Section b.i) on a shaker at 37 °C.

Spot 10-μl drops of each separately (controls) and together (mating mixture) on nutrient agar plates and incubate overnight at 37 °C.

Purify His^+ transconjugants on minimal agar plates (Section b.i.) using a platinum wire with a ball-point to obtain isolated colonies.

Inoculate 1-ml volumes of nutrient broth in small test-tubes with isolated colonies of test and control strains using toothpicks, and incubate on a shaker at 37 °C for 2–4 h.

Add 0.2 of the cultures to 4 ml of NFDM (Section b.ii) in 7-ml bijou bottles and perform quantitative acetylene reduction tests as described in Section b.ii. Express the levels of nitrogenase activity in Nif^-/Nif^- merodiploids as a percentage of those obtained for the corresponding Nif^- recipients carrying pRD1 (Nif^+). Lack of complementation is indicated by similar levels of activity in merodiploids and parental haploids.

Plate method. The transfer of the *E. coli rec*A56 mutation to *K. pneumoniae* is described by MacNeil *et al.* (1978b). This transfer was facilitated by the presence of an insertion mutation, *srl*: : Tn10, closely linked to *rec*A56, which permitted selection for tetracycline resistance followed by a screen for UV sensitivity. The *rec*A56 mutation can now be transduced into other *K. pneumoniae* strains using the procedures described in Section b.v. For plate complementation tests using Nif^- derivatives of pRD1, merodiploids can be constructed on NFDM plates.

Patch colonies of the donor strains (Sm^s) on duplicate nutrient agar plates (20–30 patches per plate) and incubate overnight at 37 °C. Include a strain carrying pRD1 or derivative as a positive control.

Spread 0.1 ml of fresh overnight cultures of recipients (His^-, Nif^-, Rec^- Sm^r) in nutrient broth on NFDM agar plates containing streptomycin (250 μg ml^{-1}). These plates select His^+Sm^r transconjugants. As a test for transdominant effects of mutant plasmids, include a plate with a $His^-Nif^+Rec^-Sm^r$ recipient for every plate of donors.

Replicate one donor plate on to the plates containing the recipients and the second on NFDM plates containing amino acids (20 μg ml^{-1}) required by donors. The latter serve as controls for the Nif phenotype of the donors.

Incubate the plates anaerobically at 30 °C for 6–7 days.

Confluent growth of corresponding patches on both mating plates and the absence of growth on the plate containing the donors alone indicate complementation.

Growth of Nif^-/Nif^+ heterogenotes and its absence in the corresponding Nif^-/Nif^- heterogenotes and Nif^- haploid controls indicate failure of the two mutations to complement.

When Nif^- derivatives of pMF100 (Merrick *et al.*, 1978) are used for complementation analysis, a similar procedure is used except that the matings are carried out on minimal agar plates selective for His^+Sm^r transconjugants by mixing 10-μl drops of donor and recipient (10 per plate) and incubating the plates overnight at 37 °C. (Also spot donors and recipients alone on minimal agar plates supplemented as required.) The spots of confluent growth are then replicated on to NFDM plates and incubated as above. This modification is necessary because the transfer frequency of pMF100 is approximately 10-fold lower than that of pRD1 (Filser and Cannon, unpublished work).

(xi) Fine structure mapping of *nif* mutations

MacNeil *et al.* (1978b) reported a plate method by which *nif* deletion mutants characterized by complementation analysis were used to map point and insertions mutations similarly characterized. Plasmid-borne *nif* point or insertion mutations can be mapped with chromosomal deletions, and chromosomal point and insertion mutations can be mapped with *nif* deletions on plasmids.

The procedure for mapping is similar to that described in Section b.x for plate complementation. Heterogenotes (Nif^-/Nif^V, Rec^+) are constructed by transferring the donor plasmids in plate matings and the resulting patches are observed for the presence or absence of Nif^+ recombinants. Confluent patches indicate either a high frequency of recombination or complementation.

(xii) SDS–polyacrylamide gel electrophoresis of pulse-labelled, cell-free extracts

Pulse labelling of nif *repressed and derepressed cultures*

Method. Suitable strains for a typical labelling experiment are:

UNF107 (total *nif* deletion)	−label under derepressing conditions.	controls
UNF107(pRD1) (Nif^+)	−label under repressing and derepressing conditions.	controls
UNF107(pRD194, *nif*J : : Tn7)	−label under derepressing conditions.	

Inoculate 2 ml of Luria broth (Section b.v) with fresh colonies of strains to be labelled and grow on a shaker at 37 °C for 3 h.

Dilute 0.2 ml in 4 ml of NFDM (Section b.ii) $+(NH_4)_2SO_4$ (10 mM) + amino acids (50 μg ml^{-1}) as required, contained in 7-ml bijou bottles. Set up each culture in triplicate—one set for each of three labelling times. Grow overnight on a shaker at 30 °C.

Centrifuge the bottles (3000 *g*, 10 min, room temperature) and pour off the supernatant; shake each bottle to remove the last drops.

Add 4 ml of NFDM + amino acids (50 μg ml^{-1}) as required [$+(NH_4)_2SO_4$ (10 mM) for repressed cultures only] to the bottles, replace the caps with Suba Seal closures (Section b.ii), equilibrate the gas pressure with a hypodermic needle and Whirlimix to re-suspend the pellets.

Incubate the bottles with vigorous shaking at 28 °C. This is the zero time for derepression.

Flush 7-ml bijou bottles (one for each culture to be labelled) with argon, replace the caps with Suba Seal closures, pressing the Suba Seal just far enough into the bottle to minimize gas exchange and facilitate easy removal. Using a mechanical pipette, add 50 μl of ^{14}C-labelled amino acids (50 μCi ml^{-1}, 58 mCi $matom^{-1}$) to each bottle by removing the Suba Seal and replacing it as quickly as possible. Ideally, two people should work together on this step. Insert the Suba Seal fully into position, equilibrate the gas pressure and place the bottles on a shaker at 28–30 °C.

After 2 h of incubation, transfer 1 ml of each of a set of cultures (T2) to a corresponding set of labelling bottles using 2-ml sterile syringes and needles previously flushed with nitrogen. Shake for 10 min.

During the 10-min labelling period, add 1 ml of acetylene to the unused portions of the T2 cultures and continue incubation on a shaker at 28 °C. After 40 min measure the ethylene present in 0.5 ml of the gas phase using a gas chromatograph. The level of ethylene in the Nif^+ NFDM culture will indicate the degree of *nif* derepression.

After 10 min quench the labelling by adding 1 ml of casamino acids (1 mg ml^{-1}) in saline phosphate buffer (Section b.i) to each bottle and centrifuge at 3000 *g* for 10 min.

Re-suspend the pellets in 0.5 ml of casamino acids (1 mg ml^{-1}) in saline phosphate buffer and transfer to small polypropylene vials (e.g. 1.5-ml Eppendorf tubes) suitable for further centrifugation as above and storage of pellets in liquid nitrogen.

Label 1-ml aliquots of the second set of cultures (T4) after 4 h, the third set (T6) after 6 h, and measure acetylene reduction as described for T2 cultures.

To lyse the cells, re-suspend in 200 μl of water, add 200 μl of double strength (2×) 'sample buffer' (as below), and place in boiling water for 4 min.

Cool to room temperature and Whirlimix carefully—just enough to reduce viscosity so that the lysate flows freely as the vial is inverted.

Centrifuge at 10000 *g* for 15 min to remove debris.

Transfer extract to clean vials and store at −20°C or in liquid nitrogen.

Preparation and running of acrylamide slab gels. A Laemmli (Laemmli, 1970) and a modified Ortec (Instruction Manual, Ortec 4200 Electrophoresis System, Ortec, Oak Ridge, Tenn., U.S.A.) gel system are described. Both systems consist of a 5% stacking gel and a separating gel. The modified Ortec gel gives better resolution of proteins between 30 and 10 × 10^3 daltons from *K. pneumoniae* grown in NFDM. The separation of nitrogenase polypeptides is affected by the chemical grade of SDS used—see Kennedy *et al.* (1976) for details. Gel dimensions are 16 × 16 × 0.15 cm.

Solutions (Modified Ortec gels)

A. Separating gel (10%)*:

Stock solution	*Volume (ml) for 30 ml of gel*	*Final concentration*
1.5 M Tris–SO_4, pH 9.0	7.5	0.375 M
10% SDS	0.3	0.1%
30% acrylamide, 0.8% bisacrylamide	10.0	10% acrylamide
TEMED	0.02	—
10% $(NH_4)_2S_2O_8$ (freshly prepared)	0.3	0.1%
Water	11.9	—

B. Stacking gel (5%):

Stock solution	*Volume (ml) for 10 ml of gel*	*Final concentration*
1.5 M Tris–SO_4, pH 9.0	0.834	0.125 M
10% SDS	0.1	0.1%
30% acrylamide, 0.8% bisacrylamide	1.67	5% acrylamide
TEMED	6.7 μl	—
10% $(NH_4)_2S_2O_8$	0.1	0.1%
Water	7.3	—

*To detect polypeptides of molecular weight less than 30 × 10^3 daltons, use 11% or 12.5% gels.

C. Sample buffer:

Stock solution	*Volume (ml) for 15 ml of 2×*	*Final concentration*
1.5 M Tris–SO_4, pH 9.0	2.0	0.1 M
10% SDS	6.0	2% w/v
Glycerol	3.0	10% v/v
Mercaptoethanol	1.5	5% v/v
0.2% bromophenol blue	0.3	0.002%
Water	2.2	

D. Reservoir buffer (final concentration) (pH = 9.0):

Component	*Amount (g) for 2 l of 10×*
0.064 M Trisma base	155
0.0177 M boric acid	21.85
0.1% SDS	20

Solutions for Laemmli gels

A. Separating gel (10%):

Stock solution	*Volume (ml) for 30 ml of gel*	*Final concentration*
1.5 M Tris–Cl, pH 8.8	7.5	0.375 M
10% SDS	0.3	0.1%
30% acrylamide 0.8% bisacrylamide	10.0	10% acrylamide
TEMED	0.02	—
10% $(NH_4)_2S_2O_8$	0.3	0.1%
Water	11.9	

B. Stacking gel:

Stock solution	*Volume (ml) for 10 ml of gel*	*Final concentration*
0.5 M Tris–Cl, pH 6.8	1.25	0.0625 M
10% SDS	0.1	0.1%
30% acrylamide 0.8% bisacrylamide	1.67	5% acrylamide
TEMED	6.7 μl	—
10% $(NH_4)_2S_2O_8$	0.3	0.1%
Water	6.87	—

C. Sample buffer:

Stock solution	*Volume (ml) for 15 ml of 2×*	*Final concentration*
1.0 M Tris–Cl, pH 6.8	3.0	0.1 M
10% SDS	6.0	2% w/v
Glycerol	3.0	10% v/v
Mercaptoethanol	1.5	5% v/v
0.2% bromophenol blue	0.3	0.002%
Water	1.2	—

D. Reservoir buffer (final concentration):

Component	*Amount (g) for 2 l of 10×*
0.025 M Trisma base	60.55
0.192 M glycine	288.384
0.1% SDS	20.0

Method. Use very clean glass plates, combs, and spacers.

Assemble glass plates and spacers and clamp with six bulldog clamps—two on each side and two on the base. The spacers should be approximately 2 mm in from the edge.

Seal the outside edges with 2% agar and place vertically on the bench.

Prepare the separating gel by mixing the components in the order listed above in a 150-ml beaker.

Mix gently and pour. The upper edge of the separating gel should be approximately 2 cm below the lower end of the loading slots.

Overlay with water delivered from an atomiser and leave for approximately 4 h to set.

Pour off the water and wash several times with water; remove any remaining water with a piece of blotting paper. Take care not to touch the gel!

Pour the stacking gel prepared by mixing the components in the order listed above in a 100-ml beaker.

Slowly insert a comb and avoid trapping any air bubbles.

Remove the comb gently when gel has set (approximately 2 h).

Remove the bulldog clamps, place the gel apparatus flat on the bench and remove the bottom spacer while applying gentle hand pressure on the plates to prevent separation.

Insert the gel apparatus in an electrophoresis box, secure with screw clamps, and seal the interface with 2% agar to prevent leakage of reservoir buffer.

Fill the sample slots with reservoir buffer and make sure that the slot spacers are straight and separate.

Fill the top and bottom reservoirs with appropriate buffer. At this stage the gels can be kept for up to 24 h before use.

Load the samples using a 100- or 50-μl syringe fitted with a piece of fine PTFE tubing; use a different piece for each sample. Load 5 μl of extracts from derepressed cultures and 2.5 μl of extracts from repressed cultures.

Connect the electrodes to the power pack and switch on. With a constant current of 25 mA the dye will reach the end of the gel in approximately 2.5 h.

After electrophoresis remove the gel apparatus from the box, separate the glass plates carefully, separate the stacking gel from the separating gel with a scalpel, and discard the former. Wear gloves!

Holding the gel in the hand, run distilled water over it and then place it in a 2-l beaker containing approximately 200 ml of 20% ethanol. Rest the gel on a platform made from a Petri dish with holes to prevent floating. Stir the solution gently with a stirring bar placed underneath the platform.

After fixing for 1 h, place the gel on a piece of 3-mm Whatman paper with dimensions approximately 2 cm greater than those of the gel, dry under vacuum in a gel drier.

Expose the gel to X-ray film (e.g. Kodak XR-1) for approximately 7 days.

Fluorography

Fix the gel in 20% ethanol as above.

Transfer to 500 ml of dimethyl sulphoxide (DMSO). Leave for 30 min.

Replace the DMSO with a further 500 ml of DMSO. (The DMSO washes can be used several times.)

Place the gel in 200 ml of 20% 2,5-diphenyloxazole (PPO) in DMSO for 3 h—the gel will shrink.

Place the gel in 500 ml of water for 30 min, change the water and leave for a further 30 min. The PPO will precipitate and give a white colour to the gel, which will also re-swell.

Dry the gel as described above.

Expose to a pre-fogged X-ray film at −70 °C.

(xiii) Preparation of cell extracts from small culture volumes for assaying nitrogenase components

Inoculate 2 ml of nutrient broth (Section b.i) with fresh colonies of mutant and Nif^+ strains. Grow on a shaker at 37 °C for 3 h.

Dilute 2 ml in 50 ml of NFDM (Section b.ii) +amino acids (50 μg ml^{-1}) as required $+(NH_4)_2SO_4$ (10 mM), contained in a 100-ml flask with a screw-cap. Grow on a shaker at 28 °C overnight with caps tightly fitted.

Transfer the cultures to argon-flushed centrifuge tubes and centrifuge at 10000 *g* for 10 min at room temperature.

Discard the supernatant, flush the tubes again with argon or nitrogen and re-suspend the bacteria in 10 ml of NFDM +amino acids (50 μg ml^{-1}) as required. Transfer the culture to a 100-ml flask, add NFDM +amino acids to a final volume of 50 ml, flask, and cap with Suba Seal closures.

Incubate on a shaker at 28 °C. This is the zero time for derepression. Use a Nif^+ and a total *nif* deletion strain as positive and negative controls.

After 4.5 h add 1 ml of acetylene to the positive control flask, continue incubation for a further 30 min, and check for *nif* derepression by measuring the ethylene produced.

The positive control should be derepressed after 5–6 h.

Harvest cells in argon-flushed centrifuge tubes at 10000 *g* for 10 min at room temperature.

Re-suspend the cells in 10 ml of a solution of Tris–Cl, pH 8.0 (10 mM), Na_2EDTA (100 mM), NaCl (150 mM), and $Na_2S_2O_4$ (100 μg ml^{-1}), and transfer to polycarbonate conical centrifuge tubes (15 ml) previously flushed with argon or nitrogen and capped with Suba Seal closures. Centrifuge at 10000 *g* at 15–20 °C for 10 min.

Re-suspend the cells in 4.5 ml of a solution of sucrose (25% w/v), Tris–Cl, pH 8.0 (50 mM), and $Na_2S_2O_4$ (100 μg ml^{-1}).

Add 0.5 ml of a solution of lysozyme (10 mg ml^{-1}), Tris–Cl, pH 8.0 (250 mM), $Na_2S_2O_4$ (100 μg ml^{-1}), and Na_2EDTA (5 mM), and incubate at room temperature for 30 min.

Centrifuge for 6–10 min at 5–6 × 10^3 r.p.m. and room temperature.

Gently layer 5 ml of a solution of sucrose (25% w/v), Tris–Cl, pH 8.0 (50 mM), and $Na_2S_2O_4$ (100μg ml^{-1}) over the pellets to dilute any remaining EDTA.

Centrifuge for 6–10 min at 5–6 × 10^3 r.p.m. at room temperature. Discard the supernatant.

Re-suspend the cells in 1.0 ml of a solution of Tris–Cl, pH 7.4 (50 mM), $MgCl_2$ (10 mM), $Na_2S_2O_4$ (200 μg ml^{-1}), and dithiothreitol (100 μg ml^{-1}). Add 0.1 ml of Triton X-100 (10%) +$Na_2S_2O_4$ (100 μg ml^{-1}) and Whirlimix gently. The extracts are now ready to be assayed for nitrogenase activity and that of its component proteins as described by Eady in Chapter II.6. Store the extracts in liquid nitrogen.

(c) ISOLATION OF RHIZOBIUM SYMBIOTIC (*sym*) MUTANTS

(i) NG dose–response curve

Media and growth conditions are the main differences between the procedure for determining the toxic effect of NG on Rhizobium and that for *K. pneumoniae*.

Materials

Rhizobium minimal medium (RMM), based on that of Schwinghamer (1969):

K_2HPO_4/KH_2PO_4 buffer, pH 7.2	2.5 mM
$(NH_4)_2SO_4$	15.0 mM
$MgSO_4 \cdot 7H_2O$	1.0 mM
NaCl	0.2 mM
$Ca(NO_3)_2 \cdot 4H_2O$	0.01 mM
Trace element solution (as below)	0.1% v/v
Glucose or mannitol	0.2% w/v
Vitamins (as required)	1 mg l^{-1}

Notes. Yeast extract (200μg ml^{-1}) will satisfy the vitamin requirements of most strains. For slow-growing strains use sodium glutamate (0.12% w/v) instead of $(NH_4)_2SO_4$. To decrease slime production use sodium succinate (0.14% w/v) as a carbon source.

RMA is RMM + agar (12 g l^{-1}).

TY medium:

Bactotryptone (Difco)	5 g l^{-1}
Bacto-yeast extract (Difco)	3 g l^{-1}
$CaCl_2 \cdot 6H_2O$	1.3 g l^{-1}

For TY agar add agar (12 g l^{-1}).

NG solution: 10 mg ml^{-1} in acetone, freshly prepared.

Saline: 0.85% NaCl.

Trace elements solution:

H_3BO_3	1.0 g l^{-1}
$ZnSO_4 \cdot 7H_2O$	1.0 g l^{-1}
$CuSO_4 \cdot 5H_2O$	0.5 g l^{-1}
$MnCl \cdot 4H_2O$	0.5 g l^{-1}
$NaMoO_4 \cdot 2H_2O$	0.1 g l^{-1}
FeNaEDTA	1.0 g l^{-1}

Method. Add 5 ml of a late log-phase culture in RMM +yeast extract (200 μg ml^{-1}) or TY medium to 45 ml of fresh medium in a 250-ml flask.

Grow to a population density of $3–5 \times 10^8$ bacteria ml^{-1} on a shaker at 30°C. The time required varies with the strain used (usually 4–6 h).

Add 1 ml of NG solution (10 mg ml^{-1} in acetone). This is the zero time.

Disperse 5-ml aliquots into eight flasks (25 ml) and continue shaking at 30°C.

Take one culture for each time interval tested (0, 10, 20, 30, 40, 60, 90, and 120 min) and determine the surviving fraction of bacteria as described in Section b.i for *K. pneumoniae*, except that saline instead of saline phosphate buffer and RMM +yeast extract (200 μg ml^{-1}) or TY agar plates instead of nutrient agar plates should be used.

Incubate the plates at 30°C for 2–3 days and construct a dose–response curve as described in Section b.i.

(ii) NG mutagenesis and isolation of *sym* mutants

Add 1 ml of a late log-phase culture in RMM +yeast extract (200 μg ml^{-1}) to 9 ml of fresh medium and incubate on a shaker at 30 °C until the population density has reached 3–5 × 10^8 bacteria ml^{-1}.

Add 0.2 ml of NG solution (10 mg ml^{-1} in acetone) and continue incubation on a shaker at 30 °C for the time required to decrease the viable count to 20–50% (obtained from the dose–response curve).

Collect cells on a membrane filter or harvest by centrifugation, wash with 10 ml of saline (Section c.i) and re-suspend in 50 ml of RMM + yeast extract (200 μg ml^{-1}).

Dispense 4-ml aliquots into 12 flasks (25 ml) and incubate on a shaker at 30 °C for 24–36 h. (The culture is divided to avoid characterizing sibs.)

Spread 0.1, 0.2, and 0.3 ml of a 10^{-6} dilution of each culture in saline on RMA plates. Prepare a plate of the parental strain similarly to compare colony growth rate.

Colonies that grow at a rate comparable to that of wild type are suitable for characterization of their symbiotic properties.

Prepare stocks on TY agar slants of a large number of such colonies (100–200 from each of the 10 cultures) and patch samples (e.g. 10) of each culture on TY agar plates. Incubate stock cultures and plates at 30 °C for 2–3 days. The patches are then ready for nodulation testing as described by Gibson (Chapter II.4).

(iii) Isolation of *sym*::Tn5 plasmid mutants of *Rhizobium leguminosarum*

The procedure is based on that of Beringer *et al.* (1978) and Johnson *et al.* (1978) and with the appropriate modifications in cultural conditions and the availability of suitably marked strains it should be applicable to all Rhizobium species. Transposition-directed mutagenesis is mediated by the P1-type plasmid pJB4JI, which is transferable from *E. coli* to Rhizobium but not stably maintained in the latter. If Km^r progeny are selected, Tn5, which is carried on the plasmid, can be stably inherited by transposition into a resident replicon. The following protocol is designed to isolated Tn5 insertion mutations in *sym* genes on the *R. leguminosarum* plasmid pRL1JI.

Bacterial strains	Genotype or phenotype
R. leguminosarum	
6015 (Johnston *et al.*, 1978)	*phe trp str inf*
1063 (Beringer *et al.*, 1978)	*ura trp rif*
E. coli	
JC5466 (Cannon *et al.*, 1976)	*trp rec*A56 *his rps*E
Plasmids	
RP4 (Dixon *et al.*, 1976)	Ap Km Tc IncPl Tra$^+$

pRL1JI (Johnston *et al.*, 1978) is a conjugative *R. leguminosarum* plasmid carrying gene(s) for the production of a bacteriocin.

pJR4JI (Beringer *et al.*, 1978) is pPH1JI::Mu::Tn5. pPH1JI is a P1-type plasmid which confers gentamicin and low-level streptomycin and spectinomycin resistance.

Materials

TY Sm Km agar contains streptomycin (250 μg ml^{-1}) and kanamycin (25 μg ml^{-1}).

TY Rif Km agar contains rifampicin (20 μg ml^{-1}) and kanamycin (25 μg ml^{-1}).

Soft TY agar is TY medium + agar (6 g l^{-1}).

Method. Grow cultures of *R. leguminosarum* 6015 (pRLlJI) on TY slants in Universal bottles at 30°C for 2 days.

Wash the bacteria off the slants with 5 ml of 20% v/v glycerol and store at −20°C until required.

Grow a late log-phase culture (1×10^9 bacteria ml^{-1}) of JC5466(pJB4JI) in nutrient broth on a shaker at 37°C.

Mix 1 ml of donor and recipient cultures (each containing 1×10^9 bacteria ml^{-1}), add 8 ml of distilled water, and collect bacteria on a Millipore membrane (pore size 0.45 μm).

Place the membranes on the surface of TY plates and incubate for about 20 h at 28°C.

Suspend the bacteria in 5 ml of distilled water and plate 0.3 ml on 16 TY Sm Km agar plates.

Colonies should appear after incubation for 3 days at 28°C. The frequency of Km^r transconjugants varies considerably among strains.

Patch 1000 colonies (taken from more than one mating if necessary) on TY Sm Km plates and, after incubation for 2 days at 28°C, replicate patches on to lawns of a rifampicin-resistant *E. coli* and *R. leguminosarum* 1063 on TY Rif Km plates. Use *R. leguminosarum* 6015(RP4) as a control for plasmid transfer to *E. coli* and *R. leguminosarum* 1063.

Incubate mating plates at 28°C for 3 days.

Clone donor strains which transfer Km to *R. leguminosarum* 1063 but not to *E. coli* and make stocks of them. These carry presumptive Rhizobium plasmid: : Tn5 insertions and are suitable for further characterization.

Check these strains for the production of the pRL1JI specific bacteriocin as follows. Patch colonies onto TY plates and after incubation for 2 days at 30 °C kill the bacteria by exposure to chloroform vapour for 5 min. Then overlay the plates with 3 ml of soft TY agar containing *R. leguminosarum* 1063 at a population density of $1–2 \times 10^8$ bacteria ml^{-1}. A halo effect around the patches indicates bacteriocin production.

Check the symbiotic phenotypes of these strains in plant tests as described by Gibson in Chapter II.4.

(d) ISOLATION OF AZOTOBACTER *nif* MUTANTS

Nif^- mutants of *Azotobacter vinelandii* have been isolated following NG mutagenesis (Fisher and Brill, 1969). The protocol described below was therefore designed to isolate NG-generated *nif* mutants. However, it should be borne in mind that the high frequency of multiple mutations caused by this mutagen can give rise to problems during subsequent genetic analysis.

(i) NG dose–response curve

A culture of Azobacter grown on N_2 or NH_4^+ can be used for mutagenesis. The doubling time of NH_4^+-grown batch cultures is approximately 25% less than that of N_2-grown cultures (R. Robson, personal communication).

Materials

AzMM (Azotobacter minimal medium). This is a modification of Burk's medium (Newton *et al.*, 1953):

K_2HPO_4/KH_2PO_4 buffer, pH 7.2*	5 mM
CH_3COONH_4	15 mM
$MgSO_4 \cdot 7H_2O$	1 mM
NaCl	1 mM
$CaCl_2 \cdot 2H_2O$	0.1 mM
$FeSO_4 \cdot 7H_2O$	0.02 mM
$NaMoO_4 \cdot 2H_2O$	0.01 mM
Sucrose or mannitol	2% w/v

*Autoclave 1 M phosphate buffer (pH 7.4) separately and add an appropriate volume to the medium before use. Autoclaving decreases the pH to 7.2. Solidify with 1.5% agar.

NDAzM is AzMM without CH_3COONH_4. Use agar (1.5%) with a low N content (e.g. Serva agar) to solidify.

AzNA (Azotobacter nutrient agar) (J. Postgate, personal communication). This is nutrient broth (Oxoid No 2) and AzMM mixed in a 1 : 1 ratio and solidified with 1.5% agar (Difco bactoagar).

Method. Add 5 ml of a late log-phase culture in AzMM to 45 ml of fresh medium in a 250-ml flask.

Incubate at 30 °C without shaking for 1 h and then with shaking until a population density of 3–5 × 10^8 bacteria ml^{-1} is reached (approximately 5 h).

Add 1 ml of NG solution (10 mg ml^{-1} in acetone). This is the zero time.

Dispense 5-ml aliquots into eight flasks (25 ml) and continue shaking at 30 °C.

Take one culture for each time interval tested (0, 10, 20, 30, 40, 60, 90, and 120 min) and determine the surviving fraction of bacteria as described in Section b.i for *K. pneumoniae*, except that sucrose free AzMM should be used as diluent instead of saline phosphate buffer and AzNA instead of NA plates.

Incubate the plates at 30 °C for 2–3 days and construct a dose–response curve as described in Section b.i.

(ii) NG mutagenesis and isolation of *nif* mutants

Add 1 ml of a late log-phase culture in AzMM to 9 ml of fresh medium in a 50-ml flask and grow on a shaker at 30 °C to a density of 3–5 × 10^8 bacteria ml^{-1}.

Add 0.2 ml of NG solution (10 mg ml^{-1} in acetone) and continue incubation on a shaker at 30 °C for the time required to decrease the viable count to 20–50% (obtained from the dose–response curve).

Collect cells on a membrane filter (pore size 0.45 μm) or harvest by centrifugation, wash with 10 ml of sucrose-free AzMM, and re-suspend in 50 ml of AzMM.

Dispense 4-ml aliquots into 12 flasks (25 ml) and incubate on a shaker at 30 °C for 24–36 h.

Spread 0.1, 0.2, and 0.3 ml of a 10^{-6} dilution of each culture in sucrose-free AzMM on AzNA plates. Incubate the plates at 30 °C for 3 days.

Replicate colonies on to NDAzM and AzMM plates and incubate at 30 °C for 4–5 days. Colonies which grow poorly on the former but at a rate comparable to that of the parental strain on the latter plates are presumptive *nif* mutants.

(e) PURIFICATION OF PLASMID DNA

(i) Plasmids from *K. pneumoniae* and *E. coli*

The procedure described for the purification of super-coiled plasmid DNA in the molecular weight range 2.0–120 × 10^6 daltons is based on that of Humphreys *et al.* (1975).

Materials

MM-CAS-YE is minimal medium (Section b.i) supplemented with casamino acids (2 mg ml^{-1}) and yeast extract (1 mg ml^{-1})

Brij58/DOC 'Lytic mix':

Brij58 (Sigma)	1% (w/v)
Sodium deoxycholate (Sigma)	0.4% (w/v)
Na_2EDTA	60 mM
Tris–Cl, pH 8.0	50 mM

TEN buffer:

Tris–Cl, pH 8.0	50 mM
Na_2EDTA	5 mM
NaCl	50 mM

TE buffer:

Tris–Cl, pH 8.0	10 mM
Na_2EDTA	0.25 mM

Method. Grow a 30-ml culture of a plasmid bearing strain to late log-phase (1×10^9 bacteria ml^{-1}) in a medium selective for the plasmid.

For large (>30×10^6 daltons) plasmids, inoculate 3 l of MM-CAS-YE with the 30-ml culture and grow overnight with aeration at 37°C. If the plasmid being isolated is unstable impose selection at this stage also.

Harvest cells at 12000 *g* and 4°C for 15 min, wash with 160 ml (4×40 ml in 50-ml tubes) of saline phosphate buffer (Section b.i), and suspend in a solution of sucrose (25%), Tris–Cl (50 mM), pH 8.0, to a final volume of 64 ml. Keep the cells chilled on ice.

Dispense 4-ml aliquots into 16 centrifuge tubes (50 ml) standing in ice.

Add 0.8 ml of lysozyme solution (5 mg ml^{-1} in 0.25 M Tris–Cl, pH 8.0) to each tube and leave on ice for 5 min.

Add 1.6 ml of Na_2EDTA (0.25 M, pH 8.0) and leave on ice for a further 10 min.

Add 6.4 ml of Brij58/DOC 'lytic mix' and leave on ice for a further 20 min. Invert the tubes gently every 5 min.

Centrifuge the crude lysates at 28–30 × 10^3 *g* and 4°C for 1 h.

Collect the supernatant in a chilled flask and measure the volume: expect 190–200 ml.

Add 1/10 volume of 5 M NaCl and dispense 20-ml aliquots into centrifuge tubes containing 2 g of polyethylene glycol 6000 (PEG-6000).

Leave at room temperature and occasionally invert the tubes gently until the PEG-6000 has dissolved.

Place tubes in ice and leave standing overnight.

Gently centrifuge the tubes in a bench-top centrifuge (*ca.* 600 *g* for 4 min) to compact the precipitate.

Discard the supernatant and add 1.5 ml of TEN buffer to each pellet. Keep the tubes on ice for approximately 2 h to allow the pellets to soften.

Combine the contents of the tubes and dilute to 12 ml with TEN buffer.

Add 2 ml of ethidium bromide solution (5 mg ml^{-1}, 10 mM in Tris–Cl, pH 8.0) and 13 g of CsCl. Invert gently to dissolve and leave the tubes standing on ice overnight.

Centrifuge the tubes at 8000 *g* for 15 min and gently remove the supernatant with a 10-ml pipette fitted with a propipette.

Adjust the refractive index with water to 1.3900, dispense into two 12-ml centrifuge tubes, overlay with mineral oil, and centrifuge at 200000 *g* and 10°C for 36–48 h.

After centrifugation, two DNA bands should fluoresce strongly in UV light. The lower band is the supercoiled DNA.

Fractionate the gradients dropwise by piercing the bottoms of the tubes and collect the lower bands.

Extract with equal volumes of isoamyl alcohol until the pink fluorescence of the aqueous phase disappears.

Transfer to a sterile dialysis bag and dialyse against ten consecutive 1-l volumes of TE buffer.

Dilute 20 μl of the DNA solution in 0.58 ml of TE buffer to obtain a UV spectrum and determine the DNA concentration (1 absorbance unit cm^{-1} at 260 nm corresponds to 50 μg ml^{-1} of DNA).

Using this procedure, the author has obtained DNA yields of pRDl (mol. wt. 101×10^6 daltons) between 150 and 200 μg from 3 l cultures.

This procedure can also be used to purify large quantities of small amplifiable plasmids with the following modifications.

Inoculate 1 l of MM-CAS-YE, containing an antibiotic selective for the plasmid if necessary, with 10 ml of late log-phase culture and grow with aeration at 37°C to a cell density of 2×10^8 ml^{-1}.

Add 5 ml of chloramphenicol solution (40 mg ml^{-1} in ethanol) to give a final concentration of ca. 200 μg ml^{-1} and continue overnight incubation as above.

Harvest and wash the cells as described above and suspend in 16 instead of 64 ml of 25% sucrose, 50 mM Tris–Cl, pH 8.0.

Dispense 4-ml aliquots into four centrifuge tubes and continue as above except that 3 ml instead of 1.5 ml of TEN buffer should be added to the DNA–PEG pellets.

(ii) Large plasmids from *Rhizobium* spp.

The procedure described (M. Nuti, personal communication) is based on those of Ledeboer *et al.* (1976) and Nuti *et al.* (1977). It has been used successfully to purify plasmids in the molecular weight range $90–360 \times 10^6$

daltons from *R. trifolii*, *R. leguminosarum*, *R. meliloti*, *R. phaseoli*, and fast-growing *R. japonicum* strains.

Materials

TY medium: see Section c.i.

TE buffer:

Tris–Cl, pH 8.0	50 mM
Na_2EDTA	20 mM

TES buffer:

Tris–Cl, pH 8.0	10 mM
Na_2EDTA	50 mM
NaCl	50 mM

Pronase E (Merck) solution: 5 mg ml^{-1} in TE buffer. Prepare freshly and incubate at 37 °C for 90 min before use.

Vibromixer (A.F. Für Chemie Apparatenbau, Switzerland).

Method. Inoculate 3 l of TY medium with 60 ml of a late log-phase culture of a Rhizobium strain in the same medium and grow with aeration at 30 °C to mid-log phase (3–4 × 10^8 bacteria ml^{-1}).

Harvest cells at 12000 *g* and 4 °C for 15 min, wash with 160 ml (4 × 40 ml in 50-ml tubes) of TE buffer, and suspend in 1 l of TES buffer.

Add 60 ml of a pronase E solution (5 mg ml^{-1} in TE buffer, see above) and 30 ml of SDS (20% w/v). The cells should lyse after incubation for 10 min at 37 °C.

Shear 250-ml aliquots of the lysate in 400-ml beakers for approximately 10 s with a Vibromixer. From this point treat each of the four aliquots separately.

Increase the pH to 12.3–12.4 by adding 3 M NaOH while the sheared lysate is stirred at approximately 150 r.p.m. with a magnetic stirrer, and leave at room temperature for 10 min.

Add sufficient 2 M Tris–Cl (pH 7.0) to decrease the pH to 8.7.

Add solid NaCl to give a final concentration of 3% w/v and shake gently to dissolve.

Add 1 volume of distilled phenol saturated with a solution of NaCl (3%) and shake gently for several seconds.

Centrifuge at approximately 1000 *g* for 10 min and transfer the aqueous phase to a sterile flask. Make up to 1 l with chloroform–isoamyl alcohol (24:1), shake gently, and centrifuge as above.

Transfer the aqueous phase to a sterile flask and add 1/10 volume of a 3 M sodium acetate solution and 2 volumes of ethanol (kept at −20 °C). Mix gently, dispense into four centrifuge tubes (250–300 ml) and leave overnight at −20 °C. Collect the precipitate by centrifugation at approximately 10000 *g* for 10 min.

Pour off the bulk of the supernatant and drain the remainder by standing the tubes inverted on a paper towel for approximately 30 min.

Add 0.5 ml of TES buffer to each tube and dissolve the pellets by gentle hand shaking.

Combine the 16 DNA solutions (4 from each of the 4 original aliquots of lysate) and make up the volume to 12 ml with TES buffer.

Add 2 ml of ethidium bromide solution (5 mg ml^{-1} in 10 mM Tris–Cl, pH 8.0) and 13 g of CsCl.

Adjust the refractive index with water to 1.3900, dispense into two 12-ml centrifuge tubes, overlay with mineral oil, and centrifuge at 200000 *g* and 10 °C for 36–48 h.

After centrifugation, collect the supercoiled DNA band and purify as described in Section e.i.

(f) DETECTION OF PLASMIDS IN RHIZOBIUM BY AGAROSE GEL ELECTROPHORESIS

The procedure described (M. Nuti, personal communication) is based on that of Casse *et al.* (1979). It can be used to screen up to 20 strains simultaneously for the presence of plasmids, to determine their molecular weights, and to detect the presence of more than one plasmid per strain (Prakash *et al.*, 1978).

Materials

TY medium: see Section c.i.

TE buffer:

Tris–Cl, pH 8.0	50 mM
Na_2EDTA	20 mM

TES buffer: see Section e.ii.

TBE buffer:

Trisma base (Sigma)	89 mM
Boric acid	89 mM
Na_2EDTA	2.5 mM

Sample buffer:

Ficoll-400	100 mg ml^{-1} of TBE buffer
SDS	10 mg ml^{-1} of TBE buffer
Bromophenol blue	0.02 mg ml^{-1} of TBE buffer

Beckman buffer No. 3010: prepare according to the manufacturer's instructions.

Agarose: Sigma No. A-6877.

Capillary matting.

Method. Grow 25-ml cultures of Rhizobium strains in TY medium to late log phase (approximately 1×10^9 bacteria ml^{-1}) on a shaker at 30 °C.

Harvest the cells by centrifugation (12000 *g*, 4 °C, 10 min) and wash with an equal volume of TE buffer in weighed tubes.

Re-suspend the cells in TE buffer (0.5 ml per 100 mg wet weight of cells). At this point the cells can be stored at −20°C or used immediately.

Standardize a pH meter to pH 12.45 at 25°C with Beckman buffer No. 3010.

Prepare a lysing mixture by dissolving 1 g of SDS in 95 ml of 50 mM Na_2EDTA (pH 8.0) and adjusting the pH to 12.45 with 3 M NaOH.

Transfer 0.5 ml of cell suspensions to 50-ml sterile beakers and add 9.5 ml of freshly prepared lysing mixture.

Stir for 1.5 min with a magnetic stirrer at 60 r.p.m.

Incubate in a water-bath at 34°C for 25 min.

Add 0.6 ml of a 2 M solution of Tris–Cl (pH 7.0) and stir for 1.5 min as above.

Add 320 mg of solid NaCl—the mixture should become cloudy after a few minutes.

Add 10 ml of distilled phenol previously saturated with a solution of NaCl (3% w/v).

Centrifuge at 2000 *g* and 4°C for 10 min.

Draw the aqueous phase (approximately 8.0 ml) into the wide end of a sterile Pasteur pipette and transfer to a sterile centrifuge tube.

Add 1/10 volume of 3 M sodium acetate (pH 8.0) followed by 2 volumes of ethanol previously kept at −20°C and leave overnight at −20°C.

Centrifuge at 12000 *g* and 0°C for 10 min, pour off the bulk of the supernatant, and drain the remainder by standing the tubes inverted on a paper towel for approximately 30 min.

Dissolve the pellet in 100–200 μl of TES buffer and remove the residual ethanol by placing in a vacuum for 2–3 min. The DNA is now ready for electrophoresis.

Prepare an agarose slab gel (0.7% agarose in TBE buffer) as described in Section g.ii.

Load 1/4–1/3 volume of each DNA solution mixed with 1/5 volume of electrophoresis sample buffer per gel lane. Use sheets of capillary matting or Whatman 3MM paper to make contact between the reservoir buffer (TBE) and the gel ends and electrophorese at 5–10 V per cm of gel length for 8 h.

For molecular weight markers use similarly prepared plasmid DNAs of known molecular weights.

Stain and photograph the gel as described in Section g.ii below.

(g) CLONING THE *nif* STRUCTURAL GENES FOR *K. pneumoniae* NITROGENASE

The structural genes for *K. pneumoniae* nitrogenase are carried on an EcoRl restriction fragment (4.5×10^6 daltons) of pRD1 (Cannon, 1978; Cannon *et al.*, 1979). Details of the protocols used to clone this fragment on the gene vector pACYC184 (Chang and Cohen, 1978) are given below. With the appropriate modifications these protocols are suitable for cloning DNA fragments in general.

Bacterial strains	*Genotype or phenotype*
CK2601 is a His^{+} revertant of CK260 (Kennedy, 1977)	*nif*H2260 *hsd*R1 *rps*L4

Plasmids

pRD1 (Section b.iii)	
pACYC184 (Chang and Cohen, 1978)	Tc Cm

Materials

EcoR1 reaction solution:

Tris–Cl, pH 7.5	80 mM
$MgCl_2$	10 mM
NaCl	50 mM

T4 polynucleotide ligase reaction buffer:

Tris–Cl, pH 7.5	66 mM
$MgCl_2$	6.6 mM

ATP: 100 and 20 mM.

Dithiothreitol (DTT): 1 M and 200 mM.

TAE buffer:

Trisma base	40 mM
CH_3COONa	5 mM
Na_2EDTA	1 mM

Adjust pH to 7.8 with glacial acetic acid.

Sample buffer:

Ficoll-400	100 mg ml^{-1} of TAE
SDS	10 mg ml^{-1} of TAE
Bromophenol blue	0.02 mg ml^{-1} of TAE

TMC:

Tris–Cl, pH 7.5	10 mM
$MgCl_2$	10 mM
$CaCl_2 \cdot 2H_2O$	30 mM

T medium:

K_2HPO_4/KH_2PO_4 buffer, pH 7.2	5.0 mM
$(NH_4)_2SO_4$	15.0 mM
$MgSO_4 \cdot 7H_2O$	1.0 mM
NaCl	0.2 mM
$CaCl_2$	0.1 mM
Glucose	0.4% w/v
Casamino acids	2 mg ml^{-1}
Yeast extract	1 mg ml^{-1}

For growth of auxotrophs, supplement with the appropriate amino acids (40 μg ml^{-1}).

(i) DNA restriction and ligation

Perform reactions in Eppendorf tubes (1.5 ml) as outlined below.

Tube No.	1	2
Volume of pRD1 DNA (200 μg ml^{-1})	4	4 μl (≡0.8 μg)
Volume of pACY184 DNA (100 μg ml^{-1})	3	3 μl (≡0.3 μg)
EcoR1 reaction solution (10×)	1	1 μl
EcoR1 (10^3 units ml^{-1})	2	2 μl
Total volume	10	10 μl
Final concentration of pRD1 DNA	80	80 μg ml^{-1}
Final concentration of pACY184 DNA	30	30 μg ml^{-1}
Centrifuge: 600*g*, 1 min	√	√
Incubate: 37 °C, 1 h	√	√
Tubes at 65 °C for 10 min	√	√
Tubes at room temperature for 5 min	√	√
Remove 5 μl from tube for gel analysis	—	√
Transfer tubes to ice	√	√
600 mM Tris–Cl, pH 7.5	1	0.5 μl
Tubes on ice for 3 h	√	√
ATP (20 mM)	1	0.5 μl
DTT (200 mM)	1	0.5 μl
T4 ligase (10^3 units ml^{-1})	4	2 μl
Total volume	17	8.5 μl
Centrifuge: 600 g, 4 °C, 1 min	√	√
Incubate: 13 °C, 3 h	√	√
Tubes on ice	√	√

Use contents of tube 2 for agarose gel analysis

T4 ligation buffer	370 μl
ATP (100 mM)	4 μl
DTT (1 M)	4 μl
T4 ligase (10^3 units ml^{-1})	5 μl
Total volume	400 μl
Incubate: 13 °C overnight	√

Use 200 μl for transformation and store the remaining 200 μl at −20 °C.

(ii) Agarose gel electrophoresis

Horizontal or vertical slab gels or tube gels may be used. The procedure described is for horizontal slab gels which measure 18 × 11.6 × 0.5 cm.

To prepare a 0.8% agarose gel (120-ml slab gel) add 120 ml of TAE buffer to 0.96 g of agarose and autoclave for 5 min. When the agarose has cooled to 55–65 °C pour into a gel mould fitted with a slot-forming comb and leave to set for at least 1 h.

Prior to loading of DNA samples add 0.2–0.5 volume of sample buffer to each DNA solution and heat at 65 °C for 5 min, then cool to room temperature.

Fill the gel slots with TAE buffer before underlaying the samples.

If submerged gels are used, add enough TAE buffer to the gel box so that the level of the buffer is *ca.* 2 mm above the surface of the gel.

Electrophorese overnight at a constant voltage of 40V.

Stain the gel for 30 min in ethidium bromide (0.5 μg ml^{-1} in TAE buffer), view with a Transilluminator (260 or 360 nm, UV Products Inc., San Gabriel, Calif., U.S.A.), and photograph.

For photographs, a Polaroid or a 35-mm camera with an orange (Kodak No. 22) or red filter (Kodak 25A) and Ilford HP5 or Kodak Tri-X pan films (10 s for *f* 2.8, 20 s and 40 s for *f* 5.6) may be used.

(iii) Transformation of *K. pneumoniae* CK2601 with ligated DNA

Grow an overnight culture of CK2601 in T medium on a shaker at 30 °C.

Inoculate 1.5 ml in to 75 ml of T medium and grow to a cell density of 2×10^8 ml^{-1} on a shaker at 37 °C. Chill the culture on ice.

Harvest cells by centrifugation at 12000 *g* and 4 °C for 5 min.

Suspend the cells in 10 ml of chilled 30 mM $CaCl_2$ and leave on ice for 20 min.

Harvest cells by centrifugation at 12000 *g* and 4 °C for 5 min.

Suspend in 0.25 ml of chilled 30 mM $CaCl_2$. The cells are now ready for transformation by the following procedure.

Tube No.	1	2
Cells	100	100 μl
DNA (0.6 μg)	200	—
TMC	—	200 μl
$CaCl_2$ (1 M)	6	—

Leave the tubes on ice for 1 h.

Heat pulse at 42 °C for 2 min and return the tubes to ice for a few minutes.

Add the tube contents to two separate flasks containing 5 ml of Luria broth (Section b.v) and incubate on a shaker at 37 °C for 90 min.

(iv) Ampicillin/D-cycloserine enrichment of Tc^rCm^s transformants

Add 15 μl of a Tc solution (5 mg ml^{-1}) to both flasks (final concentration 15 μg ml^{-1}).

Incubate for 5 h on a shaker at 37 °C.

Harvest cells in a bench centrifuge, remove the supernatant (including the last drop), re-suspend the cells in 5 ml of Luria broth (Section b.v) containing Cm (25 μg ml^{-1}) and incubate the cultures on a shaker at 37 °C for 1 h.

Add 0.5 ml of ampicillin solution (5 mg ml^{-1}) and 0.05 ml of D-cycloserine solution (1 M) to give final concentrations of 500 μg ml^{-1} and 10 mM, respectively.

Incubate for 2 h on a shaker at 37 °C.

Wash three times with saline phosphate buffer (Section b.i).

Suspend the cells in 5 ml of Luria broth (Section b.v) + Tc (15 μg ml^{-1}) and incubate for 2 h on a shaker at 37 °C.

Plate 0.1 ml of transformation culture on each of 20 plates of nutrient agar + Tc (15 μg ml^{-1}) and 0.1 ml of control culture on five similar plates.

Incubate the plates overnight at 37 °C. Expect 100–200 colonies per plate from the transformed culture.

Patch 1000 colonies on nutrient agar + Cm (25 μg ml^{-1}), NFDM agar and nutrient agar + Tc (15 μg ml^{-1}). Incubate the nutrient agar plates overnight at 37 °C and the NFDM plates anaerobically at 30 °C for 6–7 days.

Less than 10% of transformants screened should be Cm^r and 1–2% of Cm^s transformants should give rise to patches with Nif^+ colonies. These can be presumed to have the desired plasmid and 10 should be screened for its presence as follows.

Streak bacteria from patches on the nutrient agar Tc plates which correspond to Nif^+ patches on NFDM agar to obtain isolated colonies.

Suspend a large colony from each culture in 100 μl of TAE buffer in small tubes and add 25 μl of sample buffer containing 50 instead of 10 mg ml^{-1} of SDS. Place the tubes in a water-bath at 65 °C for 30 min. Whirlimix for a few seconds and, while hot, load 50 μl of each on an agarose gel (0.8%) for electrophoresis as described above. Use pMB9 DNA for molecular weight markers (Ausubel *et al.*, 1977).

A plasmid of 6.9×10^6 daltons should be detected in transformants which give rise to Nif^+ colonies.

(h) USE OF THE SOUTHERN TECHNIQUE TO LOCATE *nif* GENES IN RHIZOBIUM

Statistical analysis of the amino acid composition of nitrogenase proteins (Eady, 1977) to obtain SΔQ values has suggested a high probable degree of sequence homology between the corresponding DNAs of *R. japonicum* and *K. pneumoniae*. TheDNA which encodes nitrogenase in these two organisms should therefore hybridize.

Nuti *et al.* (1979) have shown that a DNA restriction fragment which encodes the nitrogenase polypeptides of *K. pneumoniae* (Cannon *et al.*, 1979) hybridized to plasmid DNA sequences from *R. leguminosarum*. The procedure used involved the preparation of Southern filters (Southern, 1975) carrying restriction fragments of plasmid DNA from *R. leguminosarum*, synthesis of radioactive copy DNA of pSA30 which carries the structural genes for *K.*

pneumoniae nitrogenase, and hybridization of the radioactive probe to restriction fragments bound to the filters. Details for the procedure are given below.

Materials

Plasmids pACYC184, pSA30, and pCM1: see Section b and Cannon *et al.* (1979).

[α-^{2}P]dCTP: specific activity >350 Ci mmol^{-1}.

dNTPs: 1 mM stock solutions in water. Store at −20°C.

TEN buffer:

Tris–Cl, pH 7.4	10 mM
Na_2EDTA	1 mM
NaCl	5 mM

SSC:

$Na_3C_6H_5O_7 \cdot 2H_2O$	0.015 M
NaCl	0.15 M

Prepare a 20× stock solution (pH 7.0).

Denaturation solution:

NaOH	0.5 M
NaCl	1.5 M

Renaturation solution:

Tris–Cl,pH 7.0	0.5 M
NaCl	3.0 M

TMN-G buffer:

Tris–Cl, pH 7.4	10 mM
$MgCl_2$	10 mM
NaCl	10 mM
Gelatin	100 μg ml^{-1}

Gelatin stock solution: 10 mg ml^{-2} in water, freshly prepared.

Homes–Bonner phenol–chloroform–isoamyl alcohol (25:24:1): use distilled phenol saturated with TEN buffer.

DNA polymerase I.

Deoxyribonuclease I (grade 1, Boehringer Mannheim): 50 μg ml^{-1} of TMN-G–glycerol (1:1). Store at −20°C and dilute 1/100 in TMN-G before use.

Denhardt's solution (DS):

Bovine serum albumin (fraction V, Sigma)	0.2 mg ml^{-1}
Ficoll-70	0.2 mg ml^{-1}
Polyvinylpyrrolidone	0.2 mg ml^{-1}

Prepare a 20× stock solution and store at −20°C.

Calf thymus DNA (Sigma type I): 2.5 mg ml^{-1} in TEN buffer. Sonicate for 1 min.

Nitrocellulose filter-paper (Schleicher & Schüll, type BA85).

Plastic bag for hybridization: boilable cooking bags (8 × 12 in) by Sears (U.S.A.), catalogue no. 6546, or equivalent.

(i) ^{32}P Labelling of DNA by 'nick translation'

Place 220 μl of a [^{32}P]dCTP solution (220 μCi, 627 pmol) in an Eppendorf tube and place under vacuum until the solvent has competely evaporated.

Make the additions listed in the following order:

Distilled water	327 μl
TEN buffer (10×)	40 μl
pSA30 DNA (220 μg ml^{-1})	7 μl (≡1.5 μg)
dATP (1 mM)	2 μl (≡2 nmol)
dGTP (1 mM)	2 μl (≡2 nmol)
dTTP (1 mM)	2 μl (≡2 nmol)
$MgCl_2$ (1 M)	4 μl
Mercaptoethanol (1 M)	4 μl
Gelatin (10 mg ml^{-1})	4 μl
DNase I (0.5 μg ml^{-1})	4 μl
DNA polymerase I	4 μl (≡16 units)
Total volume	400 μl

Incubate the tube at 13 °C for 2 h.

Extract with 400 μl of Homes–Bonner phenol–chloroform–isoamyl alcohol (25:24:1).

While the reaction is incubating, pour a Sephadex G-50 column (bed volume 5 ml) in a disposable 5-ml syringe (10 ml of a 5% G-50 slurry will give a 5-ml column).

Load the reaction mixture on the column and elute with TEN buffer. The flow-rate should be approximately 6 drops min^{-1}.

Collect 30 fractions (5 drops each, approximately 300 μl).

Measure the radioactivity (Cerenkov radiation) of 5-μl aliquots using the tritium channel of a liquid scintillation counter (this gives approximately 25% counting efficiency).

Two peaks of radioactivity should be obtained—the first to be eluted is the labelled DNA and the second is the unincorporated label.

Pool the fractions (usually 4 = volume of approximately 1.2 ml) of the DNA peak and dispense 200 μl aliquots in Eppendorf tubes.

The labelled DNA is now ready for hybridization. Store at −20 °C until required.

Also prepare ^{32}P-labelled copy DNA of pACYC184 by a similar procedure. pACYC184 was the gene vector used in the construction of pSA30 and should be used as a negative control probe.

(ii) Preparation of Southern transfers

The procedure used is based on that of Southern (1975).

Digest Rhizobium plasmid DNA with restriction endonucleases as described in Section g.i; use enzymes with six base specificities. Digest pCM1 DNA with EcoR1 and pSA30 DNA with EcoR1 and with EcoR1 + HindIII to obtain the following molecular weight markers:

(1) pSA30 × EcoRl	(2) pSA30 × (EcoR1 + HindIII)	(3) pCM1 × EcoR1
2.65×10^6 daltons	0.96×10^6 daltons	1.29×10^6 daltons
4.5×10^6 daltons	1.69×10^6 daltons	1.54×10^6 daltons
	2.01×10^6 daltons	3.77×10^6 daltons
	2.34×10^6 daltons	4.27×10^6 daltons
		5.43×10^6 daltons

Combine the three digested DNAs and use an aliquot containing 100, 200, and 400 pg of (1), (2), and (3), respectively, for electrophoresis in a 0.8% agarose slab gel (18 × 11.6 × 0.5 cm) as described in Section g.ii. Load 2 μg of digested Rhizobium plasmid DNAs per lane on the same gel. Prepare duplicate gels, one for each of the two probes, pSA30 and pACYC184 DNAs.

Stain the gel with ethidium bromide and photograph as described in Section g.ii.

Leave the gel exposed to UV light for approximately 5 min. This causes nicking of the DNA, which reduces the size of single strands after denaturation and thus facilitates transfer to the nitrocellulose filters.

Place the gel in a dish containing approximately 500 ml of denaturation solution and shake very gently for 40 min.

Wash four times with distilled water.

Add approximately 500 ml of neutralization buffer and continue gentle shaking for a further 40 min. The gel should be handled carefully since it becomes brittle after the above treatment.

Fill the buffer compartments of the horizontal gel box used for agarose gel electrophoresis with 20 × SSC.

Wet two sheets (12.5 × 30 cm) of Whatman 3 MM paper with 20× SSC and place on the gel platform of the electrophoresis box with the ends dipping into the SSC.

Wet three sheets (12.5 × 20 cm) of Whatman 3 MM paper with 20× SSC.

Place one sheet on the top surface of the gel and the remaining two over the sheets in the gel box.

Invert the gel with the sheet attached and place it carefully over the sheets in the gel box. Avoid air bubbles between the sheets.

Wet a sheet (11.5 × 16 cm) of nitrocellulose filter-paper with 2× SSC and place on the top surface of the inverted gel. Avoid air bubbles!

Trim the gel around the nitrocellulose sheet, removing the gel sample wells so that the sheet fits neatly over the gel. Mark the filter by cutting off a corner.

Wet a sheet (11.2 × 15.7 cm) of Whatman 3 MM paper with 2× SSC and place it neatly over the nitrocellulose sheet.

Place 10 dry Whatman 3 MM sheets (11.2 × 15.7 cm) neatly over the wet sheet, followed by 12–16 paper towels trimmed to approximately 11.2 × 15.7 cm, and finally place a glass plate (approximately 12 × 18 × 0.4 cm) on top to keep the layers compact. Cover with a clear plastic wrap to minimize evaporation.

Leave for a minimum of 5 h for DNA transfer to occur, preferably overnight.

Transfer the nitrocellulose sheet to a dish containing 300 ml of 2× SSC and shake gently for 30 min.

Place the sheet on a paper towel and leave overnight to dry. Then transfer the sheet and towel to an oven at 80°C for 2.5 h.

The nitrocellulose sheet is then ready for hybridization and can be stored at room temperature until required.

(iii) DNA hybridization on nitrocellulose sheets

Pre-treatment of nitrocellulose sheets

This reduces non-specific binding of the label.

Place the sheet in a plastic bag (see Materials) and add 20 ml of filtered distilled water.

Remove most of the water with a pipette and squeeze out the remainder by placing the bag on the bench with a paper towel at the open end and rolling a 10-ml pipette over it.

Add 20 ml of a solution made up to 6× SSC and 1× DS.

Remove most of the air from the bag before sealing and incubate overnight at 65 °C in a water-bath with gentle shaking.

Denaturation of radioactive DNA probe

To a conical polycarbonate tube (10 ml) add:

DNA probe (prepared as above)	200 μl
Filtered distilled water	2.75 ml
SDS (20%)	125 μl
DNA (2.5 mg ml^{-1}, calf thymus)	200 μl

Boil for 6 min and then cool to 70 °C. Add 1.5 ml of 20× SSC and 0.25 ml of 20× DS. The probe is now ready for hybridization.

Hybridization

Remove most of the membrane pre-treatment solution from the plastic bag and squeeze out the remainder as described above.

Add the probe (*ca*. 5.0 ml) to the bag, remove air, and re-seal. Incubate in a water-bath at 65 °C for 24 h with gentle shaking. Place a weight around the edges of the bag to keep it submerged.

Remove the probe from the bag and discard it.
Wash the filter as follows:

15 min at 65 °C in 10 ml of 1× DS + 6× SSC + 0.5% SDS;
1 h at 65 °C in 100 ml of the same solution;
1 h at 65 °C in 100 ml of 2× SSC + 0.5% SDS.

Repeat last wash four times.
Air-dry the nitrocellulose sheet for at least 1 h at room temperature.
Expose X-ray film to the membrane for 5–7 days and develop. The exposure time can be decreased to 3–4 days by using intensifying screens (Swanstom and Shank, 1978).

(i) ACKNOWLEDGEMENTS

I thank F. Ausubel, J. Beringer, R. Dixon, M. Filser, C. Kennedy, M. Merrick, M. Nuti, J. Postgate, G. Riedel, and R. Robson for useful discussions, and M. Cannon and C. Kennedy for constructive criticism of the manuscript.

Section III

METHODS FOR FIELD USE

Methods for Evaluating Biological Nitrogen Fixation
Edited by F. J. Bergersen

J. Brockwell
Division of Plant Industry,
CSIRO, Canberra, Australia.

1

Experiments with Crop and Pasture Legumes—Principles and Practice

'There is nothing grateful but the earth; you cannot do too much for it: it will continue to repay tenfold the pains and labour bestowed upon it'.

Lord Ravensworth

(a) INTRODUCTION

Field testing of legume host/*Rhizobium* strain associations is an essential final step in programmes designed to evaluate nitrogen-fixing capacity, strain competitiveness with other strains of *Rhizobium*, and long-term persistence in the face of competition from the established microflora. Experiments designed to assess the need for inoculation of legumes sown into particular soils are also best conducted in the field. Light-room and glasshouse experiments, of course, are invaluable for making preliminary measurements of symbiotic characteristics, but, however sophisticated these experiments may be, they cannot simulate the complexities of the soil environment. Nor are glasshouse trials using soil in pots a satisfactory solution; disturbance of the soil in transferring it from field to pots invariably leads to substantial changes in its microbiological and nutritional status. Examples of these complexities include soil acidity and soil moisture conditions which prevent survival of seed inocula, materials toxic to rhizobia associated with plant residues, microbial antagonism which may prevent multiplication of the inocula in the rhizosphere, calcium ion concentration which may influence infection of the legume roots by the rhizobia, naturally occurring strains which compete with the inocula in nodule formation, low or high soil temperatures which may delay the formation of nodules and the onset of nitrogen fixation, and mineral nutrient deficiencies which interfere with nodule function. Success of a strain of *Rhizobium* in establishing itself as a permanent member of the natural microflora, and fixing nitrogen in association with its host under these conditions, is more meaningful than success in the controlled environment cabinet or glasshouse.

Field experiments of any sort are always difficult to conduct and usually represent a large investment in time and sometimes money. In field work dealing with *Rhizobium* strains, there is the further difficulty that living bacteria are part of the experimental system. For instance, extreme caution is necessary to avoid mixing of inocula or accidentally transferring them from plot to plot. In addition, techniques must be available to enable the experimenter to recognize whether nodulation has been due to the applied rhizobia or to naturally occurring strains. Moreover, field experiments are subject to the vagaries of weather and innumerable other unpredictables. This author will always remember with awe the astonishing notation, referring to missing plots in a field experiment reported in a journal of tropical agriculture, 'trampled by elephants'.

Because of the resources involved, the failure of a field experiment is usually more costly than a corresponding loss of a laboratory or glasshouse experiment. Accordingly, field experiments should be meticulously planned. It is desirable to define the characteristics of the naturally occurring populations of rhizobia at the chosen site, to sow high quality seed, to inoculate that seed efficiently with known numbers of viable rhizobia that can be recognized later when they are reisolated from the experiment, to use adequate controls, and to choose an experimental design and a field location that will meet the objectives of the experiment and the demands of statistical analyses. These matters are dealt with in this chapter.

(b) NATURAL POPULATIONS OF RHIZOBIA IN SOILS

The Leguminosae is one of the largest families of flowering plants. The nodulating habit among legumes is very widespread (Allen and Allen, 1961), although non-nodulating species occur in the subfamilies Caesalpinioideae and Mimosoideae. Legume root nodule bacteria are widely distributed also, a result of the natural distribution of the host plants and through the cultivation of leguminous crop and pasture plants. Although some soils still exist that are devoid of rhizobia, most sites chosen for field experiments will contain populations of one or more *Rhizobium* spp. A knowledge of the characteristics of such populations is necessary to enable the experimenter to assess what influence they may have upon his procedures and/or results.

Rhizobial characteristics most likely to affect field experiments are the size of the population in the soil and its symbiotic characteristics. Both of these factors have a bearing on strain competition and the success of the inoculum in forming nodules on the test plant. In practice, one observes only the competition which occurs between strains of the same species. For instance, if one intends to conduct an inoculation experiment with *Trifolium* spp., only the soil population of *Rhizobium trifolii* is of practical consequence; likewise, only the characteristics of *R. japonicum* populations are of significance when one is experimenting with soybeans [*Glycine max* (L.) Merr.]. Moreover, it seems generally true that effective strains of rhizobia compete more intensively with inocula than do less effective strains (Robinson, 1969a; Marques Pinto *et al.*, 1974; Diatloff and Brockwell, 1976), although exceptions are known to occur (Nicol and Thornton, 1941; Škrdleta and Pelikan, 1973). Thus, some preliminary estimate of the competition from soil rhizobia likely to be encountered by field inocula can be obtained by enumeration of the population of the relevant *Rhizobium* sp. in the soil and assessment of its effectiveness with the intended host plant.

(i) Soil sampling

The area to be sampled is divided by eye into a grid of 20 sections. A subsample of soil, roughly cylindrical in shape, 8–10 cm deep and 3–4 cm in diameter, is taken from each section, using a pre-sterilized coring implement or a long-nosed trowel, and placed in a polyethylene bag. In order to prevent cross-contamination, if another area is to be sampled, a second pre-sterilized sampling tool should be used or the first one washed in 95% ethanol or methylated spirits and sterilized by flaming. Because of the dynamic nature of populations of rhizobia in soil, soil samples should be examined immediately following collection. If this is not feasible, the samples should be stored at 4°C but for no longer than absolutely necessary.

In the laboratory, the subsamples are thoroughly mixed on a clean surface, quartered, mixed again, and so on until a small composite sample, as homogeneous as feasible, is obtained. A coarse sieve may be required to remove stones but fine sieving is neither necessary nor desirable because of the risk that particles of organic matter harbouring colonies of rhizobia may be lost. If desired, soil moisture content may be determined (Chapter II.4) in order to express population figures on a dry weight basis.

(ii) Enumeration

Except for two unconfirmed reports (Graham, 1969; Nutman, 1973), no selective medium for rhizobia in soil has proved successful. In any case, there seems little prospect of distinguishing culturally and reliably between rhizobia from the various nodulation groups. Therefore, bacteriologically controlled plant-infection tests are usually used. The latter technique is described in detail below and selective media are dealt with only briefly.

Plant-infection tests

Plant-infection tests have been in use for over 50 years (Wilson, 1926). Basically, the method consists of inoculating test plants, grown aseptically, with aliquots from a dilution series of the sample being examined. The number of rhizobia in the sample can be calculated from the proportion of test plants forming nodules at each dilution. Numerous versions of the test now exist (Tuzimura and Watanabe, 1961; Date and Vincent, 1962; Brockwell, 1963; Ham and Frederick, 1966; Thompson and Vincent, 1967; Elliott and Blaylock, 1971; Weaver and Frederick, 1972; Brockwell *et al.*, 1975a). Two of the tests are described below, one suitable for test plants which can be grown within tubes and the other for larger plants.

In tube culture. From considerations of asepsis, convenience of handling, and economy of space, it is preferable to grow the test plants within glass tubes closed with cotton-wool plugs. Experience in this laboratory with plant-infection tests for counting populations of rhizobia in soil has shown that, when test plants are cultured in agar, as in Thornton (1930a) tubes, prolific growth of algae and saprophytic fungi sometimes interferes with nodulation. Effective control can be exercized by the use of the antibiotic mycostatin at a rate of 100 parts per 10^6 in the agar medium (Robinson, 1968). Control of soil-borne fungi and algae can also be obtained by growing the test plants on a vermiculite substrate (Brockwell *et al.*, 1975a). Moreover, certain species, suitable for use as test plants, grow better in vermiculite culture than in an agar medium (Brockwell and Hely, 1966). For these reasons, it seems desirable whenever possible to grow test plants in test-tubes containing vermiculite moistened with a plant nutrient solution (Brockwell, 1963). It is important to note that vermiculite often requires pre-treatment to remove impurities. A suitable method of doing this and details of tube preparation are presented in Chapter II.4.

Many small-seeded legume species grow best in tubes and accordingly are sought after as test plants. Where the legume under investigation is large-seeded and, therefore, inappropriate for tube culture, it is often possible to find a symbiotically related species with small seeds for plant-infection tests, e.g. the use of *Ornithopus sativus* Brot. for counting populations of *Rhizobium lupini* appropriate to *Lupinus* spp. A list of some small-seeded legumes which can be grown in test tubes as test plants for plant-infection tests for enumerating various *Rhizobium* spp. is given in Table 1.

Seed preparation. Clean, undamaged seeds, selected for uniform size by weighing, hand-sorting, or sieving, are rinsed briefly in 95% ethanol and then immersed for 5 min in 0.1% $HgCl_2$ solution. They are washed thoroughly in at least ten changes of sterile distilled water to remove all traces of $HgCl_2$. The seeds are allowed to stand in the final change of water for several hours until fully imbibed. Seeds are then spread on to 1.5% water agar in Petri dishes which are incubated at an appropriate temperature (usually 25 °C) in an inverted position to provide uniform seedlings with straight roots. Seeds (e.g. soybean) which imbibe during immersion in $HgCl_2$ solution often produce deformed seedlings; such samples are usually sterilized satisfactorily merely by immersion for 30 s in 95% ethanol. Other hard-seeded species (e.g. *Medicago*) sometimes do not imbibe even after soaking for many hours. These lines should be lightly scarified before commencing, or may be sterilized by immersion for 5 min in concentrated H_2SO_4 (which acts simultaneously as an agent for scarification and sterilization), and then rinsed very quickly in ten changes of sterile water and allowed to imbibe as before. It is important that the H_2SO_4 be freshly opened; the presence of absorbed water is harmful to seeds. Some species (e.g. *Trifolium*) appear to germinate more evenly when the Petri dishes containing the newly sown seeds are refrigerated at 4 °C for 48 h before transfer to an incubator. These techniques almost invariably ensure freedom of the sterilized seed from cells of rhizobia and nodulation of uninoculated seedlings is very rare. Total freedom from all microorganisms is difficult to attain (Ash and Allen, 1948) and, where it is required for specialized experiments, more drastic techniques, including removal of the seed coat, will be necessary.

Table 1. List of some small-seeded legumes which can be grown in test-tubes as test plants for plant-infection tests for enumerating various *Rhizobium* spp.

Rhizobium sp.	Test plant	Minimum desirable test-tube dimensions (cm)
R. meliloti	*Medicago sativa* L., *M. arabica* (L.) Huds., *M. falcata* L., *M. lupulina* L., *M. minima* (L.) Bart., *M. praecox* DC., *M. rigidula* (L.) All., *Melilotus alba* Desr., *Mel. indica* (L.) All., *Mel. officinalis* (L.) Lam., *Trigonella suavissima* Lindl.	15.0 × 1.85
	Medicago littoralis Rhode, *M. orbicularis* (L.) Bart., *M. polymorpha* L., *M. rugosa* Desr., *M. scutellata* (L.) Mill., *M. tornata* (L.) Mill., *Trigonella foenum-graecum* L.	15.0 × 2.5
R. trifolii	*Trifolium alexandrinum* L., *T. dubium* Sibth., *T. fragiferum* L., *T. glomeratum* L., *T. hybridum* L., *T. incarnatum* L., *T. pratense* L., *T. repens* L., *T. resupinatum* L., *T. subterraneum* L.	15.0 × 1.85
R. leguminosarum	*Vicia atropurpurea* Desf., *V. hirsuta* (L.) S.F. Gray, *V. sativa* L.	15.0 × 2.5
R. phaseoli	None available*	—
R. lupini	*Ornithopus sativus* Brot., *O. compressus* L.	15.0 × 1.85
R. japonicum	*Glycine ussuriensis* Regel and Maack	20.0 × 3.0
Rhizobium sp. (for cowpeas and symbiotically related species)	*Macroptilium atropurpureum* (DC.) Urb., *M. lathyroides* (L.) Urb., *Teramnus uncinatus* (L.) Sw.	15.0 × 2.5
Rhizobium sp. (for *Lotus* spp.)	*Lotus corniculatus* L., *L. hispidus* Desf., *L. pedunculatus* Cav.	15.0 × 1.85
Rhizobium sp. (for chickpea, *Cicer arietinum* L.)	None available*	—
Rhizobium sp. [for *Leucaena leucocephala* (Lam.) de Wit]	*Desmanthus virgatus* (L.) Willd.†	15.0 × 2.50

This list is not exhaustive.
*Suitable test plants are currently being sought in this laboratory.
†P. Davis (Chitedze Research Station, Lilongwe, Malawi), personal communication, 1978.

When the roots are 1–2 cm long, the seedlings are transferred aseptically into the test-tubes, one per tube. With agar culture, the seedling is placed pointing down with the root, particularly the growing tip, in contact with the moist agar surface. Care is needed as the root is delicate and easily damaged. A pair of long, fine forceps, lightly sprung, is ideal for this manipulation although some operators prefer an inoculating needle with the tip formed into the shape of a shepherd's crook. With vermiculite culture, the seedling is buried vertically, most conveniently against the wall of the tube, with the seed coat approximately 2 mm beneath the surface of the vermiculite. After planting, it is necessary to press the vermiculite lightly with a sterile, flat-ended tamp in order to provide a firm seed bed and uniform emergence of the seedlings. The tubes should then be irrigated (but not flooded, and taking care to avoid washing the seedling off the agar surface) with sterile, distilled water or dilute seedling solution and placed in a shaded glasshouse (Hely, 1959) or a lighted growth room.

Preparation of dilutions and inoculation of test plants. The method for counting rhizobia described in detail in this section is the fivefold serial dilution plant-infection test of Brockwell (1963).

Materials

1. Twenty-five healthy test plants, 3–5 days old; weak plants are discarded.
2. Measured amounts of diluting solution, sterilized by autoclaving in screw-capped bottles: (a) a single 90-ml volume in a 300-ml bottle; (b) six 4-ml volumes in 30-ml bottles. The diluting solution may be dilute seedling solution or physiological saline, but preferably not water because of its capacity for osmotic damage to rhizobia.
3. Pre-sterilized, straight-sided pipettes plugged with cotton-wool, calibrated to 1 ml, and with a wide tip for rapid delivery.

Procedure. Ten grams of the soil sample are suspended in the 90-ml volume of diluting solution and shaken at 200–300 cycles per minute for 10 min on a wrist-action shaker. Then 1 ml of this suspension is pipetted into a 4-ml amount of diluting solution and shaken for a further 5 min. (When diluting, it is good practice to suck the liquid being sampled up and down the pipette six times before transferring the subsample to the next dilution level; this helps to keep the suspension in motion during sampling and reduces the risk of non-uniform distribution of rhizobia within the pipette due to any tendency of bacterial cells to attach themselves to glass surfaces.) From this, using a fresh pipette each time, four 1-ml portions are used to inoculate four test plants, the fifth millilitre being used to make the next fivefold dilution in the same way, but without shaking. The series is repeated six times, thus using 25 test plants (5 plants are inoculated with the last dilution). This sequence of operations is illustrated in Figure 1. The test plants are grown for 4 weeks after inoculation, being watered if necessary during that time, and are then examined for the presence of nodules. From the proportion of plants forming nodules at each dilution level, the most probable number of rhizobia in the

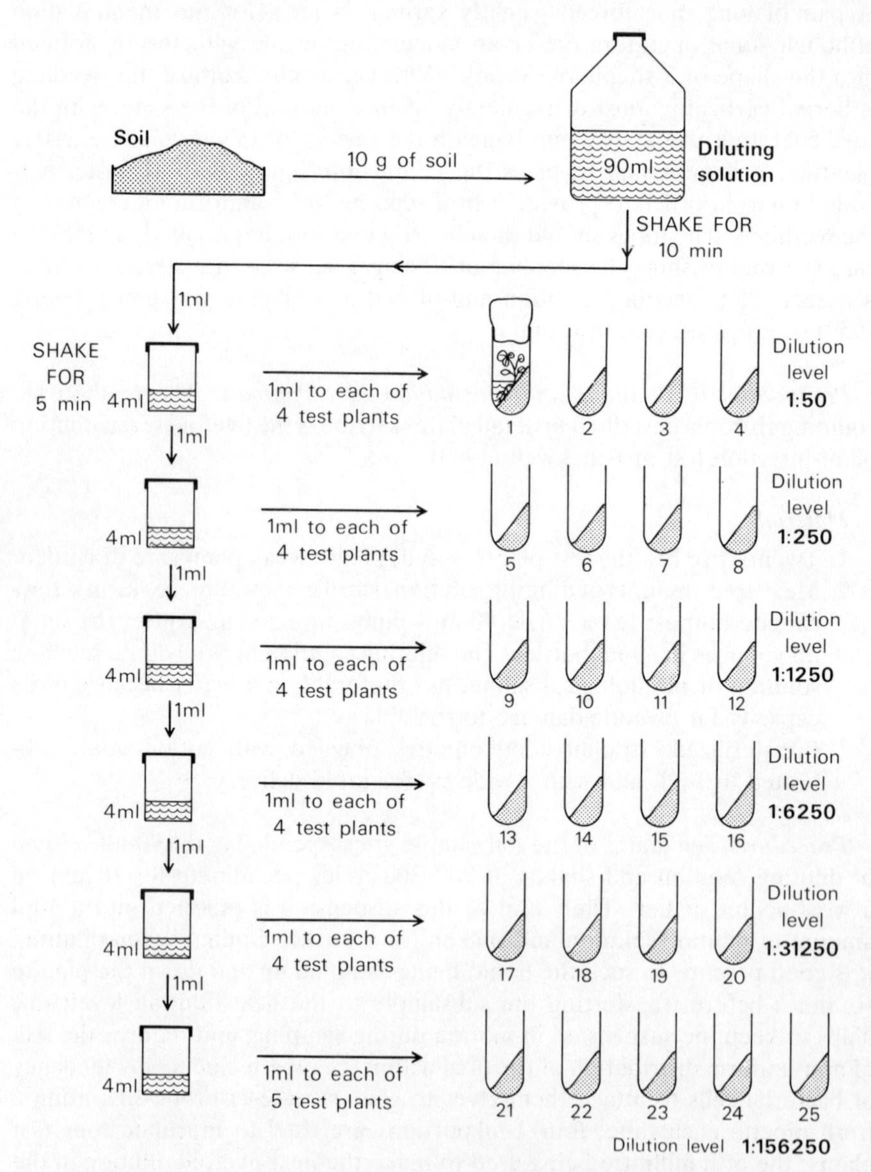

Figure 1. Sequence of operations in performing a five-fold serial dilution plant-infection test to enumerate soil populations of *Rhizobium* spp

sample is calculated using a modified version of Fisher and Yates (1963) tables—see Table 2. This method provides for a range of 1–25 200 rhizobia per millilitre of the initial soil suspension, the 95% fiducial limits being approximately +(estimate × 2.6) and −(estimate × 1/2.6). Where higher numbers are expected, additional tenfold dilutions may be inserted in the procedure before the first fivefold dilution. Where lower numbers are expected, inoculation of the test plants may be commenced using the initial 10:90 soil–diluting solution suspension, thus providing for a range of 0.2–5040 rhizobia per millilitre of the initial soil suspension; the method does not lend itself to lower numbers.

Table 2. Most probable number of nodule bacteria calculated from the distribution of positive (nodulated) test plants in a plant-infection test based on a fivefold dilution series. Initial dilution—soil : diluting solution = 10:90. (After Brockwell *et al.*, 1975a)

No. of positive (nodulated) test plants (of 4) resulting from inoculation with 1-ml aliquots						Most probable no. of nodule bacteria in original sample before any dilution	
Level of dilution of original sample						Estimate	Confidence limits (95%)
1:50	1:250	1:1250	1:6250	1:31250	1:156250*		
1	0	0	0	0	0	1.1×10^1	$0.2–7.9 \times 10^1$
2	0	0	0	0	0	2.6×10^1	$0.6–10.1 \times 10^1$
3	0	0	0	0	0	4.6×10^1	$1.5–14.1 \times 10^1$
4	0	0	0	0	0	8.0×10^1	$3.0–21.5 \times 10^1$
0	1	0	0	0	0	1.0×10^1	$0.1–7.7 \times 10^1$
1	1	0	0	0	0	2.3×10^1	$0.6–9.6 \times 10^1$
2	1	0	0	0	0	4.0×10^1	$1.2–12.8 \times 10^1$
3	1	0	0	0	0	6.5×10^1	$2.3–18.0 \times 10^1$
0	2	0	0	0	0	2.1×10^1	$0.5–9.2 \times 10^1$
1	2	0	0	0	0	3.5×10^1	$1.1–11.9 \times 10^1$
2	2	0	0	0	0	5.5×10^1	$1.9–16.0 \times 10^1$
3	2	0	0	0	0	8.7×10^1	$3.3–23.0 \times 10^1$
0	3	0	0	0	0	3.0×10^1	$0.9–10.6 \times 10^1$
1	3	0	0	0	0	4.9×10^1	$1.6–14.6 \times 10^1$
2	3	0	0	0	0	7.2×10^1	$2.7–19.6 \times 10^1$
3	3	0	0	0	0	11.3×10^1	$4.4–29.2 \times 10^1$
4	1	0	0	0	0	11.4×10^1	$4.4–29.5 \times 10^1$
4	2	0	0	0	0	16.2×10^1	$6.2–42.4 \times 10^1$
4	3	0	0	0	0	24.2×10^1	$9.0–64.9 \times 10^1$
4	4	0	0	0	0	40.4×10^1	$15.3–106.6 \times 10^1$
4	0	1	0	0	0	10.8×10^1	$4.2–28.1 \times 10^1$
4	1	1	0	0	0	15.1×10^1	$5.8–39.2 \times 10^1$
4	2	1	0	0	0	21.5×10^1	$8.1–57.4 \times 10^1$

Table 2. (*cont.*)

No. of positive (nodulated) test plants (of 4) resulting from inoculation with 1-ml aliquots						Most probable no. of nodule bacteria in original sample before any dilution	
Level of dilution of original sample						Estimate	Confidence limits (95%)
1:50	1:250	1:1250	1:6250	1:31250	1:156250*		
4	3	1	0	0	0	32.8×10^1	$12.2–87.9 \times 10^1$
4	0	2	0	0	0	14.1×10^1	$5.4–36.6 \times 10^1$
4	1	2	0	0	0	19.6×10^1	$7.4–51.9 \times 10^1$
4	2	2	0	0	0	28.3×10^1	$10.5–76.1 \times 10^1$
4	3	2	0	0	0	43.6×10^1	$16.6–114.2 \times 10^1$
4	0	3	0	0	0	18.1×10^1	$6.9–47.7 \times 10^1$
4	1	3	0	0	0	25.2×10^1	$9.4–67.6 \times 10^1$
4	2	3	0	0	0	36.4×10^1	$13.7–96.8 \times 10^1$
4	3	3	0	0	0	56.5×10^1	$21.9–146.0 \times 10^1$
4	4	1	0	0	0	5.7×10^2	$2.2–14.7 \times 10^2$
4	4	2	0	0	0	8.1×10^2	$3.1–21.2 \times 10^2$
4	4	3	0	0	0	12.1×10^2	$4.5–32.4 \times 10^2$
4	4	4	0	0	0	20.2×10^2	$7.6–53.3 \times 10^2$
4	4	0	1	0	0	5.4×10^2	$2.1–14.0 \times 10^2$
4	4	1	1	0	0	7.5×10^2	$2.9–19.6 \times 10^2$
4	4	2	1	0	0	10.8×10^2	$4.0–28.7 \times 10^2$
4	4	3	1	0	0	16.4×10^2	$6.1–43.9 \times 10^2$
4	4	0	2	0	0	7.1×10^2	$2.7–18.3 \times 10^2$
4	4	1	2	0	0	9.8×10^2	$3.7–26.0 \times 10^2$
4	4	2	2	0	0	14.1×10^2	$5.3–38.1 \times 10^2$
4	4	3	2	0	0	21.8×10^2	$8.3–57.1 \times 10^2$
4	4	0	3	0	0	9.1×10^2	$3.4–23.8 \times 10^2$
4	4	1	3	0	0	12.6×10^2	$4.7–33.8 \times 10^2$
4	4	2	3	0	0	18.2×10^2	$6.9–48.4 \times 10^2$
4	4	3	3	0	0	28.2×10^2	$10.9–73.0 \times 10^2$
4	4	4	1	0	0	2.9×10^3	$1.1–7.3 \times 10^3$
4	4	4	2	0	0	4.1×10^3	$1.6–10.6 \times 10^3$
4	4	4	3	0	0	6.0×10^3	$2.3–16.2 \times 10^3$
4	4	4	4	0	0	10.1×10^3	$3.8–26.6 \times 10^3$
4	4	4	0	1	0	2.7×10^3	$1.0–7.0 \times 10^3$
4	4	4	1	1	0	3.8×10^3	$1.5–9.8 \times 10^3$
4	4	4	2	1	0	5.4×10^3	$2.0–14.4 \times 10^3$
4	4	4	3	1	0	8.2×10^3	$3.1–22.0 \times 10^3$
4	4	4	0	2	0	3.5×10^3	$1.4–9.2 \times 10^3$
4	4	4	1	2	0	4.9×10^3	$1.8–13.0 \times 10^3$
4	4	4	2	2	0	7.1×10^3	$2.6–19.0 \times 10^3$
4	4	4	3	2	0	10.9×10^3	$4.2–28.6 \times 10^3$
4	4	4	0	3	0	4.5×10^3	$1.7–11.9 \times 10^3$
4	4	4	1	3	0	6.3×10^3	$2.3–16.9 \times 10^3$

Table 2. (*cont.*)

No. of positive (nodulated) test plants (of 4) resulting from inoculation with 1-ml aliquots						Most probable no. of nodule bacteria in original sample before any dilution	
Level of dilution of original sample						Estimate	Confidence limits (95%)
1:50	1:250	1:1250	1:6250	1:31250	1:156250*		
4	4	4	2	3	0	9.1×10^3	$3.4–24.2 \times 10^3$
4	4	4	3	3	0	14.1×10^3	$5.4–36.7 \times 10^3$
4	4	4	4	1	0	14.3×10^3	$5.5–36.9 \times 10^3$
4	4	4	4	2	0	20.3×10^3	$7.8–53.0 \times 10^3$
4	4	4	4	3	0	30.2×10^3	$11.2–81.3 \times 10^3$
4	4	4	4	4	0	50.5×10^3	$19.0–133.8 \times 10^3$
4	4	4	4	0	1	13.5×10^3	$5.2–35.3 \times 10^3$
4	4	4	4	1	1	18.8×10^3	$7.2–49.0 \times 10^3$
4	4	4	4	2	1	26.9×10^3	$10.1–71.8 \times 10^3$
4	4	4	4	3	1	41.0×10^3	$15.3–110.2 \times 10^3$
4	4	4	4	0	2	17.7×10^3	$6.8–45.9 \times 10^3$
4	4	4	4	1	2	24.5×10^3	$9.2–65.0 \times 10^3$
4	4	4	4	2	2	35.3×10^3	$13.1–95.4 \times 10^3$
4	4	4	4	3	2	54.4×10^3	$20.6–143.8 \times 10^3$
4	4	4	4	0	3	22.6×10^3	$8.6–59.7 \times 10^3$
4	4	4	4	1	3	31.4×10^3	$11.7–84.7 \times 10^3$
4	4	4	4	2	3	45.5×10^3	$17.0–121.4 \times 10^3$
4	4	4	4	3	3	70.6×10^3	$27.1–184.2 \times 10^3$
4	4	4	4	4	1	7.1×10^4	$2.7–18.6 \times 10^4$
4	4	4	4	4	2	10.1×10^4	$3.8–27.0 \times 10^4$
4	4	4	4	4	3	15.1×10^4	$5.4–42.6 \times 10^4$
4	4	4	4	4	4	25.2×10^4	$8.6–74.0 \times 10^4$
4	4	4	4	4	5	$>35.5 \times 10^4$	

*Five test plants inoculated with 1-ml aliquots from this dilution level.

A tenfold serial dilution plant infection test can be used when a lower accuracy is required. The method of preparation of the material for examination is similar to that used for fivefold dilutions. From each of six tenfold (1 ml into 9 ml) serial dilutions (not counting the initial dilution—soil: diluting solution = 10:90), three 1-ml portions are used to inoculate three test plants; 18 test plants are required for each series. The most probable number of nodule bacteria in the original sample is calculated from the proportion of test plants forming nodules at each dilution level (see Table 3). It gives a range of 4–1 099 000 organisms per millilitre of the initial soil suspension, the 95% fiducial limits being approximately +(estimate × 3.8) and −(estimate × 1/3.8).

Table 3. Most probable number of nodule bacteria calculated from the distribution of positive (nodulated) test plants in a plant-infection test based on a tenfold dilution series. Initial dilution—soil : diluting solution = 10 : 90. (After Brockwell *et al.*, 1975a)

No. of positive (nodulated) test plants (of 3) resulting from inoculation with 1-ml aliquots						Most probable no. of nodule bacteria in original sample before any dilution	
Level of dilution of original sample						Estimate	Confidence limits (95%)
1 : 10^2	1 : 10^3	1 : 10^4	1 : 10^5	1 : 10^6	1 : 10^7		
1	0	0	0	0	0	0.4×10^2	$0.1–2.5 \times 10^2$
2	0	0	0	0	0	0.9×10^2	$0.2–3.7 \times 10^2$
2	1	0	0	0	0	1.5×10^2	$0.4–5.2 \times 10^2$
3	0	0	0	0	0	2.3×10^2	$0.7–8.0 \times 10^2$
3	1	0	0	0	0	4.2×10^2	$1.0–17.2 \times 10^2$
3	2	0	0	0	0	9.2×10^2	$2.3–36.7 \times 10^2$
3	2	1	0	0	0	14.7×10^2	$4.1–52.1 \times 10^2$
3	3	0	0	0	0	23.0×10^2	$6.6–80.3 \times 10^2$
3	3	1	0	0	0	42.4×10^2	$10.4–172.5 \times 10^2$
3	3	2	0	0	0	91.8×10^2	$22.9–367.2 \times 10^2$
3	3	2	1	0	0	14.7×10^3	$4.1–52.1 \times 10^3$
3	3	3	0	0	0	23.0×10^3	$6.6–80.4 \times 10^3$
3	3	3	1	0	0	42.4×10^3	$10.4–172.7 \times 10^3$
3	3	3	2	0	0	91.9×10^3	$23.0–367.7 \times 10^3$
3	3	3	2	1	0	14.7×10^4	$4.1–52.2 \times 10^4$
3	3	3	3	0	0	23.0×10^4	$6.6–80.8 \times 10^4$
3	3	3	3	1	0	42.7×10^4	$10.4–175.2 \times 10^4$
3	3	3	3	2	0	93.3×10^4	$23.2–375.1 \times 10^4$
3	3	3	3	2	1	14.9×10^5	$4.2–53.6 \times 10^5$
3	3	3	3	3	0	24.0×10^5	$6.7–85.7 \times 10^5$
3	3	3	3	3	1	46.2×10^5	$10.4–205.0 \times 10^5$
3	3	3	3	3	2	109.9×10^5	$25.7–470.5 \times 10^5$
3	3	3	3	3	3	$>240.0 \times 10^5$	

Appraisal of plant-infection tests in tubes. Experience with these techniques in a number of laboratories has shown that the results obtained are readily reproducible. Thus, replicate counts of a single sample are unnecessary; it is better to count replicate samples.

When the test plants are grown in agar, the plant-infection test, as described here, is an accurate method for estimating most probable numbers of *Rhizobium* spp. in either pure culture or in the presence of soil (Brockwell, 1963), and gives results closely similar to those obtainable by direct counting. The rhizosphere influence of the test plant apparently extends throughout the agar medium and enables a single added rhizobial cell to initiate nodulation.

In vermiculite culture, on the other hand, the rhizosphere effect appears to be restricted and a proportion of the introduced rhizobial cells apparently do not reach the rhizosphere. This leads to an underestimation of the population of *Rhizobium* sp. being counted. Working with *R. trifolii*, Brockwell (1963) found this underestimation to be 39%, but it varies according to the test species and to the moisture status at which the vermiculite substrate is maintained (Grassia and Brockwell, 1978). Nonetheless, since the underestimation remains constant for each particular set of experimental conditions, the results obtained are reproducible and may, therefore, be regarded as relative. If quantitative estimates of population size are required, it is necessary to conduct preliminary experiments to determine the factor by which numbers are underestimated in a standard plant-infection test for each particular group of rhizobia (Grassia and Brockwell, 1978).

A complication that sometimes arises with plant-infection tests is the occurrence of 'skips', i.e. low dilution—negative (test plant without nodules), higher dilution—positive (with nodules) (Halvorsen and Ziegler, 1933). This presents a difficulty in using the most-probable-number tables which do not provide for such non-random occurrences. These 'skips' are indicative of unfavourable conditions for nodulation within the tube, due either to failure of the rhizobia to survive or multiply, to poor health of the test plant resulting from poor environmental conditions, or to interference from algae or saprophytic fungi. 'Skips' usually occur in counting small populations of rhizobia when it is necessary to introduce relatively large amounts of soil (>0.1 g) into the tube and they are more common on agar culture than in vermiculite. It is better to use an alternative method for estimating the size of small populations (see below).

Plant-infection tests in larger assemblies. For enumerating *Rhizobium* spp. such as *R. phaseoli* for which no satisfactory small-seeded test plant is available, Leonard (1943) jar assemblies are suitable. Their preparation is described in detail in Chapter II.4. Vermiculite makes an ideal substrate for the test plants.

If desired, the fivefold (25 test plants) or tenfold (18 test plants) dilution series using tube culture, described above, can be adapted for larger assemblies. However, because Leonard jars are larger, and their preparation is more time consuming, it is usual to have fewer test plant units. For instance, Date and Vincent (1962) propose four tenfold dilutions with test plants in duplicate [8 test plants, 95% fiducial limits approximately $+(\text{estimate} \times 6.0)$ and $-(\text{estimate} \times 1/6.0)$] or quadruplicate [16 test plants, 95% fiducial limits approximately $+(\text{estimate} \times 3.5)$ and $-(\text{estimate} \times 1/3.5)$]. This arrangement provides for a range of 0.6–690 rhizobia per millilitre of the lowest dilution

(Table 4). If the rhizobial population of the sample under examination cannot be estimated within these limits, additional tenfold dilutions and test plants can be included.

Table 4. Determination of most probable number (MPN) of rhizobia using a plant-infection test with a tenfold dilution series and test plants in duplicate or quadruplicate. (After Date and Vincent, 1962)

No. of positive units of 4 dilutions when tested in:		MPN per ml of the lowest dilution*
Duplicate	Quadruplicate	
8	16	>700
	15	
7	14	690
	13	340
6	12	180
	11	100
5	10	59
	9	31
4	8	17
	7	10
3	6	5.8
	5	3.1
2	4	1.7
	3	1.0
1	2	0.58
0	1,0	<0.5

*Calculated from Fisher and Yates (1963).

Seed preparation and preparation of dilutions are as described for plant infection tests in tube culture. There are, however, two modifications to that technique that apply to inoculation of test plants. Firstly, inoculation commences with 1-ml portions of the initial tenfold dilution of sample in diluting solution; secondly, all 1-ml portions used for inoculating test plants should be suspended in 10 ml of sterile seedling solution (not saline) and this suspension poured into the vermiculite around the root of the test plant. The reason for doing this is to reduce the slight risk that a 1-ml portion may become 'lost' when inoculating a small plant growing in a relatively large volume of substrate. Finally, the vermiculite surface is covered to a depth of 2 cm with a layer of sterilized dry gravel, waxed sand, or polyethylene mulch, to prevent contamination from the air, and the jars put into a growth room or glasshouse. Date and Vincent (1962) claim to have experienced no difficulty in maintaining uninoculated controls free from rhizobia and nodulation under dusty conditions. Four weeks after inoculation, the test plants may be removed from the jars and examined for the presence of nodules. The most probable number of rhizobia can then be calculated by reference to Table 4.

Counting samples with very low populations of rhizobia. The test-tube culture version of the plant infection test becomes inappropriate once soil populations of rhizobia fall below approximately 2.0 per gram. It is then necessary to use a modification of the Leonard jar technique above. The assemblies are sterilized without the vermiculite substrate. Increasing quantities of the soil under examination are mixed with suitable amounts of sterile, dry vermiculite (0.5, 1.0, 2.0, 4.0, 8.0, 16, 32, 64 g make a satisfactory progression), taking care to prevent contamination with rhizobia from outside sources. The mixtures are then packed into the jars, irrigated with seedling solution, planted with pre-germinated seedlings of the chosen test plant, covered with half a Petri dish, and placed into a growth room or glasshouse. When the seedlings have emerged sufficiently, the Petri dish is removed and the surface of the substrate covered with an anti-contamination layer.

The plants should be grown for at least 10 weeks to allow a sufficient period for the roots to explore most, if not all, of the substrate and encourage nodulation even when a single rhizobial cell is located in a 'remote corner' of the Leonard jar. During that time it will probably be necessary to replenish the seedling solution in the reservoir section of the assembly at least once. The opportunities for contamination increase greatly during such a long growing period and it is necessary, therefore, to have uninoculated control plants (two for each set of eight inoculated plants) as part of the experiment. Should the controls bear nodules when the test plants are harvested, the experiment must be treated as suspect. At best, results obtained using this method can only be regarded as approximate.

Direct Counts

Direct counts of legume nodule bacteria are feasible when the population of rhizobia of the sample equals or exceeds that of other microorganisms, for instance, with heavily inoculated legume seed.

Plate counts. Glassware and procedure for diluting should meet the standards of the American Public Health Association (1955). The sample is diluted as described above and 1-ml portions of an appropriate dilution (calculated to contain 30–300 cells of rhizobia per millilitre) pipetted into Petri dishes and mixed with 14 ml of cooled, molten yeast extract mannitol agar (Fred *et al.*, 1932) containing cycloheximide (50 $\mu g\ ml^{-1}$). At this concentration, cycloheximide does not influence the growth of *Rhizobium* in any detectable way, and minimizes the growth of fungal species. The addition of aqueous congo red (0.25% w/v) to the medium (1 $ml\ l^{-1}$) will assist in distinguishing colonies of rhizobia (Hahn, 1966). Petri dishes are incubated for 5 days (for fast growers) or 10–14 days (for slow growers) at 26°C in an inverted position. The number of colonies present may then be counted with the aid of magnification on a dark-field colony counter. Purchase and Nutman (1957) have shown that total (haemocytometer) counts and viable (plate) counts can be closely comparable.

Selective medium for rhizobia. Graham (1969) described a medium selective for strains of *R. leguminosarum*, *R. meliloti*, and *Rhizobium* sp. with the composition mannitol 5.0, lactose 5.0, K_2HPO_4 0.5, NaCl 0.2, $CaCl_2 \cdot 2H_2O$ 0.2, $MgSO_4 \cdot 7H_2O$ 0.1, $FeCl_3 \cdot 6H_2O$ 0.1, yeast extract 0.5, and agar 20.0 g l^{-1}. After autoclaving the above ingredients, the pH is adjusted to 7.0 aseptically. Cycloheximide (200 mg), pentachloronitrobenzine (100 mg), sodium benzylpenicillin (25 mg), chloromycetin (10 mg), sulphathiazole (25 mg), and neomycin (2.5 mg), sterilized by passage through a millipore filter, are then added. Congo red (1 ml of a 1% solution) may also be added if desired. The medium should be poured immediately following addition of the antibiotic drugs because some of them are heat-labile and may lose activity if the medium is allowed to set and then re-melted.

Graham (1969) states that this medium can be used directly to count populations of rhizobia in soils growing leguminous species. Reports of the use of the medium are insufficient in number to allow proper appraisal of its value.

(iii) Assessment of symbiotic properties

It has been asserted above (Section b, preamble) that, in practice, one observes only competition in nodule formation which occurs between strains of *Rhizobium* of the same species, and that effective naturally occurring strains are generally, although not always, more competitive with effective inocula than ineffective naturalized strains. Therefore, the interpretation of results of field inoculation experiments is aided by assessment of the effectiveness of naturally occurring strains. By and large, methods for measuring nitrogen fixation in *Rhizobium* strain/host legume associations have been thoroughly dealt with in Chapter II.4. However, some aspects of the subject deserve consideration here.

Choice of test host

There is evidence that legumes, especially *Trifolium* spp., tend to select effective strains of rhizobia for nodulation from a population of mixed effectiveness (Singer *et al.*, 1964; Robinson, 1969a, b; Masterson and Sherwood, 1974). It becomes important, therefore, to choose the correct test host for assessing the effectiveness of soil populations of rhizobia (Sherwood and Masterson, 1974). Generally, the plant line intended for the field experiment should be used for evaluating symbiotic effectiveness of the naturally occurring field population.

When counting field populations of rhizobia prior to field experimentation, employing the same plant line for both plant-infection tests and for the experiment itself, the enumeration test plants can be used to make a preliminary evaluation of the effectiveness of the naturally occurring rhizobia. The positive (nodulated) test plants are assessed visually or by measuring their oven dry weight or total nitrogen content in comparison with both uninoculated control plants and plants inoculated with an effective inoculum strain. It must be stressed, however, that this evaluation is only qualitative and approximate. Other soil microorganisms in the soil dilutions may interfere with the expression of nitrogen fixation; also, the response of test plants to nodulation, especially at low dilution levels, may or may not represent a synthesis of the effectiveness of several different strains of rhizobia (Mytton and de Felice, 1977).

Choice of field isolates

The method of selecting the rhizobial isolate (the term 'isolate' is defined below in Section d.i) for testing is equally important for similar reasons. For instance, Lowe and Holding (1970) found that some naturally occurring *R. trifolii* were more effective on indigenous white clover than they were on commercial lines. Accordingly, nodulated plants of the line intended for the field experiment should be used for isolation. Assuming such plants are not already growing at or near the experimental site, nodules may be taken either from nodulated test plants used for counting the soil population of rhizobia or from plants grown (as trap-hosts) in the soil for the particular purpose of providing nodules for isolation. Methods employed for isolation are described below (Section d.ii). Ten isolates are a minimum sample for estimating the effectiveness of the naturally occurring strains likely to compete most intensively in the field with inocula. The results should be treated with some caution even when larger samples of isolates are examined. The properties of soil populations of rhizobia are dynamic and significant changes in symbiotic effectiveness occur from time to time and from place to place even within a short distance (Gibson *et al.*, 1975). It is probably fair to say that there is no entirely satisfactory way to evaluate accurately the nitrogen-fixing capacity of a mixed population of rhizobia.

Symbiotic properties for strain identification

Three symbiotic criteria, nodulation, nitrogen fixation, and level of effectiveness, can be employed for strain identification. A few examples are given below, but the use of these criteria is not, of course, limited to the species mentioned and the principles are applicable to very many host plant/*Rhizobium* strain combinations.

Nodulation. Lotononis bainesii Baker is a leguminous herb whose curious red-pigmented, highly specific rhizobia almost certainly do not occur naturally outside the habitat of the host 'in south central Africa in the interior near the Tropic of Capricorn' (Oliver, 1868). Norris (1958) found no sign of nodulation when *L. bainesii* was inoculated with 59 strains of *Rhizobium* spp. derived from 31 species in 18 diverse genera. Thus, field reisolations of a strain of *L. bainesii* rhizobia introduced into coastal Queensland, for instance, can be recognized by the mere fact of ability to nodulate and fix nitrogen with *L. bainesii* (Diatloff 1977). The distinctive red pigmentation of colonies of this *Rhizobium* sp. can also be used as a means of its identification.

Nitrogen fixation. Many strains of rhizobia possess some symbiotic feature(s), which distinguish them from other strains and which can be employed for strain identification. For instance, Caldwell (1966) found that ineffective nodulation in soybean cv. Hardee was conditioned by a single dominant gene 'Rj_2' originating in the parent line CNS. The ineffectiveness manifested itself in associations between Hardee and *R. japonicum* strain CB1809 and other strains of the 122 serogroup (Caldwell *et al.*, 1966). Analogous situations have been described by Sloger (1969), Vest (1970), and Vest and Caldwell (1972). This aberrancy can be used to identify field isolates of CB1809. Test plants of Hardee and another soybean cultivar not possessing the Rj_2 gene, say Williams, are inoculated with the isolates. Strain CB1809 (and other strains of serogroup 122) will be ineffective with Hardee but effective with Williams; other strains will be effective with both plant lines (Table 5).

Another example in which two symbiotic features can be utilized for strain identification is *R. trifolii* NA30. This former Australian inoculant strain (Date, 1969) has the slightly unusual characteristic (Strong, 1937) of effectiveness with both subterranean clover (*Trifolium subterraneum* L.) and white clover (*Trifolium repens* L.). In addition, it is unusual in that it is ineffective with subterranean clover cv. Howard (Gibson, 1964). An appropriate symbiotic test employing both these features is illustrated in Table 5.

Two *R. meliloti* strains, once used in Australian inoculants, represent a third example. Strains U45 and SU47 differ from other *R. meliloti* in that they fix nitrogen with spotted medic [*Medicago arabica* (L.) Huds.] and barrel medic (*Medicago truncatula* Gaertn.) but not with King Island melilot [*Melilotus indica* (L.) All.] (Brockwell and Hely, 1966). These characteristics have been used to identify field reisolates (Brockwell, 1971; see also Table 5).

Level of effectiveness. The symbiotic character, level of effectiveness, has

Table 5. Examples of the use of nitrogen-fixing ability as a characteristic for the identification of strains of *Rhizobium*

Strain(s) of *Rhizobium*	Unusual feature(s)	Reference	Symbiotic test (+ = effective, − = ineffective)				
			Test plant	Strain response			
R. japonicum CB1809	Ineffective with *G. max* cv. Hardee	Diatloff and Brockwell (1976)		CB1809	Unknown strains		
			G. max cv. Hardee	−	+		
			G. max cv. Williams	+	+		
R. trifolii NA30	(i) Effective with both *T. subterraneum* and *T. repens*	Gibson (1964)		NA30	Unknown A	Unknown B	Unknown C
			T. subterraneum cv. Howard	−	+	+	−
	(ii) Ineffective with *T. subterraneum* cv. Howard		*T. subterraneum* cv. Tallarook	+	+	+	−
			T. repens	+	+	−	+
R. meliloti U45 and SU47	Effective with both *M. arabica* and *M. truncatula* but ineffective with *Mel. indica*	Brockwell and Hely (1966)		U45 & SU47	Unknown X	Unknown Y	Unknown Z
			M. truncatula	+	+	+	−
			M. arabica	+	−	+	−
			Mel. indica	−	−	+	−

been little used for strain identification. One such attempt, however, has been described by Brockwell (1971) for distinguishing between field reisolates of effective inoculum strains for the strain specific species, *Medicago rugosa* Desr. (Paragosa medic) (Brockwell and Hely, 1966) and naturally occurring *R. meliloti. M. rugosa* test plants were grown in 15 × 2.5 cm specimen tubes filled with vermiculite moistened with nitrogen-free nutrient solution. In order to set standards, 20 plants were inoculated with each inoculum strain. Two plants were inoculated with each isolate from the field. Twenty-eight days after inoculation, a visual rating for vigour of growth was made on a 0–4 basis (Brockwell and Hely, 1961). The mean rating for each inoculum strain was compared with the mean rating for each isolate obtained from the homologous inoculation treatment in the field. If the difference between the two ratings was less than one rating unit (e.g. 3.7 for the inoculum strain, mean of 20 plants; and 2.8 for the isolate from the field plot inoculated with that strain, mean of two plants), the isolate was regarded as symbiotically similar to the inoculum.

Appraisal of symbiotic properties as a means of strain identification

The precision of these methods of strain identification appears to depend on the degree of specificity of the *Rhizobium* strain with the legume test plant. Where the symbiotic property is almost unique as with highly specific host/strain combinations such as *Lotononis bainesii* and its rhizobia, results are accurate and consistently reliable (Diatloff, 1977). Where the property is merely uncommon as with *Trifolium subterraneum* cv. Howard and *R. trifolii* strain NA30, the method is useful but less reliable. For instance, the data in Table 6 illustrate a failure of symbiotic identification to distinguish between NA30 and unrelated naturally occurring strains. It was necessary to use alternative means of identification, *viz*. serology (Dudman and Brockwell, 1968; see also Chapter II.9) to resolve the confusion. Identification of *Rhizobium* strains based on level of effectiveness with a particular host is only a qualitative method and even less reliable. Positive results can only be described as 'symbiotic similarity'.

Undoubtedly, symbiotic identification has a place in ecological studies of the rhizobia. However, the method works better in bacteriologically controlled systems, such as test-tube culture, where the properties of competing strains are known, than with samples from the field, where some elements of the naturally occurring population may be confused symbiotically with inoculum strains. In appraising several methods for identification of *Rhizobium* strains, Brockwell *et al.* (1978) concluded that symbiotic markers, even highly specific ones, were tedious to use; less specific ones, requiring the use of more than one host species, were very time consuming indeed.

Table 6. Serological and symbiotic characteristics of 27 *Rhizobium trifolii* strains isolated from *Trifolium subterraneum* inoculated with NA30 and sown at Murrumbateman, New South Wales. (After Brockwell *et al*., 1978)

No. of field isolates	Serological identity with NA30	Effectiveness of *T. subterraneum*	
		cv. Howard	cv. Bacchus Marsh*
10	+	−	+
10	−	+	+
2	−	−	−
5	−	−	+
Control = NA30	+	−	+

*Symbiotically similar to *T. subterraneum* cv. Tallarook.

(c) HOST PLANTS FOR FIELD EXPERIMENTS

Whatever the nature of an experiment, the importance of using high-quality seed that is true to type cannot be overemphasized. Some means of obtaining first-class seed and maintaining it in that condition are discussed in this section. Barton's *Bibliography of Seeds* (1967) is a prolific source of general information.

(i) Source of seed

Seed for experimental purposes can be obtained by production of one's own requirements, purchase of commercial lines from seed suppliers, collection of undomesticated lines from the field, or by acquisitions from colleagues, other laboratories, or botanic gardens.

Production of one's own requirements

One usually reliable source of seed is that which the experimenter has grown himself. Many workers produce their own seed to ensure continuity of supply for long-term experiments: commercial lines are apt to go out of fashion (e.g. soybean varieties); there is often no commercial source of undomesticated species needed for special purposes (e.g. *Glycine ussuriensis* Regel and Maack for enumerating *R. japonicum*). Seed of many self-fertilized species can be produced in pots of soil in a glasshouse or cold-frame. Other species may need to be grown in field plots. In this case, most economic use of limited seed supplies is obtained by 'starting' the seedlings in a glasshouse and, when they are well established, transplanting them into the field as spaced plants. The soil should be pre-treated with a pre-emergence herbicide. A layer of black polyethylene sheeting spread between the plants will control most subsequent weed growth but the plots should also be hand-weeded to eliminate any possibility of weed seeds contaminating the harvest. Plants may need protection from insect attack.

Purchase of commercial lines

When ordering seed from commercial sources, it is important to specify that the seed be certified. Many countries, or states within countries, operate seed certification schemes. One of these is described in detail by Cowan (1972). The Organization for Economic Cooperation and Development (OECD) also has an international certification programme for herbage and forage crops, cereals, tree seeds, and certain other seeds (Cowan, 1972). Each certifying agency has standards that define minimum requirements for certified seed as determined by seed inspection. The standards usually require that seeds have a germination capacity of at least 80% and maximum tolerances are given for impurities, such as weed seeds, seeds of other kinds, and inert matter, which may be different for each crop. The label on a seed container (Figure 2) is of prime significance since it indicates to the buyer that the seed lot has met the requirements of certification. The label describes the species and variety or cultivar and provides an identification number relating to the grower, seed lot, and point of origin. Nevertheless, weed seeds, especially those of related species and varieties, can occur in nuisance quantities in certified seed, more commonly with pasture species. Accordingly, field plots sown with seed from commercial sources should be examined at intervals and 'strangers' removed.

Collection of undomesticated lines

For ecological studies of naturally occurring rhizobia, it is often more convenient to collect seed of undomesticated species oneself than to rely on other sources. Where possible, seed should be collected from a single representative plant or group of plants growing in close proximity. For reasons of ecotypic difference, it is never desirable to mix seed samples collected in different localities or from markedly different habitats. Field data should be recorded and a herbarium specimen or photograph made. Smith (1971) gives details about collecting plants and recording field data. Corby (1970) advises the making of a reference collection of legume specimens. Seeds should be collected when they are ripe and kept in paper (not plastic) envelopes until thoroughly dry. Many collectors like to spray their seed with an insecticide. If practicable, this practice should be avoided with seeds intended for rhizobial studies as rhizobia are very often incompatible with insecticides (e.g. Brockwell and Robinson, 1976). Gunn (1972) describes seed collection procedures in great detail.

The collection of soil samples or nodulated root material for isolation of rhizobia can often be done most conveniently at the same time as seed is collected (see Section d.i).

Acquisition from diverse sources

It is frequently possible to obtain small amounts of seed from colleagues or other laboratories at home or abroad. Plant breeders, in particular, often maintain stocks of seed of excellent quality. Some countries have National Seed Storage Laboratories [e.g. U.S.A. at Fort Collins, Colo. (James, 1972); Japan at Hiratsuka, Kanagawa (Ito, 1972)] which operate seed exchange services or are able to advise about sources and availability. Botanic Gardens may also be prepared to supply samples of seed. With the latter organizations, however, it is desirable to check the authenticity of any seed received. It

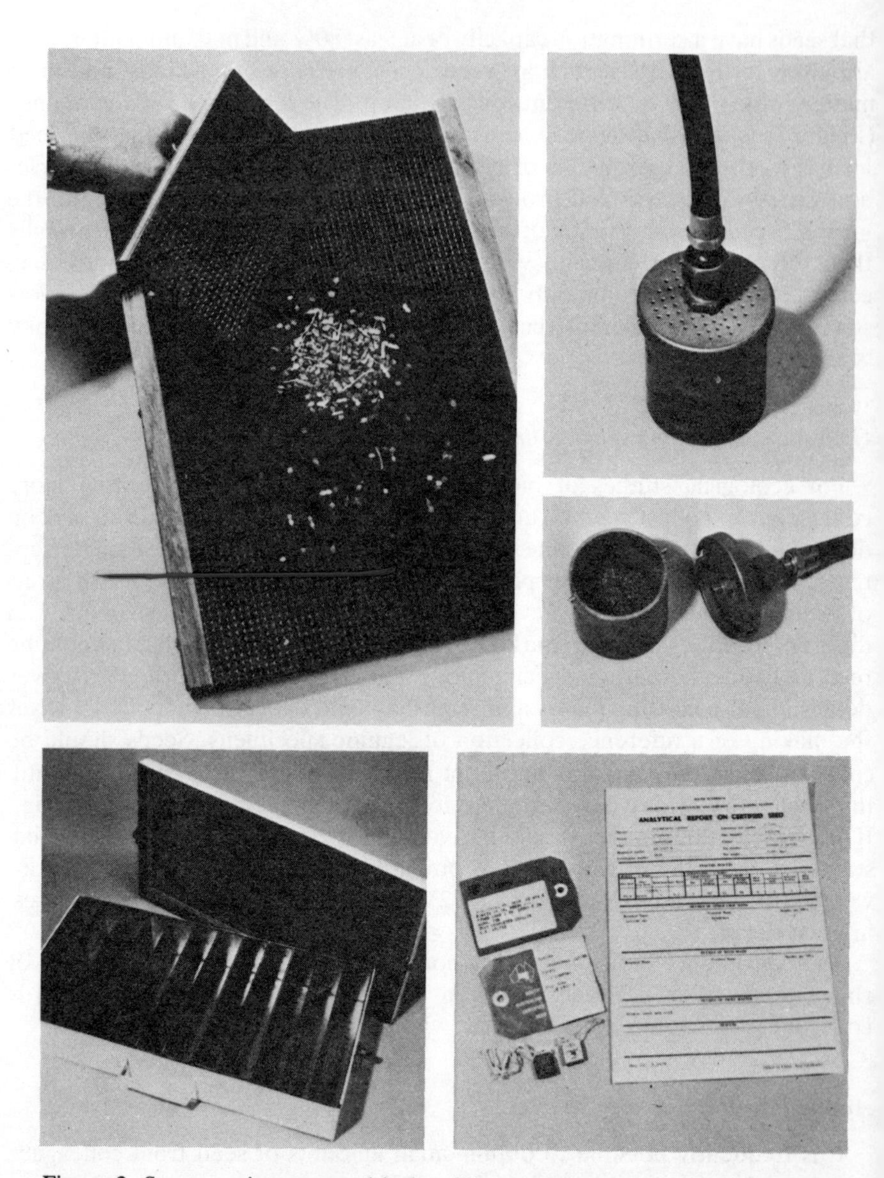

Figure 2. Some equipment used in handling seed: top left, rubbing board for breaking open hard pods; bottom left, seed storage box (divided into compartments) with snap-on, gasketed lid; top right and centre right, two views of a seed scarifier (operated by compressed air); bottom right, documentation for seed certification, showing label (back and front views), validation tag (back and front views), and certificate

sometimes happens that Gardens which are handicapped by manpower shortages have trouble in maintaining their nurseries in a weed-free condition. For example, in this laboratory, when acquiring seed from Botanic Gardens, we have encountered samples of *Trifolium ambiguum* Bieb. (Caucasian clover) heavily contaminated with the morphologically similar *Trifolium repens*.

Many countries, e.g. New Zealand and Australia, have stringent quarantine regulations applying to the importation of plant material. It is advisable, therefore, when importing seeds into such countries first to contact plant quarantine authorities for information about how best to address incoming samples in order to expedite quarantine formalities. The sender should be given this information. Likewise, when dispatching seeds overseas, the package must be labelled and addressed in a manner that meets the customs and quarantine requirements of the recipient country.

(ii) Cleaning, storing, and testing seed

Acquisition of high-quality authenticated seed is a sufficiently time-consuming operation to make it well worthwhile maintaining one's stocks in first class condition.

Seed cleaning

Newly harvested seed must be cleaned to separate sound seed from trash and broken or light seed. Only seeds of high density should be kept for long storage as the evidence indicates that seeds of greatest longevity are those with the greatest density (Harrington, 1972). Trash, such as plant refuse and soil, and broken seeds may harbour insects and fungi. Proper cleaning and handling procedures for seeds are well documented by Harmond *et al.* (1968). Particular care may be necessary with some samples, especially large-seeded species, to avoid damage during mechanical harvesting, cleaning, and dehulling operations. A hard, dimpled rubber, hand-operated 'rubbing' board for breaking open hard pods is shown in Figure 2.

Seed storage

It has long been known that certain hard-seeded species, particularly in the Leguminosae, can survive for very many years (Ewart, 1908). However, storage becomes more complex as severity of climate and length of storage increase.

In most cases legume seeds are dry at harvest. However, those that reach physiological maturity in times or places of high relative humidity will need further drying in a warm, dry atmosphere. Too high a temperature or too rapid drying may decrease germination and damage seed. Naturally, seed intended for storage must be free from insects. If it is necessary to use an insecticide, this should be clearly indicated on the storage container, otherwise problems due to incompatibility between rhizobia and insecticide may arise during experimentation.

The two main factors influencing seed longevity are seed moisture and storage temperature. Two generally valid rules apply (Harrington, 1959):

1. For each 1% increase in seed moisture (between about 5 and 14%) the life of the seed is halved.
2. For each 5 °C increase in temperature (between about 0 and 50 °C) the life of the seed is halved.

Figure 3 demonstrates some hazards associated with storage conditions outside the optima.

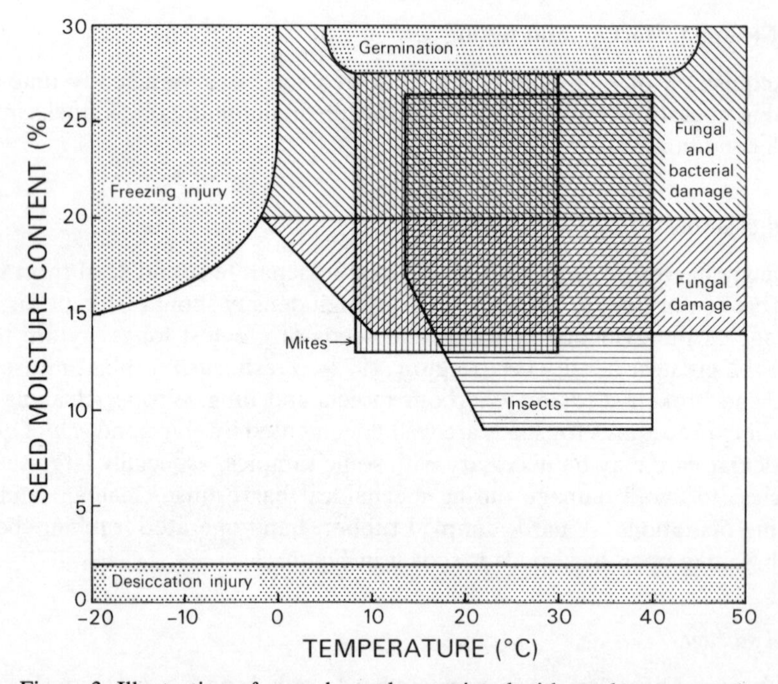

Figure 3. Illustration of some hazards associated with seed storage under less than optimum conditions (adapted from Roberts, 1972)

Therefore, seed should be stored in dry, cool conditions (below 45% relative humidity, 20–25 °C; Harrington, 1972). Seeds, dried to a low moisture content, can be stored in clearly labelled glass jars with gasketed screw-caps. Alternatively, small bags or packets of seed can be stored in metal boxes with gasketed snap-on lids containing a desiccant (see Figure 2). Silica gel (100 g per kilogram of seed), treated with cobalt chloride as an indicator of moisture content, is an ideal desiccant. When dry it is blue. When the indicator turns pink, the silica gel is reactivated by drying in an oven at 160 °C, cooled in a sealed container, and returned to the metal box. Exhaustive accounts of all aspects of seed storage, including major storage facilities, have been presented by Roberts (1972) and Harrington (1972).

Seed testing

Justice (1972) deals comprehensively with the essentials of seed testing. International rules for seed testing have been set out in great detail (Anon., 1966). This section, however, confines itself to a description of a few methods of testing germination capacity.

Knowledge of the germinability of experimental seed is essential for planning plant densities in field and pot experiments and numerical requirements for special laboratory work. Germination tests are usually made on moistened filter-paper, paper towels, or blotters. These substrates should never be so wet that a film of water is formed around the seeds. The substrate is placed on a tray and the seeds are spread out upon it. As a general rule, the distance between seeds should be at least twice their diameter or width. The tray is then put in a germination chamber of any construction that will provide a humid atmosphere to minimize drying of the substrate. For most legumes, germination tests may be conducted in the dark at a constant temperature of 20–25 °C. The substrate may need moistening from time to time. Germination is defined (Anon., 1940) as follows: 'A seed shall be considered to have germinated when it has developed into a normal seedling. Broken seedlings and weak, malformed, and obviously abnormal seedlings shall not be considered to have germinated'. The experimenter may wish to modify this definition to suit his own particular needs. For instance, he may have a requirement for seedlings which develop straight radicles of a minimum length within a certain time.

In this laboratory, we prefer to conduct germination tests on discs of moistened filter paper (Whatman No. 2 or similar quality) in Petri dishes. These assemblies are sterilized by autoclaving before use. This procedure minimizes fungal growth (except for seed-borne fungi) which is often troublesome in germination tests and the Petri dish constitutes its own germination chamber.

A method of testing not included in the international rules for seed testing (Anon., 1966) is the use of an agar substrate for testing the germination of surface-sterilized seed. Seeds, sterilized as described in Chapter II.4, are set out aseptically on water agar (1.5% w/v) in Petri dishes and a few drops of sterile water placed on the surface of the agar. The Petri dish is incubated in the dark at 20–25 °C in an inverted position, which results in seedling with straight radicles. All except large seeds adhere by water tension to the agar surface.

In planning sowing rates for field experiments, it is often useful to conduct germination tests in pots of soil taken from the experimental site. The seed is sown at normal field sowing depth, the soil watered to 80–90% field capacity and maintained at that moisture content, and the pots are placed in a glasshouse at an appropriate temperature until germination is complete. This method is suitable also for testing the germination capacity of coated or pelleted seed.

Dormancy and hard-seededness sometimes interfere with the expression of germination capacity. Some species have very little or no period of seed dormancy, others a well defined period. Requirements for breaking dormancy vary. For example, subterranean clover requires a low temperature (Hills 1944) and alyce clover [*Alysicarpus vaginalis* (L.) DC.] a high temperature (Drake 1947). Scarification by hand or mechanically (see Figure 2) may be needed for some species, usually perennials, to permit imbibition of seeds with hard, nearly impermeable testas. Scarification should be performed only on seed intended for immediate use as the operation greatly diminishes seed viability in storage.

(iii) Importance of nomenclature

Although the importance of full and correct identification of experimental material should be self-evident, it is an unfortunate fact that, from time to time, one encounters a seed sample that is not fully identified or even wrongly named.

The Leguminosae is an immense family. Estimates of its size vary from 500–550 genera and 12000–13000 species to 700 genera containing 14000 species (Allen and Allen, 1961). Taxonomists sometimes disagree about schematic patterns within the family but the following is an accepted heirarchal arrangement: family Leguminosae, subfamilies 3 (4), tribes, (subtribes), genera, sections, (subsections), species, (subspecies), varieties. Within species there often exist cultivars (of domesticated plants) and ecotypes.

The proper way botanically to refer to a plant is by its generic name (with a capital letter) and then its specific name (lower-case letters); thus, *Trifolium* (genus) *repens* (species) for white clover, or *Medicago polymorpha* var. *vulgaris* (variety) for common burr medic, or *Glycine max* cv. Williams (cultivar) for Williams soybean. In first referring to a plant species by its botanical name, it is proper also to use the authority (generally an abbreviation of the name of a prominent taxonomic botanist) designated according to the rules of nomenclature; thus, *Trifolium repens* L. (Linnaeus), or *Medicago polymorpha* L. var. *vulgaris* (Benth.) Shin., or *Glycine max* (L.) Merr. (A recent, undesirable development has been the abandonment by several otherwise reputable publications of this proper practice of using authorities for botanical names). In second and subsequent references to a species, the authority is unnecessary and it is usual to abbreviate the generic name to an initial and to omit the varietal name; thus, *T. repens, M. polymorpha, G. max,* unless confusion could arise in so doing.

The use of common names for plants can also lead to confusion. Sometimes the same common name is used for different species (e.g. hop clover for *Trifolium campestre* Shreb. and for *Trifolium dubium* Sibth.), more usually several different, often local, common names for the same species (e.g. Cooper clover, Darling clover, Cuttaburra clover, Menindee clover, channel clover for *Trigonella suavissima* Lindl.). Publications of 'Standardized Plant Names' (e.g. Anon., 1953) may help to resolve some of these problems. It is never advisable to use the common name alone for labelling stored material. In other situations, the use of the common name is acceptable. For instance, it is obviously convenient, in reports and papers, to use the term lucerne (or alfalfa) for *Medicago sativa* L., and it is legitimate to do so provided the first reference to the common name is followed by the botanical name; thus, lucerne (*Medicago sativa* L.), or *vice versa*, *Medicago sativa* L. (lucerne).

Even more precise identification is mandatory with continuing experiments in order to ensure that the same plant line (and preferably the same population) is used throughout. With domesticated plants, seed certification of commercial cultivars is generally a guarantee of authenticity but the seedsman's lot number should be noted also. With seeds acquired from colleagues or other laboratories, it is always worthwhile to record collection or plant introduction numbers. With undomesticated plants, the point of origin is essential information since major ecotypic differences are common. For instance, the range of ecotypes of the so-called *Psoralea eriantha* Benth. group in the leguminous genus *Psoralea* display striking differences in habit type (De Lacy and Britten, 1977).

Systematic revision of genera, or groups of genera, is constantly taking place and often leads to changes in botanical nomenclature. Two major revisions of recent years, by Heyn (1963) of the genus *Medicago* and Verdcourt (1970a,b) of several tropical genera, have resulted in many alterations. A list of some of these changes and others is shown in Table 7. Clearly, it is desirable to use the most recent 'accepted' name to describe a plant but it is not always so clear to the non-taxonomist if and when the new name has become accepted by the taxonomic community. Any doubts should be resolved by the use of both botanical names with their appropriate authorities. It is sufficiently difficult to keep abreast of taxonomic revisions that we consider it worthwhile in this laboratory to maintain a card index to cross-reference nomenclatural changes.

Table 7. List of some changes in the nomenclature of some legumes

Preferred name	Name previously used
Austrodolichos errabundus (Scott) Verdc.	*Vigna canescens* C. T. White
Dipogon lignosus (L.) Verdc.	*Dolichos lignosus* L.
Dolichos sericeus E. Mey. subsp. *formosus* Verdc.	*Dolichos formosus* A. Rich
Dolichos trilobus L.	*Dolichos falcatus* Willd.
Glycine wightii (Wight et Arn.) Verdc.	*Glycine javanica* L.
Lablab purpureus (L.) Sweet	*Dolichos lablab* L.
Lotus cruentus Court	*Lotus coccineus* Schlecht.
Lotus pedunculatus Cav.*, †	*Lotus uliginosus* Schkuhr[2]
Macroptilium atropurpureum (DC.) Urb.	*Phaseolus atropurpureus* DC.
Macroptilium bracteatum (L.) Urb.	*Phaseolus bracteatus* L.
Macroptilium lathyroides (L.) Urb.	*Phaseolus lathyroides* L.
Macroptilium martii (Benth.) Urb.	*Phaseolus martii* Benth.
Macrotyloma africanum (Wilczek) Verdc.	*Dolichos africanus* Wilczek
Macrotyloma axillare (E. Mey.) Verdc.	*Dolichos axillaris* E. Mey.
Macrotyloma uniflorum (Lam.) Verdc.	*Dolichos uniflorus* Lam. (*D. biflorus* auctt. non L.)
Medicago arabica (L.) Huds.	*Medicago maculata* Sibth.
Medicago polymorpha L. var. *brevispina* (Benth.) Heyn‡	*Medicago denticulata* Willd. var. *confinis* Batt.
Medicago polymorpha L. var. *polymorpha*‡	*Medicago denticulata* Willd. *lappacea* (Lam.) Benth.
Medicago polymorpha L. var. *vulgaris* (Benth.) Shin.‡	*Medicago denticulata* Willd. var. *vulgaris* Benth.
Medicago truncatula Gaertn.	*Medicago tribuloides* Desr.
Pseudovigna argentea (Willd.) Verdc.	*Dolichos argenteus* Willd.
Trifolium ambiguum Bieb.	*Trifolium ambiguum* (M.) Bieb.
Trifolium campestre Shreb.	*Trifolium procumbens* auctt.
Trifolium dubium Shreb.	*Trifolium minus* Sm.
Trifolium resupinatum L.	*Trifolium suaveolens* Willd.
Vigna angularis (Willd.) Ohwi & Ohashi	*Phaseolus angularis* Willd.
Vigna radiata (L.) Wilczek§	*Phaseolus aureus* Roxb.§
Vigna mungo (L.) Hepper	*Phaseolus mungo* L.
Vigna umbellata (Thunb.) Ohwi & Ohashi	*Phaseolus calcaratus* Roxb.
Vigna unguiculata Walp.	*Vigna sinensis* (L.) Endl. ex Hassk.

**Lotus pedunculatus* and *L. uliginosus* are now considered by some authorities to be different species.
†A name used even earlier was *Lotus major* Sm. non Scop.
‡Names used even earlier included *Medicago hispida* Gaertn.—diverse varietal names.
§A name used even earlier was *Phaseolus radiatus* L.

(iv) Special hosts for specific purposes

The legume/*Rhizobium* association is a relationship of such complexity, so widespread, and involving so many diverse lines of plants and strains of rhizobia that curious expressions of response to the symbiosis occur with some frequency. With a little ingenuity, these phenomena can often be put to experimental use. One such curiosity, already described for the bacteria, is the distinctive red pigmentation in culture of colonies of the *Rhizobium* sp. appropriate to *Lotononis bainesii* which constitutes a ready means of its identification. Several examples of the utility of special-purpose plant lines are presented in this section. The observant worker will occasionally encounter other examples which may make suitable tools for specific experimental tasks.

Small-seeded types

The usefulness of small-seeded species, which will grow satisfactorily in test tubes, as test plants for serial dilution plant-infection tests for enumerating populations of rhizobia has been dealt with above (Section b.ii).

Anthocyanin-rich types

The anthocyanins, characterized by red or blue pigmentation, occur throughout the plant kingdom and may be found in solution in all plant organs (Blank, 1947). Leaf pigmentation is often most pronounced under conditions of high light intensity and low nitrogen supply. There exist anthocyanin-rich lines of some legume species, whose leaves under normal field conditions are strikingly pigmented red, although they produce normally coloured green leaves when adequately supplied with available nitrogen in the reduced light intensity of a glasshouse. This change in leaf colour also occurs when the plant is nodulated by an effective nitrogen-fixing strain of *Rhizobium*. The change in leaf colour from red to green may therefore be used to distinguish between effective and ineffective strains of nodule bacteria. Experiments have indicated that estimates of effectiveness by colour may be made at a much earlier stage than from dry weight (or nitrogen) determinations. Examples of such plants include the red leaf lines of *Trifolium subterraneum* and *Medicago truncatula* (Brockwell, 1956, 1958). Their utilization and efficacy for prompt assessment of strain effectiveness in *R. trifolii* and *R. meliloti* respectively is illustrated in Table 8.

Distinctive nodules

The formation of distinctive nodules by a particular legume/*Rhizobium* association can be a most convenient character as it permits a large number of nodules to be classified with a minimum of effort.

Table 8. Comparison of mean dry weight and leaf colour values as estimates of effectiveness in red leaf lines of *Medicago truncatula* and *Trifolium subterraneum* inoculated with effective and ineffective strains of *Rhizobium*. (After Brockwell, 1956, 1958)

Host/*Rhizobium* association	Effectiveness	Time from inoculation									
		19 days		23 days		26 days		32 days		33 days	
		Dry wt. (mg)	Leaf colour (0–4)	Dry wt. (mg)	Leaf colour (0–4)	Dry wt. (mg)	Leaf colour (0–4)	Dry wt. (mg)	Leaf colour (0–4)	Dry wt. (mg)	Leaf colour (0–4)
M. truncatula ×	Effective	8.56	1.1			11.16	0.7			16.0	0.0
R. meliloti	Ineffective	7.94	2.0			9.71	2.4			12.6	3.3
T. subterraneum	Effective			6.25	1.3			11.0	0.7		
× *R. trifolii*	Ineffective			5.25	3.6			7.75	3.8		
No. of replicates required for significant differences between effective and ineffective associations											
M. truncatula ×	$P < 0.05$	15	10			4	3			1	1
R. meliloti	$P < 0.01$	26	16			7	5			2	1
T. subterraneum	$P < 0.05$			11	1			3	1		
× *R. trifolii*	$P < 0.01$			19	2			5	1		

Effective and ineffective nodules. For some purposes, it may be sufficient merely to distinguish between effective and ineffective nodules. Even with small, very young nodules, which are indistinguishable externally, the presence of leghaemoglobin (nodule internal tissue pigmented pink) will help to identify effective nodules.

Distinctive nodule appearance. Occasionally, the nodules formed by a particular host/strain association in a particular laboratory or field environment are so different in form from nodules formed by other associations that their appearance can be used confidently for identification purposes. An example for *T. subterraneum* × *R. trifolii* in the field is given by Hely (1965).

Pigmented nodules. There are several reports of the formation of black nodules by some tropical legumes when inoculated with certain strains of *Rhizobium* spp. (see Table 9). The black coloration appears to be due to the presence of black granules in the central nodule tissue usually confined to the uninfected cells and absent from the cells containing bacteroids (Diatloff, 1972). The characteristic provides a reliable and very convenient marker for identification of these particular strains. A similarly useful character may be the formation of red nodules by *Lupinus arboreus* Sims in association with certain *R. lupini* strains (Sprent and Silvester, 1971).

Chlorosis factor

Rhizobial-induced leaf chlorosis, in Lee soybeans inoculated with *R. japonicum* strain 76, was first observed by Erdman *et al.* (1957a); the effect was *Rhizobium* strain specific. Although other examples have been reported since (Table 9), the phenomenon is sufficiently uncommon to permit its use in a moderately direct, though somewhat specialized, technique for strain recognition (Johnson and Means, 1964).

Non-nodulating lines

Lines of soybean exist which do not nodulate in the presence of strains of *R. japonicum* which form effective nodules on most other soybean lines (Williams and Lynch, 1954). Such plants are useful as controls in field experiments designed to evaluate the effectiveness of populations of *R. japonicum* in the soil.

Host plants having aberrant nodulation

Phenomena such as those described above are examples of aberrant symbiotic behaviour which can be utilized in ecological or agronomic experimentation. However, they also illustrate a type of hazard that occurs with a low but inconvenient frequency in practical legume inoculation.

Table 9. Some tropical legume/*Rhizobium* spp. strain associations reported to form pigmented nodules (A) and some soybean/*R. japonicum* associations reported to lead to rhizobial-induced plant chlorosis (B) and some others where chlorosis is lacking or negligible (C)

Effect	Host	Strain	Source
A	*Lablab purpureus* (L.) Sweet	CB756 (black nodules)	Cloonan (1963)
A	*Macrotyloma uniflorum* (Lam.) Verdc.	CB756 (black nodules)	Cloonan (1963)
A	*Centrosema pubescens* Benth.	CB756 (black nodules)	Döbereiner (1965)
A	*Macroptilium atropurpureum* (DC.) Urb.	Q9888* (black nodules)	Diatloff (1972)
A	*Lupinus arboreus* Sims	PDD4142 (red nodules)	Sprent and Silvester (1971)
B	*Glycine max* (L.) Merr. cv. Lee, Roanoke, Ogden	Strains 76 and 77	Erdman *et al.* (1957b)
C	*G. max* cv. Blackhawk, Harosoy, Improved Pelican	Strains 76 and 77	Erdman *et al.* (1957b)
C	*G. max* cv. Lee, Roanoke, Ogden	Strain 31	Erdman *et al.* (1957b)

*Temporary accession number, Queensland Department of Primary Industries (Plant Pathology Branch).

Examples. There are three pertinent instances from Australian inoculant experience:

1. In the late 1950s, the subterranean clover cultivar Howard, newly bred for resistance to the virus disease clover stunt, was found to nodulate ineffectively with the normally effective *R. trifolii* strain NA30 then used in Australian clover inoculants (Gibson, 1964).
2. Similarly, 10 years later, it was found that the widely adapted subterranean clover cultivar Woogenellup fixed nitrogen only poorly with the important clover inoculant strain TA1 (Gibson, 1968).
3. More recently, the very valuable soybean inoculant strain *R. japonicum* CB1809 was found to be ineffective or unable to form nodules on the soybean cultivar Hardee and closely related lines (Diatloff and Brockwell, 1976).

In each case the problem was solved by switching to alternative inoculant strains.

Precautions. Nonetheless, the fact that plant lines, having aberrant nodulation behaviour with commercial inoculants, were released for use exemplifies the danger in assuming that a new cultivar will respond to inoculation in the same way as symbiotically related plants. It is obvious that all new legume material should be examined for symbiotic aberrancies prior to commercialization. Indeed, a strong case can be made for conducting such examinations before committing resources to breeding work and agronomic testing.

Testing for symbiotic aberrancies. A very simple glasshouse experiment to detect nodulation aberrancies in new plant lines can be made using two replicates of three inoculation treatments:

1. *Rhizobium* strain standard for symbiotically related plants.
2. Alternative *Rhizobium* strain with similar host specificity.
3. Uninoculated control.

Failure to respond to inoculation or gross differences in response between the two strains should be regarded as aberrant behaviour and, if necessary, investigated further.

(d) RHIZOBIUM STRAINS AND INOCULATION

This section covers the isolation of rhizobia from various sources, some details of culture collections and catalogues, criteria for selecting improved rhizobia, and inoculation methods. Other aspects of this general subject can be found in Chapter III.2.

(i) Isolation of rhizobia

The term 'isolate' is defined as 'a culture prepared from a single nodule'. Strains of rhizobia are easily obtained from nodules using very simple equipment. Similar principles apply whether the isolation is to be made from a fresh nodule, a nodule preserved by desiccation, decaying nodules, dried root material, or from soil, or whether a single isolation is being attempted or isolations are to be made from several hundred nodules.

From fresh nodules

Wherever possible firm, young nodules should be used. Whatever their age, nodules invariably have other microorganisms, as well as rhizobia, on their surfaces. Older nodules, particularly ones that are cracked or otherwise damaged, may also have non-rhizobia within them. Surface sterilization is necessary to remove these other microorganisms. The severity of nodule surface sterilization is always a compromise between destroying the unwanted organisms and preserving the rhizobia. Therefore, with small, firm, young nodules, 15 s in the sterilant [preferably 0.1% mercury (II) chloride] is sufficient, whereas large, old nodules may require an exposure of 2 min. Soft or decaying nodules cannot be dealt with in this way and an alternative procedure must be used (see immediately below).

Materials and equipment. Isolation, of course, is a procedure that must be handled aseptically. Ideally, it should be done in a 'sterile' room, a glove box, or a laminar flow cabinet, but the laboratory bench is almost as good provided that the atmosphere is reasonably dust-free, air movement negligible, and the bench surface clean. The bench top can be cleansed and sterilized simultaneously by swabbing with 95% ethanol or a mild phenolic solution.

The requirements for isolating from a single nodule are shown in Figure 4. They are: nodulated roots, a piece of glass tubing (1.0 × 4.0 cm) covered at one end with a piece of muslin or fine nylon mesh (sterilization cylinder), specimen tubes (2.5 × 5.0 cm) for sterilizing and washing, 95% ethanol, 0.1% $HgCl_2$, sterile distilled (or tap) water, sterile physiological saline (0.85%) or dilute seedling solution (Chapter II.4), a glass rod (0.5 × 12.0 cm) with a flat end, a sterile, empty Petri dish, an inoculating loop made from fine Nichrome wire, and poured plates (see Chapter II.1) of yeast extract mannitol (YEM) agar. The glassware is pre-sterilized by autoclaving or, for the glass tubing, specimen tubes, and glass rod, by rinsing with 95% ethanol.

Procedure. The sequence of operations for isolating from a single nodule is illustrated in Figure 4. The roots (1) are thoroughly washed under the tap and a nodule (2) is removed, with a piece of root attached in the case of small nodules for ease of handling, placed in the sterilization cylinder (3) and, if necessary, again washed in running tap water. The cylinder, with its nodule, is dipped briefly into 95% ethanol (4) (wetting agent and partial sterilant), then into 0.1% $HgCl_2$ (5) for from 15 s (for young, small, clean nodules) up

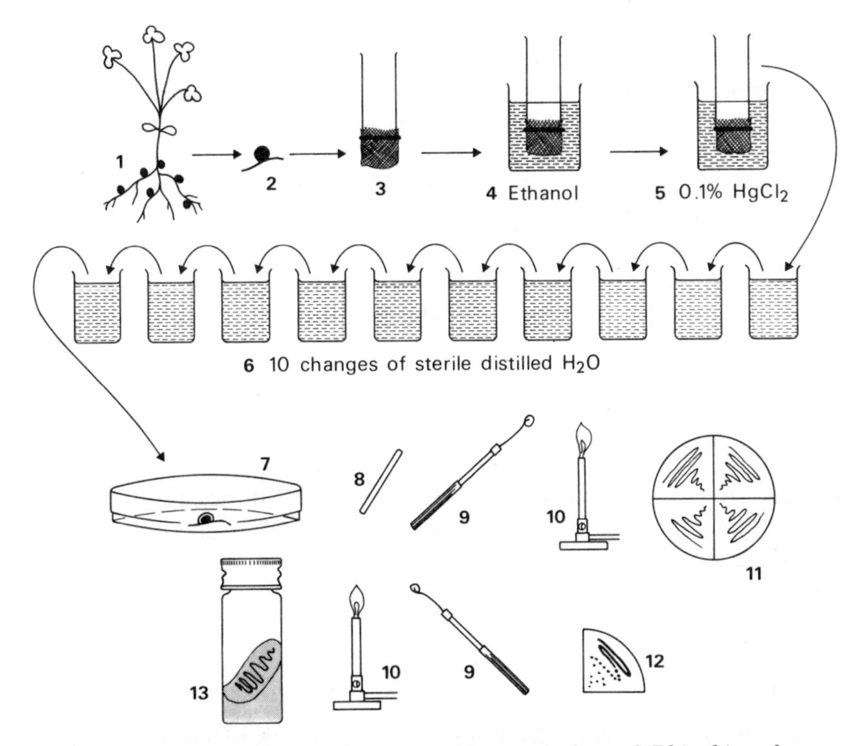

Figure 4. Sequence of operations in making an isolate of *Rhizobium* from a single nodule. 1, Nodulated roots; 2, nodule with piece of root attached; 3, sterilization cylinder; 4, 95% ethanol; 5, 0.1% $HgCl_2$; 6, 10 changes of sterile distilled water; 7, surface-sterilized nodule in sterile Petri dish preparatory to squashing; 8, glass squashing rod; 9, loop of Nichrome wire; 10, Bunsen burner flame for sterilizing loop; 11, Petri dish of YEM agar, divided into four segments, showing streak lines; 12, growth of rhizobia along streak line in a Petri dish segment, showing individual colonies; 13, single colony from Petri dish streaked on YEM agar in Macartney bottle

to 2 min (for older, larger, irregularly shaped nodules), then successively into ten changes of sterile water (6). Such thorough washing is necessary to remove as much $HgCl_2$ as possible, since even trace amounts of residual mercury are highly toxic to the rhizobia released when the nodule is crushed. The surface-sterilized nodule is then placed in a sterile Petri dish (7), a drop of saline (or seedling solution) added, and the nodule squashed with the flat end of the glass rod (8). A loopful (9) of the squashings is then streaked across the surface of the plate of YEM agar (11). To conserve materials, the Petri dishes may be divided into four segments by marking the bottom of the dish, and each dish used for isolating from four nodules (11). If the nodule is old, or the host plant has been grown in soil, or if the dust level in the laboratory is high, it is advisable to incorporate 0.002% of the antifungal antibiotic cycloheximide (actidione) into the medium to reduce the incidence of fungal contamination. The plate is incubated in an inverted position at 26 °C until colonies of rhizobia appear along the streak line (12). This will take between 3–4 days for *R. trifolii* and 14 days for very slow-growing strains of *Rhizobium* sp. A single colony is then picked off the agar surface and transferred to a slant of YEM agar (13). Although the operator will quickly become familiar with the appearance of rhizobial colonies, which may be watery, translucent, or whitely opaque (or even pink in the case of isolates from *Lotononis* spp.), it is always necessary to verify the identification (see Authentication of Isolates, below).

From desiccated nodules

Nodules collected in the field and preserved by desiccation should be reconstituted by soaking overnight in sterile water. Then the procedure for isolation is exactly the same as for fresh nodules. The period of immersion in $HgCl_2$ for surface sterilization is 1–2 min.

From soil

Direct isolation of rhizobia from soil is rarely possible. Thus, it is necessary either to sow surface-sterilized seed into the soil or to add a sample of the soil to seedlings grown from surface-sterilized seed under bacteriologically controlled conditions (Chapter II.4). When nodules form, cultures are isolated from them as described above. Naturally, the strains of rhizobia obtained in this way will be determined by the specificity of the plant (trap-host) used for the purpose.

From dried root material

Legume roots, even without nodules, are often a potent source of rhizobia. Dried roots should be reconstituted in sterile water overnight, homogenized in a micro-blender or with a mortar and pestle, and then used to inoculate a seedling grown under bacteriologically controlled conditions. The cultures are isolated from nodules which form as described above.

From decaying nodules

It is rarely possible to isolate directly from aged, soft, or decaying nodules or nodules that have been attacked by insects (e.g. Diatloff, 1965) because they cannot be surface sterilized properly. Such nodules should be homogenized and handled as for dried root material.

Alternative procedures

The procedure described for isolation is a basic one and is subject to numerous acceptable modifications. For instance:

1. A siphon device, instead of successive transfers to containers of sterile distilled water, can be used for washing nodules free of sterilant.
2. Hydrogen peroxide (5%) is an alternative to $HgCl_2$ for surface sterilizing small, young, firm, clean nodules.
3. Large nodules may be dissected prior to squashing and streaking.
4. Large, clean nodules can be stabbed with a sterile needle and the needle used to inoculate (by stab or streak) the YEM agar plate (Franco and Vincent, 1976).
5. Sophisticated apparatus can be constructed to permit surface sterilization of up to 60 nodules simultaneously (Gault *et al.*, 1973).

Authentication of isolates

An isolate cannot properly be regarded as a species of *Rhizobium* until its identity has been confirmed. This is usually done by demonstrating its ability to form nodules on an appropriate host plant growing under bacteriological control. Sometimes, doubts about the authenticity of a strain within a characterized collection, owing to variation in cultural or symbiotic characteristics or loss of invasiveness, may be resolved by examination of serological properties which are both specific and stable. Information on serological techniques is contained in Chapter II.9.

Maintenance of isolates

It is best to choose an isolated colony from the YEM isolation plate to make a culture. If there is any doubt about the purity of the isolate, it should be re-streaked on YEM agar and another isolated (clean) colony chosen. The colony is then grown on a YEM agar slant in a cotton-wool-plugged tube or screw-capped bottle and mass-transferred to fresh slopes three or four times during the next 2 months to allow it to adapt to growth conditions on an artificial medium. It should then be stored appropriately in order to maintain it in a viable and biologically stable state. There are several methods for storing cultures (Vincent, 1970; see also Chapter III.2), the best of which is lyophilization.

Culture records

It is, of course, most important that cultures be clearly and unambiguously labelled. The use of waterproof, glass-fast ink is advisable. Culture records may be kept on cards. The type of information recorded is illustrated in Figure 5. With large collections of cultures of *Rhizobium* spp., especially those where exchange between laboratories is frequent, it is often worthwhile to assemble a catalogue. Where computer facilities are available, an information storage and retrieval system relating to the culture collection can be devised (e.g. Simpson *et al.*, 1971).

(ii) Acquisition of *Rhizobium* strains

A collection of strains of *Rhizobium* spp. is usually built upon isolates made in one's own laboratory since such strains will normally represent the populations of rhizobia found in the soils or suitable for the legumes being studied. However, most collections also acquire strains from elsewhere for comparative purposes, because of special characteristics such as good nitrogen-fixing ability, or because suitable strains for inoculating a particular legume are not available locally.

Field collection of nodules, root material, and soil

Nodules are the most common and reliable source of field isolates of *Rhizobium* spp. However, at times when nodules are not present, root material or soil will often yield rhizobia.

Nodules. Nodules freshly collected in the field deteriorate very rapidly and they should be kept moist, packed in soil or between sheets of damp blotting paper, and isolations made from them within 24 h. The period between collection and isolation can be extended slightly by refrigerating the nodules, but this is not really satisfactory because of the development in storage of 'spreading' bacteria which are difficult to destroy by surface sterilization and whose rapid growth on YEM agar interferes with isolation.

Nodule collecting tubes. It has been found (Hely, 1958) that significant numbers of nodule bacteria survive in legume nodules collected in a fresh state and dried quickly and completely without heating. Moreover, they remain viable for several months if they are kept relatively cool and in a desiccated state. This can be done by drying nodulated roots between news-

CSIRO (CANBERRA) COLLECTION OF RHIZOBIUM SPP.

GROUP: Rhizobium trifolii CC No.: TA1 SYNONYMS Rothamsted 221; SU659; NA14-1; CC248.

ORIGIN: Host – T. subterraneum Locality – Bridport, Tasmania Date of Isolation – unknown.
Soil – Sandy loam Elevation – 20 m. Climate – Maritime
Rainfall – 975 mm Distribution – Autumn, winter, spring
Sender – H.V. Rees, Tasmanian Dep. Agric., Launceston, Tas. Date of Acquisition – Sept. 1955

HOW MAINTAINED: Freeze dried ; YMA agar in screw-cap Macartney bottles

EFFECTIVE FOR: T. repens, T. subterranean, T. incarnatum, T. fragiferum, T. pratense, etc.

INTERMEDIATE FOR: ——

INEFFECTIVE FOR: T. ambiguum, T. semipilosum

MUTANTS: TA1 str 1 (streptomycin resistant)

COMMENTS: Australian inoculant strain; very wide host range; poor nodulation of T. subterraneum cv. Woogenellup in the field.

REFERENCES: Gibson, A.H., Aust. J. Biol Sci. 19: 499, 1966; Gibson, A.H., Aust. J. Agric. Res. 19: 907, 1968; Gibson, A.H., Aust. J. Agric. Res. 19: 891, 1968; Brockwell, J., J. Aust. Inst. Agric. Sci. 34: 224, 1968; Dudman, W.F., Aust. J. Agric. Res. 19: 739, 1968; Date, R.A., J. Aust. Inst. Agric. Sci. 35: 27, 1969; Marques Pinto, C., Aust. J. Agric. Res. 25: 317, 1974; Chatel, D.L., Soil Biol. Biochem. 5: 443, 1973.

Figure 5. Illustration of culture record information

paper or blotting paper as one would collect and preserve a plant specimen. In this laboratory, however, we prefer to use a small, air-tight, polyethylene tube (see Figure 6) containing calcium chloride as a desiccant both to dry the nodules and to store them in a dry state. The nodules should be superficially cleaned and blotted dry before being put into the tube, which is then labelled

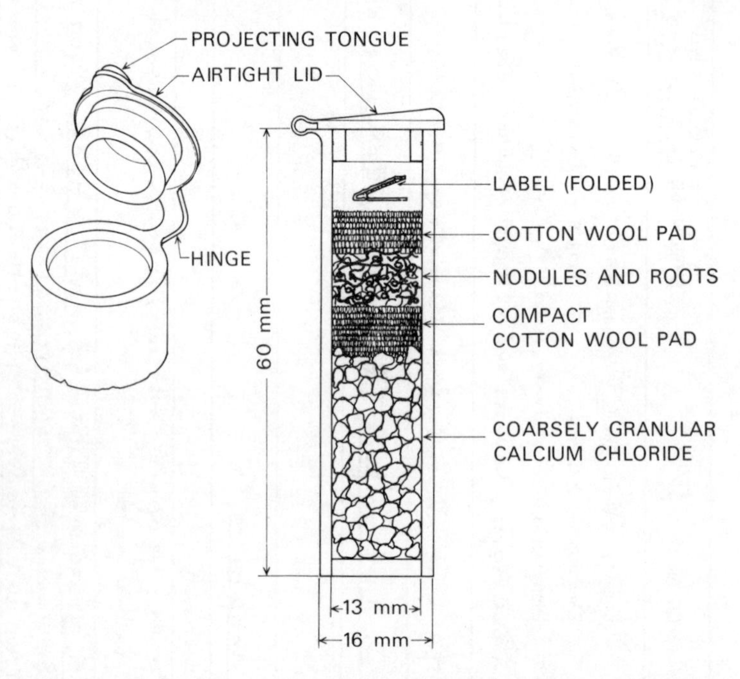

Figure 6. Polyethylene tube used to desiccate, store, and transport legume nodules

inside and outside with indelible ink. Being so light, these tubes are very convenient to transport. The nodules are reconstituted in water before isolation.

Root material and soil. There is much virtue, in plant exploration expeditions, in making provision to collect nodules simultaneously with the collection of leguminous plants. This is particularly important in plant introduction programmes dealing with species for which no suitable rhizobia occur naturally in the country for which the introduction is intended (e.g. *Lotononis* spp. for Australia). Unfortunately, the best time to collect seed is not usually a good time to collect nodules. So, root material or soil must often be used as a source of rhizobia. The small quantities needed can be collected conveniently in nodule collection tubes (see above). Isolation of rhizobia from root material and soil is described in Section d.i.

Mobile laboratories. Some institutions make use of elaborately modified motor vans, trailers, or caravans to conduct experiments of various sorts in the field. Certain work on diazotrophs (e.g. collection and maintenance of living plant material; excavation of soil cores; preliminary gassing of plants or soil for the estimation of dinitrogen fixation by acetylene reduction—Chapter II.3) can usefully be conducted in mobile laboratories. By and large, however, they are not satisfactory for bacteriological work such as the isolation of rhizobia; inevitable dustiness results in unacceptable levels of contamination.

Culture collections

There are a number of important collections of *Rhizobium* strains around the world and some of them are listed in Table 10. They will normally send strains upon request, either lyophilized in ampoules or growing on an agar medium in screw-capped bottles. Usually the culture is supplied without cost to the recipient but sometimes a nominal charge is made to cover packaging and postage.

Catalogues

Most of the larger *Rhizobium* culture collections publish their own catalogues (e.g. Rothamsted Experimental Station, 1972; University of Hawaii, 1977) but they are not always readily available. The International Biological Programme has published a catalogue (*IBP World Catalogue of Rhizobium Collections*, Allen *et al.* 1973) listing about 3000 strains of *Rhizobium* forming nodules on about 430 leguminous species, from 59 collections in 29 countries. It is a most valuable document. Another major catalogue (*World Directory of Collections of Cultures of Microorganisms*, Martin and Skerman, 1972), giving a comprehensive listing of culture collections (including collections of *Rhizobium* strains) throughout the world, has been published under the auspices of UNESCO. This information is also available from the World Data Center for Microorganisms and the World Federation of Culture Collections (see Chapter II.1).

Microbiological Resource Centers (MIRCENs)

The United Nations Environment Programme has commenced the establishment of MIRCENs in various parts of the world. The functions of some of these centres will be to maintain and distribute cultures of strains of

Table 10. Some *Rhizobium* culture collections

Country	Institute*	Address
Australia	A.I.R.C.S., Department of Agriculture	Horticultural Research Station, P.O. Box 720, Gosford, N.S.W. 2250
Australia	Division of Plant Industry, CSIRO	P.O. Box 1600, Canberra City, A.C.T. 2601.
Australia	Division of Tropical Crops and Pastures, CSIRO	The Cunningham Laboratory, Mill Road, St. Lucia, Queensland, 4067.
Brazil	I.P.E.A.C.S.	km 47, via Campo Grande, ZC-26, GB.
Canada	Department of Microbiology, University of Guelph	Guelph, Ontario.
Colombia	C.I.A.T.	Apartado Aereo 67–13, Cali.
Czechoslovakia	Department of Microbiology, Central Research Institute of Plant Production	16106 Praha 6, Ruzyně 507.
France	Station de Recherches de Microbiologie des Sols	7 rue Sully, 21 Dijon.
India	I.C.R.I.S.A.T.	1-11-256 Begumpet, Hyderabad 500016.
Japan	Department of Soils and Fertilizers, National Institute of Agriculture Science	Nishigahara, Kita-Ku, Tokyo.
New Zealand	Plant Diseases Division, D.S.I.R.	Private Bag, Auckland.
Nigeria	I.A.T.A.	Oyo Road, P.M.B. 5320, Ibadan.
Papua New Guinea	Department of Agriculture, Stock and Fisheries	Port Moresby.
Roumania	Department of Microbiology, Research Institute of Cereal and Technical Plants	Bulev Ion-Ionescu dela Brad No. 8, Bucuresti, Baneasa.

South Africa	Plant Protection Research Institute	Private Bag 134, Pretoria.
Sweden	Department of Microbiology, Agricultural College	Ultuna S-75007, Uppsala 7.
U.K.	Department of Agricultural Botany, University College of Wales	Penglais, Aberystwyth.
U.K.	Department of Soil Microbiology, Rothamsted Experimental Station	Harpenden, Herts.
U.S.A.	Crops Research Division, Agricultural Research Service, U.S.D.A.	Beltsville, Md. 20705.
U.S.A.	Department of Bacteriology, University of Wisconsin	Madison, Wisc. 53706.
U.S.A.	NifTAL Project, University of Hawaii	P.O. Box 0, Paia, Maui, Hawaii 96779.
Zimbabwe	Grasslands Research Station	Private Bag 701, Marandellas.

*Correspondence should be addressed to the 'Head of the Collection' or to the 'Curator'.

Rhizobium spp. At the time of writing (1978), two such MIRCENs have been constituted:

1. (Dr. J. R. Jardim Freire), Seccão de Microbiologia Agricola, Secretaria da Agricultura, Porto Alegre RS, Brazil.
2. (Dr. S. O. Keya), Department of Soil Science, University of Nairobi, P.O. Box 30197, Kenya.

(iii) Selecting improved rhizobia

At one time nitrogen fixation was the only criterion seriously considered for a good inoculant strain, but now there are many (Brockwell *et al.*, 1968; Chatel *et al.*, 1968; Diatloff, 1970; Herridge, 1974):

1. Competitive ability with other strains for infection sites on the roots of the host legume.
2. Nitrogen-fixing ability over a range of environmental conditions.
3. Nodule-forming and nitrogen-fixing abilities in the presence of soil nitrogen.
4. Ability to multiply in broth and survive in peat.
5. Ability to survive when incorporated in seed pellets.
6. Persistence in the soil.
7. Ability to migrate from the initial site of inoculation.
8. Ability to colonize soil away from the influence of host roots.
9. Ability to survive adverse physical conditions such as desiccation, heat, or freezing.
10. Strain stability during storage and growth.

Procedures for evaluating some of these characters are given in Chapters II.4 and III.2. However, while it is certainly feasible to examine a few strains in such depth, the total concept of multiple desirable characteristics is idealistic. The prospect of testing many strains so thoroughly is so formidible a task as to be unrealistic. The pragmatic approach is to presume that the strain that performs best in the field is the most desirable strain for field inoculation. This presumption underlines the significance of strain testing in field situations.

In choosing prospective inoculant strains, it is noteworthy that the rhizobia most competitive in nodule formation and persistent in a particular field environment are often those isolated from similar environments (Chowdhury, 1965; Chatel and Greenwood, 1973; Chatel and Parker, 1973). It is especially important, of course, that inoculant strains remain stable during storage. Loss or partial loss of symbiotic capacity unfortunately is relatively common (Labandera and Vincent, 1975). It occurs most commonly with strains maintained on agar culture. To avoid this situation, therefore, it is well worthwhile to preserve valuable inoculant strains by lyophilization in a large number of ampoules which can be opened from time to time as the need occurs.

The subject of selecting improved rhizobia is dealt with more thoroughly in Chapter III.2.

(iv) Inoculation methods

Methods that can be used for inoculating legume seed are very numerous and often determined by the aims of the experiment itself. Some general principles apply whatever the technique chosen:

1. If uninoculated control treatments are included, they should always be handled before treatments in which the seed is inoculated. This reduces the risk of contamination and subsequent nodulation of the controls.
2. Inoculation levels should always be as high as feasible within the objectives of the experiment. The presence of large numbers of inoculant rhizobia reduces the scope for contamination and nodulation by naturally occurring rhizobia or strains from other inoculation treatments.
3. Rhizobia are mobile organisms and are easily transferred by water movement, human or animal agency, or accidentally from treatment to treatment or plot to plot. Awareness of this characteristic, therefore, is needed in siting the experiment, fencing to exclude animals, weeding it, and even in walking across it.
4. Rhizobia are incompatible with many pesticides applied as seed dressings.

Methods of inoculation may be classed as direct, where the inoculant is in contact with the seed, or indirect, where the inoculant is 'implanted' alongside or beneath the seed but separate from it. Either liquid- or solid-based inoculants may be used.

Direct application—liquid cultures

Cultures grown in broth or agar media are used almost exclusively for inoculation of experimental material grown under bacteriological control or in pots in a glasshouse (Chapter II.4). They have the advantages that preparation and handling are simple matters and purity easily confirmed. Inoculation is done by soaking the seed in a culture suspension before sowing or by applying it to the newly sown seeds or to the roots of the young seedlings. Because of their very high mortality rate, rhizobia in liquid culture are unsuitable for inoculating field experiments unless the weather is cool and soil moisture ideal for immediate germination.

Direct application—peat cultures

Legume inoculants based on solid carriers, mainly soil and peat but also other materials, have advantages over other forms of inoculant in terms of preparation, packaging, storage, and distribution (Date and Roughley, 1977; see also Chapter III.2). Also, the survival of rhizobia applied to seed is superior and nodulation in the field is generally better when solid-based inoculants, especially peat cultures, are applied to the seed than when other forms of inoculant are used (Brockwell and Whalley, 1962; Burton and Curley, 1965; Brockwell, 1977). There are several methods of application.

Dusting. Peat inoculant is mixed with dry seed immediately before sowing. Some culture adheres to the seed by lodging in the micropyle and in scratches and irregularities on the testa, and by electrostatic attraction. However, much of it falls off, especially during passage of the seed through planting machinery. Dusting is the simplest method of inoculation but also the least efficient.

Slurry inoculation. To increase the amount of inoculant adhering to the seed, peat culture is applied as a water suspension; alternatively, the inoculant is mixed with moistened seed. The seed must be dried before sowing (not in direct sunlight), but as it dries a proportion of the inoculant falls off. Better attachment of the culture to the seed can be obtained by using an adhesive in the slurry; e.g. a 10% solution of household sugar. Gum arabic and substituted cellulose compounds are more tenacious adhesives. There seems little to choose between them. One cellulose derivative used in early work, methyl ethyl cellulose (Cellofas A), slightly increased the inoculant mortality rate (Brockwell, 1962). Particular caution must be exercized to avoid adhesives to which preservatives, lethal to rhizobia, have been added. The celluloses are cheaper than gum arabic. Polyvinyl alcohols and derivatives have been used as inoculant adhesives, but some of these compounds are toxic to rhizobia.

Seed pelleting. Coating legume seeds with lime to promote nodulation apparantly had its origin in Maine, U.S.A., nearly 40 years ago (Snieszko, 1941). The development of the process as a technique for establishing pasture species on acid soils was begun in Australia in the 1950s (Loneragan *et al.*, 1955). Its use is now widespread extending beyond mere protection against soil acidity (Brockwell, 1977). Advantages include protection of the inoculant rhizobia against unfavourable physical and chemical conditions in soil, competition from the soil microflora, the effects of acid fertilizers, and against seed-harvesting ants. Pelleting makes aerial sowing of inoculated seed a practical proposition and ensures better survival of the rhizobia when delays between inoculation and sowing are unavoidable.

The steps and rates of materials for pelleting seed are given in Table 11. Small batches of seed pellets for experimental purposes can be prepared using standard laboratory utensils. Larger quantities are best made in motor-driven rotating drums. Seed can also be satisfactorily pelleted in a bucket or by

Table 11. Summary of the steps and rates of materials recommended for pelleting seed. (After Roughley *et al.*, 1966)

Ingredient	Quantity	Operation
Gum arabic	60 g	Dissolve gum arabic in hot water. Cool. Chill overnight in a refrigerator. Add inoculum to gum solution and mix thoroughly.
Drinking water	200 ml	
Inoculum	Appropriate amount	
Seed—		
small, e.g. white clover	5 kg	Add the gum–inoculum mixture to the seed and mix until seed is evenly coated.
medium, e.g. subterranean clover	10 kg	
large, e.g. vetch (*Vicia* spp.)	20 kg	
very large, e.g. soybean	40 kg	
Fine lime	4 kg	Add lime all at once and mix rapidly for 1–1.5 min only. Remove excess of lime, if necessary, by sieving.

agitation back and forth on a tarpaulin. Some pelleting equipment is illustrated in Figure 7.

Materials used for pellet coatings include calcium carbonate in many forms, dolomite, various grades of gypsum, bentonite and other clay minerals, rock phosphate and other phosphorus compounds, inert materials such as titanium dioxide and talc, soil and humus, and activated charcoal. The main requirements for a good coating material are that it should be relatively close to neutrality and finely ground (90% at least passing 300 mesh). Adhesives used for attaching the coating material to seed include synthetic glues, glues of vegetable or animal origin, gelatin, various sugars, and honey. The adhesive should have sufficient tenacity so that the coating material does not slough off, but should not be so tenacious that the cotyledons are damaged during germination. Adhesives *must* be free of preservatives.

Indirect application

Alternative forms of inoculation may be preferable to direct application; for example, in the following circumstances:

1. With some species, notably soybean and subterranean clover, which have the habit of lifting the seed coat out of the ground during emergence of the cotyledons; the rhizobia clinging to the seed coat are removed from the soil.
2. Where pre-emergence diseases or insect attack constitute problems it may be necessary to use dressings of pesticides incompatible with rhizobia.
3. With very small-seeded legumes or where naturally occuring rhizobia present strong competition, it is desirable to apply inoculant at a higher rate than is possible with seed inoculation.
4. The seed coats of some legumes contain rhizobicidal materials.

Solid inoculant. The use of solid inoculant was first suggested by Bonnier (1960) and *Rhizobium* granules, or soil implant inoculant as it is sometimes called, was first described by Fraser (1966). Solid inoculant is made by coating solid granulated material with peat inoculant in an adhesive. Suitable adhesives include a 25% aqueous solution of gum arabic (no preservatives) or a 3% solution of methyl cellulose. Tenacity of the adhesive solution can be improved by chilling overnight. Peat inoculant is thoroughly stirred into the adhesive and this suspension poured on to the beads and mixed together until all beads appear evenly coated. The beads should be dried by spreading in a thin layer. When dry and any lumps have been broken up, the material is ready for use. While any inert, particulate, freely flowing material, e.g. marble

Figure 7. Some equipment used in seed pelleting: top left, bowl from a small cream separator (geared down to 37 r.p.m.) handles up to 1 kg of seed; top right, pelleting on a tarpaulin; bottom left, concrete truck for preparing very large quantities; centre right, dough mix (geared down to 28 r.p.m.) for batches up to 50 kg; bottom right, concrete mixer (paddles removed) makes excellent seed pellets. (Reproduced by permission of John Wiley and Sons Inc., New York)

chips or coarse-graded sand, can be used for preparing solid inoculant, in this laboratory we prefer polyethylene beads.

For large-scale experiments, solid inoculant can be applied through the fertilizer box or insecticide attachment of a seed drill. For smaller experiments, it is 'implanted' by hand in the row alongside the seed. Solid inoculant is particularly apt for experiments on rates of inoculant application or for the inoculation of numerous small samples of legume seed, e.g. plant breeders' lines, where conventional inoculation of each sample separately would be tedious and time consuming. It has also been used successfully to 'reinoculate' (by 'sod-seeding') fields where young lucerne was uninoculated and suffering severely from nitrogen deficiency.

Liquid inoculation. Excellent nodulation can be obtained by spraying inoculum into the row beside or beneath the seed. A peat culture of rhizobia (culture grown on broth or agar is not suitable because of poor survival) is mixed into a paste with water, diluted to a slurry, then added to a water-filled tank prior to spray application. Any spray equipment is satisfactory provided it has not been used previously for toxic chemicals.

A liquid inoculant injection system that can be mounted on a seed drill is illustrated in Figure 8. A gear pump is favoured to provide pressure but alternative types may be suitable. However, piston pumps may be damaged by the small amount of mineral grit contained in peat inoculant; low-output pumps, e.g. diaphragm pumps, may not generate sufficient bypass flow back into the tank to maintain dispersion of the peat. Nozzles should be checked for flow at frequent intervals; however, blockages usually clear readily when the pump is operating at about 140 kPa.

(e) AGRONOMIC ASPECTS

While there is virtually no limit to the number of types of field experiments possible, descriptions in this chapter will be limited to designs for strain testing, for ecological studies, and for assessment of the nitrogen-fixing capacity of naturally occurring strains and available soil nitrogen. The use of a lysimeter for measuring nitrogen budgets under field conditions is dealt with in Chapter III.4.

To minimize variation due to factors outside the experimental treatments, a careful approach is necessary. This involves a sound knowledge of the agronomy of the test plant, careful selection of an experimental site, proper replication and randomization of treatments, correct choice of plot size and shape, consideration of edge effects, and conscientious maintenance of the experiment.

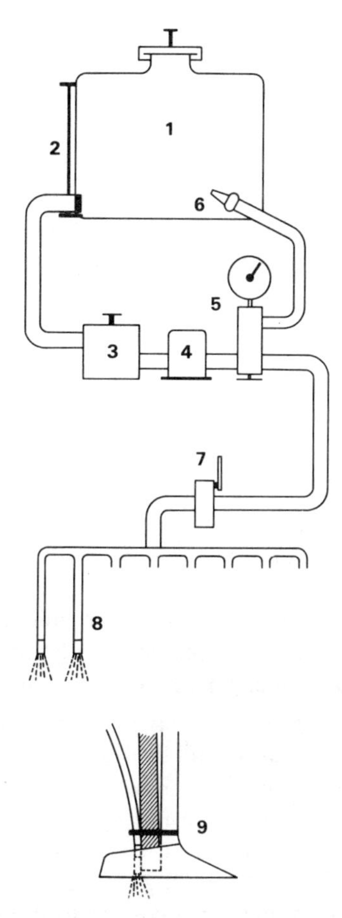

Figure 8. Diagrammatic representation of a liquid inoculant injection system. 1, tank, with sufficient capacity to match delivery time of seed and fertilizer; 2, liquid level indicator, desirable but not essential; 3, filter, peat inoculant contains small particles and fibres that may block nozzles; 4, gear pump, operated from tractor power take-off; 5, pressure relief valve, to indicate flow and provide bypass; 6, bypass outlet, to agitate inoculant suspension in tank; 7, quick-action gate valve, to control flow to nozzles and to maintain bypass flow for agitation within tank; 8, 9, nozzles, best attached immediately behind the boot of the drill tyne. (After Brockwell *et al.*, 1977. Reproduced by permission of the Irrigation Research and Extension Committee, Griffith, N.S.W., Australia)

(i) Experimental site selection

Selection of a site for strain experiments with *Rhizobium* spp. is determined by practical considerations. Sometimes the site will merely be one requiring investigation, but more commonly it will be chosen because it meets the requirements of the experimental conditions. For instance, the ideal site for strain testing is one where soil nitrogen is low and naturally occurring rhizobia are absent or few in number. On the other hand, experiments testing strain competitiveness and persistence require a soil with an established rhizobial population. The site should be nearly level and homogeneous in aspect. The soil must be uniform and the area big enough to provide adequate buffer zones between plots, drainage ditches to prevent cross-contamination by movement of surface water, fencing against animals, and, in large experiments, to manoeuvre machinery.

(ii) Land preparation

A major consideration in land preparation is that cultivation often leads to sufficient mineralization of soil nitrogen to obscure or delay inoculation effects. This can be averted:

1. By growing a cereal crop (or two) on the land in anticipation of its use for nodulation experiments in order to reduce levels of available soil nitrogen.
2. By incorporation of heavy dressings of organic matter low in nitrogen, such as straw chaff, several weeks before planting.
3. By minimal mechanical tillage, such as sod-seeding or by chemical tillage.

When weed growth is abundant, it may be necessary to use a 'knockdown' herbicide (e.g. glyphosate: rate of application 6–9 l ha^{-1}, depending on the species of weeds to be controlled; 36% w/v active ingredient) before cultivation. The land should be cultivated using equipment appropriate to the area of the site; a rotary hoe with a 30-cm cut is ideal for small areas. Preliminary cultivation should be followed by raking or harrowing to remove old root material. Further cultivation may be necessary to prepare a fine seed bed. It is always worthwhile to incorporate a pre-emergence herbicide (e.g. trifluralin: rate of application 1.0–3.5 l ha^{-1}, depending on the crop to be grown; 40% w/v active ingredient) during the final cultivation.

(iii) Application of fertilizer

The nitrogen-fixing capacity of a host legume/*Rhizobium* strain association cannot be fully expressed if the other elements essential for plant growth are deficient in the soil. It is therefore necessary to ameliorate mineral deficiencies before commencing experimentation. Sometimes the requirements of a particular soil are known in advance. Most often, however, it is convenient to apply a basal dressing of complete fertilizer containing (per hectare):

Limestone (finely ground), 250 kg (may not be necessary for legumes adapted to acid soils);

Superphosphate, 1250 kg (may be reduced for some tropical legume species);
Molybdenum trioxide, 80 g;
Sodium borate (borax), 4 kg;
Potassium sulphate, 125 kg;
Copper sulphate, 8 kg;
Zinc sulphate, 8 kg;
Manganese sulphate, 16 kg;
Magnesium sulphate, 62.5 kg.

The design of a preliminary field experiment to determine the nutrient deficiencies of a soil is shown in Figure 9a. The design can be modified for pot experiments.

The fertilizer may be spread over the whole of the plot area and incorporated, if necessary, during cultivation. This practice sometimes stimulates weed growth. Alternatively, the fertilizer can be placed in the row beneath the seed, but contact between fertilizer and inoculated seed must be avoided since it will damage the rhizobia and may reduce nodulation.

Since most inoculation experiments are intended to measure symbiotic nitrogen fixation, nitrogenous fertilizer is normally omitted. When it is used for control treatments or other purposes, ammonium sulphate, calcium ammonium nitrate, or sodium nitrate are good sources of nitrogen. Urea should be avoided as losses due to soil urease activity may be considerable. Fritted, slow-release, commercial nitrogenous fertilizers are also very satisfactory, particularly in soils with a low colloidal content.

(iv) Experimental design for assessing inoculation effects

A simple design for a strain-testing experiment on a single legume species is illustrated in Figure 9b. It provides for three inoculation treatments plus an uninoculated control with the treatments replicated four times and arranged in randomized blocks. Each plot is represented in Figure 9b by a single line irrespective of its dimensions. Four replicates is usually sufficient for statistical purposes (Section e.viii) unless the site chosen for the experiment is particularly heterogeneous.

Plot size and row spacing

The size and spacing of plots will be chosen on the basis on the growth habit of the legumes under test. For most pasture species, a single row of closely planted (2–5 cm), uniformly spaced plants makes a satisfactory plot. Some workers with spreading-type legumes prefer to use single plants in wider spacing, say 25–30 cm apart. Wider spacings still may be desirable for experiments that run for more than 1 year. In order to reduce the risk of cross-

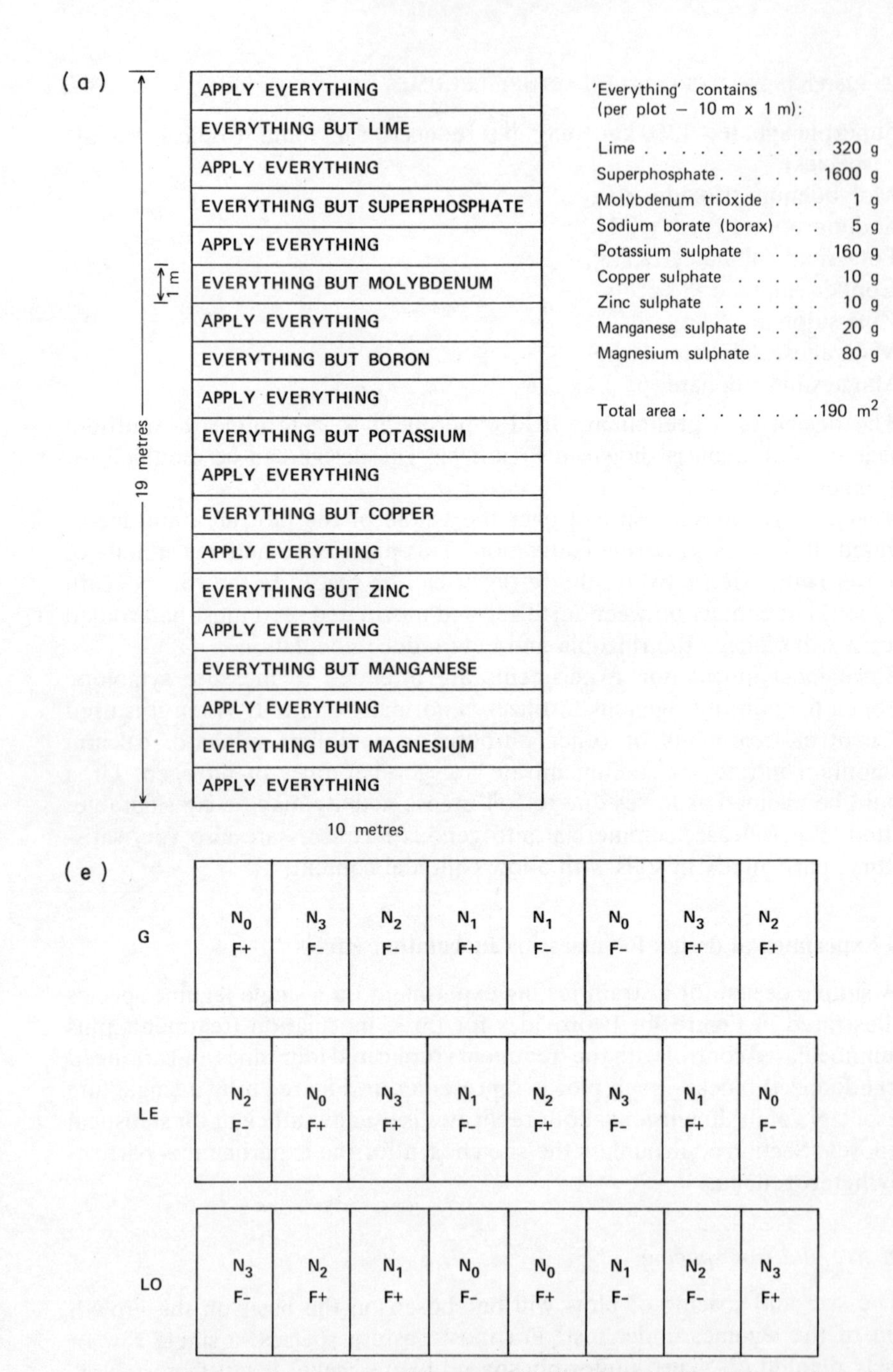

Figure 9. Experimental designs: (a) for determining soil nutrient deficiencies (after Anon., 1952); (b) for strain testing experiments; (c) plot arrangement to reduce influence of edge effects; (d) for measuring rate of spread of inoculum in soil (inoculated seed sown along unbroken lines,

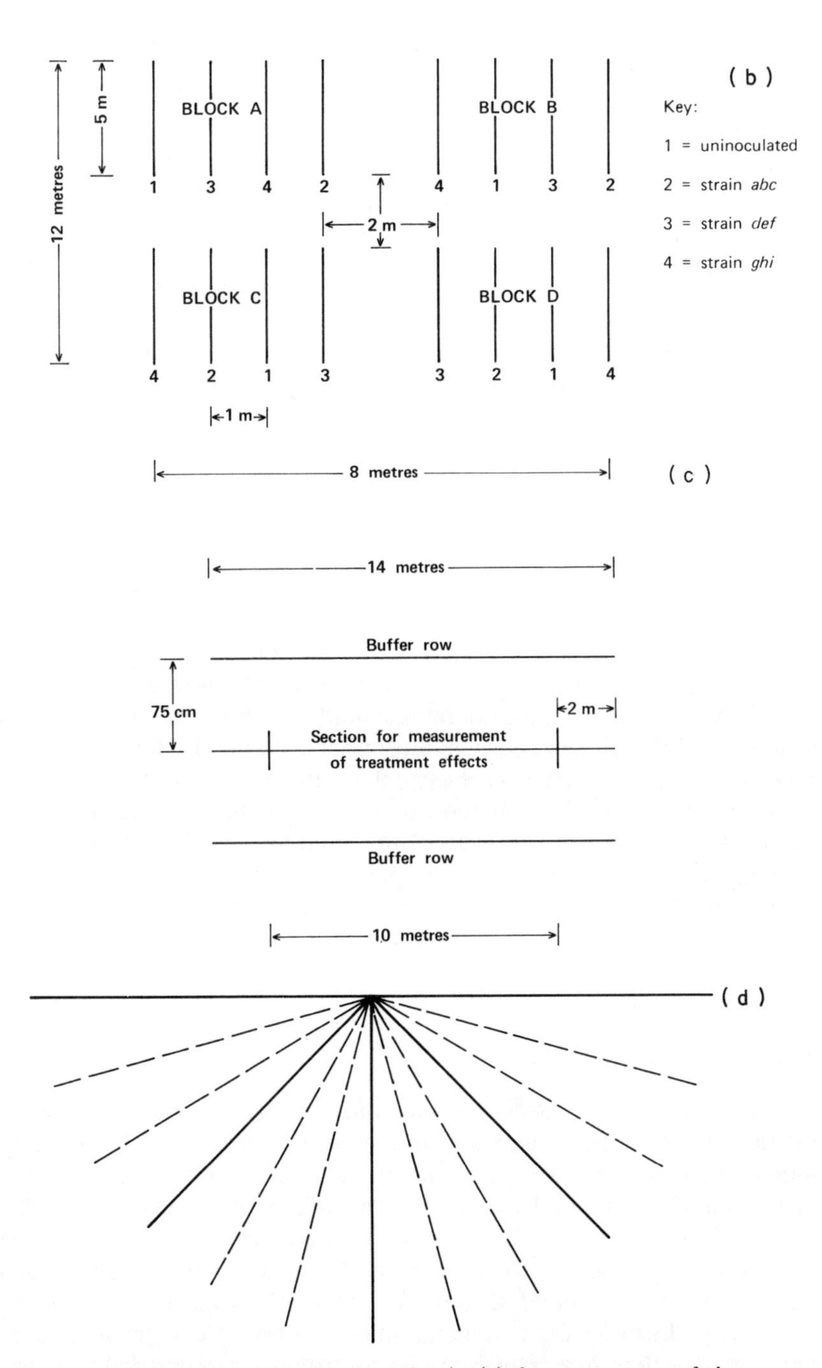

uninoculated seed sown along broken lines); (e) for assessment of the nitrogen-fixing capacity of naturally occurring strains of *Rhizobium* and available soil nitrogen (after Bell and Nutman, 1971)

contamination, the distance between rows should be as wide as practicable and never less than 50 cm. If swards are used, plot sizes should not be less than 1.5 × 1.5 m and may need to be up to 5 × 5 m for spreading species such as *Macroptilium atropurpurem* (DC.) Urb. and *Desmodium intortum* (Mill.) Urb. with wide buffer zones, at least 1.5 m, between plots.

With crop species, edge effects seem to be more prevalent than with pasture species. Therefore, it is advisable to have a minimum of three rows per plot, using only the middle section of the centre row for measurement of treatment effects. An example of this arrangement for a single plot from a soybean experiment is shown in Figure 9c.

Sowing seed

Sowing rates to achieve the desired plant density should be calculated following preliminary tests to determine seed viability (Section c.ii). Wherever feasible, seed should be inoculated immediately before sowing. Uninoculated treatments should always be sown first. In experiments on rates of inoculation, the low-rate treatments should be sown before the high-rate treatments. With hand-sown experiments, the operator should wash his hands in a mild phenolic solution or 70% ethanol between different inoculation treatments. Seed drills for sowing large experiments are extremely difficult to decontaminate properly between treatments. One method that is reasonably satisfactory is to clear the seed box and seed lines with compressed air and then sterilize them by spraying with 95% ethanol. A hand-operated aseptic seeding device for sowing inoculated legume seeds has been described by Ireland (1969). Seed must always be covered with soil immediately it is sown.

Control treatments

The criteria used to assess responses to inoculation must be considered in relation to data from uninoculated control treatments. Minus-nitrogen, uninoculated controls are necessary to judge the nodulating and nitrogen-fixing abilities of naturally occurring rhizobia at the experimental site and should always be included as one of the treatments in a field experiment. Plus-nitrogen, uninoculated controls can be used to determine the ability of the test plant to respond to nitrogen under the experimental conditions. However, difficulties exist in maintaining a constant and realistic supply of nitrogen and, therefore, plus-nitrogen controls are rarely strictly comparable with inoculated treatments. They are often excluded from field experiments (see later comments; Section e.vi).

Results

The criteria used most frequently to evaluate inoculation treatments are nodulation, yield, and nitrogen uptake.

Nodulation. Nodulation is usually assessed twice during the experiment, once sufficiently soon after germination to detect differences in the rate (or earliness) of nodulation and again later, shortly before flowering, when the plants should be fully nodulated. A number of plants (not less than ten) are carefully dug up and their root systems washed free of soil with water running over a sieve. Nodule number, nodule mass, and nodule volume have all been used as criteria of 'nodulation response to inoculation', but each is extremely laborious and time consuming, tends to be inaccurate, and may be misleading. In this laboratory, we prefer a scoring system to classify nodulation (Table 12), taking into account nodule number, distribution, and effectiveness. It may need some modification, depending on the nodulating habits of the legume under consideration. The reliability of this method is indicated by the fact that a high correlation is often found between nodule score and growth indices such as foliage wet weight, foliage dry matter, and foliage nitrogen content. The identification of the rhizobia forming the nodules is dealt with below.

Table 12. Classification of nodulation. (After Corbin and Brockwell, 1977)

	Distribution and number of effective nodules*	
Nodule score	Crown†	Elsewhere
0	0	0
½	0	1–4
1	0	5–9
1½	0	>10
2	Few	0
2½	Few	Few
3	Many	0
4	Many	Few
5	Many	Many

*Effectiveness judged on basis of nodule size and internal pigmentation; ineffective nodules not considered.
†Crown regarded as top 5 cm of root system.

Plant yield. Yield may be measured at any number of stages during plant growth to suit the convenience of the experimenter or the objectives of the experiment. In the vegetative stage, foliage dry matter production is the most reliable index of total nitrogen uptake. If fresh weight is to be measured, it must be done immediately the plants are cut becauses losses due to transpiration and bleeding sap occur very rapidly. A clock-face spring balance is adequate for field weighing especially when the samples are bulky. Samples for dry-matter measurements should be loosely packed in open-weave bags and dried in a forced-draught oven at 100°C where dry matter determination is the main objective, or at 60°C where chemical analyses are intended. Ideally, drying should commence within 4 h of cutting; otherwise, the samples should be packed in polyethylene bags to prevent water loss or maintained in cold storage. In the absence of these drying facilities, air drying can be accomplished by hanging the bags in a dry, well ventilated place until they attain constant weight. A knife-edge, single-pan balance is most satisfactory for weighing dried samples.

Harvesting small swards is best done by cutting all the growth above 3 cm from quadrats selected at random within each plot. Shears, hand-operated or mechanical, are more adaptable than other devices for the purpose. It is often desirable with mixed swards to measure proportion of species on a weight basis. A subsample is selected from each main sample for hand sorting, and the various components (e.g. legume, grass, weeds, dead vegetative matter, inert material) dried and weighed separately. Visual estimates of growth which are often suitable for rather sparse pastures are not reliable for dense swards.

The subject of harvesting procedures is dealt with very fully by Shaw *et al.*, (1976).

With crop species, seed yield is the ultimate criterion of response. Various machines are available for harvesting seed from field experiments. Factors to be considered in selecting a suitable one are the cost, ease of maintenance, and ease of cleaning. Some machines cut and thresh in one operation, others thresh material cut by hand. After harvesting, seed should be dried (to approximately 12% moisture content) if necessary as described earlier in this chapter (Section c.ii). Yield can be measured as total weight of seed, mean seed number per pod (hand harvesting necessary), proportion of fully mature, sound seeds (hand sorting required), and weight per hundred seeds (an indication of whether response is due to increased seed weight and/or increased seed number). With some species, oil content may be calculated (British Standard Method No. 4289).

Nitrogen uptake. Total foliage nitrogen is a popular index of nitrogen fixation. A method for determining nitrogen, following Kjeldahl digestion of plant material, is described in Chapter II.2. The dried sample, or a subsample of it, is ground in a standard laboratory mill. Care must be taken before grinding to brush the plant material free of any surface contamination such as soil. Ground material is stored in glass or plastic bottles with airtight lids to await digestion.

An approximate figure for crude protein percentage can be obtained by multiplying total nitrogen content (%) by a factor of 6.25. This is an over-simplification and may be theoretically objectionable. More accurate means of determining percent protein are also set out in Chapter II.2.

Acetylene-reducing ability (Hardy *et al.*, 1973; Chapter II.3) is also frequently used to estimate nitrogen fixation, but it must be recognized that this technique measures a rate of nitrogenase activity at a particular time (itself often a most valuable criterion) rather than total nitrogen fixation over a period.

Marked rhizobial strains to identify inocula

Techniques for identifying strains of diazotrophs, including rhizobia, have been dealt with in some detail in Chapter II.9. Many of them are applicable to distinguishing between field reisolates of inoculum strains and rhizobia that occur naturally in the soil. The most popular involve serological typing and drug resistance.

Isolation of rhizobia from nodules on experimental plants in the field is done as described earlier in this chapter (Section d.i). Several methods of nodule sampling exist:

1. All nodules removed from ten random plants and bulked together from which a subsample is then taken randomly.
2. One prominent nodule, selected on the basis of size, shape, and colour and usually in the crown region, taken from a number of random plants.
3. A given number, or all, of the nodules selected randomly from a few random plants.

The proportion of inoculant strains detectable appears to be consistent whatever the method chosen (Gibson *et al.*, 1976a). Isolations are made on an appropriate medium when testing reisolates for drug resistance. For some serological tests, standard yeast extract mannitol (YEM) agar medium in Petri dishes is used for isolation and the growth of the isolates increased on YEM agar slants. For other serological tests, this step can be omitted. For immunofluorescense, enough antigen is obtainable, even from small nodules, merely by squashing them to extract the 'nodule juice'. Nodules, 2–3 mm in diameter, provide sufficient antigen material for gel immunodiffusion. Steaming the nodules for 15 min before extracting the antigen or sonication for 2 min afterwards often improves precipitin band formation and, therefore, differentiation.

For some suitable host/*Rhizobium* associations, the use of nodule morphology, especially the pigmented nodule characteristics, makes field identification of the organisms responsible for nodule formation a simple matter.

Strain identification using symbiotic properties (symbiotaxonomy)

The use of symbiotic properties for strain identification involves inoculating each isolate on to several different host species and measuring the responses. Identical strains would be expected to produce similar symbiotic response levels on all the hosts. While the method is tedious and time consuming, recent evidence from this laboratory suggests that it is an extremely delicate technique for differentiation and that it may be apt for distinguishing between symbiotic mutants that arise in the field but which retain the serological and/or drug resistance characters of the parents that would otherwise be used for their identification.

This technique is also adaptable for examining symbiotic relationships between host plants. For instance, several plant lines may be inoculated with a number of strains of the appropriate *Rhizobium* sp. and response to inoculation measured. These data are then used to calculate correlation coefficients for the relationships between all pairs of host species for their effectiveness with each of the isolates (Brockwell and Katznelson, 1976). Pairs of host plants whose symbiotic behaviour is highly correlated can be said to be symbiotically related. These symbiotic relationships appear to have affinities with phylogenetic relationships. This approach has been termed symbiotaxonomy (Norris, 1965).

(v) Experimental designs for ecological studies

Designs for two types of experiment are described in this section, firstly for measuring competitiveness in nodule formation and persistence and secondly for measuring spread of rhizobia in soil.

Competitiveness and persistence experiments

It is a common requirement of ecological studies to compare the competitiveness of two or more strains in mixed inocula or of inoculum strains with naturally occurring rhizobia. A design identical with that described for assessing inoculation effects (Section e.iv; Figure 9b) is ideal for competition studies. Row spacing should be as wide as feasible, each treatment should be sown with particular care, and the experimental area must be effectively fenced against ingress of stock or animal pests because cross-contamination is an especially serious hazard in this type of experiment. Preliminary tests must be made to ensure that the identifying markers chosen are adequate to distinguish one strain from another and from the naturally occurring population of soil rhizobia. Results are normally expressed as the proportion of nodules occupied by the inoculum strain(s) using the techniques for identifying rhizobia described earlier in this chapter (Section e.iv) and in Chapter II.9.

An identical experiment conducted over a longer period can be used for persistence studies. It may be necessary to re-sow the original plots if annual test species do not regenerate naturally. This must be done with great care to avoid movement of soil from one plot to another. It is inevitable, however, in long-term experiments that some plot-to-plot transfer of rhizobia will occur as a result of blowing dust or water movement through the soil and over its surface. Provided that adequate precautions have been taken against contamination, this should be regarded as normal and as a measure of persistence and migratory ability of the strains of rhizobia involved.

Spread of rhizobia

Figure 9d illustrates a half-wheel design for a simple experiment to measure the rate of spread of inoculum. Inoculated seed is sown in rows along the spokes represented by the unbroken lines. Plants grown from uninoculated seed sown (at the same time or later) in the other rows (broken lines) are sampled from time to time and the rhizobia responsible for forming nodules on them are identified. The time taken for inoculum strains to move across from inoculated to uninoculated rows and the minimum distances traversed in so doing are used as indices of rate of spread.

(vi) Experimental designs for assessment of the nitrogen-fixing capacity of naturally occurring strains and available soil nitrogen

The simplest possible design for assessing the nitrogen-fixing capacity of naturally occurring strains and available soil nitrogen in a particular field situation using a single legume species is the one illustrated in Figure 9b. The four treatments are:

1. No inoculation, no applied nitrogen.
2. No inoculation, plus applied nitrogen.
3. Plus inoculation, no applied nitrogen.
4. Plus inoculation, plus applied nitrogen.

Interpretation of results is relatively straightforward. For instance, if test plants respond to applied nitrogen (i.e. treatment $2 > 1$), the soil has some degree of nitrogen deficiency; if test plants respond to inoculation ($3 > 1$), the naturally occurring populations of rhizobia are inadequate for the legume under examination; and so on.

An experiment of this design presents difficulties in interpretation only when there is a response to nitrogen but not to inoculation. This may be caused by a nutrient deficiency which interferes with nodulation or nodule function but not with uptake of mineral nitrogen, or by failure to produce effective nodulation. A basal application of complete fertilizer (minus nitrogen) disposes of nutrient deficiency, but plus and minus fertilizer treatments would show whether nutrient deficiency is a problem. The second possibility can be partly resolved by inspection of plant nodulation (presence and type).

A more complex design with additional inoculation and applied nitrogen treatments and the inclusion of fertilizer treatments has been described by Bell and Nutman (1971). A modification of their design is shown in Figure 9e. The treatments are:

1. Inoculation:
   a. legume uninoculated (LO);
   b. legume effectively inoculated (LE);
   c. non-legume or non-nodulating legume (G).
2. Applied nitrogen:
   a. no applied nitrogen (N_0);
   b, c, and d, three levels of applied nitrogen (N_1, N_2, N_3).
3. Fertilizer:
   a. minus fertilizer (F−);
   b. plus fertilizer (F+).

The choice of a test legume is a matter for decision by the experimenter on the basis of local conditions and the objectives of the test. The choice of a non-legume control poses different problems. The purpose of its inclusion is to measure available soil nitrogen. The selected species should have a growth pattern, rate of establishment, nutrient requirements, and ability to extract soil nitrogen similar to the test legume. This is likely to prove very difficult to accomplish. A non-nodulating variety of legume (Section c.iv), if one is available, could be used in place of the non-legume control. Even so, the two plant lines would not be strictly comparable if their agronomic vigour differed.

An applied nitrogen treatment is necessary in this type of trial to confirm that the test plant is able to respond to nitrogen under the conditions of the experiment. Various levels of nitrogen application are intended to define the optimum rate of application in the absence of symbiotic nitrogen fixation. However, the method of application of four different levels of nitrogen also raises difficulties. If it is applied at sowing beneath the seed, seedlings in the no-nitrogen treatments are at a disadvantage until their own nitrogen fixation commences; it if is applied on the soil surface, weed growth may be stimulated; if heavy dressings are applied all at once, some nitrogen may be lost due to leaching or denitrification; periodic application becomes physically difficult when crop species with a bushy habit (e.g. soybeans) are sown in rows.

Nonetheless, the more complex of these two experimental designs is capable of producing a great deal more information than the simple one and may be well worthwhile conducting, provided that experienced workers are available to do it. With inexperienced personnel, however, the simple design is probably desirable. If necessary, additional information on the responses of the non-legume to applied nitrogen, or on responses of nitrogen fixation by the legumes to other mineral nutrients can be obtained from equally simple subsidiary experiments.

(vii) Precautions with field experiments

Besides meticulous planning, there are two areas in field experimentation with rhizobia where the exercize of caution will pay dividends in terms of meaningful results: microbiological hygiene and awareness of operator and observer effects.

Hygiene and contamination

This chapter has repeatedly emphasized the importance of microbiological hygiene and the hazards of contamination. The point is valid and worth reiterating. *Rhizobium* spp. are mobile organisms, all too readily transferable from treatment to treatment or plot to plot. Contamination of uninoculated controls may easily render dubious the results from otherwise well planned, well conducted experiments. Amongst other precautions, it is always worthwhile to handle uninoculated treatments before inoculated ones, to wash one's hands and to sterilize one's equipment between one treatment and the next, to exercize vigilance to walk between field plots and never across them, and to ensure that drainage water in irrigated experiments is so diverted that it does not flow from one plot to another.

Operator and observer effects

It is important that the experimenter should be aware of operator and observer effects. Personal idiosyncrasies often result in different operators performing the same task in ways that are only slightly different but sufficiently so to influence results. For instance, in hand sowing, one operator may cover the seeds a little more deeply than another, thereby influencing rate of emergence. It follows, therefore, that different operators should not sow different treatments, rather, for instance, that one operator should sow replicates one and two in each treatment and the second should sow replicates three and four. Similarly, in making observations of a qualitative nature, one observer may see things somewhat differently from another. Thus, for example, when making visual ratings on plot growth or extent of nodulation, it is better for one observer to do the whole task than to share it with another. If the job is too big for one observer, then the work load should be shared on the basis of replicates, not treatments.

(viii) Statistics—analysis of variance

Basic statistical procedures can be applied to experimental data without great mathematical knowledge. This section presents a brief outline for the conduct of analyses of variance on data from experiments of three different designs.

Variation is a feature of all experimental work. The objective of statistical analyses is to separate the variation due to experimental treatments from the variation due to soil, climate, and other intrinsic factors. When the variation due to treatments has been determined, logical deductions can be made about the experimental results.

Randomized complete block design

The randomized block design can be used for large numbers of treatments and is usually regarded as the design which gives greatest accuracy.

Let us assume a field experiment on soybeans inoculated with strain a (treatment A), strain b (B), or uninoculated (C). Each treatment is replicated four times and the replicates arranged in four blocks. The yields are:

Block I	A (50 kg)	Block III	A (45 kg)
	C (35 kg)		B (48 kg)
	B (40 kg)		C (41 kg)
Block II	B (45 kg)	Block IV	C (36 kg)
	A (40 kg)		A (55 kg)
	C (40 kg)		B (46 kg)

These treatment yields may be described as variates. The total number (n) of variates in the experiment above is 12 and there are 11 degrees of freedom. The concept of degrees of freedom may be clarified by a hypothetical example. The total yield from the experiment is 533 kg apportioned amongst the 12 plots. If we assign whatever yield we wish to 11 of the plots, the twelfth one *must* receive what is left. Therefore, in assigning the total yield, we have 11 degrees of freedom. Hence, degrees of freedom are usually given by $n-1$. In making comparisons between the three treatments ($n=3$), there are $n-1$, or 2, degrees of freedom; for blocks ($n=4$), 3 degrees of freedom. The remaining 6 degrees of freedom in the experiment are associated with error, being variation between treatments within blocks.

The data are now tabulated (Table 13). The steps in calculation are as follows:

1. Correction factor (CF).

$$CF = \frac{(\text{experimental total})^2}{n \text{ for the experiment}} = \frac{533^2}{12} = 23674$$

Table 13. Yield of soybean seed. (Example of hypothetical data from a randomized complete block design)

	Inoculation treatments				
Blocks	A	B	C	Total	Mean
I	50	40	35	125	41.7
II	52	45	40	137	45.7
III	45	48	41	134	44.7
IV	55	46	36	137	45.7
Mean	50.5	44.8	38.0	533 (Total)	

2. Total sums of squares.
   Each item is squared and summed and the correction factor (*CF*) subtracted from the result, thus:

$$50^2 + 40^2 + 35^2 \ldots + 46^2 + 36^2 - CF = 24101 - 23674 = 427$$

3. Block sums of squares.
   Each block total is squared, summed, the total divided by the number of treatments, and the correction factor (*CF*) subtracted from the result, thus:

$$\frac{125^2 + 137^2 + 134^2 + 137^2}{3} - 23674\,(CF) = 32$$

4. Treatment sums of squares.
   Each treatment total is squared, summed, the total divided by the number of blocks, and the correction factor (*CF*) subtracted from the result, thus:

$$\frac{202^2 + 179^2 + 152^2}{4} - 23674\,(CF) = 313$$

The analysis of variance (Table 14) has used the means of blocks and treatments to estimate the variance (amount of variation) associated with them. The remaining variance is due to differences between treatments within blocks, i.e. error associated with soil, climate, and other factors. The F value (variance ratio) may, or may not, indicate that the mean square from one of the sources of variation is significantly greater than that for error. The significance of the F value is determined by the use of tables (Snedecor and Cochran, 1967). In the example given, the amount of variation associated with blocks is not significantly greater than that associated with error, but the amount of variation associated with treatments is significantly greater than that for error at the 1% level; i.e. the probability of such a result occurring by chance is less than 1 in 100. Significance at the 5% level is the generally accepted minimum for biological data.

The analysis so far has shown that the experimental layout was good; i.e. there was no significant difference between blocks. The coefficient of variation (CV) is now calculated as an indication of the overall efficiency of the experiment:

$$CV = \frac{\sqrt{\text{error } MS}}{\text{experimental mean}} \times 100$$

$$CV = \frac{\sqrt{13.7}}{533/12} \times 100 = 8.3\%$$

Table 14. Analysis of variance using hypothetical data from a randomized complete block design

Source of variation	Degrees of freedom (df)	Sums of squares (SS)	mean squares ($MS = SS/df$)	Variance ratio (F) (MS/error MS)
Blocks	3	32	10.7	0.78 (NS)*
Treatments	2	313	156.5	11.42†
Error	6	82	13.7	
Total	11	427		

*NS = not significant.
†Significant at 1% level ($P < 0.01$).

A CV value below 15% is evidence that the experiment was properly conducted.

The analysis of variance has shown that significant treatment differences exist but not where they exist. To determine this, least significant difference (LSD) is used. This test, however, cannot validly be applied unless the F-test has indicated the existence of a significant difference. The LSD value is calculated as:

$$LSD = t\sqrt{\frac{\text{error } MS \times 2}{\text{no. of replicates}}}$$

where t is obtained (using error df) from tables of t values (Snedecor and Cochran, 1967) for various levels of significance (5%, 1%, 0.1%).

Thus:

$$LSD = 2.45\sqrt{\frac{13.7 \times 2}{4}} = 6.41\ (P < 0.05)$$

or

$$LSD = 3.71\sqrt{\frac{13.7 \times 2}{4}} = 9.71\ (P < 0.01)$$

Since the difference between the means of treatment A and B does exceed 6.41, they are not significantly different, but they are both significantly greater than treatment C (A = B > C; $P < 0.05$) at the 5% level of effectiveness; A is significantly different from C at the 1% level. Where large numbers of means are involved, it may be necessary to use a range test (e.g. Duncan's multiple range test) in which the means are arranged in order of rank, and values obtained from tables are used to separate them. Groupings of treatment means may suggest certain natural comparisons. It is legitimate to test such comparisons using a decomposition of treatment sums of squares similar to the type of decomposition found in the analysis of a factorial experiment (see below, also Table 17). Doubts as to the appropriate test should be resolved by consulting a statistician.

Completely randomized design

The completely randomized design has the advantages of being easy to set out and it maximizes the number of degrees of freedom for error. However, it is usually suited only for small numbers of treatments and for uniform experimental material.

If for some reason, instead of being arranged in blocks, the experiment has been laid out as follows:

A A B C B A C C B A C B

there would no longer be any blocks from which to remove error and the analysis of variance would be as shown in Table 15.

Table 15. Analysis of variance (using the same hypothetical data as in Tables 13 and 14) from a completely randomized design

Source of variation	Degrees of freedom (*df*)	Sums of squares (*SS*)	mean square (*MS* = *SS*/*df*)	Variance ratio (*F*) (treatment *MS*/error *MS*)
Treatments	2	313	156.5	12.32*
Error	9	114	12.7	
Total	11	427		

*Significant at 1% level ($P < 0.01$).

The variance previously associated with blocks is now associated with error. However, in this instance, the error has not increased sufficiently to make the treatment effect non-significant, as might have been the case if the blocks had previously removed a large amount of variation (perhaps because of fertility differences across the site). Calculating LSD:

$$LSD = t\sqrt{\frac{\text{error } MS \times 2}{\text{no. of replicates}}}$$

$$= 2.26\sqrt{\frac{12.7 \times 2}{4}}$$

$$= 5.70\ (P < 0.05)$$

Thus, A > B > C.

Factorial experiments

It is often desirable to examine more than one set of treatments (or factors). The factorial experiment is not an experimental design, but contains all combinations of several levels of several factors. The simplest example is a 2 × 2 factorial, e.g. two soybean cultivars (C_1 and C_2) and two inoculation treatments, say CB1809 (E) and uninoculated (U). This can be arranged in a randomized complete block design, with yields in brackets, as follows:

Block I	C_1 E (60 kg)	Block III	C_1 E (58 kg)
	C_2 E (39 kg)		C_2 U (36 kg)
	C_2 U (45 kg)		C_1 U (41 kg)
	C_1 U (42 kg)		C_2 E (44 kg)
Block II	C_2 U (35 kg)	Block IV	C_2 E (42 kg)
	C_1 E (59 kg)		C_1 U (40 kg)
	C_2 E (37 kg)		C_1 E (61 kg)
	C_1 U (38 kg)		C_2 U (39 kg)

Table 16. Yield of soybean seed. (Example of hypothetical data from a factorial experiment)

	Treatments					
	Cultivar 1 (C_1)		Cultivar 2 (C_2)			
Block	E	U	E	U	Total	Mean
I	60	42	39	45	186	46.5
II	59	38	37	35	169	42.3
III	58	41	44	36	179	44.8
IV	61	40	42	39	182	45.5
Total	238	161	162	155	716	
Mean	59.5	40.3	40.5	38.8		

The data are tabulated (Table 16) and the analysis proceeds as before, except that there are now two main effects in the treatments (cultivar and inoculation) plus an interaction between the two main effects. The interaction tests whether the cultivars have responded in approximately the same way to inoculation treatment. The steps in calculation are as follows (Table 17):

Table 17. Analysis of variance using hypothetical data from a factorial experiment

Source of variation	Degrees of freedom (*df*)	Sums of squares (*SS*)	Mean square (*MS* = *SS*/*df*)	Variance ratio (*F*) (*MS*/error *MS*)
Blocks	3	40	13.3	1.17 (NS)*
Cultivars (C)	1	420	420	36.8†
Inoculation (I)	1	441	441	38.7†
C × I	1	307	307	26.9†
Treatment	3	1168		
Error	9	103	11.4	
Total	15	1271		

*NS = not significant.
†Significant at 1% level ($P < 0.01$).

1. Correction factor (*CF*).

$$CF = \frac{(\text{experimental total})^2}{n \text{ for the experiment}} = \frac{716^2}{16} = 32041$$

2. Total sums of squares.
Each item is squared and summed and the correction factor (*CF*) subtracted from the result, thus:

$$60^2 + 42^2 + 38^2 \ldots + 42^2 + 39^2 - CF = 33312 - 32041 = 1271$$

3. Block sums of squares.
Each block total is squared, summed, divided by the number of treatments, and the correction factor (*CF*) subtracted from the result, thus:

$$\frac{185^2 + 169^2 + 179^2 + 182^2}{4} - \text{CF} = 32081 - 32041 = 40$$

4. Treatments sums of squares.
Each treatment total is squared, summed, the total divided by the number of blocks, and the correction factor (*CF*) subtracted from the result, thus:

$$\frac{238^2 + 161^2 + 162^2 + 155^2}{4} - \text{CF} = 33209 - 32041 = 1168$$

In addition, the treatments sums of squares must be partitioned between the main effects and the interaction.

5. Cultivar sums of squares.
   The totals for each pair of treatments are squared, summed, divided by the number of plots of each cultivar (irrespective of inoculation treatment), and the correction factor (*CF*) subtracted from the result, thus:

$$\frac{(238 + 161)^2 + (162 + 155)^2}{8} - \text{CF} = 32461 - 32041 = 420$$

6. Inoculation sums of squares.
   The totals for each pair of treatments are squared, summed, divided by the number of plots of each inoculation treatment (irrespective of cultivar), and the correction factor (*CF*) subtracted from the result, thus:

$$\frac{(238 + 162)^2 + (161 + 155)^2}{8} - \text{CF} = 32482 - 32041 = 441$$

7. Interaction sums of squares.
   Sum of squares for cultivars and inoculation treatments are subtracted from treatments sums of squares, thus:

$$1168 - (420 + 441) = 307$$

The coefficient of variation and least differences for significance may be calculated as before. The analysis of variance shows that there is not significance differences between blocks but that the cultivars and inoculation treatments are significantly different. In addition, there is a highly significant interaction of cultivars with inoculation which arises because cultivar 1 responded to inoculation whereas cultivar 2 did not.

For more complicated experimental designs and procedures and for other types of analyses, e.g. linear correlation and regression analysis, a statistician should be consulted before the experiment is commenced.

(f) CONCLUDING REMARKS AND ACKNOWLEDGEMENTS

It has not been possible in the space available to cover field experiments for symbiotic nitrogen fixation research as fully as the subject deserves. The reader is referred to Shaw and Bryan (1976), Andrew and Fergus (1976), Norris and Date (1976), Shaw *et al.* (1976), Lovett (1976) and Vincent (1970) for more details. I thank Ms L. O'Brien and Mrs A. Stafford for the illustrations, Mr C. J. Totterdell for the photographs, and Miss L. Main for typing the manuscript.

NOTE ADDED IN PROOF. Then botanical name *Glycine wightii* (Wightn et Arn.) Verdc. (Table 7, page 446) has recently been altered to *Neonotonia wightii* (Arn.) Lackey [Lackey, J. A. (1977), *Phytologia* **37**, 209]. This change and perhaps others, has been noted in Hartley, W. (1979), *A checklist of economic plants in Australia,* CSIRO, Melbourne, 214 pp.

Methods for Evaluating Biological Nitrogen Fixation
Edited by F. J. Bergersen

J. A. Thompson
*New South Wales Department of Agriculture,
Gosford, N.S.W., Australia.*

2

Production and Quality Control of Legume Inoculants

(a) INTRODUCTION

Production of legume inoculants on a commercial scale commenced with the applications for patents in the U.K. and the U.S.A. by Nobbe and Hiltner in 1895 (Fred *et al.*, 1932). In the U.S.A. the greatest expansion was between 1929 and 1940 (Burton, 1967). The product name Nitragin used by the company of that name in the U.S.A. was in fact the trade name for all preparations put out under the Nobbe and Hiltner process (Fred *et al.*, 1932). The U.S.A. has probably remained the largest producer in the world, but commercial production is now carried on in all continents, and a new impetus has arisen with the development of grain crops in both the developed and developing countries.

The initial pre-eminence of the U.S.A. is reflected in the descriptions by Fred *et al.* (1932) and Burton (1967). Many of the developed countries commenced production of inocula at official institutions early in the century, e.g. Canada in 1905 (Newbould, 1951), Sweden in 1914 (Ljunggren, personal communication), and Australia in 1914 (Roughley, 1962). In 1932 Fred *et al.* listed 10 commercial manufacturers (9 in Europe and 1 in New Zealand).

Periodically, particular local requirements have precipitated considerable activity in inoculant production and associated research, e.g. polder reclamation in The Netherlands for over 20 years from the 1930s (van Shreven *et al.*, 1953; van Shreven, 1958). Although Canada currently does not support any commercial inoculant production there was considerable study of inoculant production up to 1954 (e.g. Hedlin and Newton, 1948; Newbould, 1951; Spencer and Newton, 1953; Gunning and Jordan, 1954).

Australian involvement in commercial production has coincided with the post World War II boom in pasture development in areas lacking suitable rhizobia and has been fully described, particularly by members of testing authorities (Vincent, 1965, 1968, 1970, 1977; Date, 1969, 1970; Roughley, 1970, 1976; Date and Roughley, 1977).

In Uruguay and Argentina the development of commercial legume inoculants has also followed post-War emphasis on pastures, whilst Brazil's production has been particularly geared to an enormous increase in soybean production. The present situation regarding availability of inoculants in Latin America has been described comprehensively by Batthyany (1977).

The current inoculant production and usage has been recently described for New Zealand by MacKinnon *et al.* (1977), South Africa by van Rensburg and Strijdom (1974) and Rhodesia by Corby (1967).

The problems of starting production of inoculants in developing countries are well described by Persaud (1977), who works in Guyana and by Shaw *et al.* (1972) and Elmes (1975) in Papua New Guinea.

The earliest documented production of inoculants in India was in 1934, and when the state of the industry was recently described by Sahni (1977), it probably had more manufacturers than any other country.

With minimal publicity, inoculant production is also well developed in many European countries. Irradiated peat inoculants are produced commercially in the U.S.S.R. (Roughley, personal communication) and France (Obaton, personal communication). Inoculants based on autoclaved carrier have been produced by the Swedish College of Agriculture at Uppsala since 1914 (Ljunggren, personal communication).

The desirability of control measures to test legume inoculants was recognized early in the history of the U.S. industry. The U.S. Federal Government was responsible for testing cultures and issuing licences in the 1930s and individual states had legislation as early as 1912 to control quality (Fred *et al.*, 1932). However, by the late 1940s regulatory control by Federal authorities was considered no longer necessary (Burton *et al.*, 1972). In recent years the only authority to publicise results of independent tests of commercial inoculants has been the state of Indiana (e.g. Schall *et al.*, 1975).

In a number of countries, government and University institutions supply inocula for commercial purposes, although the fact that the source of an inoculant is a reputable institution is not *per se* a guarantee of its value. The scale of individual private enterprises, however, is often such that fully trained personnel cannot be employed and quality control is more difficult. It was the latter situation which led to the formation of the Australian testing authority U-DALS (University–Department of Agriculture Laboratory Service) in 1956 and its current successor AIRCS (Australian Inoculants Research and Control Service) in 1971.

While it is probable that legislation is available in many countries to allow checking of the quality of the product on sale, only a few countries have established regulatory authorities which control the release of the product in the market. Canada has recently invoked legislative powers to set standards and control quality (Anon., 1979a and b). The Australian body developed out of co-operation between manufacturers, scientists, and State Departments of Agriculture, and functions without legislative backing. The Uruguayan system was established on the Australian model but with legislative powers. The control body of South Africa functions in a very similar fashion to that of Australia (e.g. van Rensburg and Strijdom, 1974). India has recently prepared a very detailed set of standards (Anon., 1977b), although the mechanics of control are not yet clear. In New Zealand the Inoculant and Coated Seed Testing Service (ICSTS) recently commenced operation (Anon., 1979c).

The need for legume inoculants in most countries may be satisfied by a range of options ranging from purchase of the prepared product from another country on the one hand to a full local programme of inoculant development, including selection of local strains and use of local materials, on the other. The degree of commitment is clearly governed by the size of the potential market, and availability of funds, technical expertise, and suitable raw materials.

(b) TYPES OF INOCULANT PRODUCED

The listing of inoculants is not exhaustive because a number of modifications to techniques of preparation are directly related to particular usage especially in relation to pre-inoculated seed. Whilst some of the processes and products are the subject of patents, others involve procedures for which no details are available.

The continuing trend towards international access to patent information makes it not only very desirable but relatively practicable for any potential manufacturer to ensure that his product does not infringe patent rights.

(i) Agar cultures

The simplest form of inoculant is a culture grown directly on an agar or gelatin surface in a glass container. The technique is still used for small-scale production of strains of inocula in most laboratories. The basic medium recommended corresponds essentially to that described by Fred *et al.* (1932).

Advantages

1. Inoculation of each bottle merely requires normal aseptic subculturing procedure.
2. Growth is visible and many contaminants are readily detected.
3. No special carrier materials are required beyond those necessary for laboratory cultures of rhizobia.

Disadvantages

1. The numbers of rhizobia per unit volume are low compared with cultures prepared by other means.
2. The inoculant has a relatively short shelf life.
3. The glass container is liable to breakage.
4. Survival of the inoculant on the seed is inferior to that of a peat-based inoculum (e.g. Vincent, 1965).
5. On some media, cultures tend to produce considerable amounts of gum, so that suspension prior to application is not readily achieved.

(ii) Broth cultures

Growth of rhizobia in liquid medium is also a standard laboratory practice, only requiring a change of scale to be used for commercial inoculant production. The media are developed from standard laboratory media. The major specialized technology arises from the need to introduce a starter aseptically into the main fermenter.

Advantages

1. The product is readily usable as a liquid inoculum directly to the soil, without the problems of dispersal or distribution which can occur with suspensions of peat cultures.
2. For special purposes the cells can be readily concentrated by centrifugation, minimizing transport costs.

Disadvantages

1. As a seed inoculum, survival of the bacteria is no better than those from agar preparation and inferior to those from peat or similar carriers.
2. The strong containers for transport may need to be returnable unless an inexpensive disposable container can be used. Survival during transport may also require refrigeration.

In recent times the most publicized large-scale production of such inoculant has been that prepared in modified milk cans in the Netherlands (van Schreven *et al.*, 1954).

(iii) Dried cultures

A specialized product patented by Scott and Bumgarner (1965) involves suspension of a broth in oil, bubbling of air to remove water, and subsequent centrifugation and separation of the oil and rhizobia. The dried rhizobia are finely ground and mixed with a carrier such as talc or kaolin to provide an inoculum which is used in the dry state.

Fraser (1975) described a technique of growing *Rhizobium* on the surface of gypsum ($CaSO_4$) granules which were then allowed to dry out. The granules were designed for use as an inoculant separate from the seed.

Nilsson (1957) used water-free sodium sulphate to dry the cells of a broth culture by formation of water of crystallization to give $Na_2SO_4 \cdot 10H_2O$. The resultant cake was then ground to a fine powder.

(iv) Freeze-dried cultures

The lyophilization process, used extensively in the preparation of bacteria, food and medicinal products, has been successfully applied to preparations of *Rhizobium* cultures in the U.S.A. (Appleman and Sears, 1944) and Australia (McLeod and Roughley, 1961). However, it has not proved to be a popular commercial process for this purpose. The process involves removal of water from a frozen broth by sublimation under vacuum.

Advantages

1. Small volume of highly concentrated cells.
2. Long shelf-life, especially at higher temperatures.
3. Lack of opportunity for growth of contaminants.

Disadvantages

1. Requirement for expensive sophisticated equipment.
2. Poor survival on seed (Vincent, 1965).

(v) Powdered carrier inoculants

The majority of legume inoculants currently produced in the world utilize powdered organic carrier materials. In spite of the wide range of alternatives which have been tested, peat is still unchallenged as a carrier (Strijdom and Deschodt, 1976). It undoubtedly has the desirable attributes of high moisture-holding capacity and can commonly be used without additives except, at times, $CaCO_3$.

It is relatively easy to devise a substrate from a variety of materials which will support satisfactory growth and survival of rhizobia (Strijdom and Deschodt, 1976) and the search for new carriers has revealed suitable materials which are usually cheap and readily available at the local level. However, the greatest attribute of peat, *viz.* its protective effect on rhizobia used as seed inocula, has been rarely used as a criterion for evaluation of alternative substances.

Alternative carriers which have shown some promise are coal (Strijdom and Deschodt, 1976; Roughley, 1976), charcoal alone (Newbould, 1951), or with composted straw (Wu and Kuo, 1969, quoted by Date and Roughley, 1977), mixtures of soil and compost (Afify *et al.*, 1968), and/or ground plant material such as those formulations popular in the Netherlands (e.g. van Schreven *et al.*, 1954). In Sweden the current formulation containing soil, peat, composted bark, and wheat husks has been little changed, except for relative quantities, since 1925 (Ljunggren, personal communication). The emphasis on ground fibrous plant material has naturally occurred in tropical areas whenever peats are rare, e.g. cellulose powder (Pugashetti *et al.*, 1971), bagasse (Leiderman, 1971), coir dust (John, 1966), and composted corn cobs (Corby, 1976). Filter mud, a waste product of sugar cane mills has shown promise (Philpotts, 1976). The majority of Indian inoculant manufacturers use lignite (Sahni, 1977), described as a low-grade coke by Tilak and Subba Rao (1978) in the most recent and comprehensive of the Indian studies of carrier materials. Inorganic materials such as bentonite and talc have also been studied (Date and Roughley, 1977) and commercial products are available in the U.S.A. based on vermiculite (Schall *et al.*, 1975). Recently Dommergues *et al.* (1979) have successfully prepared a polyacrylamide carrier.

(c) PRODUCTION TECHNIQUES

(i) Media

The essential components for culture media for *Rhizobium* spp. are readily available commercially. The standard medium (Fred *et al.*, 1932) is based on yeast extract as the nitrogen and/or growth factor source with a suitable carbon source, and minerals. Considerable latitude is allowable in formulating a medium (e.g. Burton, 1967) and the operator must experiment with the available materials. The components of the medium of Fred *et al.* (1932) were as follows:

	$g\ l^{-1}$
K_2HPO_4	0.5
$MgSO_4 \cdot 7H_2O$	0.2
NaCl	0.1
$CaCO_3$	3.0
Mannitol	10.0
Yeast water	100 ml
(10% bakers yeast extract)	

In various media, the concentrations of inorganic salts has been varied at least twofold (e.g. Vincent, 1970), and at best are somewhat arbitrary, especially as the yeast source may contain at least some of the mineral requirement. Ertola *et al.* (1969) found that the addition of potassium nitrate increased growth and maintained the pH near neutrality.

In practice, yeast water is commonly replaced by concentrated yeast extract, and care must be taken in the light of the deleterious effect of amino-acid supplementation (Strijdom and Allen, 1966). A safe level is of the order of $3\ g\ l^{-1}$, although some published media contain up to $10\ g\ l^{-1}$ (Vincent, 1970). A number of workers have demonstrated deleterious effects of concentrations between 3.5 and $10\ g\ l^{-1}$ on viable numbers, nitrogen fixation, nodulating ability or cell morphology (Staphorst and Strijdom, 1972; Date, 1972; Skinner *et al.*, 1977).

Mannitol as a carbon source has commonly been replaced, because of cost, by sucrose or glucose. Most rhizobia utilize both mono- and disaccharides, although the fast growers use a wider range than the slow growers (Graham and Parker, 1964). Glycerol has been used commercially for *R. japonicum* and gelatin or arabinose are preferred by some slow growers. It is probable that most sugars are not fully utilized. Not only will the excess be wasted, but it may contribute to unnecessary multiplication of contaminating organisms already present in unsterilized carriers, following impregnation with broth. Molasses, malt extract and soybean extracts have been used successfully in commercial production units in India (Nandi and Sinha, 1974).

(ii) Starter cultures

Suspended agar cultures may serve as the direct starter inoculum by addition to sterile packaged carrier containing nutrients which allow multiplication within the carrier. Alternatively, agar cultures may be used as starters for broth cultures. Starter cultures should:

1. be in the log phase of growth to minimize the lag period in the new medium;
2. be in a form, or amenable to ready preparation in a form, which allows for ready use as an inoculum for the next stage;
3. be sufficiently well grown to provide an inoculum of up to 1% of the total broth to be prepared.

Rhizobia are comparatively slow growing, some are very slow growing, and commercial media are unselective. Consequently, contamination is a great hazard in inoculant manufacture. Inoculation of the fermenter with the starter is the point at which the greatest risks of contamination occur and special efforts are justified to minimize these risks. Liquid starters are best prepared in Buchner flasks with sterile rubber tubing already connected to the side-arm for direct connection to the inoculation port of the fermenter. If starter is to be grown on solid medium this also can be prepared in a Buchner flask connected to a second flask containing fluid for suspending the culture. The whole unit can be autoclaved together. After growth of the culture the suspending fluid is carefully decanted, the cultures are suspended, and then introduced to the inoculation port via the rubber tubing.

(iii) Fermentation vessels

Construction and handling

Any fermentation vessel must be readily sterilized, allow access for inoculation, and provide aeration of the culture. Access is also necessary for ready cleaning. The simplest form is a glass bottle or flask which is either aerated by air bubbles or by shaking. In the latter case, no introduction of air is necessary, but some degree of air exchange is preferable. Cotton-wool plugs can be used, provided that they are not wetted by the culture. In 1976 most of the 40 Indian manufacturers of inocula were using 1–2-l glass bottles or flasks on shakers (Sahni, 1977).

Sizes of small fermenters can vary from a few litres to a working maximum of approximately 80 l, containing 50 l of broth. With an adequate autoclave these can be sterilized complete with medium and all accessories. Above this size, sterilization of the unit and medium requires pressure gasketing and steam for sterilization. Few inoculant manufacturers could justify purchase of large industrial fermenters for inoculant production alone, although they are a clearly suitable means of growing rhizobia.

The essentials of a small fermenter unit for inoculant production have been illustrated by van Schreven (1958) and Date (1974) and are described in general terms in Chapter II.1. Preferably the unit should be constructed entirely of stainless steel. Mild steel is satisfactory, especially when coated with epoxy finishes, but rusting will result in short life. Cocks should be of brass which can be flamed to high temperatures. A suitable fermenter should have at the top:

1. air inlet port, pipe to the base of the unit, and a metal or glass sparger for generating small bubbles;
2. air outlet port;
3. inoculation port.

In addition, it is convenient to include a sampling cock at the base.

For autoclaving the medium should not be filled to more than 80% its total volume. The air supply should pass through filters packed with cotton-wool or glass-wool. Unless an oil-free compressor is used, an oil trap should be fitted in the line before the two large filters which are necessary. Attached to the air inlet of the fermentation unit should be a final filter, which is sterilized with the unit. This may also be packed with glass-wool or replaced with a flat millipore type filter. The air outlet should preferably already be vented through a filter to ensure sterility during cooling. The air outlet should be left open to allow equalization of pressure but the air inlet which reaches to the bottom of the fermenter must be closed during autoclaving to prevent loss of medium.

For inoculation with the starter the inoculation port must be thoroughly sterilized by flaming, allowed to cool, briefly flamed again and the starter introduced, preferably via rubber tubing already connected to the culture flask. Risks of contamination may be further reduced by cutting the rubber tubing to provide a sterile end to the tube.

Commencement of aeration before inoculation ensures a positive air pressure at all outlets, again minimizing contamination. This is particularly important if frothing occurs. A spray trap between the air outlet and the filter will prevent the filter becoming moist after aeration has commenced.

Aeration

Air flow-rates as high as $120\,l\,l^{-1}\,h^{-1}$ are commonly recommended for some microorganisms (e.g. Calam, 1969), and $5\,l\,l^{-1}\,h^{-1}$ has been found satisfactory with small-scale commercial units (van Shreven *et al.*, 1953; Roughley, 1970). The pressure necessary ($0.7\,kg\,cm^{-2}$) may require a reducing diaphragm on the compressor outlet (Date, 1974). Fine pore size spargers (*ca.* 5 μm) ensure better solution of oxygen but impellers are not recommended because of the difficulty of maintaining a bacteriological seal on the shaft (Date, 1974).

Frothing is not unusual but it is generally a characteristic of fast growing rather than slow-growing rhizobia. For this reason, fermenters may be filled with more medium for slow growers (approximately 80% full) than fast growers (approximately 60% full). As frothing can be controlled to some extent by close monitoring and control of maximum numbers (Pulsford, Agricultural Laboratories, Australia, personal communication), any commercial anti-frothing agent should be tested for compatibility with rhizobia before use. It is also advisable to ensure that components of the media do not induce frothing, e.g. some dried yeast media induce frothing immediately they are aerated.

Incubation temperature

Incubation temperatures for commercial production are usually 26–30 °C with all strains, although Burton (1967) prefers 30–32 °C with 35 °C for *Rhizobium meliloti*.

Inoculum level and incubation time

Inoculum levels are commonly of the order of 0.1–1.0%, providing 10^6–10^7 rhizobia per millilitre of culture. Fast-growing strains (e.g. *R. meliloti*, *R. trifolii*) generally reach maximum viable numbers in 2–3 days while 'slow' growers may require 5–8 days.

The mean generation time (MGT) will be affected by the stage of growth of the starter, and the resultant lag phase, temperature of incubation, availability of nutrients, aeration, and of course the size of the inoculum.

The following minimum batch times will be necessary for a starter containing 10^9 rhizobia ml^{-1}, to provide a finished broth of 5×10^9 ml^{-1}:

	Inoculum volume (%)	
	0.1	1.0
Fast growers (MGT 2–4 h)	25–52 h	18–36 h
Slow growers (MGT 6–12 h)	78–156 h	54–108 h

Whilst it is practicable to maximize the rate of achievement of maximum bacterial numbers, the organization of an inoculant production system may not benefit from the most rapid growth conditions. For example, it may be desirable to use a low-percentage inoculum to extend the period of growth, thus allowing for development of any contaminants (which commonly grow more rapidly than *Rhizobium*) and allowing the manufacturer to complete checks for their presence before the broth is ready for harvest. In all operations, however, it is essential to maximize the proportion of living bacteria available at the point of harvest. Manipulation of nutrient sources and aeration can also affect this during growth. Storage of finished broths should be kept to a minimum but, if it is unavoidable, cultures should be held at 4 °C.

(iv) Carriers

General requirements

It is necessary to test any proposed carrier thoroughly. Interactions between rhizobia, carrier, method of treatment, and storage period are common and even peats from the same area can vary in suitability (e.g. Roughley and Vincent, 1967).

Carrier materials for inoculants should meet the following requirements:

1. the material should be ground sufficiently finely to allow thorough mixing with other components and to be compatible with its final use;
2. pH should be readily adjustable to 6.5–7.0;
3. good moisture-holding capacity is desirable and is probably one of the major reasons for the popularity of peat;
4. the carrier should be sterilizable to favour survival of the inoculant;
5. the carrier should be free of toxic materials.

Sources and characteristics of peat

Apart from amenability to modification by drying, grinding, and mixing with other materials, there are no clear criteria for choice of suitable peats. Many of the world's inocula are based on peat with a high percentage of organic matter, although this in not essential. Australian inoculant peats are about 65% organic matter (Stephens, 1943) and Wisconsin peat contains 86% (Burton, 1967). The ultimate test must be the suitability for multiplication and survival of the range of rhizobia to be grown. Strains of *R. meliloti* and *R. trifolii* differed in their tolerance of sodium and chloride ions when salt contamination affected the quality of Australian peats (Steinborn and Roughley, 1974).

In examining peats and other organic carriers for suitability, the moisture content must be defined. Moisture percentage is generally expressed in terms of wet weight of peat, so that when the weight of water equals the weight of dry peat, the moisture content is 50%. However, expression in terms only of percentage can be misleading when carriers of different moisture-holding capacity are compared, and water potential (i.e. suction or negative pressure) is a more precise criterion. This may be expressed in a number of units [1 bar = 0.987 atmosphere = 1022 cm water = 10^5 pascals; pF = 3 + log (-bars)]. When moisture potential is related to percentage moisture content the resultant graph provides the moisture characteristic curve of the material and allows realistic comparisons of materials with different moisture-holding capacities. Examples of pF values in an Australian peat illustrate the range of significant values. A pF value of 4.88 (moisture content of 30%) adversely affected growth of two of three strains, pF 4.15–3.42 was optimal for the three strains in unsterile peat (Roughley and Vincent, 1967), while for the sterile peat the optimum was 4.15–2.69 (Roughley, 1968). Determination of moisture potential involves use of specialized equipment, but, provided that reference materials can be calibrated against this equipment, a simple procedure, described by Fawcett and Collis-George (1967), can be applied. Calibrated filter paper is allowed to equilibrate with the moistened material and a moisture characteristic curve can then be derived. Different carrier materials can then be compared in terms of their moisture potential.

Milling and drying of peat

Although peat can often be successfully air-dried sufficiently to allow milling, heating may be used to assist the drying process. However, Roughley and Vincent (1967) found that, with Australian peats, heating to 135 and 160 °C caused changes which were lethal to subsequently added rhizobia; maxima of 80–100 °C were not harmful. They ascribed the harmful effect to production of inhibitory substances in the peat. In contrast, Burton (1967) found that flash-drying of peat in a rolling drum with an air inlet temperature of 650 °C produced a satisfactory peat. It is evident that a distinction must be made between drying temperature and product temperature (which would be expected to approximate the wet bulb temperature until drying is complete). The apparent absence of toxic materials makes it probable that the temperature of Burton's peat did not exceed 100 °C. However, it is clear that particular care should be taken with drying procedures and that oven drying should be avoided.

Air drying should be used where practicable. Drying with heat should be effected at the lowest possible temperatures, preferably below 100 °C. Australian and South African peats are commonly ground following air drying (Roughley and Vincent, 1967; Strijdom and Deschodt, 1976).

Although the required particle size is dependent on final use, peat is commonly milled to pass at least a 0.25-mm sieve (e.g. Burton, 1967), some require 50% to pass a 0.075-mm sieve (Strijdom and Deschodt, 1976), and Australian peats generally pass this standard. Not only does grinding to 0.075 mm improve adhesion of peat to dry seed, but Australian peat was found to be more subject to caking when moistened, if ground to only 0.15 mm (Roughley, personal communication). Care must be exercized with very fine grinding of a number of materials because of the risk of spontaneous combustion. A criterion of acceptability could be the bulk density of the final material. A very 'fluffy' product may require an unacceptably large volume per unit weight. Milling to a fine particle size can also be difficult if the material is very fibrous.

Sterilization of carriers

The decision to use a sterile carrier makes it desirable to place the carrier in its final package and to sterilize it before inoculation. Sterilization methods for packaged carrier usually involve either autoclaving or gamma irradiation. The alternative method of fumigation requires access of gases to the carrier, long exposure to ensure adequate diffusion, freedom from residues of the fumigant, and measures to prevent recontamination. Fumigation of South African peats with ethylene oxide and methyl bromide (Deschodt and Strijdom, 1974) gave poorer subsequent *Rhizobium* survival than autoclaving, although the harmful effect was not due to residual fumigant. Some organisms also survived the fumigation treatment.

Although Roughley and Vincent (1967) claim that the temperature of peat needs to be kept below 100 °C during drying, there is no published evidence that prior autoclaving at 121 °C is harmful to the *Rhizobium* subsequently added to these peat products. However, it is particularly desirable that tests of suitability of carrier materials should include confirmation of compatibility with sterilization procedure and with additives. Sahni (1977) found that the amount of $CaCO_3$ required by lignite for pH adjustment was dependent on whether or not the material was sterilized.

Even if facilities are available, gamma irradiation is expensive and steam sterilization requires 'retortable (autoclavable) pouches', which also may be difficult to obtain in some countries.

Dosage of gamma irradiation used in Australia are of the order of 5 Mrad, which does not produce a sterile product but ensures very low numbers of surviving contaminants, which remain in a minority even after storage of the inoculant for 12 months. As absolute sterilization by irradiation may not be economically feasible, it is not practicable for treatment of enriched, high-moisture-content carriers, which rely on considerable multiplication after inoculation.

Inoculation of sterilized carriers

Although powdered carriers are commonly used, two different approaches to inoculation are adopted. The method typified by van Schreven (1970), and commonly used in Europe, is to add a small inoculum of rhizobia to a sterilized carrier containing growth-promoting constituents and held essentially at its final moisture content, in its final distribution pack. Increases in numbers of rhizobia of the order of 100- to 1000-fold may thus occur within the carrier before distribution. Development of inoculants on an experimental basis frequently leads to the use of techniques closely following normal laboratory procedures. Thus, van Schreven *et al.* (1954) used a sterile needle inserted between the cotton-wool plug and the neck of the glass container of sterile carrier, and aseptically siphoned a small volume of culture from a flask. It was thus not necessary to remove the cotton-wool plug and an important source of contamination was eliminated. Subsequent mixing was achieved by periodical turning of the container, which was only half filled.

The alternative technique, more common in the U.S.A. and Australia, is for a relatively large inoculum of well grown broth (approximating 30–50% of the total final weight) to be added to the relatively dry carrier powder (*ca.* 10% moisture). The method is less dependent on multiplication so that it is also suitable for use with unsterile carriers, in which multiplication of contaminants inevitably follows addition of the medium containing the rhizobia.

Electrically operated automatic media dispensers are suitable for adding larger volumes of culture because the syringes and attached tubes and needles can all be removed for sterilization. Such units are also amenable to dispensing mixed, measured quantities of broths from separate containers (e.g. containing different strains). In Australia, where techniques require relatively large volumes of broth to be added to peat prepacked in polyethylene, such equipment is used to inject broth directly through the wall of the polyethylene bag. The surface of the bag is sterilized around the point of injection and the small hole is immediately covered with a self-adhesive label. The inoculum and carrier are mixed by manipulation immediately after sealing.

Inoculation of unsterilized carriers

Unsterilized carriers are normally held in the dried form after grinding so that the natural populations of organisms have no opportunity to multiply before the *Rhizobium* broth is added.

Mixing with inoculum may be achieved by spraying or pouring the broth on to the powdered carrier while it is being agitated in a ribbon or paddle-type batch mixer. A concrete mixer is adequate for the task. The proportions of broth and peat are governed by the nature and moisture-holding capacity of the carrier but it is generally desirable to add broth to the point where the carrier remains friable without forming balls (Date, 1974). In the U.S.A. and Australia, the broth to carrier ratio is commonly 1:1.5 to 1:2 (Roughley and Vincent, 1967; Burton, 1967). It has generally been considered essential to use broths of the highest quality, so that minimal multiplication is necessary for the rhizobia to dominate the other organisms. However, a recent study of data from 277 commercial batches of unsterilized peat inoculant produced in Australia (prior to the change to sterile peat) revealed that when broth counts exceeded $5 \times 10^8\ ml^{-1}$ the final number of rhizobia in peat cultures bore no relationship to inoculum size (Roughley and Thompson, 1978). No data were available for inocula less than $5 \times 10^8\ ml^{-1}$.

After mixing, the carrier should be covered to prevent desiccation and held during a 'curing' period of up to 1 week. During this period growth of the inoculum can occur; moisture levels equilibrate, and any 'heat of wetting' is dissipated. Heat of wetting results from water additions to particles whose heavily bound water has been removed by high-energy inputs. Thus there is a positive relationship between temperature of drying and heat of wetting (Roughley and Vincent, 1967). If peat is dried at 135 °C, temperature rises of 20 °C can be measured, but with recommended drying temperatures (<100 °C), the rise is of the order of only 5 °C (Roughley, personal communication). With unsterile peat, which is normally held in trays during curing, it is desirable to restrict the depth to 7–8 cm to minimize the temperature increase. With sterile peats dried at below 100 °C no special precautions are used to prevent temperature increases, although the packets could be readily spread out.

Following the 'curing' period, during which some multiplication will generally occur, the inoculant should be passed through a coarse seive or hammer mill to remove lumps. If the final inoculant is to contain more than one strain, the separate batches of inoculated carrier should be mixed at this stage. Ideally, the finished inoculant should then be packaged ready for testing and sale. Storage in bulk may be a practical alternative only if suitable storage conditions can be provided (i.e. low temperature without loss of moisture).

Cultures based on unsterilized peat and subjected to a 'maturation' period of 28 days after packaging have better survival rates on seed than peats 7 days old (Burton, 1976). Similar benefits to survival on seed were found with sterilized peat held for 14 days (Thompson, unpublished work).

(v) Packaging

Materials

Glass. Bottles have traditionally been used for agar cultures and retain a place for special orders and low volumes of inoculant prepared as pure cultures on solid medium. A bottle of capacity 250 ml, containing 70 ml of yeast–mannitol agar, provides sufficient rhizobia for approximately 10 kg of small seed or 25 kg of soybeans. Survival in culture is poor so storage should be restricted to a few weeks. Although gas exchange is prevented, screw-caps are the most suitable method of sealing to facilitate handling. Van Schreven (1958) used bottles for inocula made with peat–soil mixtures with cotton-wool and a Cellophane cover to prevent moisture loss. For many years Czechoslovakian peat humus cultures were packaged in glass bottles.

A freeze-dried culture previously available in Australia was marketed in vaccine bottles under nitrogen with a sealed vaccine cap.

Agar cultures are space-consuming and glass containers are subject to breakage, but they retain the advantage of being autoclavable and transparent so that growth can be observed.

Metal. Metal cans were used for many early preparations in the U.S.A. (Fred *et al.* 1932) and normally contained soil or peat cultures. Inoculant production in Sweden has continued since 1914 to be based on a metal can of capacity 150 ml containing sufficient inoculant for 0.5 ha (Ljunggren, personal communication). The junction of the cap is covered with a fixed label so that air exchange is minimal. Although cans are more resistant to breakage than bottles and are also autoclavable, they have not retained their popularity, possibly in part because of the large space requirement for storage before and after filling.

Plastic. The development of the plastics industry has changed the packaging of inocula and many other goods. The majority of the world's inoculant production is marketed in plastic pouches.

A choice of plastic material for pouches involves balancing the requirements for gas exchange, moisture retention, etc., with strength and resistance to temperature. Polyethylene allows a high gas exchange, relatively low moisture transmission, and is heat-sealable. It is strong enough for normal handling but is readily punctured and usually cannot be autoclaved.

For sterilization by autoclaving a number of choices are available. The melting point of polyethylene rises with increasing density so that a suitable material can be selected to resist autoclaving. However, a major problem of obtaining a 'retortable pouch' is that temperature exchange between the pouch and atmosphere is too slow during cooling and the higher temperature of the pouch can cause its swelling and bursting. This may be overcome in part by choosing a polyethylene of heavier gauge or strengthening by lamination with polyester or nylon. Although the latter materials are stronger, they suffer from poor gas exchange and high moisture loss. Foil laminates with polyethylene are, of course, completely gas and moisture proof.

If autoclaving is carried out on a package which is not sealed but only closed (e.g. by an elastic ring), the heat-sealing properties of polyethylene are not necessary and other materials can be considered. An excellent outline of the properties of the above materials is provided by Pinner (1967).

In Australia, low-density polyethylene of thickness 0.038–0.051 mm has proved to be satisfactory, although moisture loss remains higher than desirable. The peat-filled packs are sterilized by gamma irradiation after heat-sealing. Inoculation is effected by injection through the wall, which is subsequently re-sealed.

Rigid plastic containers of various shapes have been utilized in recent years by a number of manufacturers in the U.S.A, generally for smaller volumes of specialist inocula. One manufacturer provides 190 g of frozen concentrate in a plastic beaker with a tight-fitting lid. This is sufficient for 4 ha.

The most common Australian polyethylene pack contains 250 g of moist peat, sufficient to treat 100 kg of soybean or 50 kg of lucerne. A South African product, packaged in the same fashion, contains 230 g, sufficient to treat 90 kg of soybeans or 11 kg of lucerne. An Indian product containing 200 g is recommended for use on seed for 1 acre.

Effects of packaging on survival of Rhizobia

Packaging techniques reflect a range of influences: the form of the product and method of its production, need for resistance of the package to sterilization or distribution stresses, the availability of suitable containers, and the overall scale of the inoculant production process. However, all these techniques must finally be judged in terms of the survival of the rhizobia in the prepared inoculant—in this regard the available data are frequently not in agreement and a compromise may be necessary.

Moisture. With any carrier-based inoculum there is an optimum moisture content which ideally should be maintained for the life of the culture. If aeration is not considered necessary (see below), it is practicable to seal any container and prevent moisture loss. All of the early commercially prepared inocula in the U.S.A. (Fred *et al.*, 1932) were packaged in either tin cans with lids or screw-caps, or bottles with stoppers or screw-caps.

The change to pliable bags, normally polyethylene, has been a relatively recent development, reflecting needs for ease of handling, the availability of a wide range of synthetic materials, and the desire for aeration of the culture. The result has frequently been loss of moisture and there is considerable evidence of its adverse effects on survival. For example, Vincent (1958) found that a moisture loss of 24% per week at 5 °C gave a weekly death rate of 0.085, which was reduced to 0.001 if the moisture loss was 0.7% per week. Also, there is an interaction between rhizobia and contaminants at particular moisture levels, so that in Australian unsterilized peats the optimum moisture content range is 40–50% (pF 4.15—3.12), while with sterile peat the optimum is 40–60% (pF 4.15–2.69) (Roughley, 1968).

Moisture loss may also be confounded with increased concentration of harmful soluble salts (Steinborn and Roughley, 1974).

Aeration. The conflicting data on the need for aeration seem in part to be a reflection of the wide range of materials used in the studies (e.g. van Schreven *et al.*, 1954; Hedlin and Newton, 1948; Roughley, 1968). Whilst Roughley's (1968) data support the need for aeration with Australian peats, and have greatly influenced the choice of package material, they illustrate the need for comparative studies of package types with any carrier. There are certainly no generally accepted principles regarding aeration.

With sterilized carriers it is essential, however, to ensure that any aeration is achieved without contamination. Thus, the use of pin holes in polyethylene is not recommended and it is also important to take care with any additional packaging involving stapling, which could damage the bag.

Temperature. With agar-based inocula, the death rate is more than halved by reducing the storage temperature from 25 to 5 °C (Vincent, 1958). With peats after maturation at 26—30 °C, low-temperature (4 °C) storage is generally more favourable for survival than higher temperatures, including those at which growth would normally occur, e.g. 26 °C. Temperature effects are frequently confounded with effects of moisture loss. By preventing moisture loss, Roughley (1968) found that there was little or no death during 26 weeks with sterile Australian peats stored at 4 and 26 °C. More recent studies, with normal moisture loss, have been less predictable. While the survival of fast-growing strains of *Rhizobium* was favoured by storage at 4° C, with little decline over 12 months, the numbers at 26 °C commonly declined, although generally within acceptable limits. Conversely, the survival of slow-growing strains was superior at 26 °C with marked reduction occurring at 4 °C (Thompson, unpublished data).

(vi) Labelling

Ideally, each separate packet of inoculant should contain the following information:

1. legume hosts for which the contents are suitable;
2. quantities of seed or areas to be treated;
3. batch number of inoculant;
4. expiry date;
5. instructions for storage and use;
6. any certification by a controlling body;
7. extent of legal responsibility of the manufacturer.

The flat plastic package used for impregnated peat lends itself readily to provision of all this information, although commonly the polyethylene package is enclosed in a further polyethylene, cardboard, or even foil package. Under these circumstances, instructions, and other information, may be provided on a separate sheet packed within the cover.

(vii) Distribution

The essential requirements for distribution are that the inoculant should not be subjected to excessive temperatures; ideally it should be held at 4 °C throughout transport and storage. Although this may not generally be practicable during transport, precautions should be taken to minimize exposure by using rapid transport, preferably at night. Inocula in bulk should be transported in strong cartons or boxes. It is essential that any products normally marketed in the frozen form should reach their destination without thawing so that particular packaging requiring insulation and coolant is essential. Storage at less than 0 °C of any product other than those prepared in the frozen form is not recommended. In Australia it has generally been the practice of manufacturers to replace stocks unsold at expiry date with current material, to discourage sale of inferior products.

(d) CONTROL OF THE QUALITY OF INOCULANTS

(i) Organization

Primarily, the quality of any product is the responsibility of the manufacturer. His product may in addition be subject to official regulations. The powers provided under legislation may be periodically or only occasionally invoked by government agencies and the degree of inspection, evaluation, or penalty can vary widely. Many biological products, particularly food and drug items, may be subject to particularly stringent external controls of this nature. If failure to pass regulatory requirements involves loss of the product or restrictions on its use or distribution, it is clearly essential for the producer to adopt rigid control of the quality of the product. In the absence of any official regulation the manufacturer may adopt very low standards, which are readily attained and require minimal implementation.

The powers of external control bodies vary widely between countries. The various control measures in the U.S.A. since the 1930s served to protect the farmer from worthless products, but were not designed to be, nor did they function as, real measures of inoculant quality (Burton, 1967). The Indiana State Chemist (Schall *et al.*, 1975), which is the only authority currently publishing data on inoculant quality in the U.S.A., presents results of tests in qualitative terms only. However, in specifying brand names and manufacturing companies they are more explicit than any other testing authority in the world.

India (Anon., 1977b) has set explicit standards of both production and quality testing. Attainment of the standards allows the use of the ISI Certification Mark. However, the manufacturer is free to choose his own strains of rhizobia.

Probably the most centralized quality control system is that of the Australian Inoculants Research and Control Service (AIRCS)—previously the University–Department of Agriculture Laboratory Service (U-DALS). It has no official regulatory powers. The history of the development of AIRCS is worth noting in relation to the possible development of inoculant industries in other countries. Government Laboratories vacated the commercial inoculant production field in 1954 in favour of private companies. The absence of suitable naturally occuring rhizobia from large areas being sown with legume-based pastures highlighted the many failures which then became evident (Waters, 1954), and Vincent (1954) initiated positive moves to secure improved quality. It is significant that initially Vincent (1954) advocated setting up a licensing authority and quantitative and qualitative standards which he stated 'should be the definite responsibility of the manufacturer'. Poor quality inoculants continued to appear and an informal meeting attended by all active manufacturers and interested scientists resolved to establish mutually acceptable standards of quality and to ensure that only mother cultures from a central collection should be used. Manufacturers provided financial support to the control laboratory, which tested the final products. Quality of the final products was still often poor so a further meeting of the same group of people resulted in the introduction of a system of progressive control in which the initial broth, the freshly manufactured inoculant and the product on sale were all tested. 'This marked the beginning of an effective control programme in Australia' (Date, 1969). The principles of the organization are still retained, the standards have been raised and procedures have been modified continually as the need arose.

The functions of U-DALS expressed by Date (1969) remain the basic functions of AIRCS today:

1. selection, testing, and maintenance of suitable rhizobial strains;
2. control of the quality of legume inoculants;
3. advice to and research for manufacturers, distributors, and users of inoculants, on the problems of production, handling, and application that affect the quality and efficiency of inoculant cultures.

In 1978 AIRCS was also required by its funding authority to examine standards for pre-inoculated seed.

The original U-DALS organization was funded in part by the manufacturers with a large proportion of the funds being raised by the University of Sydney. In 1971 AIRCS was formed, with major financial assistance from the governments of all States of Australia. Contributions from a State are calculated on the basis of the proportion of the total inoculant used in that State. Financial support by manufacturers was discontinued, although charges are made for particular services outside the normal framework of local inoculant production.

Compliance with the standards set by AIRCS is still undertaken voluntarily by manufacturers.

(ii) Facilities required

Suitably qualified personnel are essential. Ideally the laboratory should be in charge of a professional microbiologist with at least one technically trained assistant. Suitable premises need to cater for contaminant-free handling of cultures. If a suitable room is not available it may be desirable to provide a suitable aseptic cabinet (e.g. a laminar flow cabinet). Normal aseptic procedures in a clean closed room should be adequate for control procedures.

Preparation of antisera requires suitable facilities for housing of small animals (preferably rabbits). The requirements for facilities and correct handling procedures are outlined by Kingham (1971). (N.B. In certain countries the performance of many of the procedures necessary requires the possession of a valid animal licence).

Controlled-environment facilities are essential for growth of plants for routine tests of infective ability and nitrogen fixation. These commonly need to provide conditions to suit both temperate (e.g. 15 °C night, 20 °C day) and tropical or sub-tropical species (e.g. 20–25 °C night, 25–30 °C day), although strict adherence to a temperature regime is less important than avoidance of excessively high temperatures. Glasshouses or shade houses are suitable for growth of plants in pots or similar open units, but the particular specialized assemblies used for testing of infectiveness or effectiveness of *Rhizobium* spp. (e.g. closed test-tubes or the modified Leonard jar; see Gibson, Chapter II.4) may require special precautions. Temperatures within enclosed test-tubes in a glasshouse can readily reach 35 °C in an ambient temperature of 25 °C. Shading the outside of the glasshouse with louvres (Hely, 1959) or blinds (Norris and Date, 1976) can allow good control in hot environments. Even a metal frame covered with a shade cloth protecting a small area within a glasshouse can effectively prevent heating of tubes above the ambient glasshouse temperature. Alternatively, tubes may also be housed in controlled-temperature water-baths. Controlled-environment rooms should provide adequate light, air circulation, and temperature control, and the day-length should be controlled by time switches. Ballasts for fluorescent lights are sources of heat which must be housed outside the room. 'Warm white' or 'cool white' fluorescent tubes have a limited light spectrum which needs to be supplemented at the red end of the spectrum by either incandescent bulbs or Grolux fluorescent tubes. The latter are not readily available in many countries and, even with the full expected life, replacements are frequent and expensive. Controlled environment cabinets have built-in refrigeration and are expensive in terms of cost of unit space compared with controlled-environment rooms.

It is useful to have field-testing facilities nearby, but any full-scale programme will require access to a number of sites. The special requirements of strain evaluation and the need to avoid cross-contamination between treatments, render it essential for any test area to be under the full control of the testing body, and normally involves fencing.

Equipment requirements are not elaborate but an autoclave and dry sterilizer are essential and incubators and refrigerators are highly desirable.

A shaker or blender is necessary for suspension of inoculant or inoculated seeds in water for quantitative evaluation. While wrist-action shakers are commonly used it has been found in this laboratory that the Stomacher Lab-Blender (A. J. Seward, Bury St Edmunds, Suffolk, U.K.) is faster (15–20 s per sample) and provides a superior suspension (see Figure 1). The essential

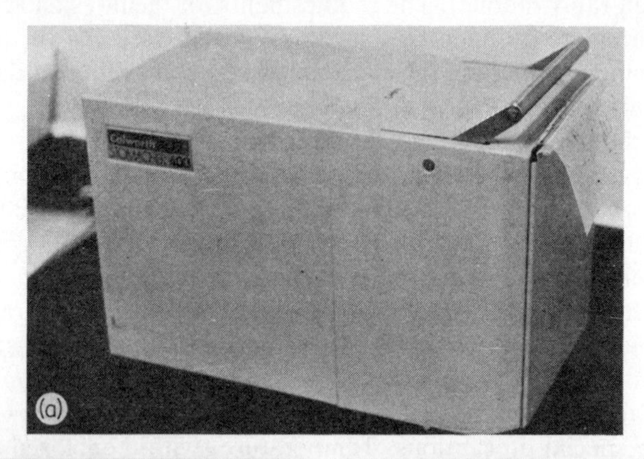

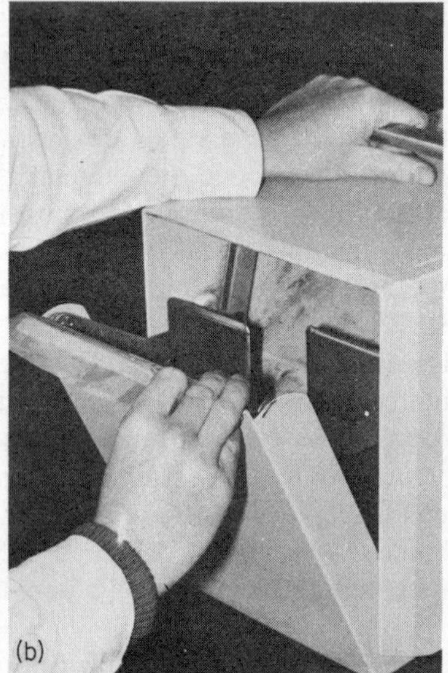

Figure 1. The Stomacher Lab-Blender in use for dispersion of inoculant or inoculated seeds for counting of rhizobia. (a) During operation, the material being examined is contained in a plastic bag, whose top is seen protruding at left. (b) Placing the plastic bag in the chamber before locking the door closed with the handle (upper right). The agitator paddles can be seen within the chamber. (c) Removal of the sample after blending

feature of the Stomacher is a pair of flat, vertical paddles alternatively moving horizontally towards a vertical surface on which is held a closed plastic bag containing the diluent and suspension material. The paddles approach within about 3 mm of the vertical surface but the mixing action is caused by the pounding action of the paddles producing a sponging and shearing action on the sample. When rhizobia are being suspended from seed, possible puncturing of the bag is avoided by mounting it between thin layers of foam plastic.

A microscope is necessary for examination of stained cells and preferably should be fitted with phase contrast equipment for total counts with a bacterial counting chamber (e.g. Petroff–Hausser).

For serological (agglutination) tests, suitable racks and a controlled temperature water bath are necessary.

A freeze-drier provides the most satisfactory method for long term storage of *Rhizobium* strains and an ampoule tester and an ampoule constrictor are valuable extras (all available from Edwards High Vacuum, Crawley, Sussex, U.K.).

(iii) Methods

A detailed description of the methods below has been presented by Vincent (1970) and in Chapter III.1. The outline given here is restricted to the particular requirements of inoculant quality tests.

Counting of Rhizobia

Total cell count. A Petroff–Hausser or similar counting chamber is used because its shallow depth (0.002 cm) requires less microscope adjustment than a haemocytometer slide of depth 0.01 cm. A microscope with phase contrast illumination and a 20–40× objective is necessary.

The slide is divided into clearly marked squares of measured area so that conversion factors, provided with the slide, allow calculation of numbers per unit volume. Populations up to 10^8 ml^{-1} can be counted without dilution and the lower limit for a reliable estimation is about 10^7 ml^{-1}.

Plate count. The plate count is used for estimating numbers of rhizobia in broth cultures and inocula prepared from sterilized carriers. Dispersed suspensions are diluted serially in 10-fold steps to a level where 30–300 cells are expected to be in the sample aliquot. The materials required are:

1. *Diluent.* The proposed water supply should be checked for suitability as a diluent. Protective salts may be necessary additions for good survival of rhizobia. Sterile diluent (9 ml) is prepared in bottles containing 99 ml, or in stoppered, capped, or cotton-wool-plugged tubes.

2. *Pipettes.* Straight-sided 'blow-out' pipettes of capacity 1 ml are best, although a volume of 2 ml may be useful where additional 1-ml aliquots are necessary. A fresh pipette is used for each dilution step; 1-ml rubber bulbs may be used to avoid sucking by mouth.
3. *Yeast–mannitol agar medium and Petri dishes (9 cm diameter).* Yeast-extract–mannitol medium (Section c.i) is usually used but $CaCO_3$ is omitted to avoid clouding the plates. Congo red (10 ml of a 0.25% aqueous solution per litre of medium) may be included to indicate possible contaminants. Rhizobia absorb less dye than most other bacteria. For poured plates, the medium is held at 50 °C in a water-bath until 10–15 ml is added to the plate and mixed with a 1.0-ml aliquot of diluted sample. Alternatively, 0.2 ml of aliquot may be spread evenly over the surface of a pre-poured plate.

In quality control work it is necessary only to prepare a dilution series for counting at pre-determined levels. An example of the procedure to examine, with poured plates, a broth expected to contain at least 10^9 rhizobia ml^{-1} is shown below. A fresh pipette is used for each transfer, avoiding contact between the pipette and the contents of the next dilution tube. Six pipettes would be used in the example. The contents of bottles and tubes are mixed by shaking or by sucking and expelling the suspension with the pipette. Duplicate plates are prepared for each dilution cultured.

Dilution	
	Broth (minimum expected 10^9 ml^{-1})
	↓ 1 ml via pipette
10^{-2}	99 ml dilution bottle (minimum expected 10^7 ml^{-1})
	↓ 1 ml via pipette
10^{-4}	99 ml dilution bottle (minimum expected 10^5 ml^{-1})
	↓ 1 ml via pipette
10^{-6}	99 ml dilution bottle (minimum expected 10^3 ml^{-1})
	↓ 1 ml via pipette
10^{-7}	9 ml dilution tube (minimum expected 10^2 ml^{-1})
	→1 ml → { duplicate Petri dishes
	→1 ml → { for addition of molten agar
10^{-8}	9 ml dilution tube (minimum expected 10^1 ml^{-1})
	→1 ml → { duplicate Petri dishes
	→1 ml → { for addition of molten agar

The estimate of numbers from such a series is derived from a mean of the numbers of colonies on duplicate plates on which 30–300 colonies develop. Use of spread plates, which receive only 0.2-ml aliquots, reduces the series by one 10-fold dilution, as the minimum expected number at dilution 10^{-6} is 200.

Plant infection count. There are no reliable culture tests for identification of legume root nodule bacteria. Thus, the plant infection dilution count must be used to estimate their numbers where other organisms are present in a suspension or culture. The technique is described by Brockwell in Chapter III.1.

Under the particular conditions of quality control, where it is necessary only to determine whether a population reaches a certain minimum number, tests at four selected levels of a 10-fold dilution series are adequate and five levels will provide a precise estimate considerably above the minimum. Thus, the tables need only cover four levels and adjustment can be made for the primary dilution. Table 1 has been prepared on the basis of three plants per dilution, to illustrate the minor differences between two popular sets of tables, but alternative numbers can be readily calculated from Fisher and Yates (1963).

Table 1. Estimates of Rhizobium numbers obtained by two methods of calculation from 10-fold serial dilutions with three tubes at each level

No. of positive tubes				Estimates of no. in aliquot of lowest dilution		
Relative dilution						
10^0	10^{-1}	10^{-2}	10^{-3}	MPN estimate from Brockwell *et al.* (1975a)	Total no. of positives	Estimate from Fisher and Yates (1963)
3	3	3	3	⩾2300	12	⩾1726
3	3	3	2	919	11	861
3	3	3	1	424	10	424
3	3	3	0	230	9	180
3	3	2	1	147	9	180
3	3	2	0	91.8	8	88
3	3	1	0	42.4	7	38
3	3	0	0	23.0	6	17
3	2	1	0	14.7	6	17
3	2	0	0	9.2	5	8.6
3	1	0	0	4.2	4	3.8
3	0	0	0	2.3	3	1.7
2	1	0	0	1.5	3	1.7
2	0	0	0	0.9	2	0.9
1	0	0	0	0.4	1	0.4
Approximate range Factor for 95% fiducial limits (×, ÷)				4.1		4.8

If the standard required of a peat is 10^9 g^{-1} and each plant receives a 1-ml aliquot, plants should be tested at dilutions 10^{-7}, 10^{-8}, 10^{-9}, and 10^{-10}. Thus in the following example:

Dilutions tested	10^{-7}	10^{-8}	10^{-9}	10^{-10}
No. of positive tubes	3	3	3	1

the estimate is 424× the primary dilution, in this case 10^{-7}; *i.e.* $424 \times 10^7 = 4.24 \times 10^9$. If all plants were nodulated the peat would contain at least 1.726 (or 2.3) $\times 10^{10}$. Use of 0.2-ml aliquots would make it preferable to test at lower dilution. In this example:

Dilutions tested	10^{-6}	10^{-7}	10^{-8}	10^{-9}
No. of positive tubes	3	3	3	1

the estimate is $424 \times 10^6 \times 5$ (because only 1/5 ml used) $= 2.12 \times 10^9$ ml^{-1}.

The materials required are the same as for the plate count except for growing plants aseptically (see Chapter III.1). For estimating numbers of rhizobia it is necessary only to use a test host on which nodules can be formed by the *Rhizobium* of interest; it is not necessary that they fix nitrogen. It is desirable to use as small a seed as possible to minimize the need for growing plants in large assemblies. The following is the list of useful test hosts commonly employed in this laboratory (exceptions are tested only on the correct host):

1. *Rhizobium trifolii–Trifolium repens* (white clover) (exception *T. semipilosum*).
2. *Rhizobium meliloti–Medicago sativa* (lucerne).
   N.B. This is a particularly complex group (Brockwell and Hely, 1966) so that exceptions are likely, e.g. *M. rugosa.*
3. *Rhizobium leguminosarum–Vicia dasycarpa* (woolly pod vetch).
4. *Rhizobium lupini–Ornithopus sativus* (Serradella).
5. *Rhizobium* spp. (cowpea group)–*Macroptilium atropurpureus* (siratro).
6. *Rhizobium japonicum* (soybean group)–*Glycine ussuriensis* (wild soybean).

Van Rensburg and Strijdom (1974) found that siratro was adequate as a test host for the presence of *Rhizobium lupini*, *R. japonicum*, and *R. phaseoli*, but the results with the latter two have been inconsistent in Australian tests (Vincent, 1970; Thompson, unpublished work).

Serological identification

Although a wide variety of methods are available (Chapter II.9) the simplest and most rapid procedure is somatic agglutination.

Materials

1. *Antiserum* is commonly stored in volumes of a few millilitres and held frozen without additives. The stock is normally diluted to 1:100 or 1:200, although the titre of rabbit antiserum is commonly at least 1:1600.
2. *Antigen* suspension must be cloudy (i,e, at least 10^7 ml^{-1}) so that a positive reaction will be clearly visible
3. *Saline*: 0.85% NaCl.
4. *Agglutination* tubes, capacity 1 ml. The Dreyer pattern is to be preferred, although Durham tubes are satisfactory alternatives.
5. *Water-bath* (52 or 37°C).

Procedure

1. Mix equal parts of suspension and saline in a capped test-tube and hold in boiling water-bath for 30 min to inactivate flagellar reaction.
2. Using a Pasteur pipette, mix 18 drops of boiled antigen with 2 drops of antiserum in one tube, and 18 drops of boiled antigen with 2 drops of saline in a second tube (control).
3. Place in a rack in the water-bath with the water level below the level of the reactants to promote mixing by convection.
4. The somatic agglutination should be visible by 4 h at 52°C, but longer times are often necessary. The reaction commences with a granular appearance which should proceed to full settling out. Auto-agglutination of the saline control necessitates repetition of the test, and possibly reduction of the concentration of saline to 0.5%.

Testing for contaminants

Gram stain. This is the standard procedure for testing agar and broth cultures for presence of Gram + contaminants, spore formers, and cells with distinctly different morphology (the procedure is outlined by Vincent, 1970). In broth cultures allowances may need to be made for a few dead Gram + cells carried over from the autoclaved medium components and some yeast cells are commonly visible in medium based on yeast–water.

Glucose–peptone test. Glucose–peptone does not favour growth of most rhizobia but many contaminants readily grow and produce pH changes. The medium consists of glucose 5 g, peptone 10 g, agar 15 g, water 1 l. Bromocresol purple (1.0% in ethanol) (10 ml) is added to the melted agar before dispensing into 28-ml McCartney bottles or 15 × 150 mm test-tubes for sterilization and sloping. A loopful of culture streaked on a slope and incubated at 28–30°C should be examined after one and two days. Marked growth, especially if associated with change of pH, indicates gross contamination. Some strains of rhizobia show slight growth but generally without appreciable change in pH.

(iv) Strains of Rhizobium for inoculants

Single strain or multistrain?

Inoculants are generally used for more than one cultivar of a legume species (e.g. soybeans) or for more than one legume species (e.g. clovers), and even for a number of legume genera covering a number of legume families or sub-families (e.g. cowpea cross-inoculation group). However, mixing of species of *Rhizobium* (e.g. *R. meliloti* and *R. trifolii*), which do not normally cross-infect, is not favoured, largely because it reduces the number of organisms available for each host. More importantly, it is essential that an inoculant should never contain a strain which will form ineffective nodules with any of the hosts for which it is recommended.

The advantages of single strain inoculants are:

1. Whether the broth or the final inoculant receives equal numbers of cells of more than one strain, subsequent multiplication can result in dominance by one strain (Marshall, 1956). Strains within a *Rhizobium* species can differ in ability to survive in peat (Thompson, unpublished data). Use of mixtures of *Rhizobium* species is even more likely to lead to differential death rates because of differing resistance to adverse conditions (e.g. salt concentrations; Steinborn and Roughley, 1974).
2. Any change in host requirements can be catered for by developing a separate inoculant which can be clearly and specifically labelled.
3. Any unfavourable variation showing up in a strain (e.g. loss of effectiveness or infectiveness) is readily evident if the host is grown in a low nitrogen environment with adequate controls, i.e. the deficiency is not masked by other strains in the inoculum.
4. If quality control is applied only to the final product it is much simpler to check the identity of the single strain.

The disadvantages are:

1. Strict adherence to a policy of selection of the single best strain for each variety or species of host can result in the need for a larger range of inoculant types. This can lead to organizational problems at both manufacturing and retailing levels.
2. Loss of effectiveness or infectiveness will result in complete failure of inoculation while multistrain inoculants will provide an infective alternative.

In view of our relative ignorance of factors governing rhizosphere colonization, infective processes and selection by the host plant, the proponents of mixed-strain inoculants can justifiably claim that the inevitable host × strain × site interactions will be best catered for by a mixture of effective strains of which at least one may reasonably be expected to form an association. However, even if all added strains have survived (Marshall,

1956) the published evidence is even ambivalent on the ability of a host to select the most effective strain (Vincent and Waters, 1953; Robinson, 1969), and the competitive ability of strains in soil is even less understood. Thus, recently separate research groups (Gibson *et al.*, 1976a; Roughley *et al.*, 1976) found that *Rhizobium trifolii* strain WU95 was consistently competitive with other strains for nodulation of subterranean clover, but the other strains were ranked differently by the two groups of workers.

It is perhaps significant that the greatest emphasis on single-strain inoculants has been in Australia which has thorough quality control, thus reducing the risk of undetected failure of the strain.

Sources of strains

A collection of strains is normally built up with a combination of field isolates and accessions from collections of other workers. Some of the principles and procedures have most recently been discussed by Date (1976) and Norris and Date (1976).

Other collections. Many strains in use for culture of legume inocula have originally been obtained directly from collections of other workers. The possible sources have been discussed by Dalton and Brockwell in Chapters II.1 and III.1, respectively. It is important that such accessions should retain their original collection number, even though a local renumbering system is normally necessary for storage. There is clearly an ethical requirement on the part of the recipient to advise the custodian of a collection if he proposes to use a strain for commercial inoculant production.

Field isolates. Initiation of a programme to produce legume inoculants commonly results from the need to successfully inoculate species introduced into a new environment. If this need has been demonstrated by poor nodulation of test plants it is unlikely that the natural population of rhizobia will yield a suitable isolate. However, good nodulation of isolated uninoculated plant hosts may indicate the presence of a small population of rhizobia which may be potentially useful as an inoculant. Guidelines have been set out by Brockwell in Chapter III.1.

Criteria for selection of inoculant strains

The number of criteria to be considered in selection of strains for legume inocula have steadily increased at the same time as improved understanding of the legume symbiosis and of the ecology of *Rhizobium*. The most obvious are:

1. ability to nodulate the legumes for which it is recommended;

2. effectiveness in nitrogen fixation in the nodules so formed;
3. suitability for inoculant production;
4. usefulness under field conditions.

The first two criteria may be tested under controlled conditions; certain environmental factors, especially temperature, can modify the symbiotic response. There is little point in pursuing tests of strains which are poor performers over the normal temperature range in (1) and (2) when alternative good performers are available.

At the manufacturing level it is sometimes found that strains differ in their reaction to normal growing procedures and use of alternatives may be justified to avoid change of growing procedures.

Usefulness under field conditions embraces a wide range of attributes which have been discussed (Chapter III.1). However, these may not only be difficult to test for technical reasons of availability of adequate test criteria but also for logistical reasons, particularly as inoculants are used over a wide range of environments. Thus, the essential test necessary is to ensure that the strain can form nodules and fix nitrogen in normal field situations at least as well as alternative strains. It is particularly important that such ability is demonstrated with the full range of proposed hosts.

Evaluation under controlled conditions

This involves use of a *Rhizobium*-free medium in a container sufficiently large to allow good differentiation of growth of nodulated and unnodulated plants. Although use of nitrogen-free nutrient medium is common, there is evidence that a more realistic evaluation of strain performance will be obtained in the presence of a small quantity of added nitrogen (Gibson, 1976b). Such conditions are more akin to the normal field situation so that choice of strains favoured by the presence of nitrogen should provide more generally useful inoculant strains. Optimal conditions for plant growth should be chosen. For nitrogen controls, combined nitrogen should be applied immediately after nodules are formed on the inoculated treatments.

Because of the need for prevention of cross contamination, the most commonly used assemblies have been sterilized, often enclosed, containers e.g. tubes or modified Leonard jars, which have been discussed.

Evaluation under field conditions

These tests are much more time, labour, and space consuming because of soil variation within sites, the need to sow treatments sufficiently well spaced to minimize cross contamination, and the possible presence of soil nitrogen which may result in some growth of uninoculated controls, so that differentiation between treatments is delayed. The reduced differentiation between treatments can also be the result of the presence of naturally-occurring rhizobia and in fact the uninoculated controls may be as effectively nodulated as the treatments (see Chapter III.1). In this case it may be necessary to carry out serological identification of nodules to determine whether the inoculant strain has in fact formed the nodules (Chapter II.9). Inoculation rates should be normal and nitrogen-free fertilizers added to ensure adequate nitrogen fixation and plant growth.

Strains for forage legumes. It is common practice to compare inoculant treatments in the field by sowing in rows and the practice of removing individual plants to examine nodulation has tended to focus attention on the individual plant as a source of yield data. However, with clovers sown in rows and swards Thompson *et al.* (1974) found that the only reliable guide to proportion of plants nodulated was obtained by using plants where there was no possibility of plant-to-plant cross-infection. Further, yield was best measured on a unit area or row length basis.

Strains for grain legumes. Compared with the controlled environment, time to harvest may be of less concern and the pressure of available soil nitrogen may favour the use of a criterion based on the longest period of differentiation. On the other hand, the common cessation of fixation in grain legumes prior to pod fill may result in interactions during the period between pod fill and harvest, e.g. plants well provided with nitrogen by an efficient strain may suffer more from a post-flowering moisture stress than smaller plants with less nitrogen. Further, such stress may be evident in grain quality which is not readily measured in simple terms. If labour and facilities permit, measurements of yield of both dry matter and grain are desirable.

A particular problem is encountered with canning peas which have highly critical harvest dates for desirable quality. Generally, harvest needs to be before maximum yield is reached.

(v) Maintenance of stock cultures

The principles of accession, maintenance, storage and collection have been presented in Chapters II.1 and III.1.

A stock culture collection is an essential part of legume inoculant control, whether for one small manufacturer or a number of users. It is expensive, time consuming and demanding of careful manipulations and good records, but it is the basis of the whole operation.

The essential features of a good collection for servicing inoculant quality control are that it provides:

1. strains of proven ability for the legumes of interest;
2. 'back-up' strains also of proven ability;
3. strains of current anticipated and potential usefulness.

Agar cultures. The most convenient form of culture for storage is the slope of yeast–mannitol agar in a cotton-wool-plugged test-tube, but these cultures are most likely to be unstable genetically (Vincent, 1970). On this medium, organisms survive for several months and even up to 2 years at low temperature. The largest problem is moisture loss, which can be reduced by using screw-capped tubes or McCartney bottles, or covering the cotton-wool plug with Parafilm M (American Can Co., Neenah, Wisc., U.S.A.) or with sterile liquid paraffin. Low-temperature storage is favoured.

Freeze-dried (lyophilized) cultures

Freeze-drying is the preferred method of storage and has been described by Dalton in Chapter II.1. It requires practiced technique to produce good results consistently. Because of the long storage practicable (many years), cultures are thus stored with minimal transfers from the original and should go into a freeze-dried collection as soon as they are received and authenticated as *Rhizobium*. Because of the low volume requirement it is practicable to minimize the slow death rate by storage under refrigeration. Sufficient units should be prepared to meet several years' requirements.

(vi) Mother cultures

Mother cultures provided to the manufacturing laboratory for production of inocula must always be readily available in an immediately usable form (normally in agar tubes), should not vary, and should originate from the same stock culture.

It is simple to ensure ready availability, but the greatest problem is to ensure that genetic instability is not expressed during transfers. There should be minimal culture transfers between the original stored culture and the mother culture and the authentication tests should be made at the latest possible transfer.

A balance needs to be struck between availability and the limited safe storage life of such cultures. As inoculants are normally prepared on a seasonal basis, annual production of mother cultures followed by refrigerated storage is a reasonable compromise.

The simplest procedure with full checking is shown in Figure 2A. All transfers are by mass streaking on agar medium and all tests are made on a single reference or stock culture which is the source of inoculum for all mother cultures. The method does not provide any information on the range of variability within the culture.

An alternative procedure (Figure 2B) tests each mother culture for ability to produce effective nodules. Any not fixing nitrogen effectively are discarded. Clearly, incidence of such failure should precipitate more intensive studies of the original culture.

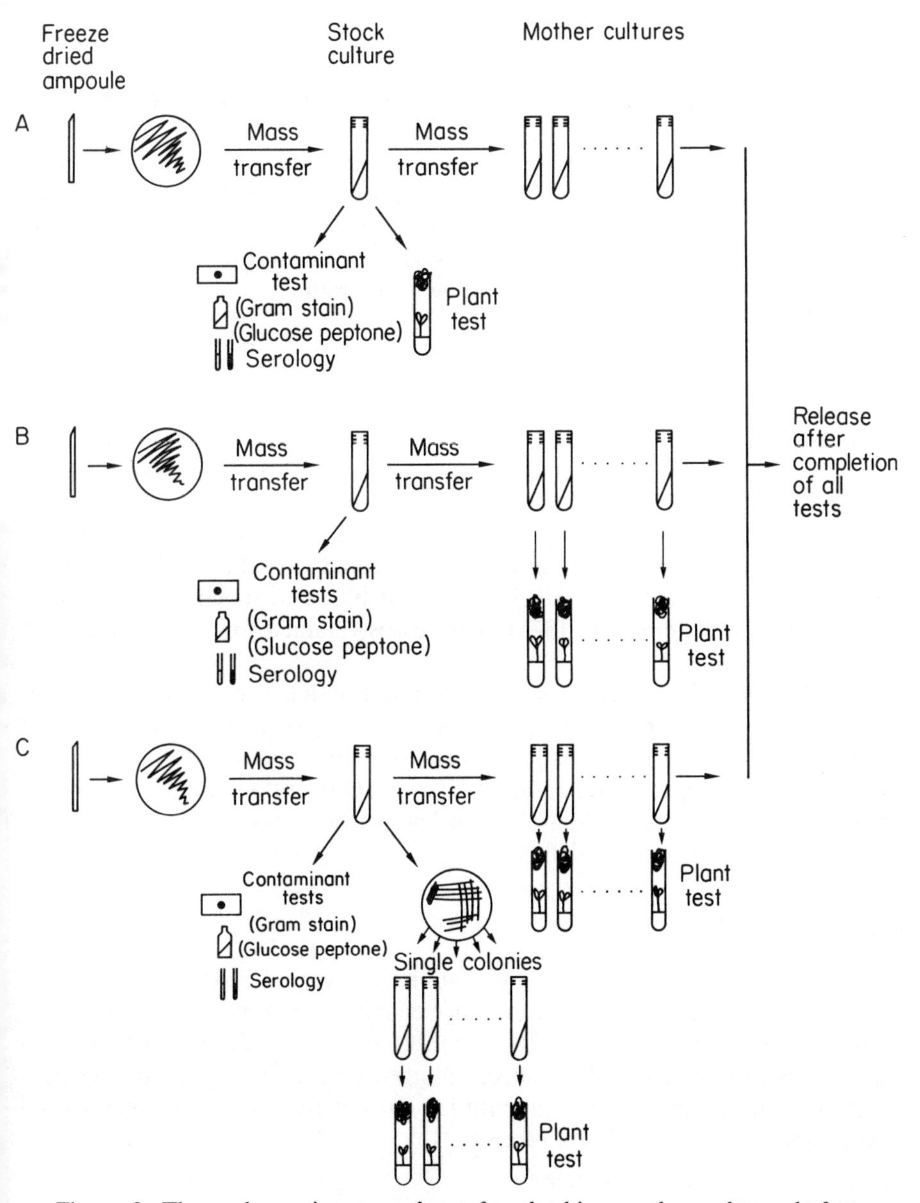

Figure 2. Three alternative procedures for checking mother cultures before release for inoculant production

The genetic instability of *Rhizobium* justifies the incorporation into procedures of a test of single colony isolates from the stock culture, when preparing mother cultures (Figure 2C). Each of 10 single colony isolates is tested on a number of replicated plants. If a single colony performs poorly, compared with the mass culture, it is re-tested and, if confirmed to be inferior, an effective single colony isolate should replace the original stock. Such re-isolation should be recorded by re-numbering.

Once the stock culture is authenticated, the mass transfer technique is used for production of any subsequent sub-cultures. Those for use as mother cultures are then re-tested on one plant to confirm ability to form nodules. The major criticism of the latter techniques is that the logistics of the operations can become prohibitive for large numbers of strains.

(vii) Quality control procedures

Most of the following control procedures have been followed at some stage by U-DALS and AIRCS in Australia in its quality control programmes (Vincent, 1970; Date, 1970). However, a recent study of the available data revealed that numbers in broth above 5×10^8 ml^{-1} were not always reflected in higher numbers in peat culture. It was concluded that such tests were unnecessary except during the development phase of inoculant production (Roughley and Thompson, 1978). The test is retained for broth cultures of any new strains.

The testing of inocula purchased from retail outlets is considered essential to provide a full quality testing service, even when inocula are controlled during production. The major problem is to obtain a representative sampling because of the large number of outlets. Testing at the retail outlet is commonly the first point of examination and can lead to more thorough initial testing if defective inoculants are detected.

Broth culture stage

Sampling. Broth samples (10 ml is sufficient) should be drawn aseptically from the fermenter at the time at which maximum numbers of live cells are expected. Separate samples must be collected for each strain and forwarded as rapidly as possible to the control laboratory in screw-capped bottles protected by insulation and packed with ice but not frozen. The sample should be examined immediately on receipt.

Priority tests

1. Serological identity is tested by agglutination as above.
2. The Gram stain should be made on undiluted broth.
3. Glucose–peptone agar is also streaked with undiluted broth.
4. The total count of rhizobia is obtained with the Petroff–Hausser chamber but gross contamination with morphologically distinct organisms may

also be observed. Commonly, at least half the cells counted are dead at this stage of the broth development.

The above tests can provide presumptive evidence of a pass or failure within 24 h so that a decision can be made whether to use the broth for the next stage of manufacture. The following tests are also initiated on receipt. Final results of slow-growers may need up to 10 days.

5. For detecting presence of contaminants, a sample of undiluted broth is placed on yeast–mannitol agar. This can reveal contaminants which have been detected in the above tests.
6. A viable count of rhizobia is obtained by normal plate count.

Optional tests

1. Measurement of pH on receipt is used primarily as a guide to possible contamination depending on the *Rhizobium* species. Commonly, *R. meliloti* strains produce a pH of 5.4 but cultures of strains for *Lotononis* can exceed pH 8.
2. The plant dilution count can be combined with the normal plate count. A delay of 3–4 weeks will be necessary before reading results but a plant test made at this stage replaces the plant count which would otherwise be necessary on the final product.

Inoculant at manufacture

Sampling. Samples should be taken from each batch and forwarded to the control laboratory as soon as preparation, or maturation, is complete. Control of temperature during transport is probably less critical than for broths, because the final product is probably more stable numerically, but refrigeration is preferred and certainly high temperatures must be avoided. In Australia six packets are collected per batch of inoculant but the variation between packets is small. Five of the six must reach the standard. Where greater variation exists heavier sampling intensity may be necessary.

Priority tests following full broth tests

1. For the viable count with sterilized carrier, the normal plate count is adequate and provides information on the presence of contaminants. Autoclaved carriers should be absolutely free of contaminants but gamma irradiation may not provide full sterilization. The requirement is therefore that there are no contaminating organisms at the lowest dilution examined (normally 10^{-6} in the AIRCS laboratory). With unsterilized carrier the plant dilution count is essential for estimates of *Rhizobium* population, although the plate count may provide useful information on relative number of contaminants.

2. The serological identity of the rhizobia is tested on cells obtained by suspension of colonies from plate count. When more than one strain is used it will be necessary to grow test material from colony picks. This is simplified if colony characteristics allow clear selection.

Priority tests without prior broth test: sterilized carrier

1. For the viable count, the normal plate count is all that is necessary as presence of contaminants should result in rejection.
2. At least one packet should be subjected to a full plant dilution count procedure as confirmation that an effective symbiosis is produced by the majority of the rhizobia. Provided some plants nodulate effectively, an inoculant should not be failed on the basis of one such test but nodulation failure at higher dilutions alerts the operator to possible problems requiring investigation.
3. Determination of serological identity is relevant only if the testing authority provides strains.
4. The Gram stain is made on a mass streak of colonies from a low dilution of the viable count.

Priority tests without prior broth test: unsterilized carrier

1. For a viable count, the plant dilution test is essential.
2. No separate test of infectivity is necessary because information is obtained from viable count.
3. Serological identity is best tested from nodules obtained in above count but is only relevant if testing authority provides strains.

Inoculant from retail outlets

Priority tests on previously tested inoculants

1. A check on labelling should show that batch number, expiry date, and hosts must agree with previous records.
2. The viable count is made by plate count or plant count depending on the sterility of the carrier.
3. Serological identity is especially important with sterile carriers where the plant count is not necessary for estimation. The serological check therefore becomes the only proof of strain suitability.

Optional tests on previously tested inoculants

1. The moisture content is useful for record purposes.
2. The plant test is clearly the only definite test for inoculating ability.

Priority tests on previously untested inoculants. If the testing authority has had no jurisdiction over the strains used by the manufacturer, there is no point in attempting to identify the strains available in the inoculum, even if sterilized carrier is used.

1. For the viable count and test of infectivity, the plant infection dilution count is essential, although a plate count will be a useful indicator of sterility.
2. An effectiveness test should be made but it may be combined with the plant count and the effectiveness of the association measured by growth of the host, provided the assemblies allow differentiation of treatments from the uninoculated controls.

(viii) Standards

Broth culture stage

Irrespective of the carrier, and of the number of strains in the inoculant, any broth or surface-grown culture of rhizobia must be a single strain and be free of contaminating organisms. It should provide as high a population as possible. The Australian standard has been 5×10^8 ml^{-1} for some years for all except *Lotononis* (3×10^8 ml^{-1}).

Inoculant at manufacture

Unsterilized carrier. The standards set for impregnated carrier arise largely from the levels which can be achieved by competent manufacturers. There is little point in setting unattainable standards but low numbers of rhizobia may be offset in part by modifying the total quantity of seed or area to be inoculated. Thus a working minimum of 100 rhizobia per seed of small legumes was adopted early in the history of Australian inoculant control and the rate of application of the inoculant to the seed chosen accordingly. While unsterilized peat was used in Australian inoculants a standard of 10^7–10^8 rhizobia g^{-1} was considered adequate for 2 months' expiry and greater than 10^8 g^{-1} was allowed 6 months' expiry. South Africa also sets a standard of 10^8 g^{-1} (van Rensburg and Strijdom, 1974).

Sterilized carrier. The introduction of sterilized peat in Australia allowed the standard for all inoculants (except *Lotononis*, 5×10^8 g^{-1}) to be raised to 10^9 g^{-1}. Such peats are also required to be free of contaminants at the lowest dilution tested (commonly 10^{-6}). Standards set in New Zealand (Anon., 1979c) and Canada (Anon., 1979a) are similar.

Inoculant from retail outlets

The minimum standard set for inoculants on sale will ultimately be based on measured survival and will obviously be dependent on time to expiry. For some inoculants stored in the frozen state, or perhaps freeze-dried preparations, it is possible that minimal decline may occur. The carrier-based inocula in Australia have generally been allowed a 10-fold drop before expiry. Thus, on present standards the fresh peat requirements of 10^9 g^{-1} allows for a count at expiry of 10^8 g^{-1}.

Expiry periods

A balance must be struck between reasonable storage life, which can be ascertained only by measurement of survival, and commercial requirements for distribution and ready availability. A minimum period for the latter is probably 2 months if transport is adequate, but there seems to be little justification for a maximum exceeding 12 months.

In Australia, manufacturers are allowed a maximum expiry date which is 12 months after the date of commencement of the tests. The manufacturer may store the product at 4 °C for any period up to expiry but is restricted to a maximum of 6 months' expiry period after release from store. Thus marked expiry dates can fall anywhere within 6–12 months from the testing date. Before the maximum expiry date, sample packets of inoculant can be resubmitted for test, and are subjected to the same tests as at manufacture. If they pass they are allowed a further 12 months' expiry provided they also contain more than a specified moisture percentage (currently 40% compared with approximately 50% for fresh peat).

Calculation of realistic standards

If a standard considerably below these is set (e.g. 10^6 g^{-1}) the attainment of even 100 $seed^{-1}$ on alfalfa would require a fresh 250-g pack of inoculant to be applied to 5 kg of seed. With a decline to one tenth of these numbers during its life to expiry, the 250-g pack could then only be applied to 0.5 kg of seed. It is doubtful whether such a quantity of peat inoculum could be successfully attached to the seed.

Implementation of standards

It is essential that adequate records are kept by both control body and manufacturers and the need is most obvious when more than one stage of manufacture is checked.

Broth samples normally need to be checked rapidly to determine whether the next step in manufacture should proceed. Manufactured inoculants should be accepted for testing only if the broth has passed the previous test. Within certain defined limits it may be practicable to re-test a stored inoculant for extension of expiry. The results of all tests should be available at the control laboratory and at the manufacturing plant.

Regulatory powers

Procedures and standards such as those described above were developed in Australia during a long and close association between manufacturers, the control body, and its advisory committee. As a result, there is no dissent regarding the application of standards and the control authority has had no need for regulatory powers. It has clearly been in the interests of the manufacturers to retain the right to quote official approval of the product. As the main funding authority for AIRCS, the State Departments of Agriculture are kept fully informed on the standard of the commercial products. The one State which has legislative authority to confiscate unsatisfactory material, Queensland, has not found it necessary to invoke its powers.

The less the control authority is involved in tests of the process during preparation the more arbitrary its standards are likely to be, and because the contact between the producer and the controlling authority is reduced the implementation of standards may be more dependent on regulation.

(e) QUALITY CONTROL OF PRE-INOCULATED SEED

Although inoculation is only a further step in the use of inoculants, pre-inoculated seed (i.e. seed inoculated before sale) has become a marketable inoculant commodity in its own right. Methods of production have been discussed by Brockwell in Chapter III.1, but surveillance of the quality of these products may be a reasonable extension of the duties of an inoculant quality control laboratory. Even in Australia, where control of the quality of inoculants has been established for many years, the recent advent of pre-inoculated seed on the local market has not been accompanied by a smooth transition of the principles of quality control. Potentially, the products available are likely to be more variable because the absence of need for special expertise and equipment can attract a large number of operators. The generally shorter life of inoculant on seed compared with that in the stored packet renders quantitative standards more critical, necessitates shorter storage periods, and necessitates adequate labelling to prevent claims that old material retains viable rhizobia. A particular problem commonly associated with a number of these products is the use of secret processes which preclude adequate discussion with the control authority. Patenting can also involve restriction of use and a degree of inflexibility in outlook. The net result of these problems in Australia has been variable numbers of rhizobia on pre-inoculated seed even though inoculants meeting the same standards are common to all producers.

This discussion is limited to evaluation of the final product in terms of the identity of the bacteria used, their number and the inoculation produced when the seeds are sown.

(i) Methods and control procedures

Identification of Rhizobia

If the control of quality of inoculated seed is part of a full programme of inoculant testing it is essential to confirm that the strain in use is the correct recommendation. This is best achieved by isolation of colonies (3–4) from a plate count.

The alternative is by examination of nodules (3–4) obtained from the grow-out test. To avoid the problems of re-isolation and testing by agglutination, direct identification by fluorescence microscopy (Trinick, 1969) is possible if facilities are available.

If more than one strain is present, large numbers of colonies or nodules may need to be tested to ensure that all strains are identified. Alternatively, recognition of one indicator strain may be considered to be sufficient evidence of the use of a known, tested inoculant.

Counting of Rhizobia

As contaminating organisms are commonly present, the plant dilution count is essential. Samples of 100–200 seeds are counted and weighed and the inoculant recovered by suspension with a Stomacher Blender or wrist-action shaker. For large seed a wrist-action or reciprocating shaker is necessary or, if the inoculant procedure is considered to warrant it, maceration may be desirable. With shakers, inclusion of glass beads with the seed may be found to improve the recovery of rhizobia.

Plate counts may be obtained if contaminating organisms are absent or in small numbers. Spread plates should be used to facilitate identification of contaminants and congo red medium is preferred, but even if the proportion of contaminants is low, counts are unlikely to be reliable because of competitive organisms.

Grow-out Tests

The most direct test of the quality of the inoculum on such seed should clearly be the ability of all the seeds of a sample to become effectively nodulated under the most adverse conditions likely to be encountered. With suitable field sites successful nodulation would satisfy all the criteria apart from a check on strain identity.

A control programme, however, frequently involves testing of seed under controlled and generally favourable conditions, usually in containers of *Rhizobium*-free medium. This technique has been used to differentiate between coating materials in inoculation of clover seed and Schall *et al.* (1975) routinely use grow-out tests in sand in crocks as a sole evaluation technique for tests of pre-inoculated seed in the U.S.A.

Despite the simplicity of the 'grow-out' test, its significance is often overrated because the essential principles of the test are ignored. In its simplest form, where conditions for nodulation are optimal, so that one *Rhizobium* in a unit can lead to nodulation of its host, the following relationship, described by Fisher (1925), holds:

$$\text{Percentage of fertile samples} = e^{-m}$$

where m is the mean number of organisms per sample.

Fisher's (1925) table 8 provides the following data on this basis:

Fertile samples (%)	10	30	50	70	90
Mean no. of organisms	0.1054	0.3567	0.6932	1.2040	2.3026

It can be calculated that a mean of one per seed will nodulate 67% of seeds and that only 6 rhizobia per seed are needed to nodulate 99.5% of all seeds. Reasonable agreement with this expected result has been found (Thompson, unpublished work) with *Trifolium subterraneum* under favourable controlled conditions, as follows:

Mean no. of rhizobia per seed	Expected nodulation (%)	Observed nodulation (%)
0	0	0
0.25	23	16
0.38	32	61
1.70	82	79
1000	100	100

However, when less favourable conditions are encountered, the relationship is not maintained. In the laboratory the use of tubes as 'grow-out' containers have been suspect because of anaerobic conditions (Ham and Frederick, 1966) and interactions between seed-coating materials and environmental factors (Norris, 1971). Porter and Scott (1977) were also critical of the lack of uniformity of tubes filled with a sand–vermiculite blend with resultant variation in moisture-holding capacity.

With field tests, where conditions are often unfavourable, a positive relationship may exist between percentage plants nodulated by the inoculum and mean inoculum size but the curve is often laterally displaced 100–1000-fold, i.e. upwards of 1000 rhizobia may be required to produce the 67% nodulation which could theoretically be expected from a mean inoculum of one organism. Competition from native rhizobia and lack of knowledge of the surviving populations are clearly contributing factors to this discrepancy but even under laboratory conditions there is a possibility of non-random occurrence of rhizobia on seeds. With one pre-inoculation process tested immediately after inoculation, 75–95% of the bacteria were recovered from 5–15% of the seeds (Cooper, 1962).

A further contributory factor to confusion regarding the 'grow-out test has arisen from the practice of evaluating results from seeds sown close enough to risk cross-infection. While the current procedures for testing pre-inoculated seed by the Indiana State Authorities are used to differentiate between satisfactory and unsatisfactory batches (Schall *et al.*, 1975) the practice of growing a number of seeds in one crock increases the chances of a *Rhizobium*-free seedling becoming nodulated from a neighbouring plant. Thus, Ham and Frederick (1966) found that all the plants in a container were nodulated when only one of the 16 seeds was inoculated and no direct watering used. Similar results were obtained by Burton *et al.* (1972).

The following conclusions are therefore drawn:

1. It is essential to sow seeds as individuals in individual containers or, if in the field, to space seeds adequately to prevent cross-infection.
2. To ensure that the test assemblies are satisfactory for nodulation of inoculated seed, controls should be prepared by inoculating parallel series of plates and plants in the same assemblies as for the test, with a dilution series of *Rhizobium* diluted to the point of extinction. Failure of assemblies to detect one or a few rhizobia visible on parallel plates clearly invalidates the test assembly.
3. Distribution of rhizobia may not be random between seeds either as the result of inoculation procedures or differential death rates.
4. The predictive field value of the laboratory grow-out test is minimal because successful nodulation can be obtained with so very few rhizobia. However, a low-percentage nodulation in the laboratory must be considered as indicative of unsatisfactory performance under field situations. Thus, the test is only of value as an indicator of failure.

Procedure for laboratory grow-out tests

1. Prepare the required number of sterile units, each sufficient to allow the growth from a single seed to a point where nodules form and can be seen to benefit the host. The units may range from plastic pouches (Weaver and Frederick, 1972), cotton-wool-plugged test-tubes (15 × 150 mm) containing plant nutrient agar for small seeds, such as *Trifolium repens*, and even *T. subterraneum*, to larger tubes or small

pots of sterile sand or similar medium for larger seeds. At least 30 seeds should be tested per treatment.
2. Place the seed for test on a *Rhizobium*-free surface, e.g. a Petri dish or notepaper.
3. Dip forceps in alcohol, flame, and allow to cool. Several pairs of forceps can be flamed together.
4. Sow one seed with each pair of forceps and return the forceps to alcohol. This single use prevents cross-contamination by the forceps.
5. After growth for approximately 4 weeks under favourable conditions, record plants as nodulated, not nodulated, or abnormal and/or non-germinating. Express percentage nodulation in terms of the germinating seedlings.

Alternative interpretations of grow-out test

The grow-out test has been primarily interpreted in qualitative terms, i.e. presence or absence of nodules. However, Burton *et al.* (1972) have presented data which showed a differential rate of nodule formation depending directly on *Rhizobium* populations. Thus, at 14 days 90% of soybeans in Leonard jars contained 3 or more tap root nodules with an inoculum of 10^6/seed, while only 10% did so with an inoculum of 100/seed. It is conceivable that the use of sufficient counted populations as control treatments, and intensive observations of these *vis-à-vis* the test seedlings, may allow some degree of quantification of the *Rhizobium* population. It is suggested, however, that such estimates are better made by the alternative of direct counting techniques.

(ii) Standards

Numbers of Rhizobium per seed

There are few data available on which to base a set of standards. New Zealand government agencies currently require that white clover and lucerne seed available for purchase by the government should carry 300 viable rhizobia per seed after storage for 28 days at 20 °C (Anon, 1979c). Canadian authorities have adopted standards as follows: small seeds (e.g. clover) 1000 per seed; intermediate seeds (e.g. sanfoin) 10,000 per seed; and large seed (e.g. soybean) 100,000 per seed (Anon., 1979b).

In Australia it has generally been accepted that pre-inoculated seed should be required to meet the same standard up to its expiry date as that of seed freshly inoculated with a satisfactory culture. For some time the requirement for an inoculant at expiry was that it provided at least 100 rhizobia per seed of white clover (Vincent, 1965). Subsequent improvements in inoculant standards raised this figure to 300 (e.g. Date, 1970) and currently, with an inoculant containing the minimum number of rhizobia (10^8 g^{-1}), 600 rhizobia per seed can be applied. With fresh inoculant commonly containing up to 5×10^9 g^{-1}, these figures can be 50-fold higher at inoculation. It is important to note that the discussion is largely centred on small pasture seed because they carry the least rhizobia and in fact are the most common seed to be pre-inoculated.

The numbers suggested by Thompson *et al.* (1975) on the basis of the use of the standard Australian 250-g inoculant pack were:

Trifolium repens	600
Trifolium pratense / *Medicago* spp.	2500
Trifolium subterraneum	3300

These differ to a small extent from those quoted by Date and Roughley (1977) only because of differences in the calculation of the quantities of seeds inoculated by 70-g and 250-g packs. While there has been criticism of the quality of commercially produced, pre-inoculated seed in Australia (Brockwell *et al.*, 1975b), such standards can be achieved. It is probable that cool (4 °C) storage will be necessary to ensure a useful commercial life for the inoculant on the seed.

Percentage of nodulating plants

Clearly, the majority of the seeds must be capable of producing nodulated plants. Schall *et al.* (1975) nominate 90% as satisfactory, while Burton *et al.* (1972) advocate 80% as a minimum for a pass. In so far as it is difficult to extrapolate from percentage nodulation to any other attribute, there seems little point in adopting a 'fair' classification such as that of Schall *et al.* (1975), who use 67–89% in this class (E. D. Schall, personal communication).

Expiry

The wide variety of products and usages makes it difficult to nominate an expiry period. In fact, Thompson *et al.* (1975) recommended that Australian manufacturers should prominently display date of manufacture, adopt a 1-month expiry until further information on survival was obtained, and ensure that, after 1 month, retailers sold the product as uninoculated seed requiring re-inoculation. The period must be kept to a practical minimum to ensure maximum numbers. One month would seem to be a reasonable minimum storage period to allow for transport, while 3–4 months should be sufficient to allow for clearance of stocks. Seasonal sowing requirements generally do not exceed these periods.

(iii) Packaging

As many pre-inoculation processes involve some re-drying after treatment, or are already dry on application, moisture loss is not of such concern as with the inoculant itself. It is therefore a reasonable principle to use the same type of packaging as for seed. It is particularly important, however, that labelling should clearly indicate the fact that the pre-inoculated seed should be stored under temperature conditions similar to inoculants and ideally at 4 °C. Further, clear labelling should indicate date of expiry and/or, because current knowledge of the useful life of most processes is minimal, the date of preparation. The information should also include advice on the need to re-inoculate the seed after the expiry period.

Methods for Evaluating Biological Nitrogen Fixation
Edited by F. J. Bergersen

Johanna Döbereiner
EMBRAPA/SNLCS-CNPq,
Rio de Janeiro, Brazil.

3

Forage Grasses and Grain Crops

(a) INTRODUCTION

In contrast with legumes, methods for the study of the recently discovered N_2-fixing grass and cereal associations are not yet well defined. Nitrogen-fixation rates are usually much lower and the systems are less tolerant of changes in the environment. Perhaps this is because no specifically adapted root structures seem to exist which can replace the sophisticated nodule environment which encloses the symbiotic bacteria in legume nodules. Most information available to date refers to the associations involving *Azospirillum* spp. (Döbereiner and Day, 1976; Patriquin and Döbereiner, 1978; Burris *et al.*, 1978). These organisms, like *Rhizobium* spp., are typical aerobes when supplied with combined N. However, cultures grown with N_2 as sole N source tolerate only very low oxygen concentrations (2.5 μM in solution, equivalent to a gas-phase pO_2 of 0.002 atm; Döbereiner, 1977). Most rapid growth and N_2 fixation are achieved with ample oxygen supply, provided that the respiration requirements are not exceeded. In culture this can be achieved by various methods which will be discussed in this chapter. *In vivo* the oxygen supply seems delicately balanced. The best model system so far seems to be a colony in which cells near the centre adjust their respiration rates to the rate of O_2 diffusion. *Rhizobium* was first shown to fix N_2 *in vitro* in a similar system (Pagan *et al.*, 1975). Colonies of *Rhizobium* and of *K. pneumoniae* can fix N_2 under air but immediately cease to do so if only the border of the colony is touched (Wilcockson and Werner, 1976).

These observations are of major importance in all aspects of methodology to be used in the study of the *Azospirillum* associations, and also are relevant to other heterotrophic diazotrophs which associate with grasses, if high energy conversion efficiencies of the systems are envisaged.

(b) DETECTION OF NITROGEN FIXATION AND ASSESSMENT

Since the first reports on N_2 fixation in grasses (Rinaudo *et al.*, 1971; Döbereiner *et al.*, 1972a,b; Patriquin and Knowles, 1972), obtained with the C_2H_2 reduction method, a large number of data have accumulated (Neyra and Döbereiner, 1977). It seems now easy to demonstrate nitrogenase activity in a large variety of grasses, especially under tropical conditions and in wet soils and also in temperate regions, but many questions still exist as to methods which permit quantification. Earlier reports indicate that N balance studies in pots (Moore, 1966; Döbereiner and Day, 1975), in the field (Parker, 1957; Day *et al.*, 1975), and in test-tubes (Rinaudo *et al.*, 1971) by direct measurement of N increments are possible, if a sufficiently long time is allowed.

Nitrogen fixation rates in tropical forage grasses (De-Polli *et al.*, 1977; De-Polli, 1975) and sugar cane (Ruschel and Vose, 1977) have been shown to be high enough to be measured after one or two days by $^{15}N_2$ incorporation. The C_2H_2 reduction method, with its simpler procedures and greater sensitivity, has been widely used and yields valuable results as long as they are properly interpreted.

(i) Excised root assay by C_2H_2 reduction

For qualitative and semiquantitative assessment of N_2 fixation in grasses, the very sensitive C_2H_2 reduction test seemed a promising solution. Unlike legume root nodules, however, grass roots removed from the soil reduce C_2H_2 only after an 8–12 h lag phase (Rinaudo *et al.*, 1971; Döbereiner *et al.*, 1972b; Yoshida and Ancajas, 1973; Döbereiner and Day, 1976). The reason for this lag has not been clarified and it is probable that more than one factor interferes in the rather delicate grass systems, when roots are removed from the soil and the micro-environment is destroyed. As long as this question remains unsolved this method should not be used for extrapolations to quantitative estimates.

The material required for the excised root assay is cheap and simple. For small plants assay vials of various kinds can be used; glass or plastic bottles with a volume of 200–300 ml, fitted with rubber stoppers or screw-caps and with wide openings, are most convenient. In our laboratory, baby feeding bottles with inverted teats are used for most experiments in which large numbers of bottles are required. For larger samples, polyethylene bags or larger jam-jars can be used. A large number of disposable gas-tight syringes are essential for proper timing.

A gas chromatograph with a flame-ionization detector as described in Chapter II.3, Section b, is required. Once serious work with the C_2H_2 reduction method is envisaged, a gas chromatograph with two columns should be available continuously. In addition, a gas chromatograph with thermodifferential detection with a column of molecular sieve (5 Å) to measure O_2 concentration is very useful. It permits the pO_2 to be monitored during the assay. Leakage of O_2 into bottles is much faster than that of C_2H_2 out of the bottles, and roots use up O_2, especially if nitrogenase activity is high.

Field plants should be harvested on a sunny afternoon to allow maximum accumulation of photosynthates. Plants are removed with a hoe and immediately placed in a bucket containing distilled water. Tap water often contains chlorine, which can affect microbial activity. The plant top and excess soil are removed and the entire root system is placed in the bottles or jars filled with distilled water. With this method a delay of up to 2 h has no effect on the ultimate activity, and it is therefore possible to harvest an entire experiment (up to 100 samples) at a time and then process them in the laboratory. There, the water is replaced with N_2 from a cylinder and 5% v/v of air (1% of O_2) injected for pre-incubation overnight at room temperature. In the morning, 12% v/v of C_2H_2 is injected and the pO_2 in the bottles is adjusted to 1% v/v, if possible, based on gas chromatographic measurements.

Rates of C_2H_2 reduction measured over 2–3 h are adequate to detect activities of significance and longer incubation periods increase hazards of artifacts. Because activities on grass roots are often small, it is important to use proper controls for C_2H_4 contamination in the C_2H_2. After assaying, the roots must be carefully washed before weighing and nitrogenase activities are expressed as nmoles or μmoles of C_2H_4 per gram of dry roots per hour. For calculations, see Chapter II.3.

Recently we observed that the lag period of maize roots was shortened to 7 h, and more reproducible results were obtained, when entire root systems attached to the stem are assayed instead of root pieces and when the results are based on an entire plant instead of root weight basis.

Whatever the assay vessel is, it is important that the ratio of free volume to root dry weight should be at least 500. If more roots are placed in the vessels, rapid changes in gas composition can occur, which can result in fermentation processes and consequent multiplication of many kinds of microorganisms (e.g. Okon *et al.*, 1977).

Interpretation of results obtained with the excised root assay must consider the restrictions discussed in the introduction to this section. Comparisons of excised root assays with intact core assays should be made for every specific environment and system. In grasses and small grains, statistically significant correlations of these two assay methods were obtained (Döbereiner, 1977; Nery *et al.*, 1977; Boddey *et al.*, 1978) and comparable results were also reported by Weier and Date (personal communication), Witty (1977), and Patriquin and Denike (1978).

Difficulties in removing large cores without seriously disturbing the system make such comparisons difficult if large grains (e.g. maize) or grasses (e.g. sugar cane) are to be assayed (Tjepkema and Van Berkum, 1977). Gas diffusion is another problem with large cores (Patriquin and Denike, 1978; Witty, 1977).

The excised root assay has been widely criticized (Barber *et al.*, 1978; Koch, 1977; Tjepkema and Van Berkum, 1977; Eskew and Ting, 1977) because multiplication of diazotrophic bacteria has been observed during the pre-incubation period, which seems to invalidate the method. The above-mentioned correlations with intact core assays, with small grasses and grains, however, indicate at least relative estimates by this method. Also, no nitrogenase activity can be discovered in roots of young maize plants (Pereira *et al.*, 1978a; Bülow and Döbereiner, 1975) even after prolonged pre-incubation, although numbers of diazotrophs are as high as during reproductive growth (10^7 or more) (Magalhaes *et al.*, 1979) and roots excised from very active grass cores show zero nitrogenase activity during the pre-incubation period, just like those of excised maize roots (De-Polli, 1975). These observations indicate that the lag before onset of C_2H_2 reduction in excised roots has more complicated causes than just multiplication of bacteria.

We have therefore recommended the excised root assay for qualitative or semiquantitative estimates and many repeatable and sensible results have been obtained with it, when the described methods are rigorously observed.

(ii) Intact core assays

For quantitative estimates of N_2 fixation in grasses, intact soil–plant systems seem most reliable. Nitrogen fixation as low as $10\,g\,ha^{-1}$ per day can be measured with satisfactory precision by incorporation of $^{15}N_2$ and still smaller values by the C_2H_2 reduction method. Several methods have been proposed (Balandreau, 1975a; Dart *et al.*, 1972; De-Polli *et al.*, 1977; Witty and Day, 1977; Lee *et al.*, 1977) where the natural soil structure is maintained and therefore microsites with specific pO_2 regulation systems should be expected to remain intact. Such cores show linear C_2H_2 reduction rates during short incubation periods (Balandreau, 1975a), but over longer periods are not linear because there is a pronounced day–night cycle in grasses (Balandreau *et al.*, 1974; Döbereiner and Day, 1975). The day–night variations in nitrogenase activity are caused by photosynthate availability, but also by changes in insolation, temperature, and soil and air humidity. Attempts at quantitative estimates must therefore include measurements during at least 24 h. This causes difficulties of maintaining natural gas compositions and light energy input in closed systems. A second major limitation of intact core assays is the various diffusion difficulties (Day *et al.*, 1975; Lee *et al.*, 1977) which seem more serious in short-term assays. Not only diffusion of C_2H_2 or labelled N_2 into the system and liberation of C_2H_4 have to be considered, but also O_2 and possibly CO_2.

$^{15}N_2$ incorporation in cores

Exposure of soil–plant cores to $^{15}N_2$ permits the measurement of actual N_2 fixation and also observations on the incorporation into plant tissue of the fixed nitrogen. This is best achieved by exposing the plants to $^{15}N_2$ during 12–24 h and analysing them after increasing additional growth periods. To give good precision it is desirable that highly labelled gas (>50% $^{15}N_2$) is used in such experiments. In the apparatus developed by De-Polli (1975) and De-Polli *et al.*, 1977), soil–plant cores taken in the field in steel cylinders of 9 cm diameter are transfered to cleaned oil cans with the same diameter. Active cores are selected by C_2H_2 reduction assays (see below) and placed into closed, gas-tight vessels, provided with a gas sampling port. The gas phase is first replaced with helium or argon by several evacuations and refillings and then adjusted with $^{15}N_2$ (70–90% enrichment) to a pN_2 of 0.4 atm; O_2 and CO_2 are added according to the experimental conditions. Adequate light and temperature controls are essential and CO_2 and O_2, which are monitored by gas chromatography, should be adjusted whenever necessary. An example of such an assembly which permits the complete recovery of the $^{15}N_2$ is shown in Figure 1. Plants should be harvested after a maximum of 24 h in such

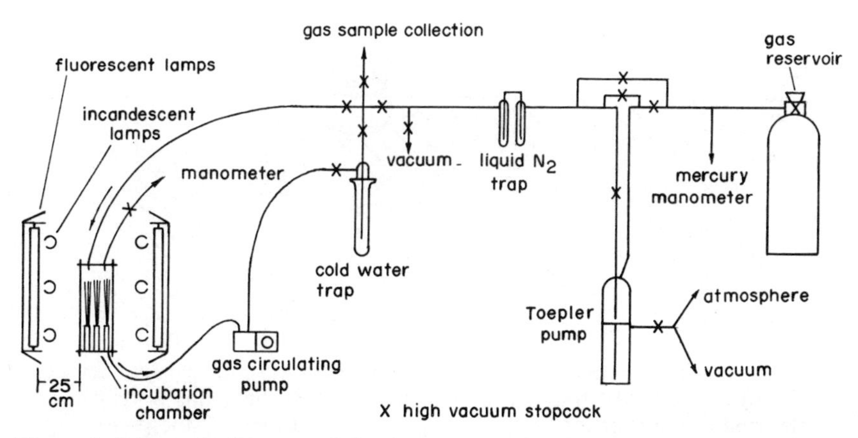

Figure 1. Schematic diagram of simple apparatus for exposure of plants to $^{15}N_2$ (from De-Polli *et al.*, 1977. Reproduced by permission of Pergamon Press)

systems. They are washed, dried, and ground and then analysed for total N and ^{15}N enrichment (see Chapter II.2).

More precise results can be obtained with gas lysimeter assemblies, such as the apparatus proposed by Witty and Day (1978).

Cores assayed by C_2H_2 reduction

Various types of assay methods have been proposed for intact soil–plant core measurements *in situ* by the C_2H_2 reduction method, which is cheap and very sensitive. Open cores as proposed by Balandreau (1975a) and Dart *et al.* (1972) consist of a steel tube of 30 cm diameter, which is hammered or rammed into the soil around the plant to be assayed (5–7 cm deep), and the soil is heavily watered around the core. Undisturbed water represents a surprisingly good barrier against diffusion, even for the very soluble C_2H_2

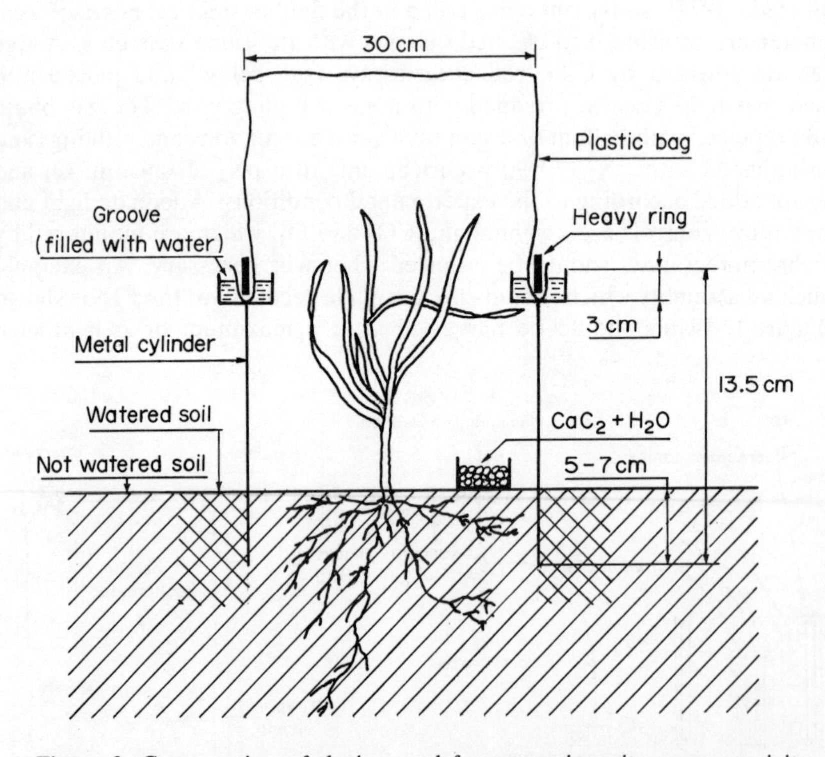

Figure 2. Cross-section of device used for measuring nitrogenase activity (C_2H_2) in the field (from Balandreau *et al.*, 1978). Sampling ports can be attached in the cylinder or by glueing a piece of rubber to the plastic bag, to allow passage of a hypodermic needle. (Reproduced by permission of Plenum Publishing Corporation)

and still better for C_2H_4. A clear plastic or glass dome is fitted to the steel cylinder with a water seal and 15–20% v/v of C_2H_2 is produced within the assembly by adding water to a calculated amount of calcium carbide just before closing it. Direct filling with a 20% C_2H_2–air mixture using two flow meters is another possibility. Rubber sampling ports permit easy repeated samplings. Based on this earlier device, which can be used for legumes and small grasses, Balandreau *et al.*, (1978) proposed a modified assembly for assaying large maize plants in the field (Figure 2).

Open cylinders are assumed to leak at a constant rate. To be able to correct for the leakage, propane is used as an internal standard (Chapter III.4). Longer incubation periods with open cores lead to errors, especially in light soils. Cores with heavy soils, on the other hand, are unsuitable for most types of C_2H_2 reduction assays because of serious diffusion problems. Witty (1977) compared C_2H_2 diffusion into grass and legume systems and concluded that the site of nitrogen fixation in legumes has ready access to O_2 and therefore also C_2H_2, while the site of nitrogen fixation in grasses, in order to function *in situ*, may be in niches where O_2 access is limited and therefore might also be more difficult to reach for the C_2H_2.

Closed cores avoid leakage of C_2H_2 and therefore permit incubations over longer periods. For larger plants, the same steel tubes as for open cores can be used but they are introduced completely into the soil and removed with as much as possible of the roots. The tubes are sealed to a plate to close the bottom, using any commercial adhesive. It should be kept in mind, however, that such a core includes only the main roots and disturbance of the root system can seriously affect the delicate balance of oxygen access and photosynthate supply, and perhaps cause changes in mineral composition of roots and soil. For these reasons we think that reliable intact field assays with larger plants such as maize have so far not been achieved with the open core method. Small grains (wheat, rice, etc.) and forage grasses are best assayed in steel tubes (10 cm diameter), which can be hammered into the soil even when it is reasonably compact and hard. The cores are removed from the field and placed into polyethylene, Saran, or other plastic bags (See Chapter II.3) which can be closed on top or around the stems (Figure 3). Various means of closing the bags have been proposed; we find closing around the stems with strings or rubber bands and filling with a soft agar solution ($7\ g\ l^{-1}$) to be most convenient. For gas exchanges and sampling, various injection ports may be used (Chapter II.3 and II.4).

Such cores are best placed on a glasshouse bench for 24 h before closing the bag, to allow the system to equilibrate. Changes in the delicate oxygen equilibrium are unavoidable when the cores are removed from the field and no serious modification should be expected by this method if glasshouse temperatures and the humidity of the cores are maintained comparable to field conditions. If potential N_2 fixation rather than actual fixation is to be

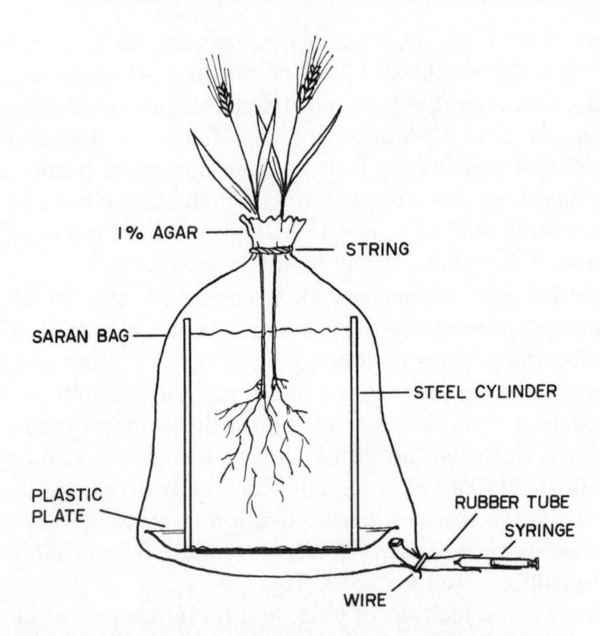

Figure 3. Schematic diagram of an assembly which permits measurements of nitrogenase activity (C_2H_2) in intact soil plant cores with the leaves exposed to open air (Döbereiner, unpublished work)

measured, application of fertilizers (P, K, and Mo) and optimal soil moisture 1 week before assay, either in the field or in the cores, seem reasonable. Weier and Date (personal communication) and Pereira *et al.* (1978b) cut the forage grasses, applied K, P, and Mo, and assayed after re-growth only. Care must be taken with this method if algal N_2 fixation is to be avoided. All cores, incubated in closed systems in the glasshouse or in the field, must be carefully shielded from direct insolation because temperatures can rise within a few minutes to more than 50 °C within the bags. If the leaves are left outside, the core is covered with aluminium foil. If the leaves are inside the assay vessel this problem is more serious if incubation longer than 1–2 h is envisaged, because natural light conditions are important to ensure normal photosynthesis.

As in the *in situ* assays, equilibration of the gas mixtures for at least 1 h is essential and the most reliable rates are obtained between 1 and 2 h after injection of C_2H_2. If propane is not used, the C_2H_2 peak can be used as an internal standard to measure the free gas volume and check for leakage. Then additional blanks must be used, consisting of a vessel with a known volume containing one non-N_2-fixing core (soil without plants). The C_2H_2 to C_2H_4 ratio is slightly different in such standards than in an empty vessel, due to the difference in solubility of the two gases. This small difference is important if low nitrogenase activities are to be measured in large gas volumes. At least three successive measurements of C_2H_2 reduction should be made in order to check for continuing increases.

Special precautions are recommended when assaying flooded rice in cores. The flooding water obstructs diffusion of C_2H_2 into the system and of C_2H_4 out of the system. The flood water contains algae, which confound the measurements. Gases diffuse through the rice stem, which leads to losses if the leaves are left outside the core. In fact, most of the gas exchange seems to occur through the plant (Balandreau *et al.*, 1975b; Lee *et al.*, 1975) and is dependent on open stomata. This means that under low air humidity and at night there is little gas exchange with the roots in these systems.

Taking most of these peculiarities into account, Lee *et al.* (1975, 1977) developed the following method for open, *in situ* rice cores. Square biscuit cans (30 × 30 cm) are used, which can be pressed into the soft mud by hand. The flooding water containing the algal component is pumped with a hand pump into a polyethylene bag and assayed separately after incubation with C_2H_2 in the light with vigorous shaking. Clean water is used to replace the flood water and a bamboo stick is placed within the core. The core is covered with a polyethylene bag, which is fastened to the can by a tightly fitting steel ring. The bag is collapsed and then air and C_2H_2 are passed in simultaneously at a ratio of 8:2 by means of two flow meters attached to the cylinder outlets. The equipment can be mounted on a trolley and rolled to the experiment site. Before each gas sampling the flood water is stirred with the bamboo stick to release C_2H_4.

(c) IDENTIFICATION OF RESPONSIBLE DIAZOTROPHIC BACTERIA

Many reports (e.g. Nelson *et al.*, 1976; Barber *et al.*, 1978; Balandreau, 1975a; Watanabe and Lee, 1977; Barber and Evans, 1976) have described the isolation of facultative diazotrophic bacteria from roots and soil, but none of them shows selective stimulation of the bacteria by the plant or any relation with nitrogenase activity on field-grown roots. So far, the only bacteria for which plant effects and correlations of root nitrogenase activity with bacterial activity or numbers have been shown are the heterotrophic aerobes *Azotobacter paspali* (Döbereiner, 1970), *Beijerinckia* spp. (Döbereiner. 1961), *Derxia* sp. (Campelo and Döbereiner, 1970), *Azospirillum* spp. (Döbereiner and Day, 1976; Bülow and Döbereiner, 1975; Vlassak and Reynders, 1977), and *Bacillus* sp. (Neal and Larson, 1976). We shall therefore concentrate on the methodology for the isolation of these organisms only. More detailed descriptions of culture techniques for other diazotrophs are discussed in Chapter II.1.

(i) Specific methods for isolation, identification, and counts

Methods for the isolation of heteroptroph aerobic or microaerophilic diazotrophs should always take advantage of the highly selective character of dinitrogen fixation and therefore nitrogen-free or poor media are most suitable.

The following media are recommended (always prepared in the order stated):

1. Nitrogren-free sucrose medium for *Azotobacter* and *Derxia* (LG):

Sucrose	20 g
K_2HPO_4	0.05 g
KH_2PO_4	0.15 g
$CaCl_2$	0.01 g
$MgSO_4 \cdot 7H_2O$	0.20 g
$Na_2MoO_4 \cdot 2H_2O$	0.002 g
$FeCl_3$	0.01 g
Bromothymol blue (0.5% in ethanol)	2.0 ml
$CaCO_3$	1.00 g
Agar	15 g
Water	to 1000 ml

Adjust the pH to 6.8 (green colour).

For the isolation of *Derxia* sp., replace the cane sugar with starch or glucose, omit $CaCO_3$, and add 0.01 g of $NaHCO_3$ (Campelo and Döbereiner, 1970).

2. Silica gel plates for *Beijerinckia* and *Azotobacter*. Prepare three stock solutions:

A. HCl, $d = 1.100$

B. $Na_2O_5 \cdot SiO_2$ (water-glass), $d = 1.060$
Exact densities are essential; commercial water-glass solidifies faster but chemically pure $NaSiO_3$ can also be used.

C. Winogradsky's salt solution:

KH_2PO_4	5 g
$MgSO_4 \cdot 7H_2O$	2.5 g
NaCl	2.5 g
$Na_2MoO_4 \cdot 2H_2O$	0.005 g
$MnSO_4 \cdot H_2O$	0.05 g
$Fe_2(SO_4)_3$	0.05 g
Water	to 1000 ml

Adjust the pH to 6.5 (also for *Beijerinckia*).

Silica gel plates are much easier to prepare than is usually thought if stock solutions are kept at hand and a suitable sink for dialysing them in running water is available.

A 500-ml volume of solution B is poured into 503 ml of solution A with constant mixing and then distributed immediately into Petri dishes (30 ml). These are allowed to solidify for 24–48 h, then placed in running water for 3 days until the $AgNO_3$ test for chloride is negative (no white precipitate is formed when one drop of a 1% $AgNO_3$ solution is placed on one of the plates). Such plates can be stored in a closed vessel for many months. On the day they are to be used, they are sterilized by immersion in boiling distilled water, which also removes any residual chloride. Glucose (10% w/v) is added to solution C and 2-ml portions are heated to boiling in test-tubes and poured over the plates, which are allowed to dry in a clean drying oven.

Rapidly air-dried soil is sieved (1 mm) and 20–100 mg are spread uniformly over the plates. Numbers of colonies on such plates correspond to numbers of microcolonies present in the soil and are usually 100–1000 times lower than dilution counts.

3. Nitrogen-free malate medium for *Azospirillum* (NFb):

Malic acid	5.0 g
K_2HPO_4	0.5 g
$MgSO_4 \cdot 7H_2O$	0.2 g
NaCl	0.1 g
$CaCl_2$	0.02 g
$Na_2MoO_4 \cdot 2H_2O$	0.002 g
$MnSO_4 \cdot H_2O$	0.01 g
FeEDTA (1.64% w/v, aqueous)	4.0 ml
Bromothymol blue (0.5% w/v in ethanol)	3.0 ml
KOH	4.5 g
Biotin	0.1 mg
Water	to 1000 ml

Adjust the pH to 6.8 (green colour) with NaOH and only then add the agar.

For semi-solid enrichment media add 1.75 g of agar and distribute 5-ml portions in 10-ml serum vials. Other vials or test-tubes can be used but the surface area to depth ratio of the medium should not exceed unity. For MPN counts, the amount of malic acid is reduced to 0.5 g and 0.5 g of cane sugar is added. The amount of KOH is then reduced to 0.4 g. For agar plates add 20 mg of yeast extract and 15 g of agar.

4. Potato infusion for non-selective growth and purity checks (BMS):

Potatoes	200 g
Malic acid	2.5 g
KOH	2.0 g
Cane sugar	2.5 g
Biotin	0.1 mg

Cook washed potatoes for 30 min and then filter through cotton. Prepare potassium malate by dissolving 2.5 g of malate in 50 ml of water, adding two drops of bromothymol blue (0.5% in ethanol) and 2.0 g of KOH. Adjust the pH to a green colour. Add malate, sugar, and biotin to the potato filtrate and dilute to 1000 ml with water.

Semi-solid growth medium is obtained by adding 1.75 g of agar and solid medium by adding 15 g of agar. Most heterotrophic diazotrophs show excellent growth on this medium and very characteristic colonies. It is a convenient medium for purity checks because fungi and most other bacteria grow on it. There is no nitrogenase activity during growth on this medium.

5. Hino and Wilson's (1958) medium for diazotrophic *Bacillus*. Two solutions are sterilized separately and then mixed:

A. Sucrose	20 g
$MgSO_4 \cdot 7H_2O$	0.5 g
NaCl	0.01 g
$FeSO_4 \cdot 7H_2O$	0.015 g
$Na_2MoO_4 \cdot 2H_2O$	0.005 g
$CaCO_3$	10 g
Water	500 ml
B. *p*-Aminobenzoic acid	10 μg
Biotin	5 μg
KH_2PO_4	0.13 g
K_2HPO_4	0.17 g
Water	500 ml

$CaCO_3$ can be replaced with 0.05 g of $CaCl_2$ if absorbance readings are desired, but the pH then drops very rapidly.

Azotobacter *spp*

Most aerobic or microaerophilic diazotrophs are very characteristic and can be identified by colony type, cell shape, and nitrogenase activity. The root-associated diazotrophs are also very easy to isolate in pure culture, possibly because they produce antibiotics or bacteriocins (Thompson, 1976) which inhibit growth of other organism within colonies, as frequently observed with *Azotobacter chroococcum.*

Azotobacter paspali (Döbereiner, 1966) is the most specific of all diazotrophic bacteria, even more specific than *Rhizobium*. It has been found in 98% of a large number (252) of root or soil samples collected in various countries from *Paspalum notatum* cv. batatais, which is the only ecotype among 31 examined which associates with this bacteria. To isolate these bacteria, therefore, roots of this grass or soil are collected in the field, the root surface soil or rhizosphere soil is then sieved through a fine sieve, mixed, and about 100 mg are spread uniformly over agar plates with nitrogen-free sucrose medium (LG) and incubated at 35 °C. Colonies which appear after 24 h on such plates are not *A. paspali* but most probably *A. chroococcum* or *A. vinelandii.* Colonies appearing after 48 h, and which are denser, smaller, and yellowish due to assimilation and acidification of the bromothymol blue indicator, are most probably *A. paspali.* To test for nitrogen fixation, such colonies can be placed intact into small vials and exposed to 10% C_2H_2 in air. Usually hundreds of colonies grow on such plates inoculated with root surface soil of *Paspalum notatum*. They are streaked out twice on LG medium and then checked for purity on BMS agar. Microscopic observations show very motile (peritrichous flagellation) rods of width 2 μm and length 10–15 μm, which become still longer but then divide into almost cocoid forms of the same width.

Comparative counts can be performed by the same method as isolation where the numbers of colonies on the plates correspond to micro-colonies per gram of soil. For this purpose, the amount of soil used should be reduced to 10–20 mg per plate. For counts of cell numbers, dilutions in mineral salt solution (LG medium minus sucrose) are mixed with 2 ml of soft medium (7.5 g l^{-1} of agar at 45 °C) and poured on the top of solidified LG plates.

Beijerinckia *spp*

Two species are frequently found in tropical soils, especially under sugar cane or forage grasses: *B. indica* (Derx) and *B. fluminensis* (Döbereiner and Ruschel, 1958). Isolation of *Beijerinckia* spp. is best achieved on silica gel plates which, owing to their completely mineral composition, do not permit significant growth of any non-diazotrophic organism. Owing to the prolonged incubation period necessary for growth of *Beijerinckia*, agar media are less suitable. Becking (1961) suggested shallow-layer liquid media with sucrose in Petri dishes as the enrichment medium, where *Beijerinckia* growth becomes apparent after more than 1 week by transforming the medium into a sticky tenacious gum. This process, however, does not permit even semi-quantitative estimates. Isolation and estimates of occurrence are therefore obtained best with silica gel plates (medium SG). Soil samples are mixed and sieved (1 mm) and 200–500 mg spread over the surface of the plates (*Beijerinckia* is usually less abundant than *Azotobacter paspali* and therefore larger volumes of soil are recommended). Glucose is used because colonies spread less than with sucrose, which is the favoured sugar. Although *Beijerinckia* is known to be acid tolerant, silica gel plates of pH 6.5 give the highest counts, probably owing to less growth of contaminants, which are mostly yeasts. The plates are incubated at 30 °C.

Beijerinckia colonies start to grow 4–10 days after inoculation. *B. indica* forms small, raised, white colonies, which become rapidly larger and very tenacious. *B. fluminensis* colonies are small, lightly beige, and dry, much like a few grains of sand. They do not continue to grow and are not sticky, as is *B. indica. Beijerinckia* cells are very characteristic under the microscope. In wet mounts, medium-sized rods (1 × 3 μm) with two very refractive fat gobules, one on each extremity, are observed (stain with Sudan black). *B. fluminensis* cells are packed in zoogloea like clusters surrounded by a 'membrane' visible under the light microscope.

For isolation, entire colonies are dispersed in test-tubes with sand and sterile water (colonies become soft after 2 h in water) and streaked out on LG medium without $CaCO_3$. *Beijerinckia* colonies can be tested for nitrogenase activity by placing intact colonies into small vials with C_2H_2 (20%). Quantitative C_2H_2 reduction assays are difficult because of the very high C_2H_2 concentration (more than 70%) necessary to saturate the enzyme of this organisms (Spiff and Odu, 1973; Döbereiner, unpublished work). Relative estimates of occurrence can be made on silica gel plates. Counts of individual bacteria by plating or MPN methods do not give reliable results because of the difficulty of dispersing naturally occurring micro-colonies in soil.

Derxia gummosa

Isolation is best achieved from roots of rice or other plants grown in heavy, poorly drained soils with much organic matter. Plates with LG medium containing starch are inoculated with small, washed root pieces or soil grains which are partially introduced into the agar, and incubated at 35–37 °C. Yellowish brown, smooth or plicated, round colonies start to grow after 5–7 days on the cut ends of the roots or below the soil grains. The colonies grow fast and become very large (0.5–1 cm) and high owing to the extremely tough gum (harder than that of *Beijerinckia*) so that only entire colonies can be removed.

Isolation again is easy; entire colonies are placed in sterile water with sand for about 2 h, to soften the consistency, then dispersed by shaking, and streaked out on agar plates (LG with glucose instead of starch). It is very characteristic of *Derxia* that two types of colonies form on N-free agar plates: small colonies which are only lightly beige and do not fix N_2 and large brown colonies which fix N_2. When ammonium acetate or some other suitable nitrogen source is added or when the plates are incubated under reduced pO_2 (0.05 atm O_2), all colonies are uniformly large and brown. This has been explained by the lack of an oxygen protection mechanism for nitrogenase. Only colonies which have reached a certain size are able to fix N_2 in air (Hill, 1971). Colonies on plates inoculated with soil or roots seem to have enough combined N to start *Derxia* growing and all colonies become large and brown. *Derxia* cells are 1–1.2 by 3–6 μm and contain large refractive lipid bodies.

Almost *Spirillum*-like motility can be observed in peptone broth culture where few lipid bodies are formed. Only one species has so far been properly described (Jensen *et al.*, 1960). Counts are still more difficult than those of *Beijerinckia* and estimates of the percentage of root pieces or soil grains containing *Derxia* seem the best expression for relative evaluations.

Azospirillum *spp*

This organism was described as *Spirillum lipoferum* in 1925 by Beijerinck, who was unable to show nitrogen fixation in pure culture. Becking (1963) isolated it again and showed $^{15}N_2$ incorporation. After nitrogen fixation in root associations with this organisms had been shown (Döbereiner and Day, 1976), it became widely known. More recent detailed studies (Krieg, 1977; Tarrand *et al.*, 1978) showed the need of reclassification of this organism into two species, *Azospirillum lipoferum* and *A. brasilense*.

Methods for the isolation of this organism must consider the strictly aerobic metabolism on the one hand and the high sensitivity due to lack of an oxygen protection mechanism for nitrogenase on the other. Semi-solid media, where the organism selects optimal pO_2 within the oxygen diffusion gradient, seem most convenient. Only in such media is it possible to make use of the highly selective N-free environment. Based on this principle, a very simple and selective method has now been developed by which hundreds of strains have been isolated and identified in our laboratory: 10-ml serum vials with 5 ml of semi-solid NFb medium are inoculated with one grain of soil or 0.5-cm pieces of washed or surface-sterilized roots (0.5–2 min sterilization with 1% chloramine T followed by washings with 50 mM phosphate buffer, pH 6.8 and water are most convenient). After incubation for 40–48 h, thin initially veil-like pellicles form, which originate on the root extremities and within 1 day move to the surface. About 1 mm below the surface thin but very dense, white, undulated pellicles are formed, which can be recognized with certainty once the researcher has some practice. These pellicles are checked for N_2 fixation by replacing the cotton plug with a rubber closure and measuring C_2H_2 reduction for 30 min. It is important to measure nitrogenase activity while the lower part of the medium is still green. A high pH inhibits it. The vials must be treated very carefully to avoid shaking and disturbance of the pellicle, which result in immediate loss of nitrogenase activity. *Azospirillum* enrichment cultures reduce at least 30–100 nmol h^{-1} of C_2H_2 and less active cultures are best discarded.

From the most active cultures a portion of the pellicle is transferred to new semi-solid NFb medium and after 24 h the veil-like pellicle is streaked out into agar plates with NFb medium containing 20 mg of yeast extract. The yeast is needed for growth on plates because colonies on the surface do not fix N_2 and therefore cannot grow without combined N. Colonies which appear after 1 or 2 days on such plates are probably not *Azospirillum*. The plates are best incubated for 1 week, when *Azospirillum* colonies are more characteristic, e.g. white, small dry and often slightly merged into the agar. Single colonies are transferred into semi-solid medium again (without yeast) and checked again for C_2H_2 reduction. If they reduce C_2H_2 and the colonies were well separated, almost always upon one purification on potato infusion, pure cultures of *Azospirillum* can be obtained. They are then transferred to stock media, such as nutrient agar, and the species and subspecies are identified according to Table 1. For a detailed description of these organisms, see Tarrand *et al.* (1978).

Both species contain denitrifying (nir^+) and non-denitrifying (nir^-) strains. All strains reduce NO_3^- to NO_2^- (nr^+) and the key enzyme for denitrification is a dissimilatory nitrite reductase (nir). The nir test is best performed in semisolid NFb medium containing 5 mM NH_4NO_3. Cultures are incubated for 24 to 36 h (36°C) and when the pellicle has just reached the top of the medium but is still green at the bottom, the culture is mixed throughly. Abundant gas bubbles after 1 h are proof of denitrification. *A. brasilense* nir^+ strains usually produce gas faster and more abundantly, than *A. lipoferum* nir^+ strains. Nir^- strains produce no gas, even after several hours.

Counts of *Azospirillum* in soils or roots are best performed by the most probable number method. Adequate dilutions (10^{-2}–10^{-8}) are made in mineral salt solution (NFb without malate) and 0.1-ml portions are placed in the center of the vials with semi-solid NFb medium in which 0.5% malate is replaced with 0.025% malate and 0.025% cane sugar. The 0.1 ml of inoculant should not be dispersed because out-growth of *Azospirillum* from single cells is difficult. Probably for this reason a mixture of malate with sucrose yields higher counts than malate alone. Other bacteria, even if they cannot fix N_2, start respiring near the *Azospirillum* cell, thus facilitating its initial multiplication. The addition of sugar to *Azospirillum* counting media has the additional advantage that in the semi-solid medium most other N_2-fixing bacteria can grow and will be estimated. The presence of *Azospirillum* or other diazotrophs in the various dilutions is determined by C_2H_2 reduction tests. The use of malate (blue colour of the medium) or light turbidity in N-free medium (NFb) are not proof of the presence of diazotrophic bacteria. The presence of a dense white pellicle is good evidence for *Azospirillum* growth.

Characteristic cell forms should be checked microscopically or by isolation from the highest dilutions which reduce C_2H_2. In order to make sure that C_2H_2 reduction is determined during active growth, C_2H_2 reduction tests must be performed every 2 days on the vials with apparent growth because cultures past the log phase do not fix N_2. After 1 week, one last C_2H_2 reduction test should be made, including the last dilution with apparent growth and one higher dilution. Another possibility is to incubate all dilutions for 1 week and replicate then the three highest dilutions with apparent growth plus one

Table 1. Characteristics for the identification of *Azospirillum* spp

	A. lipoferum	*A. brasilense*
Semi-solid NFb medium with glucose or α-keto glutaric acid as sole carbon source[a] incubated at 37°C	Heavy growth on surface after 72 h	No growth or fine pellicle much below surface after 72 h
Semi-solid NFb medium without biotin	No growth after several successive transfers[b]	Normal growth
Cell form in 48-h old[c] cultures in semi-solid NFb medium with 0.005% yeast extract	Pleomorph large S-shaped or helical cells (1.4–1.7 × 5–30 μm), slow or non-motile with prominent PHB granules	Short, slightly curved, very motile rods (1.0 × 2.1–3.8 μm) with prominent PHB granules. Retain their cell form and characteristic motility for many days

[a] The bromothymol blue indicator must be omitted because ethanol can be used as carbon source.
[b] Large inoculants (0.1 ml) should be used to avoid misjudgement due to starter problems. Instead of successive transfers the inoculant can be centrifuged and washed twice and resuspended in biotin free medium.
[c] 24-h old cultures of both species are alike and very motile spinning around their own axis.

further dilution, into new medium in which all dilutions can be assayed for C_2H_2 reduction after 48 h.

Plate counts can be made only with cultures where *Azospirillum* is the most numerous component in the total microbial population. In this instance plating on potato agar, on which *Azospirillum* colonies are characteristically more or less pinkish, usually wrinkled, and partially inserted into the agar, seems convenient.

Bacillus spp

Facultative N_2-fixing *Bacillus* spp. have been reported to establish in the rhizosphere of certain wheat lines (Rovira, 1963; Neal and Larson, 1976). Isolation of these organisms must be performed under anaerobic incubation (see Chapter II.1). Soil or root macerate dilutions (10^{-3}–10^{-6}) in mineral salt solution are inoculated into low-N medium (Hino and Wilson, 1958) with $CaCO_3$ replaced with 0.05 g of $CaCl_2$ and 0.25 g of ascorbic acid added as reducing agent (Neal and Larson, 1976). Roll-tubes with the increasing dilutions are incubated for 5 days at 28 °C and then C_2H_2 is injected. From nitrogenase-positive tubes, cultures are picked and streaked on reduced low-N medium in pre-rolled tubes. Repeated re-isolation of large, opaque colonies with undulate edges yields pure cultures of a Gram-positive *Bacillus* with sub-terminal spores and high anaerobic nitrogenase activity. Aerobic growth of these organisms occurs only with combined N.

(ii) Assessment of plant–bacteria interactions

It is now widely accepted that asymbiotic nitrogen fixation in soil is negligible owing to lack of available carbon substrates. Except in a few exceptional environments (Ruschel and Vose, 1977; Döbereiner *et al.*, 1972a), nitrogen fixation in association with non-nodulated plants is localized on or in the roots and is negligible in the absence of plants (Rinaudo *et al.*, 1971; Day *et al.*, 1976; Witty, 1977). In order to identify the bacteria responsible for N_2 fixation in certain systems it seems essential to identify interactions of plants with diazotrophic bacteria. One approach is the demonstration of selective effects of certain vegetations. The effect of sugar cane on the occurrence of *Beijerinckia* (Döbereiner, 1961), that of *Paspalum notatum* on *Azotobacter paspali* (Döbereiner, 1970), and that of *Panicum maximum* on *Azospirillum* (Döbereiner *et al.*, 1976) are examples. Rhizosphere effects (comparison of numbers of diazotrophs within and outside the root zone) have been widely used. Examples are *Azotobacter* and *Beijerinckia* in the rhizosphere of legumes (Vancura *et al.*, 1965), *Beijerinckia* in the rhizosphere of sugar cane (Döbereiner, 1961), rice (Balandreau, 1975b), and forage grasses (Ruschel and Döbereiner, 1965), *Bacillus* spp. in the rhizosphere of certain wheat lines

(Neal and Larson, 1976), and *Azospirillum* in the endorhizosphere of grasses (Döbereiner, 1978; Patriquin and Döbereiner, 1978).

To assess vegetation or rhizosphere effects it is more important to use a simple and rapid method, which may give only relative results, than elaborate, precise counting methods, because it is essential that large numbers of field samples can be analysed and compared. Soil or root samples should consist of at least six subsamples which are mixed together before assay samples are weighed. Six to ten such composite samples are necessary for each treatment if reliable results are to be obtained. So, for example, if sugar cane vegetation effects are to be studied, from each site six composite soil samples are collected within the sugar cane field and six in adjacent comparable fields with other vegetation. Sugar cane rhizosphere effects can be studied in soil samples taken in between rows which are compared with samples taken below the plant (rhizosphere) or from the root surface (rhizoplane).

In addition to the classical concept of rhizosphere, which includes the soil zone around the roots and the root surface, internal colonization of roots (inter- and intracellular infections) has recently been termed endorhizosphere colonization (Patriquin and Döbereiner, 1978). For the study of such endorhizosphere associations, washed roots are compared with surface-sterilized roots after increasing periods of exposure to disinfectants. Note that in all these studies comparative (relative) values are obtained and absolute numbers are not necessary.

If various diazotrophic bacteria occur in large numbers in the ecto- or endorhizosphere (as, for example, *Azospirillum* and facultative diazotrophs in maize roots), a comparison of root nitrogenase activity with enrichment culture activities in various media is recommended. The medium which gives significant correlation of these two variables can then be considered to enrich for the organisms which are responsible for nitrogen fixation on or in the roots (Döbereiner and Day, 1976; Bülow and Döbereiner, 1975).

To make the understanding of this important aspect easier, we shall describe in an example how to proceed to identify diazotrophs responsible for nitrogen fixation in a plant association.

We assume that intact core assays of various sites show high nitrogenase activity with a certain plant. The cores are then taken apart, the soil is mixed and a 2-g sample taken, which is spread on paper for 1–2 h to dry. The same is done with one third of the roots from each core. The remaining roots are washed with distilled water and half of them immersed for 30 s in a 1% chloramine T solution, washed for 30 s with sterile 0.05 M phosphate buffer (pH 6.8) and then several times with sterile water. About 5-mm long pieces of the washed and surface-sterilized roots are then used to inoculate six replicates each of the following media: (a) semi-solid NFb medium; (b) Hino and Wilson's medium containing 0.15% w/v of agar to make it semi-solid and incubating with rubber stoppers under N_2; (c) silica gel plates impregnated with N-free salt solution and glucose; (d) agar plates of LG medium supplemented with starch. The root pieces are immersed and crushed with forceps within the semi-solid media (a and b) or half immersed in the agar plates (d). One root piece is used for each semi-solid medium vial but about ten pieces can be accommodated on a plate. The soil is passed through a 1-mm sieve (sterilized with a flame before each new example) and then 20–100 mg are spread over the plates (c and d) or small grains attached to a moistened loop inserted into the semi-solid media without mixing (d).

It is important to identify the diazotrophs in the first enrichment culture, because prolonged further enrichments will select for specific organisms which might have nothing to do with root or soil nitrogenase activity. Therefore, attempts should be made to identify at least the genus of the predominant diazotrophs in the first enrichment media.

The following criteria can be used.

Azotobacter: After 24–48 h, C_2H_2-reducing, soft, milky colonies appear on media c and d. Large, very soft colonies appearing after 24 h are mostly *A. chroococcum* or *A. vinelandii.* Microscopic examination shows large (3×3–$6\ \mu m$) coccoid or oval cells in pairs. Smaller denser and yellow colonies appearing after 48 h are *A. paspali.* Cells are smaller (2×2–$3\ \mu m$) and usually mostly coccoid in young colonies on agar.

Azospirillum: Thin pellicle starting at the root ends or soil grain appear after 24–40 h in medium a. After 48 h a dense white pellicle and high nitrogenase activity (more than 30 nmol h^{-1} of C_2H_4 per vial); characteristic cell form with spirillum-like motility. *A. lipoferum* usually can be recognized in these vials by larger, less motile, and polymorph forms, once the medium turns alkaline. *A. brasilense* cells remain very motile and unchanged in cell form.

Beijerinckia: Characteristic sticky, C_2H_2-reducing colonies appear after 4–10 days on medium c. Cells $1 \times 3\ \mu m$ with one fat globule on each end.

Derxia: Characteristically brown, tenacious colonies appear after 5–7 days, below the soil crumbs or root pieces on medium d. Cells are 2×4–$6\ \mu m$, non-motile, containing refractive globules.

Bacillus and Clostridium: Turbidity occurs in medium b with C_2H_2 reduction after 1 week. Microscopic examination shows typical spore-forming Gram-positive cells.

Facultative diazotrophs: Turbidity and gas formation occur in medium d, with low nitrogenase activity (usually less than 10 nmol h^{-1} per vial) and non-characteristic small rods.

Once the major groups of diazotrophs have been identified, they can be correlated with root nitrogenase activity. For this, roots are cut into 0.5-cm long pieces and about 100 such pieces are incubated for 24 h in 1-ml disposable syringes, which are filled with a gas mixture of 1% O_2, 15% C_2H_2, and 84% N_2, and closed by inserting the needles into rubber stoppers. C_2H_4 formation is assayed by directly injecting part of the gas into a gas chromatograph and classifying the 100 root pieces into activity groups. The same root pieces are then placed in the media which have been previously selected as most suitable for the responsible diazotrophs and the number of colonies or nitrogenase activity after 40 h in semi-solid media are correlated with root activities. Significant correlations have been found with grass roots and *Azospirillum* or *A. paspali* incidence (Döbereiner and Day, 1976; Bülow and Döbereiner, 1975).

Methods for Evaluating Biological Nitrogen Fixation
Edited by F. J. Bergersen

R. Knowles
Department of Microbiology,
Macdonald Campus of
McGill University, Quebec, Canada.

4

Nitrogen Fixation in Natural Plant Communities and Soils

(a) INTRODUCTION

It is appropriate in introducing this chapter to point out that the methods used in natural systems are, of course, identical with those employed in many agricultural systems. However, three comments may be made.

Firstly, the great variety of natural systems (terrestrial or aquatic, small or large size of associated macroorganisms) clearly requires the use of varied methods. Direct measurement of activities of forests, for example, is not possible. Gas diffusion rates and nutrient exchange rates vary greatly and gas sampling procedures must be modified to suit the system under study.

Secondly, the low activities usually observed result in the accuracy and precision of estimates being serious problems in both direct assays and long-term balance studies. In the latter estimates the quantitation of other nitrogen inputs and outputs becomes an important requirement.

Thirdly, there are as yet no generally accepted procedures which are valid under all (or any?) conditions and therefore one must compromise (for example, in establishing the length of assay period) and whenever possible attempt to compare two or more corroborative methods.

(b) ACETYLENE REDUCTION AND $^{15}N_2$ ASSAYS

Low activities magnify some of the problems inherent in all kinds of direct ($^{15}N_2$) or indirect (C_2H_2) measurements. In both it is very important in the assay to attempt to match the conditions obtaining in nature with respect to O_2 concentration, temperature, and absence of disturbance. The higher sensitivity of the C_2H_2 reduction assay compared with the $^{15}N_2$ assay suggests that C_2H_2 assays can be shorter. Nevertheless, exposure to C_2H_2 of more than a few hours is frequently necessary. In such longer assays it is usually difficult to control the changes in microbial populations, nutrient concentrations, and atmospheric gas concentrations, which can make the subsequent interpretation of data difficult.

(i) Closed systems

Relatively sophisticated apparatus has been described with which small soil–plant ecosystems may be enclosed and maintained at constant temperature, humidity, and O_2 and CO_2 concentrations. An example is the 'gas lysimeter' of Ross *et al.* (1964), which permits long exposures to $^{15}N_2$ to be made (Stefanson, 1973). However, shorter experimental assays will be emphasized in this chapter.

Isolated components

Excised roots, rhizomes, leaves, litter, and small soil or sediment samples, as well as intact small animals (termites, aphids, sea urchins, ship worms, etc.) can be enclosed in suitable containers for assay by either C_2H_2 reduction or ^{15}N procedures. The methods to be used and the hazards to be avoided are described in Chapter III. 3, Section b.i, dealing primarily with the study of excised roots of agricultural crops. Details of analyses are given in Chapters II.2 and II.3.

Materials required

1. Suitable sample containers of appropriate size:

   Serum bottles and serum stoppers (with crimping tool if necessary).

Flasks or jars and serum stoppers. Large flasks may be closed with a large rubber stopper containing a small serum-stoppered glass tube. Serum stoppers may also be inserted through holes drilled in the lids of screw-capped preserving jars.
Plastic bags of low-permeability Saran-type material. This can be heat-sealed if it is laminated with polyethylene. The bag should be closed with a serum-stoppered glass tube or it may have on its surface a bead of silicone rubber, etc., through which hypodermic needles may be inserted.

2. Equipment and instruments for handling samples: spatula or small core sampler for soils; truncated disposable syringe for transferring sediment; forceps, knife and/or scissors for excising plant components; plastic glove bag and supply of N_2 if it is necessary to prevent exposure of samples to O_2.
3. Source of either acetylene (cylinder of CaC_2-generated; see Chapter II.3, Section c.i) or $^{15}N_2$ (see Chapter II. 2, Section b). See also Section b.iii for sources for field use. Syringes of appropriate size are required for addition of C_2H_2 or $^{15}N_2$.
4. Source of internal standard gas if desired (see Section b.iii) and syringe reserved for its use.
5. Gas sampling syringes. These are most conveniently 1-ml disposable plastic syringes and needles with or without PTFE valves. Glass gas syringes with or without valves may also be used, but are more expensive. If storage of the gas sample is necessary, refer to Section b.iii.
6. For $^{15}N_2$ assays. A supply of Ar or He as well as O_2 and an electrical or hand-operated vacuum pump will be required.

Methods. Transfer the sample to the container chosen for assay. Materials such as excised roots may be transferred intact or cut into short lengths. They may be surface sterilized by appropriate treatment with chloramine T, hypochlorite, or silver nitrate. It is recommended that, particularly for very biologically active materials (plants or animals), the ratio of gas phase volume (millilitres) to sample fresh weight (grams) should be at least 300 so as to minimize rapid changes in the composition of the atmosphere during the assay. The exact weight of sample introduced may be more conveniently determined after the assay by difference weighing of the fresh or oven-dried sample in the tared containers.

Add supplement solutions. Adjustment of soil or sediment moisture contents, the provision of a suitable aqueous phase for aquatic plant parts, or the addition of required concentrations of organic carbon or combined nitrogen may be carried out at this stage.

Close the container by inserting the appropriate stopper or by sealing the plastic bag.

Establish the appropriate atmosphere for the assay. For C_2H_2 reduction assays evacuation of the system should be avoided if possible, but if reduced O_2 concentrations are desired, either flushing with an N_2–O_2 mixture or partial evacuation and back-filling with N_2 may be necessary. For $^{15}N_2$ assays it is necessary to remove all air N_2 and to replace it with highly enriched $^{15}N_2$, and therefore more thorough evacuation and back-filling with Ar or He are necessary. For more details, see Section b.iii.

Introduce C_2H_2 or $^{15}N_2$. Remove by syringe a volume of atmosphere equal to that of the gas to be introduced. Then introduce the C_2H_2 to give a partial pressure in the range of 0.05–0.1 atm [a greater concentration may be required to saturate nitrogenase in some systems containing *Beijerinckia* (Spiff and Odu, 1973) or in some sediments], or the $^{15}N_2$ to give a partial pressure of preferably at least 0.4 atm.

Introduce internal standard gas if desired (see Section b.iii).

Incubate *in situ* or at controlled temperature in the laboratory. Light and/or dark conditions may be used, depending on objectives.

Sample the gas phase at intervals by syringe (0.1–1.0 ml) for gas chromatographic (GC) analysis. The frequency of sampling depends on the activity of the sample and the appropriateness of the assay conditions. Although assays of 1–4 h are most desirable (van Berkum and Sloger, 1979), low activity may require the taking of several samples over longer periods. Lag periods of 10–14 h are frequently observed, but the cause is not always clear. They may be due to disturbance and/or exposure to O_2 followed by re-equilibration, but they may also be due to proliferation of N_2 fixers when a source of combined nitrogen at low concentration is present (Barber *et al.*, 1976; van Berkum, personal communication). If the chamber has developed a significant positive pressure it is necessary to sample with a syringe fitted with a PTFE mini-valve to permit quantitative transfer of a known sample volume to the gas chromatograph. If storage of samples is necessary, see Section b.iii.

For $^{15}N_2$ assays the ^{15}N atom-% excess of the experimental atmosphere and of the N_2-fixing sample should be determined at appropriate intervals.

Core samples

The use of core samples of soils and sediments, with or without intact plants, has the advantage of involving much less disturbance than is inherent in the study of isolated components. This is particularly true if the sample is retained within the core cylinder. Thus, cores frequently (but not always) show no lag before the onset of activity, but on the other hand the problem of gas diffusion (C_2H_2, C_2H_4, $^{15}N_2$) into or out of the sample may be increased, resulting in a possible underestimation of rates.

The inherent variation associated with relatively small diameter cores up to about 5 cm (Knowles and O'Toole, 1975) is reduced when larger cores (diameter greater than 10 cm) are used (Day *et al.*, 1975).

Because of the large volumes involved in core studies, $^{15}N_2$ assays may be prohibitively expensive.

Materials required

1. Core sampling tubes, stainless-steel (or other material) cylinders at least 10 cm diameter, and of the required depth (usually 15–20 cm) should have the lower edge sharpened unless it is possible to use the cylinder as a liner inside a mechanized core sampling device.
2. Rigid bottom plate. For some experiments it may be desirable to close the bottom end of the core with a sheet of plastic or metal sealed on with a suitable cement such as Terostat, butyl rubber, or silicone rubber.
3. Assay chambers. These must be large enough to accept the intact core, or in cases where a bottom plate has been applied they may be attached at the top end with a suitable seal. The chambers may be made of gas-tight rigid metal, glass (e.g. Mason preserving jar), or plastic with a rubber serum stopper inserted through the lid, or they may be non-rigid Saran-type plastic bags sealed as in (2) above.
4. Source of either C_2H_2 (cylinder or CaC_2-generated; see Chapter II.3, Section b.i) or $^{15}N_2$ (see Chapter II.2, Section b). See also Section b.iii for sources for field use. A syringe of appropriate size will be required for addition of C_2H_2 or $^{15}N_2$. Note that gas syringes up to 1 l in volume are available.
5. Source of internal standard gas if desired (see Section b.iii) and syringe reserve for its use.
6. Gas sampling syringes. These are most conveniently 1-ml disposable plastic syringes and needles with or without PTFE valves. Glass gas syringes with or without valves may also be used, but are much more expensive. If storage of the gas sample is necessary, refer to Section b.iii.
7. For $^{15}N_2$ assays. A supply of Ar or He as well as O_2, and an electrical or hand-operated vacuum pump will be required.

Method. When collecting core samples, it may be desirable either to sample individual plants of interest or to sample randomly within more or less uniform types of vegetation cover. For some systems (e.g. salt marsh; Patriquin and Denike, 1978) it may be more practicable to sample large ‘soil slices’ cut by spade, instead of using a corer device.

Enclose the core by attaching the bottom plate and head chamber or by placing the entire core in a separate closed container.

Establish the appropriate atmosphere. For C_2H_2 reduction assays the ambient air is probably adequate, and evacuation should certainly be avoided. However, if a reduced O_2 concentration is desired, either flushing with an N_2–O_2 mixture or partial evacuation and back-filling with N_2 may be necessary. For $^{15}N_2$ assays it is necessary to remove all air N_2 and to replace it with highly enriched $^{15}N_2$. Therefore, more thorough evacuation and back-filling with Ar or He (or Ar–O_2 mixture, etc.) is necessary. For more details, see Section b.iii.

Introduce C_2H_2 or $^{15}N_2$. Remove by syringe a volume of atmosphere equal to that of the gas to be introduced. Then introduce the C_2H_2 to give a partial pressure in the range 0.05–0.1 atm [a greater concentration may be required to saturate nitrogenase in some systems containing *Beijerinckia* (Spiff and Odu, 1973) or in some sediments], or $^{15}N_2$ to give a partial pressure of at least 0.4 atm.

Introduce internal standard gas if desired (see Section b.iii).

Incubate *in situ* or at controlled temperature in the laboratory. Light and/or dark conditions may be used, depending on objectives.

Sample the gas phase at intervals by syringe (0.1–1.0 ml) for GC analysis. The frequency of sampling depends on the activities observed and the presence or absence of a lag phase. The lag phase is less common in cores than in isolated components since disturbance is minimized. Representative time courses should be studied and rates calculated for the period 2–6, 10–24, or 10–48 h, whichever is considered most appropriate. If storage of samples is necessary, see Section b.iii.

For $^{15}N_2$ assays the ^{15}N atom-% excess of the experimental atmosphere and of the N_2-fixing core material should be determined at appropriate intervals. For the atmosphere analysis care must be taken to freeze out all CO_2 and N_2O before analysis to avoid interference with the 28, 29, and 30 ion species by fragments from molecules other than N_2. All three ion species must be determined since the N_2 molecules in the gas phase come from two different populations (residual occluded air N_2 in the sample and the added $^{15}N_2$), with the result that distribution of ^{15}N atoms amongst the three ion species is not in equilibrium (see Chapter II.2, Section e.vii). The resulting enrichment of ^{15}N in the sample material is determined as described in Chapter II.2, Sections b.iii, and e.vi.

Representation of data.

In many cases, presentation of data on an area or volume basis is adequate and they may be calculated from the known dimensions of the core. The data may also be converted to a soil or root fresh weight or dry weight basis by carrying out appropriate determinations at the termination of the assay.

(ii) *In Situ* open-system assays

Non-disturbing C_2H_2 reduction assays for N_2 fixation in various systems were first used by Balandreau and Dommergues (1971) and, although beset by various problems in different laboratories, do appear to yield useful data if appropriate precautions are taken. In principle, an enclosure is inserted into the soil surface, C_2H_2 and a suitable internal standard gas are introduced and gas samples are analysed at appropriate time intervals. The C_2H_2 diffuses to the C_2H_2-reducing sites at a rate dependent on the water and root content of the system. Some C_2H_2 may diffuse out of the enclosed soil column around the bottom of the enclosure if high porosity soil is involved. Ethylene produced at the active sites diffuses up into the headspace of the enclosure and its concentration is measured by GC. The concentration of the internal standard gas is assumed to provide an estimate of the total gas phase volume within which the produced C_2H_4 equilibrates. Thus, the total amount of C_2H_4 produced by the enclosed area can be calculated.

The following account is derived mainly from Balandreau and Dommergues (1971, 1973), Balandreau (1975a), Lee and Watanabe (1977), Lee *et al.* (1977), Patriquin and Denike (1978), and Patriquin and Keddy (1978).

Materials required

1. Cylindrical enclosures of appropriate size (10–30 cm diameter and 15–30 cm high, depending on test site and water content, etc.) made from acrylic plastic, galvanized steel, or stainless steel. Each may have at its upper rim a trough in which an upper enclosure component can be inserted in a water seal (see, for example, Chapter III.3, Figure 2).
2. Enclosure covers. They may be rigid (sheets or cylinders of acrylic plastic) or flexible (sheets or bags of low-permeability Saran-type plastic, laminated with polyethylene if heat sealing is necessary). Each cover is attached with the water seal as in Chapter III.3, Figure 2, or sealed to the lower cylinder using silicone rubber, butyl rubber, Terostat, or a similar sealant. Each cover is fitted with one or two rubber septa, stoppers, or beads of silicone rubber through which syringe needles may be inserted.
3. Source of C_2H_2. Calcium carbide or C_2H_2 cylinder. A syringe of capacity at least 500 ml will be required if cylinder C_2H_2 is used.
4. Source of internal standard gas (see Section b.iii) and a syringe reserved for its use.
5. Thermometer for measurement of temperature inside the enclosure.
6. Gas sampling syringes. These are most conveniently 1-ml disposable plastic syringes and needles with or without PTFE valves. Glass gas syringes with or without valves may also be used but are much more expensive and no more reliable. If storage of the gas sample is necessary, refer to Section b.iii.

Method. Locate the desired test site and insert the lower cylinder (considerable force may be required, involving a stout cover and heavy sledgehammer, etc.). The selection of individual plants, often easy in agricultural systems, may be difficult or impossible with natural vegetation where random positions within a community type may be advisable. In diurnal studies, assays may be initiated successively at different times of day.

Introduce the thermometer if desired and a beaker or flask of CaC_2 sufficient to generate 0.1 atm of C_2H_2 if C_2H_2 is to be generated in this fashion.

Apply the upper part of the enclosure and seal with water or other suitable material.

Introduce C_2H_2 either by injecting the appropriate amount of water on to the CaC_2 or by injecting the appropriate volume of C_2H_2 from a cylinder. (Before adding C_2H_2 a volume of atmosphere equal to that of the C_2H_2 to be introduced may be withdrawn, but this is not always done.)

Introduce propane or other internal standard gas (see Section b.iii).

Sample the gas phase at intervals by syringe (0.1–1.0 ml) for GC analysis. The frequency of sampling depends on the system under study, but it is recommended that samples be taken after 0.5 or 1.0, 2.0, 5.0, 24, and 48 h so that the time course may be observed. Short assays of 2–5 h duration are preferable to minimize changes in atmosphere and loss of C_2H_2 from the system. If storage of samples is necessary, see Section b.iii.

Representation of data

An example of data calculations involving an internal standard gas is given in Section b.iii. The calculated rates of C_2H_4 production may be placed most simply on an area basis but for certain applications the root mass may be excavated and dried, and rates placed on a root dryweight basis.

Notes

(a) After the introduction of a gas such as C_2H_2, C_2H_4, C_2H_6, or C_3H_8 into *in situ* enclosures there is initially a very rapid drop in concentration, followed by a slower decrease (Witty, 1979a). This is thought to be caused by rapid diffusion into the occupied volume, followed by the slower processes of solution in the liquid phase and adsorption on soil particles. The latter is reported (Witt and Weber, 1975) to occur at relative rates as follows: $C_2H_4 = C_3H_6 \gg CH_4 \gg C_2H_6 \gg C_3H_8$, suggesting that C_3H_8 is not a desirable internal standard gas. Nevertheless, the C_2H_4 to C_3H_8 ratios within enclosures frequently remain relatively constant (Balandreau and Dommergues, 1973; Patriquin and Denike, 1978), a point which it is desirable to check in *in situ* studies of this kind.

(b) In wet or waterlogged systems, diffusion is greatly restricted and may require an increase in C_2H_2 concentration up to 0.2 atm. Furthermore, solution effects are more pronounced and the C_2H_4 produced may not equilibrate with the gas phase unless the liquid is stirred with a stick placed within the flexible plastic enclosure (Lee and Watanabe, 1977).

(c) The use of rigid enclosures over long time intervals leads to a pumping effect on the soil atmosphere due to pressure differences caused by temperature changes and the introduction (and/or removal) of gas. These effects are minimized in non-rigid enclosures, although pumping due to changes in atmospheric pressure will occur (for a discussion of such effects, see Lemon, 1978).

(iii) Technical notes

Atmospheres for C_2H_2 reduction assay

Saturation of the N_2-fixing sites is achieved easily with 0.05 atm pC_2H_2 in soils having moderate porosity, but 0.2 atm may be required in water-saturated soil or sediment systems despite the high solubility of C_2H_2 in water. Furthermore, some bacteria (e.g. *Beijerinckia*; Spiff and Odu, 1973) are reported to require 0.2 atm of C_2H_2 for saturation even in pure culture. It is clear, therefore, that it exceedingly difficult to match, in C_2H_2 reduction assay, the availability of the substrate C_2H_2 at the nitrogenase sites with the availability of N_2 under natural conditions.

Inactivation of nitrogenase activity by O_2, though unlikely in little disturbed systems, possibly occurs in cases where isolated components (e.g. roots) are removed from O_2-deficient environments. In such cases exposure to O_2 should be reduced to a minimum and pO_2 during assay adjusted to a value in the range 0.1–0.05 atm. Clearly, evacuation and back-filling with O_2-containing mixtures should be avoided.

Internal standards

For many studies C_2H_2 itself may act as an adequate internal standard. However, in the presence of a significant aqueous phase, and also in *in situ* assays, it may be desirable to use as an internal standard some other gas having solubility and diffusion properties closer to those of the C_2H_4 to be determined. Candidate hydrocarbons, some of which have been used for this purpose, are methane (Pedersen *et al.*, 1978), ethane (Hanson, 1977), propylene, and propane (Balandreau and Dommergues, 1971). Table 1 shows some properties of these gases in comparison with C_2H_2 and C_2H_4. They clearly differ, not only in water solubility but also in the order of elution from commonly used Porapaks, and moreover it is reported (Witt and Weber, 1975), that their adsorption on soil decreases in the following order: ethane =

Table 1. Some properties of potential internal standard gases compared with those of acetylene and ethylene

Property	Methane (CH_4)	Acetylene (C_2H_2)	Ethylene (C_2H_4)	Ethane (C_2H_6)	Propylene (C_3H_6)	Propane (C_3H_8)
Solubility in water at 20°C*	0.036	1.05	0.122	0.0496	0.22	0.0394
Order of elution on:						
Porapak R	1	3	2	3	4	5
Porapak N	1	4	2	3	5	5
Porapak Q	1	2	3	4	5	6

*Solubility data are from Stephen and Stephen (1963); units are cm^3 of gas corrected to S.T.P. per cm^3 of water at 20°C. Other solubility data are found in Wilhelm *et al.* (1977).

propylene ⩾ methane ⩾ ethane ⩾ propane. The choice of an internal standard must be made carefully with due consideration of the possibility of metabolism by the sample and the compatibility with the GC system being used.

Calculations may be made as follows (e.g. Balandreau and Dommergues, 1973). Assume that propane is the internal standard; e_1 and e_2 (mol ml^{-1}) are the C_2H_4 concentrations of samples taken at times t_1 and t_2, p_1 and p_2 (mol ml^{-1}) are the C_3H_8 concentrations of samples taken at times t_1 and t_2, P (mol) is the amount of C_3H_8 injected into the enclosure, a (m^2) is the area of surface enclosed, t_1 and t_2 are the sampling times in hours from initiation of the assay, and V_1 and V_2 (ml) are the volumes occupied by the gases at times t_1 and t_2. Then, the rate of production of C_2H_4, R (mol m^{-2} h^{-1}), is given by

$$R = \frac{e_2V_2 - e_1V_1}{a(t_2 - t_1)}$$

Now, $V_1 = P/p_1$ and $V_2 = P/p_2$. Therefore,

$$R = \frac{\dfrac{e_2P}{p_2} - \dfrac{e_1P}{p_1}}{a(t_2 - t_1)}$$

$$= \frac{P}{a(t_2 - t_1)}\left(\frac{e_2}{p_2} - \frac{e_1}{p_1}\right)$$

The relative sensitivity of the gas chromatograph to C_2H_4 and C_3H_8 can be determined by injecting equal molar quantities (within the range covered by routine analyses) of the two gases.

If e'/p' is the ratio of C_2H_4 to C_3H_8 peak heights (including attenuation) obtained in such a calibration comparison, and h_{e_1}/h_{p_1} and $h_{e_2/}h_{p_2}$ are the ratios of the C_2H_4 and C_3H_8 peak heights obtained in the samples taken at times t_1 and t_2, then

$$R = \frac{P}{a(t_2 - t_1)}\left(\frac{p'}{e'}\cdot\frac{h_{e_2}}{h_{p_2}} - \frac{p'}{e'}\cdot\frac{h_{e_1}}{h_{p_1}}\right)$$

$$= \frac{Pp'}{ae'(t_2 - t_1)}\left(\frac{h_{e_2}}{h_{p_2}} - \frac{h_{e_1}}{h_{p_1}}\right)$$

Thus, for any particular experiment in which the amount of propane injected and the relative GC sensitivity are constant, only the ratios of C_2H_4 to C_3H_8 peak heights in the gas samples are needed. Furthermore, the analyses are independent of the volume of the system as well as the size of the sample and the amount injected for GC analysis.

A more elaborate calculation was presented by Patriquin and Keddy (1978), taking into account initial rapid and subsequent slow phases of dilution of C_2H_4 and internal standard in the occupied volume in enclosures over relatively long time periods. A theoretical problem (which may be potential rather than real) is the fact that C_2H_4 and the internal standard gas are actually diffusing in opposite directions.

Atmospheres for $^{15}N_2$ assay

To maximize the sensitivity of measurement, it is necessary to use the highest possible percentage of $^{15}N_2$ and to remove as completely as possible all air N_2. Thus, flushing with Ar or He or repeated partial evacuation and back-filling with Ar or He is necessary, followed by addition of the appropriate amount of O_2. Ideally, the N_2 partial pressure should be in the range 0.8–1.0 and certainly at least 0.4 in order to saturate N_2-fixing sites. For reasons of economics a compromise may have to be made between the final pN_2 and the final atom-% ^{15}N obtainable in the atmosphere.

Gas supplies for field use

It is often convenient to use small gas cylinders (lecture bottles or larger) in field studies. However, as mentioned earlier, C_2H_2 may also be generated from CaC_2 in appropriately equipped flasks (Chapter II.3, Section b.i).

Gases can also be transported in, for example, rubber football bladders fitted with serum stoppers (Waughman, 1971).

Isotopic nitrogen is most conveniently prepared in the laboratory from $(^{15}NH_4)_2SO_4$ (Chapter II.2, Section e.iii) and, if necessary, transported to the field in large (e.g. 50-ml) glass syringes sealed with vacuum grease or silicone rubber to reduce leakage.

Storage of gas samples

For *in situ* assays and for intensive sampling regimes it is necessary to store small gas samples prior to GC analysis. This may be done in three main ways.

(a) Samples may be taken in disposable glass syringes (e.g. Glaspak) having a plastic plunger but a glass barrel to reduce leakage. The syringe can be fitted with a small valve made of PTFE (e.g. Mininert) or stainless steel (BD), or the needle can be plunged into a rubber stopper. Leakage is further reduced by storing the syringes under water.

(b) Pre-evacuated tubes (e.g. Vacutainers; Schell and Alexander, 1970) or small serum bottles or vials may be used as storage containers. However, the residual vacuum varies greatly in such containers and therefore use of the double-ended needle and simple equilibration of the atmospheres in the assay and storage containers should be avoided. It is better to inject a known volume of sample into the storage tube and then withdraw a known volume for GC analysis. The use of an internal standard simplifies the calculations and reduces errors. Again, leakage is reduced by storage under water.

(c) Samples may also be stored by injection into small serum tubes or bottles filled with water, the displaced water exiting via a second needle inserted through the septum. Depending on the amount of water remaining in the storage container a correction may need to be made for the solubility of the components of interest.

Statistical treatment of data

It was recently pointed out that the distribution of biological parameters in nature is frequently not normal and that in the case of several sets of C_2H_2 reduction activity data there was a log–log relation between mean and variance (Roger *et al.*, 1977). Thus, the commonly used Student *t* test would not be applicable without an appropriate log transformation of the data.

(iv) Controls and miscellaneous potential problems

No-acetylene control

To test for and quantitate endogenous C_2H_4 production (i.e. not nitrogenase-mediated) a control is usually recommended in which C_2H_2 is not introduced. Not only is C_2H_4 produced by certain fungi (Young *et al.*, 1951), bacteria (Primrose, 1976), and anaerobic soils (Smith and Restall, 1971; Goodlass and Smith, 1978), but it can be released from unsterilized (Kavanagh and Postgate, 1970) or autoclaved (Jacobsen and McGlasson, 1970) serum stoppers or by reaction between the trichloroacetic acid used to stop certain assays and the rubber or plastic liners of serum caps (Thake and Rawle, 1972). Such a control, however, shows only net C_2H_4 production in the absence of C_2H_2 and, as will be discussed below, net C_2H_4 metabolism may be changed in the presence of C_2H_2.

Ethylene control

A control in which only C_2H_4 is introduced in low concentration serves to show the existence of net C_2H_4 metabolism (De Bont, 1976; De Bont and Mulder, 1976; Yoshida and Suzuki, 1975). It was suggested that since C_2H_2 inhibits the oxidation of C_2H_4, such a metabolism of C_2H_4 would not interfere with the C_2H_2 reduction assay (De Bont, 1976). However, it appears that systems may exist in which, in the absence of C_2H_2, both production and further metabolism of C_2H_4 occur such that net accumulation of C_2H_4 does not occur in the control. In such a system the presence of C_2H_2 blocks the metabolism of C_2H_4 so that endogenously produced C_2H_4 then accumulates and is included as C_2H_4 formed by reduction of C_2H_2:

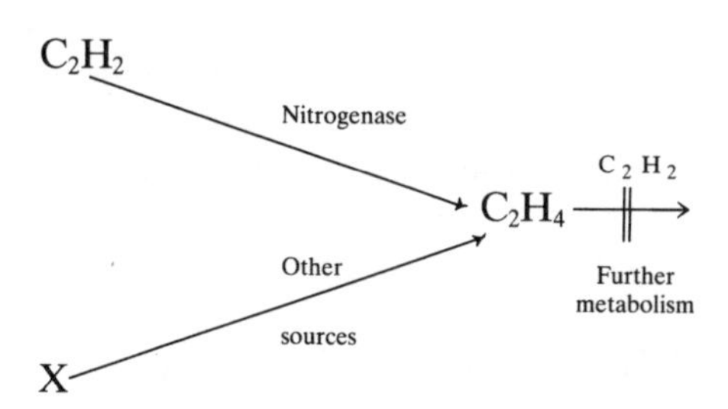

Witty (1979b) showed that such a phenomenon occurred by demonstrating that in the presence of $^{14}C_2H_2$ a significant proportion (up to 50%) of the C_2H_4 produced was not labelled, even though in a control without C_2H_2 no C_2H_4 was produced.

The use of a C_2H_4-only control may detect C_2H_4-metabolizing potential, but the possibility remains that C_2H_2-reducing activities may be seriously overestimated, particularly in some low-activity systems.

No-$^{15}N_2$ control

In $^{15}N_2$ assays it is necessary to include a control not exposed to $^{15}N_2$ but otherwise treated in exactly the same way as the exposed samples. In this way slight variations in $\delta^{15}N$ amongst natural materials are corrected for and the ^{15}N excesses observed can be assumed to be due to N_2 fixation (see Chapter II.2).

Acetylene metabolism

Although the metabolism of C_2H_2 does occur (Kanner and Bartha, 1979), it is mainly after long-term exposure to the gas (De Bont, 1976) and it is extremely unlikely that any significant change in pC_2H_2 could result from this process.

Ethane production

Small amounts of low molecular weight hydrocarbons are sometimes produced by soils and sediments. It is therefore essential that such products, particularly C_2H_6, should be resolved from C_2H_4 by whatever GC system is employed.

Saturation of nitrogenase with substrate

Reference was made in Section b.iii to the difficulty in matching substrate availabilities (C_2H_2 and N_2) at the sites of N_2-fixing activity. It is clear that in some systems (e.g. water-saturated), air N_2 does not normally saturate the sites (Rice and Paul, 1971), whereas pC_2H_2 values of 0.1–0.2 or so may indeed cause saturation of the sites and therefore lead to an overestimation of activities (or a high value for the C_2H_4 to $^{15}N_2$ molar ratio).

Side-effects of C_2H_2

The absence of ATP- and nitrogenase-dependent H_2 evolution in the presence of C_2H_2 can lead to (a) an overestimation of N_2 fixation using the C_2H_2 reduction assay, (b) increased sensitivity to O_2 in some microaerophilic N_2-fixing bacteria, and (c) the absence of H_2 accumulation in the C_2H_2 reduction assay which could otherwise occur under N_2 with eventual inhibition of nitrogenase by H_2 (Dixon, 1972; Knowles, 1976).

Since C_2H_2 is not a physiological substrate for nitrogenase, in nitrogen-deficient environments the presence of C_2H_2 leads to the imposition of nitrogen limitation such that growth of N_2 fixers is prevented (Brouzes and Knowles, 1973).

Apart from the imposition of nitrogen limitation mentioned above, C_2H_2 inhibits even NH_4^+-dependent proliferation of some N_2-fixing bacteria (e.g. *Clostridium*; Brouzes and Knowles, 1971) but not others (e.g. *Azotobacter*; Knowles *et al.*, 1973).

Other side-effects of C_2H_2 are its inhibition of the following reactions: N_2O reduction by denitrifying bacteria (Balderston *et al.*, 1976; Yoshinari and Knowles, 1976); NH_4^+ oxidation by *Nitrosomonas europaea* (Hynes and Knowles, 1978); CH_4 production by methanogenic bacteria (Oremland and Taylor, 1975; Raimbault, 1975); and CH_4 oxidation by the methylotrophic bacteria (De Bont and Mulder, 1974, 1976). The last effect precludes the use of C_2H_2 reduction assays for determining N_2-fixing activity which is supported by CH_4.

It is clear that these various effects of C_2H_2 on several important natural processes can make relatively difficult the interpretation of C_2H_2 reduction data from natural habitats.

(v) Simultaneous measurement of N_2 fixation and denitrification

Utilization of the C_2H_2 *inhibition of* N_2O *reduction*

In the presence of concentrations of C_2H_2 routinely used for N_2 fixation measurements, the reduction of N_2O by denitrifying bacteria is completely inhibited (Yoshinari *et al.*, 1977). Thus, the measurement of the accumulation of N_2O in the system can provide an estimate of the total rate of denitrification:

$$NO_3^- \rightarrow NO_2^- \rightarrow (NO) \rightarrow N_2O \xrightarrow[]{C_2H_2 \; \nparallel} N_2$$

It is conceivable that both N_2 fixation and denitrification may occur (but in different microsites) in some systems and therefore the determination of both C_2H_4 and N_2O produced in the presence of C_2H_2 may be desirable.

Analysis of gas samples for C_2H_2, C_2H_4, *and* N_2O

The hydrogen flame-ionization detectors used for C_2H_2–C_2H_4 measurements do not detect N_2O and therefore other detectors must be used. In order of

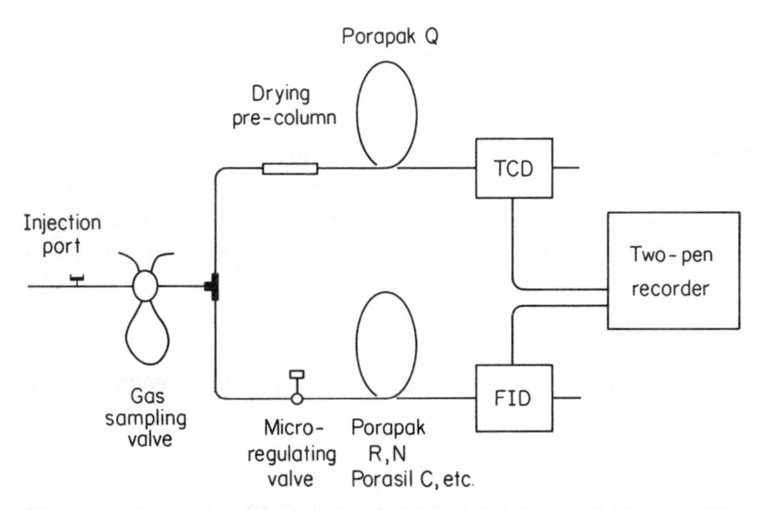

Figure 1. A simple GC system for single injection analysis of C_2H_2, C_2H_4, other hydrocarbons, CO_2, and N_2O

increasing sensitivity to N_2O, these are: thermal conductivity (hot wire and thermistor types), ultrasonic, helium ionization, and electron capture (Knowles, 1980). A convenient method for single-injection GC analysis of mixtures containing C_2H_2, C_2H_4, and N_2O is as follows (Nelson and Knowles, 1978).

A narrow-bore T (e.g. 1/16-in Swagelok fitting) is installed downstream from the injection port and/or gas sampling valve (Figure 1). One arm of the T is connected to a suitable column for N_2O separation (e.g. 2–3 m × 3 mm column of 80–100-mesh Porapak Q) and thence to a thermal conductivity detector (TCD) or a more sensitive detector. The other arm of the T is connected via a narrow-bore micro-regulating valve to a suitable column for hydrocarbon separation (e.g. 2 m × 3 mm column of 80–100-mesh Porasil C/methyl isocyanate) and thence to a hydrogen flame-ionization detector (FID). The flow-rate of the helium carrier gas is adjusted by means of the micro-regulating valve to give approximately 16 ml min^{-1} to the FID and 46 ml min^{-1} to the TCD, thus enhancing the sensitivity of detection of N_2O. The column temperature is maintained at 50 °C. The FID and TCD outputs are connected to a dual-pen recorder.

If problems are encountered with water peaks when using the TCD, a small (e.g. 80 × 6 mm) glass pre-column of indicating Drierite is inserted before the Porapak Q so that the sorbent can be replaced at appropriate intervals.

(c) OTHER ^{15}N-BASED TECHNIQUES

Several other methods involving ^{15}N have been reported for agricultural legumes. Although the methods appear to be useful for such systems there is no good evidence that they are valid for plants lacking root nodules, even in agricultural systems, and less so in natural undisturbed plant communities. Nevertheless, it seems desirable to discuss the broad principles here.

Most of the following methods require comparative measurements of a non-N_2-fixing plant to provide an estimate of available soil nitrogen. In legume experiments this can be a non-nodulating line or it can be a grass. In experiments on plants without root nodules the choice of an appropriate non-N_2-fixing comparison plant may be virtually impossible, as will be discussed below.

(i) Isotope dilution in the presence of a ^{15}N-enriched inorganic or organic N pool

Inorganic ^{15}N *pool*

As applied to studies of legumes, methods of this type involve the addition to the legume crop of not more than about 30 kg of inorganic ^{15}N per hectare [usually $(NH_4)_2SO_4$ or $Ca(NO_3)_2$ containing up to 30 atom-% of ^{15}N] so as not to suppress N_2 fixation. To estimate the availability of soil N a similar or greater addition of the ^{15}N compound is made to a control or comparison plant not fixing N_2, such as a non-nodulating line of the same legume (Ham, 1977) a cereal (Fried and Broeshart, 1975) or a grass (Goh *et al.*, 1978; Edmeades and Goh, 1979; Williams *et al.*, 1977). Herein lies the main difficulty in applying this method to non-root-nodulated plant N_2 fixation. Activities of the putative N_2-fixing systems are low and not necessarily significantly greater than those associated with any control which might be chosen. Furthermore, activity may be greatly affected by the addition of inorganic N. Thus, calculations based on the difference in A values (Fried and Broeshart, 1975) are probably not valid for other than nodulated systems.

To carry out an experiment of this type in which inorganic ^{15}N is added to the soil one must either (1) determine the ^{15}N enrichment of the available inorganic N pool by chemical methods or (2) grow a selected non-N_2-fixing plant and assume that the enrichment of its assimilated N is equal to that assimilated by the N_2-fixing test plant. Determinations of the first type are not adequately precise for a study of this kind. In the second procedure, the assumption that two different plants utilize exactly the same available soil N pool cannot be validated when one (or even both) plants fix N_2. However, in the event that a plant comparison is to be attempted, let N_t = total N contained in the N_2-fixing plant sample, E_f = ^{15}N enrichment in the N_2-fixing plant sample, and E_c = ^{15}N enrichment in the control 'non-fixing' plant sample. Then, the amount of N_2 fixed by the N_2-fixing plants, N_f, is given by

$$N_f = \frac{E_c - E_f}{E_c} \cdot N_t$$

Organic ^{15}N *pool*

This is a similar method, but one in which added inorganic ^{15}N is incorporated into the soil organic N fraction by adding sufficient carbohydrate to stimulate heterotrophic assimilation. It is claimed (Legg and Sloger, 1975) that the soil biomass and metabolite nitrogen thus formed is indistinguishable from the native soil organic N reservoirs.

In this case, the direct determination of the enrichment of the available soil N fraction is even more fraught with difficulty than that in the previous section. Thus, a suitable control plant, not fixing N_2, must be sought. In legume studies, non-nodulating lines (Legg and Sloger, 1975) and tall fescue grass (Legg, 1978) have been used. For studies of non-symbiotic N_2-fixing systems, the choice of an appropriate control plant as an index of soil ^{15}N availability will be exceedingly difficult.

Such experiments are best conducted in lysimeters (up to 1 m in diameter and 1 m or so in depth) in which homogeneous soil can be placed (see Section d.ii for further information on lysimeters). The soil is supplemented with not more than 30 kg ha^{-1} of N in the form of $(NH_4)_2SO_4$ containing 1–5 atom-% of ^{15}N. Sufficient sucrose (and possibly cellulose) is then mixed with the soil at 1-week intervals until no soil inorganic N is detectable. This reduces the possibility of a suppression of N_2 fixation. After a further few weeks of incubation, the test and control plants are planted. At an appropriate time the plants are analysed for total nitrogen and ^{15}N enrichment and the calculations are performed as in the previous section.

It is clear that the isotope dilution experiments described in these two sections cannot in any sense be applied to undisturbed natural systems.

(ii) Measurements of ^{15}N natural abundance

Soil nitrogen frequently contains a slightly higher percentage of ^{15}N than does nitrogen in the atmosphere. Furthermore, in most biochemical reactions, through isotope discrimination, the lighter of two isotopes is favoured slightly over the heavier. Thus, during N_2 fixation these two phenomena result in the fixed nitrogen having a very slightly lower ^{15}N content than does the nitrogen derived from the soil. Studies of the $\delta_A\,^{15}N$ values (parts per thousand difference in % ^{15}N compared with an appropriate air standard) show that N_2-fixing legumes have generally low $\delta^{15}N$ values and it is suggested that it may be possible to calculate the fraction of the plant nitrogen derived from fixation (Amarger *et al.*, 1977; Bardin *et al.*, 1977; Delwiche and Steyn, 1970; Mariotti, 1977). Rennie *et al.* (1976), for example, reported the following representative $\delta^{15}N$ values: faba bean 1.0–1.6, barley 3.7–4.2, soil 7.4, NH_4^+ fertilizer 1.3. It has been suggested that such considerations could be applied to a non-symbiotic or associative N_2-fixing sugar system (Ruschel and Vose, 1977). The calculation involved would be similar to that described in the previous sections. For example, if δ_A (fixer) = $\delta^{15}N$ value of N_2-fixing plant (e.g. sugar cane) N compared with atmosphere N, and δ_A (control) = $\delta^{15}N$ value of control non-fixing plant (such as lettuce) compared with atmosphere N (this is assumed to be equal to the $\delta_A{}^{15}N$ of that fraction of the soil N which is available to both plants), then the fraction of the plant N which is derived from fixation, F, is given by

$$F = \frac{\delta_A\,(\text{control}) - \delta_A\,(\text{fixer})}{\delta_A\,(\text{control})} = 1 - \frac{\delta_A\,(\text{fixer})}{\delta_A\,(\text{control})}$$

and if N_t = total N contained in the N_2-fixing plant sample, the total amount of N_2 fixed by the N_2-fixing plant sample, N_f, is given by

$$N_f = N_t\left[1 - \frac{\delta_A\,(\text{fixer})}{\delta_A\,(\text{control})}\right]$$

It must be said, however, that the hazards with such estimates arise from problems in precision (Shearer *et al.*, 1974a), the complexity of the isotope discrimination process occurring in the soil (Knowles, 1979; Shearer *et al.*, 1974b), and the difficulty in the choice of control plant. The last aspect is illustrated by the report of Bardin *et al.* (1977) that both *Calluna* and *Pinus* have lower $\delta^{15}N$ values than those of the wild legume *Sarothamnus* (broom), and that mycorrhizal pine has a $\delta^{15}N$ slightly different from that of its non-mycorrhizal counterpart. Furthermore, Delwiche *et al.* (1979) reported that not all suspected N_2-fixing plants were depleted in ^{15}N.

For the present it must be concluded, with Bremner (1975), that 'there is no evidence to suggest that determinations of the nitrogen isotope ratios of natural materials are likely to have more than qualitative value for research on nitrogen fixation', and this appears to be especially so for non-nodulated plants.

(iii) ^{15}N-depleted materials and ^{13}N

There are no reports of ^{15}N-depleted compounds being used for N_2 fixation research. Such materials are not satisfactory for studies in which great dilution of the tracer occurs and they require the use of a mass spectrometer with very great precision. It is therefore very unlikely that such materials can be used for the study of N_2 fixation, particularly in low-activity natural systems.

In view of the short half-life of ^{13}N (just over 10 min) and the problems associated with its generation and use (all other ^{13}N-labelled contaminants and other contaminating radioactive isotopes must be rigorously removed), it is unlikely that it can be used for the study of natural plants and soils. Should appropriate techniques become available, however, its very great sensitivity would be a major point in its favour.

(d) NITROGEN BALANCE TECHNIQUE

(i) General principles—relation to other N-cycle processes

Precise estimates of N_2 fixation are exceedingly difficult to obtain by means of overall nitrogen balance studies. The major reason for this difficulty is that the proper interpretation of any nitrogen balance data requires the precise quantitation of as many as possible of the nitrogen inputs and outputs of the system, as well as the precise quantitation of any accretion (or the reverse) by the system. Table 2 briefly summarizes the inputs and outputs which must be considered and the living and non-living components of the system in which estimates of changes in total nitrogen must be made.

Discussions of nitrogen balance studies with particular reference to forests were presented by Ovington (1962) and Richards (1964), and Moore (1966) reviewed most of the earlier attempts to measure non-symbiotic N_2 fixation. Many of these attempts utilized a nitrogen balance approach but with different degrees of precision and comprehensiveness.

Table 2. Factors to be considered in nitrogen balance studies

Nitrogen inputs	Nitrogen outputs	Accretion
N_2 fixation: asymbiotic, symbiotic	Denitrification	Mineral soil or sediment
Precipitation: NH_4^+, NO_3^- organic, particulate	Abiological (NO, etc.)	Organic matter layer
Dry sorption of NH_3	Volatile NH_3	Biomass: small plants, shrubs, trees, animals
Dust, droppings	Fires (N_2, NO, N_2O)	
Animals	Leaching	
Weathering	Erosion	
	Cropping, harvesting	
	Animals (game, etc.)	

The extent to which it is possible to quantitate inputs and outputs depends on the experimental approach adopted. The four broadly different methods discussed in the following sections (lysimeter, successive cropping, total system analysis, and watershed study) allow difficulties to be solved in different ways. At the same time, they imply differing degrees of disturbance and, as in the case of the first two, may not be applicable to natural systems. They are not appropriate for short-term studies. On the contrary, most published studies continued for periods of from 10 to 90 years.

(ii) Lysimeters

A lysimeter is a mass of soil in a cylindrical container from which the leachate can be collected. The major advantage of this approach is that it allows relatively precise measurements of leaching or drainage losses, which are not possible with other methods except watershed studies. Probably the major disadvantage is the disturbance involved and the necessity to re-establish vegetation. Thus, lysimeters have been used most in continuously cropped agricultural systems and are not entirely suited to natural systems. Nitrogen gains are most likely to occur in continuously vegetated systems.

Materials required

1. Cylindrical containers from 60 cm to 3 m in diameter and 1 m or more in depth. The bottom should be a shallow funnel so that the leachate can be collected via a plastic tube and reservoir bottle protected from evaporation losses. The inside of the lysimeter should preferably be coated with plastic or corrosion-resistant paint and a thin layer of fine gravel and plastic screening placed in the bottom.
2. Rain and snow collectors should be available to permit frequent collection and analysis of precipitation and dry fallout.

Method. The normal soil profile is reconstituted as closely as possible within the container using individual horizons which have each been screened and homogenized to facilitate sampling for chemical analysis. The soil should be compacted to close to its original bulk density—this is aided by application of water. An alternative method is the preparation of a 'monolith' type of lysimeter in which a large mass of undisturbed profile is inserted into the cylinder. Difficulties associated with the monolith are the probability of water channelling and the poor replication.

The loaded lysimeter is allowed to remain fallow or vegetated for at least 1 year to permit equilibration and settling to occur.

All inputs and outputs are then quantitated by at least weekly collection, measurement, and analysis of precipitation and leachate. In addition, the nitrogen contained in any added chemicals or plant material and in any plant material or soil removed is quantitated. All components can be calculated uniformly on a per hectare basis.

It seems clear that such experiments should continue for at least several years so that an average annual balance can be calculated.

The above input/output data may be supplemented with core analyses of the soil profile to detect possible increases or decreases in total nitrogen in the soil.

Comments

Chapman *et al.* (1949) discussed the various sources of error in lysimeter studies, and Pratt *et al.* (1960) presented data from a 20-year experiment in California. From their results and other considerations it is questionable whether significant measurements can be made of the low rates of N_2 fixation now considered to occur in natural systems. There is no doubt that they can detect large volatile losses arising from denitrification of applied fertilizer.

The fact that the soil mass in a lysimeter is detached from the natural water-table means that its water content probably remains higher than that in a natural profile. This may be a serious disadvantage if it results in increased denitrification, since it would introduce another unknown factor in any attempt to extrapolate data to natural conditions.

(iii) Successive cropping

The classical example of this type of study is the Broadbalk Experiment at Rothamsted Experimental Station, U.K., where certain plots have been cropped to wheat and barley continuously for 90 years. In the absence of fertilizer addition, the soil carbon and nitrogen content has remained relatively constant over a long period of time (Jenkinson, 1973). Measurement of precipitation inputs and plant material outputs permits the derivation of a long-term balance which, in the case of the Rothamsted experiment, suggests an average N_2 fixation rate of 30 kg ha^{-1} yr^{-1} (Jenkinson, 1973).

The application of this method to natural communities is of doubtful merit.

(iv) Total system analysis

Evidence of N_2 fixation has been obtained by the determination of the total nitrogen accumulated in the non-living and biomass components of grassland (Parker, 1957) and of shrub and forest communities (Jenkinson, 1970; Moore, 1966; Ovington, 1962; Richards, 1964). An estimate of precipitation and other inputs is required but unfortunately it is not possible to estimate such outputs as leaching and denitrification losses. The measurement of accretion is made by determining total nitrogen, (1) in communities of several different ages (Ovington, 1957b), as has been done in studies of contributions by root-nodulated non-legumes such as alder (Crocker and Dickson, 1957; Dickson and Crocker, 1953), or (2) in a community of a single age but in comparison with a control system. A suitable control may be a previous total analysis of the unvegetated or un-regenerated area such as was used by Jenkinson (1970) in his study of the development of the Broadbalk and Geescroft Wilderness sites at Rothamsted, or it may be a simultaneous comparative total analysis of an open (unforested) area such as was used by Ovington (1956a, b, 1957a) in his studies of the development of certain tree plantations in England.

A major problem for this kind of study is that for very many natural communities, appropriate control sites or analyses are not available, nor are records and analyses of precipitation obtainable over long time spans.

In what follows it will be assumed that a woodland or forest area is to be studied. Appropriate modifications would be required in other systems.

Materials required

1. Facilities for the dimension analysis of large numbers of trees. Appropriate data for conversion of dimensions to dry weight (e.g. Bunce, 1968; Whitaker and Woodwell, 1968).

2. Facilities for handling, sampling, drying, milling, and analysing representative samples of soil, litter, and vegetation.

Method. Within an area representative of the study forest, at least 100 random trees are selected for complete dimension analysis (Bunce, 1968; Whitaker and Woodwell, 1968). When the dimension analysis is completed, the study may proceed in two different ways.

(a) The dimensions may be converted to dry weights by calculation (Bunce, 1968; Whitaker and Woodwell, 1968), and representative samples of root, bole, and crown taken for analysis (Bormann *et al.*, 1977; Jenkinson, 1970). The total nitrogen in the trees may then be calculated from the dry weight and percentage nitrogen data.

(b) Alternatively, based on the dimension analysis, a small number (one to three) of average trees are sampled systematically. The total tree fresh weight is determined in the field and samples are taken at 2- or 3-m intervals along the bole and in the crown for dry weight and total Kjeldahl nitrogen determination. In this way the total nitrogen in the average tree is calculated, and from the stocking density the stand nitrogen in kilograms per hectare is obtained (Ovington, 1957a).

Herbaceous vegetation is sampled completely within random metre-square or other appropriate-sized quadrats. Dry weight (80 °C) and total nitrogen are obtained.

In sites where an appreciable organic layer accumulates, this must be sampled separately. Since variability is frequently high, a metal square (of the order of 25 cm per side) is placed randomly and the organic matter sampled within it. If no separate estimate of variation is required, 5–10 such samples may be bulked and the resulting composite sampled for dry weight and total nitrogen determination (Ovington, 1954).

In some studies, it is assumed that the change in the nitrogen content of the mineral soil is zero (Borman *et al.*, 1977). However, it is known that whereas the vegetation and organic layer both tend to increase in total nitrogen content during stand development, the mineral soil nitrogen may in fact decrease (Ovington, 1957b, 1962). It seems important, therefore, to sample and analyse the mineral soil to some specified depth (e.g. to 40 cm below the organic layer), usually through the analysis of replicated core samples.

Jenkinson (1970) pointed out that during the development of woodland from previously cultivated land the soil expands and its bulk density decreases. Thus, it is desirable to sample to an 'equivalent depth', namely, that which contains an equal weight of ignited fine soil per hectare. This notion assumes that nothing leaving a residue on ignition was gained or lost during the study period (Jenkinson, 1970).

(v) Watershed study

In the previous two sections (Sections d.iii and d.iv) an important unknown value in the nitrogen balance is the loss of nitrogen in groundwater drainage.

An advantage of the so-called watershed study is the ability to quantitate drainage output. Suitable watersheds are preferably small and, if possible, replicated so that some measure of variation can be obtained and differential treatments imposed. The solid geology should be such that all of the drainage water outflow is restricted to one stream so that this may be monitored continuously with V-shaped or flume-type weirs (Likens *et al.*, 1977).

Agricultural or other types of drainage basins may be amenable to the watershed approach. One of the best described studies, however, and one in which an estimate of N_2 fixation has been forthcoming, is the Hubbard Brook Experimental Forest in New Hampshire. This study is summarized by Likens *et al.* (1977), who provide in their book a useful discussion of the earlier original reports.

Precipitation inputs are quantitated (Fisher *et al.*, 1968) by methods already mentioned in Sections d.ii and d.iv. Accretion (or otherwise) of nitrogen in the ecosystem itself is quantitated by methods described in the previous section (Section d.iv). Drainage or leaching losses are calculated from (a) the hydrological stream flow data derived from continuous water-level recorders at the V-notch weir and (b) the nitrogen analyses carried out on stream water samples collected a short distance upstream from the weir installation at weekly intervals (Fisher *et al.*, 1968).

Denitrification remains a complete unknown in all of the above nitrogen balance methods. Thus for the watershed (Bormann *et al.*, 1977):

Net N_2 fixation = actual N_2 fixation − denitrification

= (annual N accretion of the system + annual hydrological export of N) − (annual precipitation N input)

(e) IDENTIFICATION OF LIMITING FACTORS

Because of the problems in achieving valid reproducible measurements of N_2 fixation rates in natural samples, and because of the low activities frequently observed, the identification of the factors(s) which may be limiting in any particular system is extremely difficult. Furthermore, the large number and variety of potentially limiting factors (Table 3) make any analysis of their participation in the control of N_2-fixing activity a formidable task. A discussion of the influence of such factors on N_2 fixation (Knowles, 1976) is beyond the scope of this chapter, although reference has already been made to several of them in earlier sections.

The measurement of N_2 fixation in closed systems (Section b.i) permits the control of certain parameters and the manipulation of others. Thus, it is possible to investigate the influence of changes in such factors as energy availability (Balandreau *et al.*, 1971; O'Toole and Knowles, 1973b), air

Table 3. Potentially limiting factors in N_2-fixing microbial and plant–microbial systems

Factors	Comments
Microbial population	Appropriate N_2 fixers required
Energy availability:	
light intensity	Controls net assimilation rate
photosynthetic area	Controls total assimilation
fruit development	Provides competing energy sink
organic carbon	Required for heterotrophs
Water availability:	
air	Low humidity causes stomatal closure and stops photosynthesis
soil	Affects plant and microbial activity, affects aeration status
Oxygen	Required for aerobes and microaerophiles, inhibits anaerobes and microaerophiles
Substrate availability	Diffusion resistance may limit N_2 and C_2H_2
Inhibitors:	
combined N	NH_4^+ and NO_3^- repress nitrogenase except in derepressed mutants
hydrogen	Could inhibit N_2 fixation but not C_2H_2 reduction
Temperature:	
air	Affects photosynthesis, translocation and mobilization of organic carbon reserves
soil	May be frequently limiting
Other nutrients	P and Mo most likely to be critical

humidity (Balandreau *et al.*, 1975), oxygen (O'Toole and Knowles, 1973a), substrate N_2 or C_2H_2 availability (Brouzes and Knowles, 1973; Rice and Paul, 1971), and combined nitrogen (Knowles and Denike, 1974; Yoshida *et al.*, 1973). Such experiments, however, are particularly difficult to design for plant–soil systems.

An alternative approach which makes use of the relatively non-disturbing *in situ* assay (Section b.ii) has been proposed by Balandreau (1975) (see also Balandreau *et al.*, 1978). A large number of *in situ* activity measurements are made on a particular plant–soil system along with simultaneous measurements of potential limiting factors such as total leaf dry weight, light intensity, air and soil moisture, and air and soil (5 and 15 cm depth) temperature. The observed activities (nmol C_2H_4 plant^{-1} h^{-1}) calculated as a percentage of the maximum activity observed are then plotted on the ordinate as a function of each factor in turn as the abscissa. Thus, for each factor a scatter of points is obtained below an upper boundary line which appears to be frequently but not always linear and which mostly has a positive slope.

It is suggested that, for any particular scatter diagram, points which lie on or close to the upper boundary line represent those instants in time for which the factor concerned was limiting. Points lying below the boundary represent occasions when some other factor was limiting (Balandreau *et al.*, 1978). When all environmental factors are non-limiting there remains the intrinsic limitation of the particular plant–soil system under study. Balandreau *et al.* (1978) suggest that this intrinsic factor is given by the maximum activity obtainable per unit of photosynthetic area. It is derived from the upper boundary line in the scatter plots of nmol C_2H_4 plant^{-1} h^{-1} against green leaf dry weight or

$$K = \frac{\text{nmol } C_2H_4 \text{ plant}^{-1}\ \text{h}^{-1}}{\text{g leaf dry weight}} = \text{nmol } C_2H_4\ (\text{g leaf dry weight})^{-1}\ \text{h}^{-1}$$

K 'reflects the ability of the plant to divert a part of its assimilates towards nitrogen- fixing organisms, and the efficiency of utilization of these assimilates after exudation in the soil under study' (Balandreau *et al.*, 1978). K values will, of course, differ between soils but, for any particular soil, different plant species and genotypes within species will presumably have different K values. This suggests that K values could be a useful index for plant breeders interested in the possibility of enhancing biological N_2 fixation.

Methods for Evaluating Biological Nitrogen Fixation
Edited by F. J. Bergersen

W.D.P. Stewart
*Department of Biological Sciences,
The University, Dundee, Scotland.*

5

Systems Involving Blue-green Algae (Cyanobacteria)

(a) INTRODUCTION

Blue-green algae (Cyanobacteria) are photosynthetic O_2-evolving prokaryotes, many species of which fix N_2. Evidence suggesting that such forms could fix N_2 was obtained as early as 1889 (Frank, 1889; Prantl, 1889) and unequivocal proof of the capacity of some species to fix N_2 was provided by Drewes (1928). Their possible role in helping to maintain the nitrogenous fertility of tropical rice fields was appreciated by De (1936, 1939), but despite this it is only recently that real interest in their potential in this connection has been forthcoming from agriculturalists, physiologists, biochemists, molecular biologists, and geneticists. This chapter considers information on some of the more practical aspects of their biology and ecology. Recent reviews on the group include those of Carr and Whitton (1973), Fogg *et al.* (1973), Stewart (1973, 1978), Mague (1977), Haselkorn (1978), and Stewart *et al.* (1979).

(b) CLASSIFICATION AND SYSTEMATICS

Photosynthetic O_2-evolving prokaryotes have, until recently, been regarded as algae and their classification and systematics has been based on the *International Code of Botanical Nomenclature* (Stafleu *et al.*, 1972). Systematic and taxonomic treatises on blue-green algae which are based on this code of nomenclature and taxonomy include those of Bornet and Flahault (1886–1888), Gomont (1892), Frémy (1929, 1934), Geitler (1932, 1936, 1942), and Desikachary (1959). Such works are recommended especially to field workers interested in classifying their organisms in a way which any other scientist can recognize. It is now clear, nevertheless, that such taxonomic works are fraught with error, owing among other things, to the morphological plasticity of the organisms, to their small size and lack of morphological markers, to the fact that living holotypes are not available, and because morphologically similar forms may be very different biochemically.

There is considerable evidence that these organisms may be better classified as bacteria (see Stanier, 1977; Stanier and Cohen-Bazire, 1977); indeed, the similarities to bacteria have been recognized for over a century (Cohn, 1853, 1871–1872). Stanier *et al.* (1978) proposed that from January 1st, 1979, blue-green algae should be formally reclassified as cyanobacteria and that their taxonomy from then onwards should be in accordance with the *International Code of Nomenclature of Bacteria* (La Page *et al.*, 1975). There has been criticism of this proposal, not so much on scientific grounds, although that has occurred (Bourrelly, 1979), but particularly because of what appears to some to be an unseemly haste to implement the proposal. At the time of writing (1979) the matter has not been resolved. I believe that on balance the organisms are more akin to bacteria than to algae; that being so, I shall call them cyanobacteria in the remainder of this chapter.

What has been useful has been a comparative study by Rippka *et al.* (1979) of 178 strains, representing all the major groups of cyanobacteria, which will serve as a beginning for any new taxonomic treatment based on the *International Code of Nomenclature of Bacteria.* Such strains have been deposited with the American Type Culture Collection (12301 Parklawn Drive, Rockville, Md. 2085, U.S.A.), although a recent request by my group for six of the strains from the collection elicited only two strains; the reason why the others could not be supplied was not given. Using characters which are readily determinable in cultured material, Rippka *et al.* (1979) have recognized 22 genera and 3 groups which they place in five sections. Until cultural holotypes become available, and accepted, reference strains have been designated for the genera described. Only strain numbers have been given to these cultures, although according to the *International Code of Nomenclature of Bacteria* (LaPage *et al.*, 1975) such strains lack nomenclatural standing. The most important thing is that it should be possible, in time, for herbarium specimens, descriptions, and illustrations to be replaced by a culture which subsequently becomes the holotype, and a more satisfactory taxonomy and systematics may eventually emerge.

The major sub-groups of cyanobacteria, based on the paper by Rippka *et al.* (1979), are shown in Table 1 and the basis on which the genera are differentiated is outlined in Table 2. Strain differences are based on characters

Table 1. The major sub-groups of Cyanobacteria (Rippka *et al.*, 1979)

Unicellular forms (the cells occur singly or form colonial aggregates held together by additional outer cell wall layers)	
Reproduction by binary fission or by budding	Section I
Reproduction by multiple fission giving rise to small daughter cells (baeocytes) or by both multiple fission and binary fission	Section II
Filamentous forms (the cells are arranged as a chain of cells (the trichome) which grows by intercalary cell division)	
Reproduction by random breakage of trichome, by the formation of hormogonia, and (Sections IV and V only) sometimes by the germination of akinetes	
Trichome always composed of vegetative cells only; division in one plane only	Section III
In the absence of combined nitrogen the trichome contains heterocysts; some also produce akinetes	
Division in one plane	Section IV
Division in more than one plane	Section V

such as their ability to grow photoheterotrophically, their capacity to synthesize nitrogenase aerobically or anaerobically, the presence or absence of c-phycocyanin, whether they are marine or freshwater in origin, their vitamin B_{12} requirement, and mode of cell division, etc., as well as on gross morphology, cell shape and size, etc.

Table 2. Genera of Cyanobacteria based on the criteria of Rippka *et al.* (1979)

Section I	Unicellular forms which reproduce by binary fission or by budding	
	Reproduction by binary fission, thylakoids absent, division in one plane, sheath present	*Gloeobacter*
	Reproduction by binary fission, thylakoids present, division in one plane, sheath present	*Gloeothece*
	Reproduction by binary fission, thylakoids present, division in one plane, sheath absent	*Synechococcus*
	Reproduction by binary fission thylakoids present, division in two or three planes, sheath present	*Gloeocapsa*
	Reproduction by binary fission, thylakoids present, division in two or three planes, sheath absent	*Synechocystis*
	Reproduction by budding, thylakoids present, division in one plane	*Chamaesiphon*
Section II	Unicellular cyanobacteria which reproduce by multiple fission	
	Reproduction by multiple fission only, motile baeocytes without fibrous outer wall layer	*Dermocarpa*
	Reproduction by multiple fission only, immotile baeocytes with fibrous outer wall layer	*Xenococcus*
	Reproduction by both binary fission and multiple fission; binary fission yields pear shaped structure of one or two basal cells and one apical cell; subsequent multiple fission of apical cell yields motile baeocytes without fibrous outer wall layer	*Dermocarpella*
	Reproduction by both binary fission and multiple fission; binary fission yields cubical cellular aggregates; subsequent multiple fission yields motile baeocytes without fibrous outer wall layer	*Myxosarcina*
	As *Myxosarcina* but baeocytes immotile with fibrous outer wall layer	*Chroococcidiopsis*
	Reproduction by both binary fission and multiple fission; binary fission yields irregular cellular aggregates (pseudofilamentous); subsequent multiple fission yields motile baeocytes without fibrous outer wall layer	Pleurocapsa group

Table 2 (continued)

Section III	Filamentous non-heterocystous cyanobacteria which divide in one plane only	
	Trichome helical, composed of isodiametric, cylindrical or disc-shaped cells, little or no constriction between adjacent cells; reproduction probably by transcellular trichome breakage; trichome motile, either not ensheathed or thinly sheathed	*Spirulina*
	Trichome straight, composed of disc-shaped cells which are not separated by deep constrictions; reproduction is by transcellular trichome breakage; trichome motile, either not ensheathed or thinly sheathed	*Oscillatoria*
	Trichome straight, composed of disc-shaped cells which are not separated by deep constrictions; reproduction is by transcellular trichome breakage; trichome immotile, enclosed by heavy sheath; motility restricted to sheathless or thinly sheathed hormogonia	LPP group A*
	Trichome straight, composed of isodiametric or cylindrical cells; reproduction is by transcellular or intercellular trichome breakage; trichome motile, no sheath; cells contain polar gas vacuoles and are separated by deep constrictions	*Pseudanabaena*
	Trichome straight, composed of isodiametric or cylindrical cells; variable degree of constriction between adjacent cells; reproduction by transcellular or intercellular trichome breakage; sheath present, or absent; no gas vacuoles	LPP group B*
Section IV	Filamentous heterocystous cyanobacteria which divide in one plane only	
	Reproduction by random trichome breakage, and (in some) by germination of akinetes, to produce trichomes indistinguishable from the mature vegetative trichomes; heterocysts intercalary or terminal; position of akinetes (if produced) variable; vegetative cells spherical, ovoid, or cylindrical	*Anabaena*
	Reproduction by random trichome breakage, and (in some) by germination of akinetes, to produce trichomes indistinguishable from the mature vegetative trichomes; heterocysts intercalary or terminal; position of akinetes (if produced) variable; vegetative cells disc-shaped	*Nodularia*
	Reproduction by random trichome breakage, and (in some) by germination of akinetes, to produce trichomes indistinguishable from the	

Table 2 (continued)

	mature vegetative trichomes; heterocysts exclusively terminal, formed at both ends of the trichome; akinetes always adjacent to heterocysts; vegetative cells isodiametric or cylindrical	*Cylindrospermum*
	Reproduction by random trichome breakage, and (in some) by germination of akinetes, to produce trichomes indistinguishable from the mature vegetative trichomes; reproduction also by formation of hormogonia distinguishable from mature trichomes by the absence of heterocysts and by one or more of rapid gliding motility, smaller cell size, cell shape, and gas vacuolation; hormogonia give rise to young filaments with a terminal heterocyst at both ends of the cellular chain; vegetative cells spherical, ovoid or cylindrical; akinetes (if produced) not adjacent to heterocysts, often formed in chains	*Nostoc*
	Reproduction by random trichome breakage, and (in some) by germination of akinetes, to produce trichomes indistinguishable from the mature vegetative trichomes; reproduction also by formation of hormogonia distinguishable from mature trichomes by the absence of heterocysts and by one or more of rapid gliding motility, smaller cell size, cell shape, and gas vacuolation; hormogonia give rise to young filaments with a terminal heterocyst at only one end of the cellular chain; mature trichome composed of cells of even width, heterocysts are predominantly intercalary; vegetative cells disc-shaped, isodiametric or cylindrical	*Scytonema*
	Reproduction by random trichome breakage, and (in some) by germination of akinetes, to produce trichomes indistinguishable from the mature vegetative trichomes; reproduction also by formation of hormogonia distinguishable from mature trichomes by the absence of heterocysts and by one or more of rapid gliding motility, smaller cell size, cell shape, and gas vacuolation; hormogonia give rise to young filaments with a terminal heterocyst at only one end of the cellular chain; mature trichome tapers from base, which bears terminal heterocyst, to apex; vegetative cells disc-shaped, isodiametric or cylindrical	*Calothrix*
Section V	Filamentous heterocystous cyanobacteria which divide in more than one plane	
	Reproduction by random trichome breakage, by	

Table 2 (continued)

formation of hormogonia and (if produced) by germination of akinetes; hormogonia composed of small cylindrical cells which enlarge and become spherical; heterocysts develop in terminal and intercalary positions; mature trichome cells divide in more than one plane; associated detachment of groups of cells leads to irregular *Gloeocapsa*-like aggregates containing terminal heterocysts; hormogonia produced within such aggregates	*Chlorogloeopsis*
Reproduction by random trichome breakage, by formation of hormogonia and (if produced) by germination of akinetes; hormogonia composed of small cylindrical cells which enlarge and become rounded; heterocysts develop almost exclusively in an intercalary position; mature trichome cells divide in more than one plane to produce a partly multiseriate trichome with lateral uniseriate branches; heterocysts in the primary trichome predominantly terminal or lateral; hormogonia produced from the ends of trichomes or from lateral branches	*Fischerella*

*LPP = *Lyngbya–Phormidium–Plectonema* group.

(c) THE RANGE OF N_2-FIXING CYANOBACTERIA

About 125 strains of free-living cyanobacteria have been shown to fix N_2 in axenic culture (Table 3). These belong to all major typological groups, although the extent to which nitrogenase occurs and the conditions under which it functions vary from group to group. This enormous increase in numbers has resulted from the use of the acetylene reduction technique (Stewart *et al.*, 1967, 1979) and from the appreciation that, like the photosynthetic bacteria, certain strains can fix N_2 under anaerobic conditions but not in air (Stewart and Lex, 1970; Rippka and Waterbury, 1977; Stewart *et al.*, 1979).

(i) Heterocystous filamentous types

The most important group of N_2-fixing cyanobacteria are the heterocystous forms which can fix N_2 aerobically and anaerobically. Heterocysts are distinctive empty-looking cells with a thick cell envelope, whose main function is to fix N_2. In N_2-grown cultures they represent about 3–7% of the total cells present. This proportion increases to about double this in nitrogen-starved

Table 3. Nitrogen-fixing Cyanobacteria

Group	Genus	Total tested	Aerobic nitrogenase	Anaerobic/ microaerobic nitrogenase	Assay condition*
Chroococcacean	*Aphanothece*	1	1	1	T.N.
	Gloeothece†	5	5	5	C_2H_2
	Synechococcus	27	0	3	C_2H_2
Pleurocapsalean	*Dermocarpa*	6	0	2	C_2H_2
	Xenococcus	3	0	1	C_2H_2
	Myxosarcina	2	0	1	C_2H_2
	Chroococcidiopsis	8	0	8	C_2H_2
	Pleurocapsa	12	0	7	C_2H_2
Non-heterocystous‡ filamentous forms	*Oscillatoria*	9	0	5	C_2H_2
	Pseudanabaena	8	0	4	C_2H_2
	Lyngbya–Plectonema–Phormidium	25	0	16	C_2H_2, $^{15}N_2$, T.N.
Heterocystous filamentous forms	*Anabaena*	15	15	15	C_2H_2, $^{15}N_2$, T.N.
	Anabaenopsis	2	2	2	C_2H_2, $^{15}N_2$, T.N.
	Aulosira	1	1	1	T.N.
	Calothrix	4	4	4	C_2H_2, $^{15}N_2$, T.N.
	Cylindrospermum	5	5	5	C_2H_2, T.N.
	Fischerella	2	2	2	T.N.
	Hapalosiphon	1	1	1	T.N.
	Mastigocladus	1	1	1	T.N.
	Nostoc	13	13	13	C_2H_2, $^{15}N_2$, T.N.
	Scytonema	3	3	3	T.N.
	Stigonema	1	1	1	T.N.
	Tolypothrix	2	2	2	T.N.
	Westiella	1	1	1	T.N.
	Westiellopsis	1	1	1	$^{15}N_2$, T.N.

*Certain, or all of the cyanobacteria have been tested by these methods; T.N. = total nitrogen.
†Includes strains previously designated as N_2-fixing *Gloeocapsa* strains.
‡The data given here are those of Rippka and Waterbury (1977), but the exact numbers of strains tested and shown to have nitrogenase may be larger since various earlier workers (Stewart and Lex, 1970; Stewart, 1970; Stewart *et al.*, 1977) had examined and obtained positive results with strains which may or may not correspond to those tested by Rippka and Waterbury (1977). (From Stewart *et al.*, 1979.)

cultures [grown in the absence of N (Kulasooriya *et al.*, 1972), or in Mo^+-deficient cultures (Fay and Vasconcelos, 1974), etc.] In cultures grown on high levels of combined nitrogen heterocysts and nitrogenase are not synthesized (Fogg, 1949; Stewart *et al.*, 1968; Wilcox, 1970; Bone, 1971; Kulasooriya *et al.*, 1972). There is often a good correlation between heterocyst frequency and the rate of nitrogenase activity in metabolically active cultures of aerobically grown cyanobacteria (Stewart *et al.*, 1968), although this correlation obviously varies depending on the metabolic state of the culture. The presence of heterocysts is an immediate visual marker that the organism has the potential to fix N_2 aerobically and anaerobically. Non-N_2-fixing mutants of heterocystous forms are extremely rare, except when selected for in laboratory cultures. Presumably such mutants do not compete in nature successfully.

Heterocysts appear to be the only N_2-fixing cells in aerobic cultures of those cyanobacteria which possess them (Fay *et al.*, 1968; Stewart *et al.*, 1969; Van Gorkom and Donze, 1971; Fleming and Haselkorn, 1973; Tel-Or and Stewart, 1976; Peterson and Burris, 1976b; Peterson and Wolk, 1978). Enzymatic evidence for the presence of nitrogenase activity in heterocysts only has depended on techniques for separating the heterocysts and comparing their metabolic activity with that of vegetative cell fractions. The currently most popular methods of preparing metabolically active heterocysts are based, often with minor modifications, on the techniques of Peterson and Burris (1976b), Thomas *et al.* (1977), and Peterson and Wolk (1978). The presence of nitrogenase is detected enzymatically using the acetylene reduction technique (Stewart *et al.*, 1969) or by detecting the nitrogenase proteins electrophoretically (Peterson and Wolk, 1978). The success of each method depends to some extent on the organism used.

Within heterocysts the nitrogenase is protected from O_2 damage (the Mo–Fe protein and especially the Fe protein are both irreversibly damaged by O_2) (Haystead *et al.*, 1970; Tsai and Mortenson, 1978). Protection from O_2 is brought about in several ways. The heterocyst envelope appears to have restricted permeability to O_2 and gases may enter the heterocyst mainly from the vegetative cells via the heterocyst pore. Heterocysts also lack a capacity to evolve O_2, owing mainly to a lack of photosystem II accessory pigments and a deficiency of Mn^{2+} (Thomas, 1970; Tel-Or and Stewart, 1977). They also show high rates of dark respiratory activity (Fay and Walsby, 1966) and an oxy-hydrogen reaction which may scavenge O_2 and generate ATP (Bothe *et al.*, 1978; Peterson and Burris, 1978). Thus, in heterocysts O_2 input is limited by diffusion, there is no O_2 evolution, and there is substantial O_2 uptake.

A third major factor facilitating nitrogenase activity in heterocysts is that heterocysts show little photofixation of CO_2 and thus within heterocysts the ATP and reductant generated can be used for N_2 fixation without competition from CO_2 fixation. In addition to the modifications associated with photosystem II activity, the key carboxylating enzyme of the Calvin cycle, ribulose 1,5-bisphosphate carboxylase (RuBP Case), is deficient in heterocysts (Winkenbach and Wolk, 1973; Stewart and Codd, 1975; Codd and Stewart, 1977a), as is phosphoribulokinase which catalyses RuBP formation (Codd *et al.*, 1980). Phosphoglycollate phosphatase and glycollate dehydrogenase, which metabolize 2-phosphoglycollate produced as a result of oxygenase activity, are also deficient in heterocysts (Codd *et al.*, 1980). Since cyanobacterial RuBP carboxylase may also act as an oxygenase (Codd and Stewart, 1977b; Okabe *et al.*, 1979) there can be no light-dependent scavenging of O_2 (photorespiration) in heterocysts as a result of the activity of this enzyme.

Fourthly, heterocysts have evolved a specialized mechanism for the removal of fixed nitrogen which, otherwise, may lead to nitrogenase inactivation and repression of nitrogenase synthesis (Bone, 1972; Stewart *et al.*, 1975). Evidence for inactivation of nitrogenase comes from the finding that on adding NH_4^+ to N_2-grown cultures the loss of activity is more rapid than can be accounted for by simple dilution of the enzyme as a result of continued growth and protein turnover, without nitrogenase synthesis. Removal of NH_4^+ from heterocysts is brought about mainly by glutamine synthetase (GS), which is present in higher concentrations in heterocysts than it is in vegetative cells (Dharmawardene *et al.*, 1973; Thomas *et al.*, 1977). It has a low K_m for NH_4^+ (1 mM) and can efficiently scavenge NH_4^+ from around the N_2-fixing site (Stewart *et al.*, 1975). Other primary NH_4^+-assimilating enzymes such as glutamic acid dehydrogenase and alanine dehydrogenase are either not present in heterocysts and/or have K_m values for NH_4^+ which make them less satisfactory than GS as the primary NH_4^+-assimilating enzyme. The ATP requirement for GS (and for nitrogenase) can be met by photophosphorylation and to a lesser extent by oxidative phosphorylation and an oxy-hydrogen reaction. Substrate-level phosphorylation appears to be unimportant in cyanobacteria (Bottomley and Stewart, 1977; Peterson and Burris, 1978).

Glutamate synthase (GOGAT), which is ferredoxin-dependent in cyanobacteria (Lea and Miflin, 1975) and which is coupled with GS in the GS–GOGAT pathway in vegetative cells, is undetectable in heterocysts, and tracer studies suggest that the glutamine produced is exported from the heterocysts via the pores to the adjacent vegetative cells (Stewart *et al.*, 1975; Thomas *et al.*, 1977) and that some of the glutamate produced there through the activity of GOGAT is transported back to the heterocysts to provide substrate for GS (Thomas *et al.*, 1977), the remainder being used in general amino acid biosynthesis.

The GS of *Anabaena* spp. (Stacey *et al.*, 1977; Sampaio *et al.*, 1979) and *Nostoc* (Sampaio *et al.*, 1979) have been purified to homogeneity. The molecular weight of the enzyme from all three species is approximately 600000, and the *Anabaena* and *Nostoc* enzymes, at least, are composed of 12 identical subunits arranged in two superimposed hexagonal rings. *In vitro*, the enzyme dissociates and loses activity in the absence of stabilizing ligands and re-association and re-activation occur on adding 2-mercaptoethanol and substrates, particularly glutamate (Sampaio *et al.*, 1979). *In vivo* the *A. cylindrica* enzyme is deactivated on darkening the culture, but the enzyme from *Nostoc* and from *Plectonema boryanum*, both of which can grow heterotrophically, is not (Rowell *et al.*, 1978, 1979).

Fifthly, heterocysts, being unable to fix CO_2 photosynthetically, receive fixed carbon compounds from the vegetative cells via the pores. This has been shown by autoradiography using ^{14}C as tracer (Wolk, 1968; Stewart *et al.*, 1969). There is some dispute about the nature of the compound which is mainly transported but it is a dissaccharide (possibly maltose, according Juttner and Carr, 1976). Heterocysts show high activities of the enzymes of the major route of dark carbon dissimilation in autotrophs—the oxidative pentose phosphate pathway (Winkenbach and Wolk, 1973; Lex and Carr, 1974; Apte *et al.*, 1978)—and whereas this pathway operates in vegetative cells mainly in the dark, in heterocysts it also functions in the light (Apte *et al.*, 1978). Apte *et al.* (1978) have also shown that a major route of electron transfer from NADPH generated in the oxidative pentose phosphate pathway to ferredoxin is via ferredoxin–$NADP^+$ oxidoreductase, an enzyme which may otherwise have been functionless in heterocysts in the absence of RuBP carboxylase. In heterocysts this enzyme can transfer electrons from NADPH to ferredoxin both in the light and in the dark, although it may function in this direction in vegetative cells only in the dark. The factors which permit such an uphill transfer of electrons are NADPH to $NADP^+$ ratios or reduction charge values of 0.3 or higher *in vitro* (Apte *et al.*, 1978) and possibly membrane potentials (Haaker *et al.*, 1980). Ferredoxin–$NADP^+$ oxidoreductase from *Anabaena cylindrica* has been purified in this laboratory (Rowell *et al.*, 1980) and multiple forms of the catalytically active enzyme have been found, as in higher plant chloroplasts (Fredricks and Gehl, 1976; Gozzer *et al.*, 1977). The enzyme can act in chloroplasts as a NADPH-specific diaphorase (Avron and Jagendorf, 1956), as a pyridine nucleotide transhydrogenase (Keister *et al.*, 1960), and as a NADPH-dependent cytochrome *f* reductase (Zanetti and Forti, 1966). The functions of the multiple forms in cyanobacteria remain to be determined.

(ii) Non-heterocystous filamentous types

Various non-heterocystous filamentous cyanobacteria have an active nitrogenase under microaerobic or anaerobic conditions only. This was first shown by Stewart and Lex (1970) using *Plectonema boryanum* strain 594, a finding which also emphasized the similarity between such cyanobacteria and certain photosynthetic bacteria which can grow aerobically on combined nitrogen but which fix N_2 only anaerobically (see Pfennig, 1977). Since then about half the non-heterocystous filamentous cyanobacteria tested have been shown to have an active nitrogenase, but under anaerobic conditions only (Rippka and Waterbury, 1977; Stewart *et al.*, 1979). The genera with N_2-fixing representatives include the *Lyngbya–Plectonema–Phormidium* group, *Oscillatoria*, and *Pseudanabaena*, while genera tested and so far without N_2-fixing representatives are *Gloeobacter*, *Gloeocapsa* (*sensu* Rippka *et al.*, 1979), *Synechocystis*, *Chamaesiphon*, and *Spirulina* (the only non-heterocystous filamentous genus yet tested and shown to have no N_2-fixing strains).

In our laboratory we have consistently been able to obtain C_2H_2 reduction and $^{15}N_2$ incorporation by incubating the algae under a gas phase of Ar–CO_2 (99.96 : 0.04 v/v) for 48 h to deplete the organisms of nitrogen and to maintain an anaerobic gas phase (Stewart and Lex, 1970). However, Rippka and Waterbury (1977), in their survey of non-heterocystous forms, included 1×10^{-5}M DCMU [3-(3,4-dichlorophenyl)-1,1-dimethylurea] in their assays, to prevent O_2 production as a result of the photolysis of water (Cobb and Myers, 1962; Cox, 1966; Cox and Fay, 1969; Bothe, 1970). It will be of interest to determine whether any, or many, non-heterocystous filamentous cyanobacteria can fix N_2 aerobically. The finding that when soil samples, for example, are plated on to medium free of combined nitrogen, heterocystous rather than non-heterocystous forms develop suggests that aerobic N_2-fixing non-heterocystous filamentous forms are rare. However, there are various early unsubstantiated reports of fixation, presumably under aerobic conditions by non-heterocystous filamentous forms including strains of *Oscillatoria* (Copeland, 1932; Moyse *et al.*, 1957), *Phormidium* and *Spirulina* (Copeland, 1932), *Plectonema* (Watanabe, 1959a), and *Lyngbya* (Van Baalen, 1962). The marine genus *Trichodesmium* also fixes N_2 (Dugdale *et al.*, 1961, 1964; Goering *et al.*, 1966; Bunt *et al.*, 1970; Carpenter, 1973; Mague *et al.*, 1974; Carpenter and McCarthy, 1975), possibly only when aggregated in colonies (Carpenter and Price, 1976). Recently, Pearson *et al.* (1979) obtained evidence of C_2H_2 reduction under aerobic conditions by *Microcoleus chthonoplastes* contaminated by two non-N_2-fixing heterotrophic bacteria. Schloesing and Laurent (1892), however, obtained no evidence of fixation by *Microcoleus vaginatus* although heterocystous forms fixed nitrogen under similar conditions. It is remarkable that many of the non-heterocystous filamentous forms reported to fix N_2 are marine forms, e.g. *Lyngbya aestuarii* and *Trichodesmium*; the significance of and/or reason for this remains to be determined.

(iii) Unicellular forms reproducing by binary fission or budding

There were early reports that *Gloeocapsa* strains from the Pamir mountains (Odintzova, 1941) and from American desert soils (Cameron and Fuller, 1960) fixed N_2, but little attention was paid to such findings until Wyatt and

Silvey (1969), using pure cultures, obtained C_2H_2 reduction by a strain of *Gloeocapsa alpicola* (*Gloeothece* sp.). Supporting evidence for nitrogenase activity in this genus then followed using axenic (Rippka *et al.*, 1971) and non-axenic (Gallon *et al.*, 1972, 1973, 1975ab, 1978) cultures.

There is now evidence based on axenic cultures that three genera of unicellular cyanobacteria which divide by binary fission have N_2-fixing representatives. These are: (a) *Aphanothece*, an Indian strain of which fixes N_2 aerobically and anaerobically (Singh, 1973); (b) *Gloeothece*, a genus which now includes all N_2-fixing strains of *Gloeocapsa*, and in which aerobic and anaerobic nitrogenase activity occurs (Rippka and Waterbury, 1977); and (c) *Synechococcus*, a genus in which 3 out of 24 strains tested fix N_2 but under anaerobic conditions only (Rippka and Waterbury, 1977). Rippka *et al.* (1979) have merged *Aphanothece* and *Synechococcus* but since there is some inconsistency in the published literature the genera are kept separate here. A problem is that the latter workers, although merging the two genera, also say that only strains of *Gloeothece* can fix N_2 aerobically and anaerobically. Thus, either *Aphanothece* (Singh, 1973) has been mis-identified and/or does not fix N_2, or aerobic nitrogenase activity is not a reliable taxonomic marker for *Gloeothece.* Irrespective of that, only 19% of all unicellular cyanobacteria so far tested (52 strains) which divide by binary fission or budding have been shown to have nitrogenase activity, and of those genera which do 17% of the strains fix aerobically and 28% fix anaerobically.

(iv) Unicellular cyanobacteria reproducing by multiple fission

This group of unicellular cyanobacteria, has six major divisions—five genera and one assemblage (see Table 2). They show a wide range of structure and development but all have G + C values within the range 38–47% and all have representatives with nitrogenase which is active only in anaerobic cultures. Of the 31 strains tested by Rippka and Waterbury (1977) none fix aerobically, but 61% fix anaerobically. Their growth and development has been detailed by Waterbury and Stanier (1978).

(v) Identification and testing for nitrogenase activity

Accurate identification is difficult and to safeguard for the future and to help resolve the existing taxonomic confusion, the following procedure is recommended:

1. Identify the organism to species level if possible using a standard phycological work and give the appropriate authorities. Then everyone knows what the organism looks like, even if the name given proves eventually to be wrong. Make drawings and take photographs of it, and note the dimensions and other characters.

2. Identify it, if possible, according to Rippka *et al*. (1979) and compare it as far as possible with a culture of the reference strain for that genus grown under identical conditions.
3. Deposit a culture with an international type culture collection for future reference (e.g. The Culture Collection of Algae and Protozoa, Cambridge; American Type Culture Collection (see Section b above); University of Texas Culture Collection, Department of Botany, University of Texas, Austin, Texas 78712, U.S.A.).
4. Present in your publication all the relevant taxonomic information and/or say where the culture or sample can be obtained.

Tests for nitrogenase activity are fairly straightforward but there are a few simple rules:

1. Use pure cultures. There is no alternative if your data on nitrogenase activity by the test organism are to be unequivocal. It is not acceptable to say that associated contaminants do not have nitrogenase activity because you may not be testing them for nitrogenase activity under the conditions which operate within a mixed culture—presumptive evidence with a mixed culture—yes; unequivocal evidence—no!
2. Tests for nitrogenase activity must be carried out under *both* aerobic and anaerobic conditions. The anaerobic conditions can be achieved by bubbling with a O_2-free gas phase and the culture medium should contain an inhibitor of photosystem II activity such as DCMU (Cox and Fay, 1969) (see Rippka and Waterbury, 1977). Rippka and Waterbury (1977) suggest the use of 1×10^{-5} M DCMU, but we find that this concentration is sometimes too high; other concentrations, which will vary with the strain, should also be tried.
3. A fairly low light intensity (*ca*. 1000 lux) should be used to prevent bleaching or the possible O_2 inactivation of nitrogenase as a result of the photoproduction of O_2 when DCMU is not added.
4. Nitrogenase activity should be monitored at intervals over a period (usually the first 72 h) after transfer to nitrogen-free medium (Stewart and Lex, 1970). There may also be a temporal separation of nitrogenase activity and photosynthetic O_2 evolution (Weare and Benemann, 1974; Gallon *et al*., 1975a) in the absence of DCMU.
5. Healthy cultures and preferably optimally growing exponential cultures must be used as inoculum and test material; the culture medium should be free of combined nitrogen.
6. The acetylene reduction test for nitrogenase activity (Stewart *et al*., 1967, 1971), although indirect, is the test of choice because of its sensitivity, and general reliability. This should be backed up where possible by tests for $^{15}N_2$ incorporation and by showing gains in combined nitrogen by the pure culture when N_2 is the sole nitrogen source (see below).

(d) CULTIVATION AND ISOLATION PROCEDURES

(i) Culture media

As discussed above, the use of axenic cultures is an essential prerequisite to any unequivocal demonstration of whether or not an organism has nitrogenase. The cultivation of cyanobacteria is fairly straightforward since all can grow in the light on simple media under aerobic conditions. Apart from a few exceptions, the golden rule is to choose a dilute inorganic medium with a neutral or alkaline pH.

Various media commonly used to grow cyanobacteria successfully are listed in Table 4.

Of the freshwater types of media, that of Allen and Arnon (1955) is long established and successful. Its main disadvantage is its relatively high phosphate concentration, which leads to precipitation on autoclaving unless the medium is cooled slowly and this may take overnight with large (20–50 l) batches, or unless the phosphate is added separately after cooling. The BG11 medium of Stanier *et al.* (1971) is, in general, preferable, being more dilute and with a lower phosphate concentration. Thus, precipitation is less likely, although it is always advisable to add the phosphate separately after autoclaving and cooling. Medium BG-11, with or without combined nitrogen, permits even suspension of the filaments, an essential requirement for any physiological experiment. Sutherland *et al.* (1979) found, for example, using *Nostoc* PCC 7524, that good synchrony of akinete formation occurred only in uniformly suspended cultures. Light intensity was the critical factor triggering akinete initiation and uniform dispersion of light was not achieved in cultures which clumped. BG-11 is the most successful and most widely used freshwater medium currently available and virtually any freshwater N_2-fixing cyanobacterium can be isolated using its nitrogen-free modification. There are a few reports of certain forms having lost their capacity to grow photoautotrophically and requiring an exogenous energy source, for example the *Nostoc* endophyte of cycad root nodules (Winter, 1935), and of a sole freshwater cyanobacterium (a *Dermocarpa* strain) requiring vitamin B_{12} (Rippka *et al.*, 1979), but the addition of organics to freshwater culture media should be avoided, particularly if the cultures are impure; in general they are not required, or recommended.

Successful marine media include those with a natural sea water base plus supplement, or defined artificial media (Provasoli *et al.*, 1957). Van Baalen (1962) used ASP-II of Provasoli *et al.* (1957), whereas Rippka *et al.* (1979) used medium MN (natural sea water supplemented with BG11 salts) or ASN-II, an artificial sea water medium (see Table 4). In most cases marine isolates do not grow on BG-11 medium with or without added NaCl. Unlike freshwater forms, certain marine isolates have a vitamin B_{12} requirement (Pintner and

Table 4. Composition of various standard media used for the culture of Cyanobacteria (values given are mg l^{-1})

Compound	Kratz and Myers (1955)	Allen and Arnon (1955)	Gorham *et al.* (1964), ASM	Provasoli *et al.* (1957), ASP-II	Stanier *et al.* (1971), BG11	Rippka *et al.* (1979), ASN-III	Rippka *et al.* (1979), MN
$NaNO_3$	—	—	170	50	1500	750	750
KNO_3	1000	2000	—	—	—	—	—
NaCl	—	230	—	18000	—	25000	—
KCl	—	—	—	600	—	500	—
$MgCl_2 \cdot 6H_2O$	—	—	40	—	—	2000	—
$MgSO_4 \cdot 7H_2O$	250	250	50	5000	75	3500	38
$CaCl_2 \cdot 2H_2O$	—	74	36	370	36	500	18
$Ca(NO_3)_2 \cdot 4H_2O$	25	—	—	—	—	—	—
$K_2HPO_4 \cdot 3H_2O$	1300	400	20	6	40	20	20
Na_2HPO_4	—	—	14	—	—	—	—
$FeCl_3$	—	—	0.6	2.3	—	—	—
$Fe(SO_4)_3 \cdot 6H_2O$	4	—	—	—	—	—	—
FeEDTA	—	26	—	—	—	—	—
Na_2EDTA	—	—	10	30	1	0.5	0.5
Na citrate $\cdot 2H_2O$	165	—	—	—	—	—	—
Citric acid	—	—	—	—	6	3	3
Iron(III) ammonium citrate	—	—	—	—	6	3	3
Na_2CO_3	—	—	—	—	20	20	20
H_3BO_3	2.86	2.86	2.5	34.3	2.86	2.86	2.86
$MnCl_2 \cdot 4H_2O$	1.81	—	1.4	4.3	1.81	1.81	1.81
$MnSO_4 \cdot 4H_2O$	—	2	—	—	—	—	—

MoO_3	0.02	0.15	—	—	—	—	—
$Na_2MoO_4 \cdot 2H_2O$	—	—	—	—	0.39	0.39	0.39
$ZnCl_2$	—	—	0.4	0.31	—	—	—
$ZnSO_4 \cdot 7H_2O$	0.22	0.22	—	—	0.22	0.22	0.22
$CuSO_4 \cdot 5H_2O$	0.08	0.08	—	—	0.08	0.08	0.08
$CuCl_2 \cdot 2H_2O$	—	—	0.0002	0.003	—	—	—
$CoCl_2 \cdot 6H_2O$	—	—	0.02	0.01	—	—	—
$Co(NO_3)_2 \cdot 6H_2O$	—	0.05	—	—	0.05	0.05	0.05
NH_4VO_3	—	0.02	—	—	—	—	—
$NiSO_4 \cdot 6H_2O$	—	0.04	—	—	—	—	—
$Cr_2(SO_4)_3 \cdot K_2SO_4 . 24H_2O$	—	0.19	—	—	—	—	—
$Na_2WO_4 \cdot 2H_2O$	—	0.02	—	—	—	—	—
$TiO(C_2O_4)_x \cdot yH_2O$	—	0.01	—	—	—	—	—
$Na_2SiO_3 \cdot 4H_2O$	—	—	—	150	—	—	—
Tris	—	—	—	1000	—	—	—
Vitamin B_{12}	—	—	—	0.002	—	—	—
Thiamine · HCl	—	—	—	0.5	—	—	—
Nicotinic acid	—	—	—	0.1	—	—	—
Capantothenate	—	—	—	0.1	—	—	—
p-Aminobenzoic acid	—	—	—	0.01	—	—	—
Biotin	—	—	—	0.001	—	—	—
Inositol	—	—	—	5	—	—	—
Folic acid	—	—	—	0.002	—	—	—
Thymine	—	—	—	30	—	—	—
Sea water	—	—	—	—	—	—	750 ml
Deionized water	1000 ml	1000 ml	1000 ml	1000 ml	1000 ml	1000 ml	250 ml
pH after autoclaving	6.9–9.0	Not given	Not given	7.6–7.8	7.1	7.5	8.3

Provasoli, 1958; Van Baalen, 1961, 1962; Rippka *et al.*, 1979) and it is advisable to routinely add this, after filter sterilization, at a concentration of 10–20 μg l^{-1}.

Both freshwater and marine isolates grow well on liquid medium or medium solidified with agar. Routinely 1% w/v agar is used but it is sometimes useful to use a softer agar (0.75%), and to use silica gel rather than agar as the base. In studies on marine cyanobacteria Stewart (1962) found silica gel prepared by the method of Smith (1951) to be a satisfactory inorganic solid base for the growth of several marine supralittoral cyanobacteria (remember when using silica gel to adjust the pH to that of the culture medium!).

(ii) Isolation procedures

These involve: (1) getting the cyanobacterium into culture; (2) freeing it from contaminating organisms; and (3) ensuring genotypic uniformity.

Initial attempts to grow the cyanobacterium usually involve taking a small sample of the raw material and adding this to an inorganic medium (with vitamin B_{12} in the case of marine forms), free from combined nitrogen in the case of heterocystous cyanobacteria, but with combined nitrogen present in the case of non-heterocystous forms. NO_3^- is preferred to NH_4^+ as a nitrogen source. The latter causes a pH drop on its assimilation, except in well buffered medium (e.g. that of Allen and Arnon, 1955), and cyanobacteria in general dislike a low pH, although any pH increase as a result of NO_3^- uptake has little adverse effect. The inoculum should be fairly large (visible to the naked eye) because small inocula often do not take for reasons which are not yet entirely understood, and organic matter in the inoculum, other than the cyanobacterium, should be minimized. If liquid medium is used, its volume should be small (5–20 ml) to enhance the chances of the inoculum taking; in all cases the light intensity should be low (probably about 1000 lux and less than 3000 lux) but continuous (various photosynthetic eukaryotes require a light–dark cycle for growth but cyanobacteria do not) and the temperature should be 30–35 °C (which selects against eukaryotes, which have temperature optima 10 °C lower). However, high temperatures are sometimes unsuccessful since they may promote bacterial growth; then a lower temperature, selected by trial and error, should be used.

When the cyanobacterium is growing it should, if in liquid culture, be inoculated on to solid medium for purification. Cyanobacteria, like other Gram-negative organisms, are highly sensitive to antibiotics such as penicillin but early attempts should be made to determine the sensitivity of the organism and its contaminants to various other antibiotics, inhibitors, etc., and add the relevant compounds at appropriate concentrations to the isolation medium. The information in Table 5 may serve as a guide to the further evaluation of various compounds.

Table 5. Minimum concentrations of antibiotics, mutagens, and inhibitors preventing growth of cyanobacteria in inorganic culture media

For further details and reference see Fogg *et al.* (1973).

Substance	Species	Concentration (mg l^{-1} except where otherwise stated)
Albamycin	*Microcystis aeruginosa*	1000
Bacitracin	*Anabaena variabilis*	100
	Microcystis aeruginosa	100
Caffeine	*Fischerella muscicola*	1940
Candicidin	*Microcystis aeruginosa*	1000
Carbenicillin*	*Nostoc* sp.	2.5
Chloromycetin	*Anabaena variabilis*	10
	Fischerella muscicola	>4000
Coumarine	*Fischerella muscicola*	> 200
Dehydrostreptomycin	*Microcystis aeruginosa*	1
Dichloronaphthoquinone	Cyanobacteria generally	0.03–0.055
Erythromycin*	*Nostoc* sp.	0.5
Gliotoxin	*Nostoc* sp.	125–250
Gramicidin	*Anabaena variabilis*	1000
Isoniazid	*Anacystis nidulans*	>10
Kanamycin*	*Nostoc* sp.	25
Lanthanum acetate	*Fischerella muscicola*	475
Maleic hydrazide	*Fischerella muscicola*	448
Manganese (II) chloride	*Fischerella muscicola*	1500
Mitomycin	*Anacystis nidulans*	5
Neomycin	*Nostoc* sp.	4
	Phormidium sp.	4
	Anacystis nidulans	>100
	Microcystis aeruginosa	1
Patulin	*Microcystis aeruginosa*	10
Penicillin	*Anabaena variabilis*	0.1
	Microcystis aeruginosa	2
	Microcystis aeruginosa	1
	Anacystis nidulans	0.2
	Phormidium muscicola	0.1
	*Nostoc** sp.	100
Phenylethanol*	*Nostoc* sp.	0.1%
Polymyxin A	*Nostoc* sp.	10–20
	Phormidium sp.	20–40
Polymyxin B	*Anabaena variabilis*	5
	Cylindrospermum licheniforme	2
	Microcystis aeruginosa	2
	Microcystis aeruginosa	10
	Anacystis nidulans	16
	Chlorogloea fritschii	80

Table 5. (continued)

Substance	Species	Concentration (mg l^{-1} except where otherwise stated)
Proflavin	*Anacystis nidulans*	>4
Rifampicin*	*Nostoc* sp.	10
Ristocetin	*Microcystis aeruginosa*	100
Streptomycin	*Anabaena variabilis*	0.1
	Cylindrospermum licheniforme	2
	Microcystis aeruginosa	2
	Microcystis aeruginosa	1
	Nostoc sp.	2
	Phormidium sp.	2
	Anacystis nidulans	0.05
Terramycin	*Anabaena variabilis*	10
	Microcystis aeruginosa	2
	Nostoc sp.	233
	Phormidium sp.	58–117
Tetracycline*	*Nostoc* sp.	10
Thioglycollic acid	*Fischerella muscicola*	>92
Tyrothricin	*Microcystis aeruginosa*	10
Uranyl nitrate	*Fischerella muscicola*	500
Usnic acid	*Microcystis aeruginosa*	1000

*J. Reaston, *personal communication.*

On a solid medium it is useful to provide unidirectional light so that motile forms can migrate towards the light, hopefully leaving the heterotrophic bacteria behind. A successful treatment used by Bunt (1961) to purify motile filamentous cyanobacteria was to inoculate on to the surface of a solid medium, then to pour a second agar layer on top, place a shallow layer of culture solution on top of this, and then to illuminate from above. He found that the motile cyanobacterial filaments grew through the upper agar layer towards the light to give pure filaments in the upper liquid layer. These were subcultured. Irrespective of the precise treatments used, an important part of any purification procedure is usually repeated subculturing from the least contaminated area (usually the edge) of the cyanobacterial growth.

Irradiation of cyanobacterial cultures with ultraviolet light is often successful in freeing cyanobacteria from contaminants, the cyanobacteria being most resistant to irradiation (Gerloff *et al.*, 1950). The usual technique is to take a dilute homogeneous suspension of exponentially growing cyanobacteria, add this to a sterile silica-glass flask and shake it vigorously whilst irradiating it using an ultraviolet lamp placed a set distance away. Samples of cyanobacterial suspension are withdrawn at intervals during irradiation and plated on to sterile medium. Samples from the longest irradiation period which allows cyanobacterial development are sometimes found to be free of contaminants. Gamma irradiation, although used (Kraus, 1966), is less successful than ultraviolet treatment. The precise method which proves to be successful in isolating pure cultures of cyanobacteria depends very much on the cyanobacterial strain and on the contaminants present. Various combinations of the above treatments will probably have to be tried—be patient, it could take a long time! The possibility of cyanobacteria being contaminated by phages is also real, particularly with cyanobacteria of the LPP group, and less frequently with unicellular and heterocystous genera (Stewart and Daft, 1977). Such phages are often lytic forms, although temperate phages also occur. It is difficult to free cultures from phages although resistant strains develop spontaneously with time (see Barnet *et al.*, 1980); usually it is best to discard phage-contaminated cultures.

(iii) Tests for purity

The most critical test is microscopic examination using a good microscope fitted with dark ground illumination. The absence of contaminating microorganisms should also be tested for by streaking on to cyanobacterial culture media supplemented with nutrient agar, or any other bacteriological or fungal medium which contains readily utilizable energy (0.25% glucose) and nitrogen (0.025% casamino acids) sources. However, all microbiological culture media are selective to some extent and the absence of organisms on test media is no guarantee that they are not there, or important. Hence the importance of critical microscopical examination as well is stressed.

(iv) Genotypic uniformity

With unicellular forms this can be achieved using simple dilution and plating techniques and homogeneous cell suspensions, discrete colonies which arise having originated from single cells. With filamentous forms, spore suspensions serve the same purpose, and spores have the added advantage of being particularly resistant to environmental extremes (except high temperatures) and thus can be readily purified from contaminants. Filamentous forms may also be disrupted to form a suspension of unicells or short filaments before plating, and hormogonia likewise give rise to colonies derived from genotypically similar cells. Such unicells, short filaments, and hormogonia can be handled by conventional micromanipulation techniques, and when subcultured the chance of growth is enhanced if they are transferred to very small amounts of liquid, *e.g.* single drops in sterile cavity slides.

(v) Stock cultures

These should be maintained on solid medium (1–2% agar) at a low light intensity (100–300 lux, Osram fluorescent white tubes) at room temperature

(25 °C). Maintenance in sealed 30-ml McCartney bottles for many months can be achieved without subculturing, or opening, if a 16 : 8 light–dark regime, and a small inoculum is used. Cyanobacteria can also be freeze-dried (Watanabe, 1959*b*; Holm-Hansen, 1964) or stored in a deep-freeze (Whitton, 1962), but they often do not survive storage at 0–4 °C, even although some field populations of free-living cyanobacteria (Fogg and Stewart, 1968) and lichens containing them (Englund and Meyerson, 1974; Kallio and Kallio, 1978) grow and fix N_2 at such temperatures.

(vi) Batch cultures

Cyanobacteria grow readily in batch cultures with the only factors of importance being containers in which sterile conditions can be easily maintained, an adequate inoculum to ensure that the culture takes, a sufficient supply of CO_2 for optimum photosynthesis, optimum illumination which varies with the culture density, aeration to remove excess O_2 and reduce photorespiration (see Lex *et al.*, 1972), optimum temperature (30–35 °C), and a good culture medium.

In our laboratory we use two main types of batch culture. Conical (Erlenmeyer) flasks are used for up to 5 l of suspension. The flasks are stoppered with cotton-wool and the stoppers covered with aluminium foil and shaken on a rotary shaker at 80–100 r.p.m. at 3000 lux (at the surface of the flasks) and 25–35 °C, depending on the strain. The inoculum is added as a homogeneous suspension of exponentially growing cells to a density at which the greenish tinge of the cyanobacteria can just be seen in the inoculated flask. As mentioned above, it is important to add an adequate inoculum. Eberly (1967) has found that the length of the lag phase may decrease with increase in inoculum size. Such cultures are used to maintain stocks, to provide inocula for larger batch cultures, or are used for experimentation. They are usually harvested towards the end of the exponential growth phase.

When material is required in large quantities for enzyme extraction, 10–40-l batch cultures are used. Although various specialized flasks are available for growing cyanobacteria, e.g. Walsby flasks (Walsby, 1967), we use standard commercially available vessels: either 10–20-l flat-bottomed round flasks (containing 7–15 l of medium) or 15–60-l reagent bottles (containing 10–40 l of medium). These are stirred magnetically (be careful not to use worn magnetic followers, and to use a fairly weak magnet otherwise the cyanobacterial cells will be ground between the follower and the base of the vessel) and bubbled with an appropriate sterile gas phase (usually air or air +5% CO_2) at a rate of 1–2 l min^{-1}. The inoculum is again added until a tinge of green can just be seen in the culture; the initial light intensity is 3000 lux. With an increase in culture density, the light intensity may be increased to 10 000 lux at the surface of the vessel. Cells are again harvested in the late exponential growth phase.

The physiology, biochemistry, molecular biology, structure, and ultrastructure of cyanobacteria change during growth in batch culture. Thus, if several batches of material are to be used it is important to ensure that an identical inoculum is used, that the growth conditions are identical, and that the cyanobacterium is harvested at the same time during the growth phase. The variation which can occur is exemplified by our studies on *Nostoc* 7524 (Sutherland *et al.*, 1979), which showed that the length of the log phase and the time at which akinetes first differentiate varies with the light intensity used.

(vii) Continuous cultures

Because of the difficulties which are sometimes inherent in achieving good replication with batch cultures, the use of continuous cultures is to be preferred, provided that large amounts of cyanobacterial cells are not required. Cyanobacteria grow very well in continuous culture and evenly dispersed suspensions of unicells or single filaments can be readily achieved with many strains. In our laboratory we maintain continuous cultures in 600-ml volumes. In general, the precautions regarding gas phase, sterility, light, and other environmental conditions are the same as for batch cultures. With *Anabaena cylindrica* and a light intensity of 3000 lux, 25°C and a 600-ml volume of culture at a standard absorbance at 665 nm of 2.0, exponentially growing cells with a doubling time of 20 h can be maintained indefinitely. Other organisms have considerably faster doubling times, especially at elevated temperatures (30–35°C). Continuous cultures serve as the most satisfactory source material for experimental work or for inocula for batch cultures.

(viii) Synchronous cultures

Prolonged synchrony of cyanobacterial cultures cannot be routinely obtained, although some progress has been made with the unicellular *Anacystis nidulans* (Lorenzen and Venkataraman, 1969, 1972; Herdman *et al.*, 1970; Lindsey *et al.*, 1971; Lorenzen and Kaushik, 1976). Lorenzen and Venkataraman (1969) achieved partial synchrony using a combination of temperature and light treatments, and Herdman *et al.* (1970) could sustain synchrony for three generations in cultures which were darkened and starved of CO_2 until the nucleic acid content of the cells stabilized, and which were then re-illuminated and supplied with CO_2 to induce synchronous growth. With filamentous types complete synchrony cannot yet be achieved, but it is possible to synchronize the production of particular cell types, e.g. akinetes (Sutherland *et al.*, 1979). It is also possible to synchronize akinete germination up to the stage of first heterocyst production. By selecting particular environmental conditions it is also possible to sustain the cyanobacterium in one particular phase of its developmental cycle. For example, Robinson and Miller (1970) found that in green light, or white light from fluorescent lamps, *Nostoc commune* remained in the aseriate stage, while red light resulted in hormogone formation. Robinson and Miller (1970), like Lazaroff (1966), suggest that *Nostoc commune* may possess a phytochrome-like pigment with photoconvertible forms which have absorption maxima in the red and green regions of the spectrum.

(e) MEASUREMENT OF NITROGENASE ACTIVITY

Several methods for demonstrating that a cyanobacterium fixes N_2 or has an active nitrogenase are available.

(i) Measuring increases in total N when N_2 is the sole source of N

This conventional method of testing for N_2 fixation is unfashionable but is cheap and satisfactory when properly used. It is, however, time consuming and not very sensitive. Thus, fairly large increases in total nitrogen are required to minimize experimental error (gains of 5–10 $\mu g\ ml^{-1}$ of N in inorganic medium, and considerably higher values in medium of high organic content; Wilson, 1958). The most usual system is to introduce scrubbed air (i.e. air which has been slowly bubbled through 1% $NaHCO_3$ to remove oxides of nitrogen and 25% v/v H_2SO_4 to remove traces of ammonia, and then re-moistened by passage through distilled water) to cultures grown on medium free of combined nitrogen, to note the nitrogen content of the cyanobacterium and medium at the start of the experiment and again at the end (a growth period of 1–3 weeks is usual) and to attribute any gains in combined nitrogen over two control series (one containing an autoclaved inoculum and one without an inoculum) to N_2 fixation by the cyanobacterium if the cultures have remained pure (this must be tested for) (see Fogg, 1942).

(ii) Use of nitrogen isotopes

Two isotopes of nitrogen are available: the stable isotope ^{15}N, which is present to the extent of 0.360–0.370 atoms of ^{15}N per 100 atoms of ^{14}N, and the radioactive isotope ^{13}N, which is a γ-emitter and which has a half-life of 10.05 min. Isotope techniques are detailed and referenced in Chapter II.2. Cyanobacteria, like other N_2-fixing prokaryotes, do not discriminate between ^{14}N and ^{15}N when fixing N_2 and an enrichment of 0.015 atom-% ^{15}N is usually accepted as reliable evidence of $^{15}N_2$ incorporation, although much lower enrichments can be reliably detected by experienced workers using good mass spectrometers. It may be noted that poor methodology (e.g. allowing traces of CO_2 and/or water into the gas sample), leads to false positives by increasing the mass 29 peak due to the production within the instrument of, for example,

CO ($^{13}C^{16}O$). The sensitivity of the ^{15}N technique can also be increased in short-term experiments by measuring ^{15}N incorporation into acid-soluble nitrogen only, since most of the newly fixed ^{15}N accumulates as ammonia (probably NH_4^+), free amino acids, and amides.

Tracer studies using $^{13}N_2$ incorporation as a measure of N_2 fixation have been used for almost 20 years (Nicholas *et al.*, 1961). Now, with improvements in technology, the method affords an extremely sensitive, albeit very expensive, technique for demonstrating N_2 fixation and for following the fate of the nitrogen fixed (Thomas *et al.*, 1975, 1977; Wolk *et al.*, 1976; Meeks *et al.*, 1977).

The acetylene reduction technique is now the preferred assay for measuring nitrogenase activity, being extremely sensitive, simple, and reproducible. It depends on the fact that nitrogenase reduces C_2H_2 to C_2H_4 (Dilworth, 1966; Schöllhorn and Burris, 1966) and was first used with cyanobacteria by Stewart *et al.* (1967, 1968). The technique is detailed in Chapter II.3. With cyanobacteria it is usually sufficient to use a gas phase of 10% C_2H_2 in air and, after the desired exposure period, to withdraw a sample of the gas phase into a syringe and to store this by plunging the needle into a rubber stopper until the gas phase is assayed for C_2H_4 production by chromatography. The ratio of the rate of C_2H_2 reduction to the rate of N_2 fixation depends on the extent to which electrons are lost through the nitrogenase acting as an ATP-dependent hydrogenase (see Bothe *et al.*, 1978; Peterson and Burris, 1978), but is usually about 4:1. Stewart *et al.* (1967) recommended calibration against $^{15}N_2$ uptake, and this should be done wherever possible.

Other techniques of measuring N_2 fixation by cyanobacteria, including the measurement of changes in N_2 to Ar ratios on exposing cyanobacteria to a gas phase containing both gases, have been used but have not found widespread application.

(f) THE ROLE OF CYANOBACTERIA IN RICE CULTIVATION

Free-living cyanobacteria are common and often abundant in rice paddy soils with a neutral or alkaline pH. Their possible importance as contributors of combined nitrogen was realized by the Indian worker De (1936, 1939) and their ecology was studied in some detail by Singh and his students particularly in the alkaline Upper Pradesh and Bihar regions of India (see Singh, 1961; Venkataraman, 1972). With the subsequent introduction of field assays to measure the nitrogenase activity of cyanobacteria using the acetylene reduction technique (Stewart *et al.*, 1967, 1968) it has proved possible, for the first time, to provide an insight into the range of organisms capable of fixing N_2 there and to quantify their input of combined nitrogen. The cyanobacterial flora of rice paddy soils is diverse and because many ideas were formulated during a period when it was thought that only heterocystous cyanobacteria fixed N_2, the conclusions on the importance of the species detected relative to the amounts of nitrogen which may be fixed require re-assessment.

There is substantial variation in cyanobacterial species, biomass, and in abundance in paddy soils (see Singh, 1961; Fogg *et al.*, 1973; Pantastico and Suayan, 1973; Renaut, *et al.*, 1975; Watanabe and Cholitkul, 1979; Roger and Reynaud, 1979). Examples from two areas, India and Senegal, illustrate this. In India, for example, cyanobacteria may represent about 70% of the total algal species (Pandey, 1965) and in the alkaline areas N_2-fixing forms appear to constitute a major part of the flora (Singh, 1942, 1961). In terms of seasonal periodicity, Singh (1961) noted that an extensive growth of cyanobacteria develops during the early part of the rainy season, which lasts from June until mid-October. This flora is dominated by species of *Aphanothece*, *Microcoleus*, *Aulosira*, *Scytonema*, *Nostoc* and *Cylindrospermum.* At the end of July the cyanobacterial growth is mixed into the upper soil layers (puddled), after which the rice is transplanted. Extensive growths of *Microcoleus*, *Anabaena*, *Cylindrospermum*, *Tolypothrix* and *Fischerella*, with *Aulosira fertilissima*, then develop, particularly from mid-September until mid-December when the soils dry out. It is notable that although *Aulosira fertilissima*, a heterocystous N_2-fixing species, was considered to be particularly important in contributing fixed nitrogen to the paddy soil, it has since become evident that virtually all the dominant cyanobacteria recorded during the seasonal cycle are in fact N_2-fixing species, since many of the non-heterocystous forms also fix N_2 anaerobically (see Stewart *et al.*, 1979, for further discussion of this point).

In equatorial Senegal, the seasonal pattern is quite different (Roger and Reynaud, 1976, 1977, 1978). There, after rain and during the early parts of the rice cultivation cycle, the microbial flora becomes dominated by diatoms and unicellular chlorophytes. Highest biomasses are recorded from the time of tillering to the time of development of panicles when filamentous chlorophytes and non-heterocystous cyanobacteria predominate if the soil flora is protected from high light (>80 000 lux at 1300 h). After panicle development the total soil flora then declines, except when the cover remains high, and then heterocystous and non-heterocystous cyanobacteria dominate; if the shading is reduced chlorophytes and non-heterocystous cyanobacteria predominate. That is, in areas of extremly high light intensities cyanobacteria compete poorly, and chlorophytes predominate. Reynaud and Roger (1978) showed this by placing screens giving 100%, 60%, 22%, and 7% of incident light over submerged unplanted soils and found that after 30 days cyanobacterial growth and nitrogenase activity were highest in the most shaded soil. Roger and Reynaud (1976) also found in Senegal soils, where the dry period lasts for 8 months, that at the end of this period spores of heterocystous cyanobacteria constituted more than 95% of the flora. Roger and Reynaud (1977) calculated a cyanobacterial biomass of up to 6 tonnes (fresh weight)

ha^{-1}. This compares with up to 24 tonnes ha^{-1} for the Philippines (Watanabe *et al.*, 1977*a*). It is not surprizing, in view of this low value and the high ratio of eukaryotic algae to cyanobacteria in the Senegal flora, that N_2-fixation rates there are lower than in most other areas (see below).

(i) Measurement of cyanobacterial abundance

It is difficult to measure cyanobacterial abundance in soils accurately because the soil microflora shows great heterogeneity in space and time and because in terms of measuring abundance even in one sample there is no completely satisfactory method. In terms of sampling strategy, both an adequate sampling density and rigorous statistical treatment of the data are necessary. Such aspects are considered fully by Reynaud and Roger (1978).

Three techniques are commonly used to measure cyanobacterial abundance within a sample. Dilution and plating techniques are used most frequently for qualitative studies now that good culture media on which most cyanobacteria can be isolated are available. Reynaud and Roger (1979) have used the following technique with some success. They first prepare core samples (each 2 cm in diameter and including the top 1 cm of soil and the surface water) at a density of 1600 samples per hectare. The organisms present in these are then cultured on agar plates using medium selecting for eukaryotic algae (BG11 medium with 15 p.p.m. of bacitracin), prokaryotic algae (BG11 medium with 20 p.p.m. of cycloheximide) and N_2-fixing cyanobacteria (nitrogen-free BG11 medium with 20 p.p.m. of cycloheximide). The biomass (volume) of each 'unit', that is, cell, filament, or colony, depending on species, of the most important cyanobacteria is then obtained by making 100 measurements of algae present in freshly collected core samples. The values obtained, together with counts of each unit, give the biomass of each organism present. The disadvantage of this method is that plating techniques do not ensure that the flora which develops on the plates has a similar composition to that in the soil. Also remember that aerobic incubation on medium free of combined nitrogen will not select for anaerobic N_2-fixing forms such as *Plectonema*. Plating techniques are usually used as an adjunct to the measurement of nitrogenase activity rather than as an attempt to obtain an indirect measure of nitrogenase activity, and as such are useful.

The use of pigment analyses as a measure of cyanobacterial abundance in soils was suggested by Singh (1961). It involves extracting the pigments in 80% acetone, and using the absorbance at 490 nm (myxoxanthin) as a measure of cyanobacterial biomass and the absorbance at 665 nm (chlorophyll *a*) as a measure of total algae plus cyanobacteria, and other plant fragments. The method gives no indication of species composition. Although techniques of pigment extraction are generally satisfactory with laboratory populations, they do not work well, or at all, with soil populations where organic materials such as humic acids and chlorophyll degradation products are also extracted. In general, the method is not recommended with field samples.

The most satisfactory method in our hands, although the most time consuming, conventional, and boring, is direct examination and counting by light microscopy. Tchan (1953) and Drew and Anderson (1977) used fluorescence microscopy to distinguish pigmented cells, but usually it has to be combined with conventional microscopy. The method can be used quantitatively by taking known volumes of soil (say 2 cm^3), adding this to a stout 30-ml bottle containing some glass beads, the diameter of which will vary with soil type used, usually 2–5 mm in diameter, and 10 cm^3 of water, and shaking mechanically or by hand until a homogeneous suspension is obtained. The cyanobacteria and algae in known volumes of such a suspension are counted using a haemocytometer, and the volume of each counting unit (cell, short filament, spore, etc.) also determined. In this way the biomass of different algal groups as well as species can be determined. Lund (1947) used a slide technique whereby the slides were placed in, or on, the soil and the algae which developed on the slides were then examined microscopically.

(ii) Amounts of nitrogen fixed and methods for measuring N_2 fixation by cyanobacteria in rice fields

The general consensus is that anything from 7 to 75 kg ha^{-1} yr^{-1} of N may be fixed by free-living cyanobacteria in rice paddy soils (see Singh, 1961; Fogg *et al.*, 1973; Watanabe and Cholitkul, 1979), with the amounts fixed varying substantially with geographical location and soil type (compare, e.g. the data of Roger and Reynaud, 1979, with those of Watanabe and Cholitkul, 1979). However, despite the various estimates and techniques used (see De and Mandel, 1956; Subrahmanyan *et al.*, 1965; Aboul-Fadl *et al.*, 1967; MacRae and Castro, 1967; Sankaram *et al.*, 1967; Venkataraman and Goyal, 1968, 1969*a*, *b*; Yoshida *et al.*, 1973; Watanabe and Cholitkul, 1979), quantitative measurements of the amounts of nitrogen fixed on a seasonal basis are difficult. One problem is the fact that, in additon to cyanobacteria, various other N_2-fixing diazotrophs including species of *Azotobacter, Arthrobacter, Beijerinckia, Clostridium, Enterobacter, Pseudomonas, Spirillum,* photosynthetic bacteria, and the *Azolla–Anabaena* symbiosis may occur in rice fields.

The three most satisfactory techniques are probably the following: (1) comparison of crop yields in the presence and absence of added cyanobacteria with yields obtained when chemical nitrogen fertilizer is available; (2) use of ^{15}N as a tracer in field studies; (3) the C_2H_2 reduction assay.

The comparison of crop yield in the presence and absence of cyanobacteria involves the setting up of a series of pot experiments, or preferably of field plots, and comparing yields obtained under different treatments. The treatments should include as a minimum: (1) untreated soils; (2) added chemical nitrogen fertilizer, e.g. $(NH_4)_2SO_4$, urea; (3) added nutrients other than nitrogen, e.g. PO_4^{3-}; trace elements such as Mo; (4) an inoculum of one or more cyanobacteria; and (5) several different rice varieties. The effects of these treatments in various combinations should be studied. Plant yield is probably best measured as grain yield or as total dry weight or shoot dry weight. Two general points are important in such experiments. Firstly, the duration of the experiment is critical. It is dangerous and erroneous to carry out experiments lasting a few weeks only because although inorganic nitrogen is assimilated immediately by the crop plant, the bulk of the nitrogen fixed by cyanobacteria is probably released only on death and decay of the cyanobacteria, or as extracellular organic nitrogen, and then it has to be mineralized. Thus, the beneficial effects of cyanobacterial N_2 fixation on soil fertility may take some time to appear, yet only in a few instances have long-term experiments been carried out. Indeed, the slow release of biologically fixed nitrogen is one of its advantages in paddy soils (see below). Secondly, the pH of the soil is important. Cyanobacteria, in general, do not flourish in acidic conditions; if the pH is not above 6.5, and preferably 7.0 or higher, cyanobacterial growth will be poor, little nitrogen will be fixed, and the response of the crop will be low. Stewart *et al.* (1979) summarized some of the experiments which have been carried out. One example will suffice. Subrahmanyan *et al.* (1965), using rice varieties Ptb10 and T141 and field plots, obtained the results summarized in Table 6. These show that substantial gains in crop yield can be achieved using cyanobacteria as a source of nitrogen fertilizer, especially when non-nitrogenous fertilizers are also provided.

Table 6. Effect of cyanobacteria on grain yield of paddy (from Subramanyan *et al.*, 1965)
Data are for field experiments. Each value is the mean of data from three growing seasons (1962–63), in each of which average values were obtained from four replicate samples.

	Grain yield	
	kg ha^{-1}	Increase (%)
Control	1536	0
Control + cyanobacteria*	1810	18
Control + $(NH_4)_2SO_4$†	2095	36
Control + fertilizer mixture‡	2244	46
Control + cyanobacteria + fertilizer mixture	2799	82

*200 g (dry wt) ha^{-1} of mixture of *Nostoc, Anabaena, Scytonema,* and *Tolypothrix*.
†20 kg ha^{-1} of N.
‡1000 kg ha^{-1} of lime +20 kg ha^{-1} of P_2O_5 +0.28 kg ha^{-1} of Na molybdate.

^{15}N has not been used to any extent in field studies on N_2 fixation, although MacRae and Castro (1967) measured ^{15}N assimilation by rice plants via cyanobacteria in the glasshouse. Their technique was to add soils to tubes, flood each soil to give 5 cm of free water above it, and place the tubes in a Mason jar, which also contained 20% KOH to remove excess CO_2. $^{15}N_2$ was generated from $(NH_4)_2SO_4$ by hypobromite treatment and the initial gas phase was 0.10 atm O_2, 0.25 atm N_2 enriched with 96.8% ^{15}N, and 0.65 atm He. The incubation period was 28 days in the light or in the dark, after which total soil nitrogen was determined by Kjeldahl digestion and distillation and the NH_4^+ in the distillate was then analysed for ^{15}N enrichment. The calculated amounts of nitrogen fixed were equivalent to 10–55 kg ha^{-1} yr^{-1} of N according to MacRae and Castro (1967) and 40–80 kg ha^{-1} yr^{-1} of N according to Yoshida *et al.* (1973). An alternative use of ^{15}N is to use an isotope dilution technique (see Chapter III.4), but in general this is not particularly useful with cyanobacteria.

The C_2H_2 reduction technique has been used extensively in field studies on N_2 fixation by cyanobacteria and other prokaryotes in rice paddy soils (Rinaudo *et al.*, 1971; Dommergues *et al.*, 1972; Kalininskaya *et al.*, 1973; Balandreau and Dommergues, 1973; Yoshida and Ancajas, 1973; Balandreau *et al.*, 1974; Balandreau, 1975*b*; Matsuguchi and Shimomura, 1977; Matsuguchi *et al.*, 1975, 1976, 1978; Renaut *et al.*, 1975; Alimagno and Yoshida, 1977; Boddey *et al.*, 1978; Watanabe *et al.*, 1978*a, b*; Dommergues and Rinaudo, 1979). Such assays should be carried out in the field because of the difficulty in replicating field conditions in the laboratory and because the *in situ* conditions are so critical to the rates of activity which may pertain with autotrophs, heterotrophs, photoheterotrophs, aerobes, anaerobes, facultative anaerobes, forms capable of using reduced sulphur or carbon compounds as reductant supply for photosynthesis, etc., all being present *in situ* in various microniches. Perhaps the most satisfactory use of the technique is that of Balandreau (1975*b*) for rice rhizosphere microorganisms in general. The equipment is illustrated in Chapter III.3, Figure 2. The technique can be modified to measure cyanobacterial activity using the technique of Watanabe *et al.* (1978*b*), who calculated C_2H_2 reduction rates before and after removing cyanobacterial growths from the test area. By difference, they concluded that in the Philippines nitrogenase activity by cyanobacteria and any attached diazotrophs contributed about 0.5 kg ha^{-1} day^{-1} of N. The C_2H_2 reduction assay has revolutionized our understanding of biological N_2 fixation in that it provides a simple, extremely sensitive, and inexpensive method of indirectly measuring nitrogenase activity, but the findings must be interpreted cautiously (see above).

Several other points are important when attempting to measure nitrogenase activity in paddy soils.

Diffusion

One of the major problems with *in situ* C_2H_2 reduction tests (and $^{15}N_2$ experiments) is ensuring adequate gaseous exchange, for example C_2H_2 input and C_2H_4 output. Fortunately, C_2H_2 is very soluble in water (0.08 ml ml^{-1} at pC_2H_2 0.1 atm and 30°C) and C_2H_4 is less soluble. Nevertheless, in soil core assays, various workers have emphasized the problems which slow diffusion can cause and have suggested techniques applicable to their particular systems which can be used to overcome this (Rice and Paul, 1971; Flett *et al.*, 1976; Lee and Watanabe, 1977; Kuwatsuka and Nakano, 1978; Matsuguchi *et al.*, 1978; Zuberer and Silver, 1978). Suggested solutions include evacuation of the samples before adding C_2H_2, stirring the samples, or vigorously shaking the samples, but such techniques may destroy the soil micro-habitats which are so important for particular types of N_2-fixing organisms to be active, and should be avoided if possible. One of the simplest ways of doing so is to use small cores, and numerous replicates, although this too has disadvantages since the smaller the core the larger the proportion of disturbed relative to undisturbed soil. In soil core assays the cored soil should be incubated immediately with C_2H_2 under field conditions. In the canopy technique of Balandreau (1975*b*) (Chapter III.3, Figure 2), which in many respects is superior to core assays, adequate diffusion into the rice rhizosphere, into the flood water layer, and into the upper parts of the anaerobic layer does occur, but obviously the bulk of the lower soil layers are C_2H_2-limited (Lee and Watanabe, 1977; Watanabe *et al.*, 1978*a*).

Quantity of C_2H_2 which should be added

Algal nitrogenase is saturated at low pC_2H_2 (0.05–0.1 atm). In short-term experiments the addition of higher concentrations of C_2H_2 is not inhibitory (see, for example, Stewart *et al.*, 1968) and we have routinely used 20% C_2H_2 in air in short-term incubations ($<$4 h) in various freshwater habitats (Stewart *et al.*, 1971). C_2H_2 toxicity may occur in the longer term, and this may be partially due to a conformational change of the nitrogenase enzyme in the presence of C_2H_2 (David and Fay, 1977; David *et al.*, 1978). The advantages of better diffusion at 20% C_2H_2 have to be balanced by the possibility of underestimating nitrogenase activity due to a possible conformational change. Despite the results of David and Fay (1977) and David *et al.* (1978), various other workers have actually found a stimulation of C_2H_2 reduction with time, possibly because of derepression of nitrogenase synthesis through nitrogen depletion in the absence of N_2 fixation (see e.g. Stewart *et al.*, 1967).

C_2H_2 can be transported in small cylinders, in football bladders (Burris, 1974), in sealed bottles of various types, or in a variety of other containers (Chapter II.3), or can be generated from calcium carbide by adding water

(Burris, 1974; Balandreau, 1975*b*). The technique which proves to be most satisfactory depends, among other things, on the ease with which the C_2H_2 can be transported to the sampling site.

Length of the incubation period

This is critical and, in general, the shorter the incubation, concomitant with maximum diffusion of C_2H_2 in and C_2H_4 out of the soil, the better. This should be determined by experimentation. Long incubation periods should be avoided, if possible.

Sampling for ethylene production

Ethylene may be sampled in various ways. If the test material is not to be used further for tests on nitrogenase activity, it is best to try to obtain representative gas samples by mixing the gas phase using a syringe as a pump prior to sampling, or by stirring the sample, if possible, using a magnetic stirrer (Matsuguchi *et al.*, 1978). Gas samples taken by syringe can be stored in these by sealing the needles in rubber bungs, or they can be added to pre-evacuated tubes, which may be sealed with silicone rubber and submerged in water (or water saturated with NaCl; Watanabe and Cholitkul, 1979) to minimize gas leakage. Stutz and Bliss (1973) used vials containing KOH, which were evacuated to 0.5 atm before adding the test gas sample, and they sealed the vials using finger-nail polish.

Calculation of amounts of N_2 fixed using C_2H_2 reduction data

The C_2H_2 reduction assay can be used indirectly to calculate rates of nitrogenase activity if the ratio of the rate of C_2H_2 reduction to the rate of N_2 reduction is known accurately. The theoretical ratio, assuming complete electron transfer to the substrates, is 3:1, and in most critical tests where diffusion has been adequately considered ratios of 3:1 to 4:1 have usually been obtained. For example, Stewart *et al.* (1968) obtained ratios ranging from 2.8:1 with *Anabaena cylindrica* to 3.6:1 with *Nostoc muscorum*. However, there are two complicating factors. Firstly, in cyanobacteria, as in other N_2-fixing organisms (Dixon, 1968, 1972; Schubert and Evans, 1976; Argüeso *et al.*, 1978), nitrogenase may also act as a hydrogenase with electrons which are not used up in substrate reduction being evolved as H_2. When N_2 is the substrate H_2 is evolved, but with C_2H_2 as substrate there is little H_2 production. Thus a C_2H_2 reduction to N_2 reduction ratio of 3:1 overestimates N_2 fixation since it implies that all of the electrons are being used for N_2 reduction when in fact they are not. Further, the extent of electron wastage in this way varies depending on the physiological state of the organism, the conditions under which it is grown, and on the presence of an efficient uptake hydrogenase to recycle H_2 (see above). It is therefore important to test the system used to determine the pertaining C_2H_2 reduction to N_2 reduction ratio. If that cannot be done, then a ratio of 4:1 is usually more likely than 3:1.

Another complicating factor in field studies is that C_2H_4 may be produced on the degradation of organic matter. This C_2H_4 is oxidized and lost under aerobic conditions (but not anaerobic conditions). C_2H_2, however, prevents C_2H_4 oxidation in air so that endogenous C_2H_4 may accumulate, relative to the control without C_2H_2 and C_2H_2 reduction is overestimated. Witty (1978) concluded on the basis of $^{14}C_2H_2$ tests that in ecosystems where the N_2-fixation rate is less than 100 g ha^{-1} day^{-1} of N endogenous C_2H_4 production in the presence of C_2H_2 could overestimate the rate of C_2H_2 reduction. Such rates are commonly obtained in paddy soils. Despite these problems, the C_2H_2 reduction technique is extremely valuable and its importance should not be under-emphasized. It is the most useful technique for routine studies on the nitrogenase activity of paddy soils so long as its limitations are appreciated.

(iii) Potential of N_2 fixation by free-living cyanobacteria in rice fields

It may be possible to exploit cyanobacteria in two main ways: (1) to make use of the existing natural populations of the paddy soil, and (2) to inoculate paddy soils with efficient N_2-fixing strains.

Use of existing natural populations of cyanobacteria in paddy soils

Most paddy soils have a natural population of cyanobacteria which thus provides, free of cost, a potential source of combined nitrogen, if it can be harnessed. Singh (1961) exploited such natural floras to fertilize areas of alkaline Usar land in North India. His technique was to divide the land into plots, each of less than half an acre, and to build round each of these an earth embankment about 1.5 ft high. During the rainy season such areas became waterlogged, rich cyanobacterial growths, particularly of *Nostoc commune* and *Aulosira fertilissima*, developed, and the fertility of the land was increased after several years in this way. Of the factors which are important in the exploitation of cyanobacteria, four may be considered.

Availability of water. The ability of cyanobacteria to resist desiccation is an important factor contributing to their success in paddy soils. This resistance is conferred by their basic protein structure, their gel-like protoplasm and thick mucilaginous sheaths (Figures 1 and 2), and the facts that they commonly grow in colonies, bundles, or globose aggregates, many are motile so that they can move to less desiccated microniches within the soil, and various heterocystous forms possess spores (akinetes) (see Fogg *et al.*, 1973). Such cyanobacteria lose moisture extremely slowly and absorb it very rapidly by passive processes (an advantage since the cyanobacteria may be dormant when the rains commence after months of dryness). The water absorbed also makes nutrients available to the cyanobacteria. We have shown that Nigerian soil cyanobacteria, mainly *Scytonema* and *Nostoc* strains, and *Stigonema* from Brazil (Stewart *et al.*, 1978), after being desiccated for several months,

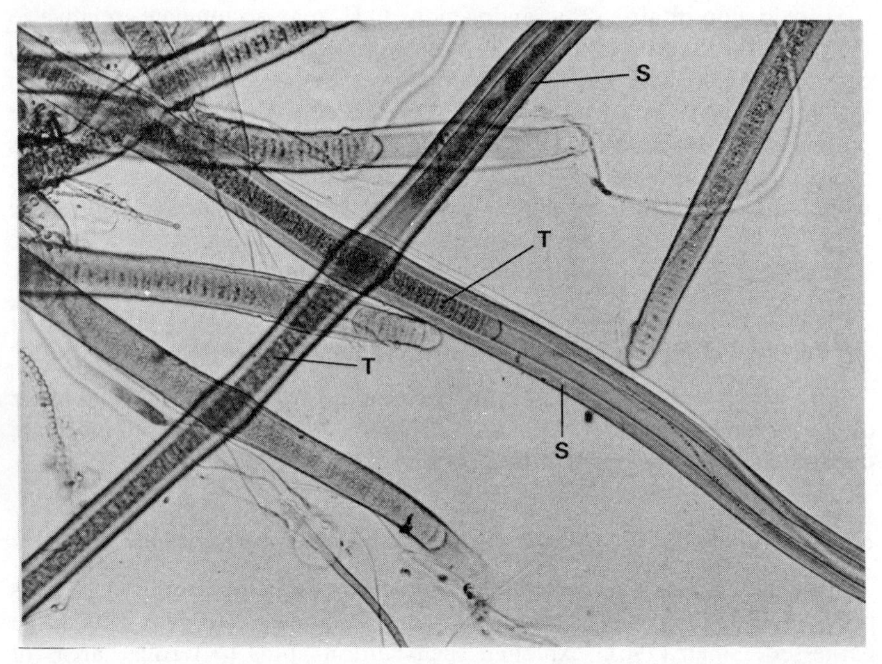

Figure 1. Light micrograph of nitrogen-fixing *Scytonema* sp. showing thick sheath (S) surrounding the trichome (T) (× 680)

commenced fixing N_2 within 24h of re-wetting; the *Stigonema* samples absorbed 5 times their own weight of water within 2 min of moistening. Indeed, Cameron and Blank (1966) have shown that *Nostoc commune* could be revived from a 107-year-old herbarium specimen by moistening. It is important in rice paddy soils to conserve water for as long as possible using simple irrigation channels, built-up banks, etc., and to ensure that the other conditions for nitrogenase activity are optimized as far as possible.

Importance of neutral or alkaline pH. Good cyanobacterial growth necessitates a pH above 6.5, the pH optima of most cyanobacteria being above 7 and sometimes as high as 10.0 (Fogg *et al.*, 1973). However, there is some evidence that in tropical rice fields, such as those of Senegal (Roger and

Reynaud, 1979), N_2-fixing *Nostoc* and *Anabaena* species occur most abundantly in soils of pH 5.9, and that other species are most abundant at pH 6.3 We have also detected substantial nitrogenase activity by *Scytonema* and *Stigonema* species even at pH 4 (although these were short-term C_2H_2 reduction assays and growth was not measured). At pH 4, nitrogenase activity was 25% of the optimum (at pH 8.0) for *Scytonema* and was about 33% of the

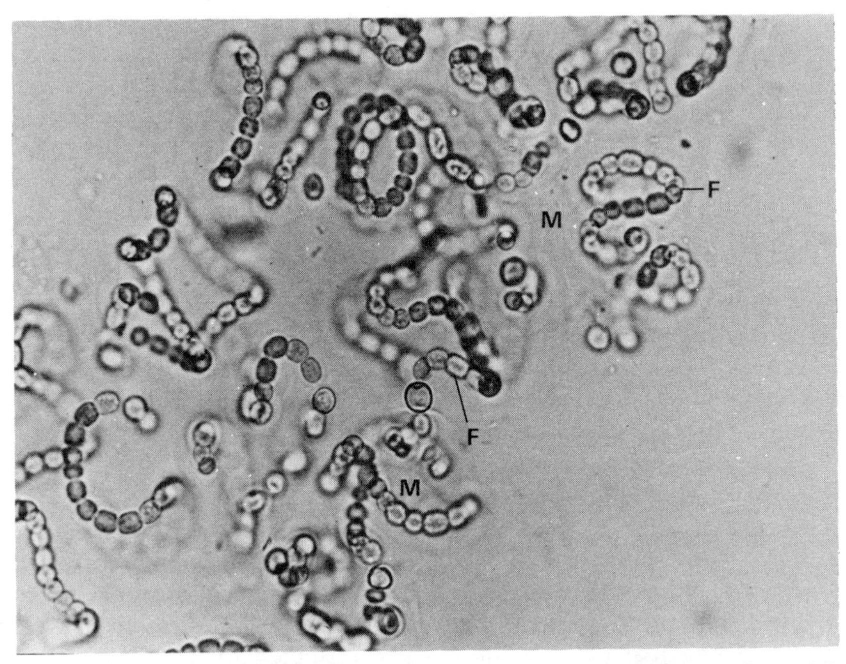

Figure 2. Light micrograph of nitrogen-fixing *Nostoc* sp. with cyanobacterial filaments (F) embedded in thick mucilage (M) (×680)

optimum (at pH 8.0) for *Stigonema* (Stewart *et al.*, 1978). Despite the growth and nitrogenase activity which may occur under acidic conditions it is good practice, if the pH is not neutral or alkaline, to increase it to 7–8 by liming, if it is hoped to use cyanobacteria as an important source of fertilizer nitrogen. It is their inability, in general, to fix N_2 well under acidic conditions which most seriously limits their use in many paddy soils.

Importance of non-nitrogenous fertilizer. Although most attention has been paid to the shortage of available combined nitrogen in paddy soils, it is important that, in order to maximize the advantages of N_2-fixing cyanobacteria, other nutrients do not limit their growth. Of these, molybdenum, which is a component of nitrogenase, phosphorus, which is often limiting in tropical soils, and calcium, which increases the soil pH, are often the most important (e.g. Subrahmanyan *et al.*, 1965; Aboul-Fadl *et al.*, 1967; Sankaram *et al.*, 1967; Venkataraman and Goyal, 1969*a, b*).

Possible complementary use of synthetic fertilizer. Venkataraman (1972, 1979) has advocated the simultaneous addition of cyanobacteria and chemical nitrogenous fertilizer to improve crop yield. The advantages, if any, of using this dual supplement depend on a variety of factors. Venkataraman (1979) considers, on the basis of numerous Indian experiments, that cyanobacterial N_2 fixation alone may contribute 25–30 kg ha^{-1} yr^{-1} of N and that on average the addition of cyanobacteria along with fertilizer nitrogen may reduce the requirement for fertilizer nitrogen by about a third.

Several points are relevant here. Firstly, the effect of combined nitrogen is to inhibit nitrogenase activity and synthesis in cyanobacteria, just as in other organisms, if the concentrations are high enough (Fogg, 1949; Stewart *et al.*, 1968, 1975; Bone, 1971; Fogg *et al.*, 1973). Secondly studies using axenic cultures and/or continuous cultures show that at similar concentrations NH_4^+-N is more inhibitory than NO_3^--N or urea. Thirdly, in the long term, nitrogenous fertilizer, irrespective of the form supplied, if not lost to the atmosphere by leaching, or assimilated, may be converted into NH_4^+. Venkataraman (1979) observed simultaneous assimilation of NH_4^+ and N_2 fixation at NH_4^+ concentrations as high as 40 p.p.m. This is unusual since axenic cultures show inhibition of nitrogenase activity and synthesis at much lower levels (Fogg *et al.*, 1973; Stewart *et al.*, 1975). Venkataraman (1975, 1979) also suggests that cyanobacteria may produce growth-promoting substances for the rice plant, as well as fix N_2, but there is no good evidence that growth-promoting substances are important or beneficial, and I sometimes have an intuitive feeling that some of the results attributed to the production of growth-promoting substances by cyanobacteria may simply be due to them increasing the water-holding capacity of the soil, thus preventing desiccation, particularly in pot experiments. Nevertheless, Venkataraman (1979) has obtained results showing that, after four crops, the addition of cyanobacteria alone increased the grain yield by 12.7%, the addition of cyanobacteria + 25 kg of N increased productivity by 30.3%, and that cyanobacteria + 50 kg of N increased productivity by 54.9%. In these and other experiments the addition of large amounts of fertilizer nitrogen (or indeed cyanobacteria) is justified only if nitrogen is the limiting factor (see below).

Possible beneficial effects of using a cyanobacterial inoculum

The idea of producing a cyanobacterial inoculum which can use light energy and air to fix N_2 at field temperature is exciting, but we are some way from doing this routinely yet and the techniques for applying free-living cyanobacteria to field crops today are probably less advanced than techniques of rhizobial inoculation of legumes half a century ago. Two facets are particularly relevant if we are to make progress in this area.

Strain selection. Despite the fact that it has proved possible to select efficient rhizobial strains which have greatly benefitted legume growth, it is another matter entirely to select superior N_2-fixing cyanobacteria which will not be closeted within a root nodule but which will have to grow and compete effectively in the soil as well as fix large amounts of N_2. The difficulties in this respect should not be underestimated. I have discussed elsewhere (Stewart *et al.*, 1979) some of the characteristics which may be beneficial in a good N_2-fixing strain. In summary these are:

1. Organisms should be selected which grow rapidly in the field and withstand environmental extremes. Different strains will probably have to be selected for particular areas. For example, acidophilic N_2-fixing species would be particularly beneficial in South American paddy soils.
2. The organisms should be able to fix N_2 aerobically, anaerobically, and microaerobically, as do heterocystous cyanobacteria and a few unicellular strains.
3. They should be able to grow photoautotrophically, photoheterotrophically, and chemoheterotrophically.
4. There should be little reductant wastage as H_2; thus the cyanobacteria should, if they evolve H_2 via nitrogenase, have a good uptake hydrogenase.
5. They should have a non-respressible nitrogenase which can thus fix N_2 in the presence of fertilizer nitrogen.
6. There may also be advantages if the cyanobacteria are able to liberate extracellular NH_4^+ in excess of their requirements for growth. One possible disadvantage of this is that if the cyanobacteria are photoautotrophs they will be growing predominantly in the upper oxygenated layers and there, nitrification leading to NO_3^- formation followed by denitrification may lead to losses in combined nitrogen (the same will happen, of course, if fertilizer NH_4^+ or NO_3^- is applied). It is possible that the liberation of extracellular NH_4^+ may not outweigh the advantages which the decay of the cyanobacteria in the lower deoxygenated soil layers provides. Further research in this area is required.

Growing an inoculum. Watanabe (1961), who carried out extensive studies on cyanobacteria in Japan and South-East Asia for about 20 years, investigated the possibility of mass culturing *Tolypothrix tenuis* as an inoculum for the rice crop. Unfortunately, Watanabe's studies were carried out over a period when cheap chemical nitrogen fertilizer was readily available (1950–70) and he received little encouragement. He started with an inoculum in small culture flasks, increased the yield using large illuminated tanks in the laboratory (this gave a yield of 0.2 g l^{-1} day^{-1}), then transferred the cyanobacteria outdoors, growing them in poly(vinyl chloride)bags. In this way he grew about 7 tons of cyanobacteria per year. He also grew the cyanobacteria on moist, porous volcanic gravel (it could be stored in air-tight bags in this way for up to 3 years without loss of activity) and used this at a concentration of 1–5 kg (dry weight) ha^{-1} of cyanobacteria to inoculate the rice fields at the time when the rice was transplanted. Watanabe (1961) considered the addition of cyanobacteria to be equivalent to adding 29 kg ha^{-1} yr^{-1} of N and the increases obtained were on average 2% in the first, 8% in the second, 15.1% in the third, 19.5% in the fourth, and 10.6% in the fifth year. Such a technique, however, had major inherent disadvantages, in addition to the fairly low increases in rice yield obtained. These included low yields of *Tolypothrix*, the high capital cost of preparing the cyanobacteria in tank culture in the laboratory, and transportation costs, bearing in mind that 1–5 kg (dry weight) of cyanobacterium plus associated porous gravel was used for field inoculation.

If the idea of inoculating rice fields is to be successful, it will almost certainly be essential for the farmers, alone or in groups, to produce their own cyanobacterial biomass, possibly using small culture ponds near the rice fields. Venkataraman (1972) has attempted to do this using an open-air soil-culture method in which the inoculum is grown in shallow rectangular ponds, harvested, dried together with soil, and the soil then broadcast on to the field. Venkataraman (1979) has produced some cost–benefit analysis values and calculates that whereas it costs (presumably in 1978) $12.20 for 25 kg of chemical nitrogen fertilizer in India, the equivalent cost of a cyanobacterial inoculum required to produce this amount of nitrogen commercially is $3.65, and that it costs virtually nothing if the farmers prepare the inoculum themselves. The values, and the efficiency with which it can currently be done, are perhaps a little optimistic, but the system can surely be improved in years to come. According to Venkataraman (1979), attempts to produce inocula of cyanobacteria are being initiated in Nepal, Bangladesh, Burma, and Sri Lanka. It seems that for too long the possible use of cyanobacterial inoculation of rice crops has been tinkered with and it is important to set up an effective international research programme in which efficient strains of N_2-fixing cyanobacteria for use in diverse paddy soils can be selected and tested, and in which efficient, economically viable techniques for preparing inocula and for their application can be developed for use by farmers.

Grazing of cyanobacteria

A major disadvantage of using cyanobacteria as the sole source of nitrogen in paddy soils is that they are subject to grazing by animals. Extensive growths of daphnids may occur and devour cyanobacterial populations within 1–2 weeks (Venkataraman, 1961; Watanabe, 1961). Such pests can be controlled using pesticides. Watanabe (1967), in summarizing earlier work, reports that attack by daphnids can be prevented by using pesticides such as Folidol or Parathion, that Parathion is also effective against *Bulimus, Branchinella, Leptestheria,* and other small animals (Hirano *et al.*, 1955), and that pentachlorophenol (PCP) at 100 p.p.m. inhibits chlorophytes but not cyanobacteria. It has to be remembered, however, that various N_2-fixing cyanobacteria are susceptible to pesticides. Thus, DaSilva *et al.* (1975) observed that immediately on adding pesticides to soils there was an initial decrease in algal growth, and thereafter only some algae recovered; among those which did not recover were various N_2-fixing cyanobacteria. Herbicides such as MCPA, 2,4-D, TCA and trifluralin also inhibit cyanobacterial growth, as shown by Cullimore and McCann (1977). It will probably be important to add efficient pesticides to cyanobacterial inocula for use in the field. Herbivorous fish may also graze on cyanobacteria (e.g. Moriarty *et al.*, 1973), and this offers the possibility of secondary production from cyanobacteria, if such fish are produced simultaneously in the rice paddy soils. This is an important aspect worth further study, especially as the fish excrement may serve as a nitrogen source for the growth of rice.

(g) SYMBIOTIC CYANOBACTERIA

Cyanobacteria form symbiotic associations with a very small number of eukaryotic plant genera (Table 7).

Table 7. Symbiotic associations of N_2-fixing cyanobacteria

Eukaryote group	Eukaryote genus	Cyanobiont
Algae	*Rhizosolenia* spp.	*Calothrix* (?)
Fungi	Various ascomycetes	Usually *Nostoc*
Bryophytes	*Anthoceros, Blasia, Cavicularia,* etc.	*Nostoc*
Pteridophytes	*Azolla*	*Anabaena*
Cycads	*Encephalartos Macrozamia, Zamia,* etc.	*Nostoc*
Angiosperms	*Gunnera*	*Nostoc*

(i) Associations with algae

The most studied example is the association between the endophytic rivulariacean cyanobacterium *Richelia intracellularis* (a short *Calothrix*-type filament with basal heterocysts) and the marine diatom *Rhizosolenia* (to identify diatoms, see Hustedt, 1930, 1959 and Hendey, 1964). Venrick (1974) has shown that *R. intracellularis* may reach concentrations of 10^3–10^4 filaments l^{-1} in localized patches of the Northern Pacific during June–September and there C_2H_2 reducing activity equivalent to up to 5 ng l^{-1} h^{-1} of N_2 have been recorded (Mague *et al.*, 1974). In the Northern Pacific, biological N_2 fixation contributes about 3% of the total daily fixed nitrogen requirement for phytoplankton growth. The *Richelia–Rhizosolenia* association, like *Trichodesmium*, with which it often occurs, has not yet been examined for nitrogenase activity in bacteria-free culture, probably because many marine microorganisms are notoriously difficult to grow (but see Guillard and Ryther, 1962; McLachlan, 1964; Provasoli *et al.*, 1957).

An association between *Enteromorpha linza* and *Calothrix braunii* has been found in intertidal regions of north-west France (Lami and Meslin, 1959). The *Calothrix*, with distinct heterocysts, is located in the air spaces of the *Enteromorpha* filaments. The association has not yet been tested for N_2 fixation but is probably active. A *Dichothrix* species which occurs as an epiphyte on pelagic *Sargassum* also fixes N_2 (Carpenter and Cox, 1974). There is no evidence that any of the associations between cyanobacteria and aquatic animals fix N_2. Possibly their main functional role in such ecosystems is to recycle nitrogen released by the animal.

(ii) Associations with fungi (lichens)

Cyanobacteria occur in symbiosis in about 8% of the 17000 species of lichens (see Duncan, 1970, for a key to lichen taxonomy, and Ahmadjian, 1967, for information on cyanobiont taxonomy). There are two types of thalli. In those in which a cyanobacterium (or a eukaryotic alga) is the sole photosynthetic partner (homoisomerous types), the latter occurs in a distinct layer near the thallus surface and usually represents about 6% of the total lichen nitrogen. In the second type (heteroisomerous forms), the main photosynthetic partner is a eukaryotic alga (usually a chlorophyte) and the cyanobacterium may occur in discrete packets, associated with fungal hyphae, called cephalodia. These cephalodia may be superficial, as in *Peltigera aphthosa* (where they represent about 2.6% of the thallus biomass), or may be internal, as in *Lobaria pulmonaria.* In a few, the cyanobiont forms a layer which is distinct from that of the primary phycobiont. Typical lichens have not yet been re-synthesized from the isolated components, but a partial re-synthesis has been reported (Ahmadjian, 1967). Koch's postulates have not been satisfied although it is possible to isolate the same cyanobacterium many times from the lichen. Isolation can be achieved by surface sterilizing lichen discs, transferring these to inorganic medium supplemented with combined nitrogen and actidione (100 μg ml^{-1}) on moist filter-paper in Petri dishes or on medium solidified with 1% agar. Under such conditions, in the light, the symbiosis tends to break down and the cyanobacteria grow out. Eukaryotic phycobionts can be isolated by gently disrupting the thallus in a homogenizer, differentially centrifuging the homogenate and transferring the algae to nitrogen-containing medium plus penicillin (20 $\mu g\ ml^{-1}$) and polymyxin B sulphate (10 $\mu g\ ml^{-1}$).

Lichens, in general, are particularly susceptible to atmospheric pollution and some may be used as indicators of pollution (Ferry *et al.*, 1973). Thus, many do not keep well in the laboratory. This may be because environmental conditions which are satisfactory for the maintenance of both symbionts must occur and possibly because, in symbiosis, membrane permeability of the components is altered. For experimental work we use field material which is freshly collected, freed from adhering material, washed, and incubated under standard conditions for up to 48 h before use. Lichen thalli should not be left water-saturated in the laboratory at high temperatures (>20 °C) for long (>4 h), as this stimulates the growth of contaminating bacteria, particularly anaerobic forms. Cengia-Sambo (1926) suggested that *Azotobacter* was the agent of N_2 fixation in lichens, but this is not so (Bond and Scott, 1955). The youngest parts of lichen thalli are most metabolically active and should be used. To avoid variability of the material, use sets of discs taken with a cork-borer from the younger parts of the thalli, when possible.

(iii) Associations with bryophytes

N_2-fixing cyanobacteria may form associations with the moss *Sphagnum* and with various thallose liverworts. *Sphagnum* species (for keys see Duncan, 1962; Hill, 1978) are important components of the flora of temperate and subtemperate regions of both hemispheres. They are characterized by having discrete empty-looking hyaline cells. West and Fritsch (1927) reported on *Nostoc* and *Anabaena* occurring endophytically in *Sphagnum* and on *Hapalosiphon* being associated with submerged sphagna in the U.K. Stewart (1966) noted *Hapalosiphon* within hyaline cells. Granhall and Von Hofsten (1976) studied a *Nostoc–Sphagnum* association in detail. The associations, which were found on least acidic sites, were of *Nostoc* growing intracellularly within the hyaline cells of *S. lindbergii* and *S. riparium.* No cyanobacteria occurred in *S. balticum, S. fuscum,* or *S. annulatum* (*S. yensenii*). According to Sonesson (1973), sphagna with cyanobacteria grow faster than those without cyanobacteria. Some (Granhall and Selander, 1973; Alexander, 1974; Granhall and Lid-Torsvik, 1975) consider that N_2-fixing heterocystous cyanobacteria in association with *Sphagnum* and *Drepanocladus* may contribute significantly to the nitrogen economy of sub-arctic mires.

Nostoc species occur symbiotically with certain thallose liverworts, as Leitgeb (1878) first noted with *Anthoceros.* The liverworts include *Anthoceros, Blasia,* and *Cavicularia.* N_2 fixation by the symbiotic cyanobacterium in each has been clearly established (Bond and Scott, 1955; Watanabe and Kiyohara, 1963; Rodgers and Stewart, 1977; Stewart and Rogers, 1977). *Anthoceros* (Figure 3) is easily identified by having elongated sporophytes (hence the common name hornworts) (for details of the genus, see Proskauer, 1948, 1960, 1967, 1969; Rodgers and Stewart, 1977). *Blasia* (Figure 4) is a member of the Jungermanniales or leafy liverworts, although the plant itself is more thallose than most leafy liverworts (see MacVicar, 1926, and Watson, 1963,

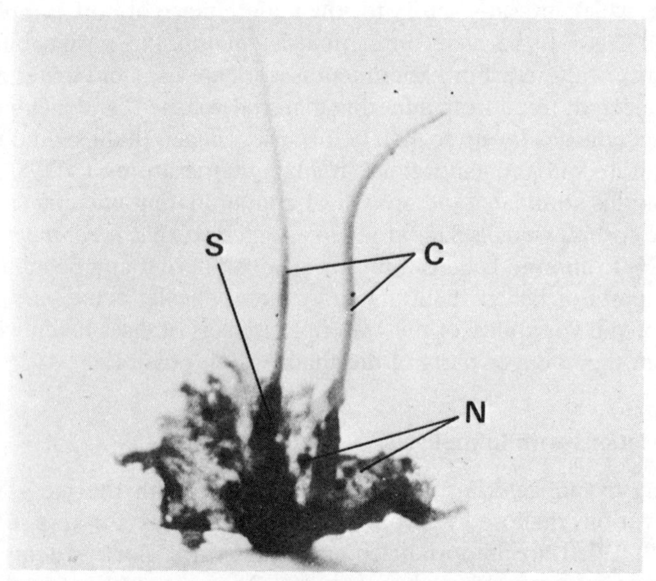

Figure 3. The liverwort *Anthoceros punctatus* with gametophyte containing *Nostoc* colonies (N) and sporophyte (S) with capsules (C) (×5) (from Rodgers and Stewart, 1977)

for further details of the genus). For information on *Cavicularia*, see Watanabe and Kiyohara (1963). The cyanobacteria develop in mucilage-filled cavities on the undersurfaces of the gametophyte, the sporophyte, when present, being without cyanobacteria. Colony development results from the movement of motile filaments into the cavities (the mucilage is produced by the liverwort, also being present in cavities without cyanobacteria) and there they develop as compact, easily seen colonies. There is a constant relationship during growth between colony number and thallus biomass (Rodgers and Stewart, 1977). Liverwort thalli can be propagated easily in the laboratory on Peralite moistened with nitrogen-free ASM medium, or on the same medium solidified with 1% agar. The symbiotic system grows well in the laboratory in the pH range 4–8 at 10–15 °C with a low light intensity (80 lux) and when moist but not water-saturated (Rodgers and Stewart, 1977). According to Rodgers (1978) the symbiosis does not offer the cyanobionts of *Blasia* any protection against desiccation. Cyanobacteria-free thalli are obtained by excising small portions of the thallus without cyanobacteria and propagating these, as above, but using medium containing combined nitrogen. The symbiosis can be re-established by adding the appropriate cyanobacterium, but the symbiosis is specific in that only certain *Nostoc* species develop in the cavities, and are

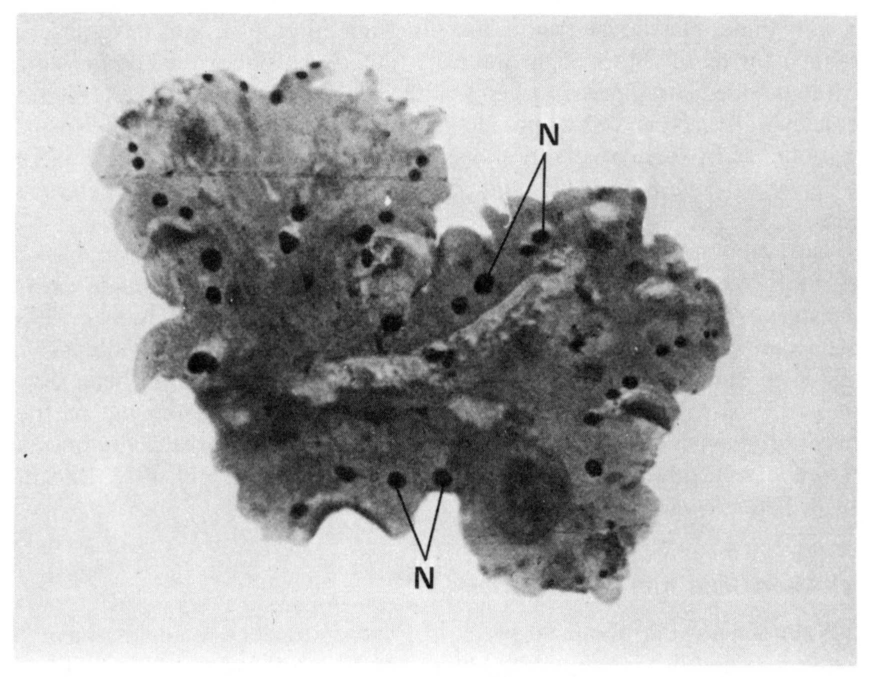

Figure 4. Thallus of the liverwort *Blasia pusilla* showing the presence of *Nostoc* colonies (N) (×11) (from Rodgers and Stewart, 1977)

found consistently in the cavities of field material (occasionally field material may contain a few diatoms and/or green algae, but such eukaryotes are probably chance contaminants; see Rodgers and Stewart, 1977).

(iv) *Anabaena–Azolla* symbiosis

Azolla is the only fern which forms a distinct association with a N_2-fixing cyanobacterium (*Anabaena azollae*). This symbiosis may serve as an important source of biologically fixed nitrogen for the growth of rice.

Azolla is a small water fern which is placed in a family of its own—

Azollaceae. The plants are heterosporous with microspores and megaspores and about 25 fossil species and 6 extant species are recognized. None of the fossils have been shown to contain a symbiotic cyanobacterium. Living species are difficult to classify in the field because the taxonomy is based largely on reproductive features and field material is often sterile. The plants have a small, pinnately branched stem with small imbricate leaves, each with an almost colourless and often submerged lower lobe and a green floating upper lobe. The leaves are arranged in two alternate rows and the endophyte occurs in cavities at the base of the dorsal lobe (Figure 5). *Azolla* can be propagated readily under glasshouse conditions, although different strains have different optima for temperature, light intensity, etc. A satisfactory medium with or without added combined nitrogen is 40% Hoaglands solution. The fern grows equally well on N_2 and on combined nitrogen. *Anabaena*-free thalli can usually be obtained by sequentially treating symbiotic plants with streptomycin (5 mg 1^{-1}), tetracycline (4 mg 1^{-1}), and penicillin (25 mg 1^{-1}) (see also Peters and Mayne, 1974).

The cyanobiont has been classified as *Anabaena azollae* (Oes, 1913; Nickell, 1958; Kawamatu, 1965*a,b*; Johnson *et al.*, 1966), but although various *Anabaena* species have been isolated (Pringsheim, 1918; Huneke, 1933; Schaede, 1947; Venkataraman, 1962; Wieringa, 1968; Becking 1976*b*; Ashton and Walmsley, 1976; Newton and Herman, 1979), none has yet been shown to satisfy Koch's postulates. Symbiosis is re-established from the partners only during a very short period of leaf development when the cavity in which the endophyte develops remains open to the exterior (see Hill, 1975; Becking, 1978; Peters *et al.*, 1978).

(v) Associations with gymnosperms

Nine genera and about 90 species of cycads (order Cycadales) have N_2-fixing root nodules inhabited by heterocystous cyanobacteria (Figure 6) (see also Allen and Allen, 1965; Grobbelaar *et al.*, 1971; Bond, 1967; Bergersen *et al.*, 1965; Halliday and Pate, 1976*a*).

The root nodules, which represent about 1.5% of the plant biomass, occur from just below the soil surface to a depth of about 50 cm, and there is controversy over whether they are normal features of the root system, or whether they develop in response to infection (see Nathanielsz and Staff, 1975). The cyanobacterium occurs in a distinct air space in the outer cortex of the mature nodule lobe (the lobe apices remain uninfected) and probably grows heterotrophically there, as it can do when isolated (although nitrogenase activity rapidly increases on re-exposure to light; Halliday and Pate, 1976*a*). The cyanobiont has been variously identified as *Anabaena cycadae* (Reinke, 1872; Horejsi, 1910; Spratt, 1915; Chaudhuri and Akhtar, 1931). *Nostoc punctiforme* (Hariot, 1892), *Nostoc commune* (Schneider, 1894), or *Nostoc*

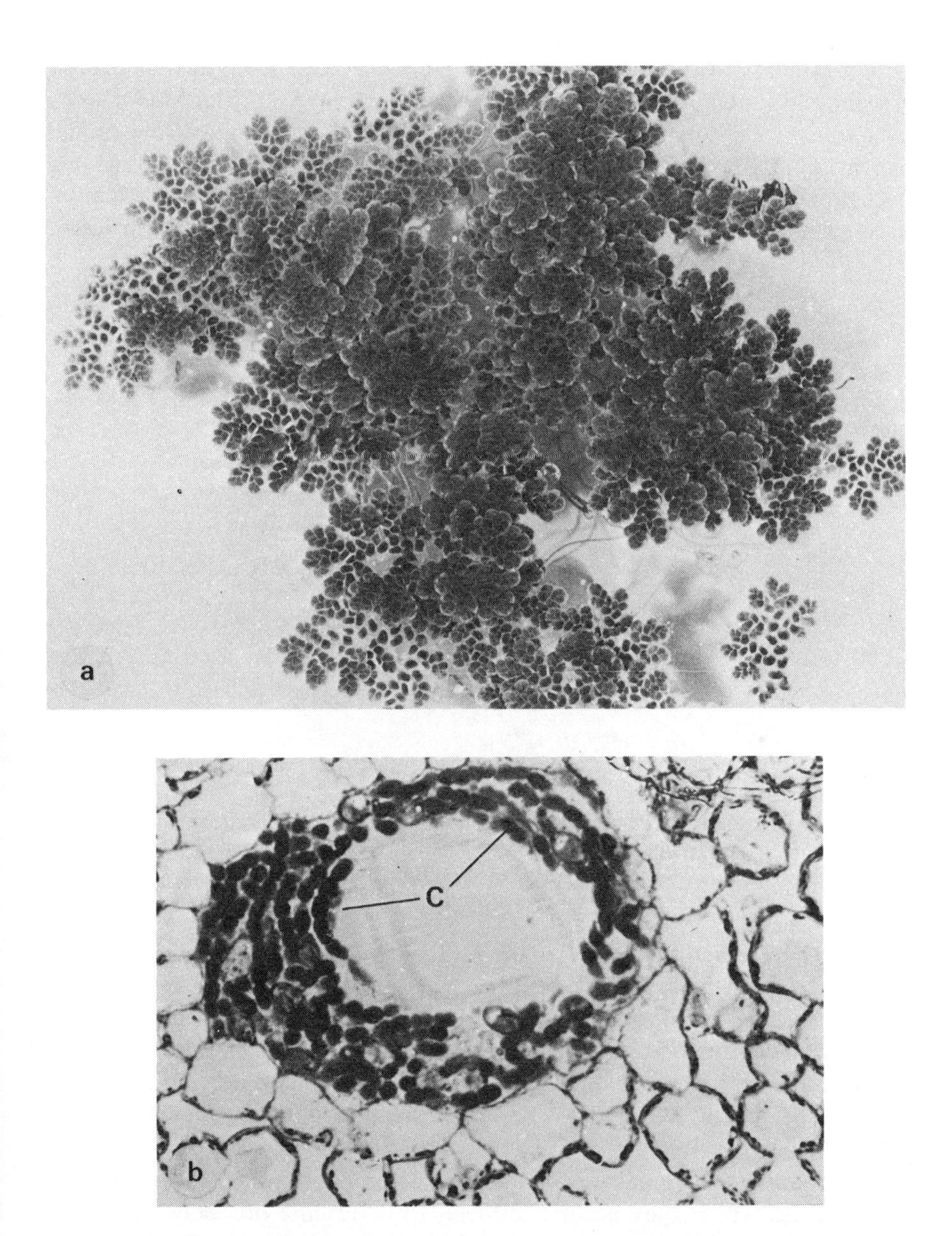

Figure 5. (a) *Azolla* plants growing in liquid culture (full scale) (b) Light micrograph of a section through the dorsal lobe of an *Azolla* leaf showing the presence of the cyanobacterium (C) *(Anabaena azollae)* within the cavity (×650)

cycadae (Watanabe and Kiyohara, 1963), and represents about 0.7% of the nodule biomass and fixes N_2 at rather similar rates to free-living cyanobacteria. An unusual finding is that a *Nostoc* strain (*Nostoc* MAC) from *Macrozamia communis* which was heterocystous and fixed N_2 when isolated (Bowyer and Skerman, 1968) lost its N_2-fixing capacity after several years in culture and became aheterocystous. It is uncertain whether this resulted from a spontaneous mutation (possibly through repeated growth on combined nitrogen),

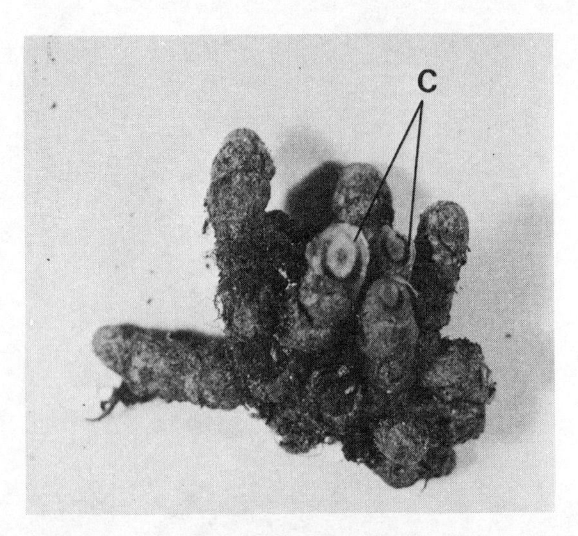

Figure 6. Root nodule of *Cycas* sp. showing the coralloid appearance of the nodule. Note the zone of cyanobacteria (C) in the outer cortex of cut nodule lobes (×4)

whether a mixed culture was initially present (highly unlikely since the clone was raised from a single filament), or whether there was misidentification, wrong labelling, or contamination during subculturing. The nitrogen fixed by the cyanobacterium is rapidly transported to the remainder of the plant with most of the ^{15}N labelling being found in citrulline, glutamine, and glutamate in the translocating strand and with little remaining in the cyanobacterial cells (Bergersen *et al.*, 1965; Halliday and Pate, 1976*a*).

The quantitative significance of cycad root nodules to the nitrogen economy of the areas where they occur is difficult to determine precisely because of the scattered distribution of the plants. Halliday and Pate (1976*a*) calculate that *Macrozamia riedlei* contributes 18.8 kg ha^{-1} yr^{-1} of N in sandy areas of Western Australia, which is a high value bearing in mind the slow growth of the plant. In other areas, for example the understorey of Australian eucalyptus forests, cycads may also be important sources of fixed nitrogen. Despite the benefits of cycads as nitrogen fixers, as a source of sago (the stems of *Cycas*; Burkill, 1966) and of starch (roots and stems of *Zamia*; Uphof, 1968), it must be remembered that the fruits of some species are poisonous.

(vi) *Gunnera*

This is the only angiosperm genus with which N_2-fixing symbiotic cyanobacteria consistently develop (Figure 7). A member of the Haloragaceae,

Figure 7. Scanning electron micrograph of a cut stem swelling of *Gunnera* sp. showing filaments of the symbiotic cyanobacterium (C) (a *Nostoc* sp.) (×1300) (photograph by Dr J. I. Sprent)

with rhubarb-like leaves, its natural distribution is the southern hemisphere and particularly moist areas such as riverbanks. The symbiotic *Nostoc* colonies are located in special mucilage-filled glands which develop on the stem at the bases of the petioles. In the glands are papillae-like outgrowths into the mucilage cavity. The *Nostoc* (*N. punctiforme* according to Winter, 1935) develops intracellularly, has a high heterocyst frequency, shows high rates of nitrogenase activity, and has various other physiological and biochemical modifications associated with its symbiotic role (see Silvester, 1976). Although Scott (1969) considered that the cyanobionts may be unimportant as N_2-fixers, Silvester and Smith (1969) showed unequivocally that the nitrogen fixed was transferred from the glands to the remainder of the plant and that during 10 weeks in an environment free of combined nitrogen, plant nitrogen increased 9-fold. Becking (1976*b*) concluded that in Indonesia *Gunnera* fixed 22–38 kg ha^{-1} yr^{-1} of N, whereas Silvester and Smith (1969) calculated an annual fixation rate in New Zealand of up to 720 kg ha^{-1} yr^{-1} of N. Such values will vary markedly depending on the density of *Gunnera* in the area.

(h) ORGANIZATION OF THE SYMBIOTIC SYSTEMS

The various symbioses involving cyanobacteria show many similar features. These have been considered in detail elsewhere (Silvester, 1976; Stewart *et al.*, 1979; Peters *et al.*, 1979) and only a resumé, together with more recent findings, need be considered here.

Firstly, although earlier workers, and indeed some recent ones (Ahmadjian, 1962; Duckett *et al.*, 1977), suggest that the associations are not specific, most recent studies suggest that they are. Thus, only heterocystous cyanobacteria occur in symbiotic association and all fix N_2. Also, specific genera are involved—*Anabaena* in *Azolla, Nostoc* in *Peltigera*, in liverworts, and in *Gunnera*, etc. Even within the one genus there are differences between the species, some forming symbiosis and others not doing so [e.g. in *Blasia pusilla* only 3 out of 16 cyanobacteria tested formed symbiotic colonies (Rodgers and Stewart, 1977), and with *Peltigera canina* haemagglutinins produced by the fungus cross-reacted fully with only 3 out of 11 cyanobacterial species tested (Lockhart *et al.*, 1978)]. Recently, we have demonstrated the presence of peritrichous fimbriae on the *Nostoc* partner of *Peltigera canina* (Dick and Stewart, 1980). Such fimbriae, which are approximately 7 nm wide and up to 3 μm long, are different from the Types 1–6 fimbriae and sex fimbriae which are distinguished in Gram-negative heterotrophic bacteria (see, Duguid 1968). The possible role of such fimbriae in the specificity of the associations will be of interest in view of their implication in the specificity of various bacteria–eukaryotic cell associations.

Secondly, in symbiotic association, the cyanobacteria change in morphology, particularly if they are associated with another photosynthetic partner. For example, they develop in distinct packets and show heterocyst frequencies of up to 20–60% (Rodgers and Stewart, 1977; Silvester and MacNamara, 1976; Englund, 1977; Peters, 1977), although it is uncertain whether all the heterocysts are metabolically active since de-differentiation of mature heterocysts does not occur. At the ultrastructural level, the vegetative cells show fewer structured granules and more carboxysomes than do free-living cyanobacteria. Polyglucoside bodies may accumulate in the light and are used up to support nitrogenase activity in the dark. Polyglucoside breakdown appears to result from the activity of glycogen phosphorylase (Lehmann and Wöber, 1978; A. N. Rai, unpublished work).

Thirdly, physiological modifications occur in symbiosis with another photosynthetic partner. For example, the *Blasia* (Stewart and Rodgers, 1977) and *Gunnera* (Silvester, 1976) cyanobionts show reduced O_2 evolution and photofixation of CO_2 in symbiosis, the pigment composition of the *Azolla* cyanobiont also changes in symbiosis and Ray *et al.* (1979) speculate that in symbiosis the phycobilin pigments may be particularly important in light-harvesting.

Fourthly, at the biochemical level the GS activity of the symbiotic cyanobacterium is very low compared with that of the free-living cyanobacterium. This may result in NH_4^+, which otherwise would have been assimilated, being released for uptake by the associated partners [by glutamic acid dehydrogenase in the mycobiont of *P. canina* (Stewart *et al.*, 1977; Stewart and Rowell, 1977) and probably by eukaryotic GS in *Azolla* (Peters *et al.*, 1979)].

(i) THE POTENTIAL OF N_2 FIXATION BY THE *AZOLLA–ANABAENA* SYMBIOSIS IN RICE FIELDS

(i) General

Azolla has been used for centuries as a source of green manure in south-east Asia. Liu (1979) reports that it was used as early as 540 B.C. as a feed for domestic animals in China, and that there are early records of its use as a fertilizer in Chinese rice paddy soils. Recently, with the appreciation that the symbiosis fixes N_2, much attention has been paid to its role as an organic source of combined nitrogen in rice soils, particularly in China, where about 1.3 million hectares of *Azolla* are grown annually (Liu, 1979), Vietnam (Dao and Tran, 1979), India (Singh, 1979), Indonesia (Becking, 1979), the Philippines (Watanabe and Cholitkul, 1979), and elsewhere (Moore, 1969), including the U.S.A. (Rains and Talley, 1979). It is a rapidly growing symbiosis which can double its nitrogen content every 1–3 days under favourable growth conditions, contains 3–5% nitrogen and, when present in paddy soils, increases the nitrogen and organic contents of the soil, improves the physical and chemical texture of the soil, and helps to prevent weed growth by shading out slower growing weeds. In chemical composition *A. pinnata* contains 10.5% ash, 3–3.36% crude fat, 24–30% crude protein, 4–5% nitrogen, 0.5–0.9% phosphorus, 0.4–1.0% calcium, 2–4.5% potassium, 0.5–0.65% magnesium, 0.11–0.16% manganese, and 0.06–0.26% iron (Singh, 1979). In agricultural practice, *A. pinnata* is the plant of choice in south-east Asia, with particular types being selected for different areas. For example, in Vietnam (Dao and

Tran, 1979), where *A. pinnata* var. *imbricata* is used, three types are used: red *Azolla*, which generally is best and can withstand both low temperatures and high salinity, purple *Azolla*, which is resistant to acidic conditions, and green *Azolla*, which resists high temperatures. In the U.S.A. both *A. mexicana* and *A. filiculoides* have been used in experiments but both have the disadvantage of being unable to withstand freezing, although they do tolerate lower temperature extremes than *A. pinnata* (see below).

(ii) Growth and physiology of *Azolla*

Azolla plants can be grown easily in the laboratory. Growth occurs best at a relative humidity above 60% and agitation of the cultures should be avoided since this leads to reduce growth and nitrogenase activity (Ashton and Walmsley, 1976). Possibly this is a reason why *Azolla* is seldom found growing in open waters, preferring the more sheltered habitat afforded to it by the rice plants. It continues to fix nitrogen in the presence of combined nitrogen, possibly because in symbiosis it possesses little GS activity (Peters *et al.*, 1979; Gadd *et al.*, 1980). This may allow it to be used together with chemical nitrogen fertilizer in rice fields and in this respect it has advantages over the use of free-living cyanobacteria where combined nitrogen inhibits N_2 fixation. It also has the advantage of being readily recognizable by the farmer.

The symbiotic system also has an advantage over free-living cyanobacteria for use in rice fields in that the *Azolla* symbiosis grows and fixes N_2 under acidic conditions (pH 5–7) whereas free-living cyanobacteria do not flourish in acidic soils.

Most workers have obtained temperature optima with *A. pinnata* of between 20 and 30°C, although Brotonegoro and Abdulkadir (1976) found increased growth up to 40°C, inhibition at 43°C and a Q_{10} between 20 and 30°C of 2. The lower temperature for growth of *A. pinnata* is 14°C (Singh, 1979) or 15°C (Dao and Tran, 1979). Watanabe *et al.* (1977*b*), who used a light–dark regime of 12:12, found equal growth of *Azolla* at average temperatures of 22, 25 and 28°C, but with decreased growth at 31°C. Night temperatures of 18°C did not affect the *Azolla.* Red *Azolla* (Dao and Tran, 1979), *A. filiculoides*, and *A. mexicana* can withstand lower temperatures, but neither of the latter two species can withstand freezing (Becking, 1979), although *A. filiculoides* can grow at 5°C (Talley *et al.*, 1977). For further details of the effects of frost and freezing on *Azolla*, see Benedict (1923), Ashton (1974), and Talley *et al.* (1977).

Growth and nitrogenase activity of *Azolla* vary with light intensity and day-length. In general, *Azolla* grows best in shaded situations among rice, although this may be a reflection of effects of disturbance rather than an effect of light *per se* (see above). Using *A. caroliniana*, Peters (1976) found that the saturating light intensity for nitrogenase activity by the isolated symbiont was 2150 lux and that by the symbiosis was 4840 lux. Becking (1976*b*), in short-term experiments, found no difference in nitrogenase activity by *Azolla* in the range 14000–27000 lux and that 50–75% shading of the plant at full sunlight for up to 10h did not affect photosynthesis or nitrogenase activity, although longer periods of shading reduced nitrogenase activity as expected.

With *Azolla*, there is evidence of complex interactions between various environmental factors on growth and nitrogenase activity. Thus, in Vietnam a high light intensity may enhance growth at pH 5, whereas at pH 6–7 it inhibits growth (Dao and Tran, 1979).

(iii) Yields of *Azolla* obtained

Azolla can double in biomass every 1–3 days. The biomass which it could produce is thus theoretically enormous. It has been calculated to fix as much as 670 kg ha^{-1} yr^{-1} of N (Becking, 1979); such very high amounts, however, have been calculated on the basis of unlimited growth for 12 months of the year and in practice this never happens. A realistic estimate of N_2 fixation by *Azolla* is that of Subudhi and Singh (1979), who calculate a daily fixation rate of 7.5 mg (g dry weight)$^{-1}$ of N.

Azolla is used in the field in two main ways. Firstly, it may be grown in rotation when rice is not cultivated. For example, in India it is grown as an annual during July–November/December (Gopal, 1967) and in Vietnam it is grown as a winter crop and is followed by spring rice in February–June, and then by summer rice in July–October/November. Singh (1979) recommends the following procedure for India. If sufficient water is available before rice planting *Azolla* should be grown as a green manure. This involves ploughing the field, levelling it, and irrigating it. When there is about 5–10 cm of water, the *Azolla* plants are added [1000 kg (fresh weight) ha^{-1}], together with 8–20 kg of P_2O_5 and 100–500 g ha^{-1} of carbofuran (see below). After 10–20 days the *Azolla* is incorporated into the soil and the rice seedlings are transplanted. Incorporation of 10 tonnes ha^{-1} of *Azolla* as green manure into the soil is as efficient as applying 30 kg ha^{-1} of fertilizer N. The grain yield increased up to 54% when such *Azolla* was grown but not incorporated. Singh (1979) recommends that when water is limiting, about 5–10% of the area should be irrigated and planted with *Azolla*. As the *Azolla* grows it should be harvested weekly and used as organic manure for the remaining area. This it can provide within 2–3 months. In China *Azolla* is grown for 15–20 days before or after rice. The entire *Azolla* plant is again mixed into the soil and provides 30–40% of the nitrogen required for the growth of one crop, increasing the rice yield by about 18% on average. The exact response varies with soil type.

Thus, in highly fertile soils the increase is 9.9–27.6%, and in low-fertility soils it may be up to 42.7%. Chinese workers recommend the use of young plants rather than old plants and also recommend turning over the *Azolla*-containing soil several times.

The alternative procedure to rotation is to grow the *Azolla* along with the rice. In India, Singh (1979) suggests that 200–1000 kg ha^{-1} of *Azolla* should be added after establishment of the rice and that the *Azolla* should be allowed to grow for 20–40 days before being incorporated into the soil. Singh (1979) noted how, after *Azolla* incorporation, the rice plants changed colour, grew better, and flowered earlier. An innovative way of growing rice and *Azolla* together has been worked out by the Chinese, who grow the *Azolla* plants and the rice in alternate rows. The rice varieties used are compact, have straight, narrow leaves to prevent shading, have a rigid stem, big ears, more grains, and are of medium tillering capacity. Liu (1979) concludes that when *Azolla* is used, rice varieties with a long growing period are superior to those with a short growing period. The rice is grown in narrow rows, two plants wide, and on either side of the row of rice a wider row of *Azolla* is planted. Using this technique the Chinese produce average rice crops of 13 200 kg ha^{-1} and 109 300 kg ha^{-1} of *Azolla*, the highest values obtained being 16 070 and 178 300 kg ha^{-1}, respectively. Occasionally *Azolla* production has been as high as 223 500 kg ha^{-1}.

Azolla can also be used as a feed for poultry (Singh and Subudhi, 1978; Subudhi and Singh, 1979*a*) and pigs (Singh, 1979), as a vegetable (Singh, 1978 quoted in Singh, 1979), as a general compost and for biogas production (Singh, 1979). It can also be used to purify water (Rains and Talley, 1979).

(j) PERSPECTIVE

The cyanobacteria, either free-living or in symbiosis with *Azolla*, can provide from sunlight and water, substantial amounts of fixed nitrogen and organic carbon for soil fertility. The extent to which such organisms will prove to be important in agricultural practice in the future will probably depend on the availability and price of chemical nitrogen fertilizer. The energy crisis of the 1970s, however, has emphasized the dangers of depending solely on fertilizer production from fossil fuels and it would be prudent, on a global basis, to diversify interest to include biological sources of fertilizer nitrogen. In the rice-growing regions of the world, cyanobacteria are the organisms of choice in any attempt to capitalize on biological N_2-fixing systems. It is known from laboratory studies that most possess nitrogenase; the environmental conditions under which they fix N_2 are known, and there are encouraging results from field trials, particularly those using *Azolla*. If science and agriculture can proceed side-by-side, the use of cyanobacteria should result in a substantial reduction in the requirements for synthetic nitrogen fertilizer in rice fields; it is doubtful, however, whether such organisms can always provide all the nitrogen required for optimum growth of the rice crop and some input of chemical nitrogen fertilizer will probably always be required.

References

Aaronson, S. (1970), *Experimental microbial ecology,* Acedemic Press, New York, p. 65.
Aboul-Fadl. M., Taha-Eid, M., Hamissa, M. R., El-Nawawy, A. S., and Shoukry, A. (1967), *J. Microbiol. UAR,* **2,** 241.
Adams, M. H. (1959), *Bacteriophages,* Interscience, New York.
Adelberg, E. A., Mandel, M., and Chen, G. C. C. (1965), *Biochem, Biophys. Res. Commun.,* **18,** 788.
Afify, M. N., Moharram, A. A., El-Nady, M. A. L., Hamdi, Y. A., El-Sherbani, M. F., and Lofti, M. (1968), *Agric. Res. Rev.,* **46,** 2.
Ahmadjian, V. (1962), *Am. J. Bot.,* **49,** 277.
Ahmadjian, V. (1967), *Phycologia,* **6,** 127.
Aho, P. E., Seidler, R. J., Evans, H. J., and Raju, P. N. (1974), *Phytopathology,* **64,** 1413.
Akkermans, A. D. L. (1971), *Nitrogen fixation and nodulation of Alnus and Hippophaë under natural conditions,* PhD Thesis, University of Leiden.
Akkermans, A. D. L., Abdulkadir, S., and Trinick, M. J. (1978a), *Nature, Lond.,* **274,** 190.
Akkermans, A. D. L., Abdulkadir, S., and Trinick, M. J. (1978b), *Pl. Soil,* **49,** 711.
Akkermans, A. D. L., and van Dijk, C. (1976), in *Symbiotic nitrogen fixation in plants* (Ed. P. S. Nutman), International Biological Programme, Vol. 7, Cambridge University Press, London, p. 511.
Akkermans, A. D. L., Straten, J. van, and Roelofsen, W. (1977), in *Recent Developments in nitrogen fixation* (Eds. W. Newton, J. R. Postgate, and C. Rodríguez-Barrueco), Academic Press, London, New York, San Francisco, p. 591.
Alexander, V. (1974), in *Soil organisms and decomposition in tundra,* (Eds. A. J. Holding and S. F. Maclean), Tundra Biome Steering Committee, Stockholm.
Alimagno, B. V. and Yoshida, T. (1977), *Pl. Soil,* **47,** 239.
Allen, M. B. and Arnon, D. I. (1955), *Pl. Physiol.,* **30,** 366.
Allen, E. K. and Allen, O. N. (1961), *Rec. Advan. Bot.,* **9,** 585.
Allen, E. K. and Allen, O. N. (1965), in *Microbiology and soil fertility.* (Eds. C. M. Gilmore and O. N. Allen), Oregon State University Press, Corvallis, p. 77.
Allen, O. N., Hamatová, E. and Skinner, F. A. (1973), *IBP World catalogue of Rhizobium collections,* IBP Central Office, London, 282 pp.

Amarger, N., Mariotti, A. and Mariotti, F. (1977), *C. R. Acad. Sci., Paris,* **284D,** 2179.

American Public Health Association (1955), *Standard methods for the examination of water, sewage and industrial wastes,* 10th Edition, American Public Health Association, New York.

Andersen, K., Shanmugam, K. T., and Valentine, R. C. (1977), in *Genetic engineering for nitrogen fixation.* (Eds. A. Hollaender *et al.),* Plenum Press, New York, p. 95.

Andrew, C. S. and Cowper, J. L. (1973), *Lab. Pract.,* **4,** 364.

Andrew, C. S. and Fergus, I. F. (1976), in *Tropical pasture research principles and methods,* CAB Bull. No. 51, Commonwealth Agricultural Bureau, Farnham Royal, U.K.

Anon. (1940), *Rules and regulations under the Federal Seed Act,* U.S.A., Service and Regulatory Announcement No. 156, p. 9.

Anon. (1952), *Rural Res. in CSIRO (Aust.),* **2,** 9.

Anon. (1953), *Standardized Plant Names,* CSIRO (Aust.) Bull. No. 272.

Anon. (1966), *Proc. Int. Seed Test. Ass.,* **31,** 1.

Anon. (1977a), *World food and nutrition study, the potential contributions of research,* National Academy of Sciences, Washington, D.C., p. 74.

Anon. (1977b), *Indian Standard Specification for Rhizobium inoculants,* Indian Standards Institution, New Delhi.

Anon. (1979a), *The method of testing legume inoculant and pre-inoculated seed products.* Canada Department of Agriculture, Ottowa, Canada, 13 pp.

Anon. (1979b), *Canada Gazette,* Part II, **113 (9),** May 9, p. 3.

Anon. (1979c), *The inoculant and coated seed testing service,* New Zealand Ministry of Agriculture and Fisheries, Seed Testing Station, Palmerston North, New Zealand.

Antonini, E. and Brunori, M. (1971), *Hemoglobin and myoglobin in their reactions with ligands,* North Holland, Amsterdam.

Appleby, C. A. (1962), *Biochim. Biophys. Acta,* **60,** 226.

Appleby, C. A. (1974), 'Leghaemoglobin', in *The biology of nitrogen fixation* (Ed. A. Quispel), North Holland, Amsterdam, p. 521.

Appleby, C. A., Nicola, N., Hurrell, J. G. R., and Leach, S. J. (1975), *Biochemistry,* **14,** 4444.

Appleby, C. A., Wittenberg, B. A., and Wittenberg, J. B. (1973), *Proc. Nat. Acad. Sci. USA,* **70,** 564.

Appleman, M. D. and Sears, D. H. (1944), *Soil Sci. Soc. Amer. Proc.,* **9,** 98.

Aprison, M. H., Magee, W. E., and Burris, R. H. (1954), *J. Biol. Chem.,* **208,** 29.

Apte, S. K., Rowell, P., and Stewart, W. D. P. (1978), *Proc. Roy. Soc. Lond., B,* **200,** 1.

Arber, W. and Linn, S. (1969), *Ann. Rev. Biochem.,* **38,** 467.

Argüeso, T. R., Hanus, J., and Evans, H. J. (1978), *Arch. Microbiol.,* **116,** 113.

Ash, C. G., and Allen, O. N. (1948), *Proc. Soil Sci. Soc. Am.,* **13,** 279.

Ashton, P. J. (1974), in *The Orange River, progress report.* (Ed. E. M. von Zinderen-Bakker), Bloemfontein, p. 123.

Ashton, P. J. and Walmsley, R. D. (1976), *Endeavour,* **36,** 39.

Askew, J. L. and Lane, C. L. (1979), in *Symbiotic nitrogen fixation in the management of temperate forests* (Eds. J. C. Gordon, C. T. Wheeler, and D. A. Perry), Forest Research Laboratory, Oregon State University, Corvallis, Oregon, p. 472.

Aston, F. W. (1927), *Proc. Roy. Soc. Lond., A,* **115,** 487.

Atkins, C. A. and Pate, J. S. (1977), *Photosynthetica,* **11,** 214.

Atkins, C. A., Pate, J. S., and Sharkey, P. J. (1975), *Pl. Physiol.,* **56,** 807.

Atwater, W. O. (1885), *Am. Chem. J.,* **6,** 365.

Atwater, W. O. (1886), *Am. Chem. J.*, **8,** 398.
Aulie, R. P. (1970), *Proc. Am. Phil. Soc.,* **114,** 435.
Austin, S. M., Galonsky, A., Bortins, J., and Wolk, C. P. (1975), *Nuc. Instrum. Methods,* **126,** 373.
Ausubel, F., Riedel, G., Cannon, F., Perkin, A., and Margolskee, R. (1977), in *Genetic engineering for nitrogen fixation* (Eds. A. Hollaender *et al.),* Plenum Press, New York, p. 111.
Averill, B. A., Bale, J. R., and Orme-Johnson, W. H. (1978), *J. Am. Chem. Soc.,* **100,** 3034.
Aviram, I., Wittenberg, B. A., and Wittenberg, J. B. (1978), *J. Biol. Chem.,* **253,** 5685.
Avron, M. and Jagendorf, A. T. (1956), *Arch. Biochem. Biophys.,* **66,** 475.
Bachhuber, M., Brill, W. J., and Howe, M. M. (1976), *J. Bacteriol.,* **128,** 749.
Bachman, B. J., Low, K. B., and Taylor, A. L. (1976), *Bacteriol. Rev.,* **40,** 116.
Baker, D. and Torrey, J. G. (1979), in *Symbiotic nitrogen fixation in the management of temperate forests* (Eds. J. C. Gordon, C. T. Wheeler, and D. A. Perry), Forrest Research Laboratory, Oregon State University, Corvallis, Oregon, p. 38.
Baker, K. (1968), *Lab. Pract.,* **17,** 817.
Baker, K. (1978), *Biotechnol. Bioeng.,* **20,** 1345.
Balandreau, J. (1975a), *Activité nitrogénasique dans la rhizosphère de quelques graminées,* Thèse de Docteur ès Sciences Naturelles, Nancy, France.
Balandreau, J. (1975b), *Rev. Ecol. Biol. Sol.,* **12,** 273.
Balandreau, J. and Dommergues, Y. (1971), *C. R. Acad. Sci., Paris,* **273,** 2020.
Balandreau, J. and Dommergues, Y. (1973), in *Modern methods in the study of microbial ecology* (Ed. T. Rosswell), *Bull. Ecol. Res. Comm. (Stockholm),* **17,** 247.
Balandreau, J., Ducerf, P., Hamad-Fares, I., Weinhard, P., Rinaudo, G., Millier, C., and Dommergues, Y. (1978), in *Limitations and potentials for biological nitrogen fixation in the tropics* (Eds. J. Döbereiner, R. H. Burris, and A. Hollaender), Plenum Press, New York, p. 275.
Ballandreau, J. P., Miller, C. R., and Dommergues, Y. R. (1974), *Appl. Microbiol.,* **27,** 662.
Balandreau, J., Rinaudo, G., Fares-Hamad, I., and Dommergues, Y. (1975), in *Nitrogen fixation by free-living micro-organisims* (Eds. W. D. P. Stewart), Cambridge University Press, London, New York, Melbourne, p. 57.
Balandreau, J. Weinhard, P., Rinaudo, G., and Dommergues, Y. (1971), *Oecol, Plant.,* **6,** 341.
Balderston, W. L., Sherr, B., and Payne, W. J. (1976), *Appl. Environ. Microbiol.,* **31,** 504.
Barber, L. E. and Evans, H. J. (1976), *Can. J. Microbiol.,* **22,** 254.
Barber, L. E., Tjepkema, J. D., and Evans, H. J. (1978), *Ecol. Bull., Stockholm,* **26,** 366.
Barber, L. E., Tjepkema, J. D., Russell, S. A., and Evans, H. J. (1976), *Appl. Environ. Microbiol.,* **32,** 108.
Bardin, R., Domenach, A. M., and Chalamet, A. (1977), *Rev. Ecol. Biol. Sol.,* **14,** 395.
Barnet, Y., Daft, M. J., and Stewart, W. D. P. (1980), in preparation.
Barrios, S., Raggio, N., and Raggio, M. (1963), *Pl. Physiol.,* **38,** 171.
Barton, L. V. (1967), *Bibliography of seeds,* Columbia University Press, New York.
Batthyany, C. (1977), in *Exploiting the legume–rhizobium symbiosis in tropical agriculture,* (Eds. J. M. Vincent, A. S. Whitney and J. Bose), University of Hawaii College of Tropical Agriculture Misc. Publ., No. 145, p. 429.

Baumgarten, J., Reh, M. and Schlegel, H. G. (1974), *Arch. Microbiol.,* **100,** 207.
Becking, J. H. (1961), *Pl. Soil,* **14,** 49.
Becking, J. H. (1963), *Antome van Leeuwenhoek, J. Microbiol, Serol.* **29,** 326.
Becking, J. H. (1970), *Int. J. Syst. Bacteriol.,* **20,** 201.
Becking, J. H. (1975), in *The development and function of roots,* (Eds. J. G. Torrey and D. T. Clarkson), Academic Press, London, p. 507.
Becking, J. H. (1976a), in *Symbiotic nitrogen fixation in plants,* (Ed. P. S.Nutman), International Biological Programme, Vol. 7, Cambridge University Press, Cambridge, London, New York, Melbourne, p. 539.
Becking, J. H. (1976b), in *Proc. 1st int. symp. on nitrogen fixation,* (Eds. W. E. Newton and C. J. Nyman), Vol. 2, Washington State University Press, Pullman, Washington, p. 556.
Becking, J. H. (1978), *Ecol. Bull., Stockholm,* **26,** 266.
Becking, J. H. (1979a), *Pl. Soil,* **51,** 289.
Becking, J. H. (1979b), in *Nitrogen and rice symposium,* IRRI, Los Baños, Laguna, Philippines, p. 345.
Bednarski, M. A. and Reporter, M. (1978), *Appl. Environ. Microbiol.,* **36,** 115.
Beechey, R. B. and Ribbons, D. W. (1972), in *Methods in microbiology* (Eds. J. R. Norris and D. W. Ribbons), Vol. 6B, Academic Press, London, New York, p. 23.
Begg, J. E. and Turner, N. C. (1976), in *Advances in agronomy,* Vol. 28, Academic Press, New York, p. 161.
Beijerinck, M. W. (1925), *Zentralbl. Bakteriol. Parasitendkd. Infektionskr. Hyg., Abt. II,* **63,** 353.
Bell, F. and Nutman, P.S. (1971), in *Nitrogen fixation in natural and agricultural habitats* (Eds. T. A. Lie and E. G. Mulder), *Pl. Soil,* Special Vol., p. 231.
Benedict, R. C. (1923), *Am. Fern. J.,* **13,** 48.
Benemann, J. R. and Weare, N. M. (1974), *Science,* **184,** 174.
Benemann, J. R. and Valentine, R. C. (1972), *Adv. Microb. Physiol.,* **8,** 59.
Benemann, J. R., Yoch, D. C., Valentine, R. C., and Arnon, D. I. (1969), *Proc. Nat. Acad. Sci. USA,* **64,** 1079.
Benson, D. R., Arp, D. J., and Burris, R. H. (1979a), in *Symbiotic nitrogen fixation in the management of temperate forests* (Eds. J. C. Gordon, C. T. Wheeler, and D. A. Perry), Forest Research Laboratory, Oregon State University, Corvallis, Oregon, p. 472.
Benson, D. R., Arp, D. J., and Burris, R. H. (1979b), *Science, N.Y.,* **205,** 688.
Benson, D. R. and Eveleigh, D. E. (1979), *Soil Biol. Biochem.,* **11,** 331.
Benveniste, R. and Davies, J. (1973), *Ann. Rev. Biochem.,* **42,** 471.
Berger, J. A., Nay, S. N., Berger, L. R., and Bohlool, B. B. (1979), *Apl. Environ. Microbiol.,* **37,** 642.
Bergersen, F. J. (1961), *Aust. J. Biol. Sci.,* **14,** 349.
Bergersen, F. J. (1963), *Aust. J. Biol. Sci.,* **16,** 669.
Bergersen, F. J. (1965), *Aust. J. Biol. Sci.,* **18,** 1.
Bergersen, F. J. (1970), *Aust. J. Biol. Sci.,* **23,** 1015.
Bergersen, F. J. (1977), in *A treatise on dinitrogen fixation, Section III, biology* (Eds. R. W. F. Hardy and W. S. Silver), Wiley, New York, London, Sydney, Toronto, p. 519.
Bergersen, F. J. and Goodchild, D. J. (1973), *Aust. J. Biol. Sci.,* **26,** 741.
Bergersen, F. J. and Hipsley, E. H. (1970), *J. Gen. Microbiol.,* **60,** 61.
Bergersen, F. J., and Kennedy, G. S., and Wittmann, W. (1965), *Aust. J. Biol. Sci.,* **18,** 1135.
Bergersen, F. J. and Turner, G. L. (1967), *Biochim, Biophys. Acta,* **141,** 507.

Bergersen, F. J. and Turner, G. L. (1970), *Biochim. Biophys. Acta,* **214,** 28.
Bergersen, F. J. and Turner, G. L. (1973), *Biochem. J.,* **131,** 61.
Bergersen, F. J. and Turner, G. L. (1975a), *J. Gen. Microbiol.,* **89,** 31.
Bergersen, F. J. and Turner. G. L. (1975b), *J. Gen. Microbiol.,* **91,** 345.
Bergersen, F. J. and Turner, F. L. (1978), *Biochim. Biophys. Acta,* **538,** 406.
Bergersen. F. J. and Turner, G. L. (1979), *Anal. Biochem.,* **96,** 165.
Bergersen, F. J., Turner, G. L., and Appleby, C. A. (1973), *Biochim. Biophys. Acta,* **292,** 271.
Bergersen, F. J., Turner, G. L., Gibson, A. H., and Dudman, W. F. (1976), *Biochim. Biophys. Acta,* **444,** 164.
Bergey's manual of determinative bacteriology (1957), (Eds. R. S. Breed, E. G. D. Murray, and N. R. Smith), 7th Edition, Williams and Wilkins, Baltimore.
Bergey's manual of determinative bacteriology, (1974), (Eds. R. E. Buchanan and N. E. Gibbons), 8th Edition, Williams and Wilkins, Baltimore.
Bergmeyer, H. U. and Bernt, E. (1974), in *Methods of enzymatic analysis* (Ed. H. U. Bergmeyer), Adademic Press, New York, London, p. 727.
Beringer, J. E., Beynon, J. L., Buchanan-Wollaston, A. V., and Johnston, A. W. B. (1978), *Nature, Lond.,* **276,** 633.
Beringer, J. E. and Hopwood, D. A. (1976), *Nature, Lond.* **264,** 291.
Berndt, H., Ostwal, K. P., Lalucat, J., Schumann, C., Mayer, F., and Schlegel, H. G. (1976), *Arch. Microbiol.,* **108,** 17.
Berndt, H., Lowe, D. J., and Yates, M. G. (1978), *Eur. J. Biochem.,* **86,** 133.
Berry, A. and Torrey, J. G. (1979), in *Symbiotic nitrogen fixation in the management of temperate forests* (Eds. J. C. Gordon, C. T. Wheeler, and D. A. Perry), Forest Research Laboratory, Oregon State University, Corvallis, Oregon, p. 69.
Bethlenfalvay, G. J. and Phillips, D. A. (1978), *Physiol. Plant.,* **42,** 119.
Beynon, J. H. (1960), *Mass spectrometry and its application to organic chemistry,* Elsevier, Amsterdam, London, New York.
Bezdicek, D. F., Evans, D. W., Abede, B., and Witters, R. E. (1978), *Agron. J.,* **70,** 865.
Biggins, J. (1967), *Pl. Physiol.,* **42,** 1442.
Biggins, D. R. and Postgate, J. R. (1969), *J. Gen. Microbiol.,* **56,** 181.
Biggins, D. R. and Postgate, J. R. (1971a), *J. Gen. Microbiol.,* **65,** 119.
Biggins, D. R. and Postgate, J. R. (1971b), *Eur. J. Biochem.,* **19,** 408.
Bishop, A. E., Guevara, J. G., Engelke, J. A., and Evans, H. J. (1976), *Pl. Physiol.,* **57,** 542.
Blank, F. (1947), *Bot. Rev.,* **13,** 241.
Bock, R. M. and Alberty, R. A. (1953), *J. Am. Chem. Soc.,* **75,** 1921.
Boddey, R. M., Quilt, P., and Ahman, N. (1978), in *Limitations and potentials for biological nitrogen fixation in the tropics* (Eds. J. Döbereiner *et al.*), Plenum Press, New York, p. 357.
Boland, M. J. and Benny, A. G. (1977), *Eur. J. Biochem.,* **79,** 355.
Boland, M. J., Fordyce, A. M., and Greenwood, R. M. (1978), *Aust. J. Pl. Physiol.,* **5,** 553.
Bond, G. (1950), *Ann. Bot.,* **15,** 95.
Bond, G. (1951), *Ann. Bot.,* **15,** 447.
Bond, G. (1956a), *J. Exp. Bot.,* **7,** 387.
Bond, G. (1956b), *New Phytol.,* **55,** 147.
Bond, G. (1957), *Ann. Bot.,* **21,** 513.
Bond. G. (1960), *J. Exp. Bot.,* **11,** 91.
Bond, G. (1961), *Z. Allg. Mikrobiol.,* **1,** 93.

Bond, G. (1963), in *Symbiotic associations* (Eds. P. S. Nutman and B. Mosse), Symp. 13, Soc. Gen. Microbiol., Cambridge University Press, London, p. 72.
Bond, G. (1967), *Ann. Rev. Pl. Physiol.*, **18,** 107.
Bond, G. (1974), in *The biology of nitrogen fixation* (Ed. A. Quispel), North-Holland, Amsterdam, p. 342.
Bond, G. (1976a), in *Symbiotic nitrogen fixation in plants* (Ed. P. S. Nutman), International Biological Programme, Vol. 7, Cambridge University Press, Cambridge, London, New York, Melbourne, p. 443.
Bond, G. (1976b), *Proc. Roy. Soc. Lond., B,* **193,** 127.
Bond, G. and Mackintosh, A. H. (1975a), *Proc. Roy. Soc. Lond., B,* **192,** 1.
Bond, G. and Mackintosh, A. H. (1975b). *Proc. Roy. Soc. Lond., B,* **190,** 199.
Bond, G. and Scott, G. D. (1955), *Ann. Bot.*, **19,** 67.
Bond, G., Fletcher, W. W., and Ferguson, T. P. (1954), *Pl. Soil,* **5,** 309.
Bond, G., MacConnell, J. T., and McCallum, A. H. (1956), *Ann Bot.*, **20,** 501.
Bone, D. H. (1971), *Arch. Mikrobiol.*, **80,** 242.
Bone, D. H. (1972), *Arch. Mikrobiol.*, **86,** 13.
Bonnier, C. (1960), *Ann. Inst. Pasteur (Paris),* **98,** 537.
Booth, C. (Ed) (1971), *Methods in microbiology,* Vol. 4, Academic Press, London, New York.
Borkowski J. D. and Johnson, M. J. (1967), *Biotechnol. Bioeng.*, **9,** 635.
Bormann, F. H., Likens, G. E., and Mellillo, J. (1977), *Science,* **196,** 981.
Bornet, E. and Flahault, C. (1886–1888), *Revision des Nostocacées Heterocystées,* Reprinted 1959, J. Cramer, Weineim.
Bothe, H. (1970), *Ber. Dt. Bot. Ges.*, **83,** 421.
Bothe, H. and Yates, M. G. (1976), *Arch. Microbiol.*, **107, 25.**
Bothe, H., Distler, E., and Eisbrenner, G. (1978), *Biochimie,* **60,** 277.
Bothe, H., Tennigkeit, J., Eisbrenner, G., and Yates, M. G. (1977), *Planta,* **133,** 237.
Bottomley, P. J. and Stewart, W. D. P. (1977), *New Phytol.*, **79,** 625.
Bourelly, P. (1979), *Rev. Algol. N. S.*, **14,** 5.
Boussingault, J. B. (1853), *C. R. Acad. Sci., Paris,* **37,** 207.
Bowyer, J. W. and Skerman, V. B. D. (1968), *J. Gen. Microbiol.*, **54,** 299.
Boyer, J. S. (1976), in *Water deficits and plant growth,* (Ed. T. T. Kozlowski), Vol. IV, Academic Press, New York, p. 153.
Boyer, P. D., Mills, R. C., and Fromm, H. J. (1959), *Arch. Biochem. Biophys.*, **81,** 249.
Bremner, J. M. (1965), in *Methods of soil analysis,* (Ed. C. A. Black), Part 2, Chapters 83–86, American Society of Agronomy, Madison, Wisc., pp. 1149, 1179, 1238, 1256.
Bremner, J. M. (1975), in *Biological nitrogen fixation in farming systems of the tropics* (Eds. A. Ayanaba and P. J. Dart), Wiley, Chichester, New York, Brisbane, Toronto, p. 335.
Bremner, J. M. and Edwards, A. P. (1965), *Soil Sci. Soc. Am. Proc.*, **29,** 504.
Bremner, J. M., Cheng, H. H. and Edwards, A. P. (1965), in *Report FAO/IAEA Tech. Mtg., Brunswick-Volkenrode,* Pergamon Press, Oxford, p. 429.
Brenchley, J. E. (1973), *J. Bacteriol.*, **114,** 666.
Brenner, M. and Niederwieser, A. (1967), *Methods Enzymol.*, **11,** 39.
Brewer, G. J. and Sing, C. F. (1971), *An introduction to isozyme technique,* Academic Press, New York.
Bridges, B. A. (1966), *Lab. Pract.*, **15,** 418.
Brill, W. J. (1974). in *The biology of nitrogen fixation* (Ed. A. Quispel), North-Holland, Amsterdam, p. 639.

Brill, W. J., Steiner, A. L., and Shah, V. K. (1974), *J. Bacteriol.,* **118,** 986.
Brockwell, J. (1956), *J. Aust. Inst. Agric. Sci.,* **22,** 260.
Brockwell, J. (1958), *J. Aust. Inst. Agric. Sci.,* **24,** 342.
Brockwell, J. (1962), *Aust. J. Agric. Res.,* **13,** 638.
Brockwell, J. (1963), *Appl. Microbiol.,* **11,** 377.
Brockwell, J. (1971), *Aust. CSIRO Div. Plant Ind. Fld. Stn. Rec.,* **10,** 51.
Brockwell, J. (1977), in *A treatise on dinitrogen fixation, Part IV, agronomy and ecology* (Eds. R. W. F. Hardy and A. H. Gibson), Wiley, New York, p. 277.
Brockwell, J., Diatloff, A., Grassia, A., and Robinson, A. C. (1975a), *Soil Biol. Biochem.,* **7,** 305.
Brockwell, J., Diatloff, A. and Schwinghamer, E. A. (1978), in *Microbial. ecology* (Eds. M. W. Loutit and J. A. R. Miles), Springer-Verlag, Berlin, Heidelberg, New York, p. 390.
Brockwell, J. and Dudman, W. F. (1968), *Aust. J. Agric. Res.,* **19,** 749.
Brockwell, J., Dudman, W. F., Gibson, A. H., Hely, F. W. and Robinson, A. C. (1968), *Trans. 9th Int. Congr. Soil Sci. Soc.,* **2,** 103.
Brockwell, J., Herridge, D. F., Roughley, R. J., Thompson, J. A., and Gault, R. R. (1975b), *Aust. J. Exp. Agric. Anim. Husb.,* **15,** 780.
Brockwell, J. and Gault, R. R. (1976), *Aust. J. Exp. Agric. Anim. Husb.,* **16,** 500.
Brockwell, J., Gault, R. R., and Chase, D. L. (1977a), *Farmers' Newsl.,* **104,** 7 (Irrigation Research and Extension Committee, Griffith, N.S.W.).
Brockwell, J. and Hely, F. W. (1961), *Aust. J. Agric. Res.,* **12,** 630.
Brockwell, J. and Hely, F. W. (1966), *Aust. J. Agric. Res.,* **17,** 885.
Brockwell, J. and Katznelson, J. (1976), *Aust. J. Agric. Res.,* **27,** 799.
Brockwell, J. and Robinson, A. C. (1976), *Aust. CSIRO Div. Plant Ind. Fld. Stn. Rec.,* **15,** 15.
Brockwell, J., Schwinghamer, E. A., and Gault, R. R. (1977b), *Soil Biol. Biochem.,* **9,** 19.
Brockwell, J. and Whalley, R. D. B. (1962), *Aust. J. Sci.,* **24.** 458.
Broida, H. P. and Chapman, M. W. (1958), *Anal. Chem.,* **30,** 2049.
Brotonegoro, S. and Abdulkadir, S. (1976), *Ann. Bogor,* **6,** 69.
Brouzes, R. and Knowles, R. (1971), *Can. J. Microbiol.,* **17,** 1483.
Brouzes, R. and Knowles, R. (1973), *Soil Biol. Biochem.,* **5,** 223.
Brown, C. M. and Dilworth, M. J. (1975), *J. Gen. Microbiol.,* **86.** 39.
Bulen, W. A. (1976), in *Proc. 1st int. symp. nitrogen fixation,* (Eds. W. E. Newton and C. J. Nyman), Vol. 1, Washington State University Press, Pullman, Washington, p. 177.
Bulen, W. A., Burns, R. C., and LeComte, J. R. (1965), *Proc. Nat. Acad. Sci. USA,* **53,** 532.
Bulen, W. A., LeComte, J. R., Burns, R. C., and Hinkson, J. (1965), in *Non-heme iron proteins: role in energy conversion.* (Ed. A. San Pietro), Antioch Press, Yellow Springs, Ohio, p. 261.
Bulen, W. A. and LeCompte, J. R. (1972), *Methods Enzymol.,* **24,** 456.
Bulow, J. F. W. von and Döbereiner, J. (1975), *Proc. Nat. Acad. Sci. USA,* **72,** 2389.
Bunce, R. G. H. (1968), *J. Ecol.,* **56,** 759.
Bunt. J. S. (1961), *Nature, Lond.,* **192,** 1275.
Bunt, J. S., Cooksey, K. E., Heeb, M. A., Lee, C. C., and Taylor, B. F. (1970), *Nature, Lond.,* **227,** 1163.
Bunting, A. H. and Horrocks, J. (1964), *Ann. Bot.,* **28,** 229.
Burkill, I. H. (1966), *A dictionary of the economic products of the Malay Peninsula,* Ministry of Agriculture, Kuala Lumpur.

Burns, R. C. (1969), *Biochim. Biophys. Acta,* **171,** 253.
Burns, R. C., Holsten, R. D., and Hardy, R. W. F. (1970), *Biochem. Biophys. Res. Commun.,* **39,** 90.
Burns, R. C. and Hardy, R. W. F. (1972), *Methods Enzymol.,* **24,** 480.
Burns, R. C. and Hardy, R. W. F. (1975), *Nitrogen fixation in bacteria and higher plants,* Springer-Verlag, Berlin, Heidelberg, New York, 189 pp.
Burris, R. H. (1941), *Science,* **94,** 238.
Burris, R. H. (1942), *J. Biol. Chem.,* **143,** 509.
Burris, R. H. (1972), *Methods Enzymol.,* **24,** 415.
Burris, R. H. (1974), in *The biology of nitrogen fixation* (Ed. A. Quispel), North Holland, Amsterdam, p. 9.
Burris, R. H., Albrecht, S. L., and Okon, Y. (1978), in *Limitations and potentials for biological nitrogen fixation in the tropics.* (Eds. J. Döbereiner, R. H. Burris, and A. Hollaender), Plenum Press, New York, p. 303.
Burris, R. H. and Miller, C. E. (1941), *Science,* **93,** 114.
Burris, R. H., Eppling, F. J., Wahlin, H. B., and Wilson, P. W. (1942), *Soil Sci. Soc. Am. Proc.,* **7,** 258.
Burris, R. H., Eppling, F. J., Wahlin, H. B., and Wilson, P. W. (1943), *J. Biol. Chem.,* **148,** 349.
Burris, R. H. and Wilson, P. W. (1957), *Methods Enzymol.,* **4,** 355.
Burris, R. H. and Orme-Johnson, W. H. (1976), in *Microbial iron metabolism; a comprehensive treatise* (Ed. J. B. Neilands), Academic Press, London, New York, p. 187.
Burton, J. C. (1967), in *Microbial technology,* (Ed. H. J. Peppler), Reinhold, New York, p. 1.
Burton, J. C. (1976), in *Symbiotic nitrogen fixation in plants,* (Ed. P. S. Nutman), Cambridge University Press, Cambridge, London, New York, Melbourne, p. 175.
Burton, J. C. and Curley, R. L. (1965), *Argon. J.,* **57,** 379.
Burton, J. C., Martinez, C. J., and Curley, R. L. (1972), *Methods of testing and suggested standards for legume inoculants and preinoculated seed,* Nitragin Sales Corp., Milwaukee, Wisc.
Buttolph. T. J. (1955), in *Radiation biology, Vol. II, Ultraviolet and related radiations* (Ed. A. Hollaender), McGraw-Hill, New York, p. 41.
Calam, C. T. (1969), in *Methods in microbiology,* (Eds. J. R. Harris and D. W. Ribbons), Vol. I, Academic Press, London, New York, p. 255.
Caldwell, B. C. (1966), *Crop Sci.,* **6,** 427.
Caldwell, B. C., Hinson, K., and Johnson, H. W. (1966), *Crop Sci.,* **6,** 495.
Callaham, D., Tredici, P. D., and Torrey, J. G. (1978), *Science,* **199,** 899.
Calvert, H. E., Chaudhary, A. H., and Lalonde, M. (1979), in *Symbiotic Nitrogen Fixation in the Management of Temperate Forests* (Eds. J. C. Gordon, C. T. Wheeler, and D. A. Perry), Forest Research Laboratory, Oregon State University, Corvallis, Oregon, p. 474.
Cameron, R. E. and Blank, G. B. (1966), Jet Propulsion Laboratory, *Technical Report No. 32-971,* Pasadena, Calif., 41 pp.
Cameron, R. E. and Fuller, W. H. (1960), *Soil Sci. Soc. Am. Proc.,* **24,** 353.
Campbell, N. E. R. and Evans, H. J. (1969), *Can. J. Microbiol.,* **15,** 1342.
Campbell, D. H., Garvey, J. S., Cremer, N. E., and Sussdorf, D. H. (1970), *Methods in immunology,* 2nd Edition, W. A. Benjamin, New York.
Campelo, A. B. and Döbereiner, J. (1970), *Pesq. Agropec. Bras.,* **5,** 327.
Cañizo, A. and Rodríguez-Barrueco, C. (1976), *N. Z. J. Bot.,* **14,** 271.
Cannon, F. (1978), in *Genetic engineering*, (Eds. H. W. Boyer and S. Nicosia), Elsevier/North Holland, Biochemical Press, Amsterdam, p. 181.

Cannon, F. C., Dixon, R. A., Postgate, J. R., and Primrose, S. B. (1974), *J. Gen. Microbiol.*, **80,** 227.
Cannon, F. C., Dixon, R. A. and Postgate, J. R. (1976), *J. Gen. Microbiol.*, **93,** 111.
Cannon, F. C. and Postgate, J. R. (1976), *Nature, Lond.*, **260,** 271.
Cannon, F. C., Reidel, G. E., and Ausubel, F. M. (1979), *Molec. Gen. Genet.*, in press.
Carnahan, J. E., Mortenson, L. E., Mower, H. F., and Castle, J. E. (1960), *Biochim. Biophys. Acta,* **44,** 1520.
Carpenter, E. J. (1973), *Deep Sea Res.*, **20,** 285.
Carpenter, E. J. and Cox, J. L. (1974), *Limnol. Oceanogr.*, **19,** 429.
Carpenter, E. J. and McCarthy, J. J. (1975), *Limnol. Oceanogr.*, **20,** 389.
Carpenter, E. J. and Price, IV, C. C. (1976), *Science,* **191,** 1278.
Carr. D. J., Carr, S. G. M., and Papst, W. R. (1979), *Aust. J. Ecol.*, in press.
Carr, N. G. and Whitton, B. A. (1973), *The biology of blue-green algae*, Blackwell, Oxford.
Carter, K. R., Jennings, N. T., Hanus, J., and Evans, H. J. (1978), *Can. J. Microbiol.*, **24,** 307.
Cartwright, P. M. (1967), *Ann. Bot.*, **31,** 309.
Casse, F., Boucher, C., Julliot, J. S., Michel, M., and Denarié, J. (1979), *J. Gen. Microbiol.*, **113,** 229.
Castric, P. A., Farnden, K. J. F., and Conn, E. E. (1972), *Arch. Biochem. Biophys.*, **152,** 62.
Cedar, H. and Schwartz, J. H. (1969), *J. Biol. Chem.*, **244,** 4112.
Cengia-Sambo, M. (1926), *Atti Soc. Ital. Nat. Museo Storia Nat., Milano*, **64,** 191.
Chaney, A. L. and Marbach, E. P. (1962), *Clin. Chem.*, **8,** 130.
Chang, A. C. G. and Cohen, S. N. (1978), *J. Bacteriol.*, **134,** 1141.
Chapman, H. D., Liebig, G. F., and Rayner, D. S. (1949), *Hilgardia.*, **19,** 57.
Chatel, D. L. and Greenwood, R. M. (1973), *Soil Biol. Biochem.*, **5,** 433.
Chatel, D. L., Greenwood, R. M., and Parker, C. A. (1968), *Trans. 9th Int. Congr. Soil Sci. Soc.*, Vol. 2, p. 65.
Chatel, D. L. and Parker, C. A. (1973), *Soil Biol. Biochem.*, **5,** 415.
Chaudhuri, H. and Akhtar, A. R. (1931), *J. Indian Bot. Soc.*, **10,** 43.
Cheng, H. H., Bremner, J. M., and Edwards, A. P. (1964), *Science*, **146,** 1574.
Cheniae, G. and Evans, H. J. (1959), *Biochim. Biophys. Acta*, **35,** 140.
Child, J. J. and La Rue, T. A. (1974), *Pl. Physiol.*, **53,** 88.
Ching, T. M., Hedtke, S., and Newcombe, W. (1977), *Pl. Physiol.*, **60,** 771.
Chowdhury, M. S. (1965), *The growth and persistence of* Rhizobium lupini *in sandy soil*, PhD Thesis, University of Western Australia.
Christeller, J. T., Laing, W. A., and Sutton, W. D. (1977), *Pl. Physiol.*, **60,** 47.
Clark, M. F. and Adams, A. N. (1977), *J. Gen. Virol.*, **34,** 475.
Cloonan, M. J. (1963), *Aust. J. Sci.*, **26,** 121.
Cloonan, M. J. and Humphrey, B. (1976), *J. Appl. Bacteriol.*, **40,** 101.
Coakley, W. T., Bater, A. J., and Lloyd, D. (1977), *Microb. Physiol.*, **16,** 279.
Cobb, H. D. and Myers, J. (1962), *Pl. Physiol.*, Suppl., **37,** VII.
Codd, G. A. and Stewart, W. D. P. (1977a), *FEMS Microbiol. Lett.*, **2,** 247.
Codd, G. A. and Stewart, W. D. P. (1977b), *Arch. Microbiol.*, **113,** 105.
Codd, G. A., Okabe, K., and Stewart, W. D. P. (1980), *Arch. Microbiol.*, submitted for publication.
Cohn, F. (1853), *Nova Acta Acad. Caesar Leop. Carol*, **24,** 103.
Cohn, F. (1871–1872), *Jb. Schles. Ges. Vaterl. Kult.*, **49,** 83.
Cole, H. A., Wimpenny, J. W. T., and Hughes, D. E. (1967), *Biochim. Biophys. Acta*, **143,** 445.

Colowick, S. P. and Kaplan, N. O. (Eds.) (1955–80), *Methods Enzymol.*, **1–69**.
Conway, E. J. (1950), *Microdiffusion analysis and volumetric error*, 3rd Edition Crosby, Lockwood and Sons, London.
Copeland, J. J. (1932), *Am. J. Bot.*, **19,** 844.
Cooper, R. (1962), *J. Appl. Bacteriol.*, **25,** 232.
Cooper, T. G. (1977), *The tools of biochemistry*, Wiley–Interscience, New York, London, p.1.
Corbin, E. J., Brockwell, J., and Gault, R. R. (1977), *Aust. J. Exp. Agric. Anim. Husb.*, **17,** 126.
Corby, H. D. L. (1967), *Proc. Grassld. Soc. S. Afr.*, **2,** 75.
Corby, H. D. L. (1976), in *Symbiotic nitrogen fixation in plants* (Ed. P. S. Nutman), Cambridge University Press, Cambridge, London, New York, Melbourne, p. 169.
Corby, H. D. L. (1970), in *A manual for the practical study of root-nodule bacteria*, IBP Handbook No. 15, Blackwell, Oxford, p. 150.
Coty, V. F. (1967), *Biotechnol. Bioeng.*, **9,** 25.
Courtois, B., Hornez, J. P., and Derieux, J. C. (1979), *Can. J. Microbiol.*, **25,** 1191.
Cowan, J. R. (1972), in *Seed biology, Vol. 3, Insects and seed collection, storage, testing and certification* (Ed. T. T. Kozlowski), Academic Press, New York, p. 371.
Cox, G. F. (1969), *Methods Enzymol.*, **13,** 47.
Cox, E. (1976), *Ann. Rev. Genet.*, **10,** 135.
Cox, R. M. (1966), *Arch. Mikrobiol.*, **53,** 263.
Cox, R. M. and Fay, P. (1969), *Proc. Roy. Soc. Lond. B*, **172,** 357.
Cramer, S. P., Hodgson, K. O., Gillum, W. O., and Mortenson, L. E. (1978a), *J. Am. Chem. Soc.*, **100,** 3398.
Cramer, S. P., Gillum, W. O., Hodgson, K. O., Mortenson, L. E., Stiefel, E. I., Chisnell, J. R., Brill, W. J., and Shah, V. K. (1978b), *J. Am. Chem. Soc.*, **100,** 3814.
Crocker, R. L. and Dickson, B. A. (1957), *J. Ecol.*, **45,** 169.
Crocker, R. L. and Major, J. (1955), *J. Ecol.*, **43,** 427.
Crone, Von der (1902), *Sber. Niederrhein. Ges. Bonn*, A167.
Crowle, A. J. (1973), *Immunodiffusion*, 2nd Edition, Academic Press, New York.
Cruickshank, R. (1965), *Medical microbiology: a guide to the laboratory diagnosis and control of infection*, 11th Edition, E. and S. Livingstone, Edinburgh.
Cullimore, D. R. and McCann, A. E. (1977), *Pl. Soil*, **46,** 499.
Daday, A., Platz, R. A., and Smith, G. D. (1977), *Appl. Environ. Microbiol.*, **34,** 478.
Dalton, H. (1974), *Crit. Rev. Microbiol.*, **3,** 183.
Dalton, H. and Mortsenson, L. E. (1972), *Bacteriol. Rev.*, **36,** 231.
Dalton, H., Morris, J. A., Ward, M. A., and Mortenson, L. E. (1971), *Eur. J. Biochem.*, **10,** 2066.
Dalton, H. and Postgate, J. R. (1969a), *J. Gen. Microbiol.*, **56,** 307.
Dalton, H. and Postgate, J. R. (1969b), *J. Gen. Microbiol.*, **54,** 463.
Dalton, H. and Whittenbury, R. (1976), *Arch. Microbiol.*, **109,** 147.
Dalton, D. A. and Zobell, D. B. (1977), *Pl. Soil*, **48,** 57.
Daniel, R. M. and Appleby, C. A. (1972), *Biochim. Biophys. Acta*, **275,** 347.
Dao, The Tuan and Tran, Quang Thuyet (1979), in *Nitrogen and rice symposium*, IRRI, Los Baños, Laguna, Philippines, p. 395.
Dart, P. J. (1977), in *A treatise on dinitrogen fixation, Section III, Biology*, (Ed. R. W. F. Hardy and W. S. Silver), Wiley–Interscience, New York, London, Sydney, Toronto, p. 367.
Dart, P. J. and Day, J. (1971), in *Biological nitrogen fixation in natural and agricultural habitats*, (Ed. T. A. Lie and E. G. Mulder), *Pl. Soil*, Special Volume, p. 167.

Dart, P. J., Day, J. M., and Harris, D. (1972), *Use of isotopes for study of fertilizer utilisation by legume crops*, International Atomic Energy Agency Technical Report 149, p. 85.
Dart, P., Day, J., Islam, R., and Döbereiner, J. (1976), in *Symbiotic nitrogen fixation in plants* (Ed. P. S. Nutman), Cambridge University Press, Cambridge, London, New York, Melbourne, p. 361.
Dasilva, E. J., Henriksson, L. E., and Henriksson, E. (1975), *Arch. Environ. Contam. Toxicol.*, **3,** 193.
Date, R. A. (1969), *J. Aust. Inst. Agric. Sci.*, **35,** 27.
Date, R. A. (1970), *Pl. Soil*, **32,** 703.
Date, R. A. (1972), *J. Appl. Bacteriol.*, **35,** 379.
Date, R. A. (1974), *Proc. Indian Nat. Sci. Acad.*, **40,** 667.
Date, R. A. (1976), in *Symbiotic nitrogen fixation in plants* (Ed. P. S. Nutman), Cambridge University Press, Cambridge, London, New York, Melbourne, p. 137.
Date, R. A. and Roughley, R. J. (1977), in *A treatise on dinitrogen fixation, Section IV, Agronomy and ecology* (Eds. R. W. F. Hardy and A. H. Gibson), Wiley, New York, London, Sydney, Toronto, p. 243.
Date, R. A. and Vincent, J. M. (1962), *Aust. J. Exp. Agric. Anim. Husb.*, **2,** 5.
David, K. A. V. and Fay, P. (1977), *Appl. Environ. Microbiol.*, **34,** 640.
David, K. A. V., Apte, S. K., and Thomas, J. (1978), *Biochem. Biophys. Res. Commun.*, **82,** 39.
Davies, S. N. (1951), *J. Gen. Microbiol.*, **5,** 807.
Davies, D. D. and Teixeira, A. N. (1975), *Phytochemistry*, **14,** 647.
Davis, J. B., Coty, V. F., and Stanley, J. P. (1964), *J. Bacteriol.*, **88,** 468.
Davis, L. C. and Orme-Johnson, W. H. (1976), *Biochim. Biophys. Acta*, **452,** 42.
Davis, L. C., Shah, V. K., and Brill, W. J. (1975), *Biochim. Biophys. Acta*, **403,** 67.
Day, E. D. (1972), *Advanced immunochemistry*, Williams and Wilkins, Baltimore.
Day, J. M., Harris, D., Dart, P. J., and Berkum, P. van (1975), in *Nitrogen fixation by free living micro-organisms* (Ed. W. D. P. Stewart), Cambridge University Press, Cambridge, London, New York, Melbourne, p. 71.
Day, J. M., Neves, M. C. P., and Döbereiner, J. (1976), *Soil Biol. Biochem.*, **7,** 107.
Dazzo, F. B. and Milam, J. R. (1975), *Soil Crop Sci. Soc. Fla. Proc.*, **35,** 121.
De, P. K. (1936), *Indian J. Agric. Sci.*, **6,** 1237.
De, P. K. (1939), *Proc. Roy. Soc. Lond. B*, **127,** 121.
De, P. K. and Mandel, L. N. (1956), *Soil Sci.*, **81,** 453.
De Bont, J. A. M. (1976), *Can. J. Microbiol.*, **22,** 1060.
De Bont, J. A. M. and Mulder, E. G. (1974), *J. Gen. Microbiol.*, **83,** 113.
De Bont, J. A. M. and Mulder, E. G. (1976), *Appl. Environ. Microbiol.*, **31,** 640.
De Bont, J. A. M. and Leijten, M. W. M. (1976), *Arch. Microbiol.*, **107,** 235.
Decker, L. E. and Rau, E. M. (1963), *Proc. Soc. Exp. Biol. Med.*, **112,** 144.
De Lacy, I. H. and Britten, E. J. (1977), *Proc. 3rd Int. Congr. Society for the Advancement of Breeding Researches in Asia and Oceania (SABRAO)*, Vol. 2, 14(b)–49.
Delwiche, C. C. and Steyn, P. L. (1970), *Environ. Sci. Technol.*, **4,** 929.
Delwiche, C. C., Zinke, P. J., and Johnson, C. M. (1965), *Pl. Physiol.*, **40,** 1045.
Delwiche, C. C., Zinke, P. J., Johnson, C. M., and Virginia, R. A. (1979), *Bot. Gaz.*, **140,** (suppl.), 565.
Dénarié, J., Rosenberg, C., Bergeron, B., Boucher, C., Michel, M., and Barate de Bertalmio, M. (1977), in *DNA insertion elements, plasmids, and episomes* (Eds. A. I. Bukhari, J. A. Shapiro, and S. L. Adhya), Cold Spring Harbor Laboratory, New York, p. 507.

De-Polli, H. (1975), *Ocorrencia de fixacão de N_2 nas gramineas tropicais Digitaria decumbens e Paspalum notatum*, MSc Thesis, ESALQ, Piracicaba, SP, Brazil.
De-Polli, H., Matsui, E., Döbereiner, J., and Salati, E. (1977), *Soil Biol. Biochem.*, **9,** 119.
Deschodt, C. C. and Strijdom, B. W. (1974), *Phytophylactica*, **6,** 229.
Desikachary, T. V. (1959), *Cyanophyta*, Indian Council of Agricultural Research, New Delhi.
Determann, H. (1968), *Gel chromatography*, Springer-Verlag, New York.
Devine, T. E. and Reisinger, W. W. (1978), *Agron. J.*, **70,** 510.
Dharmawardene, M. W. M., Haystead, A., and Stewart, W. D. P. (1973), *Arch. Mikrobiol.*, **90,** 281.
Diatloff, A. (1965), *J. Entomol. Soc. Queensl.*, **4,** 86.
Diatloff, A. (1970), *Queensl. J. Agric. Anim. Sci.*, **27,** 279.
Diatloff, A. (1972), *Aust. Plant Pathol. Soc. Newsl.*, **1,** 28.
Diatloff, A. (1977), *Soil Biol. Biochem.*, **9,** 85.
Diatloff, A. and Brockwell, J. (1976), *Aust. J. Exp. Agric. Anim. Husb.*, **16,** 514.
Dick, H. L. N. and Stewart, W. D. P. (1980), *Arch. Microbiol.*, submitted for publication.
Dickson, B. A. and Crocker, R. L. (1953), *J. Soil Sci.*, **4,** 142.
Dilworth, M. J. (1966), *Biochim. Biophys. Acta*, **127,** 285.
Dilworth, M. J. (1974), *Ann. Rev. Pl. Physiol.*, **25,** 81.
Dilworth, M. J. (1980), *Methods Enzymol.*, **69c,** in press.
Dixon, M. (1971), *Biochim. Biophys. Acta*, **226,** 241.
Dixon, R. A., Cannon, F. C., and Kondorosi, A. (1976), *Nature, Lond.*, **260,** 268.
Dixon, R., Kennedy, C., Kondorosi, A., Krishnapillai, V., and Merrick, M. (1977), *Molec. Gen. Genet.*, **157,** 189.
Dixon, R. A. and Postgate, J. R. (1976), *Nature, Lond.*, **237,** 102.
Dixon, R. O. D. (1968), *Arch. Mikrobiol.*, **62,** 272.
Dixon, R. O. D. (1972), *Arch. Mikrobiol.*, **85,** 193.
Döbereiner, J. (1961), *Pl. Soil*, **14,** 211.
Döbereiner, J. (1965), *Soil Biol. Int. News Bull.*, No. 2.
Döbereiner, J. (1966), *Pesq. Agropec. Bras.*, **1,** 357.
Döbereiner, J. (1970), *Zentralbl. Bakteriol. Parasitenkd. Infektionskr. Hyg., Abt. II*, **124,** 224.
Dőbereiner, J. (1977), in *Recent developments in nitrogen fixation* (Eds. W. Newton, J. R. Postgate, and C. Rodríguez-Barrueco), Academic Press, London, New York, San Francisco, p. 513.
Döbereiner, J. (1978), *Ecol. Bull., Stockholm*, **26,** 343.
Döbereiner, J. and Day, J. M. (1975), in *Nitrogen fixation by free living micro-organisms*, (Ed. W. D. P. Stewart), Cambridge University Press, Cambridge, London, New York, Melbourne, p. 39.
Döbereiner, J. and Day, J. M. (1976), in *Proc. 1st int. symp. nitrogen fixation* (Eds. W. E. Newton and C. J. Nyman), Washington State University Press, Pullman, Washington, p. 518.
Döbereiner, J., Day, J. M., and Dart, P. J. (1972a), *Pl. Soil*, **37,** 191.
Döbereiner, J., Day, J. M., and Dart, P. J. (1972b), *J. Gen. Microbiol.*, **71,** 103.
Döbereiner, J., Nery, M., and Marriel, I. E. (1976), *Can. J. Microbiol.*, **22,** 1464.
Döbereiner, J. and Ruschel, A. P. (1958), *Rev. Biol. (Lisboa)*, **1,** 261.
Doku, E. V. (1970), *Exp. Agric.*, **6,** 13.
Dommergues, Y. R., Balandreau, J., Rinaudo, G., and Weinhard, P. (1972), *Soil Biol. Biochem.*, **5,** 83.

Dommergues, Y. R., Diem, H. G., and Divies, C. (1979), *Appl. env. Microbiol.*, **37,** 778.
Dommergues, Y. R. and Rinaudo, G. (1979), in *Nitrogen and rice symposium*, IRRI, Los Baños, Laguna, Philippines, in press.
Drake, V. C. (1947), *Proc. Ass. Off. Seed Anal.*, **37,** 143.
Drake, J. W. (1970), *The molecular basis of mutation*, Holden-Day, San Francisco.
Drew, E. A. and Anderson, J. R. (1977), *Soil Biol. Biochem.*, **9,** 207.
Drewes, K. (1928), *Zentralbl. Bakteriol. Parasitenkd. Infektionster. Hyg., Abt. II*, **76,** 88.
Dreyer, W. J. and Bynum, E. (1967), *Methods Enzymol.*, **11,** 32.
Drozd, J. W. and Postgate, J. R. (1970a), *J. Gen. Microbiol.*, **60,** 427.
Drozd, J. W. and Postgate, J. R. (1970b), *J. Gen. Microbiol.*, **63,** 63.
Drozd, J. W., Tubb, R. S., and Postgate, J. R. (1972), *J. Gen. Microbiol.*, **73,** 221.
Drucker, D. B. (1976), in *Methods in microbiology* (Ed. J. R. Norris), Vol. 9, Academic Press, New York, p. 51.
Duckett, J. G., Prasad, A. K. S. K., Davies, D. A., and Walker, S. (1977), *New Phytol.*, **79,** 349.
Dudman, W. F. (1964), *J. Bacteriol.*, **88,** 782.
Dudman, W. F. (1971), *Appl. Microbiol.*, **21,** 973.
Dudman, W. F. (1977), in *A treatise on dinitrogen fixation, Section IV, Agronomy and ecology* (Eds. R. W. F. Hardy and A. H. Gibson), Wiley–Interscience, New York, p. 487.
Dudman, W. F. and Brockwell, J. (1968), *Aust. J. Agric. Res.*, **19,** 739.
Dugdale, R. C., Menzel, D. W., and Ryther, J. H. (1961), *Deep Sea Res.*, **7,** 298.
Dugdale, R. C., Goering, J. J., and Ryther, J. H. (1964), *Limnol. Oceanogr.*, **9,** 507.
Duguid, J. P. (1968), *Arch. Immunol. Ther. Exp.*, **16,** 175.
Duke, S. H. and Ham, G. E. (1976), *Plant Cell Physiol.*, **17,** 1037.
Dulley, J. R. (1975), *Anal. Biochem.*, **67,** 91.
Dulley, J. R. and Grieve, P. A. (1975), *Anal. Biochem.*, **64,** 136.
Dumas, J. B. (1834), *J. Pharm. Sci. Ass.*, **20,** 129.
Duncan, U. K. (1962), *Trans. Bot. Soc. Edinb.*, **39,** 290.
Duncan, U. K. (1970), *Introduction to British lichens*, T. Buncle and Co., Arbroath.
Eady, R. R. (1977), in *The evolution of metalloenzymes, metalloproteins and related materials* (Ed. G. J. Leigh), Symposium Press, London, p. 67.
Eady, R. R., Issack, R., Kennedy, C., Postgate, J. R., and Ratcliffe, H. (1978), *J. Gen. Microbiol.*, **104**, 277.
Eady, R. R. and Postgate, J. R. (1974), *Nature, Lond.*, **249,** 805.
Eady, R. R., Lowe, D. J., and Thorneley, R. N. F. (1978), *FEBS Lett.*, **95,** 211.
Eady, R. R., Smith, B. E., Cook, K. A., and Postgate, J. R. (1972), *Biochem. J.*, **128,** 655.
Eady, R. R. and Smith, B. E. (1978), in *A treatise on dinitrogen fixation, Section II, physicochemical properties of nitrogenase and its components in dinitrogen fixation* (Eds. R. W. F. Hardy, F. Bottomly, and R. C. Burns), Wiley–Interscience, New York, London, Sydney, Toronto, p. 399.
Eberly, W. R. (1967), in *Environmental requirements of blue-green algae* (Ed. A. F. Bartsch), U.S. Dept. of the Interior, Pacific Northwest Water Laboratory, Corvallis, Oregon, p. 7.
Edmeades, D. C. and Goh, K. M. (1979), *Commun. Soil Sci. Pl. Anal.*, **10,** 513.
Eigner, E. A. and Loftfield, R. B. (1974), *Methods Enzymol.*, **29,** 601.
Ellfolk, N. (1972), *Endeavour*, **31,** 139.
Ellfolk, N. and Sievers, G. (1971), *Acta Chem. Scand.*, **25,** 3532.

Elliott, W. H. (1955), *Methods Enzymol.*, **2,** 337.
Elliott, L. F. and Blaylock, J. W. (1971), *Proc. Soil Sci. Am.*, **35,** 158.
Elmes, R. P. T. (1975), *Papua N.G. Dept. Agric. Stock Fisheries Bull. No. 13*, p. 61.
Elmerich, C., Houmard, J., Sibald, L., Marheimer, I., and Charpin, N. (1978), *Molec. Gen. Genet.*, **165,** 181.
Emerich, D. W. (1978), *Diss. Abstr.*, **38,** 2480B.
Emerich, D. W. and Burris, R. H. (1978), *Methods Enzymol.*, **53,** 314.
Emerich, D. W., Ruiz-Arguesco, T., Ching, T. M., and Evans, H. J. (1979), *J. Bacteriol.*, **137,** 153.
Englund, B. (1977), *Physiol. Pl.*, **41,** 298.
Englund, B. and Meyerson, H. (1974), *Oikos*, **25,** 283.
Engvall, E. and Perlmann, P. (1972), *J. Immunol.*, **109,** 129.
Erdman, L. W., Johnson, H. W., and Clark, F. E. (1957a), *Plant Dis. Rep.*, **40,** 646.
Erdman, L. W., Johnson, H. W., and Clark, F. E. (1957b), *Agron. J.*, **49,** 267.
Ertola, R. J., Mazza, L. A., Balatti, A. P., Cuevas, C. M., and Daguerre, D. (1969), *Soil Sci.*, **108,** 373.
Eskew, D. L. and Ting, I. P. (1977), *Pl. Sci. Lett.*, **8,** 327.
Evans, E. E. (1957), in *Manual of microbiological methods*, Society of American Bacteriologists, McGraw-Hill, New York, p. 199.
Evans, H. J. and Barber, L. E. (1977), *Science*, **197,** 332.
Evans, H. J., Campbell, N. E. R., and Hill, S. (1972), *Can. J. Microbiol.*, **18,** 13.
Evans, H. J., Koch, B., and Klucas, R. (1972), *Methods Enzymol.*, **24,** 470.
Evans, L. T. and King, R. W. (1975), in *Report of the Technical Advisory committee working group on the biology of yield of grain legumes*, TAC Secretariat, FAO, Rome, 17 pp.
Evans, M. C. W. and Albrecht, S. L. (1974), *Biochem. Biophys. Res. Commun.*, **61,** 1187.
Evans, M. C. W., Telfer, A., and Smith, R. V. (1973), *Biochim. Biophys. Acta*, **310,** 344.
Ewart, A. J. (1908), *Proc. Roy. Soc. Victoria (N.S.)*, **21,** 1.
Ewing, W. H. and Fife, M. A. (1972), *Int. J. Syst. Bacteriol.*, **22,** 4.
Fåhraeus, G. (1957), *J. Gen. Microbiol.*, **16,** 374.
Farnsworth, R. B. (1976), *Nitrogen fixation in shrubs*, Report circulated privately from Dept. of Agronomy, Brigham Young University, Provo, Utah, U.S.A.
Fasold, H., Crundlack, G., and Turba, F. (1961), in *Chromatography* (Ed. E. Hettmann), Reinhold, New York, and Chapman and Hall, London, p. 378.
Fawcett, R. G. and Collis-George, N. (1967), *Aust. J. exp. Agric. Anim. Husb.*, **25,** 162.
Fawcette, C. P., Ciotti, M. M., and Kaplan, N. O. (1961), *Biochim. Biophys. Acta*, **54,** 210.
Fay, P. and Vasconcelos, L. (1974), *Arch. Microbiol.*, **99,** 221.
Fay, P. and Lang, N. J. (1971), *Proc. Roy. Soc. Lond, B*, **178,** 185.
Fay, P. and Walsby, A. E. (1966), *Nature, Lond.*, **209,** 94.
Fay, P., Stewart, W. D. P., Walsby, A. E., and Fogg, G. E. (1968), *Nature, Lond.*, **220,** 810.
Ferguson, T. P. and Bond, G. (1953), *Ann. Bot.*, **17,** 175.
Ferry, B. W., Baddeley, M. S., and Hawksworth, D. L. (1973), *Air pollution and lichens*, Athlone Press, London.
Fessenden, R. J., Knowles, R., and Brouzes, R. (1973), *Proc. Soil Sci. Soc. Am.*, **37,** 893.
Fishbeck, K., Evans, H. J., and Boersma, L. L. (1973), *Agron. J.*, **65,** 429.

Fisher, D. W., Gambell, A. W., Likens, G. E., and Bormann, F. H. (1968), *Water Resour. Res.*, **4,** 1115.
Fisher, R. A. (1925), *Statistical methods for research workers*, Oliver and Boyd, London.
Fisher, R. A. and Yates, F. (1963), *Statistical tables for biological, agricultural and medical research*, Oliver and Boyd, London.
Fisher, R. J. and Brill, W. J. (1969), *Biochim. Biophys. Acta*, **184,** 99.
Fisher, R. J. and Wilson, P. W. (1970), *Biochem. J.*, **117,** 1023.
Fleming, H. and Haselkorn, R. (1973), *Proc. Nat. Acad. Sci. U.S.A.*, **70,** 2727.
Flett, R. J., Hamilton, R. D., and Campbell, N. E. R. (1976), *Can. J. Microbiol.*, **21,** 43.
Fogg, G. E. (1942), *J. Exp. Biol.*, **19,** 78.
Fogg, G. E. (1949), *Ann. Bot.*, **8,** 241.
Fogg, G. E. and Stewart, W. D. P. (1968), *Brit. Antarct. Surv. Bull.*, **15,** 39.
Fogg, G. E., Stewart, W. D. P., Fay, P., and Walsby, A. E. (1973), *The blue-green algae*, Academic Press, London, New York.
Fottrell, P. F. and O'Hora, A. (1969), *J. Gen. Microbiol.*, **57,** 287.
Francis, G. E., Mulligan, W., and Wormall, A. (1959), *Isotopic tracers*, Athlone Press, London.
Francis, A. E. and Rippon, J. E. (1949), *J. Gen. Microbiol.*, **3,** 425.
Franco, A. A. and Vincent, J. M. (1976), *Pl. Soil*, **45,** 27.
Frank, B. (1889), *Ber. Dt. Bot. Ges.*, **7,** 34.
Fraser, M. E. (1966), *J. Appl. Bacteriol.*, **29,** 587.
Fraser, M. E. (1975), *J. Appl. Bacteriol.*, **39,** 345.
Fred, E. B., Baldwin, I. L., and McCoy, E. (1932), *Root nodule bacteria and leguminous plants*, University of Wisconsin Studies in Science, No. 5, University of Wisconsin, Madison, Wisc.
Fredricks, W. W. and Gehl, J. M. (1976), *Arch. Biochem. Biophys.*, **174,** 666.
Freese, E. (1971), in *Chemical mutagens: principles and methods for their detection* (Ed. A. Hollaender), Vol. 1, Plenum Press, New York, London, p. 1.
Frémy, P. (1929), *Archs. Bot., Caen. Mem.*, No. 2.
Frémy, P. (1934), *Mem. Soc. Nat. Sci. Nat. Math., Cherbourg*, **41,** 1.
French, J. R. J., Turner, G. L., and Bradbury, J. F. (1976), *J. Gen. Microbiol.*, **95,** 202.
Freney, J. R. and Wetselaar, R. (1967), *The determination of mineral nitrogen in soil*, CSIRO Australia, Division of Plant Industry Technical Paper No. 23.
Fried, M. and Broeshart, H. (1975), *Pl. Soil*, **43,** 707.
Fuchsman, W. H. and Appleby, C. A. (1979a), *Biochim. Biophys. Acta*, **579,** 314.
Fuchsman, W. H. and Appleby, C. A. (1979b), *Anal. Biochem.*, **18,** 1309.
Fuchsman, W. H. and Hardy, R. W. F. (1972), *Bioinorg. Chem.*, **1,** 195.
Fujita, Y. and Myers, J. (1965), *Arch. Biochem. Biophys.*, **111,** 619.
Gadd, G. M., Rowell, P., and Stewart, W. D. P. (1980), in preparation.
Gallacher, A. E. and Sprent, J. I. (1978), *J. Exp. Bot.*, **29,** 413.
Gallon, J. R., La Rue, T. A., and Kurz, W. G. W. (1972), *Can. J. Microbiol.*, **18,** 327.
Gallon, J. R., Kurz, W. G. W., and La Rue, T. A. (1973), *Can. J. Microbiol.*, **19,** 461.
Gallon, J. R., Kurz, W. G. W., and La Rue, T. A. (1975a), in *Nitrogen fixation by free living micro-organisms* (Ed. W. D. P. Stewart), Cambridge University Press, Cambridge, London, New York, Melbourne, p. 159.
Gallon, J. R., La Rue, T. A., and Kurz, W. G. W. (1975b), *Phytochemistry*, **14,** 861.

Gallon, J. R., Ul-Haque, M. I., and Chaplin, A. E. (1978), *J. Gen. Microbiol.*, **106,** 329.
Gardner, Isobel, C. (1976), in *Symbiotic nitrogen fixation in plants* (Ed. P. S. Nutman), International Biological Programme, Vol. 7, Cambridge University Press, Cambridge, London, New York, Melbourne, p. 485.
Gates, C. T. (1974), *Aust. J. Bot.*, **22,** 45.
Gault, R. R., Byrne, P. T., and Brockwell, J. (1973), *Lab. Pract.*, **22,** 292.
Gault, R. R., Dale, A. and Brockwell, J. (1980), *Lab. Pract.*, **29,** 265.
Gault, R. R., Thomas, R. C., Lemon, G. J., and Brockwell, J. (1977), *Lab. Pract.*, **26,** 403.
Geitler, L. (1932), in *Rabenhorst's kryptogamenflora von Deutschland, Österreich und der Schweiz,* Bd., **14,** (Ed. R. Kolkwitz), Akademische Verlagsgesellschaft, Leipzig. Reprinted 1971, Johnson Reprint Coorp., New York, London, 1196 pp.
Geitler, L. (1936), in *Handbuch der pflanzenanatomie* (Eds. W. Zimmermann and O. Ozenda), Vol. 6, Part 2, Bornträger, Berlin.
Geitler, L. (1942), in *Die natürlichen pflanzenfamilien* (Eds. A. Engler and K. Prantl), 2nd Edition, Vol. 1b, p. 1.
Gerloff, G. C., Fitzgerald, G. P., and Skoog, F. (1950), *Am. J. Bot.*, **37,** 216.
Gibson, A. H. (1964), *Aust. J. Agric. Res.*, **15,** 37.
Gibson, A. H. (1965), *Aust. J. Biol. Sci.*, **18,** 295.
Gibson, A. H. (1966a), *Aust. J. Biol. Sci.*, **19,** 219.
Gibson, A. H. (1966b), *Aust. J. Biol. Sci.*, **19,** 499.
Gibson, A. H. (1967a), *Aust. J. Biol. Sci.*, **20,** 1087.
Gibson, A. H. (1967b), *Aust. J. Biol. Sci.*, **20,** 1105.
Gibson, A. H. (1967c), *Aust. J. Biol. Sci.*, **20,** 837.
Gibson, A. H. (1968), *Aust. J. Agric. Res.*, **19,** 907.
Gibson, A. H. (1969), *Aust. J. Biol. Sci.*, **22,** 829.
Gibson, A. H. (1971), in *Biological nitrogen fixation in natural and agricultural habitats* (Eds. T. A. Lie and E. G. Mulder), *Pl. Soil*, Special Volume, p. 139.
Gibson, A. H. (1976a), in *Proc. 1st int. symp. nitrogen fixation* (Eds. W. E. Newton and C. J. Nyman), Washington State University Press, Pullman, Washington, p. 400.
Gibson, A. H. (1976b), in *Symbiotic nitrogen fixation in plants* (Ed. P. S. Nutman), Cambridge University Press, Cambridge, London, New York, Melbourne. p. 385.
Gibson, A. H. (1977), in *A treatise on dinitrogen fixation, Section IV, agronomy and ecology* (Eds. R. W. F. Hardy and A. H. Gibson), Wiley, New York, London, Sydney, Toronto, p. 393.
Gibson, A. H. and Brockwell, J. (1968), *Aust. J. Agric. Res.*, **19,** 891.
Gibson, A. H. and Nutman, P. S. (1960), *Ann. Bot.*, **24,** 420.
Gibson, A. H. and Pagan, J. D. (1977), *Planta (Berl.)*, **134,** 17.
Gibson, A. H., Curnow, B. C., Bergersen, F. J., Brockwell, J., and Robinson, A. C. (1975), *Soil Biol. Biochem.*, **7,** 95.
Gibson, A. H., Date, R. A., Ireland, J. A., and Brockwell, J. (1976a), *Soil Biol. Biochem.*, **8,** 395.
Gibson, A. H., Scowcroft, W. R., Child, J. J., and Pagan, J. D. (1976b), *Arch. Microbiol.*, **108,** 45.
Gibson, A. H., Scowcroft, W. R., and Pagan, J. D. (1977), in *Recent developments in nitrogen fixation,* (Eds. W. Newton, J. R. Postgate, and C. Rodríguez-Barrueco), Academic Press, London, New York, San Francisco, p. 387.
Ginsburg, A. and Stadtman, E. R. (1973), in *The enzymes of glutamine metabolism* (Ed. S. Prusiner and E. R. Stadtman), Academic Press, New York, p. 9.

Gladstones, J. S. (1970), *Fld. Crop Abstr.*, **23,** 123.
Glick, D. (Ed.) (1954–64), *Methods of biochemical analysis,* 11 Vols., Wiley–Interscience, New York, London.
Glynn, L. E. and Steward, M. W. (1978), *Immunochemistry: an advanced textbook,* Wiley, Chichester.
Goa, J. (1953), *Scand. J. Clin. Lab. Invest.,* **5,** 218.
Godfrey, C. A., Coventry, D. R., and Dilworth, M. J. (1975), in *Nitrogen fixation by free living micro-organisms* (Ed. W. D. P. Stewart), Cambridge University Press, Cambridge, London, New York, Melbourne, p. 311.
Goering, J. J., Dugdale, R. C., and Menzel, D. W. (1966), *Limnol. Oceanogr.,* **11,** 614.
Gogotov, J. N. and Schlegel, H. G. (1974), *Arch Microbiol.,* **97,** 359.
Goh, K. M., Edmeades, D. C., and Robinson, B. W. (1978), *Soil Biol. Biochem.,* **10,** 13.
Goldberg, R. B., Bender, R. A., and Streicher, S. L. (1974), *J. Bacteriol.,* **118,** 810.
Gomont, M. (1892), *Ann. Sci. Nat. (Bot.),* **15,** 263; **16,** 91.
Goodchild, D. J. (1977), *International review of cytology,* Suppl. No. 6, Academic Press, New York, p. 235.
Goodlass, G. and Smith, K. A. (1978), *Soil Biol. Biochem.,* **10,** 193.
Gopal, G. (1967), *Trop. Ecol.,* **8,** 126.
Gordon, J. C. and Wheeler, C. T. (1978), *New Phytol.,* **80,** 179.
Gorham, P. R., McLachlan, J. S., Hammer, U. T., and Kim, W. K. (1964), *Verh. Int. Ver. Theor. Angew. Limnol.,* **15,** 796.
Gozzer, C., Zanetti, G., Galliano, M., Sacchi, G. A., Michiotti, L., and Curti, B. (1977), *Biochim. Biophys. Acta,* **485,** 278.
Graham, P. H. (1964), *Antonie van Leeuwenhoek J. Microbiol. Serol.,* **30,** 68.
Graham, P. H. (1969), *Appl. Microbiol.,* **17,** 769.
Graham, P. H. and Parker, C. A. (1964), *Pl. Soil,* **20,** 383.
Granhall, U. and Lid-Torsvik, V. (1975), in *Fennoscandian tundra ecosystems* (Ed. F. E. Wilgolaski), Springer-Verlag, Berlin, p. 305.
Granhall, U. and Selander, H. (1973), *Oikos,* **24,** 8.
Granhall, U. and Von Hofsten, A. (1976), *Physiol. Pl.,* **36,** 88.
Grassia, A. and Brockwell, J. (1978), *Soil Biol. Biochem.,* **10,** 101.
Gresshoff, P. M., Skotnicki, M. L., Eadie, J. F., and Rolfe, B. G. (1977), *Plant Sci. Lett.,* **10,** 299.
Grimes, H. and Masterson, C. L. (1971), *Pl. Soil,* **35,** 289.
Grobbelaar, N., Strauss, J. M., and Groenewald, E. G. (1971), in *Biological nitrogen fixation in natural and agricultural habitats (Eds. T. A. Lie and E. G. Mulder), Pl. Soil,* Special Volume, p. 324.
Guillard, R. R. L. and Ryther, J. H. (1962), *Can. J. Microbiol.,* **8,** 229.
Gunn, C. R. (1972), in *Seed biology, Vol. 3, Insects, and seed collection, storage, testing, and certification* (Ed. T. T. Kozlowski), Academic Press, New York, p. 55.
Gunning, C. and Jordan, D. C. (1954), *Can. J. Agric. Sci.,* **34,** 255.
Gupta, B. M. and Kleckowska, J. (1962), *J. Gen. Microbiol.,* **27,** 473.
Gutteridge, C. S. and Norris, J. R. (1979), *J. Appl. Bacteriol.,* **47,** 5.
Haaker, H. and Veeger, C. (1977), *Eur. J. Biochem.,* **77,** 1.
Haaker, H., Scherings, G., and Veeger, C. (1977), in *Recent developments in nitrogen fixation* (Eds. W. Newton, J. R. Postgate, and Rodríguez-Barreuco), Academic Press, London, New York, San Francisco, p. 271.
Haaker, H., Laane, C., and Veeger, C. (1980), in Nitrogen Fixation, (Eds. W. D. P. Stewart and J. R. Gallon), Academic Press, London, in press.
Hack, H. R. B. (1971), *J. Exp. Bot.,* **22,** 323.

Hageman, R. V. and Burris, R. H. (1978), *Proc. Nat. Acad. Sci. USA,* **75,** 2699.
Hageman, R. H. and Hucklesby, D. P. (1971), *Methods Enzymol.,* **23,** 491.
Hahn, N. J. (1966), *Can. J. Microbiol.,* **12,** 725.
Halliday, J. and Pate, J. S. (1976a), *Aust. J. Pl. Physiol.,* **3,** 349.
Halliday, J. and Pate, J. S. (1976b), *J. Brit. Grassl. Soc.,* **31,** 29.
Halvorsen, H. O. and Ziegler, N. R. (1933), *J. Bacteriol.,* **25,** 101.
Ham, G. E. (1977), in *Biological nitrogen fixation in farming systems of the tropics* (Eds. A. Ayababa and P. J. Dart), Wiley, Chichester, New York, Brisbane, Toronto, p. 325.
Ham, G. E. and Frederick, L. R. (1966), *Agron. J.,* **58,** 592.
Hanson, R. B. (1977), *Appl. Environ. Microbiol.,* **33,** 596.
Hanus, F. J., Carter, K. R., and Evans, H. J. (1980), *Methods Enzymol.,* **69c,** in press.
Hardy, R. W. F. (Gen. Ed.) (1976–78), *A treatise on dinitrogen fixation, Sections 1–3,* Wiley, New York, London, Sydney, Toronto.
Hardy, R. W. F., Burns, R. C., and Holsten, R. D. (1973), *Soil Biol. Biochem.,* **5,** 47.
Hardy, R. W. F. and Havelka, U. D. (1976), in *Symbiotic nitrogen fixation in plants* (Ed. P. S. Nutman), Cambridge University Press, Cambridge, London, New York, Melbourne, p. 421.
Hardy, R. W. F. and Holsten, R. D. (1977), in *A treatise on dinitrogen fixation, Section IV, agronomy and ecology* (Eds. R. W. F. Hardy and A. H. Gibson), Wiley–Interscience, New York, London, Sydney, Toronto, p. 451.
Hardy, R. W. F., Holsten, R. D., Jackson. E. K., and Burns, R. C., (1968), *Pl. Physiol.,* **43,** 1185.
Hardy, R. W. F. and Jackson, E. K. (1967), *Fed. Proc.,* **24,** 725.
Hardy, R. W. F. and Knight, Jr., E. (1966), *Biochem. Biophys. Res. Commun.,* **23,** 409.
Hardy, R. W. F. and Knight, Jr., E. (1967), *Biochim. Biophys. Acta,* **139,** 69.
Hare, P. E. (1977), *Methods Enzymol.,* **47,** 3.
Hariot, P. (1892), *C. R. Acad. Sci., Paris,* **115,** 325.
Harmond, J. E., Brandenburg, N. R., and Klein, L. M. (1968), *US Dept. Agric., Agric. Hanb.,* No. **354,** 55 pp.
Harper, J. E. (1971), *Crop Sci.,* **11,** 347.
Harper, J. E. and Nicholas, J. C. (1976), *Physiol. Plant.,* **38,** 24.
Harper, S. H. T. and Lynch, J. M. (1973), *Lab. Pract.,* **22,** 736.
Harrington, J. F. (1959), *Proc. Short Course Seedsmen, State Coll. Miss.,* p. 89.
Harrington, J. F. (1972), in *Seed Biology, Vol. 3, insects, and seed collection, storage, testing and certification* (Ed. T. T. Kozlowski), Academic Press, New York, p. 145.
Haselkorn, R. A. (1978), *Ann. Rev. Pl. Physiol.,* **29,** 319.
Hatch, M. D. (1973), *Arch. Biochem. Biophys.,* **156,** 207.
Havelka, U. D. and Hardy, R. W. F. (1976), in *Proc. 1st int. symp. nitrogen fixation* (Eds. W. E. Newton and C. J. Nyman), Washington State University Press, Pullman, Washington, p. 456.
Hawk, P. B. (1965), in *Hawk's physiological chemistry* (Ed. B. Oser), 14th Edition, McGraw-Hill, New York, London, p. 1327.
Hayes, W. (1964), *The genetics of bacteria and their viruses. Studies in basic genetics and molecular biology,* Blackwell, Oxford.
Haystead, A., Dharmawardene, M. W. N., and Stewart, W. D. P. (1973), *Pl. Sci. Lett.* **1,** 439.
Haystead, A., King, J., and Lamb, W. I. C. (1980), *J. Exp. Bot.,* in press.
Haystead, A., Robinson, R., and Stewart, W. D. P. (1970), *Arch. Mikrobiol.,* **74,** 235.

Hebert, G. A., Pelman, P. L., and Pittman, B. (1973), *Appl. Microbiol.*, **25**, 26.
Hedlin, R. A. and Newton, J. D. (1948), *Can. J. Res.*, **26**, 174.
Heinen, W. (1971), in *Methods in Microbiology* (Eds. J. R. Norris and D. W. Ribbons), Vol. 6A, Academic Press, London, New York, p. 383.
Hellmann, H. (1956), in *Modern methods of plant analysis* (Eds. K. Paech and M. V. Tracey), Vol. 1, Springer-Verlag, Berlin, Göttingen, Heidelberg, p. 126.
Helriegel, H., and Wilfarth, H. (1888), *'Untersuchungen über die Stickstoffnahrung der Gramineen und Leguminosen'*, Beilageheft Ztschr. Ver. Rubenzucker-Industrie Deutsch. Reichs., 234 pp.
Hely, F. W. (1958), in *Plant Exploration, collection and introduction,* FAO Agric. Stud. No. 41, p. 61.
Hely, F. W. (1959), *J. Agric. Eng. Res.*, **4,** 133.
Hely, F. W. (1965), *Aust. J. Agric. Res.*, **16,** 575.
Hely, F. W., Bonnier, C., and Manil, P. (1953), *Nature (London),* **171,** 884.
Hendey, N. I. (1964, *An introductory account of the smaller algae of British coastal waters, Part V, Bacillariophyceae,* Fisheries Investigations Ser. No. 4, 317 pp.
Herbert, D., Elsworth, R. E., and Telling, R. C. (1956), *J. Gen. Microbiol.*, **14,** 601.
Herdman, M. and Carr. N. G. (1972), *J. Gen. Microbiol.*, **70,** 213.
Herdman, M., Faulkner, B. M., and Carr, N. G. (1970), *Arch. Mikrobiol.*, **73,** 238.
Herridge, D. F. (1974), *Studies of Rhizobia nodulating tropical legumes,* MSc Agric. Thesis, University of Sydney.
Herridge, D. F. (1977), *Carbon and nitrogen of two annual legumes,* PhD Thesis, University of Western Australia, p. 16.
Hevesy, G. (1948), *Cold Spring Harbor Symp. Quant. Biol.* **13,** 129.
Hewitt, E. J. and Bond, G. (1961), *Pl. Soil,* **14,** 159.
Hewitt, E. J. and Bond, G. (1966), *J. Exp. Bot.*, **17,** 480.
Heyn, C. C. (1963), *The annual species of Medicago,* Scripta Hierosolymitana Vol. XII. Magnes Press, The Hebrew University, Jerusalem.
Hill, D. J. (1975), *Planta,* **122,** 179.
Hill, M. O. (1978), in *The moss flora of Britain and Ireland* (Ed. A. J. E. Smith), Cambridge University Press, Cambridge, p. 30.
Hill, R. L. and Bradshaws, R. A. (1969), *Methods Enzymol.*, **13,** 91.
Hill, S. (1971), *J. Gen Microbiol.*, **67,** 377.
Hill, S. (1973), *Lab. Pract.*, **22,** 193.
Hill. S. (1975), *J. Gen. Microbiol.*, **91,** 207.
Hill. S. (1976), *J. Gen. Microbiol.*, **93,** 335.
Hills, K. L. (1944), *J. Counc. Sci. Ind. Res.*, **17,** 186.
Hine, P. W. and Lees, H. (1976), *Can. J. Microbiol.*, **22,** 611.
Hino, S. (1955), *J. Biochem.*, **42,** 775.
Hino, S. and Wilson, P. W. (1958), *J. Bacteriol.*, **75,** 403.
Hirano, T., Shiraishi, K., and Nakano, K. (1955), *Bull. Shikoku Agr. Exp. Stn.*, **2,** 121.
Hobart, M. J. and McConnell, I. (1975), *The immune system: a course on the molecular and cellular basis of immunity,* Blackwell, Oxford.
Hoch, G. E., Schneider, K. C., and Burris, R. H. (1960), *Biochim. Biophys. Acta,* **37,** 273.
Holme, T. and Zacharias, B. (1965), *Nature, Lond.*, **208,** 1235.
Holm-Hanson, O. (1964), *Can. J. Bot.*, **42,** 127.
Holsten, R. D., Burns, R. C., Hardy, R. W. F., and Hebert, R. R. (1971), *Nature, Lond.*, **232,** 173.
Hong, J. S. and Rabinowitz, J. C. (1970), *J. Biol. Chem.*, **245,** 6574.

Hopwood, D. A. (1970), in *Methods in microbiology* (Eds. J. R. Norris and D. W. Ribbons), Vol. 3A, Academic Press, London, p. 363.
Horejsi, J. (1910), *Rozpr. Cesk. Akad. Frantiska Josefa,* **19,** Tr ii (9), 1.
Howe, M. M. (1973), *Virology,* **54,** 93.
Howe, M. M. and Bade, E. G. (1975), *Science,* **190,** 624.
Huang, C.-Y., Boyer, J. S., and Vanderhoef, L. N. (1975), *Pl. Physiol.,* **56,** 222.
Huang, A. H. C., Liu, K. D. F., and Youle, R. J. (1976), *Pl. Physiol.,* **58,** 110.
Huang, T. C., Zumft, W. G., and Mortenson, L. E. (1973), *J. Bacteriol.,* **113,** 884.
Hubbard, R. L. and Pearson, B. O. (1958), *US For. Serv. Calif. For. Range Exp. Stn.,* Note 138.
Humphrey, B. A. and Vincent, J. M. (1973), *Microbios,* **7,** 87.
Humphrey, B. A. and Vincent, J. M. (1975), *Microbios,* **13,** 71.
Humphrey, J. H. and White, R. G. (1970), *Immunology for students of medicine,* 3rd Edition, Blackwell, Oxford.
Humphreys, E. C. (1956), in *Modern methods of plant analysis* (Eds. K. Paech and M. V. Tracey), Vol. 1, Springer-Verlag, Berlin, Göttingen, Heidelberg, New York, p. 479.
Humphreys, G. O., Willshaw, G. A., and Anderson, E. S. (1975), *Biochim. Biophys. Acta,* **282,** 457.
Huneke, A. (1933), *Beitr. Biol. Pfl.,* **20,** 315.
Hurrell, G. R., Nicola, N. A., Broughton, W. J., Dilworth, M. J., Minasian, E., and Leach, S. J. (1976), *Eur. J. Biochem.,* **66,** 389.
Hurwitz, C. and Wilson, P. W. (1940), *Ind. Eng. Chem. Anal. Ed.,* **12,** 31.
Huss-Danell, K. (1978), *Physiol. Plant.,* **43,** 372.
Husted, F. (1930), *Rabenhorst's Kryptogammenflora von Deutschland, Osterreich und der Schweiz,* Bd. 7(1), Akademische Verlagsgesellschaft, Leipzig, 8205.
Husted. F. (1959), *Rabenhorst's Kryptogammenflora von Deutschland, Österreich und der Schweiz,* Bd. 7(2), Akademische Verlagsgesellschaft, Leipzig, 8453.
Hutchinson, F. and Pollard, E. (1961), in *Mechanisms in radiobiology* (Eds. M. Errera and A. Forssberg), Vol. 1, Academic Press, New York, p. 1.
Hylemon, P. B., Wells, J. S., Kreig, N. R., and Jannasch, H. W. (1973), *Int. J. Syst. Bacteriol.,* **23,** 340.
Hynes, R. K. and Knowles, R. (1978), *FEMS Microbiol. Lett.,* **4,** 319.
Imamura, T., Riggs, A., and Gibson, Q. H. (1972), *J. Biol. Chem.,* **247,** 521.
Imshenetskii, A. A., Pariiskaya, A. N., and Gorelova, O. P. (1976), *Microbiology,* **45,** 945 (Engl. trans. of *Mikrobiologiya,* **45,** 1107).
Ireland, J. A. (1969), *Aust. J. Exp. Agric. Anim. Husb.,* **9,** 209.
Israel, D. W., Howard, R. L., Evans, H. J., and Russell, S. A. (1974),*J. Biol. Chem.,* **249,** 500.
Ito, H. (1972), in *Viability of seeds (*Ed. E. H. Roberts), Chapman and Hall, London, p. 405.
Jackson, E. K., Parshall, G. W., and Hardy, R. W. F. (1968), *J. Biol. Chem.,* **243,** 4952.
Jacobs, S. (1966), in *Methods of Biochemical analysis* (Ed. D. Glick), Vol. XIV, Interscience, New York, London, Sydney, p. 177.
Jacobsen, J. V. and McGlasson, W. B. (1970), *Pl. Physiol.,* **45,** 631.
Jakoby, W. B. (Ed.) (1971), *Methods in Enzymology,* Vol. 22.
Jakoby, W. B. (Ed.) (1974), *Methods in Enzymology,* Vol. 34.
James, E. (1972), in *Viability of seeds* (Ed. E. H. Roberts), Chapman and Hall, London, p. 397.
Jaworek, D., Gruber, W., and Bergmeyer, H. U. (1974), in *Methods of enzymatic analysis* (Ed. H. U. Bergmeyer), Academic Press, New York, London, p. 2127.

Jaworski, E. G. (1971), *Biochem. Biophys. Res. Commun.*, **43,** 1274.
Jenkinson, D. S. (1970), *Rep. Rothamsted Exp. Stn. 1970,* 113.
Jenkinson, D. S. (1973), *J. Sci. Food Agric.*, **24,** 1149.
Jensen, H. L. (1942a), *Proc. Linn. Soc. N.S.W.*, **66,** 98.
Jensen, H. L. (1942b), *Proc. Linn. Soc. N.S.W.*, **67,** 205.
Jensen, H. L., Petersen, E. J., and Bhattachary, A. (1960), *Arch. Microbiol.*, **36,** 182.
Jensen, V. and Petersen, E. J. (1955), 'Taxonomic studies on *Azotobacter chroococcum* Beij. and *Azotobacter beijerinckii* Lipman', *Den Kongelige Veterinaer- og Landbohøjskole Årsskrift (Royal Veterinary and Agricultural College Copenhagen Yearbook),* p. 107.
John, K. P. (1966), *J. Rubb. Res. Inst. Malaya,* **19,** 173.
Johnson, D. A. and Asay, K. H. (1978), *Crop Sci.*, **18,** 520.
Johnson, G. V., Mayeux, P. A., and Evans, H. J. (1966), *Pl. Physiol.*, **41,** 852.
Johnson, H. W. and Means, U. M. (1964), *Agron. J.*, **56,** 60.
Johnson, R. N., Bradbury, J. H., and Appleby, C. A. (1978), *J. Biol. Chem.*, **253,** 2148.
Johnston, A. W. B., Benyon, J. L., Buchanan-Wollaston, A. V., Setchell, S. M., Hirsch, P. R., and Beringer, J. B. (1978), *Nature, Lond.*, **276,** 634.
Jones, L. W. and Bishop, N. I. (1976), *Pl. Physiol.*, **57,** 659.
Jordan, E. B. and Bainbridge, K. T. (1936), *Phys. Rev.*, **50,** 98.
Jordan, D. C., McNicol, P. J., and Marshall, M. R. (1978), *Can. J. Microbiol.*, **24,** 643.
Joshi, H. U., Carr, A. J. H., and Jones, D. G. (1967), *J. Gen. Microbiol.*, **47,** 139.
Justice, O. L. (1972), in *Seed biology, Vol. 3, insects, and seed collection, storage, testing and certification,* (Ed. T. T. Kozlowski), Academic Press, New York, p. 301.
Jutono (1973), *IV conf. on global impacts of applied microbiology, São Paulo,* Abstr. No. 18.
Juttner, F. and Carr, N. G. (1976), in *Proc. 2nd. int symp photosynthetic prokaryotes* (Eds. G. A. Codd and W. D. P. Stewart), University of Dundee, Dundee, p. 121.
Kaback, H. R. (1971), *Methods Enzymol.*, **22,** 99.
Kabat, E. A. (1961), *Kabat and Mayer's experimental immunochemistry,* 2nd Edition, Charles C. Thomas, Springfield, Ill.
Kabat, E. A. (1968), *Structural concepts in immunology and immunochemistry,* Holt, Rinehart and Winston, New York.
Kalb, V. F., Donohue, J. T., Corrigan, M. G., and Bernlohr, R. W. (1978), *Anal. Biochem.*, **90,** 47.
Kalininskaya, T. A., Rao. V. R., Volkova, T. N., and Ippolitov, L. T. (1973), *Microbiology,* **42,** 426.
Kallio, P. and Kallio, S. (1978), *Ecol. Bull., Stockholm,* **26,** 225.
Kanner, D. and Bartha, R. (1979), *J. Bacteriol.*, **139,** 225.
Kavanagh, E. P. and Postgate, J. R. (1970), *Lab. Pract.*, **19,** 159.
Kawamatu, S. (1965a), *Cytologia,* **30,** 75.
Kawamatu, S. (1965b), *Cytologia,* **30,** 80.
Kay, D. (1972), in *Methods in Microbiology,* (Eds. J. R. Norris and D. W. Ribbons), Vol. 7A, Academic Press, London, New York, p. 191.
Keele, B. B., Hamilton, P. B., and Elkan, G. H. (1969), *J. Bacteriol.*, **97,** 1184.
Keeney, D. R. and Tedesco, M. J. (1973), *Anal. Chim. Acta,* **65,** 19.
Keister, D. L., San Pietro, A., and Stokzenbach, F. E. (1960), *J. Biol. Chem.*, **235,** 2989.
Kelly, M. (1967), *Biochem. J.*, **107,** 1.
Kelly, M., Postgate, J. R., and Richards, R. L. (1967), *Biochem, J.*, **102,** 1c.

Kennedy, C. (1977), *Molec. Gen. Genet.*, **157,** 199.
Kennedy, C., Eady, R. R., Kondorosi, E., and Klavans Rekosh, D. (1976), *Biochem. J.,* **155,** 383.
Kennedy, I. R. (1965), *Anal. Biochem.,* **11,** 105.
Kennedy, I. R. (1970), *Biochim. Biophys. Acta,* **222,** 135.
Kennedy, I. R., Rigaud, J., and Trinchant, J. C. (1975), *Biochim. Biophys. Acta,* **397,** 24.
Kenten, R. H. (1956), in *Modern methods of plant analysis* (Eds. K. Paech and M. V. Tracey), Springer-Verlag, Berlin, Göttingen, Heidelberg, p. 432.
Kidby, D. K. (1966), *Pl. Physiol.,* **41.** 1139.
Kihlman, B. A. (1966), *Actions of chemicals on dividing cells,* Prentice-Hall, Englewood Cliffs, N.J.
Kingham, W. H. (1971), in *Methods in microbiology* (Eds. J. R. Norris and D. W. Ribbons) Vol. 5A, Academic Press, London, New York, p. 282.
Kinsey, D. W. and Bottomley, R. A. (1963), *J. Inst. Brew.,* **69,** 164.
Kirkland, J. J. (Ed.) (1971), *Modern practice of liquid chromotography,* Wiley–Interscience, New York, London, Sydney, Toronto.
Kishinevsky, B. and Bar-Joseph, M. (1978), *Can. J. Microbiol.,* **24,** 1537.
Kleckner, N., Roth, J., and Botstein, D. (1977), *J. Molec. Biol.,* **116,** 125.
Kleiner, D. and Chen, C. H. (1974), *Arch. Microbiol.,* **98,** 93.
Klucas, R. (1972), *Can. J. Microbiol.,* **18,** 1845.
Knowles, R. (1976), in *Proc. 1st int. symp. nitrogen fixation* (Eds. W. E. Newton and C. J. Nyman), Washington State University Press, Pullman, Washington, p. 539.
Knowles, R. (1979), in *Soil biochemistry,* (Eds. E. A. Paul and J. Ladd), Vol. 5, Marcel Dekker, New York, in press.
Knowles, R. and Denike, D. (1974), *Soil Biol. Biochem.,* **6,** 353.
Knowles, R., Brouzes, R., and O'Toole, P. (1973), in *Modern methods in the study of microbial ecology* (Ed. T. Rosswall), *Bull. Ecol. Res. Commun. (Stockholm),* **17,** p. 255.
Knowles, R., and O'Toole, P. (1975), in *Nitrogen fixation by free-living micro-organisms* (Ed. W. D. P. Stewart), Cambridge University Press, Cambridge, London, New York, Melbourne, p. 285.
Koch, B. L. (1977), *Pl. Soil,* **47,** 703.
Koch, B., Evans, H. J., and Russell, S. (1967), *Pl. Physiol.,* **42,** 466.
Kondorosi, A., Kiss, G. B., Forrai, T., Vincze, E., and Banfalvi, Z. (1977), *Nature, Lond.,* **268,** 525.
Kratz, W. A. and Myers, J., (1955), *Am. J. Bot.,* **42,** 282.
Kraus, M. P. (1966), *Nature, Lond,* **211,** 310.
Krebber, O. (1932), *Arch. Mikrobiol.,* **3,** 588.
Krieg, N. R. (1976), *Bact. Rev.,* **40,** 55.
Krieg, N. R. (1977), in *Genetic engineering for nitrogen fixation* (Eds. A. Hollaender *et al.*), Plenum Press, New York, p. 463.
Kullasooriya, S. A., Lang, N. J., and Fay, P. (1972), *Proc. Roy. Soc. Lond., B.,* **181,** 199.
Kurz, W. G. W. and LaRue, T. A. (1977), *Can. J. Microbiol.,* **23,** 1197.
Kuwatsuka, S. and Nakano, R. (1978), *Jap. Sci. Soil Manure,* **49,** 210.
Kwapinski, J. B. G. (1965), *Methods of serological research,* Wiley, New York.
Labandera, C. A. and Vincent, J. M. (1975), *J. Appl. Bacteriol.,* **39,** 209.
Laemmli, U. K. (1970), *Nature, Lond.,* **222,** 680.
Lalonde, M. and Calvert, H. E. (1979), in *Symbiotic nitrogen fixation in the management of temperate forests* (Eds. J. C. Gordon, C. T. Wheeler, and D. A. Perry), Forest

Research Laboratory, Oregon State University, Corvallis, Oregon, p. 95.
Lalonde, M. and Fortin, J. A. (1972), *Can. J. Bot.*, **50,** 2597.
Lalonde, M. and Knowles, R. (1975), *Can. J. Microbiol.*, **21,** 1058.
Lalonde, M., Knowles, R., and Fortin, J. A. (1975), *Can. J. Microbiol.*, **21,** 1901.
Lami, R. and Meslin, R. (1959), *Bull. Lab. Dinard,* **44,** 47.
Lamprecht, W. and Trauschold, I. (1974), in *Methods of enzymatic analysis,* (Ed. H. U. Bergmeyer), Academic Press, New York, London, p. 2101.
Lane, M. D., Maruyama, J., and Easterday, R. L. (1969), *Methods in Enzymol.* **13,** 277.
LaPage, S. P., Shelton, J. E., Mitchell, T. G., and MacKenzie, A. R. (1970), in *Methods in microbiology* (Eds. J. R. Norris and D. W. Ribbons), Vol. 3A, p. 135.
LaPage, S. P., Sneath, P. H. A., Lessel, E. F., Skerman, V. B. D., Selliger, H. P. R., and Clark, W. A., (Eds.) (1975), *International code of nomenclature of bacteria,* American Society for Microbiology, Washington, D.C..
LaRue, T. A. (1977), in *A treatise on dinitrogen fixation, Section III, Biology,* (Eds. R. W. F. Hardy and W. S. Silver), Wiley, New York, London, Sydney, Toronto, p. 19.
La Rue, T. A. and Kurz, W. G. W. (1970), *Abstr. X int. congr. microbiol., Mexico City, Mexico,* p. 142.
LaRue, T. A. and Kurz, W. G. W. (1973a), *Pl. Physiol.*, **51,** 1074.
LaRue, T. A. and Kurz, W. G. W. (1973b), *Can. J. Microbiol.*, **19,** 304.
Laser, H. (1961), in *Biochemists handbook* (Ed. C. Long), Spon, London, p. 60.
Lawes, J. B. and Gilbert, J. H. (1851), *J. Roy. Agric. Soc. Engl.*, **12,** 1.
Lawes, J. B. and Gilbert, J. H. (1854), *Amounts of and methods of estimating ammonia and nitric acid in rain water,* Report to the British Association for Advancement of Science.
Lawes, J. B. and Gilbert, J. H. (1889), *Phil. Trans. Roy. Soc. Lond.*, **180,** 1.
Lawes, J. B. and Gilbert, J. H. (1890), *Proc. Roy Soc. Lond.*, **47,** 85.
Lawes, J. B. and Gilbert, J. H. (1892), *J. Roy. Agric. Soc. Engl.*, Ser. 3, **2,** 657.
Lawes, J. B., Gilbert, J. H., and Pugh, E. (1861), *Phil. Trans. Roy. Soc. Lond.*, **151,** 431.
Lawn, R. J. and Brun, W. A. (1974a), *Crop Sci.*, **14,** 11.
Lawn, R. J. and Brun, W. A. (1974b), *Crop Sci.*, **14,** 22.
Lawrence, D. B., Schoenike, R. E., Quispel, A., and Bond, G. (1967), *J. Ecol.*, **55,** 793.
Lawrie, A. C. and Wheeler, C. T. (1975), *New Phyto.*, **74,** 437.
Lazaroff, N. J. (1966), *Phycologia,* **2,** 7.
Lea, P. and Miflin, B. J. (1975), *Biochem. Soc. Trans.*, **3,** 381.
Lea, P. J., Fowden, L. and Miflin, B. J. (1978), *Phytochemistry,* **17,** 217.
Leaf, G., Gardner, I. C., and Bond, G. (1958), *J. Exp. Bot.*, **9,** 320.
Leaf, G., Gardner, I. C., and Bond, G. (1959), *Biochem. J.*, **72,** 662.
Ledeboer, A. M., Keol, A. J. M., Dons, J. J. M., Spier, F., Schilperoort, R. A., Zaenen, I., van Larebeke, N., and Schell, J. (1976), *Nucl. Acids Res.*, **3,** 449.
Lederberg, J. and Lederberg, E. M. (1952), *J. Bacteriol.*, **63,** 399.
Lechevalier, M. and Lechevalier, H. (1979), in *Symbiotic nitrogen fixation in the management of temperate forests* (Eds. J. C. Gordon, C. T. Wheeler, and D. A. Perry), Forest Research Laboratory, Oregon State University, Corvallis, Oregon, p. 111.
Lee, K.-K. and Watenabe, I. (1977), *Appl. Environ. Microbiol.*, **34,** 654.
Lee, K.-K., Alimagno, B. V., and Watanabe, I. (1975), *Non-symbiotic nitrogen fixation in paddy soil,* Saturday Seminar IRRI, Los Baños, Laguna, Philippines.

Lee, K.-K., Alimagno, B., and Yoshida, T. (1977), *Pl. Soil,* **47,** 519.
Lees, H. and Postgate, J. R. (1973), *J. Gen Microbiol.,* **75.** 161.
Legg, J. O. (1978), *11th int. congr. soil. sci., Edmonton,* Vol. 1, p. 329.
Legg. J. O. and Sloger, C. (1975), *Proc. 2nd int. conf. stable isotopes, October 1975, Chicago, Ill.*
Lehmann, M. and Wöber, G. (1978), *Pl. Cell. Environ.,* **1,** 155.
Leiderman, J. (1971), *Rev. Ind. Agric. Tucuman,* **48,** 51.
Leitgeb, H. (1878), *Sber. Akad. Wiss. Wien,* **77,** 79.
Lemon, E. (1978), in *Nitrogen in the environment* (Eds. D. R. Nielson and J. G. MacDonald), Vol. 1, Academic Press, New York, p. 493.
Leonard, L. T. (1943), *J. Bacteriol.,* **45,** 523.
Leonard, L. T. (1944), *Method of testing bacterial cultures and results of tests of commercial inoculants,* U.S.D.A. Circ. No. 703, 8 pp.
Lessler, M. A. and Brierley, G. P. (1969), *Methods Biochem. Anal.,* **17,** 1.
Lex, M. and Carr, N. G. (1974), *Arch. Microbiol.,* **101,** 161.
Lex, M., Silvester, W. B., and Stewart, W. D. P. (1972), *Proc. Roy. Soc. Lond. B.,* **180,** 87.
Lie, T. A. (1969a), *Pl. Soil,* **30,** 391.
Lie, T. A. (1969b), *Pl. Soil,* **31,** 391.
Lie, T. A. (1974), in *The biology of nitrogen fixation* (Ed. A. Quispel), North Holland, Amsterdam, p. 555.
Lie, T. A., and Brotonegoro, S. (1969), *Pl. Soil,* **30,** 339.
Lie, T. A. and Mulder, E. G. (Eds.) (1971), *Nitrogen fixation in natural and agricultural habitats, Pl. Soil,* Special Volume.
Likens, G. E., Bormann, F. H., Pierce, R. S., Eaton, J. S., and Johnson, N. M. (1977), *Biogeochemistry of a forested ecosystem,* Springer-Verlag, Berlin, Heidelberg, New York.
Lim, S. T. (1978), *Pl. Physiol.,* **62,** 609.
Lindsay, C. R. and Jordan, D. C. (1976), *Can. J. Soil. Sci.,* **56,** 495.
Lindsey, J. K., Vance, B. D., Keeter, J. S., and Scholes, V. E. (1971), *J. Phycol.,* **7,** 65.
Line, M. A. and Loutit, M. W. (1971), *J. Gen. Microbiol.,* **66,** 309.
Liu, C. C. (1979), in *Nitrogen and rice symposium,* IRRI, Los Baños, Laguna, Philippines, p. 375.
Ljones, T. and Burris, R. H. (1972), *Anal. Biochem.,* **45,** 448.
Ljones, T. (1974), in *The biology of nitrogen fixation* (Ed. A. Quispel), North-Holland, Amsterdam, p. 617.
Ljones, T. and Burris, R. H. (1978), *Biochemistry,* **17,** 1866.
Lockhart, C. M., Rowell, P., and Stewart, W. D. P. (1978), *FEMS Microbiol. Lett.,* **3,** 127.
Loneragan, J. F., Meyer, D., Fawcett, R. G., and Anderson, A. J. (1955), *J. Aust. Inst. Agric. Sci.,* **21,** 264.
Loomis, W. D. (1969), *Methods Enzymol.,* **13,** 555.
Loomis, W. D. (1974), *Methods Enzymol.,* **31,** 528.
Lorenzen, H. and Kaushik, B. D. (1976), *Ber. Dt. Bot. Ges. Bd.,* **89,** 491.
Lorenzen, H. and Venkataraman, G. S. (1969), *Arch. Mikrobiol.,* **67,** 251.
Lorenzen, H. and Venkataraman, G. S. (1972), in *Cell physiology* (Ed. D. M. Prescott), Academic Press, London, p. 373.
Loveday, J. (1963), *Aust. J. Sci.,* **26,** 90.
Lovett, J. V. (1976), *A course manual in annual crop production (maize, sorghum and soybeans),* University of New England, Armidale, NS.W.
Lowe, D. J. (1978), *Biochem. J.,* **175,** 955.

Lowe, D. J., Eady, R. R. and Thorneley, R. N. F. (1978), *Biochem. J.*, **173**, 277.
Lowe, J. F. and Holding, A. J. (1970), *Proc. symp. white clover research,* Occasional Symp. No. 6, British Grassland Society, Hurley, U.K., p. 79.
Lowe, R. H. and Evans, H. J. (1964), *Biochim, Biophys, Acta,* **85,** 377.
Lowry, O. H., Rosebrough, N. J., Farr, A. L., and Randall, R. J. (1951), *J. Biol. Chem.,* **193,** 265.
Ludden, P. W. and Burris, R. H. (1976), *Science,* **194,** 424.
Ludden, P. W. and Burris, R. H. (1978), *Biochem. J.*, **175,** 251.
Ludden, P. W., Okon, Y., and Burris, R. H. (1978), *Biochem. J.*, **173,** 1001.
Lund, J. W. G. (1947), *New Phytol.*, **46,** 35.
Lundsgaard, J. S., Grønlund, J., and Degn, H. (1978), *Biotechnol. Bioeng.*, **20,** 809.
Mackinnon, P. A., Robertson, J. G., Scott, D. J., and Hale, C. N. (1977), *N.Z. J. Exp. Agric.*, **5,** 35.
Mackintosh, M. E. (1978), *J. Gen. Microbiol.*, **105,** 215.
MacLennan, D. G. and Pirt, S. J. (1966), *J. Gen. Microbiol.*, **45,** 289.
MacNeil, D. and Brill, W. J. (1978), *J. Bacteriol.*, **136,** 247.
MacNeil, T., Brill, W. J., and Howe, M. M. (1978a), *J. Bacteriol.*, **134,** 821.
MacNeil, T., MacNeil, D., Roberts, G. P., Supiant, M. A., and Brill, W. J. (1978b), *J. Bacteriol.*, **136,** 253.
Macnicol, P. K. (1972), *Anal. Biochem.*, **45,** 624.
MacRae, I. C. (1977), *Aust. J. Biol. Sci.*, **30,** 593.
MacRae, I. C. and Castro, T. F. (1967), *Soil Sci.*, **103,** 277.
MacVicar, S. M. (1926), *The students handbook of British hepatics*, V. V. Sumfield, Eastbourne.
McCrae, R. E., Hanus, J., and Evans, H. J. (1978), *Biochem. Biophys. Res. Commun.*, **80,** 384.
McKenna, C. E., McKenna, M.-C., and Higa, M. T. (1976), *J. Am. Chem. Soc.*, **98,** 4657.
McKnight, T. (1949), *Queensl. J. Agric. Sci.*, **6,** 61.
McLachlan, J. (1964), *Can. J. Microbiol.*, **10,** 769.
McLeod, R. W. and Roughley, R. J. (1961), *Aust. J. Exp. Agric. Anim. Husb.*, **1,** 29.
McNabb, D. H., Geist, J. M., and Youngberg, C. T. (1978), *Agron. Abstr.*, 127.
McVean, D. N. (1955), *J. Ecol.*, **43,** 61.
Magalhães, F. M. M., Patriquin, D. G., and Döbereiner, J. (1979), *Rev. Bras. Biol.*, **39,** 587.
Mague, T. H. (1977), in *A treatise on dinitrogen fixation, Section IV, agronomy and ecology* (eds. R. W. F. Hardy and A. H. Gibson), Wiley. New York, p. 85.
Mague, T. H. and Burris, R. H. (1972), *New Phytol.*, **71,** 275.
Mague, T. H., Weare, N. M., and Holm-Hansen, O. (1974), *Mar. Biol.*, **24,** 109.
Mallard, T. M., Mallard, C. S., Holfeld, H. S., and La Rue, T. A. (1977), *Anal. Chem.*, **49,** 1275.
Mariotti, A. (1977), *Recherche*, **8,** 886.
Marques Pinto, C., Yao, P. Y., and Vincent, J. M. (1974), *Aust. J. Agric. Res.*, **25,** 317.
Marrack, J. R. (1963), *Brit. Med. Bull.*, **19,** 178.
Marshall, K. C. (1956), *J. Aust. Inst. Agric. Sci.*, **22,** 137.
Martin, S. M. and Skerman, V. B. D. (Eds.) (1972), *World directory of collections of cultures of microorganisms*, Wiley–Interscience, New York, London, Sydney, Toronto.
Martin, A. E. and Skyring, G. W. (1962), *Commonw. Bur. Pastures Field Crops Tech. Bull.*, No. 46.

Martin, A. E., Henzell, E. F., Ross, P. J., and Haydock, K. P. (1963), *Aust. J. Soil Res.*, **1**, 169.
Masterson, C. L. and Sherwood, M. T. (1974), *Ir. J. Agric. Res.*, **13**, 91.
Matsuguchi, T. and Shimomura, T. (1977), in *Proc. int. seminar on soil environment and fertility management in intensive agriculture*, Japanese Society for Soil Manure, Tokyo, p. 755.
Matsuguchi, T., Shimomura, T., and Lee, S. K. (1978), *Ecol. Bull., Stockholm*, **26**, 137.
Matsuguchi, T., Tangcham, B., and Patiyuth, S. (1975), *Jap. Agric. Res. Q.*, **8**, 253.
Matsuguchi, T., Tangcham, B., and Patiyuth, S. (1976), *Soil microorganisms*, **18**, 7 (in Japanese).
Mayer, S. W., Kelly, F. H., and Morton, M. E. (1955), *Anal. Chem.*, **27**, 837.
Mayr-Harting, A., Hedger, A. J., and Berkeley, R. C. W. (1972), in *Methods in microbiology* (Eds. J. R. Norris and D. W. Ribbons), Vol. 7A, Academic Press, London, p. 316.
Meade, H. M. and Signer, E. R. (1977), *Proc. Nat. Acad. Sci. USA*, **74**, 2076.
Medan, D. and Tortosa, R. D. (1976), *Boln. Soc. Argent. Bot.*, **17**, 323.
Mederski, H. J. and Streeter, J. G. (1977), *Pl. Physiol.*, **59**, 1076.
Meeks, J. C., Wolk, C. P., Thomas, J., Lockau, W., Shaffer, P. W., Austin, S. M., Wan-Shen Chien, and Galonsky, A. (1977), *J. Biol. Chem.*, **252**, 7894.
Meeks, J. C., Wolk, C. P., Lockau, W., Schilling, N., Shaffer, P. W. and Wan-Shen Chien (1978), *J. Bacteriol.*, **134**, 125.
Meister, A. (1974), in *The enzymes* (Ed. P. D. Boyer), Vol. 10, Academic Press, New York, p. 699.
Meister, A., Levintow, L., Greenfield, R. E., and Abendschein, P. A. (1955), *J. Biol. Chem.*, **215**, 441.
Merrick, M., Filser, M., Kennedy, C., and Dixon, R. (1978), *Molec. Gen. Genet.*, **165**, 103.
Meuzelaar, H. L. C., Kistemaker, P. G., and Tom, A. (1975), in *New approaches to the identification of microorganisms* (Eds. C.-G. Hedén and T. Illéni), Wiley, New York, p. 165.
Meyer, D. R. and Anderson, A. J. (1959), *Nature, Lond.*, **183**, 61.
Mian, S. and Bond, G. (1978), *New Phytol.*, **80**, 187.
Mian, S., Bond, G., and Rodriguez-Barrueco, C. (1976), *Proc. Roy. Soc. Lond. B*, **194**, 285.
Miflin, B. J. (1967), *Nature, Lond.*, **214**, 1133.
Miflin, B. J. (1974), *Pl. Physiol.*, **54**, 550.
Miflin, B. J. and Lea, P. J. (1977), *Ann. Rev. Pl. Physiol.*, **28**, 299.
Miller, J. H. (1972), *Experiments in molecular genetics*, Cold Spring Harbor Laboratory, Cold Spring Harbor, New York.
Miller, R. E. and Stadtman, E. R. (1972), *J. Biol. Chem.*, **247**, 7407.
Minchin, F. R., Neves, M. C. P., Summerfield, R. J., and Richardson, A. C. (1977), *J. Exp. Bot.*, **28**, 507.
Minchin, F. R., Summerfield, R. J., Eaglesham, A. R. J., and Stewart, K. A. (1978), *J. Agric. Sci. (Camb.)*, **90**, 355.
Mitruka, M. B. (1976), *Methods of detection and identification of bacteria*, CRC Press, Cleveland, Ohio.
Moore, A. W. (1966), *Soils Fert.*, **29**, 113.
Moore, A. W. (1969), *Bot. Rev.*, **35**, 17.
Moore, S. and Stein, W. H. (1951), *J. Biol. Chem.*, **192**, 663.
Moriarty, D. J. W., Darlington, J. P. E. C., Dunn, I. G., Moriarty, C. M., and Telvin, M. P. (1973), *Proc. Roy. Soc. Lond., B*, **184**, 299.

Morse, R. N. and Evans, L. T. (1962), *J. Agric. Eng. Res.*, **7,** 128.
Mortenson, L. E. (1972), *Methods Enzymol.*, **24,** 446.
Mortenson, L. E. and Chen, J.-S. (1974), in *Microbial iron metabolism* (Ed. J. B. Neilands), Academic Press, New York, London, p. 231.
Mortenson, L. E. and Thorneley, R. N. F. (1979), *Ann. Rev. Biochem.*, in press.
Moss, F. J. and Bush, F. (1967), *Biotechnol. Bioeng.*, **9,** 585.
Moustafa, E. and Mortenson, L. E. (1969), *Biochim. Biophys. Acta*, **172,** 106.
Moustafa, E. and Petersen, G. B. (1962), *Biochim. Biophys. Acta*, **58,** 364.
Moyse, A., Couderc, D., and Garnier, J. (1957), *Rev. Cytol. Biol. Veg.*, **18,** 293.
Muecke, P. S. and Wiskich, J. T. (1969), *Nature, Lond.*, **221,** 674.
Mulder, E. G. and Veen, W. L. van (1960), *Pl. Soil*, **13,** 265.
Mulder, E. G. and Brotonegoro, S. (1974), in *The biology of nitrogen fixation* (Ed. A. Quispel), North-Holland, Amsterdam, p. 37.
Mulongoy, K. and Elkan, G. H. (1977), *J. Bacterol.*, **131,** 179.
Münck, E., Rhodes, H., Orme-Johnson, W. H., Davis, L. C., Brill, W. J., and Shah, V. K. (1975), *Biochim. Biophys. Acta*, **400,** 32.
Munns, D. N. (1968), *Pl. Soil*, **28,** 129.
Munns, D. N. (1977a), in *A treatise on dinitrogen fixation, Section IV, agronomy and ecology* (Eds. R. W. F. Hardy and A. H. Gibson), Wiley, New York, p. 353.
Munns, D. N. (1977b), in *Exploiting the legume–rhizobium symbiosis in tropical agriculture* (Eds. J. M. Vincent, A. S. Whitney, and J. Bose), University of Hawaii College of Tropical Agriculture, Misc. Publ., No. 145, p. 211.
Munns, D. N. and Fox, R. L. (1976), *Pl. Soil*, **45,** 701.
Munns, D. N., Fox, R. L., and Koch, B. L. (1977), *Pl. Soil*, **46,** 591.
Munson, T. O. and Burris, R. H. (1969), *J. Bacteriol.*, **97,** 1093.
Murphy, P. M. and Masterson, C. L. (1970), *J. Gen. Microbiol.*, **61,** 121.
Mytton, L. R. (1976), *Ann. Appl. Biol.*, **82,** 577.
Mytton, L. R. and Felice, J. de (1977), *Ann. Appl. Biol.*, **87,** 83.
Nairn, R. C. (1976), *Fluorescent protein tracing*, 4th Edition, Churchill Livingstone, Edinburgh.
Nakos, G. and Mortenson, L. E. (1971), *Biochim. Biophys. Acta*, **229,** 431.
Nandi, P. and Sinha, N. (1974), *Proc. Indian Nat. Sci. Acad.*, **40B,** 479.
Nason, A., Lee, K.-Y., Pan, S.-S., Ketchum, P. A., Lamberti, A., and De Vries, J. (1971), *Proc. Nat. Acad. Sci. USA*, **68,** 3242.
Nathanielsz, C. P. and Staff, I. A. (1975), *Am. J. Bot.*, **62,** 232.
Neal, Jr., J. L. and Larson, R. I. (1976), *Soil Biol. Biochem.*, **8,** 151.
Neilson, A. H. and Sparell, L. (1976), *App. Environ. Microbiol.*, **32,** 197.
Nelson, A. D., Barber, L. E., Tjepkema, J., Russell, S. A., Powelson, R., Evans, H. J., and Seidler, R. J. (1976), *Can. J. Microbiol.*, **22,** 523.
Nelsen, C. E., Safir, G. R., and Hanson, A. D. (1978), *Pl. Physiol.*, **61,** 131.
Nelson, N. (1944), *J. Biol. Chem.*, **153,** 375.
Nelson, L. M. and Knowles, R. (1978), *Can. J. Microbiol.*, **24,** 1395.
Nery, M., Abrantes, G. T. V., Santos, D. dos, and Döbereiner, J. (1977), *Rev. Brasil. Ciên. Solo*, **1,** 15.
Newbould, F. H. S. (1951), *Sci. Agric.*, **31,** 463.
Newton, J. W., Wilson, P. W., and Burris, R. H. (1953), *J. Biol. Chem.*, **204,** 445.
Newton, J. W. and Herman, A. I. (1979), *Arch. Microbiol.*, **120,** 161.
Newton, W. E. and Nyman, C. J. (Eds.) (1976), *Proc. 1st int. symp. nitrogen fixation*, Washington State University Press, Pullman, Washington, 2 Vols.
Neyra, C. A. and Döbereiner, J. (1977), *Advan. Agron.*, **29,** 1.
Neyra, C. A. and Hageman, R. H. (1975), *Pl. Physiol.*, **56,** 692.

Neyra, C. A., Döbereiner, J., Lalonde, R., and Knowles, R. (1977), *Can. J. Microbiol.*, **23,** 300.
Nicholas, D. J. D. and Nason, A. (1957), *Methods Enzymol.*, **3,** 981.
Nicholas, D. J. D., Silvester, D. J., and Fowler, J. F. (1961), *Nature, Lond.*, **189,** 634.
Nicholas, J. C., Harper, J. G., and Hageman, R. H. (1976), *Pl. Physiol.*, **58,** 731.
Nickell, L. G. (1958), *Am. Fern J.*, **48,** 103.
Nicol, H. and Thornton, H. G. (1941), *Proc. Roy. Soc. Lond.*, **130,** 322.
Nicola, N. A. and Leach, S. J. (1977a), *Biochemistry*, **16,** 50.
Nicola, N. A. and Leach, S. J. (1977b), *Eur. J. Biochem.*, **78,** 133.
Nilsson, A. O. (1940), *Rev. Sci. Instrum.*, **11,** 212.
Nier, P. E. (1957), *Ann. Roy. Agric. Coll. Sweden*, **23,** 255.
Nohrstedt, H. O. (1976), *Grundfoerbaettring*, **27,** 171.
Norris, D. O. (1958), *Aust. J. Agric. Res.*, **9,** 629.
Norris, D. O. (1965), *Proc. 9th int. grassl. congr., Sao Paulo*, 1087.
Norris, D. O. (1971), *Aust. J. Exp. Agric. Anim. Husb.*, **11,** 194.
Norris, D. O. and Date, R. A. (1976), in *Tropical pasture research, principles and methods* (Eds. N. H. Shaw and W. W. Bryan), Commonw. Agric. Bur. Bull. No. 51, Alden Press, Oxford, p. 134.
Norris, J. R. (1960), *Nature, Lond.*, **185,** 634.
Norris, J. R. (Ed.) (1976), *Methods in Microbiology*, Vol. 9, Academic Press, London, New York, San Francisco.
Norris, J. R., and Ribbons, D. W. (Eds.) (1969–73), *Methods in microbiology*, Vols. 1–3 and 5–7, Academic Press, London, New York.
Novick, R. P. and Richmond, M. H. (1965), *J. Bacteriol.*, **90,** 467.
Novick, R. P., Clowes, R. C., Cohen, S. N., Curtiss, III, R., Datta, N., and Falkow, S. (1976), *Bacteriol. Rev.*, **40,** 168.
Nowotny, A. (1969), *Basic exercises in immunochemistry, a laboratory manual*, Springer-Verlag, Berlin.
Nuti, M. P., Ledeboer, A. M., Lepidi, A. A., and Schilperoort, R. A. (1977), *J. Gen. Microbiol.*, **100,** 241.
Nutman, P. S. (1959), *J. Exp. Bot.*, **10,** 250.
Nutman, P. S. (1962), *Proc. Roy. Soc. Lond., B*, **156,** 122.
Nutman, P. S. (1967), *Aust. J. Agric. Res.*, **18,** 381.
Nutman, P. S. (1973), *Rothamsted Experiment Station Report for 1972, Part 1*, Rothamsted Experiment Station, Harpenden, p. 83.
Nutman, P. S. (Ed.) (1975), *Symbiotic nitrogen fixation in plants*, Cambridge University Press, Cambridge, London, New York, Melbourne.
Oakley, C. L. (1971), in *Methods in microbiology* (Eds. J. R. Norris and D. W. Ribbons), Vol. 5a, Academic Press, New York, p. 173.
Obaton, M. (1971), *C. R. Acad. Sci., Paris*, **272,** 2630.
Odintzova, S. V. (1941), *Dokl. Akad. Nauk, SSR*, **32,** 578.
O'Donnell, M. J. and Smith, B. E. (1978), *Biochem. J.*, **173,** 831.
Oes, A. (1913), *Z. Bot.*, **5,** 145.
Ohlsson, P. I. and Paul, K. G. (1976), *Acta Chem. Scand., B*, **30,** 373.
Okabe, K., Codd, G. A., and Stewart, W. D. P. (1979), *Nature, Lond.*, **279,** 525.
Okon, Y., Albrecht, S. L., and Burris, R. H. (1976a), *J. Bacteriol.*, **128,** 592.
Okon, Y., Albrecht, S. L., and Burris, R. H. (1976b), *J. Bacteriol.*, **127,** 1248.
Okon, Y., Albrecht, S. L., and Burris, R. H. (1977), *Appl. Environ. Microbiol.*, **33,** 85.
Okon, Y., Houchins, J. P., Albrecht, S. L., and Burris, R. H. (1977), *J. Gen. Microbiol.*, **98,** 87.

Oliver, D. (1868), *Flora of tropical Africa*, L. Reeve and Co., London.
O'Neal, D. and Joy, K. W. (1973), *Arch. Biochem. Biophys.*, **159,** 113.
O'Neal, D. and Joy, K. W. (1974), *Pl. Physiol.*, **54,** 773.
Orcutt, F. S. and Fred, E. B. (1935), *J. Am. Soc. Agron.*, **27,** 550.
Oremland, R. S. and Taylor, B. F. (1975), *Appl. Microbiol.*, **30,** 707.
Orme-Johnson, W. H. and Davis, L. C. (1977), in *Iron sulphur proteins* (Ed. W. Lovenberg). Vol. 3, Academic Press, New York, London.
Orme-Johnson, W. H., Davis, L. C., Henzl, M. T., Averill, B. A., Orme-Johnson, N. R., Münck, E., and Zimmerman, R. (1977), in *Recent developments in nitrogen fixation* (Eds. W. Newton, J. R. Postgate, and C. Rodríguez-Barrueco), Academic Press, London, New York, San Francisco, p. 131.
Osborn, M. J. and Munson, R. (1974), *Methods Enzymol.*, **31A,** 642.
O'Toole, P. and Knowles, R. (1973a), *Soil Biol. Biochem.*, **5,** 783.
O'Toole, P. and Knowles, R. (1973b), *Soil Biol. Biochem.*, **5,** 789.
Ottolenghi, P. (1975), *Biochem. J.*, **151,** 61.
Ovington, J. D. (1954), *J. Ecol.*, **42,** 71.
Ovington, J. D. (1956a), *J. Ecol.*, **44,** 171.
Ovington, J. D. (1956b), *New Phytol.*, **55,** 289.
Ovington, J. D. (1957a), *New Phytol.*, **56,** 1.
Ovington, J. D. (1957b), *Ann. Bot. N.S.*, **21,** 287.
Ovington, J. D. (1962), *Advan. Ecol. Res.*, **1,** 103.
Paech, K. and Tracey, M. V. (Eds.) (1956–64), *Modern methods of plant analysis*, 7 Volumes, Springer-Verlag, Berlin, Göttingen, Heidelberg.
Pagan, J. D., Child, J. J., Scowcroft, W. R., and Gibson, A. H. (1975), *Nature, Lond.*, **256,** 406.
Pan, S. and Schmidt, E. L. (1969), *Bacteriol. Proc.*, 14.
Pandey, D. C. (1965), *Nova Hedwigia Z. Kryptogamenkd.*, **9,** 299.
Pantastico, J. B. and Suayan, Z. A. (1973), *Philipp. Agric.*, **57,** 313.
Parker, C. A. (1957), *J. Soil Sci.*, **8,** 48.
Parker, C. A. (1961), *Aust. J. Exp. Biol.*, **39,** 515.
Parkinson, D., Gray, T. R. G., and Williams, S. T. (1971), *Methods for studying the ecology of soil microorganisms*, IBP Handbook No. 19, Blackwell, Oxford, Edinburgh.
Pataki, G. (1966), *Techniques of thin layer chromatography in amino acid and peptide chemistry*, Ann Arbor Science Publishers, Ann Arbor, Mich.
Patriquin, D. G. and Denike, D. (1978), *Aquat. Bot.*, **4,** 211.
Patriquin, D. G. and Döbereiner, J. (1978), *Can. J. Microbiol.*, **24,** 734.
Patriquin, D. G. and Keddy, C. (1978), *Aquat. Bot.*, **4,** 227.
Patriquin, D. and Knowles, R. (1972), *Mar. Biol.*, **16,** 49.
Paul, E. A. (1975), in *Nitrogen fixation by free-living micro-organisms* (Ed. W. D. P. Stewart), Cambridge University Press, Cambridge, London, New York, Melbourne, p. 259.
Paul, E. A., Myers, R. J., and Rice, W. A. (1971), in *Biological nitrogen fixation in natural and agricultural habitats* (Eds. T. A. Lie and E. G. Mulder), *Pl. Soil*, Special Volume, p. 495.
Pearson, H. W., Howsley, R., Kjeldsen, C. K., and Walsby, A. E. (1979), *FEMS Microbiol. Lett.*, **5,** 163.
Pedersen, W. L., Chakrabarty, K., Klucas, R. V., and Vidaver, A. K. (1978), *Appl. Environ. Microbiol.*, **35,** 129.
Pereira, P. A. A., Bülow, J. F. W. von, and Neyra, C. A. (1978a), *Rev. Bras. Ciên. Solo*, **2,** 28.

Pereira, P. A. A., Neyra, C. A., and Döbereiner, J. (1978b), in *Limitations and potentials for biological nitrogen fixation in the tropics* (Eds. J. Döbereiner, R. H. Burris, and A. Hollaender), Plenum Press, New York, p. 367.
Perry, J. H. (1950), *Chemical engineers' handbook*, McGraw-Hill, New York, p. 1868.
Persaud, H. B. (1977), in *Exploiting the legume–rhizobium symbiosis in tropical agriculture* (Eds. J. M. Vincent, A. S. Whitney, and J. Bose), University of Hawaii, College of Tropical Agriculture, Misc. Publ., No. 145, p. 441.
Peters, G. A. (1976), in *Proc. 1st int. symp. on nitrogen fixation* (Eds. W. E. Newton and C. J. Nyman), Vol. 2, Washington State University Press, Pullman, Washington, p. 592.
Peters, G. A. (1977), in *Genetic engineering for nitrogen fixation* (Eds. A. Hollaender *et al.*), Plenum Press, New York, p. 231.
Peters, G. A. and Mayne, B. C. (1974), *Pl. Physiol.*, **53,** 813.
Peters, G. A., Toia, Jr., R. E., Raveed, D., and Levine, N. J. (1978), *New Phytol.*, **80,** 583.
Peters, G. A., Mayne, B. C., Ray, T. B., and Toia, Jr., R. E. (1979), in *Nitrogen and rice symposium*, IRRI, Los Baños, Laguna, Philippines, p. 325.
Peterson, R. B. and Burris, R. H. (1976a), *Arch. Microbiol.*, **108,** 35.
Peterson, R. B. and Burris, R. H. (1976b), *Arch. Microbiol.*, **107,** 115.
Peterson, R. B. and Burris, R. H. (1978), *Arch. Microbiol.*, **116,** 125.
Peterson, R. B. and Wolk, C. P. (1978), *Proc. Nat. Acad. Sci. USA*, **75,** 6271.
Pfennig, N. A. (1977), *Rev. Microbiol.*, **31,** 275.
Phillips, D. A. (1971), *Physiol. Plant.*, **25,** 482.
Phillips, D. A. (1974a). *Pl. Physiol.*, **53,** 67.
Phillips, D. A. (1974b), *Pl. Physiol.*, **54,** 654.
Philpotts, H. (1976), *J. appl. Bact.*, **41,** 277.
Pienkos, P. T., Shah, V. K., and Brill, W. J. (1977), *Proc. Nat. Acad. Sci. USA*, **74,** 5468.
Pike, R. M. (1967), *Bacteriol. Rev.*, **31,** 157.
Pinner, S. H. (Ed.) (1967), *Modern packaging films*, Butterworths, London.
Pintner, I. J. and Provasoli, L. (1958), *J. Gen. Microbiol.*, **18,** 190.
Pirt, S. J. (1975), *Principles of microbe and cell cultivation*, Blackwell, Oxford.
Pizelle, G. and Thiery, G.-A. (1977), *Physiol. Vég.*, **15,** 333.
Planqué, K., Kennedy, I. R., de Vries, G. E., Quispel, A., and Brussel, A. A. N. van (1977), *J. Gen. Microbiol.*, **102,** 95.
Porter, F. E. and Scott, J. M. (1977), *VI Am. rhizobium conf., Gainesville, Florida.*
Porter, F. E., Nelson, I. S., and Wold, E. K. (1966), *Crops Soils*, **18,** 10.
Postgate, J. R. (1965), *Lab. Pract.*, **14,** 1140.
Postgate, J. R. (1966), *Lab. Pract.*, **15,** 1239.
Postgate, J. R. (1970), *J. Gen. Microbiol.*, **63,** 137.
Postgate, J. R. (Ed.) (1971), *The chemistry and biochemistry of nitrogen fixation*, Plenum Press, London, New York, 326 pp.
Postgate, J. R. (1972), *Methods in microbiology*, Vol. 6B, Academic Press, New York, p. 343.
Postgate, J. (1974), in *The biology of nitrogen fixation* (Ed. A. Quispel), North Holland, Amsterdam, p. 663.
Postgate, J. R., Crumpton, J. E., and Hunter, J. R. (1961), *J. Gen. Microbiol.*, **24,** 15.
Potrikus, C. J. and Breznak, J. A. (1977), *Appl. Environ. Microbiol.*, **33,** 392.
Prakash, R. K., Hooykaas, P. J. J., Ledeboer, A. M., Kijne, J., Schilperoort, R. A., Nuti, M. P., Lepidi, A. A., Casse, F., Boucher, C., Julliot, J. S., and Denarié, J.

(1980), in *Nitrogen fixation,* (Eds. W. Newton and W. H. Orme-Johnson), University Park Press, Baltimore, p. 139.
Prantl, K (1889), *Hedwigia*, **28,** 135.
Pratt, P. F., Chapman, H. D. and Garber, M. J. (1960), *Soil Sci.*, **90,** 293.
Primrose, S. B. (1976), *J. Gen. Microbiol.*, **97,** 343.
Pringsheim, E. G. (1918), *Arch. Protistenk.*, **38,** 127.
Proksch, G. (1972), in *Isotopes and radiation in soil–plant relationships including forestry*, International Atomic Energy Agency, Vienna, p. 217.
Proskauer, J. (1948), *Ann. Bot.*, **12,** 237.
Proskauer, J. (1960), *Phytomorphology*, **10,** 1.
Proskauer, J. (1967), *Phytomorphology*, **17,** 61.
Proskauer, J. (1969), *Phytomorphology*, **19,** 52.
Provasoli, L., McLaughlin, J. J. A., and Droop, M. R. (1957), *Arch. Mikrobiol.*, **25,** 392.
Prusiner, S. and Milner, L. (1970), *Anal. Biochem.*, **37,** 429.
Pugashetti, B. K., Gopalgowda, H. S., and Patil, R. B. (1971), *Curr. Sci.*, **18,** 494.
Purchase, H. F. and Nutman, P. S. (1957), *Ann. Bot., Lond.*, **21,** 439.
Quispel, A. (1954), *Acta Bot. Neerl.*, **3,** 495.
Quispel, A. (Ed.) (1974), *The biology of nitrogen fixation*, North Holland, Amsterdam, 769 pp.
Racker, E. (1950), *Biochim. Biophys. Acta*, **4,** 211.
Raggio, M., Raggio, N., and Torrey, J. G. (1957), *Am. J. Bot.*, **44,** 325.
Raggio, M., Raggio, N., and Torrey, J. G. (1965), *Pl. Physiol.*, **40,** 601.
Raimbault, M. (1975), *Ann. Microbiol. (Inst. Pasteur)*, **126A,** 247.
Rains, D. W. and Talley, S. N. (1979), in *Nitrogen and rice symposium*, IRRI, Los Baños, Laguna, Philippines, p. 419.
Raju, P. N., Evans, H. J., and Seidler, R. J. (1972), *Proc. Nat. Acad. Sci. USA*, **69,** 3474.
Rawlins, S. L. (1976), in *Water deficits and plant growth* (Ed. T. T. Kozlowski), Vol. IV, Academic Press, New York, p. 1.
Rawlings, J., Shar, V. K., Chisnell, J. R., Brill, W. J., Zimmerman, R., Münck, E., and Orme-Johnson, W. H. (1978), *J. Biol. Chem.*, **253,** 1001.
Ray, T. B., Mayne, B. C., Peters, G. A., and Toia, R. (1979), *Pl. Physiol.*, Suppl. 623.
Rebello, J. L. and Strauss, N. (1969), *J. Bacteriol.*, **98,** 683.
Reiderer-Henderson, M. and Wilson, P. W. (1970), *J. Gen. Microbiol.*, **61,** 27.
Reinke, J. (1872), *Gott. Nachr.*, **57,** 100.
Rennie, D. A., Paul, E. A., and Johns, L. E. (1976), *Can. J. Soil Sci.*, **56,** 43.
Repaske, R. and Wilson, P. W. (1952), *J. Am. Chem. Soc.*, **74,** 3101.
Reporter, M. and Hermina, N. (1975), *Biochem. Biophys. Res. Commun.*, **64,** 1126.
Renaut, J., Sasson, A., Pearson, H. W., and Stewart, W. D. P. (1975), in *Nitrogen fixation by free-living micro-organisms* (Ed. W. D. P. Stewart), Cambridge University Press, Cambridge, London, New York, Melbourne, p. 229.
Reynaud, P. A. and Roger, P. A. (1978), *Ecol. Bull., Stockholm*, **26,** 148.
Reynaud, P. A. and Roger, P. A. (1979), *C.R. Acad. Sci., Paris, Ser. D*, **288,** 999.
Reynolds, P. H. S. and Farnden, K. J. F. (1979), *Phytochemistry*, **18,** 1625.
Rice, W. A. and Paul, E. A. (1971), *Can. J. Microbiol.*, **17,** 1049.
Richards, B. N. (1964), *Aust. For.*, **28,** 68.
Rinaudo, G., Balandreau, J., and Dommergues, Y. (1971), in *Biological nitrogen fixation in natural and agricultural habitats* (Eds. T. A. Lie and E. G. Mulder), *Pl. Soil*, Special Volume, p. 471.

Rippka, R. and Waterbury, J. B. (1977), *FEMS Microbiol. Lett.*, **2,** 83.
Rippka, R. and Stainer, R. Y. (1978), *J. Gen. Microbiol.*, **105,** 83.
Rippka, R., Neilson, A., Kunisawa, R., and Cohen-Bazire, G. (1971), *Arch. Mikrobiol.*, **76,** 341.
Rippka, R., Deruelles, J., Waterbury, J. B., Herdman, M., and Stainer, R. Y. (1979), *J. Gen. Microbiol.*, **111,** 1.
Rittenberg, D. (1946), in *Preparation and measurement of isotopic tracers* (Eds. D. W. Wilson *et al.*), J. W. Edwards, Ann Arbor, Mich.
Roberg, M. (1934), *J. Wiss. Bot.*, **79,** 472.
Roberts, E. H. (1972), in *Viability of seeds* (Ed. E. H. Roberts), Chapman and Hall, London, p. 14.
Robertson, J. G. and Taylor, M. P. (1973), *Planta*, **112,** 1.
Robertson, J. G., Farnden, K. J. F., Warburton, M. P., and Banks, J. M. (1975a), *Aust. J. Pl. Physiol.*, **2,** 265.
Robertson, J. G., Warburton, M. P., and Farnden, K. J. F. (1975b), *FEBS Lett.*, **55,** 33.
Robertson, J. G., Warburton, M. P., Lyttleton, P., Fordyce, A. M. and Bullivant, S. (1978), *J. Cell. Sci.*, **30,** 151.
Robertson, J. G. and Farnden, K. J. F. (1980), in *The biochemistry of plants—a comprehensive treatise* (Eds. P. K. Stumpf and E. E. Conn), Vol. 5, Academic Press, New York. *Amino acids and derivatives* (Ed. B. J. Miflin), Chapter 2, in press.
Robinson, A. C. (1968), *Aust. J. Exp. Agric. Anim. Husb.*, **8,** 327.
Robinson, A. C. (1969a), *Aust. J. Agric. Res.*, **20,** 827.
Robinson, A. C. (1969b), *Aust. J. Agric. Res.*, **20,** 1053.
Robinson, B. L. and Miller, J. H. (1970), *Physiol. Plant.*, **23,** 461.
Robinson, C. F. (1960), in *Physical methods in chemical analysis* (Ed. W. G. Berl), Vol. 1, Academic Press, New York, London, p. 463.
Rodgers, G. A. (1978), *Pl. Physiol.*, **44,** 407.
Rodgers, G. A. and Stewart, W. D. P. (1977), *New Phytol.*, **78,** 441.
Rodríguez-Barrueco, C. (1968), *Bot. J. Linn. Soc.*, **62,** 77,
Roger, P. and Reynaud, P. (1976), *Rev. Ecol. Biol. Sol.*, **13,** 545.
Roger, P. and Reynaud, P. (1977), *Rev. Ecol. Biol. Sol.*, **14,** 519.
Roger, P. A., Reynaud, P. A., Rinaudo, G., Ducerf, P. E., and Traore, T. M. (1977), *Cah. ORSTOM , Sér. Biol.*, **12,** 133.
Roger, P. and Reynaud, P. (1978), *Rev. Ecol. Biol. Sol.*, **15,** 219.
Roger, P. A. and Reynaud, P. A. (1979), in *Nitrogen and rice symposium*, IRRI, Los Baños, Laguna, Philippines, p. 287.
Rogers, V. E. (1969), *Aust. CSIRO Div. Pl. Ind. Fd. Stn. Rec.*, **8,** 37.
Rognes, S. E. (1970), *FEBS Lett.*, **10,** 62.
Rognes, S. E. (1975), *Phytochemistry*, **14,** 1975.
Roitt, I. M. (1974), *Essential immunology*, 2nd Edition, Blackwell, Oxford.
Rokosh, D. A., Kurz, W. G. W., and La Rue, T. A. (1973), *Anal. Biochem.*, **54,** 477.
Rosenblum, E. D. and Wilson, P. W. (1949), *J. Bacteriol.*, **57,** 413.
Ross, P. J., Martin, A. E., and Henzell, E. F. (1964), *Nature, Lond.*, **204,** 444.
Ross, P. J., Martin, A. E., and Henzell, E. F. (1968), *Trans. 9th Int. Congr. Soil Sci. Soc.*, **2,** p. 487.
Rothamsted Experimental Station (1972), *The Rothamsted collection of strains. Catalogue of strains 1972*, Rothamsted Experimental Station, Harpenden.
Roughley, R. J. (1962), *Agric. Gaz. N.S.W.*, **73,** 260.
Roughley, R. J. (1968), *J. Appl. Bacteriol.*, **31,** 259.

Roughley, R. J. (1970), *Pl. Soil*, **32,** 675.
Roughley, R. J. (1976), in *Symbiotic nitrogen fixation in plants* (Ed. P. S. Nutman), Cambridge University Press, Cambridge, London, New York, Melbourne, p. 125.
Roughley, R. J. and Thompson, J. A. (1978), *J. Appl. Bacteriol.*, **44,** 317.
Roughley, R. J. and Vincent, J. M. (1967), *J. Appl. Bacteriol.*, **30,** 362.
Roughley, R. J. and Dart, P. J. (1970), *J. Expt. Bot.*, **21,** 776.
Roughley, R. J., Blowes, W. M., and Herridge, D. F. (1976), *Soil Biol. Biochem.*, **8,** 403.
Roughley, R. J., Date, R. A., and Walker, M. H. (1966), *Agric. Gaz. N.S.W.*, **77,** 142.
Rovira, A. D. (1963), *Pl. Soil*, **19,** 304.
Rowell, P., Sampaio, M. J. A. M., and Stewart, W. D. P. (1978), *Proc. Soc. Gen. Microbiol.*, **5,** 103.
Rowell, P., Sampaio, M. J. A. M., Ladha, J. K., and Stewart, W. D. P. (1979), *Arch. Microbiol.*, **120,** 195.
Rowell, P., Diez, J., Apte, S. K., and Stewart, W. D. P. (1980), in preparation.
Ruben, S., Hassid, W. Z., and Kamen, M. D. (1940), *Science*, **91,** 578.
Ruegg, J. J. and Alston, A. M. (1978), *Aust. J. Agric. Res.*, **29,** 951.
Rundel, P. W. and Neel, J. M. (1978), *Flora*, **167,** 127.
Ruschel, A. P. and Döbereiner, J. (1965), *Proc. IX Int. Grassl. Congr. (São Paulo)*, **2,** 1103.
Ruschel, A. P. and Vose, P. B. (1977), *Bol. Sci.*, Cent. Energ. Nucl. Agric., Piracicaba, Brazil, 12 pp.
Russell, S.A. and Evans, H. J. (1970), quoted in Hardy, R. W. F., Burns, R. C., and Holsten, R. O. (1973), *Soil Biol. Biochem.*, **5,** 47.
Ryan, E., Bodley, F. and Fottrell, P. F. (1972), *Phytochemistry*, **11,** 957.
Sahni, V. P. (1977), in *Exploiting the legume–rhizobium symbiosis in tropical agriculture,* (Eds. J. M. Vincent, A. S. Whitney and J. Bose), University of Hawaii College of Tropical Agriculture, Misc. Pub., No. 145, p. 413.
Sampaio, M. J. A. M., Rowell, P., and Stewart, W. D. P. (1979), *J. Gen. Microbiol.*, **111,** 181.
Sankaram, A., Mudholkar, N. J., and Sahay, M. H. (1967), *Indian J. Microbiol.*, **7,** 57.
Schaede, R. (1947), *Planta*, **35,** 319.
Schall, E. D., Shenberger, L. C., and Swope, A. (1975), *Inspection of legume inoculants and pre-inoculated seeds*, Inspection Report No. 106, Purdue University Agricultural Experiment Station, Indiana.
Schell, D. M. and Alexander, V. (1970), *Limnol. Oceanogr.*, **15,** 961.
Schloesing, T. and Laurent, F. M. (1892), *Ann. Inst. Pasteur*, **6**, 824.
Schmidt, E. L. (1974), *Soil Sci.*, **118,** 141.
Schneider, A. (1894), *Bot. Gaz.*, **19,** 25.
Schneider, K. and Schlegel, H. G. (1977), *Arch. Microbiol.*, **112,** 229.
Schoenheimer, R. and Rittenberg, D. (1939), *J. Biol. Chem.*, **127,** 285.
Scholander, P. F., Hammel, H. T., Bradstreet, E. D., and Hemmingsen, E. A. (1965), *Science*, **148,** 339.
Scholl, R. L., Harper, J. E., and Hageman, R. H. (1974), *Pl. Physiol.,* **53,** 825.
Schöllhorn, R. and Burris, R. H. (1966), *Fed. Proc.,* **24,** 710.
Schöllhorn, R. and Burris, R. H. (1967a), *Proc. Nat. Acad. Sci. USA,* **57,** 1317.
Schöllhorn, R. and Burris, R. H. (1967b), *Proc. Nat. Acad. Sci. USA,* **58,** 213.
Schubert, K. R. and Evans, H. J. (1976), *Proc. Nat. Acad. Sci. USA,* **73,** 1207.
Schubert, K. R., Engelke, J. A., Russell, S. A., and Evans, H. J. (1977), *Pl. Physiol.,* **60,** 651.

Schwinghamer, E. A. (1965), *Aust. J. Biol. Sci.,* **18,** 333.
Schwinghamer, E. A. (1967), *Antonie van Leeuwenhoek J. Microbiol. Serol.,* **33,** 121.
Schwinghamer, E. A. (1968), *Can. J. Microbiol.,* **14,** 355.
Schwinghamer, E. A. (1969), *Can. J. Microbiol.,* **15,** 611.
Schwinghamer, E. A. (1977), in *A treatise on dinitrogen fixation, Section III, biology* (Eds. R. W. F. Hardy and W. S. Silver), Wiley, New York, p. 577.
Schwinghamer, E. A., Evans, H. J., and Dawson, M. D. (1970), *Pl. Soil,* **33,** 192.
Scott, G. D. (1969), *Plant Symbiosis,* Edward Arnold, London.
Scott, D. B., Farnden, K. J. F., and Robertson, J. G. (1976), *Nature, Lond.,* **263,** 703.
Scott, J. R. and Bumgarner, H. R. (1965), *U.S. Pat., 3 168 796.*
Scowcroft, W. R. and Gibson, A. H. (1975), *Nature, Lond.,* **253,** 351.
Seetin, M. W. and Barnes, D. K. (1977), *Crop Sci.,* **17,** 783.
Sekiguchi, T. and Nosoh, Y. (1973), *Biochem. Biophys. Res. Commun.,* **51,** 331.
Sestak, A. (1971), in *Plant photosynthetic production. Manual of methods* (Eds. A. Sestak, J. Catsky, and P. G. Jarvin), Dr. W. Junk Publishers, The Hague, p. 672.
Shah, V. K. and Brill, W. J. (1973), *Biochim. Biophys. Acta,* **305,** 445.
Shah, V. K. and Brill, W. J. (1977), *Proc. Nat. Acad. Sci. USA,* **74,** 3249.
Shah, V. K., Davis, L. C., and Brill, W. J. (1972), *Biochim. Biophys. Acta,* **256,** 498.
Shah, V. K., Chisnell, J. R., and Brill, W. J. (1978), *Biochem. Biophys. Res. Commun.,* **81,** 232.
Shanmugam, K. T., O'Gara, F., Andersen, K., and Valentine, R. C. (1978), *Ann. Rev. Pl. Physiol.,* **29,** 263.
Shapiro, B. M. and Stadtman, E. R. (1970), *Methods Enzymol.,* **17A,** 910.
Shaw, D. E., Trinick, M. J., Layton, W. A., and Cartledge, E. G. (1972), *Papua N.G. agric. J.,* **23,** 12.
Shaw, N. H. and Bryan, W. W. (Eds.) (1976), *Tropical pasture research, Principles and methods,* CAB Bull. No. 51, Commonwealth Agricultural Bureau, Farnham Royal, 454 pp.
Shaw, N. H., t'Mannetje, L., Jones, R. M., and Jones, R. J. (1976), in *Tropical pasture research, principles and methods,* CAB Bull. No. 51, Commonwealth Agricultural Bureau, Farnham Royal, p. 235.
Shearer, G. B., Kohl, D. H., and Commoner, B. (1974a), *Soil Sci.,* **118,** 308.
Shearer, G., Duffy, J., Kohl, D. H., and Commoner, B. (1974b), *Soil Sci. Soc. Am. Proc.,* **38,** 315.
Sherwood, M. T. and Masterson, C. L. (1974), *Ir J. Agric. Res.,* **13,** 101.
Shestakov, S. V. and Nguen Than Khyen (1970), *Molec. Gen. Genet.,* **107,** 372.
Sievers, G., Huhtala, M. J., and Ellfolk, N. (1978), *Acta Chem. Scand., B,* **32,** 380.
Silvester, W. B. (1976), in *Symbiotic nitrogen fixation in plants* (Ed. P. S. Nutman), Cambridge University Press, Cambridge, London, New York, Melbourne, p. 521.
Silvester, W. B. and Bennett, K. J. (1973), *Soil Biol. Biochem.,* **5,** 171.
Silvester, W. B. and McNamara, P. J. (1976), *New Phytol.,* **77,** 135.
Silvester, W. B. and Smith, D. R. (1969), *Nature, Lond.,* **224,** 1231.
Simpson, F. J., Okeren, I., and Peardon, G. (1971), *Can. J. Microbiol.,* **17,** 403.
Simpson, J. R. and Gibson, A. H. (1970), *Soil Biol. Biochem.,* **2,** 295.
Sims, A. P. and Cocking, E. C. (1958), *Nature, Lond.,* **181,** 474.
Sinclair, A. G. (1973), *N.Z. J. Agric. Res.,* **16,** 263.
Sinclair, A. G., Hannagan, R. B., Johnstone, P., and Hardacre, A. K. (1978), *N.Z. J. Exp. Agric.,* **6,** 65.
Sinclair, A. G., Hannagan, R. B., and Risk, W. H. (1976), *N.Z. J. Agric. Res.,* **19,** 451.
Singer, M., Holding, A. J., and King, J. (1964), *Trans. 8th Int. Congr. Soil Sci. Soc.,* **3,** 1021.
Singh, R. N. (1942), *Indian J. Agric. Sci.,* **12,** 743.

Singh, R. N. (1961), *Role of blue-green algae in nitrogen economy of Indian agriculture,* Indian Council of Agricultural research, New Delhi.
Singh, P. K. (1973), *Arch. Mikrobiol.,* **92,** 59.
Singh, P. K. (1979), in *Nitrogen and rice symposium,* IRRI, Los Baños, Laguna, Philippines, p. 407.
Singh, P. K. and Subudhi, B. P. R. (1978), *Ind. Farming,* **27,** 37.
Skerman, V. B. D. (1969), *Abstracts of microbiological methods,* Wiley–Interscience, New York.
Skinner, F. A., Roughley, R. J., and Chandler, M. R. (1977), *J. appl. Bact.,* **43,** 287.
Škrdleta, V. and Pelikan, J. (1973), *Zentralbl. Bakteriol. Parasitenkd. Infektionskr. Hyg., Abt. II,* **128,** 745.
Sloger, C. (1969), *Pl. Physiol.,* **44,** 1666.
Sloger, C. and Silver, W. S. (1965), in *Non-heme iron proteins: role in energy conversion* (Ed. A.San Pietro), Antioch Press, Yellow Springs, Ohio, p. 299.
Sloger, C., Bezdicek, D., Milberg, R., and Boonkerd, N. (1975), in *Nitrogen fixation by free-living micro-organisms* (Ed. W. D. P. Stewart), Cambridge University Press, Cambridge, London, New York, Melbourne, p. 271.
Smith, B. E., Lowe, D. J., and Bray, R. C. (1973), *Biochem. J.,* **135,** 331.
Smith, B. E. and Lang, G. (1974), *Biochem. J.,* **137,** 169.
Smith, B. E., Thorneley, R. N. F., Yates, M. G., Eady, R. R., and Postgate, J. R. (1976a), *Proc. 1st int. symp. nitrogen fixation* (eds. W. Newton and C. J. Nyman), Vol. 1, Washington State University Press, Pullman, Washington, p. 150.
Smith, B. E., Thorneley, R. N. F., Eady, R. R., and Mortenson, L. E. (1976b), *Biochem. J.,* **157,** 439.
Smith, B. E., Eady, R. R., Thorneley, R. N. F., Yates, M. G., and Postgate, J. R. (1977), in *Recent developments in nitrogen fixation* (Eds. W. Newton, J. R. Postgate and C. Rodríguez-Barrueco), Academic Press, London, New York, San Francisco, p. 119.
Smith, C. E., Jr. (1971), *U.S. Dept. Agric. Inf. Bull.,* No. 348.
Smith, K. A. and Restall, S. W. F. (1971), *J. Soil. Sci.,* **22,** 430.
Smith, K. C. and Hanawalt, P. C. (1969), *Molecular photobiology: inactivation and recovery,* Academic Press, London.
Smith, I. (1969), *Chromotographic and electrophoretic techniques,* Vol.s I and II, Third Edition, Heinemann, London.
Smith, J. H., Legg, J. O., and Carter, J. M. (1963), *Soil Sci.,* **96,** 313.
Smith. L. A., Hill, S., and Yates, M. G. (1976), *Nature, Lond.,* **262,** 209.
Smith, R. W. and Evans, M. C. W. (1970), *Nature, Lond.,* **225,** 1253.
Smith, W. K. (1951), *Proc. Soc. Appl. Bacteriol.,* **16,** 139.
Snedecor, G. W. and Cochran, W. G. (1967), *Statistical methods,* Sixth Edition, Iowa State University Press, Ames, Iowa.
Snieszko, S. F. (1941), *Canning Age,* **22,** 49.
Sohngen, N. L. (1906), *Zentralbl. Bakteriol. Parasitentkd. Infektionskr. Hyg., Ab. II.,* **15,** 513.
Solomons, G. L. (1969), *Materials and methods in fermentations,* Academic Press, London.
Sonesson, M. (1973), in *Progress report 1972* (Ed. M. Sonesson), Swedish IBP Tundra Project Tech. Rep., **14,** 66.
Southern, E. M. (1975), *J. Molec. Biol.,* **98,** 503.
Spackman, D. H., Stein, W. H., and Moore, S. (1958), *Anal. Chem.,* **30,** 1190.
Spackman, D. H. (1967), *Methods Enzymol.,* **11,** 3.
Spencer, J. F. T. and Newton, J. D. (1953), *Can. J. Bot.,* **31,** 253.

Spiff, E. D. and Odu, C. T. I. (1972), *Soil Biol. Biochem.*, **4**, 71.
Spiff, E. D. and Odu, C. T. I. (1973), *J. Gen. Microbiol.*, **78**, 207.
Spooner, D. F. and Sykes, G. (1972), in *Methods in Microbiology* (Eds. J. R. Norris and D. W. Ribbons), Vol. 7B, Academic Press, London, p. 211.
Spratt, E. R. (1915), *Ann. Bot.*, **29**, 619.
Sprent, J. I. (1969), *Planta (Berl.)*, **88**, 372.
Sprent, J. I. (1972), *New Phytol.*, **71**, 603.
Sprent, J. I. (1976), in *Water deficits and plant growth* (Ed. T. T. Kozlowski), Vol. IV, Academic Press, New York, p. 291.
Sprent, J. I. and Silvester, W. B. (1971), *New Phytol.*, **72**, 991.
St. John, R. T., Johnston, H. M., Seidman, C., Garfinkel, D., Gordon, J. K., Shah, V. K., and Brill, W. J. (1975), *J. Bacteriol.*, **121**, 759.
Stacey, G., Tabita, F. R., and Van Baalen, C. (1977), *J. Bacteriol.*, **132**, 596.
Stadtman, E. R. and Ginsburg, A. (1974), in *The enzymes* (Ed. P. D. Boyer), Academic Press, New York, p. 755.
Stafleu, F. A., Bonner, C. E. B., McVaugh, R., Meikle, R. D., Rollins, R. C., Roff, R., Schops, J. M., Schultz, G. M., de Vilmorin, R., and Voss, E. G. (1972), *International code of botanical nomenclature,* A. Oosthoeck, Utrecht.
Stahl, E. and Ashwood, P. (1969), *This layer chromotography,* 2nd Edition, Springer-Verlag, Berlin.
Stanford, N. P., Campelo, A. B., and Döbereiner, J. (1968), in *IVth Reun. Latino-Am. Inoculantes Legumin., Porto Alegre, Brazil,* p. 76.
Stanier, R. Y. (1977), *Carlsberg Res. Commun.*, **42**, 77.
Stanier, R. Y. and Cohen-Bazire, G. (1977), *Ann. Rev. Microbiol.*, **31**, 225.
Stanier, R. Y., Kunisawa, R., Mandel, M., and Cohen-Bazire, G. (1971), *Bacteriol. Rev.*, **35**, 171.
Stanier, R. Y., Sistrom, W. R., Hansen, T. A., Whitton, B. A., Castenholz, R. W., Pfenning, N., Gorlenko, V. N., Kondratieva, E. N., Eimhjellen, K. E., Whittenbury, R., Gherna, R. L., and Trüper, H. G. (1978), *Int. J. Syst. Bacteriol.*, **28**, 335.
Stanley, P. E. and Williams, S. G. (1969), *Anal. Biochem.*, **29**, 381.
Staphorst, J. L. and Strijdom, B. W. (1972), *Phytophylactica*, **4**, 29.
Starlinger, P. and Saedler, H. (1976), *Curr. Topics Microbiol. Immunol.*, **75**, 111.
Stefanson, R. C. (1970), *J. Agric. Eng. Res.*, **15**, 295.
Stefanson, R. C. (1973), *Soil Biol. Biochem.*, **5**, 869.
Steinborn, J. and Roughley, R. J. (1974), *J. Appl. Bacteriol.*, **37**, 93.
Stephen, H. and Stephen, T. (1963), *Solubilities of inorganic and organic compounds,* Vol. 1, Part I, Macmillan, New York.
Stephens, C. G. (1943), *Trans. Roy. Soc. S. Aust.*, **67**, 191.
Stewart, W. D. P. (1962), *Ann. Bot. Lond., N.S.*, **26**, 439.
Stewart, W. D. P. (1966), *Nitrogen fixation in plants,* Athlone Press, London.
Stewart, W. D. P. (1970), *Pl. Soil,* **32**, 555.
Stewart, W. D. P. (1973), *Ann. Rev. Microbiol.*, **27**, 283.
Stewart, W. D. P. (Ed.) (1975), *Nitrogen fixation by free-living micro-organisms,* Cambridge University Press, London, New York, Melbourne, 471 pp.
Stewart, W. D. P. (1977), in *A treatise on dinitrogen fixation, Section III, biology* (Eds. R. W. F. Hardy and W. S. Silver), Wiley, New York, p. 63.
Stewart, W. D. P. (1978), *Endeavour, N.S.*, **2**, 170.
Stewart, W. D. P. and Codd, G. A. (1975), *Brit. Phycol. J.*, **10**, 273.
Stewart, W. D. P. and Daft, M. J. (1977), in *Advances in Aquatic Microbiology* (Ed. M. R. Droop), Vol. I, Academic Press, London, p. 177.
Stewart, W. D. P. and Lex, M. (1970), *Arch. Mikrobiol.*, **73**, 250.

Stewart, W. D. P. and Pearson, M. C. (1967), *Pl. Soil,* **26,** 348.
Stewart, W. D. P. and Rodgers, G. A. (1977), *New Phytol.,* **78,** 459.
Stewart, W. D. P. and Rowell, P. (1977), *Nature, Lond.,* **265,** 371.
Stewart, W. D. P., Fitzgerald, G. P., and Burris, R. J. (1967), *Proc. Nat. Acad. Sci. USA,* **58,** 2071.
Stewart, W. D. P., Fitzgerald, G. P., and Burris, R. H. (1968), *Arch. Mikrobiol.,* **62,** 336.
Stewart, W. D. P., Haystead, A., and Pearson, H. W. (1969), *Nature, Lond.,* **224,** 226.
Stewart, W. D. P., Mague, T. M., Fitzgerald, G. P., and Burris, R. H. (1971), *New Phytol.,* **70,** 497.
Stewart, W. D. P., Haystead, A., and Dharmawardene, M. W. N. (1975), in *Nitrogen fixation by free-living micro-organisms* (Ed. W. D. P. Stewart), Cambridge University Press, Cambridge, London, New York, Melbourne, p. 129.
Stewart, W. D. P., Rowell, P., and Apte, S. K. (1977), in *Recent developments in nitrogen fixation* (Eds. W. Newton, J. R. Postgate, and C. Rodríguez-Barrueco), Academic Press, London, New York, San Francisco, p. 287.
Stewart, W. D. P., Sampaio, M. J., Isichei, A. O., and Sylvester- Bradley, R. (1978), in *Limitation and potentials for biological nitrogen fixation in the tropics* (Ed. J. Döbereiner *et al.*), Plenum Press, New York, p. 41.
Stewart, W. D. P., Rowell, P., Ladha, J. K., and Sampaio, M. J. A. M. (1979), in *Nitrogen and rice symposium,* IRRI, Los Baños, Laguna, Philippines, p. 263.
Stone, S. R., Copeland, L., and Kennedy, I. R. (1979), *Phytochemistry,* **18,** 1273.
Straatman, M. G. (1977), *Int. J. App. Radiat. Isotop.,* **28,** 13.
Streeter, J. G. (1973), *Arch. Biochem. Biophys.,* **157,** 613.
Streeter, J. G. (1977), *Pl. Physiol.,* **60,** 235.
Streicher, S. L., Gurney, E. G., and Valentine, R. C. (1971), *Proc. Natl Acad. Sci. USA,* **65,** 74.
Streicher, S., Gurney, E. G., and Valentine, R. C. (1972), *Nature, Lond.,* **239,** 495.
Streicher, S. L., Shanmugam, K. T., Ausubel, F., Morandi, D., and Goldberg, R. B. (1974), *J. Bacteriol.,* **120,** 815.
Strength, W. J., Isani, B., Linn, D. M., Williams, F. D., Vandermolen, G. E., Laughon, B. E., and Krieg, N. R. (1976), *Int. J. Syst. Bacteriol.,* **26,** 253.
Strijdom, B. W. and Allen, O. N. (1966), *Can. J. Microbiol.,* **12,** 275.
Strijdom, B. W. and Deschodt, C. C. (1976), in *Symbiotic nitrogen fixation in plants* (Ed. P. S. Nutman), Cambridge University Press, Cambridge, London, New York, Melbourne, p. 151.
Strong, J. H. (1937), *J. Counc. Sci. Ind. Res.,* **10,** 12.
Stutz, R. C. and Bliss, L. C. (1973), *Pl. Soil,* **38,** 209.
Subramanyan, R., Relwani, L. L., and Manna, G. B. (1965), *Proc. Nat. Acad. Sci. India,* **35,** 382.
Subudhi, B. P. R. and Singh, P. K. (1978a), *Poult. Sci.,* **57,** 378.
Subudhi, B. P. R. and Singh, P. K. (1979), *Biol. Plant,* **21,** 66.
Summerfield, R. J., Huxley, P. A., and Steele, W. (1974), *Fld. Crop Abstr.,* **27,** 301.
Summerfield, R. J., Wein, H. C., and Minchin, F. R. (1976), *Exp. Agric.,* **12,** 241.
Summerfield, R. J., Minchin, F. R., Roberts, E. H., and Wien, H. C. (1977), *Pl. Sci. Lett.,* **8,** 355.
Sutherland, J., Herdman, M., and Stewart, W. D. P. (1979), *J. Gen. Microbiol.,* **115,** 273.
Sutton, W. D. and Mahoney, P. (1977), *Pl. Physiol.,* **60,** 800.
Swanstrom, R. and Shank, P. R. (1978), *Anal. Biochem.,* **86,** 184.

Sweetser, P. B. (1967), *Anal. Chem.,* **39,** 979.
Swisher, R. H., Landt, M. L., and Reithel, F. J. (1977), *Biochem. J.,* **163,** 427.
Syono, K., Newcomb, W., and Torrey, J. G. (1976), *Can. J. Bot.,* **54,** 2155.
Talley, S. N., Talley, B. J., and Rains, D. W. (1977), in *Genetic engineering for nitrogen fixation* (Ed. A. Hollaender *et al.*), Plenum Press, New York, p. 259.
Tanaka, M., Haniu, M., Yasunobu, K. T., and Mortenson, L. E. (1977a), *J. Biol. Chem.,* **252,** 7081.
Tanaka, M., Haniu, M., Yasunobu, K. T., and Mortenson, L. E. (1977b), *J. Biol. Chem.,* **252,** 7089.
Tanaka, M., Haniu, M., Yasunobu, K. T., and Mortenson, L. E. (1977c), *J. Biol. Chem.,* **252,** 7093.
Tanner, J. W. and Anderson, I. C. (1963), *Canad. J. Pl. Sci.,* **43.** 542.
Tarrand, J. J., Krieg, N. R., and Döbereiner, J. (1978), *Can. J. Microbiol.,* **24,** 967.
Tchan, Y. T. (1953), *Proc. Linn. Soc. N.S.W.,* **77,** 265.
Tchan, Y. T. and de Ville, R. (1970), *Ann. Inst. Pasteur,* **118,** 665.
Tel-Or, E. and Stewart, W. D. P. (1976), *Biochim. Biophys, Acta,* **423,** 189.
Tel-Or, E. and Stewart, W. D. P. (1977), *Proc. Roy. Soc. Lond., B.,* **198,** 61.
Tel-Or, E., Luijk, L. W., and Packer, L. (1977), *FEBS Lett.,* **78,** 49.
Tel-Or, E., Luijk, W. W., and Packer, L. (1978), *Arch. Biochem. Biophys.,* **185,** 185.
Thake, B. and Rawle, P. R. (1972), *Arch. Mikrobiol.,* **85,** 39.
Thode, H. G. and Urey, H. C. (1939), *J. Chem. Phys.,* **7,** 34.
Thomas, J. (1970), *Nature, Lond.,* **228,** 181.
Thomas, J., Wolk, C. P., Shaffer, P. W., Austin, S. M., and Galonsky, A. (1975), *Biochem. Biophys. Res. Commun.,* **67,** 501.
Thomas, J., Meeks, J. C., Wolk, C. P., Shaffer, P. W., Austin, S. M., and Chien, W.-S. (1977), *J. Bacteriol.,* **129,** 1545.
Thompson, J. A., Roughley, R. J., and Herridge, D. F. (1974), *Pl. Soil,* **40,** 511.
Thompson, J. A., Brockwell, J., and Roughley, R. J. (1975), *J. Aust. Inst. Agric. Sci.,* **41,** 253.
Thompson, J. A. and Vincent, J. M. (1967), *Pl. Soil,* **26,** 72.
Thompson, J. P. (1976), *The taxonomy of the Azotobacteraceae examined by numerical methods of classification,* PhD Thesis, University of Queensland, Brisbane.
Thomson, J. J. (1907), *Phil. Mag.,* **13,** 561.
Thorneley, R. N. F. (1974), *Biochim. Biophys. Acta,* **333,** 487.
Thorneley, R. N. F. and Eady, R. R. (1974), *Biochem. J.,* **133,** 405.
Thorneley, R. N. F. and Eady, R. R. (1977), *Biochem, J.,* **167,** 457.
Thorneley, R. N. F. and Willison, K. R. (1974), *Biochem. J.,* **139,** 211.
Thorneley, R. N. F., Eady, R. R., and Yates, M. G. (1975), *Biochim. Biophys. Acta,* **403,** 269.
Thorneley, R. N. F., Yates, M. G., and Lowe, D. J. (1976), *Biochem. J.,* **155,** 137.
Thorneley, R. N. F., Eady, R. R., and Lowe, D. J. (1978), *Nature, Lond.,* **272,** 557.
Thornton, H. G. (1930a), *Ann. Bot., Lond.,* **44,** 385.
Thornton, H. G. (1930b), *Proc. Roy. Soc. Lond., B,* **106,** 110.
Tiffney, W. N., Benson, W. R., and Eveleigh, D. E. (1978), *Am. J. Bot.,* **65,** 625.
Tiffney. W., Eveleigh, D., Barrera, J., and Mitchell, S. (1979), in *Symbiotic Nitrogen fixation in the Management of temperate Forests* (Eds. J. C. Gordon, C. T. Wheeler, and D. A. Perry), Forest research Laboratory, Oregon State University, Corvallis, Oregon, p. 420.
Tilak, K. V. B. and Subba Rao, N. S. (1978), *Fert. News,* **23,** 25.
Tisdale, S. L. and Nelson, W. L. (1966), *Soil fertility and fertilizers,* 2nd Edition, Macmillan, New York.
Tjepkema, J. and Evans, H. J. (1975), *Biochem. Biophys, Res. Commun.,* **65,** 625.

Tjepkema, J. and Van Berkum, P. (1977), *Appl. Environ, Microbiol.*, **33**, 626.
Tjepkema, J. D. and Burris, R. H. (1976), *Pl. Soil*, **45**, 81.
Torrey, J. G. and Tjepkema, J. D. (1979), *Bot. Gaz.*, 140 (Suppl.), ii.
Trinick, M. J. (1969), *J. Appl. Bacteriol.*, **32**, 181.
Trinick, M. J. (1973), *Nature, Lond.*, **244**, 459.
Trinick, M. J. (1975), in *Proc. 1st int. symp nitrogen fixation* (Eds. W. E. Newton and C. J. Nyman), Vol. 2, Washington State University Press, Pullman, Washington, p. 507.
Trinick, M. J. and Galbraith, J. (1976), *Arch. Microbiol.*, **108**, 159.
Tronick, S. R., Ciardi, J. E., and Stadtman, E. R. (1973), *J. Bacteriol.*, **115**, 858.
Trotter, A. (1902), *Bull. Soc. Bot. It.*, 50.
Tsai, L. B. and Mortenson, L. E. (1978), *Biochim, Biophys. Acta*, **81**, 280.
Tso, M.-Y. and Burris, R. H. (1973), *Biochim. Biophys. Acta*, **309**, 263.
Tso, M.-Y. W., Ljones, T., and Burris, R. H. (1972), *Biochim. Biophys. Acta*, **267**, 600.
Tubb, R. S. and Postgate, J. R. (1973), *J. Gen. Microbiol.*, **79**, 103.
Turner, G. L. and Bergersen, F. J. (1969), *Biochem. J.*, **115**, 529.
Tuzimura, K. and Watanabe, I. (1961), *Soil. Sci. Pl. Nutr.*, **7**, 61.
Umbreit, W. W. and Bond, V. S. (1936), *Ind. Eng. Chem. Anal. Ed.*, **8**, 276.
Umbreit, W. W., Burris, R. H., and Stauffer, J. F. (1972), *Manometric and biochemical techniques*, Burgess Publishing Co. Minneapolis.
University of Hawaii (1977), *Catalogue of strains from the NifTAL rhizobium collection (abbreviated)*, NifTAL Project, University of Hawaii, Paia, Maui, Hawaii, 9 pp.
Upchurch, R. G. and Elkan, G. H. (1978), *Biochim. Biophys. Acta*, **538**, 244.
Uphof, J. C. (1968), *Dictionary of economic plants*, 2nd Edition, Cramer, London.
Van Baalen, C. (1961), *Science*, **133**, 1922.
Van Baalen, C. (1962), *Bot. Mar.*, **4**, 129.
Van Berkum, P. and Sloger, C. (1979), *Pl. Physiol.*, **64**, 739.
Van Dijk, C. (1978), *New Phytol.*, **81**, 601.
Van Gorkom, H. J. and Donze, M. (1971), *Nature, Lond.*, **234**, 231.
Van Rensburg, J. H. and Strijdom, B. W. (1974), *Phytophylactica*, **6**, 307.
Van Ryssen, F. W. J. and Grobbelaar, N. (1970), *S. Afr. J. Sci.*, **66**, 22.
Van Schreven, D. A. (1958), in *Nutrition of the legumes* (Ed. E. G. Hallsworth), Proc. Univ. Nottingham 5th Easter School in Agricultural Science, Butterworths, London, p. 328.
Van Schreven, D. A. (1970), *Pl. Soil*, **32**, 113.
Van Schreven, D. A., Harmsen, G. W., Lindenbergh, D. J. and Otzen, D. (1953), *Antonie van Leeuwenhoek J. Microbiol. Serol.*, **19**, 300.
Van Schreven, D. A., Otzen, D., and Lindenbergh, D. J. (1954), *Antonie van Leewenhoek J. Microbiol. Serol.*, **20**, 33.
Van Straten, J., Akkermans, A. D. L., and Roelofsen, W. (1977), *Nature, Lond.*, **266**, 257.
Van't Reit, J. and Planta, R. J. (1975), *Biochim. Biophys. Acta*, **379**, 81.
Vancura, V., Abd-el-Malek, Y., and Zayed, M. N. (1965), *Folia Microbiol., Prague*, **10**, 224.
Vainshtein, B. K., Arutunyan, É. G., Kuranova, I. P., Borisov, V. V., Sosfenov, N. I., Pavlovskii, A. G., Grebenko, A. I., Konareva, N. V., and Nekrasov, Y. V. (1977), *Dokl, Akad. NaukSSSR*, **233**, 238.
Venkataraman, G. S. (1961), *Sci. Cult.*, **27**, 9.
Venkataraman, G. S. (1962), *Indian J. Agric. Sci.*, **32**, 22.
Venkataraman, G. S. (1972), *Algal biofertilisers and rice cultivation*, Today and Tomorrow's Printers and Publishers, New Delhi.

Venkataraman, G. S. (1975), in *Nitrogen fixation by free-living micro-organisms* (Ed. W. D. P. Stewart), Cambridge University Press, Cambridge, London, New York, Melbourne, p. 207.
Venkataraman, G. S. (1979), in *Nitrogen and rice symposium,* IRRI, Los Baños, Laguna, Philippines, p. 311.
Venkataraman, G. S. and Goyal, S. K. (1968), *Soil Sci. Pl. Nutr.,* **14,** 249.
Venkataraman, G. S. and Goyal, S. K. (1969a), *Sci. Cult.,* **35,** 58.
Venkataraman, G. S. and Goyal, S. K. (1969b), Mikrobiologiya, **38,** 709.
Venrick, E. L. (1974), *Limnol. Oceanogr.,* **19,** 437.
Verdcourt, B. (1970a), *Kew Bull.,* **24,** 379.
Verdcourt, B. (1970b), *Kew Bull.,* **24,** 507.
Verma, D. P. S. and Bal, A, K. (1976), *Proc. Nat. Acad. Sci. USA,* **73,** 3843.
Verma, D. P. S., Nash, D. T., and Schulman, H. M. (1974), *Nature, Lond.,* **251,** 74.
Verma, D. P. S., Kazazian, V., Zogbi, V., and Bal. A. K., (1978), *J. Cell Biol.,* **78,** 919.
Vest, G. (1970), *Crop Sci.,* **10,** 34.
Vest, G. and Caldwell, B. C. (1972), *Crop Sci.,* **12,** 692.
Villalobo, A., Roldan, J. M., Rivas, J., and Cárdenas, J. (1977), *Arch. Microbiol.,* **112,** 127.
Vincent, J. M. (1941), *Proc. Linn. Soc. N.S.W.,* **66,** 145.
Vincent, J. M. (1942), *Proc. Linn. Soc. N.S.W.,* **67,** 82.
Vincent, J. M. (1954), *J. Aust. Inst. Agric. Sci.,* **20,** 247.
Vincent, J. M. (1958), in *Nutrition of the legumes* (Ed. E. G. Hallsworth), Proc. Univ. Nottingham 5th Easter School in Agricultural Science, Butterworths, London, P. 108.
Vincent, J. M. (1965), in *Soil nitrogen* (Eds. W. V. Bartholomew and F. E. Clark), American Society of Agronomy, Wisconsin, p. 384.
Vincent, J. M. (1968), in *Festskrift til Hans Laurits Jensen,* Gadgaard Nielsens Bogtrykkeri, Lemvig, Denmark, p. 145.
Vincent, J. M. (1970), *A manual for the practical study of root-nodule bacteria,* IBP Handbook No. 15, International Biological Programme, Blackwell, Oxford.
Vincent, J. M. (1977), in *Exploiting the legume-rhizobium symbiosis in tropical agriculture,* (Eds. J. M. Vincent, A. S. Whitney and J. Bose), University of Hawaii College of Tropical Agriculture, Misc. Pub., No. 145, p. 447.
Vincent, J. M. and Humphrey, B. A. (1970), *J. Gen. Microbiol.,* **63,** 379.
Vincent, J. M. and Waters, L. M. (1953), *J. Gen. Microbiol.,* **9,** 357.
Vincent, J. M., Humphrey, B. A., and Škrdleta, V. (1973), *Arch. Mikrobiol.,* **89,** 79.
Virtanen, A. I., Moisio, T., Allison, R. M., and Burris, R. H. (1954), *Acta Chem. Scand.,* **8,** 1730.
Vlassak, K. and Reynders, L. (1977), *Associative dinitrogen fixation in temperate regions,* IAEA Advisory Committee Group, Vienna.
Voller, A., Bidwell, D. E., and Bartlett, A. (1976), *Bull. Wld. Hlth. Org.,* **53,** 55.
Wacek, T. J. and Alm, D. (1978), *Crop Sci.,* **17,** 514.
Walker, G. A. and Mortenson, L. E. (1973), *Biochem. Biophys. Res. Commun.,* **53,** 904.
Walker, P. D., Batty, K. and Thomson, R. O. (1971), in *Methods in microgiology* (Eds. J. R. Norris and D. W. Ribbons), Vol. 5a, Academic Press, New York, p. 219.
Wallsgrove, R. M., Harel, E., Lea, P. J., and Miflin, B. J. (1977), *J. Exp. Bot.,* **28,** 588.
Walsby, A. E. (1967), *Biotechnol. Bioeng.,* **9,** 443.

Wang, C. H. (1967), in *Methods of biochemical analysis,* (Ed. D. Glick), Vol. 15, Interscience, New York, London, Sydney, p. 311.
Wang. C. H. (1972), in *Methods in Microbiology,* (Eds. J. R. Norris and D. W. Ribbons), Vol. 6B, Academic Press, London, New York, p. 185.
Wang, R., Healey, F. P., and Myers, J. (1971), *Pl. Physiol.,* **48,** 108.
Warrington, R. (1878), *J. Chem. Soc.,* 39.
Warrington, R. (1883), *On some of the changes which nitrogenous matter undergoes within the soil,* Harrison and Sons, London, 23 p.
Watanabe, A. (1959a), *J. Gen. Appl. Microbiol.,* **5,** 21.
Watanabe, A. (1959b), *J. Gen. Appl. Microbiol.,* **5,** 153.
Watanabe, A. (1961), *Stud. Rokugawa Inst. Tokyo,* **9,** 162.
Watanabe, A. (1967), *Proc. IX Int. Congr. Microbiol. Symp.,* p. 77.
Watanabe, A. and Kiyohara, T. (1963), in *Studies on microalgae and photosynthetic bacteria,* Japanese Society for Plant Physiology, University of Tokyo Press, Tokyo, p. 189.
Watanabe, I. and Cholitkul, W. (1970), in *Nitrogen and rice symposium,* IRRI, Los Baños, Laguna, Philippines, p. 223.
Watanabe, I and Lee, K.-K. (1977), in *Biological nitrogen fixation in farming systems of the tropics* (Eds. A. Ayanaba and P. J. Dart), Wiley, Chichester, New York, Brisbane, Toronto, p. 289.
Watanabe, I., Lee, K.-K., Alimagno, B. V., Sato, M., del Rosario, D. C., and de Guzman, M. R. (1977a), *IRRI Res. Pap.,* No. 3, 16 pp.
Watanabe, I., Espinas, C. R., Berja, N. S., and Alimagno. B. V. (1977b), *IRRI Res. Pap.* No. 11, 15 pp.
Watanabe, I., Lee, K.-K. and Alimagno, B. V. (1978a), *Soil Sci. Pl. Nutr.* **24,** 1.
Watanabe, I., Lee, K.-K., and de Guzman, M. R. (1978b), *Soil Sci. Pl. Nutr.* **24,** 465.
Waterbury, J. B., and Stanier, R. Y. (1978), *Bacteriol. Rev.,* **42,** 2.
Waters, L. M. (1954), *J. Aust. Inst. Agric. Sci.,* **20,** 250.
Watson, E. V. (1963), *British mosses and liverworts,* Cambridge University Press, Cambridge.
Watt, G. D. and Bulen, W. A. (1976), *Proc. 1st int. symp. nitrogen fixation* (Eds. W. Newton and C. J. Nyman), Vol. 1, Washington State University Press, Pullman, Washington, p. 248.
Watt, G. D. and Burns, A. (1977), *Biochemstry,* **16,** 264.
Waughman, G. J. (1971), *Oikos,* **22,** 111.
Waughman, G. J. (1972), *Pl. Soil,* **37,** 521.
Waughman, G. J. (1977), *J. Exp. Bot.,* **28,** 949.
Way, J. T. (1856), *J. Roy, Agric. Soc. Engl.,* **17,** 123 and 618.
Weare, N. M. and Benemann, J. R. (1974), *J. Bacteriol.,* **119,** 258.
Weaver, R. W. and Frederick. L. R. (1972), *Pl. Soil,* **36,** 219.
Weir, D. M. (1967), *Handbook of experimental immunology,* Blackwell, Oxford.
Wellman, R. P., Cook, F. D., and Krouse, H. R. (1968), *Science,* **161,** 269.
Wellner, V. P., Zoukis, M., and Meister, A. (1966), *Biochemistry,* **5,** 3509.
Weppner, W. A., Clarke, P. O., and Leach, F. R. (1977), *Can J. Microbiol.,* **23,** 1585.
West, G. S. and Fritsch, F. E. (1927), *A Treatise on the British freshwater algae,* Cambridge University Press, Cambridge, 534 pp.
Wetselaar, R., Jacobsen, P., and Chaplin, G. R. (1973), *Soil Biol. Biochem.,* **5,** 35.
Whatley, F. R. and Arnon, D. I. (1963), *Methods Enzymol.,* **6,** 308.
Wheeler, C. T. (1969), *New Phytol.,* **68,** 675.
Wheeler, C. T. (1971), *New Phytol.,* **70,** 487.

Wheeler, C. T., Cameron, E. M., and Gordon, J. C. (1978), *New Phytol.*, **80**, 175.
Wheeler, C. T. and McLaughlin, M. E. (1979), in *Symbiotic nitrogen fixation in the management of temperate forests* (Eds. J. C. Gordon, C. T. Wheeler, and D. A. Perry), Forest Research Laboratory, Oregon State University, Corvallis, Oregon, p. 124.
Whitaker, R. H. and Woodwell, G. M. (1968), *J. Ecol.*, **56**, 1.
White, F. A. (1968), *Mass spectrometry in science and technology*, Wiley, New York, London, Sydney.
Whiting, M. J. and Dilworth, M. J. (1974), *Biochim. Biophys. Acta*, **371**, 337.
Whittenbury, R. and Dalton, H. (1979), in *The Prokaryotes* (Eds. M. P. Starr, H. Stolp, G. G. Truper, A. Balows, and J. G. Schlegel), Springer-Verlag, Berlin, Heidelberg, New York, in press.
Whitternbury, R., Phillips, K. C., and Wilkinson, J. F. (1970), *J. Gen. Microbiol.*, **61**. 205.
Whitton, B. A. (1962), *Brit. Phycol. Bull.*, **2**, 177.
Wieringa, K. T. (1968), *Antonie van Leeuwenhoek J. Microbiol. Serol.*, **34**, 54.
Wieslander, A. E. and Schreiber, B. O. (1939), *Madroño*, **5**, 38.
Wilcockson, J. and Werner, D. (1976), *Ber. Dt. Bot. Ges.*, **89**. 587.
Wilcox, M. (1970), *Nature, Lond.*, **228**, 686.
Wilhelm, E., Battino, R., and Wilcock, R. J. (1977), *Chem. Rev.*, **77**, 219.
Williams, C. A. and Chase, M. W. (1967), *Methods in immunology and immunochemistry*, Vol. 1, Academic Press, New York.
Williams, C. A. and Chase, M. W. (1971), *Methods in immunology and immunochemistry*, Vol. 3, Academic Press, New York.
Williams, C. A. and Chase, M. W. (1976), *Methods in immumology and immunochemistry*, Vol. 5, Academic Press, New York.
Williams, C. H. and Twine, J. R. (1967), *CSIRO Aust. Div. Pl. Ind. Tech. Pap.*, No. 24.
Williams, L. F. and Lynch, D. L. (1954), *Agron. J.*, **46**, 28.
Williams W. A., Jones, M. B., and Delwiche, C. C. (1977), *Agron, J.*, **69**, 1023.
Wilson, G. S. and Miles, A. (1975), *Topley and Wilson's principles of bacteriology, virology and immunity*, 6th Edition, 2 Volumes, Edward Arnold, London.
Wilson, K. (1975), in *A biologist's guide to principles and techniques of practical biochemistry* (Eds. B. L. Williams and K. Wilson), Edward Arnold, London, p. 230.
Wilson, J, K. (1926), *J. Am. Soc. Agron.*, **18**, 911.
Wilson, M. H. M., Humphrey, B. A., and Vincent, J. M. (1975), *Arch. Microbiol.*, **103**, 151.
Wilson, P. W. (1940), *The biochemistry of symbiotic nitrogen fixation*, University of Wisconsin Press, Madison, Wisc., 302 pp.
Wilson, P. W. (1958), in *Encyclopaedia of plant physiology*, Vol. 8, Springler-Verlag, Berlin, p. 9.
Wilson, P. W. (1963), *Bacteriol. Rev.*, **27**, 405.
Wilson, P. W., Fred, E. B., and Salmon, M. R. (1933), *Soil Sci.*, **35**, 145.
Wilson, P. W. and Fred, E. B. (1935), *Sci. Monthly*, **41**, 240.
Winkenbach, F. and Wolk, C. P. (1973), *Pl. Physiol.*, **52**, 480.
Winter, H. C. and Burris, R. H. (1976), *Ann. Rev. Biochem.*, **45**, 409.
Winter, G. (1935), *Beitr. Biol. Pfl.*, **23**, 295.
Witt. W. W. and Weber, J. B. (1975), *Weed Sci.*, **23**, 302.
Wittenberg, J. B. (1978), *J. Biol. Chem.*, **253**, 5690.

Wittenberg, J. B., Appleby, C. A., and Wittenberg, B. A. (1972), *J. Biol. Chem.,* **247,** 527.
Wittenberg, J. B., Bergersen, F. J., Appleby, C. A., and Turner, G. L. (1974), *J. Biol. Chem.,* **249,** 4057.
Witty, J. F. (1977), *A critical analysis of the application of the acetylene reduction method in assessing nitrogen fixation by tropical crops,* MOD Research Scheme Report, No. 3041.
Witty, J. F. (1978), *Abstr. 3rd int. symp. on nitrogen fixation, Madison, Wisc.,* B50.
Witty, J. F. (1979), *New Phytol.,* in press.
Witty, J. F. (1979a), *Pl. Soil,* **52,** 165.
Witty, J. F. (1979b), *Soil Biol. Biochem.,* **11,** 209.
Witty, J. F. and Day, J. M. (1977), in *Isotopes in biological dinitrogen fixation,* International Atomic Energy Agency, Vienna, p. 135.
Wolf, J. and Barker, A. N. (1968), in *Identification methods for microbiologists* (Eds. M. B. Gibbs and D. A. Shapton), Society for Applied Bacteriology Technical Series No. 2, Part B, Academic Press, London, p. 93.
Wolk, C. P. (1968), *J. Bacteriol.,* **96,** 2138.
Wolk, C. P., Austin, S. M., Bortins, J., and Galonsky, A. (1974), *J. Cell Biol.,* **61,** 440.
Wolk, C. P., Thomas, J., Shaffer, P. W., Austin, S. M., and Galonsky, A. (1976), *J. Biol. Chem.,* **251,** 5027.
Wong, K. F. and Dennis, D. T. (1973), *Pl. Physiol.,* **51,** 322.
Wong, P. P. and Evans, H. J. (1971), *Pl. Physiol.,* **47,** 750.
Woolfolk, C. A., Shapiro, B., and Stadtman, E. R. (1966), *Arch. Biochem. Biophys.,* **116,** 177.
Worral, V. S. and Roughley, R. J. (1976), *J. Exp. Bot.,* **27,** 1233.
Wright, P. E. and Appleby, C. A. (1977), *FEBS Lett.,* **78,** 61.
Wu, T. T. (1966), *Genetics,* **54,** 405.
Wyatt, J. T. and Silvey, J. K. G. (1969), *Science,* **165,** 908.
Yao, P. Y. and Vincent, J. M. (1969), *Aust. J. Biol. Sci.,* **22,** 413.
Yates, M. G. (1972), *FEBS Lett.,* **27,** 63.
Yates, M. G. (1977), in *Recent developments in nitrogen fixation* (Eds. W. Newton, J. R. Postgate, and C. Rodríguez-Barrueco), Academic Press, London, New York, San Francisco, p. 219.
Yates, M. G. and Planque, K. (1975), *Eur. J. Biochem.,* **60,** 467.
Yates, M. G., Thorneley, R. N. F., and Lowe, D. J. (1975), *FEBS Lett.,* **60,** 89.
Yoch, D. (1972), *Biochem. Biophys. Res. Commun.,* **49,** 335.
Yoshida, T. and Ancajas, R. R. (1973), *Soil Sci. Soc. Am. Proc.,* **35,** 156.
Yoshida, T. and Suzuki, T. (1975), *Soil Sci. Plant Nutr.,* **21,** 129.
Yoshida, T., Roncal, R. A., and Bautista, E. M. (1973), *Soil Sci. Pl. Nut.* **19,** 117.
Yoshinari, T. and Knowles, R. (1976), *Biochem. Biophys. Res. Commun.,* **69,** 705.
Yoshinari, T., Hynes, R., and Knowles, R. (1977), *Soil Biol. Biochem.,* **9,** 177.
Young, R. E., Pratt, H. K., and Biale, J. B. (1951), *Pl. Physiol.,* **26,** 304.
Youngberg, C. T. and Wollum, II, A. G. (1976), *J. Soil Sci. Soc. Am.,* **40,** 109.
Zanetti, G. and Forti, G. (1966), *J. Biol. Chem.,* **241,** 279.
Ziegler, H. and Hüser, R. (1963), *Nature, Lond.,* **199,** 508.
Zarnea, G., Cracea, E., Dumitresco, S., Andreutza, C., and Herlea, V. (1966), *Soil Biol. Int. News Bull.,* **6,** 20.
Zimmermann, R., Münk, E., Brill, W. J., Shah, V. K., Henzl, M. T., Rawlings, J., and Orme-Johnson, W. H. (1978), *Biochim, Biophys, Acta,* **537,** 185.
Zobel, R. W., Tredici, D. T., and Torrey, J. G. (1976), *Pl. Physiol.,* **57,** 344.

Zuberer, D. A. and Silver, W. S. (1978), *Appl. Environ. Microbiol.*, **35,** 567.
Zumft, W. G. and Castillo. F. (1978), *Arch. Microbiol.*, **117,** 53.
Zumft, W. G. and Mortenson, L. E. (1973), *Eur. J. Biochem.*, **35,** 401.
Zumft, W. G. and Mortenson, L. E. (1975), *Biochim. Biophys. Acta,* **416,** 1.
Zumft, W. G., Mortenson, L. E., and Palmer, G. (1974), *Eur. J. Biochem.*, **46,** 525.
Zweig, G. and Whitaker, J. R. (1971), *Paper chromatography and electrophoresis,* Vols. I and II, Academic Press, New York, London.

General Index

Index of Diazotrophic Microorganisms

Index of Nodulated Legumes

(only preferred names are listed, see p. 446)

Index of N_2-fixing, Non-leguminous Plants

A. PLANTS WITH ROOT NODULES

B. NON-NODULATED PLANTS WITH ASSOCIATED OR SYMBIOTIC DIAZOTROPHS